T0185813

Infectious Disease

Series Editor

Vassil St. Georgiev
National Institute of Health Dept. Health & Human Services, Bethesda, MD, USA

The infectious Disease series provides the best resources in cutting-edge research and technology.

More information about this series at http://www.springer.com/series/7646

Manjunath P. Pai · Jennifer J. Kiser
Paul O. Gubbins · Keith A. Rodvold
Editors

Drug Interactions in Infectious Diseases: Antimicrobial Drug Interactions

Fourth Edition

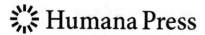 Humana Press

Editors
Manjunath P. Pai
College of Pharmacy
University of Michigan
Ann Arbor, MI, USA

Paul O. Gubbins
Division of Pharmacy Practice and
Administration
UMKC School of Pharmacy at MSU
Springfield, MO, USA

Jennifer J. Kiser
Department of Pharmaceutical Sciences
Skagg School of Pharmacy and
Pharmaceutical Sciences
University of Colorado
Aurora, CO, USA

Keith A. Rodvold
College of Pharmacy and Medicine
University of Illinois at Chicago
Chicago, IL, USA

Infectious Disease
ISBN 978-3-030-10197-8   ISBN 978-3-319-72416-4   (eBook)
https://doi.org/10.1007/978-3-319-72416-4

This Humana Press imprint is published by the registered company Springer International Publishing AG
part of Springer Nature.
The registered company address is: Gewerbestrasse 11, 6330 Cham, Switzerland

Foreword

In the 1950s and 1960s, there was euphoria that antibacterial drugs had been discovered, which seemed to have the potential to eliminate the major role infectious diseases had in reducing the quality and duration of human life. Penicillins, cephalosporins, macrolides, tetracyclines, and aminoglycosides were a small but manageable armamentarium, which seemed destined to solve many human challenges.

Since the 1960s and 1970s, we have recognized how readily most infectious agents learn to become resistant to the anti-infective agents to which they are exposed. Methicillin-resistant *Staphylococcus aureus* (MRSA), vancomycin-resistant *Enterococcus faecium* (VRE), carbapenemase-producing *Klebsiella* (KPC), azole-resistant *Candida*, and acyclovir-resistant herpes simplex have been examples of how much urgency there is to create new drugs which will have activity against organisms that have learned to evade currently available anti-infective agents.

We have also developed new classes of drugs for more recently recognized pathogens such as human immunodeficiency virus (HIV) and hepatitis C. These older and newer drugs are given to patients who are receiving a rapidly expanding armamentarium of molecules to treat their chronic and acute underlying conditions.

Healthcare providers are well aware that drugs are only effective and safe if administered with tactical and strategic planning. The right dose, given at the right time, to the right patient is a foundation for effective and safe care. However, as patients are administered more and more agents for a wide range of health challenges, interactions among drugs become more and more likely.

Every experienced clinician has anecdotes of unanticipated drug interactions that affected clinical outcome. Drug interactions can have a major negative impact on drug efficacy and can greatly enhance toxicity for the antimicrobial agent being focused on or for concurrent drugs that may be life-sustaining.

This fourth edition of *Drug Interactions in Infectious Diseases* provides healthcare providers with a unique resource for both understanding basic principles and finding important information. Volume 1 on Mechanisms and Models of Drug

Interactions and Volume 2 on Antimicrobial Drug Interactions are well organized for providers to quickly find practical information. This resource maximizes the likelihood that the healthcare team can optimize efficacy and safety in this era when patients are so often receiving multiple drugs.

Henry Masur, MD
Chief, Critical Care Medicine Department
NIH-Clinical Center
Bethesda, MD, USA

Editors' Preface

The benefits of new medical therapies in infectious diseases cannot be appreciated without understanding and mitigating risk. Drug interactions in infectious diseases are a major source of medical harm that can be prevented. Over the past two decades, we have witnessed a major expansion in our anti-infective armamentarium. This expansion has been coupled with an improved understanding of drug interaction mechanisms and scientific approaches to measure them. Our transformation of the fourth edition of this text to a two-volume series is a direct reflection of the growing knowledge in this domain. Volume 1 provides a mechanistic profile of drug interactions as well as in vitro, in vivo, in silico, and clinical methods to evaluate these interactions. Volume 2 is structured by anti-infective class to provide clinicians, researchers, and academicians a useful resource to meet their practical needs.

Given the scale of this field of study, no comprehensive reviews on antimicrobial drug-drug interactions can be easily published through journals. Software programs and deep learning algorithms that can integrate the effects of all known covariates of drug-drug interaction are in development but have as yet not entered clinical practice. Hence, clinical intuition and vigilance remain key defenses against untoward drug-drug interactions. Since the last publication in 2011, several new antimicrobials have received regulatory approval. The chapters have been updated to reflect these new additions. Three distinct chapters related to the pharmacologic management of human immunodeficiency virus- and hepatitis C virus-related infections have been added in response to recent drug approvals.

The strength of the textbook lies not only in the fact that it is a comprehensive reference book on drug interactions but it also has chapters that provide insights that are difficult to find in the medical literature. We are confident that the information provided in the detailed tables and text will increase the acumen of the practicing clinician, the academic instructor, and the infectious disease researcher.

As the editors of the fourth edition of *Drug Interactions in Infectious Diseases*, we are thrilled to deliver a text that will enhance your clinical knowledge of the complex mechanisms, risks, and consequences of drug interactions associated with antimicrobials, infection, and inflammation. The quality and depth of the information provided would not be possible without the contributions of an excellent number of authors. We are indebted to our authors for their time and diligence to ensure that this textbook remains a primary reference for those engaged in the field of infectious diseases. Finally, we thank our families for their support and encouragement throughout this endeavor.

Ann Arbor, MI, USA	Manjunath P. Pai
Aurora, CO, USA	Jennifer J. Kiser
Springfield, MO, USA	Paul O. Gubbins
Chicago, IL, USA	Keith A. Rodvold

Author's Preface

It is well known that drug interactions pose a major risk to patients. Even a cursory look at approved drug product labels for anti-infective drugs, such as HIV drugs, direct-acting antivirals for HCV, azole antifungal drugs, and anti-mycobacterial agents, reveals that drug interactions present a huge challenge for patients and their healthcare providers. However, before a drug reaches patients, drug development scientists have the opportunity to define the potential for drug interactions. The work of these scientists and the regulatory scientists responsible for drug approval results in information available to healthcare providers and patients.

Concerns related to drug interactions grow as the knowledge of pharmacology advances. The interactions may be due to CYP enzymes, non-CYP enzymes, the ever-growing list of drug transporters, changes in gastric pH, and more. It is easy to be overwhelmed by the scope of the issue. How do you develop an informative and efficient drug interaction program? What drugs are likely perpetrators or victims of interactions? Do you have to study all potential interactions? This textbook helps answer those questions. The chapters address general drug interaction concepts, specific classes of anti-infective drugs, and application of the concepts to drug development. Together, the information helps one focus on the overarching goals of a drug interaction program, determine the potential for clinically significant drug interactions, and develop management strategies for the interactions. The first goal can be divided into four questions. Does the investigational drug alter the pharmacokinetics of other drugs? Do other drugs alter the pharmacokinetics of the investigational drug? What is the magnitude of the change? Is the change clinically significant?

As indicated in the initial chapters of this book, there are many potential mechanisms for drug interactions. Also, concerns go beyond interactions between small molecules. Other considerations include interactions due to biologic products, food components, and herbal medications. However, the bulk of drug interaction evaluations involve investigation of CYP enzyme- or transporter-based interactions. Drug development programs include multiple steps to evaluate the potential for these interactions. For both CYP enzyme and transporter interactions, programs often begin with in vitro evaluations that screen for interactions. If the in vitro evaluations reveal potential interactions, additional evaluations, usually clinical studies with pharmacokinetic endpoints, follow. In some situations, model-based simulations can replace clinical studies or help refine their design [1]. Scientific quality and rigor is essential for all studies. The methods and interpretation of in vitro metabolism and transporter studies must follow best practices because the results may screen out the need for clinical evaluations [2]. Each clinical study should be designed to address the goal of the study. Some clinical studies, referred to as index studies, use perpetrators (inhibitors or inducers) or substrates (victims) with well-known pharmacokinetic and drug interaction properties [1]. Results of the index studies can be extrapolated to other drug combinations and inform the need for additional studies. The design of index studies should maximize the potential to detect an interaction. In contrast to index studies, concomitant use studies investigate drug interactions between the investigational drug and other drugs used in the target population [2]. Results of concomitant use studies provide useful information for the healthcare provider and patient.

The progression from in vitro to index and then concomitant use studies is a common drug development path. However, there are other options. In silico studies that use physiologically based pharmacokinetic (PBPK) methods may substitute for some clinical studies [1]. Instead of dedicated drug interaction studies, prospectively planned evaluations nested within a larger clinical trial may provide useful drug interaction information in the intended patient population. The nested studies often use population pharmacokinetic methods. The in silico and population PK evaluations should be carefully designed to address their specific goals.

Two draft guidance documents from the US Food and Drug Administration provide more details about in vitro and in vivo drug interaction studies: In Vitro Metabolism- and Transporter-Mediated Drug-Drug Interaction Studies Guidance for Industry [3] and Clinical Drug Interaction Studies – Study Design, Data Analysis, and Clinical Implications Guidance for Industry [4].

The progression of drug interaction evaluations that determine the presence and magnitude of pharmacokinetic changes forms the foundation for the next questions: Is the interaction clinically significant? How are clinically significant interactions managed? Thus, solid knowledge regarding general drug interaction concepts, issues related to specific classes of anti-infective drugs, and application of the concepts to drug development are essential to the development of anti-infective drugs.

Kellie Schoolar Reynolds, PharmD
Deputy Director, Division of Clinical Pharmacology IV
Office of Clinical Pharmacology, Office of Translational Sciences
Center for Drug Evaluation and Research, US Food and Drug Administration

References

1. Rekic D, Reynolds KS, Zhao P et al (2017) Clinical drug-drug interaction evaluations to inform drug use and enable drug access. J Pharm Sci 106:2214–2218
2. Yoshida K, Zhao P, Zhang L et al (2017) In vitro-in vivo extrapolation of metabolism and transporter-mediated drug-drug interactions- overview of basic prediction methods. J Pharm Sci 106:2209–2213
3. US Department of Health and Human Services, Food and Drug Administration, Center for Drug Evaluation and Research (CDER) (2017) Draft guidance for industry. In vitro metabolism- and transporter mediated drug-drug interactions. Available at https://www.fda.gov/downloads/Drugs/GuidanceComplianceRegulatoryInformation/Guidances/UCM581965.pdf. Accessed 30 Oct 2017
4. US Department of Health and Human Services, Food and Drug Administration, Center for Drug Evaluation and Research (CDER). Draft guidance for industry. Clinical drug interaction studies – study design, data analysis, and clinical implications. Available at https://www.fda.gov/downloads/Drugs/GuidanceComplianceRegulatoryInformation/Guidances/UCM292362.pdf. Accessed 30 Oct 2017

Contents

Contributors

Waheed A. Adedeji, MBBS, MSc, FMCP Department of Clinical Pharmacology, University College Hospital, Ibadan, Oyo State, Nigeria

Jarrett R. Amsden, PharmD, BCPS Department of Pharmacy Practice, Butler University College of Pharmacy & Health Sciences, Indianapolis, IN, USA

Tunde Balogun, DVM, MSc, PhD Department of Clinical Pharmacology and Therapeutics, College of Medical Sciences, University of Maiduguri, Maiduguri, Borno State, Nigeria

Lauren R. Cirrincione University of Nebraska Medical Center, College of Pharmacy, Department of Pharmacy Practice, Omaha, NE, USA

Larry H. Danziger, PharmD Department of Pharmacy, Tufts Medical Center, Clinical Pharmacist Emergency Medicine, Boston, MA, USA

Mary H.H. Ensom, Pharm D, FASHP, FCCP FCSHP FCAHS Faculty of Pharmaceutical Sciences, University of British Columbia, Vancouver, BC, Canada

Children's and Women's Health Centre of British Columbia, Vancouver, BC, Canada

Fatai A. Fehintola, MBBS, MSc, FMCP Department of Clinical Pharmacology, University College Hospital, Ibadan, Oyo State, Nigeria

Department of Pharmacology and Therapeutics, University of Ibadan, Ibadan, Oyo State, Nigeria

Douglas N. Fish, PharmD Department of Clinical Pharmacy, University of Colorado School of Pharmacy, Aurora, CO, USA

University of Colorado Hospital, Aurora, CO, USA

David R. P. Guay, PharmD, FCP, FCCP Department of Experimental & Clinical Pharmacology, College of Pharmacy, University of Minnesota, Minneapolis, MN, USA

Paul O. Gubbins, PharmD Division of Pharmacy Practice and Administration, UMKC School of Pharmacy at MSU, Springfield, MO, USA

Karolyn S. Horn, PharmD Department of Pharmacy, Tufts Medical Center, Clinical Pharmacist Emergency Medicine, Boston, MA, USA

Tony K. L. Kiang, BSc(Pharm), ACPR, PhD Faculty of Pharmacy and Pharmaceutical Sciences, University of Alberta, Edmonton, AB, Canada

Jennifer J. Kiser, PharmD Department of Pharmaceutical Sciences, Skaggs School of Pharmacy and Pharmaceutical Sciences, University of Colorado Denver, Aurora, CO, USA

Stan Louie, PharmD, FASHP University of Southern California, School of Pharmacy, Los Angeles, CA, USA

Christine E. MacBrayne, PharmD Department of Pharmaceutical Sciences, Skaggs School of Pharmacy and Pharmaceutical Sciences, University of Colorado Denver, Aurora, CO, USA

Gene D. Morse, PharmD, FCCP, BCPS School of Pharmacy and Pharmaceutical Sciences, Center for Integrated Global Biomedical Sciences, University at Buffalo, Buffalo, NY, USA

Rocsanna Namdar, PharmD, FCCP, BCPS Raymond G. Murphy Veterans Affairs Medical Center, Albuquerque, NM, USA

Manjunath P. Pai, PharmD College of Pharmacy, University of Michigan, Ann Arbor, MI, USA

Parul Patel, PharmD Clinical Pharmacology, ViiV Healthcare, Research Triangle Park, NC, USA

Charles A. Peloquin, PharmD Infectious Disease Pharmacokinetics Lab, College of Pharmacy, and Emerging Pathogens Institute, University of Florida, Gainesville, FL, USA

John C. Rotschafer, PharmD, FCCP University of Minnesota College of Pharmacy, Department of Experimental and Clinical Pharmacology, Minneapolis, MN, USA

Kimberly K. Scarsi University of Nebraska Medical Center, College of Pharmacy, Department of Pharmacy Practice, Omaha, NE, USA

Gregory M. Susla, PharmD, FCCM Medical Information, MedImmune, Gaithersburg, MD, USA

Mary A. Ullman, PharmD Regions Hospital, Department of Pharmacy, St. Paul, MN, USA

Kyle John Wilby, BSP, ACPR, PharmD College of Pharmacy, Qatar University, Doha, Qatar

Chapter 1
Beta-Lactam Antibiotics

Larry H. Danziger and Karolyn S. Horn

1.1 Beta-Lactam Antibiotics

The beta-lactam antibiotics are a large class of diverse compounds used clinically in oral, parenteral, and inhaled dosage formulation. The beta-lactam antibiotic agents have become the most widely used therapeutic class of antimicrobials because of their broad antibacterial spectrum and wide therapeutic index. Reports of drug-drug interactions with the beta-lactam antimicrobials are a relatively rare phenomenon, and when they do occur, they are generally of minor clinical significance. This chapter describes the drug-drug interactions of the beta-lactam class of antibiotics: penicillins, cephalosporins, carbapenems, and monobactams.

This chapter serves as a review and clinical assessment of the literature regarding beta-lactam drug interactions. After reading this chapter, the reader will recognize the clinical significance of drug-drug interactions associated with the beta-lactam antibiotics and understand the management of these drug-drug interactions.

1.2 Penicillin Drug Interactions

1.2.1 Acid-Suppressive Agents

The combination of various penicillins (ampicillin, amoxicillin, bacampicillin, amoxicillin/clavulanate) and H_2-receptor antagonists (cimetidine and ranitidine) and the proton pump inhibitors (omeprazole, esomeprazole, and lansoprazole) has

L. H. Danziger (✉) • K. S. Horn
Department of Pharmacy, Tufts Medical Center, Clinical Pharmacist Emergency Medicine,
Boston, MA, USA
e-mail: danziger@uic.edu

© Springer International Publishing AG, part of Springer Nature 2018
M. P. Pai et al. (eds.), *Drug Interactions in Infectious Diseases: Antimicrobial Drug Interactions*, Infectious Disease,
https://doi.org/10.1007/978-3-319-72416-4_1

1

been evaluated for effects on the bioavailability of different penicillins [1–9]. With the exception of bacampicillin, the bioavailability of the penicillins was unaffected. The area under the curve (AUC) of bacampicillin was reduced in the presence of food, ranitidine, and sodium bicarbonate [5]; however, another study did not demonstrate a difference in AUC with coadministration of omeprazole and bacampicillin [3]. The concurrent administration of most penicillins and acid-suppressive agents poses no problems except possibly with bacampicillin.

1.2.2 Allopurinol

An increased incidence of skin rash has been reported in patients receiving either ampicillin or amoxicillin concomitantly with allopurinol. In an analysis of data collected in 4686 patients receiving ampicillin, 252 of which were also receiving allopurinol, rash was reported in 5.9% of the patients receiving ampicillin compared to 13.9% of patients receiving both ampicillin and allopurinol ($p = 0.0000001$) [10]. There were no differences in age, sex, diagnosis, or admission laboratory value of blood urea nitrogen (BUN) that could be identified between the two groups. Another study identified 1324 patients who received ampicillin, with 67 also receiving allopurinol. The frequency of skin rashes was higher in the ampicillin-allopurinol combination than ampicillin alone (22.4% vs. 7.5%, $p < 0.00005$). The incidence of rash in those on allopurinol alone was 2.1% [11].

Fessel and colleagues attempted to explain why there is a higher incidence of rash in patients receiving allopurinol and ampicillin [12]. They compared the history of allergies to penicillin, allergies to other antibiotics, presence of hay fever, use of antihistamine medications, and the prevalence of asthma in 124 asymptomatic hyperuricemic individuals compared to 224 matched normouricemic controls. The following results were significant in asymptomatic hyperuricemic subjects versus the control subjects: history of penicillin allergy (14.1% versus 4.9%), hay fever (18.8% versus 8.0%), and use of antihistamine medications (9.9% versus 2.7%). The incidence of allergies to antibiotics excluding penicillin and prevalence of asthma were not significant between groups. The authors hypothesized that hyperuricemic individuals tend to have a higher frequency of allergic reactions; therefore, this altered immunologic state may explain the increased incidence of ampicillin rashes rather than an ampicillin-allopurinol interaction [12].

Fredj and colleagues reported a case of amoxicillin hypersensitivity 2 years after a drug rash with eosinophilia and systemic symptoms (DRESS) induced by allopurinol therapy [13]. Another case of erythema multiforme and allopurinol hypersensitivity syndrome in a patient on concomitant allopurinol and amoxicillin has also been reported [14].

The significance of this pharmacodynamic interaction tends to be minor. Clinicians may continue to prescribe these agents concomitantly. Patients should be monitored and counseled regarding this potential increased incidence of skin rashes when these two agents are prescribed concurrently. As with any patient who develops DRESS or another severe medication-induced rash, the clinician and patient should be cognizant of the potential for future medication-induced reactions.

1.2.3   Aminoglycosides

Penicillins and aminoglycosides are commonly used in combination to treat a variety of infections. However, concomitant use of the extended-spectrum penicillin antimicrobials may result in inactivation of the aminoglycosides. Henderson et al. reported on the in vitro inactivation of tobramycin, gentamicin, and netilmicin, when combined with azlocillin, carbenicillin, and mezlocillin in plasma samples incubated at 37° from 1 to 9 days. They noted that each of the penicillins studied decreased the concentrations of the aminoglycosides. The amount of aminoglycoside inactivation was related to temperature, contact time, and penicillin concentration [15]. Although the majority of interactions are reported in vitro, the potential for in vivo interactions is of concern, especially in those patients with end-stage renal failure [16–23].

1.2.3.1   In Vivo Aminoglycoside Inactivation

Animal models have demonstrated interactions between various penicillins and aminoglycosides. In bilaterally nephrectomized canines administered carbenicillin and tobramycin, gentamicin, and amikacin, the serum concentrations of all the aminoglycosides were decreased at 24 h and 7 days [24]. Additionally, it was noted that carbenicillin reduced serum half-lives of gentamicin and tobramycin by 40% ($P < 0.05$). In another study, a decrease in plasma concentrations as well as a variation in volume of distribution and half-life was shown with coadministration of piperacillin and netilmicin in rabbits, but renal accumulation and renal damage were similar between rabbits treated with the combination and only netilmicin [25]. McLauglin and Reeves noted that rabbits that received only gentamicin were reported to have normal gentamicin concentrations, while rabbits receiving carbenicillin and gentamicin had undetectable concentrations at 30 h [17].

Evidence for an interaction between penicillins and aminoglycosides in humans is primarily restricted to coadministration with extended-spectrum penicillins, particularly in patients with end-stage renal failure [17–23, 26, 27]. McLaughlin and Reeves reported their experience in two patients [17]. In the first patient, they reported undetectable gentamicin concentrations and clinical failure in a 12-year-old patient who received an infusion of carbenicillin and gentamicin for *Pseudomonas* bacteremia. The second patient was undergoing hemodialysis and receiving gentamicin for 8 days for the treatment of a soft tissue infection. Carbenicillin therapy was added on day 8. The authors reported that therapeutic serum concentrations for gentamicin could not be achieved, despite administration of high doses following the addition of carbenicillin. Of note, the patient received more frequent dialysis sessions during this period, which may have also contributed to subtherapeutic gentamicin concentrations. Uber et al. noted similar findings when tobramycin and piperacillin where administered concomitantly in a chronic hemodialysis patient [18]. Davies et al. evaluated gentamicin half-lives in the presence of therapeutic doses of ticarcillin or carbenicillin in eight patients with end-stage renal failure [20]. In patients receiving gentamicin concomitantly with ticarcillin, the gentamicin half-life was reduced from 31 to 22 h, whereas gentamicin half-life was reduced from 50 to 8 h in patients receiving carbenicillin and gentamicin. Blair et al. also documented a significant interaction between carbenicillin and gentamicin. The mean gentamicin serum half-life was significantly impacted by the presence of carbenicillin (18.4 ± 8.2 versus 61.6 ± 30.7 h, respectively) [22]. However, these authors reported that amikacin serum concentrations and clearance were not altered by concomitant carbenicillin administration. Lastly, Riff and Jackson reported on four patients on chronic hemodialysis receiving gentamicin and carbenicillin concomitantly, noting that the half-life of gentamicin was reduced by over 50% and that serum concentration was also reduced by 20–40% [23].

However, Halstenson et al. assessed the effect of piperacillin administration on the disposition of netilmicin and tobramycin in 12 chronic hemodialysis patients [19]. The half-life of netilmicin was not significantly altered when given concurrently with piperacillin. In comparison, the half-life of tobramycin was considerably reduced in the presence of piperacillin (59.62 ± 25.18 versus 24.71 ± 5.41 h). Lau et al. were unable to document any such drug-drug interaction between piperacillin and tobramycin in subjects with normal renal function (defined as creatinine clearances of greater than or equal to 60 mL/min) [28]. One report of healthy subjects who received 1 g of aztreonam both alone and combined with 0.5 g of amikacin showed no difference in overall exposure between monotherapy and combination therapy [29]. Hitt and colleagues reported no differences in pharmacokinetic parameters of once-daily gentamicin with the coadministration of several piperacillin-tazobactam regimens in subjects with normal renal function [30]. Similarly, Dowell et al. were unable to demonstrate differences in the pharmacokinetic parameters of tobramycin when administered alone or with piperacillin/tazobactam in subjects with moderate renal impairment (creatinine clearance between 40 and 59 mL/min), mild renal impairment (creatinine clearance between 20 and 39 mL/min), or normal renal function (creatinine clearance greater than 90 mL/min) [31].

Roberts and colleagues evaluated 18 healthy, cystic fibrosis patients administered either tobramycin alone or tobramycin plus ticarcillin. They noted that the clearance and volume of distribution of tobramycin increased by 13% ($P < 0.001$) and 14% ($P < 0.001$), respectively, with the coadministration of ticarcillin. They also noted that the concentration of tobramycin was decreased significantly (by 13%) when measured out to 330 min, in the presence of ticarcillin. The authors felt this difference was unlikely to be of any clinical significance [32].

It has been suggested that the extended-spectrum penicillins interact chemically with the aminoglycosides to form biologically inactive amides. The degree of inactivation is dependent on the specific aminoglycoside and beta-lactam used [20, 33]. In vivo inactivation of aminoglycosides occurs at such a slow rate that it appears to be clinically insignificant in patients with normal renal function [28, 33]. Some investigators have stated that this interaction could possibly be relevant for patients with renal failure who have high serum concentrations of penicillins [19, 20, 34]; therefore, close therapeutic monitoring of aminoglycosides is warranted in this specific clinical situation.

Concomitant administration of oral neomycin and penicillin VK has been reported to reduce serum concentrations of penicillin [31]. In healthy volunteers, penicillin VK concentrations decreased by over 50% following the administration of oral neomycin concomitantly with penicillin VK [31]. Due to the significant decrease in penicillin exposure, penicillin VK should not be administered to patients receiving oral neomycin.

1.2.3.2 In Vitro Aminoglycoside Inactivation

In vitro inactivation of aminoglycosides can be significant when these agents are prepared in the same intravenous mixture for administration [17, 23, 33]. Noone and Pattison showed that within 2 h of admixing at room temperature, an intravenous fluid mixture containing ampicillin (concentration equivalent to 12 g/d) and gentamicin resulted in a 50% decline in the gentamicin activity. After 24 h, no measurable gentamicin activity was noted [33]. An intravenous fluid mixture containing gentamicin and carbenicillin demonstrated a 50% reduction in activity between 8 and 12 h after admixing at room temperature. Aminoglycosides and penicillins should not be mixed together prior to infusion.

1.2.3.3 In Vitro Inactivation Aminoglycoside in Sampling Serum Concentrations

If high concentrations of penicillins are present in serum samples that are to be assayed for aminoglycoside concentrations, inactivation of the aminoglycosides by the penicillins can result in falsely decreased aminoglycoside concentrations [16, 35]. Penicillin concentration, period of time prior to sampling, and storage temperature of the sample are factors that affect the extent of inactivation [16]. When measuring aminoglycoside serum concentrations through intravenous tubing, one should flush 5–10 mL of either normal saline or 5% dextrose in water (based on drug compatibilities) through the tubing before withdrawing blood to minimize the amount of beta-lactam present in the intravenous tubing prior to sampling.

1.2.3.4 Aminoglycosides: Synergy

The concomitant use of beta-lactam and aminoglycoside antimicrobials has been described as synergistic for several Gram-positive and Gram-negative organisms [36–39]. By inhibiting the cell-wall synthesis, beta-lactams increase the porosity of the bacterial cell wall resulting in greater aminoglycoside penetration and access to target ribosomes [40].

The use of penicillin or ampicillin in combination with an aminoglycoside has been documented to be advantageous in the treatment of streptococcal and enterococcal infections [41–47]. As a result of increased efficacy with combination therapy, many severe streptococcal and enterococcal infections are routinely treated with penicillin or ampicillin plus an aminoglycoside [46].

Despite the well-documented in vitro synergy between beta-lactams and aminoglycosides, limited clinical data are available supporting superior efficacy of synergistic versus nonsynergistic combinations for the treatment of Gram-negative infections. Anderson et al. retrospectively evaluated Gram-negative bacteremias to determine if the treatment with one or two antimicrobials effected outcome and whether in vitro synergy correlated with superior efficacy [48]. Of the 173 patients treated with two drugs, the clinical response rate was 83% in patients who received synergistic versus 64% with nonsynergistic antimicrobial regimens ($p < 0.05$). The use of synergistic antimicrobial combinations (aminoglycoside plus ampicillin or carbenicillin) was associated with better clinical response in patients with neutropenia ($p < 0.001$), shock ($p < 0.001$), *Pseudomonas aeruginosa* bacteremias ($p < 0.05$), and "rapidly or ultimately fatal" conditions ($p < 0.005$). However, the data from several meta-analyses do not support the use of concomitant antimicrobial therapy for definitive treatment of Gram-negative bacterial infections [49]. In critically ill patients with severe sepsis associated with Gram-negative bacteremia, the combination of an extended-spectrum penicillin and aminoglycoside is a reasonable therapeutic approach [49].

1.2.4 Anticoagulants

1.2.4.1 Heparin

A number of case reports have suggested that parenteral penicillins in combination with heparin have caused coagulopathies [50–56] and may predispose patients to clinically significant bleeding [53–55, 57]. The exact mechanism of this interaction is unknown but may be a result of a direct effect on platelet function by penicillins, which may have an additive anticoagulant effect when combined with heparin [51, 52, 57].

Wisloff and colleagues evaluated the bleeding time of patients receiving heparin and penicillins compared to heparin alone [56]. Fifty patients were placed on heparin (5000 IU subcutaneously for 7 days) following an elective vascular surgery procedure and were also randomized to receive a combination of ampicillin and cloxacillin or no antibiotics. The patients that were receiving heparin along with the penicillins had a slightly longer bleeding time; however, this was still within an acceptable range in most cases.

Since patients receiving heparin are routinely monitored closely for coagulopathies and clinically significant bleeding, the potential interaction between these two drugs does not warrant further precautions.

1.2.4.2 Warfarin

A decreased anticoagulant effect for warfarin has been documented when given concomitantly with nafcillin [58–63], dicloxacillin [58, 64, 65], cloxacillin [66], and flucloxacillin [67, 68]. This interaction can be significant, necessitating up to a two- to fourfold increase in warfarin dose during concomitant therapy. In addition to decreasing INR levels, cloxacillin has also been described as increasing INR in a patient on chronic warfarin therapy [69].

It has been postulated that these antibiotics induce the cytochrome P450 system and may increase the metabolism of warfarin [60, 63, 70, 71]. Another possible explanation may involve the ability of these highly protein-bound agents to displace warfarin. However, Qureshi et al. performed an in vitro study and demonstrated that nafcillin did not affect the protein binding of warfarin [60]. Cropp and Busey reported that the usual onset of this interaction between nafcillin and warfarin is within 1 week after initiation of nafcillin therapy and with warfarin requirements returning to baseline usually within 4 weeks after the discontinuation of the nafcillin [72].

Krstenansky et al. studied the effect of dicloxacillin in seven patients stabilized on warfarin therapy [64]. Prothrombin times (PTs) were obtained prior to treatment and on days 1, 3, 6, and 7 of dicloxacillin administration. A decrease in the PT was observed in all patients on day 6 or 7 compared to baseline PT values. The decrease in PT ranged from 0.3 to 5.6 s (mean ± SD of −1.9 ± 1.8 s) and was statistically significant ($p < 0.05$). This interaction is described in case reports for patients being treated with dicloxacillin and warfarin [58, 69, 73]. Similar to nafcillin and warfarin, the effects of the interaction on international normalized ratio (INR) often last for up to 3 weeks after discontinuation of dicloxacillin.

Brown and colleagues presented a case report of a patient on warfarin 2.5 mg daily who developed an increased hypoprothrombinemia response after receiving high-dose intravenous penicillin (24 million units/day). Upon withdrawal of the penicillin, the patient's prothrombin time subsequently returned to his baseline [74].

Davydov et al. reported a case of a 58-year-old woman, in which warfarin interacted with amoxicillin/clavulanate resulting in an elevated international normalized ratio (INR) and hematuria [75]. More recently, amoxicillin, amoxicillin-clavulanate, and cloxacillin have all been implicated in case reports as interacting with warfarin to increase INR [76, 77]. It is important to note that some of these reports also describe clinically significant bleeding that occurred as a result of this interaction [75, 77].

Although the exact mechanism of this interaction remains unknown, it has been proposed that broad-spectrum antibiotic use may lead to a decrease in vitamin K-producing bacteria within the gastrointestinal tract. This may then result in a vitamin K-deficient state (especially in patients with low dietary intake of vitamin K) potentially leading to an increased effect of warfarin. Clinicians should be aware of the potential interaction between penicillins and oral anticoagulants and monitor the PT and INR in patients receiving these agents concurrently.

1.2.4.3   Direct Oral Anticoagulants

To date no studies have been published regarding drug interactions between the penicillin antibiotics and the direct oral anticoagulants (DOACs; dabigatran, rivaroxaban, apixaban, and edoxaban). However, Lippi et al. described that a review of the Web site of eHealthMe reported cerebral hemorrhage in two patients, rectal hemorrhage in six patients, and ten episodes of hematemesis in patients receiving amoxicillin and dabigatran concomitantly [78]. According to these reports, it would seem prudent that patients concomitantly administered amoxicillin and the DOACs require close follow-up.

1.2.4.4   Aspirin

Large doses of aspirin may increase the serum concentrations and half-lives of penicillin, oxacillin, nafcillin, cloxacillin, and dicloxacillin when administered concurrently [79, 80]. Eleven patients with arteriosclerotic disorders received penicillin G before and after high doses of aspirin (3 g/d) [79]. During aspirin administration, penicillin half-life increased from 44.5 ± 15.8 m to 72.4 ± 35.9 m ($p < 0.05$) [79]. The mechanism of this interaction remains unknown. Some have speculated that this interaction may occur as a result of aspirin displacing penicillin from protein-binding sites or of aspirin competing with penicillins for the renal tubular secretory proteins [79–83]. Avoidance of this combination is unnecessary.

1.2.5   *Beta-Adrenergic Blockers*

Coadministration of ampicillin and atenolol may lead to a decrease in the serum concentration of atenolol. In a crossover study, six healthy subjects were orally administered with 100 mg atenolol alone and with 1 g ampicillin. Atenolol pharmacokinetics were assessed after a single dose and after reaching steady state. These subjects previously received intravenous atenolol in another study, which was utilized to determine oral bioavailability in the present study. The bioavailability of atenolol was reduced from 60% (atenolol alone) to 36% (single-dose atenolol and ampicillin, $p < 0.01$) to 24% (steady-state concentrations of atenolol and ampicillin, $p < 0.01$) [84]. Other atenolol pharmacokinetic parameter values for AUC, Cmax, and mean steady-state concentrations were also significantly reduced ($p < 0.01$).

Despite the differences in atenolol serum concentration, blood pressure measurements did not differ between the groups over a 4-week treatment period.

McLean and colleagues also performed a crossover study administering oral atenolol and ampicillin to six volunteers [85]. Unlike the previous study, these investigators dosed ampicillin at clinically applicable doses of 250 mg four times a day, as well as higher doses of 1 g. The mean reduction of AUC was lower in the former dosing regimen compared to the latter one (18.2% versus 51.5%).

Although the clinical significance of this interaction is questionable, it would seem reasonable that patients should be monitored for this interaction when higher doses of ampicillin are used, especially in the presence of renal dysfunction; however, no empiric dosage alterations are recommended at this time.

1.2.6 Calcium Channel Blockers

Nifedipine appears to increase the bioavailability of amoxicillin by facilitating its active transport mechanism within the gastrointestinal tract [86]. In a randomized crossover study conducted in eight healthy volunteers, each subject received 1 g oral amoxicillin with 20 mg nifedipine or placebo. The absolute bioavailability of amoxicillin was noted to increase from 65.25% to 79.2% with the addition of nifedipine ($p < 0.01$) [86]. The AUC also increased from 29.7 ± 5.3 mg $\cdot$ h/L (amoxicillin alone) compared to 36.26 ± 6.9 mg $\cdot$ h/L (amoxicillin and nifedipine) ($p < 0.01$). Since no adverse events were associated with the alterations of these pharmacokinetic parameters, no dosage adjustments are recommended.

Nafcillin has been postulated to enhance the elimination of agents metabolized through the cytochrome P450 system [63, 70]. A crossover study was conducted to evaluate the induction potential of nafcillin on nifedipine, a substrate of the cytochrome P450 3A4 enzyme [71]. Healthy volunteers were randomized to receive 5 days of oral nafcillin (500 mg four times daily) or placebo, which was followed by a single dose of nifedipine. The subjects who received nafcillin along with nifedipine were found to have a significant reduction in the nifedipine $AUC_{0-\infty}$ (80.9 ± 32.9 $\mu g \cdot$ l/h versus 216.4 $\mu g \cdot$ l/h; $p < 0.001$) and enhanced plasma clearance (138.5 ± 42.0 l/h versus 56.5 ± 32.0 l/h; $p < 0.002$) compared to the nifedipine-placebo group. Due to the limited available data, the clinical significance of this interaction is unknown.

In an animal model, the total exposure of amlodipine was increased by 42 and 133% greater when it was coadministered with ampicillin at doses of 10 mg/kg and 20 mg/kg, respectively, compared to amlodipine alone ($p < 0.001$). The authors postulated that ampicillin (and possibly other antibiotics) increases amlodipine exposure by suppression of gut microbes responsible for metabolism of amlodipine [87].

1.2.7   Chloramphenicol

The administration of a bacteriostatic agent such as chloramphenicol may antagonize the bactericidal activity of beta-lactam antimicrobials [88, 89]. Beta-lactam
antimicrobials exhibit their bactericidal effect by binding to penicillin-binding proteins and inhibiting bacterial cell-wall synthesis. For beta-lactams to exert optimal
bactericidal effects, bacteria should be actively growing and dividing. However,
bacteriostatic agents such as chloramphenicol, which may inhibit protein synthesis,
may interfere with the bactericidal activity of penicillins.

In vitro studies have demonstrated the concomitant of penicillins and chloramphenicol to be antagonistic, particularly in the setting of chloramphenicol-resistant, ampicillin-susceptible strains [88, 90, 91]. Although Solberg and Andersen
did not find antagonism with the combination of penicillin and chloramphenicol
against meningococci in vitro [92], Asmar and Dajani found that the combination
of ampicillin and chloramphenicol could be synergistic if the chloramphenicol was
bactericidal against the organism, but antagonistic if chloramphenicol was bacteriostatic [93]. Human data do not support these findings [94, 95]. Patients with
gonococcal infections treated with a combination of penicillin and chloramphenicol had better clinical outcomes than patients treated with penicillin alone [94].
Superior outcomes were also reported among patients infected with typhoid fever
who were treated with chloramphenicol plus ampicillin compared to chloramphenicol alone [95].

Relevant clinical information is limited for this drug-drug interaction. Since the
in vivo and in vitro data concerning this interaction are contradictory, it seems prudent to avoid the concurrent use of these antimicrobials.

1.2.8   Chloroquine

Investigators conducted a study in healthy volunteers to evaluate the coadministration of chloroquine and ampicillin on the pharmacokinetics of ampicillin [96].
Ampicillin pharmacokinetics alone or in the presence of chloroquine was determined by characterizing the drug's renal elimination. The mean percent of dose
excreted was 29% for ampicillin alone versus 19% for the ampicillin/chloroquine
combination ($p < 0.005$). The coadministration of ampicillin and chloroquine
resulted in a significant reduction in ampicillin bioavailability, but not in time of
maximal excretion [96]. Based on limited data, coadministration of these agents
may lead to a reduction in ampicillin concentrations. The antimicrobial effects of
the combination of chloroquine with either penicillin G or penicillin VK on
Staphylococcus aureus were evaluated [97]. Antagonistic microbiological effects
were observed with chloroquine when combined with these two penicillins. The
clinical significance of these interactions remains unknown, but avoidance of the
combination appears to be unnecessary.

1.2.9 Ciprofloxacin

Several in vitro studies examining penicillins (piperacillin, azlocillin) coadminis-tered with ciprofloxacin found an antagonistic effect when the concentrations of the antibiotics were at less than one-fourth of the MIC for the organism studied [98, 99]. Smith and Eng found no synergy or antagonism with the combination of ampicillin and ciprofloxacin against *Enterococcus faecalis* [100]. Fuursted and Gerner-Smith found a synergistic effect with the combination of piperacillin and ciprofloxacin against *Pseudomonas aeruginosa* [101]. Overbeek and colleagues did not find syn-ergism or antagonism with the combination of ciprofloxacin and azlocillin against *Acinetobacter anitratum* [102]. Smith and colleagues found synergy with the com-bination of levofloxacin and ampicillin against selected isolates of *E. faecium* [103].

Interactions between the penicillins and fluoroquinolones have been rarely docu-mented in humans [104, 105]. Barriere and colleagues assessed the effect of the concurrent administration of ciprofloxacin and azlocillin in a crossover trial [104]. Six subjects were administered single doses of ciprofloxacin and azlocillin alone and in combination. Similar pharmacokinetic profiles were noted with azlocillin; however, when coadministered with azlocillin, a statistically significant reduction in total clearance and renal clearance of ciprofloxacin was noted. Orlando and col-leagues administered ciprofloxacin and azlocillin both in combination and alone to healthy volunteers and measured the area under the bactericidal titer curve to com-pare agents. They found an additive effect for the combination of ciprofloxacin and azlocillin for *Escherichia coli*, *Staphylococcus aureus*, and *Klebsiella pneumoniae*, but not for *Serratia marcescens* [106]. Based on limited data, coadministration of these agents need not be avoided.

1.2.10 Contraceptives: Oral Estrogen

Several case reports of breakthrough bleeding and pregnancies have been reported in patients receiving oral contraceptives and antibiotics concomitantly [107–111]. It has been postulated that antibiotics interfere with the enterohepatic circulation of oral estrogens, resulting in subtherapeutic estrogen concentrations [109–111]. After oral estrogens are absorbed, they undergo hepatic metabolism to glucuronide and sulfate conjugates and are excreted into the bile. Bacteria residing in the gut hydrolyze the conjugates to active drug, which is then reabsorbed by the body [109]. The proposed mechanism of this interaction involves the ability of antibiot-ics to destroy the gut bacteria that are required to hydrolyze the conjugated estro-gen to their active form.

Studies in animal models assessing this interaction have shown mixed results [112, 113]. One investigation demonstrated no alterations in the pharmacokinetics of ethinylestradiol when administered with ampicillin [112]. Another study found differences in both AUC and plasma clearance in the group that received antibiotics compared to ethinylestradiol alone [113].

Several studies have been performed in humans to determine if the case reports and animal data represent significant findings [114–116]. Freidman and colleagues prospectively evaluated the serum concentrations of gonadotropins and other hormones in 11 volunteers receiving Demulen® (50 μg of ethinylestradiol and 1 mg of ethynodiol diacetate) plus ampicillin or placebo during two consecutive menstrual cycles [115]. Progesterone concentrations were similar between the Demulen-ampicillin and Demulen-placebo groups. Follicle-stimulating hormone and luteinizing hormone appeared to be similar between the two groups. None of the 11 patients underwent ovulation. Freidman and colleagues concluded that ampicillin should not reduce the effectiveness of Demulen. Other researchers have criticized the results of this study because of its study design which included a small number of subjects, a short duration of antimicrobial therapy, and a relatively high dose of estrogens (present in Demulen) [110].

Back and colleagues evaluated seven women receiving oral contraceptives for at least 3 months (all containing ≥30 μg of ethinyloestradiol) who presented to their clinic with an infection that required the administration of ampicillin of 8 d duration [114]. Blood samples were taken during concomitant oral estrogen and ampicillin therapy and during the next menstrual cycle without ampicillin. Six female volunteers receiving only oral contraceptives for at least 3 months were similarly evaluated for the potential drug interaction. Plasma concentrations of ethinyloestradiol, levonorgestrel, follicle-stimulating hormone, and progesterone were not significantly different between the two groups (oral contraceptive-ampicillin versus oral contraceptive alone). Despite the fact that a lower concentration of ethinyloestradiol was seen with two women on ampicillin, the authors concluded that alternative methods of protection are not necessary in most women [114].

Another study in volunteers analyzed the effect of administering ampicillin or metronidazole with an oral contraceptive preparation [116]. This summary will be limited to the group using ampicillin ($n = 6$). Subjects initially received a low-dose oral contraceptive (1 mg norethisterone acetate and 30 μg ethinylestradiol). On days 6 and 7, plasma concentrations of ethinylestradiol and norethisterone were obtained. Subsequently, subjects were administered ampicillin (500 mg twice daily orally for 5–7 days) and the contraceptive steroid. Following antibiotic treatment, serum hormones, ampicillin, and progesterone concentrations were measured in the subjects. The concentrations of norethisterone and ethinylestradiol were not altered in the presence of ampicillin, and progesterone concentrations were in the appropriate range to suppress ovulation [116].

It is difficult to determine the clinical significance of this interaction because of the small number of clinical trials, small numbers of patients, minimal number of case reports, and the limited number of oral contraceptives studied. A review article by Weisberg suggests that the possibility of a clinically significant interaction between antibiotics and oral contraceptives is likely less than 1% [117]. The author states that women with a greater extent of enterohepatic circulation, previous breakthrough bleeding, or contraceptive failure may have a higher risk for this interaction

[117]. More recently, Dickinson et al. reviewed this literature from 1969 through 1999. They concluded that although a rare occurrence, certain penicillins may affect plasma ethinylestradiol concentration in some women. Given the serious nature of an unexpected pregnancy, they advised that women should consider other protective measures while taking these antibiotics [118].

Although clinical trials have not been able to demonstrate any consistent interaction between oral contraceptives and antibiotics due to the potential risk of contraceptive failure, clinicians should still counsel patients on this potential interaction and suggest alternative method(s) of contraception if antimicrobial therapy is necessary.

1.2.11 Cyclosporine

Although nafcillin is not well-established as an inducer of the cytochrome P450 system, there have been several reports to suggest that nafcillin may reduce the serum concentrations of cyclosporine via induction of the cytochrome P450 system [119]. On two separate occasions, a 34-year-old woman, status post-renal transplant, experienced a reduction in cyclosporine serum concentration following nafcillin administration [119]. The patient received 2 g nafcillin intravenously every 6 h for a positive culture of methicillin-susceptible *S. aureus* from a perinephric abscess. Upon admission, the patient was receiving 400 mg cyclosporine daily with a corresponding trough serum concentration of 229 ng/mL. After initiation of nafcillin, her cyclosporine concentrations decreased to 119 ng/mL and 68 ng/mL on days 3 and 7 of nafcillin, respectively, despite stable daily doses of 400 mg of cyclosporine. Upon discontinuation of nafcillin, trough serum concentrations of cyclosporine increased to 141 ng/mL and 205 ng/mL on days 2 and 4 without nafcillin therapy, respectively. No change in renal or hepatic function was noted throughout this entire treatment period. The second cyclosporine-nafcillin interaction occurred when the patient was later readmitted for drainage of retroperitoneal fluid collection. The patient experienced a similar decline in cyclosporine concentrations during concomitant therapy and subsequent increases in cyclosporine concentrations following discontinuation of nafcillin. Based on the findings of this case report, cyclosporine concentrations should be closely monitored during concomitant nafcillin administration.

In addition to this case report, Jahansouz and colleagues compared nine patients immediately post lung transplant who received nafcillin for 1 week with ten patients who were post lung transplant who did not receive nafcillin. All patients received cyclosporine for immunosuppression. Despite no differences in cyclosporine levels between the two groups, the patients who received nafcillin had significantly more renal dysfunction (presumed to be cyclosporine-induced) and more viral infections than those who did not receive nafcillin. There was no difference in mortality or graft survival. The authors concluded that an alternative agent to nafcillin should be considered in lung transplant recipients receiving cyclosporine in the early post-transplant period [120].

There is not sufficient evidence or a plausible mechanism established to recommend avoidance of nafcillin and cyclosporine coadministration. However, as a precaution, cyclosporine levels should be monitored and adjusted as necessary while on nafcillin therapy.

1.2.12   Macrolides

1.2.12.1   Azithromycin

Only one in vitro study examined the coadministration of azithromycin and amoxicillin. This investigation involved the creation of a biofilm model to mimic subgingival plaque and multiple antibiotics, both alone and in combination. The combination of amoxicillin and azithromycin did not result in synergism in this model [121].

1.2.12.2   Erythromycin

The concurrent administration of erythromycin and penicillin may result in antagonism, synergy, or no effect (indifference) on the antibacterial activity of penicillin. Beta-lactams exert their cidal effects on bacteria by binding to penicillin-binding proteins and inhibiting cell-wall synthesis. For beta-lactams to exercise their optimal bactericidal activity, bacteria should be actively growing and dividing; therefore, erythromycin can interfere with the bactericidal activity of penicillin by inhibiting protein synthesis.

In vitro studies have demonstrated varied results for the concomitant administration of penicillin and erythromycin [122–140]. These differences may be due to such factors as the specific microorganism involved, susceptibility patterns to both agents, antibiotic concentrations, the inoculum effect, and time of incubation [122, 124, 126, 128, 130, 134]. Similar to the disparate results demonstrated in vitro, case reports have shown penicillin and erythromycin antagonism in the treatment of scarlatina [131] and *Streptococcus bovis* septicemia [133], whereas clinical improvement has been reported with the concurrent use of ampicillin and erythromycin in the treatment of pulmonary nocardiosis [132]. Although there has been concern about the use of the combination of beta-lactams and macrolides because of the possibility of antagonism, they have recently gained favor for the treatment of community-acquired pneumonia in the hospitalized patient. Several recently published studies found that patients with bacteremic pneumococcal pneumonia treated with a beta-lactam plus a macrolide had a lower mortality rate compared to those treated with a single agent [135–137]. As such, treatment guidelines for community-acquired pneumonia recommend a beta-lactam and macrolide as a preferred treatment option for hospitalized patients [138]. As evident from these clinical reports and in vitro testing, the antagonism risk between beta-lactams and macrolides appears to be minimal.

1.2.13 Guar Gum

Guar gum, which may be utilized as a food additive, has been reported to reduce serum concentrations of phenoxymethylpenicillin [141]. In a double-blind study, ten healthy volunteers received guar gum or placebo granules along with 3 mu of phenoxymethylpenicillin. The peak penicillin concentration decreased significantly from 7560 ± 1720 ng/mL to 5680 ± 1390 ng/mL ($p < 0.01$) when administered with placebo compared to guar gum. The $AUC_{0-6 h}$ of penicillin decreased significantly from $14,500 \pm 1860$ to $10,380 \pm 2720$ ng/mL · hr. ($p < 0.001$) when administered with guar gum. The time to peak concentration was not altered significantly. As a result of the significant decrease in the peak serum concentrations and $AUC_{0-6 h}$, phenoxymethylpenicillin should not be administered concomitantly with guar gum.

1.2.14 Interferon-Gamma

Recent data suggest that penicillin may interact with a variety of cytokines by conjugating these biological proteins [142]. Benzylpenicillin has been shown to conjugate IFN-gamma, IL-1beta, IL-2, IL-5, IL-13, and TNF-alpha; however, based on a series of in vitro experiments, benzylpenicillin only appears to alter the biologic activity of IFN-gamma [142]. Using an in vitro bioassay, Brooks and colleagues noted that benzylpenicillin inhibited the ability of IFN-gamma to induce CD54 expression on epithelial cells. Additional preclinical studies suggest that other regulatory functions of IFN-gamma may also be modulated by benzylpenicillin [142]. Because IFN-gamma promotes Th1 responses and inhibits Th2- and IgE-mediated responses, disruption of IFN-gamma activity by benzylpenicillin may result in clinically significant immunomodulatory effects, which promote allergy.

1.2.15 Khat

The chewing of khat (a natural substance obtained from shrubs grown in East Africa and Yemen) may reduce the bioavailability of ampicillin and amoxicillin [143]. In a crossover design, eight healthy adult male Yemeni subjects received ampicillin or amoxicillin under various conditions of khat chewing [143]. The urinary excretion method was utilized to determine the bioavailabilities of ampicillin and amoxicillin under the following conditions: antibiotic alone, 2 h before khat chewing, immediately prior to khat chewing, immediately prior to khat chewing with a meal, midway through khat chewing, and 2 h after khat chewing. The bioavailability of ampicillin (measured by percentage of ampicillin excreted unchanged in the urine, peak excretion, and time to peak excretion) was significantly decreased during all conditions except when administered 2 h after khat chewing. In contrast, amoxicillin's bioavailability was only affected when amoxicillin was taken midway through khat chewing. Considering the limited use of khat in the developed countries, this should not be considered a clinically relevant drug-drug interaction. However, if ampicillin and amoxicillin are administered to an individual using khat, these agents should be taken at least 2 h following khat chewing.

1.2.16   Methotrexate

Weak organic acids such as penicillins can compete with methotrexate for renal tubular secretion [144, 145] and reduce the renal elimination of methotrexate. Various studies in rabbits have demonstrated a reduction in the renal clearance of methotrexate and 7-hydroxymethotrexate [144–146]. One of the studies demonstrated nearly 50% reduction in methotrexate (MTX) clearance when piperacillin was administered 10 min before and 4 h after a single dose of MTX ($p \leq 0.05$) [145]. The AUC of MTX and its 7-hydroxymethotrexate metabolite also differed significantly from the control ($p \leq 0.05$).

Despite the rather significant results reported from animal studies, few case reports have documented this potential interaction [147–152]. Bloom and colleagues reported four cases in which the administration of various penicillins concomitantly with MTX resulted in the decreased clearance of methotrexate [148]. Methotrexate clearance before and after the addition of the following antimicrobial agents is as follows: penicillin, 2.8 L/h versus 1.8 L/h; piperacillin, 11 L/h versus 3.6 L/h; ticarcillin, 5.8 L/h versus 2.3 L/h; and dicloxacillin/indomethacin, 6.4 L/h versus 0.45 L/h, respectively. Due to reduction in clearance, these patients required an extended leucovorin rescue. Titier et al. published a case report describing severe methotrexate toxicity following the concomitant administration of high-dose methotrexate and oxacillin, which lead to a series of complications and ultimately the death of the patient [152]. Mayall and colleagues performed a retrospective chart review of patients at their institution who received low-dose methotrexate and developed neutropenia ($< 1 \times 10^9$ cell/L). Of the five patients identified, four were also being administered penicillins, which the authors postulated could have interfered with methotrexate excretion, leading to neutropenia [153]. More recently, Zarychanski and colleagues reported the interaction of piperacillin/tazobactam with methotrexate resulting in prolonged toxic concentrations of methotrexate [154]. In contrast, Herrick and colleagues reported no differences in renal clearance of methotrexate administered alone or with flucloxacillin in ten patients with rheumatoid arthritis [155].

Avoiding the concomitant use of penicillins and methotrexate is justified to avoid potential toxicity. If the concomitant administration of penicillins and methotrexate is necessary, close monitoring of methotrexate concentrations and signs of toxicity is suggested.

1.2.17 Oseltamivir

A pharmacokinetic study conducted in healthy volunteers evaluated the concurrent administration of oseltamivir (a prodrug) and amoxicillin [156]. No differences in the pharmacokinetic parameters of oseltamivir's active metabolite, Ro 64–0802, were noted when administered alone compared to coadministration with amoxicillin. Also, no pharmacokinetic differences were noted for amoxicillin with or without the administration of oseltamivir [156]. Based upon these finding, oseltamivir may be prescribed with amoxicillin.

1.2.18 Phenytoin

Highly protein-bound antibiotics such as nafcillin and oxacillin (both approximately 90% bound to plasma proteins) [80, 157] have the potential to interact with other highly protein-bound agents such as phenytoin [158, 159]. Due to drug displacement from protein-binding sites, high doses of nafcillin or oxacillin may increase unbound concentrations of phenytoin in certain patient populations [158, 159].

Dasgupta et al. conducted an in vitro study to determine the potential drug interaction between oxacillin and phenytoin [158]. Serum was collected from three separate patient populations (A, B, and C). Serum for group A was collected from healthy patients receiving phenytoin. Serums for group B and C were obtained from hypoalbuminemic and hyperuremic individuals, respectively. Subjects in these latter two groups were not receiving phenytoin; therefore, the serum was supplemented with phenytoin. Each group was tested for total and unbound phenytoin concentrations with and without 15 μg/mL or 50 μg/mL of oxacillin, which represented estimated peak oxacillin concentrations following a 500 mg oral dose and a 1 g intravenous dose, respectively. Serum from group A showed no statistical difference in unbound phenytoin concentrations with 15 μg/mL oxacillin; however, a significantly higher unbound phenytoin concentration with 50 μg/mL of oxacillin was observed when compared to serum not containing oxacillin (1.67 μg/mL versus 1.47 μg/mL) ($p < 0.05$). Serum from subjects in groups B and C also demonstrated a statistically significant increase in unbound phenytoin concentrations for both oxacillin concentrations compared to the group without oxacillin.

Dasgupta and colleagues performed another study to determine the potential effect of nafcillin on unbound phenytoin concentrations [159]. The study consisted of both in vitro and in vivo components. The authors observed both in vitro and in vivo displacement of phenytoin with the addition of nafcillin to serum. Although increases in unbound phenytoin appeared to be minor for the in vitro portion of the experiment, a significant increase in unbound phenytoin concentrations was noted in all groups compared to the control group ($p < 0.05$). Unbound phenytoin concentrations were also measured in four patients receiving phenytoin and nafcillin concurrently [159]. The investigators obtained unbound phenytoin concentrations during and after nafcillin therapy. Unbound phenytoin concentrations decreased following the discontinuation of nafcillin, although baseline phenytoin concentrations were not obtained.

Patients receiving antimicrobials with a high percentage of protein binding (90% or greater) and concomitant phenytoin should be monitored closely for signs of phenytoin toxicity. Furthermore, patients receiving high doses of any penicillin should have their unbound and total phenytoin concentrations monitored closely. Phenytoin dosage adjustments should be made according to extent of the interaction.

1.2.19 Probenecid

The interaction of probenecid and penicillins (weak organic acids) occurs primarily as a result of the inhibition of the tubular secretion of penicillin, although other mechanisms may be possible as well [160, 161]. The decrease in renal elimination results in increased penicillin serum concentrations. Studies have shown that the AUCs of amoxicillin, ampicillin, ticarcillin, and nafcillin may increase by approximately 50–100% when coadministered with probenecid [79, 161–166]. Other beta-lactams such as penicillin and dicloxacillin have also demonstrated increased serum concentrations in the presence of probenecid [79, 161, 163, 167–169]. Although probenecid significantly effects renal clearance of piperacillin/tazobactam, it does not significantly affect area under the curve or half-life of piperacillin/tazobactam [170]. An in vitro investigation with the effect of probenecid and benzylpenicillin against *Neisseria gonorrhoeae* showed enhanced effect with the combination compared to either agent alone, hinting at a mechanism beyond inhibition of tubular secretion of penicillin [171].

This drug-drug interaction may be clinically beneficial in certain situations in which higher penicillin serum concentrations are necessary especially when using oral agents. However, careful monitoring or avoidance of this combination should be considered in certain patient populations in whom drug accumulation may occur (e.g., elderly patients or patients with impaired renal function).

1.2.20 Proguanil

Babalola CP et al. conducted a study in healthy volunteers to evaluate the coadministration of proguanil and cloxacillin on the pharmacokinetics of cloxacillin [172]. Differences in pharmacokinetic parameter values for cloxacillin alone or in the presence of proguanil were determined by assaying urinary samples. Both the maximum excretion rate and total amount of excreted unchanged cloxacillin were reduced by approximately 50% when taken with proguanil compared to proguanil alone ($p < 0.0001$). No differences were noted in cloxacillin half-life or Tmax. The authors suggest that separating these two agents by 1–2 h may avoid this potential interaction.

1.2.21 Sulfonamides

The concurrent administration of penicillins and sulfonamides was evaluated in a pharmacokinetic study [80]. The unbound concentrations of penicillin G, penicillin V, nafcillin, and dicloxacillin were increased with the concurrent administration of several sulfonamides. The researcher postulated that this interaction occurred as a result of the displacement of penicillins from protein-binding sites [80]. In a separate study, Kunin reported that the coadministration of oral oxacillin and sulfonamides caused a decrease in oxacillin serum concentrations. The author postulated that perhaps the sulfonamides may cause reduced absorption of oral oxacillin; however, additional mechanisms cannot be ruled out [80]. Based on this limited clinical data, avoidance of penicillins and sulfonamides is not warranted.

1.2.22 Tetracyclines

As previously stated, the administration of a bacteriostatic agent, such as tetracycline or related compounds, may antagonize the bactericidal activity of beta-lactams. Nonetheless, both antagonism and synergy between penicillins and tetracyclines have been documented in in vitro and in vivo studies [173–178].

Lepper and Dowling reported the outcome of 57 patients diagnosed with pneumococcal meningitis who were treated with high-dose penicillin ($n = 43$) or high-dose penicillin along with the tetracycline antibiotic, aureomycin ($n = 14$) [178]. Although the severity of illness appeared similar between the treatment groups, mortality rates were significantly higher in the patients who received combination therapy compared to penicillin alone (79% versus 30%). Olsson and colleagues also noted a trend toward increased mortality in patients with pneumococcal meningitis treated with penicillin in combination with a tetracycline derivative (85%; $n = 7$) versus penicillin alone (52%; $n = 23$) or erythromycin alone (50%; $n = 6$) [179]. Strom noted that treatment of hemolytic streptococci with penicillin in combination with chlortetracycline compared to penicillin alone had similar initial clinical response but the penicillin/chlortetracycline group experienced a higher incidence of reinfection [180].

Unlike the case studies involving meningitis, Ahern and Kirby reported similar clinical outcomes in patients treated with penicillin alone versus penicillin in combination with aureomycin for pneumococci pneumonia [181]. The authors suggested that the role of rapid, bactericidal activity of penicillin is of more clinical significance in treating meningitis compared to less severe infections such as pneumonia. Ahern and Kirby stressed the importance of penicillin's role in treating meningitis due to the relatively limited phagocytic activity in the subarachnoid space compared to nonmeningeal infections such as pneumonia [181].

As a result of increasing antibiotic resistance, the eradication rates of *Helicobacter pylori* infections have been declining. This decrease in efficacy has led to the search for new therapies. Although there are precautions of using penicillins and tetracyclines concomitantly, recent research has shown that such combinations are effective in the treatment of *Helicobacter pylori* infections [182, 183].

Avoiding the combination of penicillin and tetracycline derivatives appears appropriate in severe infections requiring rapid bactericidal activity such as meningitis. In less severe infections, the use of these drugs in combination has not been documented to adversely affect outcomes.

1.2.23   *Vancomycin*

An interesting observation that has emerged in clinical practice is the possible interaction with either piperacillin or piperacillin-tazobactam when administered with vancomycin resulting in acute kidney injury (AKI). Several studies, mostly retrospective, have documented this increased incidence in AKI, in patients receiving concomitant vancomycin and piperacillin-tazobactam compared to vancomycin alone [184–187], or vancomycin administered concomitantly with other beta-lactams, such as cefepime [188–191] or meropenem [189]. This literature has predominantly evaluated the combination of piperacillin-tazobactam and vancomycin interaction. Recently, Rutter and Burgess reported the results of a large retrospective cohort study comparing the incidence of AKI in patients receiving either piperacillin-tazobactam or ampicillin-sulbactam [192]. They reported that AKI occurred at similar rates for both groups (265 patients per group: piperacillin-tazobactam 11.4% vs ampicillin-sulbactam 9.2%; $p = 0.14$). They did note, however, that adding vancomycin to piperacillin-tazobactam increased the probability of AKI compared to piperacillin-tazobactam alone (adjusted OR 1.77, 95% CI 1.26–2.46). Karino and colleagues reported several possible independent risk factors for development of AKI in patients receiving vancomycin and piperacillin-tazobactam. These included documented Gram-positive infection, the presence of sepsis, the administration of a vancomycin loading dose, and concomitant administration of other nephrotoxins [193]. In a retrospective, matched cohort study, Navalkele and colleagues found that the combination of vancomycin and piperacillin-tazobactam was an independent predictor of AKI compared to the combination of vancomycin and cefepime (hazard ratio = 4.27; 95% confidence interval, 2.73–6.68). In their patient population the onset of AKI occurred sooner in the vancomycin and piperacillin-tazobactam group (3 vs 5 days, $P = < 0.0001$) [191]. This study also reported that higher vancomycin troughs were not associated with a higher risk of AKI in these patients [191]. Lastly, studies have found no difference between intermittent or extended infusion piperacillin-tazobactam in rates of AKI [193, 194].

The combination of piperacillin-tazobactam and vancomycin resulting in an increased risk of AKI has also been reported in pediatric patients, both in a case report and two retrospective studies [195–197]. Three meta-analyses have shown an increased rate of AKI with concomitant administration of piperacillin-tazobactam and vancomycin [198–200]. Interestingly, other retrospective analyses have not found a statistically significant difference [201–204].

It is necessary to take this potential interaction into context with other patient-specific factors that increase the risk of AKI, such as critical illness, underlying disease states, and other nephrotoxic agents the patient may be receiving. Recent data seems to indicate that the addition of tazobactam is not likely the mechanism in the observed increased incidence of AKI in these patients [192]. However, prospective data are lacking evaluating the impact of beta-lactamase inhibitors alone on the incidence of AKI. Additionally, it is critical to note that these studies used differing definitions of AKI. It is also of importance to realize that there is no prospective evidence for this interaction, and no physiological mechanism has been elucidated. Nevertheless, the possible interaction between piperacillin and tazobactam resulting in an increased risk of AKI serves as a consideration for practitioners to implement into their stewardship intervention practices by discontinuing combination broad-spectrum antimicrobial therapy with piperacillin-tazobactam and vancomycin as soon as clinically possible.

1.2.24 Vecuronium

The concurrent administration of vecuronium and acylaminopenicillins has been reported to prolong muscle paralysis in both humans and animals [205–208]. Tryba reported a significant prolongation of muscle relaxation after a fixed dose of vecuronium with apalcillin, azlocillin, mezlocillin, and piperacillin in six patients. When combined with vecuronium, the mean increase in effect was 26% with apalcillin, 38% with mezlocillin, 46% with piperacillin, and 55% with azlocillin [205]. Condon et al. conducted a double-blind clinical trial to determine the ability of piperacillin or cefoxitin (control agent) to prolong the muscular blockade of vecuronium [209]. Patients were eligible for study enrollment if they were undergoing an elective operation with general anesthesia that required antibiotic prophylaxis. Patients were subsequently randomized to receive piperacillin or cefoxitin as the prophylactic antibiotic prior to the operation. All patients received vecuronium for muscle relaxation. Prolongation of neuromuscular blockade was determined before and after the administration of the antibiotic by the electromyography twitch response. Of the 27 evaluable patients enrolled in the study, five patients (two piperacillin and three cefoxitin) exhibited a nonclinically significant prolongation of neuromuscular blockade. Otherwise, the rate and extent of neuromuscular blockade was similar between groups. It appears that this interaction is clinically insignificant, although knowledge of this potential prolongation may be useful in certain surgical settings.

1.2.25   Miscellaneous Agents

The concomitant administration of penicillins and acidic drugs such as phenylbuta-
zone, sulfinpyrazone, indomethacin, and sulfaphenazole may prolong the half-life
of penicillin [79]. In this investigation, the half-life of penicillin was not noted to
change significantly with concomitant administration of chlorothiazide, sulfame-
thizole, and sulfamethoxypyridazine [79].

This is postulated to occur as a result of competition between the acidic drugs
and penicillin for renal tubular secretory proteins [79, 210]. Gothoni et al. reported
that concomitant treatment of pivampicillin and metoclopramide led to more rapid
absorption of the pivampicillin [211].

Potential drug-drug interactions between the penicillins and theophylline have
also been investigated. The coadministration of amoxicillin, ampicillin, ticarcillin/
clavulanic acid, or ampicillin/sulbactam with theophylline was not noted to alter
theophylline's properties [212–216].

Deppermann et al. assessed the effect of the coadministration of pirenzepine, an
antimuscarinic, with various antibiotics including amoxicillin in a double-blind,
randomized crossover study [4]. Coadministration of pirenzepine with amoxicillin
did not significantly alter the pharmacokinetics of amoxicillin.

Conner presented a case report, which suggested that concomitant administration of venlafaxine and amoxicillin-clavulanate might result in serotonin syndrome
[217]. The patient exhibited intense paresthesia in his fingers, tingling on his tongue,
severe abdominal cramping, profuse diarrhea, cold sweats, and uncontrollable shiv-
ering after the concomitant administration of amoxicillin-clavulanate and venlafaxine at two separate times. This patient had previously taken amoxicillin-clavulanate
without incident previously when he was not using venlafaxine without toxicity.

1.3   Cephalosporin Drug Interactions

1.3.1   Acid-Suppressive Agents

1.3.1.1   Ranitidine, Famotidine, and Omeprazole

Concomitant administration of the prodrugs, cefpodoxime proxetil, cefuroxime
axetil, and cefditoren pivoxil, with agents that increase gastric pH, such as ranitidine, results in a reduction of the antibiotic serum concentrations [5, 218]. The
bioavailability of the cefpodoxime proxetil has been reported to decrease by approximately 30–40% with concurrent administration of an H_2-receptor antagonist [218,
219]. However, no impact on the bioavailability of cefpodoxime was noted when
famotidine administration was separated from cefpodoxime by 2 h. Similarly, the
AUC of cefuroxime axetil was reduced by approximately 40% with pretreatment of
ranitidine and sodium bicarbonate [5]. The Cmax and AUC of cefditoren pivoxil

were reduced by approximately 25% with the concurrent administration of famotidine [220]. Other studies have found no significant effect on the bioavailability of cephalexin, cefaclor AF, or ceftibuten when administered concomitantly with H_2-receptor antagonists, omeprazole, or antacids [4, 221–225]. Madras-Kelly and colleagues reported that the administration of omeprazole or ranitidine with cephalexin had only a minimal effect on the pharmacokinetics of cephalexin, with the exception of a significant delay in Tmax which increased almost twofold [222]. Based on the results from these studies, concurrent administration of H_2-receptor antagonists and cefuroxime axetil, cefpodoxime proxetil, and cefditoren pivoxil should be avoided. If these agents need to be administered concurrently, the cephalosporins should be given at least 2 h after the H_2-receptor antagonist.

1.3.1.2 Acid-Neutralizing Agents: Antacids

The coadministration of antacids and certain cephalosporins including Cefaclor CD®, cefdinir, cefpodoxime, cephradine, and cefditoren may lead to decreased concentrations of the antibiotics [218–221, 223, 226]. A variety of studies have reported decreases in cephalosporin AUC and Cmax to be in the range of 20–40% for cefaclor, cefdinir, and cefpodoxime when administered with an antacid [218, 221, 223]. A minimal reduction in Cmax (14%) and AUC (11%) was noted with the concurrent administration of cefditoren with an antacid [220]. Other investigators have found no effect with cephalexin [4] or cefixime [224] when administered concomitantly with antacids. Certain cephalosporins including Cefaclor CD, cefdinir, cefpodoxime, and cefditoren should not be coadministered with antacids. If antacids are required during therapy, the cephalosporins should be separated from the antacid administration by at least 2 h.

1.3.2 Aminoglycosides

Aronoff et al. studied the in vivo interaction of ceftazidime and tobramycin in both anuric patients and normal volunteers [15]. Analysis of serum concentrations only showed minor changes in the pharmacokinetics of these agents when they were administered concurrently. However, the clearance of both ceftazidime and tobramycin was unchanged when these drugs were given concomitantly. The volume of distribution of tobramycin at steady state was increased by 20% in the normal volunteers when ceftazidime was given concurrently. The increase in distribution was accompanied by a slight decrease in ceftazidime elimination rate. The authors felt that this interaction was of little clinical significance and that no alteration in dosing was necessary when giving this combination.

1.3.3 Beta-Lactamase Inhibitors

The pharmacokinetics of the beta-lactamase inhibitors, tazobactam and ceftolozane, have been evaluated in healthy adult subjects [227]. Patients received a combination of these drugs in single doses of up to 2000 and 1000 mg of ceftolozane and tazobactam, respectively, as well as multiple doses of up to 3000 and 1500 mg of ceftolozane and tazobactam, respectively, daily. Ceftolozane was reported to have linear pharmacokinetics unaffected by the coadministration of tazobactam. Two phase I clinical trials were performed to investigate the interaction between ceftazidime and avibactam, as well as ceftazidime-avibactam and metronidazole. Healthy subjects were randomized to receive ceftazidime, avibactam, or ceftazidime-avibactam. They were given the medication once on days 1 and 4, and every 8 h on days 2 and 3. There were no differences in pharmacokinetic parameters between ceftazidime and avibactam compared to ceftazidime-avibactam or ceftazidime-avibactam and metronidazole. This indicates no drug-drug interaction exists between ceftazidime and avibactam, or ceftazidime-avibactam and metronidazole [228].

The drug-drug interaction potential of avibactam was investigated both in vitro and in a [^{14}C] mass balance study in six healthy male subjects. Avibactam was a substrate and an inhibitor of human OAT1 and OAT3 renal transporters. Given this observation was noted when avibactam concentrations were 20-fold higher than typical dosing, coupled with the rapid elimination of avibactam noted in healthy subjects, the clinical impact of this interaction is low. Nevertheless, a potentially clinically significant interaction may occur with potent inducers of OAT1 and OAT3, such as probenecid, and should be taken into consideration prior to coadministration. The authors found no interactions with the cytochrome P450 enzymes or other renal and hepatic transporters [229].

1.3.4 Calcium

Several regulatory agencies have issued warnings regarding the use of ceftriaxone concomitantly with intravenous products containing calcium [230, 231]. Reports indicate that ceftriaxone may be incompatible with calcium-containing solutions, depending on the concentrations used [232, 233]. These warnings were based originally on the reports of seven cases of neonatal or infant death and/or sudden cardiorespiratory arrest [232]. These authors reported that these patients received higher than normal ceftriaxone doses (150–200 mg/kg/day) and the use of higher concentration of calcium supplements administered via intravenous bolus [232]. In some cases these deaths were believed to have occurred as a result of the formation of these precipitates in the lungs or kidneys. Ceftriaxone being an anion, when present in high concentrations, can bind with calcium ions to form insoluble complexes that can precipitate out in various tissues [234, 235].

This warning of the concomitant use of ceftriaxone and calcium products has recently been reassessed by the FDA, resulting in the issuance of a less restrictive advisory. The most recent FDA advisory states that [236]:

- "Concomitant use of ceftriaxone and intravenous calcium-containing products is contraindicated in neonates (≤28 days of age). Ceftriaxone should not be used in neonates (≤28 days of age) if they are receiving (or are expected to receive) calcium-containing intravenous products."
- "In patients >28 days of age, ceftriaxone and calcium-containing products may be administered sequentially, provided the infusion lines are thoroughly flushed between infusions with a compatible fluid."
- "Ceftriaxone must not be administered simultaneously with intravenous calcium-containing solutions via a Y-site in any age group."
- "FDA now recommends that ceftriaxone and calcium-containing products may be used concomitantly in patients >28 days of age, using the precautionary steps above because the risk of precipitation is low in this population."

Steadman and colleagues recently reviewed the FDA Adverse Event Reporting Systems to determine the risk of serious ceftriaxone-calcium interactions in adults [237]. In these authors' opinion, their analysis of this FDA data base supported the FDA's recently revised recommendations suggesting that patients greater than 28 days of age may receive calcium and ceftriaxone sequentially. However, these authors do caution that in certain populations (such as those with intravascular depletion), the sequential administration of these two agents still warrants caution.

1.3.5   Calcium Channel Blockers

Variable data exist regarding the effects of nifedipine on cephalosporin pharmacokinetics [238, 239]. In a randomized crossover study, each healthy volunteer received cefixime with nifedipine or placebo [239]. The absolute bioavailability of cefixime was increased from 31% (cefixime alone) to 53% (cefixime and nifedipine) ($p < 0.01$). The $AUC_{0-\infty}$ also increased from 16.1 mg · h/L (cefixime alone) compared to 25.4 mg · h/L (cefixime and nifedipine) ($p < 0.01$) [239]. These investigators have also shown increased cephalexin concentrations with coadministration of nifedipine or diltiazem in an animal model [240]. The authors concluded that nifedipine can increase the absorption of these cephalosporins by enhancing the active transport mechanism in the intestine. In contrast, another study demonstrated that the pharmacokinetics of cefpodoxime did not change when coadministered with nifedipine [238]. The effects of amlodipine on cephalexin and cefuroxime axetil pharmacokinetics were investigated in healthy male subjects. There was a significant difference in the cephalexin geometric mean ratio for the AUC and the Cmax for individuals also administered amlodipine compared to those who only received cephalexin. There was no difference in pharmacokinetics seen with cefuroxime coadministration with amlodipine. The authors suggest that amlodipine may increase the bioavailability, specifically for peptidomimetic beta-lactam antibiotics [241]. Due to differences in specific antimicrobials and lack of adverse events seen with calcium channel blocker and cephalosporin combinations, no dosage changes are recommended when these agents are coadministered.

1.3.6 Cholestyramine

The coadministration of cholestyramine with cefadroxil or cephalexin has been shown to cause a delay in absorption, which is associated with a prolonged Tmax and reduction in Cmax [242, 243]. Despite these pharmacokinetic alterations, other important parameters such as AUC or amount of drug excreted in the urine were minimally affected. Although data for this interaction are limited, the clinical significance is doubtful, particularly when one considers that cholestyramine does not appear to alter cephalosporin exposure.

1.3.7 Colistin/Polymyxin

Coadministration of cephalothin and colistin has been associated with an increased incidence of nephrotoxicity compared to colistin alone [244, 245]. Koch-Weser and colleagues evaluated the incidence of adverse events related to colistin administration [244]. Renal toxicity was defined as a rise in serum creatinine (SCr) or blood urea nitrogen (BUN) that exceeded baseline (prior to colistin therapy) by 100% in the normal or by 25% in the abnormal range. The upper limit of normal range for Scr and BUN were defined as 1.5/100 mL and BUN 25 mg/100 mL. Overall, the incidence of nephrotoxic reactions occurred at a rate of 64/317 (20.2%). In patients that received administration of colistin along with cephalothin versus colistin alone, the rate of renal reaction increased to 33.3% (26/78) versus 15.9% (38/239) ($p < 0.001$).

Adler and Segel presented four case reports of acute renal failure due to colistin therapy [245]. Coadministration of cephalothin occurred in three of the four patients, while the fourth patient received cephalothin prior to colistin treatment. No other causes of renal failure could be determined in these patients.

Since both agents administered alone may have the potential to cause nephrotoxicity [245–247], it appears that the coadministration of these agents may result in an increased incidence of nephrotoxicity. The package insert for colistimethate also warns that coadministration of sodium cephalothin with colistimethate may enhance the nephrotoxicity of colistimethate and that the antimicrobials should not be administered together. The clinical impact of this interaction is limited since cephalothin is rarely used in clinical practice; however, careful monitoring of renal function is warranted if the combination is prescribed. No information exists regarding beta-lactam interactions with polymyxin B.

1.3.8 Cyclosporine

The data regarding drug interactions between cephalosporins and cyclosporine is contradictory. In rats administered cefepime and cyclosporine, both the AUC in the blood and brain increased significantly [248]. Soto and colleagues reported that two patients that had undergone renal transplants presented with significantly increased cyclosporine serum concentrations 2–3 days after the initiation of ceftriaxone 1 g twice a day [249]. These authors reported a two- to fourfold increase in cyclosporine concentration in these two patients. Cockburn reported that the concomitant use of ceftazidime or latamoxef (moxalactam) had been associated with an increase in cyclosporine concentrations [250]. Other investigators have shown no problems with the concomitant use of cyclosporine and ceftazidime. Verhagen and colleagues reported no significant impact upon renal function in 28 patients who underwent allogeneic bone marrow transplantation receiving both ceftazidime and cyclosporine for febrile neutropenia as measured by serum creatinine concentrations or creatinine clearance [251]. Since the data concerning the use of cyclosporine and cephalosporins is limited and contradictory, no firm recommendation can be made regarding their use together.

1.3.9 Contraceptives: Oral Estrogen

Refer to this topic in the discussion of penicillin.

1.3.10 Ethanol: Disulfiram-Like Reactions

Semisynthetic cephalosporins containing a methyltetrazolethiol (MTT) side chain such as cefamandole, cefoperazone, cefmenoxime, cefotetan, and moxalactam have been documented to cause disulfiram-like reactions in patients who consume ethanol during antibiotic treatment [252–254]. Cephalosporins with an MTT side chain inhibit acetaldehyde dehydrogenase, which results in the accumulation of acetaldehyde, a toxic metabolite of ethanol. Patients should be instructed not to consume alcohol during and for several days following antibiotic therapy. Refer to Chap. 10 regarding antimicrobials and food interactions for a more detailed review of this topic.

1.3.11   Iron

Coadministration of ferrous sulfate appears to cause a chelation complex and reduce the absorption of cefdinir [255]. In a randomized three-way crossover study, six healthy male subjects received the following regimens: 200 mg cefdinir alone, 200 mg cefdinir plus 1050 mg ferrous sulfate sustained release, or 200 mg cefdinir followed by 1050 mg ferrous sulfate sustained release 3 h later [255]. The $AUC_{0-12} \pm SD$ ($\mu g \cdot h/mL$) was significantly lower in the groups that received cefdinir concomitantly with ferrous sulfate (0.78 ± 0.25 $\mu g \cdot h/mL$) or at 3 h following the dose of cefdinir (6.55 ± 1.61 $\mu g \cdot h/mL$) compared to cefdinir alone (10.3 ± 1.35 $\mu g \cdot h/mL$) ($p < 0.05$). To avoid the potential for therapeutic failure of cefdinir, it should not be taken together with ferrous sulfate.

Three cases of red stools associated with cefdinir and iron-containing products have been reported in the literature [256, 257]. In all cases the discoloration of the stool was not associated with GI symptoms, and in all three instances, stool guaiac tests were negative. The reddish discoloration of the stools is thought to be due to formation of a nonabsorbable complex between iron and cefdinir or some of its breakdown products in the GI tract [257].

1.3.12   Metoclopramide

A healthy volunteer, crossover study evaluated the effect of food, metoclopramide, propantheline, and probenecid on the pharmacokinetics of cefprozil [258]. In the metoclopramide arm of the study, volunteers received cefprozil alone or cefprozil given 0.5 h after a dose of metoclopramide. Both isomers of cefprozil, cis and trans, were assayed in blood and urine. Cefprozil's isomers demonstrated a statistically significant reduction in mean residence time when administered after metoclopramide; however, there was no difference in $AUC_{0-\infty}$ or half-life of cefprozil among the treatment groups. Administration of metoclopramide prior to cefprozil did not affect its extent of absorption. Concurrent administration of these agents need not be avoided.

1.3.13   Methotrexate

Rabbits receiving concomitant infusions of methotrexate and a cephalosporin (ceftriaxone, ceftazidime, ceftizoxime, or cefoperazone) have been demonstrated to have an increased renal elimination of methotrexate and 7-hydroxymethotrexate [144].

In a case report, an 8-year-old boy receiving methotrexate for non-Hodgkin's lymphoma experienced a decrease in methotrexate clearance when methotrexate was coadministered with piperacillin [147]. The patient subsequently received methotrexate along with ceftazidime without any impact on methotrexate clearance. The differences seen in methotrexate renal elimination between cephalosporins and piperacillin may be due to the extent of tubular secretion (penicillins > cephalosporins) [144, 259]. In contrast, Tran and Herrington report a patient that received methotrexate, cefepime, and ceftriaxone, with no effect on methotrexate clearance. They suggest that cefepime and ceftriaxone do not inhibit OAT3, the main transporter involved in methotrexate clearance, which leads to a lack of interaction with coadministration of the agents [260].

There have been no documented interactions resulting in decreased renal elimination of methotrexate with the concurrent administration of cephalosporins and methotrexate in humans. However, because of the documented interaction between some penicillins and methotrexate as well as the animal data regarding some cephalosporins and methotrexate, close monitoring of methotrexate concentrations and signs of toxicity (e.g., bone marrow suppression, nephrotoxicity, mucositis) is suggested during concurrent use of cephalosporins and methotrexate.

1.3.14   Metformin

In a crossover study, healthy volunteers were randomized to receive metformin alone or metformin along with cephalexin [261]. The coadministration of metformin and cephalexin led to an increase in Cmax and AUC of metformin by approximately 30%. It appears that cephalexin interferes with renal clearance of metformin, which may be due to competition for renal transport proteins such as organic anion or cation transporter (OAT or OCT, respectively) [261, 262]. Limited data are available on the clinical significance of this interaction. Clinicians should exercise caution when using these two agents together and monitor for metformin toxicity.

1.3.15   Nonsteroidal Anti-inflammatory Drugs

Diclofenac has been reported to cause an increase in the biliary excretion of ceftriaxone [263]. A study was conducted in patients in whom a cholecystectomy was performed and a drain was placed in the common bile duct [263]. The subjects who received ceftriaxone along with diclofenac demonstrated a 320% ($p < 0.05$) increase in the amount of ceftriaxone excreted in the bile and a 56% ($p < 0.05$) reduction in the amount excreted in the urine. Due to the limited data, no therapeutic recommendations can be made.

1.3.16   Phenytoin

Highly protein-bound antibiotics such as ceftriaxone (approximately 90% bound to plasma proteins) [264] have the potential to interact with other highly protein-bound agents such as phenytoin [159]. Due to protein displacement, high doses of ceftriaxone may increase unbound concentrations of phenytoin in certain patient populations [159]. Dasgupta and colleagues performed an in vitro study to determine the effect of ceftriaxone in displacing phenytoin from protein-binding sites [159]. Estimated peak ceftriaxone concentrations (270 µmol/L and 361 µmol/L) were added to pooled sera from patients receiving phenytoin. Three groups with varying albumin concentrations were evaluated. The greatest ceftriaxone-induced displacement effect was seen in the group with the lowest albumin concentration (25 g/L). In this group, the unbound phenytoin concentrations (µmol/L) (SD) were 8.12 (0.28) for the control, 9.39 (0.12) for ceftriaxone 270 µmol/L, and 9.93 (0.36) for ceftriaxone 361 µmol/L, respectively. Although the increases appear minor, significant increases in unbound phenytoin concentrations were noted in all groups compared to the control group ($p < 0.05$). In patients receiving ceftriaxone concomitantly with phenytoin, monitoring of unbound and total serum concentrations of phenytoin in addition to watching for signs of phenytoin toxicity is warranted.

1.3.17   Oral Anticoagulants

Semisynthetic cephalosporins containing an MTT substituent at the 3-position, such as cefamandole, cefoperazone, cefmenoxime, cefotetan, and moxalactam, have been associated with the development of a hypoprothrombinemia [265]. Several case reports have implicated these agents in prolonged prothrombin time and/or bleeding episodes in patients [266–272]. Anagaran and colleagues retrospectively assessed the effect of prophylactic administration of cefamandole or vancomycin on the warfarin anticoagulation response in 60 postsurgical patients [273]. Patients who received cefamandole had a higher proportion of elevated prothrombin times compared those who received vancomycin (14 versus 1, $p < 0.05$). In another study, these same investigators characterized the effect of cefazolin, cefamandole, and vancomycin on warfarin anticoagulation in post-cardiac valve replacement patients [274]. They noted that the greatest number of patients ($n = 6$) with elevated prothrombin times received cefamandole compared to cefazolin ($n = 1$) and vancomycin ($n = 1$). In addition, cefamandole therapy was associated with a 15–20% greater change in prothrombin times compared to the cefazolin and vancomycin ($p < 0.01$). Patients who are malnourished or who have renal insufficiency may be at higher risk for this interaction [266]. There are two case reports of patients on chronic warfarin therapy who were administered ceftaroline for cellulitis and developed elevated INRs. One of the two patients also developed a clinically significant bleed [275, 276]. Saum and Balmat performed a retrospective chart review comparing INR outcomes among different agents commonly used for urinary tract infections (penicillin, ceftriaxone, ciprofloxacin, and a first-generation cephalosporin). They found patients treated with ceftriaxone had a higher peak INR (3.56, $p = 0.004$) and greater extent of change from baseline INR (+1.19, $p = 0.006$) compared to patients treated with other antibiotics [277].

The exact mechanism of the hypoprothrombinemia phenomenon is unknown, although several mechanisms have been proposed [278–281]. Clinicians are cautioned to monitor for signs and symptoms of bleeding, prothrombin time, and activated partial thromboplastin time in patients receiving cephalosporins with an MTT side chain and concomitant therapy with oral anticoagulants.

1.3.18 Probenecid

Probenecid can increase the serum concentrations of most renally eliminated *cephalosporins* [221, 258, 282–286]. Although other mechanisms may contribute, probenecid appears to inhibit tubular secretion of cephalosporins resulting in their decreased renal elimination [160, 161]. The AUCs of ceftizoxime, cefoxitin, cefaclor, and cephradine have been reported to increase by approximately 50–100% with the coadministration of probenecid [17, 160, 284]. Another study showed that probenecid prolonged the cefuroxime serum half-life by 63%, from 0.8 h to 1.3 h (P 0.05) after a 750 mg dose, and the area under the concentration-time curve (AUC) increased by 44% (P 0.05) [287]. Twenty-six patients were prospectively examined to determine the effect of probenecid on cefazolin. This study showed that probenecid 500 mg given four times a day allowed a cefazolin dosage adjustment to 2000 mg once daily, instead of three times a day, still resulting in what would be considered therapeutic serum concentrations of cefazolin [288].

Probenecid has been documented to prolong the half-life and increase the serum concentration of many other cephalosporins as well [221, 282–286, 288–298]. Certain cephalosporins such as ceforanide, ceftazidime, ceftriaxone, and moxalactam are eliminated through a different pathway, and their pharmacokinetics are not significantly altered by probenecid [282, 283, 299–304]. Caution or avoidance of this combination should be considered in certain patient populations in which drug accumulation may occur (e.g., elderly patients or patients with impaired renal function).

1.3.19 Propantheline

A healthy volunteer, crossover study evaluated the effect of food, metoclopramide, propantheline, and probenecid on the pharmacokinetics of cefprozil [258]. In the propantheline arm of the study, volunteers received cefprozil alone or cefprozil given 0.5 h after a dose of propantheline. Both isomers of cefprozil, cis and trans, were assayed in blood and urine samples. There was no difference in cefprozil $AUC_{0-\infty}$ or half-life in either treatment group. The administration of propantheline prior to cefprozil does not affect the extent of cefprozil absorption. No special precautions seem necessary for this combination.

1.3.20   Theophylline

The coadministration of cephalexin or cefaclor with theophylline has not been documented to significantly alter any pharmacokinetic parameters of theophylline [305–307]. However, Hammond and Abate reported a case of a possible interaction between theophylline and cefaclor, which resulted in theophylline toxicity [308]. It was unclear whether this was an actual drug-drug interaction or the effect of an acute viral illness on theophylline disposition. Based on these limited data, no dosage recommendation seems warranted.

1.3.21   Miscellaneous Agents

The pharmacokinetics of the combination of cefotaxime and mezlocillin were evaluated in eight healthy subjects and five patients with end-stage renal disease [309]. Simultaneous administration of mezlocillin in the volunteers resulted in a decrease of the total body clearance of cefotaxime by 42%. In the presence of end-stage renal disease, simultaneous administration of mezlocillin and cefotaxime led to an increase of the half-life of cefotaxime by roughly six times, to 5.8 h. Combined administration of cefotaxime and mezlocillin did not affect the pharmacokinetics of mezlocillin. These results suggest that lower doses of cefotaxime are probably adequate to maintain comparable cefotaxime plasma concentrations when mezlocillin is given simultaneously, in patients with normal renal function.

Older cephalosporins such as cephalothin (renamed cefalotin) and cephaloridine have been reported to cause nephrotoxicity [244, 245]. The coadministration of these older cephalosporins with other potential nephrotoxic agents including colistin [244, 245], various aminoglycosides [244–246, 310–316], and furosemide [317–320] has been associated with an increased incidence of nephrotoxicity. The clinical impact of this interaction is limited because these cephalosporins are rarely used in current clinical practice; however, careful monitoring of renal function is warranted if such combinations are prescribed. These drug-drug interactions have not been documented as a clinically significant problem for any of the newer cephalosporins [321–323].

1.4 Carbapenems

1.4.1 Aminoglycosides

As with penicillins, there are mixed reports of both inactivation and enhancement of activity with the use of carbapenems and aminoglycosides. An in vitro model of tobramycin was incubated with imipenem/cilastatin and analyzed at various time points using a fluorescence polarization assay. The authors found no degradation at typical serum concentrations of both agents [324]. In an immunocompetent guinea pig model of *Acinetobacter baumannii* pneumonia, the authors report a Cmax to MIC ratio of 8.4 for amikacin monotherapy compared to 5.4 for those treated with the combination of imipenem and amikacin. There was also a greater decrease in bacterial counts in guinea pigs treated with amikacin alone compared to imipenem and amikacin. The authors postulate that there is an in vivo chemical interaction that exists between imipenem and amikacin that diminishes the efficacy when administered together [325]. Finally, a pharmacokinetic study was performed in healthy subjects who were administered 0.5 g of imipenem both alone and then combined with 0.5 g of amikacin. The mean Cmax for imipenem increased from 7.70 to 26.00 mcg/mL when combined with amikacin. This did not change overall exposure of imipenem; therefore, the authors concluded that the clinical effects of monotherapy compared to combination therapy are unchanged [29].

1.4.2 Probenecid

Concomitant probenecid can increase the concentration of the carbapenems. It is proposed that probenecid inhibits tubular secretion of the carbapenems, resulting in their decreased renal elimination.

Of the four commercially available carbapenems in the United States, probenecid has the most impact on the renal elimination of doripenem followed by meropenem, ertapenem, and imipenem. The combination of doripenem and probenecid produced a 53% increase in half-life and 75% increase in the AUC of doripenem compared to doripenem alone [326]. Meropenem's half-life and AUC were increased by 33% and 55%, respectively, when coadministered with probenecid [291]. Ertapenem's half-life and AUC increased by 20% and 25% with the combination of ertapenem and probenecid compared to ertapenem alone [327]. Nix and colleagues also reported that the concomitant administration of probenecid and ertapenem resulted in the decreased renal clearance of unbound ertapenem by approximately 50% [328]. In contrast, imipenem's half-life and AUC only increased 6% and 13%, respectively, when coadministered with probenecid [329]. Caution and/or avoidance of this combination should be a consideration in patient populations in which drug accumulation may occur (such as elderly patients or patients with impaired renal function). The increased serum concentration noted because of this drug-drug interaction may increase the risk of central nervous system toxicity of these agents.

1.4.3   Cyclosporine

Based on case reports, cyclosporine and imipenem/cilastatin may demonstrate additive central nervous system toxicity when administered concomitantly. Bösmuller and colleagues reported five transplant patients experiencing central nervous system toxicity during administration of cyclosporine and imipenem/cilastatin [330]. None of these patients reported a history of seizures. Four of the five patients experienced a seizure despite cyclosporine concentrations within normal therapeutic range. The fifth patient experienced a myoclonia; this was associated with an elevated cyclosporine concentration of 900 ng/mL. Symptoms of central nervous toxicity occurred within 1 d in four patients, and symptoms resolved in all patients with discontinuation of imipenem/cilastatin or discontinuation, or dose reduction of cyclosporine. Zazgornik and colleagues published a case report of a 62-year-old female receiving imipenem/cilastatin and cyclosporine who developed central nervous system toxicity [331]. The patient had recently received a renal transplant secondary to interstitial nephritis and was receiving imipenem/cilastatin for a urinary tract infection. Following the second dose of imipenem/cilastatin, the patient experienced confusion, agitation, and tremors, which resulted in the discontinuation of imipenem/cilastatin. The serum cyclosporine concentration, which was obtained 4 days after imipenem/cilastatin therapy, was elevated at 1000 ng/mL compared to a previous level of 400 ng/mL. In contrast, an investigation in a rat model has demonstrated decreased cyclosporine serum concentrations when it was combined with imipenem/cilastatin [332].

Since both imipenem and cyclosporine administered alone may have the potential to cause central nervous system side effects, it is difficult to determine what role the combination of these agents may have played in these reports. Based on this limited clinical data, avoidance of imipenem and cyclosporine is not warranted.

1.4.4   Theophylline

Semel and Allen reported three cases of seizures occurring in patients receiving imipenem/cilastatin and theophylline [333]. None of the patients had a previous history of neurologic or seizure disorder. The authors concluded that the seizures could be due to both of the drugs' ability to inhibit gamma aminobutyric acid binding to receptors, thus resulting in increased excitation of the central nervous system. It is difficult to differentiate the potential for seizures between the administration of imipenem/cilastatin alone and the combination of imipenem/cilastatin and theophylline. Avoiding coadministration of theophylline and imipenem/cilastatin is not warranted.

1.4.5 Ganciclovir

Patients have experienced generalized seizures during concomitant imipenem/cilastatin and ganciclovir therapy [334, 335]. No additional information is available on these patients. Due to this limited data, it is difficult to differentiate the potential for seizures of imipenem/cilastatin alone or the combination of imipenem/cilastatin and ganciclovir. The manufacturer does not recommend coadministration of imipenem/cilastatin and ganciclovir unless the benefits outweigh the risks.

1.4.6 Valganciclovir

After oral administration of valganciclovir, it is rapidly converted to ganciclovir by intestinal and hepatic esterases. Although no in vivo drug-drug interaction studies have been conducted with valganciclovir, because of its rapid conversion to ganciclovir in the body [336], any drug-drug interaction seen with ganciclovir would be expected to occur with valganciclovir [337]. Due to the possibility of an interaction between valganciclovir and imipenem/cilastatin, the use of these drugs concomitantly should be avoided unless the benefit outweighs the risk [338].

1.4.7 Valproic Acid

The coadministration of carbapenems and valproic acid may lead to decreased concentrations of valproic acid [326]. The proposed mechanism is that carbapenems may interfere with the hydrolysis of valproic acid's glucuronide metabolite to valproic acid [326, 339–343]. A healthy volunteer study evaluated the pharmacokinetics of valproic acid and glucuronide metabolite in subjects receiving doripenem [326]. Valproic acid's Cmax, Cmin, and AUC were decreased by

44.5%, 77.7%, and 63%, respectively, when coadministered with doripenem. In contrast, an increase in the Cmax and AUC of valproic acid's glucuronide metabolite was seen when coadministered with doripenem [326]. Two retrospective studies showed decreased valproic acid concentrations of 82% and 66% in patients receiving concomitant meropenem and valproic acid compared to valproic acid alone [344, 345]. In a retrospective study examining coadministration of all carbapenems and valproic acid in 52 patients, the authors reported a decrease in valproic acid levels of 60% with 24 h of administration. Even more, 90% of patients were determined to have subtherapeutic valproic acid levels. Ertapenem and meropenem had a greater effect on valproic acid levels as compared to imipenem/cilastatin ($p < 0.005$) [346]. In both studies, the authors noted that the decrease in valproic concentrations could be seen within 24 h of concomitant administration of these two agents. Animal models have also found decreased valproic acid concentrations with the concurrent administration of imipenem [298], meropenem [347], or panipenem [348] and valproic acid. There have been several documented cases of lowering of plasma valproic acid concentrations during concomitant therapy with carbapenems [349–357]. In addition to changes in valproic acid concentrations, there have been many case reports published describing patients who experienced breakthrough seizures secondary to this drug interaction [348, 349, 358–368]. Spriet and Willems describe a patient with decompensated liver cirrhosis (Child-Pugh score 13) who was being treated with meropenem and was then started on valproic acid after experiencing a seizure. Total valproic acid levels were stable both during and after meropenem was discontinued. The authors theorized that the upregulation and inhibition of the glucuronide metabolites did not occur due to the patient's advanced liver disease [369]. Providers should avoid prescribing carbapenems in patients receiving valproic acid to prevent subtherapeutic valproic acid serum concentrations [370]. If no alternative therapy is available, close monitoring of valproic acid concentrations and dosage modifications of valproic acid is recommended [326].

1.4.8 Miscellaneous Agents

Franco-Bronson reported severe hypotension related to imipenem and haloperidol administered concomitantly in three patients. The episodes were brief and self-limiting in nature. The authors postulated that competition for protein-binding sites between these two agents might have resulted in increased free levels of haloperidol [371].

1.5 Monobactams

1.5.1 Aminoglycosides

Six healthy volunteers were administered 1 g of aztreonam alone or combined with 0.5 g amikacin [29]. The apparent serum concentrations of aztreonam achieved 0.5 and 8 h after the end of infusion were 32.68 ± 10.06 and 1.84 ± 0.75 µg/mL, respectively. After concomitant administration with amikacin, the respective concentrations were 21.63 ± 10.50 and 4.00 ± 1.71 µg/mL. This study revealed that concomitant of amikacin with aztreonam in healthy volunteers minimally effects the Cmax without impacting any other pharmacokinetic parameter.

1.5.2 Probenecid

Concomitant probenecid can increase aztreonam concentrations [372]. It is proposed that probenecid inhibits tubular secretion resulting in decreased aztreonam renal elimination. In a randomized crossover trial, six healthy men received aztreonam alone or aztreonam along with probenecid [372]. Coadministration of probenecid with aztreonam increased aztreonam concentrations from 81.7 ± 3.4 to 86.0 ± 2.2 µg/mL. This interaction seems to carry minimal clinical risk. No recommendation to avoid the concurrent administration of probenecid and aztreonam seems warranted.

1.5.3 Miscellaneous Agents

A number of other antimicrobial agents have been evaluated for the potential of drug interactions with aztreonam. Healthy subjects were given concomitant linezolid and aztreonam, in an open-label, crossover study [373]. The combined treatment compared to each drug alone resulted in an increase in the maximum serum concentration of linezolid of approximately 18% and an approximate 7% decrease in the apparent elimination rate of aztreonam. Neither of these changes are considered clinically significant. Other studies in healthy subjects administered aztreonam concomitantly with daptomycin [374], nafcillin [375], gentamicin plus metronidazole [372], or amikacin [29] showed no clinically significant drug interactions were identified in these studies.

Table 1.1 Clinical significance of beta-lactam drug interactions

	Penicillins	Cephalosporins	Carbapenems	Monobactam
Major	Contraceptives, oral estrogen Probenecid Methotrexate Warfarin	Calcium (ceftriaxone, infants) Contraceptives, oral estrogen Methotrexate Probenecid Warfarin	Cyclosporine Ganciclovir/valganciclovir Theophylline Probenecid Valproic acid	
Moderate	Cyclosporine (nafcillin) Phenytoin Tetracyclines Vancomycin Metronidazole			
Minor	Acid suppressive agents Allopurinol Aminoglycosides Aspirin Beta-adrenergic blockers Calcium channel blockers Chloramphenicol Chloroquine Ciprofloxacin Direct acting oral anticoagulants Heparin Interferon-gamma Guar gum Khat Macrolides Oseltamivir Proguanil Sulfonamides Vecuronium	Acid suppressive agents Aminoglycoside nephrotoxicity Avibactam Calcium channel blocker Cholestyramine Colistin Cyclosporine Ethanol Iron Metoclopramide Mezlocillin Nephrotoxic agents (with cefalotin, cephaloridine) Nonsteroidal anti-inflammatory drugs Phenytoin Propantheline Theophylline	Aminoglycosides Cyclosporine Haloperidol Theophylline	Probenecid

Admixtures of aztreonam and metronidazole prepared at two different concentrations revealed an incompatibility as evidenced by the appearance of a pinkish color [376]. Under acidic pH settings of these admixtures of aztreonam and metronidazole, it is suggested that the aminothiazole moiety of aztreonam is diazotized by the nitrite ion produced by metronidazole solutions. The diazotized molecule, in turn, reacts with another aztreonam molecule by diazo-coupling leading to the pink color [376]. As a result of this interaction, it has been recommended that aztreonam and metronidazole be administered separately (Table 1.1).

References

1. Rogers HJ, James CA, Morrison PJ, Bradbrook ID (1980) Effect of cimetidine on oral absorption of ampicillin and cotrimoxazole. J Antimicrob Chemother 6(2):297–300
2. Paulsen O, Höglund P, Walder M (1989) No effect of omeprazole-induced hypoacidity on the bioavailability of amoxycillin or bacampicillin. Scand J Infect Dis 21(2):219–223
3. Staniforth DH, Clarke HL, Horton R, Jackson D, Lau D (1985) Augmentin bioavailability following cimetidine, aluminum hydroxide and milk. Int J Clin Pharmacol Ther Toxicol 23(3):154–157
4. Deppermann KM, Lode H, Höffken G, Tschink G, Kalz C, Koeppe P (1989) Influence of ranitidine, pirenzepine, and aluminum magnesium hydroxide on the bioavailability of various antibiotics, including amoxicillin, cephalexin, doxycycline, and amoxicillin-clavulanic acid. Antimicrob Agents Chemother 33(11):1901–1907
5. Sommers DK, van Wyk M, Moncrieff J, Schoeman HS (1984) Influence of food and reduced gastric acidity on the bioavailability of bacampicillin and cefuroxime axetil. Br J Clin Pharmacol 18(4):535–539
6. Wittayalertpanya S, Wannachai N, Thongnopnua P, Mahachai V (2000) Effect of omeprazole on gastric mucosa and serum levels of amoxicillin in patients with non-ulcer dyspepsia. J Med Assoc Thai 83(6):611–618
7. Pommerien W, Braun M, Idström JP, Wrangstadh M, Londong W (1996) Pharmacokinetic and pharmacodynamic interactions between omeprazole and amoxicillin in helicobacter pylori-positive healthy subjects. Aliment Pharmacol Ther 10(3):295–301
8. Hassan-Alin M, Andersson T, Niazi M, Liljeblad M, Persson BA, Röhss K (2006) Studies on drug interactions between esomeprazole, amoxicillin and clarithromycin in healthy subjects. Int J Clin Pharmacol Ther 44(3):119–127
9. Mainz D, Borner K, Koeppe P, Kotwas J, Lode H (2002) Pharmacokinetics of lansoprazole, amoxicillin and clarithromycin after simultaneous and single administration. J Antimicrob Chemother 50(5):699–706
10. Jick H, Porter JB (1981) Potentiation of ampicillin skin reactions by allopurinol or hyperuricemia. J Clin Pharmacol 21(10):456–458
11. Excess of ampicillin rashes associated with allopurinol or hyperuricemia. A report from the Boston Collaborative Drug Surveillance Program, Boston University Medical Center. N Engl J Med 286(10):505–507 (1972)
12. Fessel WJ (1972) Immunologic reactivity in hyperuricemia. N Engl J Med 286(22):1218
13. Ben Fredj N, Aouam K, Chaabane A, Toumi A, Ben Rhomdhane F, Boughattas N et al (2010) Hypersensitivity to amoxicillin after drug rash with eosinophilia and systemic symptoms (DRESS) to carbamazepine and allopurinol: a possible co-sensitization. Br J Clin Pharmacol 70(2):273–276
14. Pérez A, Cabrerizo S, de Barrio M, Díaz MP, Herrero T, Tornero P et al (2001) Erythema-multiforme-like eruption from amoxicillin and allopurinol. Contact Dermatitis 44(2):113–114
15. Henderson JL, Polk RE, Kline BJ (1981) In vitro inactivation of gentamicin, tobramycin, and netilmicin by carbenicillin, azlocillin, or mezlocillin. Am J Hosp Pharm 38(8):1167–1170
16. Townsend RS (1989) In vitro inactivation of gentamicin by ampicillin. Am J Hosp Pharm 46(11):2250–2251
17. McLaughlin JE, Reeves DS (1971) Clinical and laboratory evidence for inactivation of gentamicin by carbenicillin. Lancet 1(7693):261–264
18. Uber WE, Brundage RC, White RL, Brundage DM, Bromley HR (1991) In vivo inactivation of tobramycin by piperacillin. DICP 25(4):357–359
19. Halstenson CE, Hirata CA, Heim-Duthoy KL, Abraham PA, Matzke GR (1990) Effect of concomitant administration of piperacillin on the dispositions of netilmicin and tobramycin in patients with end-stage renal disease. Antimicrob Agents Chemother 34(1):128–133
20. Davies M, Morgan JR, Anand C (1975) Interactions of carbenicillin and ticarcillin with gentamicin. Antimicrob Agents Chemother 7(4):431–434

21. Kradjan WA, Burger R (1980) In vivo inactivation of gentamicin by carbenicillin and ticarcillin. Arch Intern Med 140(12):1668–1670
22. Blair DC, Duggan DO, Schroeder ET (1982) Inactivation of amikacin and gentamicin by carbenicillin in patients with end-stage renal failure. Antimicrob Agents Chemother 22(3):376–379
23. Riff LJ, Jackson GG (1972) Laboratory and clinical conditions for gentamicin inactivation by carbenicillin. Arch Intern Med 130(6):887–891
24. Pieper JA, Vidal RA, Schentag JJ (1980) Animal model distinguishing in vitro from in vivo carbenicillin-aminoglycoside interactions. Antimicrob Agents Chemother 18(4):604–609
25. Santos Navarro M, Zarzuelo Castañeda A, López FG, Sánchez Navarro A, Arévalo M, Lanao JM (1998) Pharmacokinetic parameters of netilmicin and protective effect of piperacillin regarding nephrotoxicity caused by netilmicin. Eur J Drug Metab Pharmacokinet 23(2):143–147
26. Weibert R, Keane W, Shapiro F (1976) Carbenicillin inactivation of aminoglycosides in patients with severe renal failure. Trans Am Soc Artif Intern Organs 22:439–443
27. Eykyn S, Phillips I, Ridley M (1971) Gentamicin plus carbenicillin. Lancet 1(7698):545–546
28. Lau A, Lee M, Flascha S, Prasad R, Sharifi R (1983) Effect of piperacillin on tobramycin pharmacokinetics in patients with normal renal function. Antimicrob Agents Chemother 24(4):533–537
29. Adamis G, Papaioannou MG, Giamarellos-Bourboulis EJ, Gargalianos P, Kosmidis J, Giamarellou H (2004) Pharmacokinetic interactions of ceftazidime, imipenem and aztreonam with amikacin in healthy volunteers. Int J Antimicrob Agents 23(2):144–149
30. Hitt CM, Patel KB, Nicolau DP, Zhu Z, Nightingale CH (1997) Influence of piperacillin-tazobactam on pharmacokinetics of gentamicin given once daily. Am J Health Syst Pharm 54(23):2704–2708
31. Dowell JA, Korth-Bradley J, Milisci M, Tantillo K, Amorusi P, Tse S (2001) Evaluating possible pharmacokinetic interactions between tobramycin, piperacillin, and a combination of piperacillin and tazobactam in patients with various degrees of renal impairment. J Clin Pharmacol 41(9):979–986
32. Roberts GW, Nation RL, Jarvinen AO, Martin AJ (1993) An in vivo assessment of the tobramycin/ticarcillin interaction in cystic fibrosis patients. Br J Clin Pharmacol 36(4):372–375
33. Noone P, Pattison JR (1971) Therapeutic implications of interaction of gentamicin and penicillins. Lancet 2(7724):575–578
34. Ervin FR, Bullock WE, Nuttall CE (1976) Inactivation of gentamicin by penicillins in patients with renal failure. Antimicrob Agents Chemother 9(6):1004–1011
35. Van der Bijl P, Seifart HI (1989) In vitro evaluation of ampicillin-gentamicin interactions. J Oral Maxillofac Surg 47(5):489–494
36. Moellering RC, Wennersten C, Weinberg AN (1971) Synergy of penicillin and gentamicin against enterococci. J Infect Dis 124(Suppl):S207–S209
37. Guenthner SH, Chao HP, Wenzel RP (1986) Synergy between amikacin and ticarcillin or mezlocillin against nosocomial bloodstream isolates. J Antimicrob Chemother 18(4):550–552
38. Laverdière M, Gallimore B, Restieri C, Poonia K, Chow AW (1994) In vitro synergism of ceftriaxone combined with aminoglycosides against Pseudomonas aeruginosa. Diagn Microbiol Infect Dis 19(1):39–46
39. Marks MI, Hammerberg S, Greenstone G, Silver B (1976) Activity of newer aminoglycosides and carbenicillin, alone and in combination, against gentamicin-resistant Pseudomonas aeruginosa. Antimicrob Agents Chemother 10(3):399–401
40. Moellering RC, Wennersten C, Weinberg AN (1971) Studies on antibiotic synergism against enterococci. I. Bacteriologic studies. J Lab Clin Med 77(5):821–828
41. Weinstein AJ, Moellering RC (1973) Penicillin and gentamicin therapy for enterococcal infections. JAMA 223(9):1030–1032

42. Murray BE, Church DA, Wanger A, Zscheck K, Levison ME, Ingerman MJ et al (1986) Comparison of two beta-lactamase-producing strains of Streptococcus faecalis. Antimicrob Agents Chemother 30(6):861–864
43. Sexton DJ, Tenenbaum MJ, Wilson WR, Steckelberg JM, Tice AD, Gilbert D et al (1998) Ceftriaxone once daily for four weeks compared with ceftriaxone plus gentamicin once daily for two weeks for treatment of endocarditis due to penicillin-susceptible streptococci. Endocarditis Treatment Consortium Group. Clin Infect Dis 27(6):1470–1474
44. Shelburne SA, Greenberg SB, Aslam S, Tweardy DJ (2007) Successful ceftriaxone therapy of endocarditis due to penicillin non-susceptible viridans streptococci. J Infect 54(2):e99–101
45. Fujitani S, Rowlinson MC, George WL (2008) Penicillin G-resistant viridans group streptococcal endocarditis and interpretation of the American Heart Association's Guidelines for the Treatment of Infective Endocarditis. Clin Infect Dis 46(7):1064–1066
46. Baddour LM (1998) Infective endocarditis caused by beta-hemolytic streptococci. The Infectious Diseases Society of America's Emerging Infections Network. Clin Infect Dis 26(1):66–71
47. Smyth EG, Pallett AP, Davidson RN (1988) Group G streptococcal endocarditis: two case reports, a review of the literature and recommendations for treatment. J Infect 16(2):169–176
48. Anderson ET, Young LS, Hewitt WL (1978) Antimicrobial synergism in the therapy of gram-negative rod bacteremia. Chemotherapy 24(1):45–54
49. Tamma PD, Cosgrove SE, Maragakis LL (2012) Combination therapy for treatment of infections with gram-negative bacteria. Clin Microbiol Rev 25(3):450–470
50. Brown CH, Natelson EA, Bradshaw W, Williams TW, Alfrey CP (1974) The hemostatic defect produced by carbenicillin. N Engl J Med 291(6):265–270
51. Brown CH, Natelson EA, Bradshaw MW, Alfrey CP, Williams TW (1975) Study of the effects of ticarcillin on blood coagulation and platelet function. Antimicrob Agents Chemother 7(5):652–657
52. Brown CH, Bradshaw MJ, Natelson EA, Alfrey CP, Williams TW (1976) Defective platelet function following the administration of penicillin compounds. Blood 47(6):949–956
53. Andrassy K, Ritz E, Hasper B, Scherz M, Walter E, Storch H (1976) Penicillin-induced coagulation disorder. Lancet 2(7994):1039–1041
54. Andrassy K, Weischedel E, Ritz E, Andrassy T (1976) Bleeding in uremic patients after carbenicillin. Thromb Haemost 36(1):115–126
55. Tabernero Romo JM, Corbacho L, Sánchez S, Rodríguez JL, Macías JF, del Cañizo C et al (1979) Effects of carbenicillin on blood coagulation: a study in patients with chronic renal failure. Clin Nephrol 11(1):31–34
56. Wisløff F, Larsen JP, Dahle A, Lie M, Godal HC (1983) Effect of prophylactic high-dose treatment with ampicillin and cloxacillin on bleeding time and bleeding in patients undergoing elective vascular surgery. Scand J Haematol 31(2):97–101
57. Lurie A, Ogilvie M, Gold CH, Meyer AM, Goldberg B (1974) Carbenicillin-induced coagulopathy. S Afr Med J 48(11):457–461
58. Taylor AT, Pritchard DC, Goldstein AO, Fletcher JL (1994) Continuation of warfarin-nafcillin interaction during dicloxacillin therapy. J Fam Pract 39(2):182–185
59. Heilker GM, Fowler JW, Self TH (1994) Possible nafcillin-warfarin interaction. Arch Intern Med 154(7):822–824
60. Qureshi GD, Reinders TP, Somori GJ, Evans HJ (1984) Warfarin resistance with nafcillin therapy. Ann Intern Med 100(4):527–529
61. Shovick VA, Rihn TL (1991) Decreased hypoprothrombinemic response to warfarin secondary to the warfarin-nafcillin interaction. DICP 25(6):598–600
62. Kim KY, Frey RJ, Epplen K, Foruhari F (2007) Interaction between warfarin and nafcillin: case report and review of the literature. Pharmacotherapy 27(10):1467–1470
63. Davis RL, Berman W, Wernly JA, Kelly HW (1991) Warfarin-nafcillin interaction. J Pediatr 118(2):300–303

64. Krstenansky PM, Jones WN, Garewal HS (1987) Effect of dicloxacillin sodium on the hypo-prothrombinemic response to warfarin sodium. Clin Pharm 6(10):804–806
65. Mailloux AT, Gidal BE, Sorkness CA (1996) Potential interaction between warfarin and dicloxacillin. Ann Pharmacother 30(12):1402–1407
66. Khalili H, Nikvarz N, Najmeddin F, Dashti-Khavidaki S (2013) A probable clinically significant interaction between warfarin and cloxacillin: three case reports. Eur J Clin Pharmacol 69(3):721–724
67. Merwick A, Hannon N, Kelly PJ, O'Rourke K (2010) Warfarin-flucloxacillin interaction presenting as cardioembolic ischemic stroke. Eur J Clin Pharmacol 66(6):643–644
68. Garg A, Mohammed M (2009) Decreased INR response secondary to warfarin-flucloxacillin interaction. Ann Pharmacother 43(7):1374–1375
69. Marusic S, Gojo-Tomic N, Bacic-Vrca V, Franic M (2012) Enhanced anticoagulant effect of warfarin in a patient treated with cloxacillin. Int J Clin Pharmacol Ther 50(6):431–433
70. Rolinson GN, Sutherland R (1965) The binding of antibiotics to serum proteins. Br J Pharmacol Chemother 25(3):638–650
71. Lang CC, Jamal SK, Mohamed Z, Mustafa MR, Mustafa AM, Lee TC (2003) Evidence of an interaction between nifedipine and nafcillin in humans. Br J Clin Pharmacol 55(6):588–590
72. Cropp JS, Bussey HI (1997) A review of enzyme induction of warfarin metabolism with recommendations for patient management. Pharmacotherapy 17(5):917–928
73. Lacey CS (2004) Interaction of dicloxacillin with warfarin. Ann Pharmacother 38(5):898
74. Brown M, Korchinski E, Miller D (1979) Interaction of penicillin-G and warfarin? Can J Hosp Pharm 32:18–19
75. Davydov L, Yermolnik M, Cuni LJ (2003) Warfarin and amoxicillin/clavulanate drug interaction. Ann Pharmacother 37(3):367–370
76. Goodchild JH, Donaldson M (2013) A clinically significant drug interaction between warfarin and amoxicillin resulting in persistent postoperative bleeding in a dental patient. Gen Dent 61(4):50–54
77. Larsen TR, Gelaye A, Durando C (2014) Acute warfarin toxicity: an unanticipated consequence of amoxicillin/clavulanate administration. Am J Case Rep 15:45–48
78. Lippi G, Favaloro EJ, Mattiuzzi C (2014) Combined administration of antibiotics and direct oral anticoagulants: a renewed indication for laboratory monitoring? Semin Thromb Hemost 40(7):756–765
79. Kampmann J, Hansen JM, Siersboek-Nielsen K, Laursen H (1972) Effect of some drugs on penicillin half-life in blood. Clin Pharmacol Ther 13(4):516–519
80. Kunin CM (1966) Clinical pharmacology of the new penicillins. II. Effect of drugs which interfere with binding to serum proteins. Clin Pharmacol Ther 7(2):180–188
81. Suffness M, Rose B (1974) Potential drug interactions and adverse effects related to aspirin. Drug Intell Clin Pharm 8:694–699
82. Moskowitz B, Somani SM, McDonald RH (1973) Salicylate interaction with penicillin and secobarbital binding sites on human serum albumin. Clin Toxicol 6(2):247–256
83. Hayes AH (1981) Therapeutic implications of drug interactions with acetaminophen and aspirin. Arch Intern Med 141(3 Spec No):301–304
84. Schäfer-Korting M, Kirch W, Axthelm T, Köhler H, Mutschler E (1983) Atenolol interaction with aspirin, allopurinol, and ampicillin. Clin Pharmacol Ther 33(3):283–288
85. McLean AJ, Tonkin A, McCarthy P, Harrison P (1984) Dose-dependence of atenolol-ampicillin interaction. Br J Clin Pharmacol 18(6):969–971
86. Westphal JF, Trouvin JH, Deslandes A, Carbon C (1990) Nifedipine enhances amoxicillin absorption kinetics and bioavailability in humans. J Pharmacol Exp Ther 255(1):312–317
87. Yoo HH, Kim IS, Yoo DH, Kim DH (2016) Effects of orally administered antibiotics on the bioavailability of amlodipine: gut microbiota-mediated drug interaction. J Hypertens 34(1):156–162
88. Jawetz E, Gunnison JB, Coleman VR (1950) The combined action of penicillin with streptomycin or chloromycetin on enterococci in vitro. Science 111(2880):254–256

89. Wallace JF, Smith RH, Garcia M, Petersdorf RG (1967) Studies on the pathogenesis of meningitis. VI. Antagonism between penicillin and chloramphenicol in experimental pneumococcal meningitis. J Lab Clin Med 70(3):408–418
90. Yourassowsky E, Monsieur R (1971) Antagonism limit of penicillin G and chloramphenicol against Neisseria meningitidis. Arzneimittelforschung 21(9):1385–1387
91. Mackenzie AM, Chan FT (1986) Combined action of chloramphenicol and ampicillin on chloramphenicol-resistant Haemophilus influenzae. Antimicrob Agents Chemother 29(4):565–569
92. Solberg O, Andersen BM (1979) The interaction of penicillin and chloramphenicol against meningococci in vitro. Acta Pathol Microbiol Scand B 87B(2):103–107
93. Asmar BI, Dajani AS (1983) Ampicillin-chloramphenicol interaction against enteric Gram-negative organisms. Pediatr Infect Dis 2(1):39–42
94. Gjessing HC, Odegaard K (1967) Oral chloramphenicol alone and with intramuscular procaine penicillin in the treatment of gonorrhoea. Br J Vener Dis 43(2):133–136
95. De Ritis F, Giammanco G, Manzillo G (1972) Chloramphenicol combined with ampicillin in treatment of typhoid. Br Med J 4(5831):17–18
96. Ali HM (1985) Reduced ampicillin bioavailability following oral coadministration with chloroquine. J Antimicrob Chemother 15(6):781–784
97. Sultana N, Arayne MS (1986) Drug antibiotic interactions-antimalarials. J Pak Med Assoc 36(2):37–40
98. Pohlman JK, Knapp CC, Ludwig MD, Washington JA (1996) Timed killing kinetic studies of the interaction between ciprofloxacin and beta-lactams against gram-negative bacilli. Diagn Microbiol Infect Dis 26(1):29–33
99. Chin NX, Jules K, Neu HC (1986) Synergy of ciprofloxacin and azlocillin in vitro and in a neutropenic mouse model of infection. Eur J Clin Microbiol 5(1):23–28
100. Smith SM, Eng RH (1988) Interaction of ciprofloxacin with ampicillin and vancomycin for Streptococcus faecalis. Diagn Microbiol Infect Dis 9(4):239–243
101. Fuursted K, Gerner-Smidt P (1987) Analysis of the interaction between piperacillin and ciprofloxacin or tobramycin against thirteen strains of Pseudomonas aeruginosa, using killing curves. Acta Pathol Microbiol Immunol Scand B 95(3):193–197
102. Overbeek BP, Rozenberg-Arska M, Verhoef J (1985) Interaction between ciprofloxacin and tobramycin or azlocillin against multiresistant strains of Acinetobacter anitratum in vitro. Eur J Clin Microbiol 4(2):140–141
103. Smith CE, Foleno BE, Barrett JF, Frosco MB (1997) Assessment of the synergistic interactions of levofloxacin and ampicillin against Enterococcus faecium by the checkerboard agar dilution and time-kill methods. Diagn Microbiol Infect Dis 27(3):85–92
104. Barriere SL, Catlin DH, Orlando PL, Noe A, Frost RW (1990) Alteration in the pharmacokinetic disposition of ciprofloxacin by simultaneous administration of azlocillin. Antimicrob Agents Chemother 34(5):823–826
105. Peterson LR, Moody JA, Fasching CE, Gerding DN (1987) In vivo and in vitro activity of ciprofloxacin plus azlocillin against 12 streptococcal isolates in a neutropenic site model. Diagn Microbiol Infect Dis 7(2):127–136
106. Orlando PL, Barriere SL, Hindler JA, Frost RW (1990) Serum bactericidal activity from intravenous ciprofloxacin and azlocillin given alone and in combination to healthy subjects. Diagn Microbiol Infect Dis 13(2):93–97
107. Drug interaction with oral contraceptive steroids. Br Med J 281(6233):93–94 (1980)
108. DeSano EA, Hurley SC (1982) Possible interactions of antihistamines and antibiotics with oral contraceptive effectiveness. Fertil Steril 37(6):853–854
109. Bainton R (1986) Interaction between antibiotic therapy and contraceptive medication. Oral Surg Oral Med Oral Pathol 61(5):453–455
110. Silber TJ (1983) Apparent oral contraceptive failure associated with antibiotic administration. J Adolesc Health Care 4(4):287–289

111. Miller DM, Helms SE, Brodell RT (1994) A practical approach to antibiotic treatment in women taking oral contraceptives. J Am Acad Dermatol 30(6):1008–1011

112. Fernández N, Sierra M, Diez MJ, Terán T, Pereda P, García JJ (1997) Study of the pharmacokinetic interaction between ethinylestradiol and amoxicillin in rabbits. Contraception 55(1):47–52

113. Back DJ, Breckenbridge AM, Cross KJ, Orme ML, Thomas E (1982) An antibiotic interaction with ethinyloestradiol in the rat and rabbit. J Steroid Biochem 16(3):407–413

114. Back DJ, Breckenridge AM, MacIver M, Orme M, Rowe PH, Staiger C et al (1982) The effects of ampicillin on oral contraceptive steroids in women. Br J Clin Pharmacol 14(1):43–48

115. Friedman CI, Huneke AL, Kim MH, Powell J (1980) The effect of ampicillin on oral contraceptive effectiveness. Obstet Gynecol 55(1):33–37

116. Joshi JV, Joshi UM, Sankholi GM, Krishna U, Mandlekar A, Chowdhury V et al (1980) A study of interaction of low-dose combination oral contraceptive with Ampicillin and Metronidazole. Contraception 22(6):643–652

117. Weisberg E (1999) Interactions between oral contraceptives and antifungals/antibacterials. Is contraceptive failure the result? Clin Pharmacokinet 36(5):309–313

118. Dickinson BD, Altman RD, Nielsen NH, Sterling ML, Council on Scientific Affairs AeMA (2001) Drug interactions between oral contraceptives and antibiotics. Obstet Gynecol 98(5 Pt 1):853–860

119. Veremis SA, Maddux MS, Pollak R, Mozes MF (1987) Subtherapeutic cyclosporine concentrations during nafcillin therapy. Transplantation 43(6):913–915

120. Jahansouz F, Kriett JM, Smith CM, Jamieson SW (1993) Potentiation of cyclosporine nephrotoxicity by nafcillin in lung transplant recipients. Transplantation 55(5):1045–1048

121. Soares GM, Teles F, Starr JR, Feres M, Patel M, Martin L et al (2015) Effects of azithromycin, metronidazole, amoxicillin, and metronidazole plus amoxicillin on an in vitro polymicrobial subgingival biofilm model. Antimicrob Agents Chemother 59(5):2791–2798

122. Finland M, Bach MC, Garner C, Gold O (1974) Synergistic action of ampicillin and erythromycin against Nocardia asteroides: effect of time of incubation. Antimicrob Agents Chemother 5(3):344–353

123. Rosenfeld LS, Kirby WMM (1963) Synergism of erythromycin and penicillin against resistant staphylococci: mechanism and relation to synthetic penicilins. Antimicrob Agents Chemother:831–842. 1962

124. Waterworth P (1963) Apparent synergy between penicillin and erythromycin or fusidic acid. Clin Med (Northfield) 70:941–953

125. Oswald EJ, Reedy RJ, Wright WW (1962) Antibiotic combinations: an in vitro study of antistaphylococcal effects of erythromycin plus penicillin, streptomycin, or tetracycline. Antimicrob Agents Chemother:904–910. 1961

126. Herrell W, Balows A, Becker J (1960) Erythrocillin: a new approach to the problem of antibiotic-resistant staphylococci. Antibiotic Med Clin Ther (New York) 7:637–642

127. Manten A (1954) Synergism and antagonism between antibiotic mixtures containing erythromycin. Antibiot Chemother (Northfield) 4(12):1228–1233

128. Allen NE, Epp JK (1978) Mechanism of penicillin-erythromycin synergy on antibiotic-resistant Staphylococcus aureus. Antimicrob Agents Chemother 13(5):849–853

129. Manten A, Terra J (1964) The antagonism between penicillin and other antibiotics in relation to drug concentrtion. Chemotherapia (Basel) 64:21–29

130. Chang TW, Weinstein L (1966) Inhibitory effects of other antibiotics on bacterial morphologic changes induced by penicillin G. Nature 211(5050):763–765

131. Strom J (1961) Penicillin and erythromycin singly and in combination in scarlatina therapy and the interference between them. Antibiot Chemother (Northfield). 11:694–697

132. Bach MC, Monaco AP, Finland M (1973) Pulmonary nocardiosis. Therapy with minocycline and with erythromycin plus ampicillin. JAMA 224(10):1378–1381

133. Robinson L, Fonseca K (1982) Value of the minimum bactericidal concentration of antibiotics in the management of a case of recurrent Streptococcus bovis septicaemia. J Clin Pathol 35(8):879–880

134. Garrod L, Waterworth P (1962) Methods of testing combined antibiotic bactericidal action and the significance of the results. J Clin Pathol 15:328–338

135. Martínez JA, Horcajada JP, Almela M, Marco F, Soriano A, García E et al (2003) Addition of a macrolide to a beta-lactam-based empirical antibiotic regimen is associated with lower in-hospital mortality for patients with bacteremic pneumococcal pneumonia. Clin Infect Dis 36(4):389–395

136. Waterer GW, Somes GW, Wunderink RG (2001) Monotherapy may be suboptimal for severe bacteremic pneumococcal pneumonia. Arch Intern Med 161(15):1837–1842

137. Mufson MA, Stanek RJ (1999) Bacteremic pneumococcal pneumonia in one American City: a 20-year longitudinal study, 1978-1997. Am J Med 107(1A):34S–43S

138. Mandell LA, Wunderink RG, Anzueto A, Bartlett JG, Campbell GD, Dean NC et al (2007) Infectious Diseases Society of America/American Thoracic Society consensus guidelines on the management of community-acquired pneumonia in adults. Clin Infect Dis 44(Suppl 2):S27–S72

139. Johansen HK, Jensen TG, Dessau RB, Lundgren B, Frimodt-Moller N (2000) Antagonism between penicillin and erythromycin against Streptococcus pneumoniae in vitro and in vivo. J Antimicrob Chemother 46(6):973–980

140. Fuursted K, Knudsen JD, Petersen MB, Poulsen RL, Rehm D (1997) Comparative study of bactericidal activities, postantibiotic effects, and effects of bacterial virulence of penicillin G and six macrolides against Streptococcus pneumoniae. Antimicrob Agents Chemother 41(4):781–784

141. Huupponen R, Seppälä P, Iisalo E (1984) Effect of guar gum, a fibre preparation, on digoxin and penicillin absorption in man. Eur J Clin Pharmacol 26(2):279–281

142. Brooks BM, Thomas AL, Coleman JW (2003) Benzylpenicillin differentially conjugates to IFN-gamma, TNF-alpha, IL-1beta, IL-4 and IL-13 but selectively reduces IFN-gamma activity. Clin Exp Immunol 131(2):268–274

143. Attef OA, Ali AA, Ali HM (1997) Effect of Khat chewing on the bioavailability of ampicillin and amoxycillin. J Antimicrob Chemother 39(4):523–525

144. Iven H, Brasch H (1988) The effects of antibiotics and uricosuric drugs on the renal elimination of methotrexate and 7-hydroxymethotrexate in rabbits. Cancer Chemother Pharmacol 21(4):337–342

145. Iven H, Brasch H (1986) Influence of the antibiotics piperacillin, doxycycline, and tobramycin on the pharmacokinetics of methotrexate in rabbits. Cancer Chemother Pharmacol 17(3):218–222

146. Najjar TA, Abou-Auda HS, Ghilzai NM (1998) Influence of piperacillin on the pharmacokinetics of methotrexate and 7-hydroxymethotrexate. Cancer Chemother Pharmacol 42(5):423–428

147. Yamamoto K, Sawada Y, Matsushita Y, Moriwaki K, Bessho F, Iga T (1997) Delayed elimination of methotrexate associated with piperacillin administration. Ann Pharmacother 31(10):1261–1262

148. Bloom E, Ignoffo R, Reis C, Cadman E (1986) Delayed clearance of methotrexate associated with antibiotics and anti-inflammatory agents. Clin Res 34:560A

149. Nierenberg DW, Mamelok RD (1983) Toxic reaction to methotrexate in a patient receiving penicillin and furosemide: a possible interaction. Arch Dermatol 119(6):449–450

150. Dean R, Nachman J, Lorenzana AN (1992) Possible methotrexate-mezlocillin interaction. Am J Pediatr Hematol Oncol 14(1):88–89

151. Ronchera CL, Hernández T, Peris JE, Torres F, Granero L, Jiménez NV et al (1993) Pharmacokinetic interaction between high-dose methotrexate and amoxycillin. Ther Drug Monit 15(5):375–379

152. Titier K, Lagrange F, Péhourcq F, Moore N, Molimard M (2002) Pharmacokinetic interaction between high-dose methotrexate and oxacillin. Ther Drug Monit 24(4):570–572
153. Mayall B, Poggi G, Parkin JD (1991) Neutropenia due to low-dose methotrexate therapy for psoriasis and rheumatoid arthritis may be fatal. Med J Aust 155(7):480–484
154. Zarychanski R, Wlodarczyk K, Ariano R, Bow E (2006) Pharmacokinetic interaction between methotrexate and piperacillin/tazobactam resulting in prolonged toxic concentrations of methotrexate. J Antimicrob Chemother 58(1):228–230
155. Herrick AL, Grennan DM, Aarons L (1999) Lack of interaction between methotrexate and penicillins. Rheumatology (Oxford) 38(3):284–285
156. Hill G, Cihlar T, Oo C, Ho ES, Prior K, Wiltshire H et al (2002) The anti-influenza drug oseltamivir exhibits low potential to induce pharmacokinetic drug interactions via renal secretion-correlation of in vivo and in vitro studies. Drug Metab Dispos 30(1):13–19
157. Gravenkemper C, Bennett J, Brodie J, Kirby W (1965) Dicloxacillin. In vitro and pharmacologic comparisons with oxacillin and cloxacillin. Arch Intern Med 116:340–345
158. Dasgupta A, Sperelakis A, Mason A, Dean R (1997) Phenytoin-oxacillin interactions in normal and uremic sera. Pharmacotherapy 17(2):375–378
159. Dasgupta A, Dennen DA, Dean R, McLawhon RW (1991) Displacement of phenytoin from serum protein carriers by antibiotics: studies with ceftriaxone, nafcillin, and sulfamethoxazole. Clin Chem 37(1):98–100
160. Welling PG, Dean S, Selen A, Kendall MJ, Wise R (1979) Probenecid: an unexplained effect on cephalosporin pharmacology. Br J Clin Pharmacol 8(5):491–495
161. Gibaldi M, Schwartz MA (1968) Apparent effect of probenecid on the distribution of penicillins in man. Clin Pharmacol Ther 9(3):345–349
162. Barbhaiya R, Thin RN, Turner P, Wadsworth J (1979) Clinical pharmacological studies of amoxycillin: effect of probenecid. Br J Vener Dis 55(3):211–213
163. Ziv G, Sulman FG (1974) Effects of probenecid on the distribution, elimination, and passage into milk of benzylpenicillin, ampicillin and cloxacillin. Arch Int Pharmacodyn Ther 207(2):373–382
164. Shanson DC, McNabb R, Hajipieris P (1984) The effect of probenecid on serum amoxycillin concentrations up to 18 hours after a single 3 g oral dose of amoxycillin: possible implications for preventing endocarditis. J Antimicrob Chemother 13(6):629–632
165. Waller ES, Sharanevych MA, Yakatan GJ (1982) The effect of probenecid on nafcillin disposition. J Clin Pharmacol 22(10):482–489
166. Corvaia L, Li SC, Ioannides-Demos LL, Bowes G, Spicer WJ, Spelman DW et al (1992) A prospective study of the effects of oral probenecid on the pharmacokinetics of intravenous ticarcillin in patients with cystic fibrosis. J Antimicrob Chemother 30(6):875–878
167. Krogsgaard MR, Hansen BA, Slotsbjerg T, Jensen P (1994) Should probenecid be used to reduce the dicloxacillin dosage in orthopaedic infections? A study of the dicloxacillin-saving effect of probenecid. Pharmacol Toxicol 74(3):181–184
168. Hoffstedt B, Haidl S, Walder M (1983) Influence of probenecid on serum and subcutaneous tissue fluid concentrations of benzylpenicillin and ceftazidime in human volunteers. Eur J Clin Microbiol 2(6):604–606
169. Maeda K, Tian Y, Fujita T, Ikeda Y, Kumagai Y, Kondo T et al (2014) Inhibitory effects of p-aminohippurate and probenecid on the renal clearance of adefovir and benzylpenicillin as probe drugs for organic anion transporter (OAT) 1 and OAT3 in humans. Eur J Pharm Sci 59:94–103
170. Ganes D, Batra A, Faulkner D et al (1991) Effect of probenecid on the pharmacokinetics of piperacillin and tazobactam in healthy volunteers. Pharm Res 8:S299
171. Catlin BW (1984) Probenecid: antibacterial action against Neisseria gonorrhoeae and interaction with benzylpenicillin. Antimicrob Agents Chemother 25(6):676–682
172. Babalola CP, Iwheye GB, Olaniyi AA (2002) Effect of proguanil interaction on bioavailability of cloxacillin. J Clin Pharm Ther 27(6):461–464

173. Spek R, Jawetz E, Gunnison J (1951) Studies on antibiotic synergism and antagonism; the interference of aureomycin or terramycin with the action of penicillin in infections in mice. AMA Arch Intern Med 88(2):168–174

174. Gunnison J, Coleman V, Jawetz E (1950) Interference of aureomycin and of terramycin with action of penicillin in vitro. Proc Soc Exp Biol Med 75(2):549–552

175. Chuang YC, Liu JW, Ko WC, Lin KY, JJ W, Huang KY (1997) In vitro synergism between cefotaxime and minocycline against Vibrio vulnificus. Antimicrob Agents Chemother 41(10):2214–2217

176. Chuang YC, Ko WC, Wang ST, Liu JW, Kuo CF, JJ W et al (1998) Minocycline and cefotaxime in the treatment of experimental murine Vibrio vulnificus infection. Antimicrob Agents Chemother 42(6):1319–1322

177. Ko WC, Lee HC, Chuang YC, Ten SH, Su CY, Wu JJ (2001) In vitro and in vivo combinations of cefotaxime and minocycline against Aeromonas hydrophila. Antimicrob Agents Chemother 45(4):1281–1283

178. Lepper M, Dowling H (1951) Treatment of pneumococcic meningitis with penicillin compared with penicillin plus aureomycin; studies including observations on an apparent antagonism between penicillin and aureomycin. AMA Arch Intern Med 88(4):489–494

179. Olsson R, Kirby J, Romansky M (1961) Pneumococcal meningitis in the adult. Clinical, therapeutic, and prognostic aspects in forty-three patients. Ann Intern Med 55:545–549

180. Strom J (1955) The question of antagonism between penicillin and chlortetracycline, illustrated by therapeutical experiments in scarlatina. Antibiotic Med Clin Ther (New York). 1(1):6–12

181. Ahern J, Kirby W (1953) Lack of interference of aureomycin with penicillin in treatment of pneumococcic pneumonia. AMA Arch Intern Med 91(2):197–203

182. Song Z, Suo B, Zhang L, Zhou L (2016) Rabeprazole, minocycline, amoxicillin, and bismuth as first-line and second-line regimens for helicobacter pylori eradication. Helicobacter 21(6):462–470

183. Lee JY, Kim N, Park KS, Kim HJ, Park SM, Baik GH et al (2016) Comparison of sequential therapy and amoxicillin/tetracycline containing bismuth quadruple therapy for the first-line eradication of Helicobacter pylori: a prospective, multi-center, randomized clinical trial. BMC Gastroenterol 16(1):79

184. Burgess LD, Drew RH (2014) Comparison of the incidence of vancomycin-induced nephrotoxicity in hospitalized patients with and without concomitant piperacillin-tazobactam. Pharmacotherapy 34(7):670–676

185. Kim T, Kandiah S, Patel M, Rab S, Wong J, Xue W et al (2015) Risk factors for kidney injury during vancomycin and piperacillin/tazobactam administration, including increased odds of injury with combination therapy. BMC Res Notes 8:579

186. Rutter WC, Burgess DR, Talbert JC, Burgess DS (2017) Acute kidney injury in patients treated with vancomycin and piperacillin-tazobactam: a retrospective cohort analysis. J Hosp Med 12(2):77–82

187. Kureishi A, Jewesson PJ, Rubinger M, Cole CD, Reece DE, Phillips GL et al (1991) Double-blind comparison of teicoplanin versus vancomycin in febrile neutropenic patients receiving concomitant tobramycin and piperacillin: effect on cyclosporin A-associated nephrotoxicity. Antimicrob Agents Chemother 35(11):2246–2252

188. Gomes DM, Smotherman C, Birch A, Dupree L, Della Vecchia BJ, Kraemer DF et al (2014) Comparison of acute kidney injury during treatment with vancomycin in combination with piperacillin-tazobactam or cefepime. Pharmacotherapy 34(7):662–669

189. Peyko V, Smalley S, Cohen H (2016) Prospective comparison of acute kidney injury during treatment with the combination of piperacillin-tazobactam and vancomycin versus the combination of cefepime or meropenem and vancomycin. J Pharm Pract 30(2):209–213

190. Rutter WC, Cox JN, Martin CA, Burgess DR, Burgess DS. 2017. Nephrotoxicity during vancomycin therapy in combinationwith piperacillin-tazobactam or cefepime. Antimicrob Agents Chemother 61(2):e02089-16. https://doi.org/10.1128/AAC.02089-16

191. Navalkele B, Pogue JM, Karino S, Nishan B, Salim M, Solanki S et al (2017) Risk of acute kidney injury in patients on concomitant vancomycin and piperacillin-tazobactam compared to those on vancomycin and cefepime. Clin Infect Dis 64(2):116–123

192. Rutter WC, Burgess DS (2017) Acute kidney injury in patients treated with IV beta-lactam/beta-lactamase inhibitor combinations. Pharmacotherapy 37(5):593–598

193. Karino S, Kaye KS, Navalkele B, Nishan B, Salim M, Solanki S et al (2016) Epidemiology of acute kidney injury among patients receiving concomitant vancomycin and piperacillin-tazobactam: opportunities for antimicrobial stewardship. Antimicrob Agents Chemother 60(6):3743–3750

194. Mousavi M, Zapolskaya T, Scipione MR, Louie E, Papadopoulos J, Dubrovskaya Y (2017) Comparison of rates of nephrotoxicity associated with vancomycin in combination with piperacillin-tazobactam administered as an extended versus standard infusion. Pharmacotherapy 37(3):379–385

195. Ibach BW, Henry ED, Johnson PN (2016) Acute kidney injury in a child receiving vancomycin and piperacillin/tazobactam. J Pediatr Pharmacol Ther 21(2):169–175

196. McQueen KE, Clark DW (2016) Does combination therapy with vancomycin and piperacillin-tazobactam increase the risk of nephrotoxicity versus vancomycin alone in pediatric patients? J Pediatr Pharmacol Ther 21(4):332–338

197. Gao S, Li J, Li Z (2015) Comparison of the incidence of vancomycin-induced nephrotoxicity in hospitalized children with or without concomitant piperacillin-tazobactam. Fudan Univ J Med Sci 42:743–748

198. Hammond DA, Smith MN, Li C, Hayes SM, Lusardi K, Brandon Bookstaver P (2017) Systematic review and meta-analysis of acute kidney injury associated with concomitant vancomycin and piperacillin/tazobactam. Clin Infect Dis 64(5):666–674

199. Giuliano CA, Patel CR, Kale-Pradhan PBI (2016) The combination of piperacillin-tazobactam and vancomycin associated with development of acute kidney injury? A meta-analysis. Pharmacotherapy 36(12):1217–1228

200. Rindone JP, Mellen C, Ryba J (2016) Does piperacillin-tazobactam increase the risk of nephrotoxicity when used with vancomycin: a meta-analysis of observational trials. Curr Drug Saf 12(1):62–66

201. Moenster RP, Linneman TW, Finnegan PM, Hand S, Thomas Z, McDonald JR (2014) Acute renal failure associated with vancomycin and β-lactams for the treatment of osteomyelitis in diabetics: piperacillin-tazobactam as compared with cefepime. Clin Microbiol Infect 20(6):O384–O389

202. Hammond DA, Smith MN, Painter JT, Meena NK, Lusardi K (2016) Comparative incidence of acute kidney injury in critically ill patients receiving vancomycin with concomitant piperacillin-tazobactam or cefepime: a retrospective cohort study. Pharmacotherapy 36(5):463–471

203. Al Yami MS (2017) Comparison of the incidence of acute kidney injury during treatment with vancomycin in combination with piperacillin-tazobactam or with meropenem. J Infect Public Health 10(6):770–773

204. Davies SW, Efird JT, Guidry CA, Dietch ZC, Willis RN, Shah PM et al (2016) Top guns: the "maverick" and "goose" of empiric therapy. Surg Infect 17(1):38–47

205. Tryba M (1985) Potentiation of the effect of non-depolarizing muscle relaxants by acylaminopenicillins. Studies on the example of vecuronium. Anaesthesist 34(12):651–655

206. Singh YN, Harvey AL, Marshall IG (1978) Antibiotic-induced paralysis of the mouse phrenic nerve-hemidiaphragm preparation, and reversibility by calcium and by neostigmine. Anesthesiology 48(6):418–424

207. Harwood TN, Moorthy SS (1989) Prolonged vecuronium-induced neuromuscular blockade in children. Anesth Analg 68(4):534–536

208. Mackie K, Pavlin EG (1990) Recurrent paralysis following piperacillin administration. Anesthesiology 72(3):561–563

209. Condon RE, Munshi CA, Arfman RC (1995) Interaction of vecuronium with piperacillin or cefoxitin evaluated in a prospective, randomized, double-blind clinical trial. Am Surg 61(5):403–406
210. Nierenberg DW (1987) Drug inhibition of penicillin tubular secretion: concordance between in vitro and clinical findings. J Pharmacol Exp Ther 240(3):712–716
211. Gothoni G, Pentikäinen P, Vapaatalo HI, Hackman R, af Björksten K (1972) Absorption of antibiotics: influence of metoclopramide and atropine on serum levels of pivampicillin and tetracycline. Ann Clin Res 4(4):228–232
212. Kadlec GJ, Ha LT, Jarboe CH, Richards D, Karibo JM (1978) Effect of ampicillin on theophylline half-life in infants and young children. South Med J 71(12):1584
213. Jonkman JH, van der Boon WJ, Schoenmaker R, Holtkamp A, Hempenius J (1985) Lack of effect of amoxicillin on theophylline pharmacokinetics. Br J Clin Pharmacol 19(1):99–101
214. Matera MG, Cazzola M, Lampa E, Santangelo G, Paizis G, Vinciguerra A et al (1993) Clinical pharmacokinetics of theophylline during co-treatment with ticarcillin plus clavulanic acid in patients suffering from acute exacerbation of chronic bronchitis. J Chemother 5(4):233–236
215. Cazzola M, Santangelo G, Guidetti E, Mattina R, Caputi M, Girbino G (1991) Influence of sulbactam plus ampicillin on theophylline clearance. Int J Clin Pharmacol Res 11(1):11–15
216. Jonkman JH, van der Boon WJ, Schoenmaker R, Holtkamp AH, Hempenius J (1985) Clinical pharmacokinetics of amoxycillin and theophylline during cotreatment with both medicaments. Chemotherapy 31(5):329–335
217. Connor H (2003) Serotonin syndrome after single doses of co-amoxiclav during treatment with venlafaxine. J R Soc Med 96(5):233–234
218. Hughes GS, Heald DL, Barker KB, Patel RK, Spillers CR, Watts KC et al (1989) The effects of gastric pH and food on the pharmacokinetics of a new oral cephalosporin, cefpodoxime proxetil. Clin Pharmacol Ther 46(6):674–685
219. Saathoff N, Lode H, Neider K, Depperman KM, Borner K, Koeppe P (1992) Pharmacokinetics of cefpodoxime proxetil and interactions with an antacid and an H2 receptor antagonist. Antimicrob Agents Chemother 36(4):796–800
220. Spectracef® package insert. Cornerstone BioPharma, Inc. (2007) [Press release]
221. Omnicef® package insert. Abbott Laboratories (2005)
222. Madaras-Kelly K, Michas P, George M, May MP, Adejare A (2004) A randomized crossover study investigating the influence of ranitidine or omeprazole on the pharmacokinetics of cephalexin monohydrate. J Clin Pharmacol 44(12):1391–1397
223. Satterwhite JH, Cerimele BJ, Coleman DL, Hatcher BL, Kisicki J, DeSante KA (1992) Pharmacokinetics of cefaclor AF: effects of age, antacids and H2-receptor antagonists. Postgrad Med J 68(Suppl 3):S3–S9
224. Healy DP, Sahai JV, Sterling LP, Racht EM (1989) Influence of an antacid containing aluminum and magnesium on the pharmacokinetics of cefixime. Antimicrob Agents Chemother 33(11):1994–1997
225. Radwanski E, Nomeir A, Cutler D, Affrime M, Lin CC (1998) Pharmacokinetic drug interaction study: administration of ceftibuten concurrently with the antacid mylanta double-strength liquid or with ranitidine. Am J Ther 5(2):67–72
226. Arayne MS, Sultana N, Afzal M (2007) Cephradine antacids interaction studies. Pak J Pharm Sci 20(3):179–184
227. Miller B, Hershberger E, Benziger D, Trinh M, Friedland I (2012) Pharmacokinetics and safety of intravenous ceftolozane-tazobactam in healthy adult subjects following single and multiple ascending doses. Antimicrob Agents Chemother 56(6):3086–3091
228. Das S, Li J, Armstrong J, Learoyd M, Edeki T (2015) Randomized pharmacokinetic and drug-drug interaction studies of ceftazidime, avibactam, and metronidazole in healthy subjects. Pharmacol Res Perspect 3(5):e00172
229. Vishwanathan K, Mair S, Gupta A, Atherton J, Clarkson-Jones J, Edeki T et al (2014) Assessment of the mass balance recovery and metabolite profile of avibactam in humans and in vitro drug-drug interaction potential. Drug Metab Dispos 42(5):932–942

230. Food and Drug Administration. Rocephin (ceftriaxone sodium) for injection [Press release]. Available at. http://www.fda.gov/Safety/MedWatch/SafetyInformation/SafetyAlertsforHumanMedicalProducts/ucm152863.htm. September 11, 2007

231. Health Canada. Fatal interactions ceftriaxone-calcium-notice to hospitals [Press release]. Available at: http://healthycanadians.gc.ca/recall-alert-rappel-avis/hc-sc/2008/14497a-eng.php. July 31, 2008

232. Grasberger H, Otto B, Loeschke K (2000) Ceftriaxone-associated nephrolithiasis. Ann Pharmacother 34(9):1076–1077

233. Bradley JS, Wassel RT, Lee L, Nambiar S (2009) Intravenous ceftriaxone and calcium in the neonate: assessing the risk for cardiopulmonary adverse events. Pediatrics 123(4):e609–e613

234. Avci Z, Koktener A, Uras N, Catal F, Karadag A, Tekin O et al (2004) Nephrolithiasis associated with ceftriaxone therapy: a prospective study in 51 children. Arch Dis Child 89(11):1069–1072

235. Park HZ, Lee SP, Schy AL (1991) Ceftriaxone-associated gallbladder sludge. Identification of calcium-ceftriaxone salt as a major component of gallbladder precipitate. Gastroenterology 100(6):1665–1670

236. Food and Drug Administration. Information for healthcare professionals, Ceftriaxone (marketed as Rocephin and generics) [Press release]. Available at http://www.fda.gov/Drugs/DrugSafety/PostmarketDrugSafetyInformationforPatientsandProviders/DrugSafetyInformationforHeathcareProfessionals/ucm084263.htm. April 21, 2009

237. Steadman E, Raisch DW, Bennett CL, Esterly JS, Becker T, Postelnick M et al (2010) Evaluation of a potential clinical interaction between ceftriaxone and calcium. Antimicrob Agents Chemother 54(4):1534–1540

238. Deslandes A, Camus F, Lacroix C, Carbon C, Farinotti R (1996) Effects of nifedipine and diltiazem on pharmacokinetics of cefpodoxime following its oral administration. Antimicrob Agents Chemother 40(12):2879–2881

239. Duverne C, Bouten A, Deslandes A, Westphal JF, Trouvin JH, Farinotti R et al (1992) Modification of cefixime bioavailability by nifedipine in humans: involvement of the dipeptide carrier system. Antimicrob Agents Chemother 36(11):2462–2467

240. Berlioz F, Julien S, Tsocas A, Chariot J, Carbon C, Farinotti R et al (1999) Neural modulation of cephalexin intestinal absorption through the di- and tripeptide brush border transporter of rat jejunum in vivo. J Pharmacol Exp Ther 288(3):1037–1044

241. Ding Y, Liu W, Jia Y, Lu C, Jin X, Yang J et al (2013) Effects of amlodipine on the oral bioavailability of cephalexin and cefuroxime axetil in healthy volunteers. J Clin Pharmacol 53(1):82–86

242. Marino E, Vicente M, Dominguez-Gil A (1983) Influence of cholestyramine on the pharmacokinetic parameters of cefadroxil after simultaneous administration. Int J Pharm 16:23–30

243. Parsons R, Paddock G, Hossack G (1975) Cholestyramine induced antibiotic malabsorption. In: Williams J, Geddes A (eds) Pharmacology of antibiotics. Plenum, New York/London, pp 191–198

244. Koch-Weser J, Sidel VW, Federman EB, Kanarek P, Finer DC, Eaton AE (1970) Adverse effects of sodium colistimethate. Manifestations and specific reaction rates during 317 courses of therapy. Ann Intern Med 72(6):857–868

245. Adler S, Segel DP (1971) Nonoliguric renal failure secondary to sodium colistimethate: a report of four cases. Am J Med Sci 262(2):109–114

246. Foord RD (1975) Cephaloridine, cephalothin and the kidney. J Antimicrob Chemother 1(3 Suppl):119–133

247. Coly-Mycin® M Parenteral. package insert. JHP Pharmaceuticals, LLC (2013)

248. Chang YL, Chou MH, Lin MF, Chen CF, Tsai TH (2001) Effect of cyclosporine, a P-glycoprotein inhibitor, on the pharmacokinetics of cefepime in rat blood and brain: a microdialysis study. Life Sci 69(2):191–199

249. Soto Alvarez J, Sacristán Del Castillo JA, Alsar Ortiz MJ (1991) Interaction between ciclosporin and ceftriaxone. Nephron 59(4):681–682

250. Cockburn I (1986) Cyclosporine A: a clinical evaluation of drug interactions. Transplant Proc 18(6 Suppl 5):50–55
251. Verhagen C, de Pauw BE, de Witte T, Holdrinet RS, Janssen JT, Williams KJ (1986) Ceftazidime does not enhance cyclosporin-A nephrotoxicity in febrile bone marrow transplantation patients. Blut 53(4):333–339
252. Kannangara DW, Gallagher K, Lefrock JL (1984) Disulfiram-like reactions with newer cephalosporins: cefmenoxime. Am J Med Sci 287(2):45–47
253. Foster TS, Raehl CL, Wilson HD (1980) Disulfiram-like reaction associated with a parenteral cephalosporin. Am J Hosp Pharm 37(6):858–859
254. Uri JV, Parks DB (1983) Disulfiram-like reaction to certain cephalosporins. Ther Drug Monit 5(2):219–224
255. Ueno K, Tanaka K, Tsujimura K, Morishima Y, Iwashige H, Yamazaki K et al (1993) Impairment of cefdinir absorption by iron ion. Clin Pharmacol Ther 54(5):473–475
256. Nelson JS (2000) Red stools and omnicef. J Pediatr 136(6):853–854
257. Lancaster J, Sylvia LM, Schainker E (2008) Nonbloody, red stools from coadministration of cefdinir and iron-supplemented infant formulas. Pharmacotherapy 28(5):678–681
258. Shukla UA, Pittman KA, Barbhaiya RH (1992) Pharmacokinetic interactions of cefprozil with food, propantheline, metoclopramide, and probenecid in healthy volunteers. J Clin Pharmacol 32(8):725–731
259. Neu HC (1982) The in vitro activity, human pharmacology, and clinical effectiveness of new beta-lactam antibiotics. Annu Rev Pharmacol Toxicol 22:599–642
260. Tran HX, Herrington JD (2016) Effect of ceftriaxone and cefepime on high-dose methotrexate clearance. J Oncol Pharm Pract 22(6):801–805
261. Jayasagar G, Krishna Kumar M, Chandrasekhar K, Madhusudan Rao C, Madhusudan Rao Y (2002) Effect of cephalexin on the pharmacokinetics of metformin in healthy human volunteers. Drug Metabol Drug Interact 19(1):41–48
262. Wang DS, Kusuhara H, Kato Y, Jonker JW, Schinkel AH, Sugiyama Y (2003) Involvement of organic cation transporter 1 in the lactic acidosis caused by metformin. Mol Pharmacol 63(4):844–848
263. Merle-Melet M, Bresler L, Lokiec F, Dopff C, Boissel P, Dureux JB (1992) Effects of diclofenac on ceftriaxone pharmacokinetics in humans. Antimicrob Agents Chemother 36(10):2331–2333
264. Popick AC, Crouthamel WG, Bekersky I (1987) Plasma protein binding of ceftriaxone. Xenobiotica 17(10):1139–1145
265. Andrassy K, Bechtold H, Ritz E (1985) Hypoprothrombinemia caused by cephalosporins. J Antimicrob Chemother 15(2):133–136
266. Freedy HR, Cetnarowski AB, Lumish RM, Schafer FJ (1986) Cefoperazone-induced coagulopathy. Drug Intell Clin Pharm 20(4):281–283
267. Rymer W, Greenlaw C (1980) Hypoprothrombinemia associated with cefamandole. Drug Intell Clin Pharm 14:780–783
268. Osborne JC (1985) Hypoprothrombinemia and bleeding due to cefoperazone. Ann Intern Med 102(5):721–722
269. Cristiano P (1984) Hypoprothrombinemia associated with cefoperazone treatment. Drug Intell Clin Pharm 18(4):314–316
270. Hooper CA, Haney BB, Stone HH (1980) Gastrointestinal bleeding due to vitamin K deficiency in patients on parenteral cefamandole. Lancet 1(8158):39–40
271. Pakter RL, Russell TR, Mielke CH, West D (1982) Coagulopathy associated with the use of moxalactam. JAMA 248(9):1100
272. Marier RL, Faro S, Sanders CV, Williams W, Derks F, Janney A et al (1982) Moxalactam in the therapy of serious infections. Antimicrob Agents Chemother 21(4):650–654
273. Angaran DM, Dias VC, Arom KV, Northrup WF, Kersten TG, Lindsay WG et al (1987) The comparative influence of prophylactic antibiotics on the prothrombin response to warfarin in

the postoperative prosthetic cardiac valve patient. Cefamandole, cefazolin, vancomycin. Ann Surg 206(2):155–161

274. Angaran DM, Dias VC, Arom KV, Northrup WF, Kersten TE, Lindsay WG et al (1984) The influence of prophylactic antibiotics on the warfarin anticoagulation response in the postoperative prosthetic cardiac valve patient. Cefamandole versus vancomycin. Ann Surg 199(1):107–111

275. Bohm NM, Crosby B (2012) Hemarthrosis in a patient on warfarin receiving ceftaroline: a case report and brief review of cephalosporin interactions with warfarin. Ann Pharmacother 46(7–8):e19

276. Farhat NM, Hutchinson LS, Peters M (2016) Elevated International Normalized Ratio values in a patient receiving warfarin and ceftaroline. Am J Health Syst Pharm 73(2):56–59

277. Saum LM, Balmat RP (2016) Ceftriaxone potentiates warfarin activity greater than other antibiotics in the treatment of urinary tract infections. J Pharm Pract 29(2):121–124

278. Bechtold H, Andrassy K, Jähnchen E, Koderisch J, Koderisch H, Weilemann LS et al (1984) Evidence for impaired hepatic vitamin K1 metabolism in patients treated with N-methyl-thiotetrazole cephalosporins. Thromb Haemost 51(3):358–361

279. Frick PG, Riedler G, Brögli H (1967) Dose response and minimal daily requirement for vitamin K in man. J Appl Physiol 23(3):387–389

280. Lipsky JJ (1983) N-methyl-thio-tetrazole inhibition of the gamma carboxylation of glutamic acid: possible mechanism for antibiotic-associated hypoprothrombinaemia. Lancet 2(8343):192–193

281. Lipsky JJ, Lewis JC, Novick WJ (1984) Production of hypoprothrombinemia by moxalactam and 1-methyl-5-thiotetrazole in rats. Antimicrob Agents Chemother 25(3):380–381

282. Verhagen CA, Mattie H, Van Strijen E (1994) The renal clearance of cefuroxime and ceftazidime and the effect of probenecid on their tubular excretion. Br J Clin Pharmacol 37(2):193–197

283. Lüthy R, Blaser J, Bonetti A, Simmen H, Wise R, Siegenthaler W (1981) Comparative multiple-dose pharmacokinetics of cefotaxime, moxalactam, and ceftazidime. Antimicrob Agents Chemother 20(5):567–575

284. LeBel M, Paone RP, Lewis GP (1983) Effect of probenecid on the pharmacokinetics of ceftizoxime. J Antimicrob Chemother 12(2):147–155

285. Reeves DS, Bullock DW, Bywater MJ, Holt HA, White LO, Thornhill DP (1981) The effect of probenecid on the pharmacokinetics and distribution of cefoxitin in healthy volunteers. Br J Clin Pharmacol 11(4):353–359

286. Mariño EL, Dominguez-Gil A (1981) The pharmacokinetics of cefadroxil associated with probenecid. Int J Clin Pharmacol Ther Toxicol 19(11):506–508

287. Garton AM, Rennie RP, Gilpin J, Marrelli M, Shafran SD (1997) Comparison of dose doubling with probenecid for sustaining serum cefuroxime levels. J Antimicrob Chemother 40(6):903–906

288. Spina SP, Dillon EC (2003) Effect of chronic probenecid therapy on cefazolin serum concentrations. Ann Pharmacother 37(5):621–624

289. Stoeckel K (1981) Pharmacokinetics of Rocephin, a highly active new cephalosporin with an exceptionally long biological half-life. Chemotherapy 27(Suppl 1):42–46

290. Kaplan K, Reisberg BE, Weinstein L (1967) Cephaloridine: antimicrobial activity and pharmacologic behavior. Am J Med Sci 253(6):667–674

291. Duncan WC (1974) Treatment of gonorrhea with cefazolin plus probenecid. J Infect Dis 130(4):398–401

292. Mischler TW, Sugerman AA, Willard DA, Brannick LJ, Neiss ES (1974) Influence of probenecid and food on the bioavailability of cephradine in normal male subjects. J Clin Pharmacol 14(11–12):604–611

293. Tuano SB, Brodie JL, Kirby WM (1966) Cephaloridine versus cephalothin: relation of the kidney to blood level differences after parenteral administration. Antimicrob Agents Chemother (Bethesda) 6:101–106

294. Griffith RS, Black HR, Brier GL, Wolny JD (1977) Effect of probenecid on the blood levels and urinary excretion of cefamandole. Antimicrob Agents Chemother 11(5):809–812

295. Taylor WA, Holloway WJ (1972) Cephalexin in the treatment of gonorrhea. Int J Clin Pharmacol 6(1):7–9

296. Ko H, Cathcart KS, Griffith DL, Peters GR, Adams WJ (1989) Pharmacokinetics of intravenously administered cefmetazole and cefoxitin and effects of probenecid on cefmetazole elimination. Antimicrob Agents Chemother 33(3):356–361

297. Vlasses PH, Holbrook AM, Schrogie JJ, Rogers JD, Ferguson RK, Abrams WB (1980) Effect of orally administered probenecid on the pharmacokinetics of cefoxitin. Antimicrob Agents Chemother 17(5):847–855

298. Bint AJ, Reeves DS, Holt HA (1977) Effect of probenecid on serum cefoxitin concentrations. J Antimicrob Chemother 3(6):627–628

299. Stoeckel K, Trueb V, Dubach UC, McNamara PJ (1988) Effect of probenecid on the elimination and protein binding of ceftriaxone. Eur J Clin Pharmacol 34(2):151–156

300. O'Callaghan CH, Acred P, Harper PB, Ryan DM, Kirby SM, Harding SM (1980) GR 20263, a new broad-spectrum cephalosporin with anti-pseudomonal activity. Antimicrob Agents Chemother 17(5):876–883

301. DeSante KA, Israel KS, Brier GL, Wolny JD, Hatcher BL (1982) Effect of probenecid on the pharmacokinetics of moxalactam. Antimicrob Agents Chemother 21(1):58–61

302. Patel IH, Soni PP, Carbone JJ, Audet PR, Morrison G, Gibson GA (1990) Lack of probenecid effect on nonrenal excretion of ceftriaxone in anephric patients. J Clin Pharmacol 30(5):449–453

303. Kercsmar CM, Stern RC, Reed MD, Myers CM, Murdell D, Blumer JL (1983) Ceftazidime in cystic fibrosis: pharmacokinetics and therapeutic response. J Antimicrob Chemother 12(Suppl A):289–295

304. Jovanovich JF, Saravolatz LD, Burch K, Pohlod DJ (1981) Failure of probenecid to alter the pharmacokinetics of ceforanide. Antimicrob Agents Chemother 20(4):530–532

305. Pfeifer HJ, Greenblatt DJ, Friedman P (1979) Effects of three antibiotics on theophylline kinetics. Clin Pharmacol Ther 26(1):36–40

306. Bachmann K, Schwartz J, Forney RB, Jauregui L (1986) Impact of cefaclor on the pharmacokinetics of theophylline. Ther Drug Monit 8(2):151–154

307. Jonkman JH, van der Boon WJ, Schoenmaker R, Holtkamp A, Hempenius J (1986) Clinical pharmacokinetics of theophylline during co-treatment with cefaclor. Int J Clin Pharmacol Ther Toxicol 24(2):88–92

308. Hammond D, Abate MA (1989) Theophylline toxicity, acute illness, and cefaclor administration. DICP 23(4):339–340

309. Rodondi LC, Flaherty JF, Schoenfeld P, Barriere SL, Gambertoglio JG (1989) Influence of coadministration on the pharmacokinetics of mezlocillin and cefotaxime in healthy volunteers and in patients with renal failure. Clin Pharmacol Ther 45(5):527–534

310. Plager JE (1976) Association of renal injury with combined cephalothin-gentamicin therapy among patients severely ill with malignant disease. Cancer 37(4):1937–1943

311. Wade JC, Smith CR, Petty BG, Lipsky JJ, Conrad G, Ellner J et al (1978) Cephalothin plus an aminoglycoside is more nephrotoxic than methicillin plus an aminoglycoside. Lancet 2(8090):604–606

312. Gurwich EL, Sula J, Hoy RH (1978) Gentamicin-cephalothin drug reaction. Am J Hosp Pharm 35(11):1402–1403

313. Hansen MM, Kaaber K (1977) Nephrotoxicity in combined cephalothin and gentamicin therapy. Acta Med Scand 201(5):463–467

314. Cabanillas F, Burgos RC, Rodríguez C, Baldizón C (1975) Nephrotoxicity of combined cephalothin-gentamicin regimen. Arch Intern Med 135(6):850–852

315. Fillastre JP, Laumonier R, Humbert G, Dubois D, Metayer J, Delpech A et al (1973) Acute renal failure associated with combined gentamicin and cephalothin therapy. Br Med J 2(5863):396–397

316. Bobrow SN, Jaffe E, Young RC (1972) Anuria and acute tubular necrosis associated with gentamicin and cephalothin. JAMA 222(12):1546–1547
317. Kleinknecht D, Jungers P, Fillastre JP (1974) Letter: nephrotoxicity of cephaloridine. Ann Intern Med 80(3):421–422
318. Dodds MG, Foord RD (1970) Enhancement by potent diuretics of renal tubular necrosis induced by cephaloridine. Br J Pharmacol 40(2):227–236
319. Norrby R, Stenqvist K, Elgefors B (1976) Interaction between cephaloridine and furosemide in man. Scand J Infect Dis 8(3):209–212
320. Simpson IJ (1971) Nephrotoxicity and acute renal failure associated with cephalothin and cephaloridine. N Z Med J 74(474):312–315
321. Trollfors B, Norrby R, Kristianson K (1978) Effects on renal function of treatment with cefoxitin sodium alone or in combination with furosemide. J Antimicrob Chemother 4(B):85–89
322. Korn A, Eichler HG, Gasic S (1986) A drug interaction study of ceftriaxone and frusemide in healthy volunteers. Int J Clin Pharmacol Ther Toxicol 24(5):262–264
323. Walstad RA, Dahl K, Hellum KB, Thurmann-Nielsen E (1988) The pharmacokinetics of ceftazidime in patients with impaired renal function and concurrent frusemide therapy. Eur J Clin Pharmacol 35(3):273–279
324. Ariano RE, Kassum DA, Meatherall RC, Patrick WD (1992) Lack of in vitro inactivation of tobramycin by imipenem/cilastatin. Ann Pharmacother 26(9):1075–1077
325. Bernabeu-Wittel M, Pichardo C, García-Curiel A, Pachón-Ibáñez ME, Ibáñez-Martínez J, Jiménez-Mejías ME et al (2005) Pharmacokinetic/pharmacodynamic assessment of the in-vivo efficacy of imipenem alone or in combination with amikacin for the treatment of experimental multiresistant Acinetobacter baumannii pneumonia. Clin Microbiol Infect 11(4):319–325
326. Doribax ® package insert. Shionogi & Co. Ltd. (2014)
327. Invanz® package insert. Merck & Co (2012)
328. Nix DE, Majumdar AK, DiNubile MJ (2004) Pharmacokinetics and pharmacodynamics of ertapenem: an overview for clinicians. J Antimicrob Chemother 53(Suppl 2):ii23–ii28
329. Norrby SR, Alestig K, Ferber F, Huber JL, Jones KH, Kahan FM et al (1983) Pharmacokinetics and tolerance of N-formimidoyl thienamycin (MK0787) in humans. Antimicrob Agents Chemother 23(2):293–299
330. Bösmüller C, Steurer W, Königsrainer A, Willeit J, Margreiter R (1991) Increased risk of central nervous system toxicity in patients treated with ciclosporin and imipenem/cilastatin. Nephron 58(3):362–364
331. Zazgornik J, Schein W, Heimberger K, Shaheen FA, Stockenhuber F (1986) Potentiation of neurotoxic side effects by coadministration of imipenem to cyclosporine therapy in a kidney transplant recipient – synergism of side effects or drug interaction? Clin Nephrol 26(5):265–266
332. Mraz W, Sido B, Knedel M, Hammer C (1987) Concomitant immunosuppressive and antibiotic therapy – reduction of cyclosporine A blood levels due to treatment with imipenem/cilastatin. Transplant Proc 19(5):4017–4020
333. Semel JD, Allen N (1991) Seizures in patients simultaneously receiving theophylline and imipenem or ciprofloxacin or metronidazole. South Med J 84(4):465–468
334. Primaxin® package insert. Merck & Co. (2016)
335. Tseng A, Foisy M (1996) The role of ganciclovir for the management of cytomegalovirus retinitis in HIV patients: pharmacological review and update on new developments. Can J Infect Dis 7(3):183–194
336. Jung D, Dorr A (1999) Single-dose pharmacokinetics of valganciclovir in HIV- and CMV-seropositive subjects. J Clin Pharmacol 39(8):800–804
337. Valcyte® package insert. Genentech USA, Inc. (2015)
338. Valcyte (Film-Coated Tablets) Roche Products Ltd. UK Summary of Product Characteristics (2015)

339. Suzuki E, Yamamura N, Ogura Y, Nakai D, Kubota K, Kobayashi N et al (2010) Identification of valproic acid glucuronide hydrolase as a key enzyme for the interaction of valproic acid with carbapenem antibiotics. Drug Metab Dispos 38(9):1538–1544

340. Masuo Y, Ito K, Yamamoto T, Hisaka A, Honma M, Suzuki H (2010) Characterization of inhibitory effect of carbapenem antibiotics on the deconjugation of valproic acid glucuronide. Drug Metab Dispos 38(10):1828–1835

341. Nakamura Y, Nakahira K, Mizutani T (2008) Decreased valproate level caused by VPA-glucuronidase inhibition by carbapenem antibiotics. Drug Metab Lett 2(4):280–285

342. Mori H, Takahashi K, Mizutani T (2007) Interaction between valproic acid and carbapenem antibiotics. Drug Metab Rev 39(4):647–657

343. Nakajima Y, Mizobuchi M, Nakamura M, Takagi H, Inagaki H, Kominami G et al (2004) Mechanism of the drug interaction between valproic acid and carbapenem antibiotics in monkeys and rats. Drug Metab Dispos 32(12):1383–1391

344. Haroutiunian S, Ratz Y, Rabinovich B, Adam M, Hoffman A (2009) Valproic acid plasma concentration decreases in a dose-independent manner following administration of meropenem: a retrospective study. J Clin Pharmacol 49(11):1363–1369

345. Spriet I, Goyens J, Meersseman W, Wilmer A, Willems L, Van Paesschen W (2007) Interaction between valproate and meropenem: a retrospective study. Ann Pharmacother 41(7):1130–1136

346. CC W, Pai TY, Hsiao FY, Shen LJ, FL W (2016) The effect of different carbapenem antibiotics (ertapenem, imipenem/cilastatin, and meropenem) on serum valproic acid concentrations. Ther Drug Monit 38(5):587–592

347. Yokogawa K, Iwashita S, Kubota A, Sasaki Y, Ishizaki J, Kawahara M et al (2001) Effect of meropenem on disposition kinetics of valproate and its metabolites in rabbits. Pharm Res 18(9):1320–1326

348. Yamamura N, Imura K, Naganuma H, Nishimura K (1999) Panipenem, a carbapenem antibiotic, enhances the glucuronidation of intravenously administered valproic acid in rats. Drug Metab Dispos 27(6):724–730

349. De Turck BJ, Diltoer MW, Cornelis PJ, Maes V, Spapen HD, Camu F et al (1998) Lowering of plasma valproic acid concentrations during concomitant therapy with meropenem and amikacin. J Antimicrob Chemother 42(4):563–564

350. Yoon H, Kim DH (2013) Unusual drug reaction between valproate sodium and meropenem. Int J Clin Pharm 35(3):316–318

351. Park MK, Lim KS, Kim TE, Han HK, Yi SJ, Shin KH et al (2012) Reduced valproic acid serum concentrations due to drug interactions with carbapenem antibiotics: overview of 6 cases. Ther Drug Monit 34(5):599–603

352. Hellwig TR, Onisk ML, Chapman BA (2011) Potential interaction between valproic acid and doripenem. Curr Drug Saf 6(1):54–58

353. Muzyk AJ, Candeloro CL, Christopher EJ (2010) Drug interaction between carbapenems and extended-release divalproex sodium in a patient with schizoaffective disorder. Gen Hosp Psychiatry 32(5):560.e1–3

354. Clause D, Decleire PY, Vanbinst R, Soyer A, Hantson P (2005) Pharmacokinetic interaction between valproic acid and meropenem. Intensive Care Med 31(9):1293–1294

355. Nacarkucuk E, Saglam H, Okan M (2004) Meropenem decreases serum level of valproic acid. Pediatr Neurol 31(3):232–234

356. Miranda Herrero MC, Alcaraz Romero AJ, Escudero Vilaplana V, Fernández Lafever SN, Fernández-Llamazares CM, Barredo Valderrama E et al (2015) Pharmacological interaction between valproic acid and carbapenem: what about levels in pediatrics? Eur J Paediatr Neurol 19(2):155–161

357. Paulzen M, Eap CB, Gründer G, Kuzin M (2016) Pharmacokinetic interaction between valproic acid, meropenem, and risperidone. J Clin Psychopharmacol 36(1):90–92

358. Suntimaleeworakul W, Patharachayakul S, Chusri S (2012) Drug interaction between valproic acid and meropenem: a case report. J Med Assoc Thail 95(2):293–295

359. Spriet I, Meersseman W, De Troy E, Wilmer A, Casteels M, Willems L (2007) Meropenem-valproic acid interaction in patients with cefepime-associated status epilepticus. Am J Health Syst Pharm 64(1):54–58
360. Taha FA, Hammond DN, Sheth RD (2013) Seizures from valproate-carbapenem interaction. Pediatr Neurol 49(4):279–281
361. Nagai K, Shimizu T, Togo A, Takeya M, Yokomizo Y, Sakata Y et al (1997) Decrease in serum levels of valproic acid during treatment with a new carbapenem, panipenem/betamipron. J Antimicrob Chemother 39(2):295–296
362. Coves-Orts FJ, Borrás-Blasco J, Navarro-Ruiz A, Murcia-López A, Palacios-Ortega F (2005) Acute seizures due to a probable interaction between valproic acid and meropenem. Ann Pharmacother 39(3):533–537
363. Santucci M, Parmeggiani A, Riva R (2005) Seizure worsening caused by decreased serum valproate during meropenem therapy. J Child Neurol 20(5):456–457
364. Fudio S, Carcas A, Piñana E, Ortega R (2006) Epileptic seizures caused by low valproic acid levels from an interaction with meropenem. J Clin Pharm Ther 31(4):393–396
365. Lunde JL, Nelson RE, Storandt HF (2007) Acute seizures in a patient receiving divalproex sodium after starting ertapenem therapy. Pharmacotherapy 27(8):1202–1205
366. Tobin JK, Golightly LK, Kick SD, Jones MA (2009) Valproic acid-carbapenem interaction: report of six cases and a review of the literature. Drug Metabol Drug Interact 24(2–4):153–182
367. Gu J, Huang Y (2009) Effect of concomitant administration of meropenem and valproic acid in an elderly Chinese patient. Am J Geriatr Pharmacother 7(1):26–33
368. Liao FF, Huang YB, Chen CY (2010) Decrease in serum valproic acid levels during treatment with ertapenem. Am J Health Syst Pharm 67(15):1260–1264
369. Spriet I, Willems L (2011) No interaction between valproate and meropenem in a cirrhotic patient. Ann Pharmacother 45(9):1167–1168
370. Mancl EE, Gidal BE (2009) The effect of carbapenem antibiotics on plasma concentrations of valproic acid. Ann Pharmacother 43(12):2082–2087
371. Franco-Bronson K, Gajwani P (1999) Hypotension associated with intravenous haloperidol and imipenem. J Clin Psychopharmacol 19(5):480–481
372. Swabb EA, Sugerman AA, Frantz M, Platt TB, Stern M (1983) Renal handling of the mono-bactam aztreonam in healthy subjects. Clin Pharmacol Ther 33(5):609–614
373. Sisson TL, Jungbluth GL, Hopkins NK (1999) A pharmacokinetic evaluation of concomitant administration of linezolid and aztreonam. J Clin Pharmacol 39(12):1277–1282
374. Cubicin® package insert. Merck & Co. Inc. (2015)
375. Creasey WA, Adamovics J, Dhruv R, Platt TB, Sugerman AA (1984) Pharmacokinetic inter-action of aztreonam with other antibiotics. J Clin Pharmacol 24(4):174–180
376. Bell RG, Lipford LC, Massanari MJ, Riley CM (1986) Stability of intravenous admixtures of aztreonam and cefoxitin, gentamicin, metronidazole, or tobramycin. Am J Hosp Pharm 43(6):1444–1453

Chapter 2
Macrolides, Azalides, and Ketolides

Manjunath P. Pai

2.1 Introduction

Azithromycin and amoxicillin are currently the most prescribed antibiotics in the outpatient setting [1]. An estimated 60 million outpatient prescriptions are written for macrolides in the United States alone; over 90% of these are for azithromycin [1]. This high use coincides with the long history of experience and safety of these agents for respiratory tract infections. The macrolide, erythromycin, was the first major alternative for patients with hypersensitivity reactions to penicillins like amoxicillin [2]. However, this class of agent was recognized to be different very early in its development. Hinshaw noted in 1953, "The do not have the same range of bactericidal possibilities that penicillin has, and unlike penicillin they cannot be given in massive doses," referring to the gastrointestinal intolerability and hepatotoxicity observed with this class [3]. Several macrolide derivatives and the azalide, azithromycin, have been generated over the past 60 years in order to identify safer, better tolerated agents with a lower drug interaction potential [4].

Macrolides contribute to drug-drug interaction through several mechanisms but principally through inhibition of the cytochrome P450 (CYP) 3A4 isoenzyme system [5]. Specifically, coadministration with narrow therapeutic index cardiovascular drugs can lead to serious adverse events including death [5, 6]. Specifically, macrolides cause a prolongation of the QTc interval in a dose-dependent manner and can contribute to cardiac dysrhythmias especially when used with agents known to prolong the QTc interval [7]. Transport proteins like P-glycoprotein (P-gp) can lead to complimentary drug interactions with CYP3A4 [8]. Macrolides can inhibit P-gp and alter the distribution, metabolism, and elimination of substrates of these transporters [9]. It is therefore essential that clinicians gain a deeper understanding

M. P. Pai (✉)
College of Pharmacy, University of Michigan, Ann Arbor, MI, USA
e-mail: amitpai@med.umich.edu

© Springer International Publishing AG, part of Springer Nature 2018
M. P. Pai et al. (eds.), *Drug Interactions in Infectious Diseases: Antimicrobial Drug Interactions*, Infectious Disease,
https://doi.org/10.1007/978-3-319-72416-4_2

of the pharmacology of this class of antimicrobials when selecting appropriate pharmacotherapy.

Macrolides are defined by their structural characteristics as a 14-membered ring (erythromycin, clarithromycin, dirithromycin, roxithromycin), 15-membered ring (azithromycin), or 16-membered ring (spiramycin) [10]. Azithromycin is the only clinically available azalide and so classified because it contains a nitrogen atom in its macrocyclic lactone structure [10]. The rise of erythromycin resistant *Streptococcus pneumoniae* strains created an impetus for the development of ketolides [11]. The ketolides extend the microbiologic spectrum of macrolides through chemical modifications that have rendered them less prone to efflux (*mef* or *msr*)- and methylase (*erm*)-mediated mechanisms of macrolide resistance [11]. These agents replace cladinose with a 3-keto group and inclusion of a cyclic carbamate group within the lactone ring [11]. Of the four ketolide agents (modithromycin, cethromycin, telithromycin, solithromycin), only telithromycin has received regulatory approval; however the United States Food and Drug Administration (US FDA) narrowed the clinical indications of telithromycin to only mild-to-moderate pneumonia in 2007 due to concerns about hepatotoxicity [12]. Replacement of the pyridine-imidazole side chain (telithromycin) with an aminophyl-1,2,3-triazole ring was considered as a potential approach to reduce the hepatotoxicity potential of this class with the fluoroketolide agent, solithromycin [13]. Regulatory approval has been deferred for solithromycin until concerns about the hepatotoxic potential of this agent can be addressed with a higher level of statistical confidence [14].

The similarity in structure of clarithromycin and ketolides to erythromycin confers a similar drug interaction potential [4]. In contrast, the structural difference between azithromycin and these agents has been associated with a lower drug interaction potential [5]. Therefore, the current chapter focuses on the drug interaction potential of key macrolide, azalide, and ketolide (MAK) agents, namely, erythromycin, clarithromycin, azithromycin, and telithromycin. Available information regarding solithromycin is also included given that this agent unlike cethromycin demonstrated efficacy in clinical trials for community-associated pneumonia [15]. The degree of pharmacokinetic interaction of MAK agents is most often based on the changes in exposure of the substrate drug. The percentage change in the area under the plasma/serum concentration time curve (AUC) is reported in this chapter to illustrate the degree of interaction.

2.2 Basic Pharmacology

Several antimicrobials including the MAK antibiotics exert their effects by blocking ribosomal protein synthesis [16]. Macrolides bind reversibly to the 50S ribosomal subunit of sensitive microorganisms and exert a bacteriostatic effect similar to that of the lincosamides and streptogramins. Erythromycin (MW = 733.93 g/mol) specifically inhibits the translocation step by blocking the transfer of the peptide chain from the transferase site to the donor site [16]. Analysis of the crystal structure of the large ribosomal subunit (50S) form complexed with azithromycin suggests that azithromycin blocks the protein exit tunnel [17]. Resistance to macrolides and azalides occurs through four known mechanisms which include drug efflux, ribosomal protection by constitutive or inducible production of methylase enzyme, hydrolysis by esterases, and chromosomal mutations that modify the 50S ribosome [11]. Ketolides are less likely to induce production of methylase and to undergo drug efflux [11]. Dimethylation at 23S rRNA nucleotide A2058 within the ribosomal binding site does confer resistance to the ketolides [18]. The level of A2058-dimethylation correlates with reductions in ketolide sensitivity and is greatest in *S. pyogenes* strains expressing erm(B) [18]. The use of the only approved ketolide (telithromycin) is technically limited to the treatment of community-acquired pneumonia (of mild to moderate severity) due to *Streptococcus pneumonia* (including multidrug-resistant isolates), *Haemophilus influenzae*, *Moraxella catarrhalis*, *Chlamydophila pneumoniae*, or *Mycoplasma pneumoniae*, for patients 18 years and older due to its hepatotoxic potential [12]. As a result, the MAK agents continue to be prescribed for respiratory and non-respiratory infections secondary to these pathogens. Vaccination of children with the 7-valent and 13-valent pneumococcal forms is contributing to improvements in susceptibility profiles of global strains to the beta-lactams and macrolides [19].

Beyond coverage against Gram-positive pathogens, clarithromycin and azithromycin are used as prophylaxis and treatment of *Mycobacterium avium-intracellularae* in AIDS and other immunocompromised patient populations [20]. These agents are utilized to treat pathogens associated with ticks that are seeing a rise in incident infections over the past decade [21]. Clarithromycin (MW = 747.95 g/mol) has also been a key agent in the treatment of peptic ulcer disease secondary to *Helicobacter pylori* but is increasingly less effective due to emergence of resistance globally [22]. Finally, macrolides and azalides are used clinically for their non-antimicrobial properties, and non-antimicrobial macrolides are in drug development [23, 24]. Erythromycin demonstrates prokinetic effects through stimulation of the motilin receptor and is used in critically ill patients with gastroparesis [25]. Azithromycin (MW= 748.98 g/mol) has recently been demonstrated to have comparable antroduodenal effects as erythromycin [26]. Macrolides exert anti-inflammatory effects that have been exploited for chronic respiratory diseases such as diffuse panbronchiolitis and cystic fibrosis [27]. Use of chronic azithromycin has been shown to have anti-inflammatory, antisecretory, and tissue repair and healing effects. Biochemical effects include inhibition of nuclear factor kappa-B, reductions in proinflammatory cytokines, and decreases in reactive oxygen species. These biochemical effects also include cellular changes through reduced inflammatory cell migration and shifts in T-helper cells toward the type 2 helper T cell profile [28]. The MAK antimicrobials can also affect neuromuscular transmission. Telithromycin (MW = 812.00 g/mol) has been shown to inhibit postsynaptic nicotinic acetylcholine receptors, and its use is contraindicated in patients with myasthenia gravis [29]. This adverse pharmacologic effect is not seen with solithromycin (MW = 845.01 g/mol) that lacks the pyridine moiety [13]. Finally, the MAK antibiotics are associated with drug-induced QT prolongation and sudden death especially when combined with other agents implicated with QT prolongation that are metabolized by CYP3A4 [6, 30]. The affinity of clarithromycin is roughly twofold of that of erythromycin for this effect mediated by the human ether-a-go-go-related gene (HERG)-encoded potassium channels [31].

2.2.1 Absorption

The majority of MAK agents are biopharmaceutical class II compounds, having low solubility and high permeability [32]. Erythromycin is extremely sensitive to acid degradation through slow loss of cladinose from erythromycin A by gastric acid. Food significantly decreases the rate (lag time extends to 2–3 h) and extent of its absorption (50% reduction in AUC) [33]. More importantly, significant intersubject variability exists in the absorption of erythromycin [33–36]. Modification of erythromycin through esterification and enteric coating helps to reduce gastric degradation and improve bioavailability. Available formulations of erythromycin as the base, estolate, ethylsuccinate, and stearate derivatives are included in Table 2.1. Erythromycin estolate is least susceptible to acid hydrolysis, and enteric coating leads to the most predictable absorption profile [35, 37]. Erythromycin is also available as an intravenous formulation that is delivered as a lactobionate derivative. Clarithromycin is available as both an intravenous and oral formulations in several countries (Klaricid IV) but is currently only available as an immediate and extended release oral formulation (Table 2.1) in the United States. Clarithromycin undergoes

Table 2.1 Available systemic formulations of macrolides, azalides, and ketolides in the United States

Drug	Formulations
Erythromycin (base)	Injectable (100 mg/mL), injectable (200 mg/mL), capsule (250 mg), coated tablet (250 mg), timed-release tablet (250 mg), enteric-coated tablet (500 mg), delayed-release (250 mg)
Erythromycin estolate	Capsule (125 mg), suspension (125 mg/mL)
Erythromycin ethylsuccinate	Suspension (200 mg/5 mL), suspension (400 mg/5 mL), coated tablet (400 mg), tablet (400 mg)
Erythromycin lactobionate	Injectable (500 mg vial), injectable (1000 mg vial)
Erythromycin stearate	Coated tablet (250 mg), coated tablet (500 mg), tablet (500 mg)
Clarithromycin	Suspension (125 mg/5 mL), suspension (250 mg/5mL), tablet (250 mg), tablet (500 mg), extended-release tablet (500 mg)
Azithromycin	Extended-release suspension (2 g), injectable (500 mg), solution (2.5 g), solution (500 mg), suspension (500 mg), suspension (100 mg/5mL), suspension (200 mg/5mL), suspension (1 g/packet), tablet (250 mg), tablet (500 mg), tablet (600 mg)
Telithromycin	Tablet (300 mg), tablet (400 mg)

significant first-pass metabolism and has a bioavailability of approximately 50% that is not affected by the coadministration of food [38]. However, the bioavailability of the extended release formulation is significantly improved when taken with food and is the recommended approach to administration [39].

Azithromycin is available (Table 2.1) as intravenous, immediate release, and extended release oral formulations [40]. The oral bioavailability of azithromycin capsule (not commercially available) is 38% [41]. The extended release formulation is not bioequivalent to the immediate release oral suspension, and so the dosages of these agents are not interchangeable [42]. Food reduces the bioavailability of extended release oral suspension but has marginal effects on the immediate release formulations [42]. The solubility of azithromycin is higher in its amorphous instead of the dihydrate-crystalline form, but these alternate formulations have not been introduced to the market [43]. Improving the bioavailability of this compound can have a major impact on the global supply chain of this agent given that high doses are needed for certain clinical indications. Telithromycin is available as an oral formulation only and has a bioavailability of 57% that is not adversely affected by the coadministration of food [44]. The absolute oral bioavailability of solithromycin is 62%, and coadministration of food does not reduce its bioavailability [11].

2.2.2   Distribution

Erythromycin is approximately 70–95% protein bound, with variability noted based on the specific derivative [45]. The volume of distribution (Vd) of erythromycin is approximately 40 L in adults [46]. Clarithromycin is less protein bound (42–50%) and has a correspondingly larger apparent Vd of 243–266 L [47]. Azithromycin demonstrates concentration-dependent protein binding with a range of 7–50% and an apparent Vd that exceeds 1000 L [47]. Telithromycin in comparison has similar protein binding to clarithromycin (60–70%) and an apparent Vd of 200–250 L [4]. Solithromycin is 81% plasma protein bound with an apparent Vd of 400 L [11]. High intracellular concentrations contribute to the computation of large Vd values for these. Pulmonary tissue concentrations also exceed serum-unbound drug concentrations, with epithelial lining fluid concentrations that are 3.15- to 24.10-fold higher, and alveolar macrophage concentrations have been documented to be 84.2- to 3,234-fold higher than serum total concentrations [48]. Accumulation of azithromycin and clarithromycin in alveolar macrophages occurs in the acidic subcellular compartments of these cells such as the lysosomes [49]. The enhanced distribution of MAK into pulmonary tissue is likely mediated by active transport mechanisms as these measurements are not predicted by the free drug concentration profiles of these agents [48, 50, 51].

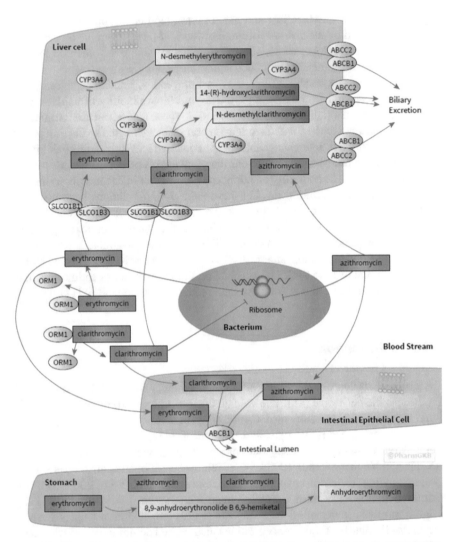

Fig. 2.1 Macrolide metabolism and transport pathways [accessible through PharmGKB]. ©PharmGKB. Permission granted by PharmGKB and Stanford University

2.2.3 Metabolism

The transport and metabolism of macrolides are illustrated in Fig. 2.1. Transport of erythromycin and clarithromycin into hepatocytes is mediated by organic anion transport polypeptide and has not been well characterized for the other agents [52]. The tertiary amine function, -N-$(CH_3)_2$, subsequently undergoes

N-demethylation by hepatic microsomes, followed by dealkylation [53]. The metabolites N-desmethylerythromycin, 14-(R)-hydroxyclarithromycin, and N-desmethylclarithromycin form stable complexes with Fe^{2+} of reduced CYP3A4 to disable enzymatic activity [54]. Of these metabolites, 14-(R)-hydroxy clarithromycin is more potent against *Haemophilus influenzae* compared to clarithromycin [55]. Erythromycin has been demonstrated to first induce CYP, which is followed by rapid complexation and inactivation by its metabolite [56]. In contrast, azithromycin induces N-demethylase but does not induce CYP, and no inactive CYP complexes are detectable despite high azithromycin accumulation in this tissue [57]. Telithromycin undergoes similar biotransformation as clarithromycin and has a N-bis-demethyl derivative that is 4–16-fold less active than the parent compound [58]. However, telithromycin has been shown to decrease the activity of both CYP1A2 and CYP3A4, while clarithromycin principally decreases the activity of CYP3A4 [59]. The mechanism-based inhibition of CYP3A by erythromycin, clarithromycin, and telithromycin suggests that several days of drug-free period may be necessary to permit generation of CYP3A4 isoenzymes to normalize the intrinsic activity of this metabolic pathway [60]. Solithromycin is metabolized by CYP3A4, n-acetyl transferase, and loss of the aminopheyl-1,2,3-triazole group [11]. The latter two metabolites are active but represent less than 6% of the systemic exposure in human mass balance studies [13]. Solithromycin auto-inhibits its CYP3A4 metabolism contributing to nonlinear clearance over time [11]. Auto-inhibition has also been shown with clarithromycin with similar nonlinearity (at higher doses) [61]. However, this effect does not contribute to further increases in exposure after 48 h of the standard 500 mg twice daily regimen of clarithromycin [61].

2.2.4 Elimination

Elimination of MAK agents is primarily through the hepatic route with unchanged drug elimination in urine contributing to 2.5% of erythromycin and 10–12% for azithromycin, telithromycin, and solithromycin [4, 47, 62]. Overall, biliary excretion accounts for the majority of the elimination of these compounds primarily through the P-gp transport. In contrast, elimination of clarithromycin through the kidney accounts for 20–40% of its excretion [63]. In addition, 10–15% of 14-hydroxy clarithromycin is excreted in urine [63]. As a consequence, the dosage of clarithromycin is recommended to be reduced by 50% in patients with renal impairment based on a creatinine clearance (CLcr) <30 mL/min. Drug interactions that increase clarithromycin exposure can necessitate a 75% reduction in the dosage when CLcr < 30mL /min [64].

2.3 Drug Interactions

The primary mechanisms associated with contraindicated and major drug interactions with MAK agents involve CYP3A4 inhibition in the small intestine and liver [65]. Interactions associated with drug transport systems such as P-gp and OATP have been increasingly recognized, but the degree of interaction is driven by the complimentary CYP3A4 inhibition [66]. Given the specific expression of human transporters within select tissues, inhibition of these systems can affect absorption, distribution, metabolism, and elimination of several compounds [67]. Inhibition of P-gp in the intestine can lead to increased bioavailability of the substrate, while inhibition of P-gp in the kidney can reduce elimination. Inhibition of P-gp in the brain leads to accumulation of substrates within this tissue. Hence, prediction of the degree of MAK drug-drug interactions on a specific substrate is complex, when the distribution, metabolism, and elimination pathways are regulated by multiple systems.

MAK can increase or decrease the bioavailability of drugs through an alteration in normal intestinal flora. Digoxin and digitoxin can be metabolized by *Eubacterium lentum* found in the gastrointestinal tract of certain individuals [68, 69]. Inhibition of *E. lentum* by erythromycin has been proposed as a mechanism leading to enhanced bioavailability of digoxin [68, 69] . Conversely, macrolides can theoretically attenuate the activity of conjugated estrogens by reducing enterohepatic recirculation of these oral contraceptive agents [70]. However, this mechanism has not been substantiated, and systematic reviews and recent studies have failed to confirm these historic findings [71–73]. Microbiome-related research is beginning to shed new light on the potential impacts of antimicrobials at a young age. Preliminary evidence from Finland demonstrates that use of macrolides in 2–7-year-olds caused a measurable shift in the microbiome and reduces bile salt hydrolases [74]. Children in this cohort were more likely to be at risk for asthma and weight gain with this early use of macrolides compared to the penicillins [74]. Modification to bile acid activity has major implications for drug that require micellar solubilization in order to be bioavailable. Further exploration of macrolide-host-microbiome-drug interactions is required.

Clarithromycin is active against bacillus Calmette-Guerin (BCG) and so can reduce the viability of this microorganism during intravesical instillation to treat bladder cancer [75]. Finally MAK have been shown to induce QTc prolongation and have the potential for a synergistic pharmacodynamic interaction with agents known to also prolong the QTc interval [7]. Multiple substrates that have the potential to increase the QTc interval are metabolized by CYP3A4 and transported by P-gp. Use of verapamil with erythromycin serves as a prime example where this interaction has been associated with a fivefold higher risk for sudden death [6, 76]. The following sections highlight the key known PK-PD interactions of MAK antimicrobials. Given the breadth of potential CYP3A4 substrates, these sections focus on interactions of MAK agents with non-antimicrobials.

2.3.1 Pharmacokinetic Drug-Drug Interactions

2.3.1.1 Absorption-Related Drug Interactions

Absorption-mediated interactions with MAK agents are primarily transport mediated. Alternate mechanisms such as the prokinetic effect of AK agents through motilin agonism are most pronounced with erythromycin and its more potent metabolite, 9-anhydro-erythromycin A-6,9-hemiketal, formed by its acid lability. The clinical implication of this prokinetic effect is limited for the MAK agents. The bioavailability of MAK is not significantly affected by the concomitant use of antacids [77]. The coadministration of antacids and azithromycin is discouraged due to the low bioavailability of this agent. The mean C_{max} of azithromycin was reduced by 24%, while the AUC was not affected when coadministered with aluminum-/magnesium-containing antacids [42, 78]. The use of an acid-suppressive agent such as cimetidine has no negative impact on azithromycin absorption [78]. Drug absorption through the intestinal barrier is regulated primarily by adenosine triphosphate (ATP)-binding cassette (ABC) drug transporters and the solute carrier (SLC) transporters [67]. P-glycoprotein (ABC transporter) is an efflux transporter found in the apical (luminal) membrane of the entire intestine from duodenum to rectum, with a high expression in the enterocytes of the small intestine. The bioavailability of digoxin (P-gp substrate) has been documented to increase with the concomitant use of clarithromycin [79, 80]. Another relevant substrate is dabigatran etexilate, an oral anticoagulant with a mean (range) bioavailability of 6.5 (2.8–12.1%) that is not metabolized by CYP and is excreted unchanged (80%) by the kidneys [81]. Coadministration of dabigatran with clarithromycin increased the mean (range) oral bioavailability to 10.1 (4.1–26.9%) noted by a 60.2% increase in the C_{max} and 49.1% increase in the AUC [82]. Of greater concern is the rise in the interindividual variability (coefficient of variation) in bioavailability of dabigatran from 38.6% to 72.5% with concomitant of clarithromycin use [82]. This degree of interaction likely extends to telithromycin and solithromycin given similar expected affinity for P-gp inhibition compared to erythromycin and azithromycin. This expectation is further demonstrated through a 15-year population-based nested case-control study that documented a fourfold higher associated risk for digoxin toxicity with clarithromycin compared to either erythromycin or azithromycin [83].

Uptake transport into enterocytes is mediated by OATP and PEPT transporters [84]. In addition, select OATP type transporters are specifically expressed in the liver (OATP1B1, OATP1B3) [84]. The IC_{50} for inhibition of OATP1B3-mediated uptake of pravastatin was determined to be 11 μM, 32 μM, and 34 μM for telithromycin, clarithromycin, and erythromycin, respectively [85]. Clarithromycin inhibits the hepatic uptake of HMG-CoA reductase inhibitors such as pitavastatin, pravastatin, and rosuvastatin, leading to an increase in the bioavailability of these agents [86, 87]. A 200% increase in exposure of pravastatin with clarithromycin is attributed to this mechanism as pravastatin is not a substrate of CYP3A4 [88]. Limiting the daily dose of pravastatin to 40 mg/day is recommended if it is used concomitantly with clarithromycin. However, the impact of OAT1B1 inhibition is limited for other substrates such as ambrisentan where only a 41% increase in AUC is observed with concomitant clarithromycin use [89].

2.3.1.2 Distribution-Related Drug Interactions

Other than the intestine and liver, P-gp is also expressed in various organs such as the brain, kidney, placenta, adrenal, testes, and retina [90]. However, the relative expression and influence of P-gp vary by the tissue type. Inhibition of P-gp has been shown to increase distribution of radio-labeled verapamil to a greater extent in the retina compared to the brain [91]. The potential for increased exposure of drugs that are substrates of P-gp is unlikely to increase substantially in restricted distribution compartments (brain, eyes, testes) when MAK agents are coadministered. Use of clarithromycin has not been shown to alter the brain distribution of verapamil because a secondary transport mechanism mediated by a proton antiporter may overcome this inhibited pathway [92, 93]. Data on altered distribution due to plasma protein-binding displacement with the use of MAK agents are limited. The mean [95% CI] ratio of the unbound fraction of quinine in patients with severe liver dysfunction who received erythromycin was 1.76 [1.42, 2.11]. In comparison the mean [95% CI] ratio of the unbound fraction of quinine in healthy volunteers who received erythromycin was 1.41 [1.28, 1.55] [94]. The effects of MAK agents on distribution-mediated interactions are likely to be limited. However, tissue-specific toxicities can be attributed in part to altered distribution when drugs with narrow therapeutic indices are used with MAK agents as discussed in the next section.

2.3.1.3 Metabolism-Related Drug Interactions

Clarithromycin, telithromycin, and solithromycin are strong inhibitors, erythromycin is a moderate inhibitor, and azithromycin is not an inhibitor of CYP3A4 [11, 95]. Strong inhibitors are classified as agents that have the ability to raise substrate AUC values at least fivefold, while moderate inhibitors have the potential of raising AUC values by two- to less than fivefold [66]. Evaluation of the CYP3A4 drug interaction potential in healthy volunteers has historically included use of ketoconazole as the index agent [66]. Regulatory agencies like the FDA recommended against the use of ketoconazole, and consideration of clarithromycin or itraconazole in healthy volunteer studies due to the potential for liver injury [96, 97]. The selection of ritonavir or cobicistat has been suggested as better alternative for healthy volunteer studies, while other studies suggest that the risk of hepatotoxicity does not outweigh the scientific benefit of retaining ketoconazole as the index inhibitor [96–98]. These studies highlight various transport mechanisms impacted by CYP3A4 inhibitors that truly limit clinical prediction and generalizations about the specific drug-drug interactions of these inhibitors. As a result, the interactions of key drug classes with MAK agents where documented serious adverse events have occurred are listed in Table 2.2 and detailed as follows. Please note that a detailed summary

Table 2.2 Pharmacokinetic alterations of key drugs when combined with the macrolides and ketolides

Drugs	Erythromycin	Clarithromycin	Telithromycin
Anticancer			
Cabazitaxel	↑ 25% AUC[a]	↑ 50% AUC[a]	↑ 50% AUC[a]
Vinblastine	↑ 25% AUC[a]	↑ 50% AUC[a]	↑ 50% AUC[a]
Vinorelbine	↑ 25% AUC	↑ 50% AUC[a]	↑ 50% AUC[a]
Anticoagulant			
Edoxaban	↑ 185% AUC	↑ 185% AUC	↑ 185% AUC
Rivaroxaban	↑ 34% AUC	↑ 54% AUC	↑ 54% AUC
Dabigatran	↑ 30% AUC	↑ 60% AUC	↑ 60% AUC
Antidiabetic			
Glyburide	↑ 18% C_{max}	↑ 18% C_{max}[a]	↑ 18% C_{max}[a]
Glibenclamide	↑ 25% C_{max}[a]	↑ 25% C_{max}	↑ 25% C_{max}[a]
Repaglinide	↑ 40% AUC[a]	↑ 40% AUC	↑ 40% AUC[a]
Saxagliptin	↑ 100%[a] AUC	↑ 250%[a] AUC	↑ 250%[a] AUC
Antimigraine			
Dihydroergocryptine	↑ 1650% AUC	↑ 1650% AUC[a]	↑ 1650% AUC[a]
Eletriptan	↑ 400% AUC	↑ 600% AUC[a]	↑ 600% AUC[a]
Benzodiazepines			
Alprazolam	↑ 60% AUC	↑ 100% AUC	↑ 100% AUC
Midazolam	↑ 400% AUC	↑ 200–800% AUC	↑ 200–800% AUC
Triazolam	↑ 52% AUC	↑ 100% AUC[a]	↑ 100% AUC[a]
Calcium channel blocker			
Felodipine	↑ 250% AUC	↑800% AUC[a]	↑ 800% AUC[a]
Nifedipine	↑ 100% AUC	↑200% AUC[a]	↑200% AUC[a]
Immunosuppressants			
Cyclosporine	↑ 100% AUC	↑ 200% AUC[a]	↑ 200% AUC[a]
Everolimus	↑440% AUC	↑1500% AUC[a]	↑1500% AUC[a]
HMG-CoA inhibitors			
Atorvastatin	↑ 32.5% AUC	↑ 82% AUC	↑ 82% AUC[a]
Lovastatin	↑ 390% AUC[a]	↑ 1000% AUC[a]	↑ 890% AUC[a]
Simvastatin	↑ 390% AUC	↑ 1000% AUC	↑ 890% AUC
Pitavastatin	↑ 280% AUC	↑ 280–460% AUC[a]	↑ 280–460% AUC[a]
Pravastatin	↑ 100% AUC[a]	↑ 200% AUC	↑ 200% AUC[a]
PDE5 inhibitors			
Sildenafil	↑ 182% AUC	↑ 230% AUC	↑ 230% AUC[a]
Tadalafil	↑ 107% AUC	↑ 107–312% AUC[a]	↑ 107–312% AUC[a]
Vardenafil	↑ 400% AUC	↑ 1000% AUC[a]	↑ 1000% AUC[a]

[a]Change in area under the curve (AUC) estimated based on data available for an alternate CYP inhibitor with a comparable degree of inhibition [99]

of the interaction potential of all approved drugs that are metabolized by the CYP system and have the potential to interact with MAK agents is not feasible in this review. The following sections serve as relevant examples of the spectrum and extent of metabolism-related drug interactions secondary to MAK agents.

2.3.1.4 Anticancer Medications

Several potentially harmful interactions exist between the macrolides and anticancer medications. The degree of interaction is dependent on substrate specificity for CYP3A4. Several orally administered tyrosine kinase inhibitors interact with strong CYP3A4 inhibitors such as ketoconazole necessitating a >50% reduction in the daily dose [100]. A similar degree of interaction is expected with the use of clarithromycin or telithromycin. The potential negative pharmacokinetic effects of clarithromycin may be countered by positive chemosensitizing effects that may work synergistically with agents to induce autophagy in models of myeloma and pancreatic and non-small cell lung cancer that are presently under clinical study [101, 102]. Unfortunately, the high drug-drug interaction potential of this agent has reduced the feasibility of this clinical evaluation for non-small cell cancer but shown success for select cancers [103]. Irrespective of the potential non-antimicrobial benefits, the key concern with overexposure of select anticancer agents due to CYP3A4 inhibition is the risk for prolonged severe neutropenia. Concomitant use of clarithromycin and vinorelbine was associated with a 4.52-fold (95% CI 1.41–14.45) increased risk for grade 3/4 neutropenia [104]. A similar near fivefold risk in development of neutropenia was observed with the concomitant administration of docetaxel and clarithromycin [105]. Dose reductions to 50% or more of normal may be necessary in select cases where the coadministration of these anticancer agents and clarithromycin or telithromycin cannot be avoided.

2.3.1.5 Antidiabetic Medications

There are currently eight classes of agents that are approved for the treatment of patients with type 2 diabetes. These include the α-glucosidase inhibitors, biguanides, dipeptidyl peptidase 4 (DPP4) inhibitors, glucagonlike peptide-1 (GLP-1), meglitinides, selective sodium-glucose cotransporter-2 (SGLT-2) inhibitors, sulfonylureas, and thiazolidinediones [106]. The primary metabolic pathway for metabolism of the sulfonylureas and thiazolidinediones is via the CYP2C system. The pharmacokinetics of glyburide (sulfonylurea) was evaluated to be marginally altered in combination with erythromycin [107]. The C_{max} of glyburide was demonstrated to increase by 18%, with an associated reduction in the T_{max} from 4.9 to 3.0 h [107]. These results are thought to be a result of enhanced gastric motility by erythromycin, which increased the rate but not extent of glyburide absorption. The AUC of glyburide, a substrate of CYP2C9, was increased by a mean [90% CI] of 25% [12%, 50%] when combined with clarithromycin [107]. Despite this mild pharmacokinetic interaction, reports of severe and potentially life-threatening hypoglycemia have been documented with use of clarithromycin and sulfonylureas that may in part be transporter mediated [108, 109].

Schelleman and colleagues recently performed two case-crossover studies using US Medicaid data to evaluate if the use of oral antimicrobials increases the risk of severe hypoglycemia. An association between sulfonylurea-induced hypoglycemia and the use of fluconazole, ciprofloxacin, levofloxacin, azithromycin, clarithromycin, erythromycin, trimethoprim/sulfamethoxazole, or cephalexin was measured. Interestingly the odds ratio [95% CI] for glyburide-induced hypoglycemia and concomitant use of an antimicrobial was 2.66 [2.02, 3.49]-cephalexin, 2.65 [1.87, 3.76]-azithromycin, 3.60 [2.35, 5.50]-erythromycin, and 13.28 [10.26, 17.18]-clarithromycin [110]. Similarly the odds ratio for glipizide-induced hypoglycemia was at least 2.5-fold higher with the use of clarithromycin versus azithromycin [110]. These findings have been substantiated through an analysis of Texas Medicaid data that show that the risk associated with antimicrobial use and hypoglycemia is greatest with clarithromycin – 3.96 [2.42–6.49] [111]. Cumulatively, these data suggest that although MAK agents are less likely in theory due to a CYP3A4-mediated pathway, an interaction does exist and is likely to be exacerbated in patients with acute infections.

The pharmacokinetic interaction potential between the mitiglinides, thiazolidinediones, and MAK agents tends to be less likely with increases of AUC typically <40% [112, 113]. In contrast, the exposure of saxagliptin (DPP4 inhibitor), a CYP3A4 substrate, increased by 250% when coadministered with ketoconazole [114]. Although no cases of hypoglycemia have been reported with use of saxagliptin and CYP3A4 inhibitors, the product label recommends that the daily dosage of saxagliptin be limited to 2.5 mg once daily when used with agents like clarithromycin and telithromycin [114]. Overall, patients with diabetes receiving treatment with an oral antidiabetic agent should be counseled on the signs and symptoms of hypoglycemia, such as headache, dizziness, drowsiness, nervousness, tremor, weakness, perspiration, and palpitations when initiated on agents like clarithromycin or telithromycin.

2.3.1.6   Antimigraine Medications

The ergopeptide alkaloids and the triptans represent the two major drug classes that are used to treat migraine attacks. Chronic or high exposures to ergotamine compounds can lead to ergotism known classically as "St. Anthony's Fire" related to the convulsive and gangrenous symptoms that ensue from this toxicity [115]. The ergotamines have been used to treat migraine headaches for nearly a century but increasingly replaced by the triptans. Dihydroergotamine is metabolized by CYP3A4 and has an oral bioavailability of <1% [116]. Thus it is remarkable how the increase in oral bioavailability with the use of ergopeptides in combination with erythromycin and clarithromycin can lead to acute limb ischemia, necrosis, and gangrene [117–120]. Evaluation of the interaction potential of alpha-dihydroergocryptine and erythromycin suggests that there is a mean [95% CI] of 1650% [870%, 3150%] increase in the AUC of this ergoline [121].As a result, the use of erythromycin and clarithromycin is contraindicated with ergoline derivatives such as bromocriptine and cabergoline. These major adverse event risks are expected with telithromycin and solithromycin, and so avoidance of concomitant MAK agents with ergotamines is essential.

The currently available triptans include sumatriptan, rizatriptan, naratriptan, zolmitriptan, eletriptan, almotriptan, frovatriptan, and avitriptan that are not all metabolized by the same pathway . Most triptans are metabolized by monoamine oxidase A or alternatively by CYP1A2 [122]. Almotriptan is metabolized by CYP3A4 and CYP1A2, and its exposure has demonstrated to increase by 57% when used with the potent CYP3A4 inhibitor, ketoconazole [123]. The exception within this group is eletriptan, which is principally transported by P-gp and metabolized by CYP3A4 [124]. The exposure of eletriptan increases by 400% when coadministered with erythromycin and 600% when coadministered with ketoconazole [124]. Hence, the potential interaction between clarithromycin or telithromycin with eletriptan is significant, and concomitant use within 72 h of use of strong CYP3A4 inhibitors such as clarithromycin and telithromycin is recommended [124].

2.3.1.7 Benzodiazepines

The majority of benzodiazepine undergo CYP metabolism followed by glucuronidation [125]. The exception to these includes lorazepam that is glucuronidated by UGT2B15 and clonazepam that undergoes acetylation via NAT2 [125–127]. The three benzodiazepines influenced by CYP3A4 include alprazolam, triazolam, and midazolam [125]. Midazolam is considered an optimal CYP3A4 probe substrate and is a useful agent to characterize the degree of CYP3A4 inhibition. Erythromycin was demonstrated to increase the AUC of single-dose oral alprazolam by approximately 60% but was not demonstrated to alter psychomotor function as assessed by the Digit Symbol Substitution Test [125]. In contrast, the clearance of triazolam was demonstrated to be reduced by 52% with the concomitant use of erythromycin and was associated with psychomotor dysfunction and amnesia [125]. Prolonged sedation has been reported with the concomitant use of midazolam and erythromycin [128]. Olkkola et al. demonstrated the exposure of oral midazolam to be increased by 400%, while that of intravenous midazolam increased by 54% with use of erythromycin [129]. The dual inhibition of both intestinal and hepatic CYP3A4 contributes to the enhanced oral bioavailability of midazolam [129]. As expected, azithromycin has not demonstrated a significant interaction with midazolam [130]. In contrast, the potent inhibitors, clarithromycin and telithromycin, are expected to increase the exposure of midazolam by 200–800% with expected psychomotor adverse events [80]. This potential interaction is especially important in elderly patients and those sensitive to the effects of benzodiazepines. Although a reduction in the dose of the midazolam is recommended, use of azithromycin would be preferred over coadministration with clarithromycin or telithromycin. The isomer of zopiclone, eszopiclone, is principally metabolized by CYP3A4, and 220% increase in AUC is expected when combined with a strong inhibitor like clarithromycin [131]. Alternatively, use of sedative hypnotics such as zolpidem and zaleplon has a lower potential for interaction with agents like clarithromycin [132]

2.3.1.8   Calcineurin Inhibitors and Proliferation Signal Inhibitors

Major drug-drug interactions have been observed with the combination of calcineurin inhibitors (cyclosporine and tacrolimus) and proliferation signal inhibitors (sirolimus and everolimus) and MAK agents [133]. Overexposures to cyclosporine and tacrolimus due to this interaction have been associated with nephrotoxicity and neurotoxicity [133]. Patients that are immunosuppressed are often at an increased risk for pulmonary infection that carries the risk for an untoward reaction when a MAK agent is selected. The exposure of cyclosporine can increase by 100–500% when combined with erythromycin [134]. Initial dosage reductions of at least 50–100% may be necessary when combined with clarithromycin and telithromycin [133]. Again, as expected the interaction between azithromycin and these agents has been of marginal significance [135]. Similarly, a major interaction has been documented with the combination of tacrolimus and erythromycin or clarithromycin [136]. Again, an interaction has been documented with azithromycin and tacrolimus, but the impact of this interaction is expected to also be low [137]. Similar changes in the pharmacokinetics of everolimus have been shown in healthy volunteers, where the mean [90% CI] AUC of everolimus increased by 440% [350%, 540%] with erythromycin co-use [138]. Pea and colleagues recently illustrated a case that required a reduction in the everolimus dose to one-quarter of the normal daily dose for approximately 2 weeks upon discontinuation of clarithromycin before the everolimus dose could be administered at the baseline level [139]. Similarly, sirolimus should not be used concomitantly with strong CYP3A4 inhibitors. The combined use of sirolimus and clarithromycin has been associated with an 800% increase in the blood trough concentration of sirolimus and acute nephrotoxicity [140]. The sustained CYP3A4 inhibition induced by clarithromycin and telithromycin makes it critically important to gain a thorough medication history before these immunosuppressive agents are initiated.

2.3.1.9   Hydroxymethylglutaryl Coenzyme A (HMG-CoA) Reductase Inhibitors

Overexposure of select HMG-CoA reductase inhibitors, commonly referred to as the "statins," is associated with rhabdomyolysis and acute kidney injury. Case reports and population-based studies documenting the seriousness of this interaction are abound [141–143]. Currently available statins include atorvastatin, fluvastatin, lovastatin, pitavastatin, pravastatin, rosuvastatin, and simvastatin. One out of four adults in the United States over the age of 40 years is currently prescribed a statin that increases the probability for this major adverse drug-drug interaction event [144]. The metabolic and transport pathways of these agents are variable. Atorvastatin, lovastatin, and simvastatin are primarily metabolized by CYP3A4 and transported into hepatocytes by OAT1B1 and eliminated via P-gp through the bile [145]. Fluvastatin is primarily metabolized by CYP2C9, and rosuvastatin is primarily metabolized by CYP2C9 and CYP2C19 [145]. Pravastatin is not a substrate of CYP and instead is primarily metabolized by sulfation pathways and undergoes elimination via the kidney [145]. Pitavastatin is metabolized by CYP2C8 and CYP2C9 and forms a lactone metabolite via uridine5'-diphosphate (UDP) glucuronosyltransferase (UGT; UGT1A3 and UGT2B7) [86]. Despite the lack of CYP3A4 metabolism for select HMG-CoA, all marketed statins are substrates of OATP and are expected to interact with MAK agents to varying degrees with a corresponding risk for myopathy, rhabdomyolysis, and death [84]. In addition, multidrug resistance-associated protein (MRP) is now known to be highly expressed in skeletal muscle and contributes to reduced intracellular accumulation of agents like atorvastatin and rosuvastatin [146]. An estimated 60% of cases of statin-related rhabdomyolysis are thought to be a result of drug interactions. Simvastatin doses of 80 mg per day have been associated with a higher risk of rhabdomyolysis compared to standard doses of fluvastatin and pravastatin, which carry lower risks [147].

A 390% and 1000% increase in the AUC_{0-24} of simvastatin is expected with the concomitant use of erythromycin or clarithromycin, respectively [148]. A comparable degree of interaction between simvastatin and telithromycin is expected, while the interaction potential between azithromycin and simvastatin is low. The use of erythromycin and pitavastatin was associated with a 280% increase in the AUC of pitavastatin likely mediated by OATP1B1 inhibition [96]. These data suggest that macrolides interact with HMG-CoA reductase inhibitors via both metabolic and transporter-mediated interactions. A 50–80% reduction in the dose of other HMG-CoA inhibitors may be necessary when used with macrolide or ketolide agents. Most importantly, patients who must receive these agents concurrently should be warned to contact their healthcare provider immediately if they experience muscle pain, tenderness, or weakness.

2.3.1.10 Phosphodiesterase 5 (PDE5) Inhibitors

Sildenafil, tadalafil, vardenafil, and avanafil are approved clinically to treat erectile dysfunction and are increasingly used for non-urological conditions [149]. Avanafil is the most recent addition to this therapeutic class and has the clinical advantage of a faster rate of response (15 min). The PDE5 inhibitors are also increasingly being used to treat idiopathic pulmonary arterial hypertension [149]. Sildenafil is recognized to be a very selective substrate of CYP3A4 like buspirone, simvastatin, and midazolam. The sildenafil, tadalafil, and vardenafil AUC increases by 182–300% when coadministered with a MK [150]. In contrast, avanafil AUC increases by 1300% when coadministered with a strong CYP3A4 inhibitor that is expected with clarithromycin and telithromycin [149]. Coadministration of avanafil with clarithromycin and telithromycin is not recommended, while dosage reductions are recommended for the other agents. A lower starting dose of sildenafil (25 mg) should be considered if its concomitant use with MK agents cannot be avoided [149]. Similarly initial doses of tadalafil and vardenafil of 2.5 mg are recommended if concomitant use of CYP3A4 inhibitors cannot be used. Azithromycin does not affect the pharmacokinetics of sildenafil and is not expected to impact the other agents [149].

2.3.1.11   Miscellaneous Agents

Buspirone is a generically available anxiolytic agent that is known to be a selective substrate of CYP3A4 [151]. The plasma AUC of buspirone increased by 600% with associated drowsiness and altered psychomotor performance with erythromycin coadministration [151]. A greater than 50% reduction in the dose of buspirone is likely to be necessary if used concomitant with erythromycin, clarithromycin, or telithromycin. Seizures, drowsiness, and neutropenia have been documented in case reports of the combination of erythromycin and clozapine [152]. Clozapine is metabolized by CYP1A2, CYP3A4, and CYP2D6, and this risk for interaction with macrolides is likely most pronounced in CYP2D6 poor metabolizers. Opiates such as fentanyl and oxycodone are primarily metabolized by CYP3A4 [153, 154]. No major adverse events have been observed with the combination of clarithromycin and oxycodone, while that combination with fentanyl has been associated with serious respiratory depression [155, 156]. A less predictable interaction has been recently reported with the combination of linezolid and clarithromycin in patients with multidrug-resistant tuberculosis. A 44% reduction in linezolid clearance has been documented and thought to be P-gp mediated but not associated with any major adverse events [157]. A similar transport-mediated interaction is also suggested with the combination of montelukast and clarithromycin through OATP1B1/OAT1B3 since montelukast is a primary substrate of CYP2C8 [158]. A 245% increase in the AUC of montelukast is observed when combined with clarithromycin [159]. Overall these interactions demonstrate the importance of CYP3A4, OATP, and P-gp-mediated interactions with MK agents [160].

2.3.1.12   Elimination-Related Interactions

Serious elimination drug interactions have been documented with the combination clarithromycin and colchicine, digoxin, and the vinca alkaloids [161, 162]. The primary mechanism includes P-gp inhibition, but the intensity of interaction is also mediated by CYP3A4 inhibition. Clarithromycin and telithromycin carry the highest risk followed by erythromycin, with azithromycin exerting a minor interaction through this pathway. For example, clarithromycin increases the mean [min, max] AUC of colchicine by 281.5% [88.7%, 851.6%], and its use is contraindicated with colchicines [161]. Although not explicitly stated in the product label, use of erythromycin or telithromycin should be avoided with colchicine given the risk for life-threatening and fatal toxic reactions [161]. Azithromycin is not considered to be a concern with colchicine though one case report has documented its potential for interaction (confounded by the coadministration of cyclosporine) [160]. Rhabdomyolysis, neuromyopathy, acute kidney injury, agranulocytosis, fever, diarrhea, convulsions, alopecia, and death can occur with the coadministration of colchicine with MK agents [160, 163–165]. Importantly, patients with chronic kidney disease are at a higher risk for this interaction [166, 167].

Other well-documented examples include the interaction of MAK agents with digoxin. In a large population-based study conducted in Taiwan, the risk for hospitalization secondary to digoxin intoxication increased by 5.07-fold (95% CI 2.36, 10.89) if clarithromycin was used within 14 days of the index day [168]. Gomes and colleagues completed a 15-year population-based, case-control study and demonstrated an increased risk for hospitalization secondary to digoxin intoxication with erythromycin, clarithromycin, and azithromycin [83]. The odds ratio [95% CI] for digoxin toxicity-related hospitalization was 3.7 [1.7, 7.9], 3.7 [1.1, 12.5], and 14.8 [7.9, 27.9] with the recent exposure to erythromycin, azithromycin, and clarithromycin, respectively. The primary mechanism for this interaction is P-gp mediated though gastrointestinal antimicrobial inhibition of *E. lentum* by MAK agents has a documented role [69].The relatively low P-gp-mediated interaction potential of azithromycin is exemplified by a 60% increase in melagatran AUC, when the active form of the prodrug thrombin inhibitor ximelagatran is coadministered [169]. The vinca alkaloids such as vinblastine, vincristine, vinorelbine, and vindesine are substrates of both P-gp and CYP3A [104]. Severe adverse events have been observed with the combination of vinblastine and erythromycin when used in combination with cyclosporine [170]. The incidence of vinorelbine-associated grade 4 neutropenia was reported in 31.6% of patients treated with clarithromycin versus 2.5% of patients not treated with this agent [104]. Clarithromycin or telithromycin should be avoided in patients receiving treatment with vinca alkaloids. On a more intriguing note, clarithromycin is being repurposed as a potential adjuvant to chemotherapy against multiple myeloma [101, 102]. Although the underlying mechanisms behind this effect may be autophagy mediated, drug interactions may play a role when combined with select agents [102].

2.3.2 Pharmacodynamic Torsades de Pointes

The primary mechanism for drug-induced QTc prolongation includes inhibition of the rapid component of the delayed rectifier potassium current through a potassium channel that is regulated by the human ether-a-go-go-related gene (HERG) [171]. Potentiation of the QTc interval and development of cardiac dysrhythmias such as torsades de pointes (TdP) can have grave consequences such as sudden death. Prior to our mechanistic understanding of this phenomenon, observations of this toxicity were noted primarily with intravenous erythromycin [172]. The rate of erythromycin infusion was demonstrated to increase the QTc interval in critically ill patients and significant even at slower rates of infusion [173, 174]. Three patients developed ventricular fibrillation and one patient died [173, 174]. Oberg and Bauman retrospectively evaluated the effect of erythromycin lactobionate infusion on changes in the QTc interval and revealed a change in the QTc interval from 432 ± 39 ms to 483 ± 62 ms at baseline compared to during erythromycin therapy, respectively [175]. Only one patient developed TdP in this evaluation but brought the potential risk of intravenous erythromycin and QTc prolongation to the forefront.

Macrolides like clarithromycin inhibit HERG in a voltage- and time-dependent manner [31]. The ratio of IC_{50} to the serum C_{max} value (IC_{50}/C_{max}) has been used as a surrogate marker for the potential clinical risk for an agent known to inhibit HERG. The inhibition of HERG by agents withdrawn from the market such as ter-fenadine, astemizole, and cisapride occurs at the nanomolar concentration range with IC_{50}/C_{max} values of 0.075–5.2. Lin and colleagues evaluated 20 drugs known to induce QTc prolongation in order to identify cutoff values that could predict a higher risk of torsades de pointes [176]. Although IC_{50}/C_{max} cutoff values predictive of TdP have not been identified, values of 9.1, 12.8, and 17.5 are documented for clarithromycin, telithromycin, and erythromycin, respectively [176]. Higher IC_{50}/C_{max} values are expected with azithromycin, but despite this lower risk for QTc prolongation, reports of TdP have been recorded in the literature with this agent [177–180].

Ray and colleagues extended our knowledge of the risk of oral erythromycin and sudden death from cardiac causes through a population-based study of 1,249,943 person years of data through a Tennessee Medicaid cohort [6]. The incidence rate ratio [95% CI] for sudden death with the current use of erythromycin was 2.01 [1.08–3.75] and increased to 5.35 [1.72–16.64] with the use of a CYP3A inhibitor such as the calcium channel blocker, verapamil [6]. Although this study was not designed to evaluate the mechanistic basis for this interaction, the data suggest the necessity for caution when using macrolides with CYP3A inhibitors, especially when both agents carry the risk for dysrhythmias. A report in the New England Journal of Medicine by the same investigative group in 2012 suggested that these concerns extended to azithromycin in patients with a high baseline risk of cardio-vascular disease [30]. Relative to amoxicillin, use of azithromycin was associated with a 2.5 hazard ratio for death related to a cardiovascular event at a risk similar to levofloxacin but higher than ciprofloxacin [30]. This work along with others led to an FDA review and subsequent label revision informing healthcare providers for this risk [181]. Other groups have demonstrated that this risk is not present or lower in younger and middle-aged adults, highlighting well-characterized risk factors such as increasing age, female sex, concomitant illness, and use of concomitant agents that prolong the QTc interval [179, 182]. Table 2.3 includes a current list of agents that are contraindicated with the use of MAK agents based on a review of regulatory approved product labels.

As tabulated, azithromycin continues to not have any drug interaction-related contraindications, while telithromycin is contraindicated with cisapride and pimo-

Table 2.3 Pharmacological agents contraindicated with the use of erythromycin, clarithromycin, and telithromycin

Erythromycin	Clarithromycin	Telithromycin
1. Amifampridine	1. Alfuzosin	1. Alfuzosin
2. Amisulpride	2. Amifampridine	2. Amifampridine
3. Astemizole	3. Amisulpride	3. Amisulpride
4. Bepridil	4. Bepridil	4. Bepridil
5. Cisapride	5. Cisapride	5. Cisapride
6. Colchicine	6. Colchicine	6. Colchicine
7. Dronedarone	7. Conivaptan	7. Conivaptan
8. Ergot derivatives	8. Dronedarone	8. Dronedarone
9. Flibanserin	9. Eletriptan	9. Eletriptan
10. Fluconazole	10. Eliglustat	10. Eliglustat
11. Grepafloxacin	11. Eplerenone	11. Eplerenone
12. Levomethadyl	12. Ergot derivatives	12. Ergot derivatives
13. Lomitapide	13. Flibanserin	13. Flibanserin
14. Lovastatin	14. Fluconazole	14. Fluconazole
15. Mesoridazine	15. Isavuconazonium sulfate	15. Isavuconazonium sulfate
16. Pimozide	16. Ivabradine	16. Itraconazole
17. Piperaquine	17. Ketoconazole	17. Ivabradine
18. Posaconazole	18. Lomitapide	18. Lomitapide
19. Saquinavir	19. Lovastatin	19. Lovastatin
20. Simvastatin	20. Lurasidone	20. Lurasidone
21. Sparfloxacin	21. Maraviroc	21. Maraviroc
22. Terfenadine	22. Mesoridazine	22. Mesoridazine
23. Thioridazine	23. Naloxegol	23. Naloxegol
24. Ziprasidone	24. Nelfinavir	24. Nelfinavir
	25. Nimodipine	25. Nimodipine
	26. Pimozide	26. Pimozide
	27. Piperaquine	27. Piperaquine
	28. Posaconazole	28. Posaconazole
	29. Ranolazine	29. Saquinavir
	30. Saquinavir	30. Silodosin
	31. Silodosin	31. Simvastatin
	32. Simvastatin	32. Terfenadine
	33. Thioridazine	33. Thioridazine
	34. Tolvaptan	34. Tolvaptan
	35. Venetoclax	35. Venetoclax
	36. Ziprasidone	36. Ziprasidone

zide due to the potential risk of TdP. Clarithromycin has the longest list of contraindications and is comparable to erythromycin. Extension of the outlined list of agents contraindicated with clarithromycin to telithromycin is rational. Table 2.4 outlines a list of drugs, where the risk for combined QTc prolongation is known, probable or theoretically possible. The drugs outlined in this table demonstrate that several commercially available antiarrhythmic, antibiotic, antifungal, antihypertensive, antimalarial, antipsychotic, and anesthetic drugs have the potential to induce QTc prolongation [7, 44, 76, 176, 183, 184]. The combined risk of CYP3A4 inhibition coupled with QTc prolongation can lead to synergistic PK/PD drug interactions that can be fatal as documented with agents such as verapamil, pimozide, and cisapride

Table 2.4 Drugs with a known, probable and theoretical potential for QTc prolongation when combined with clarithromycin

Known	Probable	Theoretical		
Astemizole	Ajmaline	Artemether	Fluoxetine	Procainamide
Atazanavir	Dofetilide	Bretylium	Foscarnet	Prochlorperazine
Cisapride	Dolasetron	Telavancin	Gemifloxacin	Propafenone
Desipramine	Doxepin	Acecainide	Halofantrine	Protriptyline
Diltiazem	Dronedarone	Amiodarone	Haloperidol	Sematilide
Disopyramide	Droperidol	Amisulpride	Halothane	Sertindole
Itraconazole	Hydroquinidine	Amitriptyline	Ibutilide	Sotalol
Pimozide	Ketoconazole	Amoxapine	Imipramine	Sulfamethoxazole
Quetiapine	Mesoridazine	Aprindine	Isoflurane	Sultopride
Quinidine	Pazopanib	Arsenic Trioxide	Isradipine	Tedisamil
Risperidone	Ranolazine	Azimilide	Lidoflazine	Thioridazine
Terfenadine	Salmeterol	Bepridil	Lorcainide	Trifluoperazine
Verapamil	Voriconazole	Chloral Hydrate	Mefloquine	Trimethoprim
		Chloroquine	Moxifloxacin	Trimetrexate
		Chlorpromazine	Nortriptyline	Trimipramine
		Dibenzepin	Octreotide	Vasopressin
		Encainide	Pentamidine	Ziprasidone
		Enflurane	Pirmenol	Zolmitriptan
		Flecainide	Prajmaline	Zotepine
		Fluconazole	Probucol	

in recent years [185–187]. Finally, solithromycin has not demonstrate any QTc prolongation in a randomized crossover study in health volunteers exposed up to twice the daily dose [188]. Although this is a promising finding, post-marketing knowledge gained with the MAK agents to date suggests cautious optimism if this agent is approved for clinical use.

2.4 Summary

Macrolides have numerous antimicrobial and non-antimicrobial properties, and azithromycin represents one of the most widely prescribed antibiotics on the planet. Clarithromycin and telithromycin are important derivatives of erythromycin that have expanded the spectrum of activity of this agent and improved its tolerability. However, the drug interaction potential of these agents is greater than that of erythromycin. Azithromycin is a substantially different agent that shares pharmacologic similarities to clarithromycin and telithromycin without their major drug interaction potential. However, increasing antimicrobial resistance coupled with continued development of ketolides suggests that the benefits of this class of agents will constantly need to be weighed against their risk for drug interactions. Phase 1 drug metabolic inhibition through CYP3A4 and P-gp inhibition are the key mechanistic pathways for MAK drug-drug interactions. Given that these drug disposition pathways account for the clearance of a majority of pharmacological agents, the drug interaction potential of this class will remain high. A careful review of an individual patient's medication regimen is critical prior to their prescription.

References

1. Hicks LA, Bartoces MG, Roberts RM, Suda KJ, Hunkler RJ, Taylor TH Jr et al (2015) US outpatient antibiotic prescribing variation according to geography, patient population, and provider specialty in 2011. Clin Infect Dis 60(9):1308–1316
2. Haight TH, Finland M (1952) Laboratory and clinical studies on erythromycin. N Engl J Med 247(7):227–232
3. Hinshaw HC (1953) The treatment of infectious diseases: advances in the use of antimicrobial drugs. Calif Med 79(4):282–283
4. Zeitlinger M, Wagner CC, Heinisch B (2009) Ketolides – the modern relatives of macrolides : the pharmacokinetic perspective. Clin Pharmacokinet 48(1):23–38
5. Pai MP, Graci DM, Amsden GW (2000) Macrolide drug interactions: an update. Ann Pharmacother 34(4):495–513
6. Ray WA, Murray KT, Meredith S, Narasimhulu SS, Hall K, Stein CM (2004) Oral erythromycin and the risk of sudden death from cardiac causes. N Engl J Med 351(11):1089–1096
7. Iannini PB (2002) Cardiotoxicity of macrolides, ketolides and fluoroquinolones that prolong the QTc interval. Expert Opin Drug Saf 1(2):121–128
8. Watkins PB (1997) The barrier function of CYP3A4 and P-glycoprotein in the small bowel. Adv Drug Deliv Rev 27(2-3):161–170
9. Pal D, Mitra AK (2006) MDR- and CYP3A4-mediated drug-drug interactions. J Neuroimmune Pharmacol 1(3):323–339
10. Jelic D, Antolovic R (2016) From erythromycin to azithromycin and new potential ribosome-binding antimicrobials. Antibiotics (Basel) 5(3):pii:E29
11. Zhanel GG, Hartel E, Adam H, Zelenitsky S, Zhanel MA, Golden A et al (2016) Solithromycin: a novel fluoroketolide for the treatment of community-acquired bacterial pneumonia. Drugs 76(18):1737–1757
12. Georgopapadakou NH (2014) The wobbly status of ketolides: where do we stand? Expert Opin Investig Drugs 23(10):1313–1319
13. Fernandes P, Martens E, Bertrand D, Pereira D (2016) The solithromycin journey-It is all in the chemistry. Bioorg Med Chem 24(24):6420–6428
14. Owens B (2017) Solithromycin rejection chills antibiotic sector. Nat Biotechnol 35(3):187–188
15. Mansour H, Chahine EB, Karaoui LR, El-Lababidi RM (2013) Cethromycin: a new ketolide antibiotic. Ann Pharmacother 47(3):368–379
16. Arenz S, Ramu H, Gupta P, Berninghausen O, Beckmann R, Vazquez-Laslop N et al (2014) Molecular basis for erythromycin-dependent ribosome stalling during translation of the ErmBL leader peptide. Nat Commun 5:3501
17. Hansen JL, Ippolito JA, Ban N, Nissen P, Moore PB, Steitz TA (2002) The structures of four macrolide antibiotics bound to the large ribosomal subunit. Mol Cell 10(1):117–128
18. Douthwaite S, Jalava J, Jakobsen L (2005) Ketolide resistance in Streptococcus pyogenes correlates with the degree of rRNA dimethylation by Erm. Mol Microbiol 58(2):613–622
19. Schroeder MR, Stephens DS (2016) Macrolide Resistance in Streptococcus pneumoniae. Front Cell Infect Microbiol 6:98
20. Uthman MM, Uthman OA, Yahaya I. Interventions for the prevention of mycobacterium avium complex in adults and children with HIV. Cochrane Database Syst Rev (4):CD007191
21. Sanchez E, Vannier E, Wormser GP, Diagnosis HLT (2016) Treatment, and prevention of lyme disease, human granulocytic anaplasmosis, and babesiosis: a review. JAMA 315(16):1767–1777
22. Thung I, Aramin H, Vavinskaya V, Gupta S, Park JY, Crowe SE et al (2016) Review article: the global emergence of Helicobacter pylori antibiotic resistance. Aliment Pharmacol Ther 43(4):514–533
23. Erakovic Haber V, Bosnar M, Kragol G (2014) The design of novel classes of macrolides for neutrophil-dominated inflammatory diseases. Future Med Chem 6(6):657–674

24. Zarogoulidis P, Papanas N, Kioumis I, Chatzaki E, Maltezos E, Zarogoulidis K (2012) Macrolides: from in vitro anti-inflammatory and immunomodulatory properties to clinical practice in respiratory diseases. Eur J Clin Pharmacol 68(5):479–503

25. Camilleri M, Parkman HP, Shafi MA, Abell TL, Gerson L, American College of Gastroenterology (2013) Clinical guideline: management of gastroparesis. Am J Gastroenterol 108(1):18–37; quiz 8

26. Chini P, Toskes PP, Waseem S, Hou W, McDonald R, Moshiree B (2012) Effect of azithromycin on small bowel motility in patients with gastrointestinal dysmotility. Scand J Gastroenterol 47(4):422–427

27. Cramer CL, Patterson A, Alchakaki A, Soubani AO (2017) Immunomodulatory indications of azithromycin in respiratory disease: a concise review for the clinician. Postgrad Med 129(5):493–499

28. Kanoh S, Rubin BK (2010) Mechanisms of action and clinical application of macrolides as immunomodulatory medications. Clin Microbiol Rev 23(3):590–615

29. Liu CN, Somps CJ (2010) Telithromycin blocks neuromuscular transmission and inhibits nAChR currents in vitro. Toxicol Lett 194(3):66–69

30. Ray WA, Murray KT, Hall K, Arbogast PG, Stein CM (2012) Azithromycin and the risk of cardiovascular death. N Engl J Med 366(20):1881–1890

31. Volberg WA, Koci BJ, Su W, Lin J, Zhou J (2002) Blockade of human cardiac potassium channel human ether-a-go-go-related gene (HERG) by macrolide antibiotics. J Pharmacol Exp Ther 302(1):320–327

32. Amidon GL, Lennernas H, Shah VP, Crison JRA (1995) theoretical basis for a biopharmaceutic drug classification: the correlation of in vitro drug product dissolution and in vivo bioavailability. Pharm Res 12(3):413–420

33. Hirsch HA, Finland M (1959) Effect of food on the absorption of erythromycin propionate, erythromycin stearate and triacetyloleandomycin. Am J Med Sci 237(6):693–709

34. Mantyla R, Ailio A, Allonen H, Kanto J (1978) Bioavailability and effect of food on the gastrointestinal absorption of two erythromycin derivatives. Ann Clin Res 10(5):258–262

35. Welling PG, Huang H, Hewitt PF, Lyons LL (1978) Bioavailability of erythromycin stearate: influence of food and fluid volume. J Pharm Sci 67(6):764–766

36. Malmborg AS (1979) Effect of food on absorption of erythromycin. A study of two derivatives, the stearate and the base. J Antimicrob Chemother 5(5):591–599

37. DiSanto AR, Chodos DJ (1981) Influence of study design in assessing food effects on absorption of erythromycin base and erythromycin stearate. Antimicrob Agents Chemother 20(2):190–196

38. Chu S, Park Y, Locke C, Wilson DS, Cavanaugh JC (1992) Drug-food interaction potential of clarithromycin, a new macrolide antimicrobial. J Clin Pharmacol 32(1):32–36

39. Alkhalidi BA, Tamimi JJ, Salem II, Ibrahim H, Sallam AA (2008) Assessment of the bioequivalence of two formulations of clarithromycin extended-release 500-mg tablets under fasting and fed conditions: a single-dose, randomized, open-label, two-period, two-way crossover study in healthy Jordanian male volunteers. Clin Ther 30(10):1831–1843

40. Periti P, Mazzei T, Mini E, Novelli A (1989) Clinical pharmacokinetic properties of the macrolide antibiotics. Effects of age and various pathophysiological states (Part I). Clin Pharmacokinet 16(4):193–214

41. Ballow CH, Amsden GW (1992) Azithromycin: the first azalide antibiotic. Ann Pharmacother 26(10):1253–1261

42. Chandra R, Liu P, Breen JD, Fisher J, Xie C, LaBadie R et al (2007) Clinical pharmacokinetics and gastrointestinal tolerability of a novel extended-release microsphere formulation of azithromycin. Clin Pharmacokinet 46(3):247–259

43. Aucamp M, Odendaal R, Liebenberg W, Hamman J (2015) Amorphous azithromycin with improved aqueous solubility and intestinal membrane permeability. Drug Dev Ind Pharm 41(7):1100–1108

44. Nguyen M, Chung EP (2005) Telithromycin: the first ketolide antimicrobial. Clin Ther 27(8):1144–1163
45. Prandota J, Tillement JP, d'Athis P, Campos H, Barre J (1980) Binding of erythromycin base to human plasma proteins. J Int Med Res 8(Suppl 2):1–8
46. Welling PG, Craig WA (1978) Pharmacokinetics of intravenous erythromycin. J Pharm Sci 67(8):1057–1059
47. Piscitelli SC, Danziger LH, Rodvold KA (1992) Clarithromycin and azithromycin: new macrolide antibiotics. Clin Pharm 11(2):137–152
48. Kiem S, Schentag JJ (2008) Interpretation of antibiotic concentration ratios measured in epithelial lining fluid. Antimicrob Agents Chemother 52(1):24–36
49. Van Bambeke F, Montenez JP, Piret J, Tulkens PM, Courtoy PJ, Mingeot-Leclercq MP (1996) Interaction of the macrolide azithromycin with phospholipids. I. Inhibition of lysosomal phospholipase A1 activity. Eur J Pharmacol 314(1-2):203–214
50. Rodvold KA, Gotfried MH (2012) Still JG, Clark K, Fernandes P. Comparison of plasma, epithelial lining fluid, and alveolar macrophage concentrations of solithromycin (CEM-101) in healthy adult subjects. Antimicrob Agents Chemother 56(10):5076–5081
51. Rodvold KA, George JM, Yoo L (2011) Penetration of anti-infective agents into pulmonary epithelial lining fluid: focus on antibacterial agents. Clin Pharmacokinet 50(10):637–664
52. Lancaster CS, Bruun GH, Peer CJ, Mikkelsen TS, Corydon TJ, Gibson AA et al (2012) OATP1B1 polymorphism as a determinant of erythromycin disposition. Clin Pharmacol Ther 92(5):642–650
53. Mao JC, Tardrew PL (1965) Demethylation of erythromycins by rabbit tissues in vitro. Biochem Pharmacol 14(7):1049–1058
54. Pessayre D (1983) Effects of macrolide antibiotics on drug metabolism in rats and in humans. Int J Clin Pharmacol Res 3(6):449–458
55. Unal D, Fenercioglu A, Ozbay L, Ozkirim B, Erol D (2008) The effect of hydroxy metabolites of clarithromycin to the pharmacokinetic parameters, and determination of hydroxy metabolites ratio of clarithromycin. Eur J Drug Metab Pharmacokinet 33(4):243–246
56. Danan G, Descatoire V, Pessayre D (1981) Self-induction by erythromycin of its own transformation into a metabolite forming an inactive complex with reduced cytochrome P-450. J Pharmacol Exp Ther 218(2):509–514
57. Amacher DE, Schomaker SJ, Retsema JA (1991) Comparison of the effects of the new azalide antibiotic, azithromycin, and erythromycin estolate on rat liver cytochrome P-450. Antimicrob Agents Chemother 35(6):1186–1190
58. Evrard-Todeschi N, Gharbi-Benarous J, Gaillet C, Verdier L, Bertho G, Lang C et al (2000) Conformations in solution and bound to bacterial ribosomes of ketolides, HMR 3647 (telithromycin) and RU 72366: a new class of highly potent antibacterials. Bioorg Med Chem 8(7):1579–1597
59. Nosaka H, Nadai M, Kato M, Yasui K, Yoshizumi H, Miyoshi M et al (2006) Effect of a newly developed ketolide antibiotic, telithromycin, on metabolism of theophylline and expression of cytochrome P450 in rats. Life Sci 79(1):50–56
60. Zhou SF (2008) Drugs behave as substrates, inhibitors and inducers of human cytochrome P450 3A4. Curr Drug Metab 9(4):310–322
61. Abduljalil K, Kinzig M, Bulitta J, Horkovics-Kovats S, Sorgel F, Rodamer M et al (2009) Modeling the autoinhibition of clarithromycin metabolism during repeated oral administration. Antimicrob Agents Chemother 53(7):2892–2901
62. Jamieson BD, Ciric S, Fernandes P (2015) Safety and pharmacokinetics of solithromycin in subjects with hepatic impairment. Antimicrob Agents Chemother 59(8):4379–4386
63. Shi J, Chapel S, Montay G, Hardy P, Barrett JS, Sica D et al (2005) Effect of ketoconazole on the pharmacokinetics and safety of telithromycin and clarithromycin in older subjects with renal impairment. Int J Clin Pharmacol Ther 43(3):123–133
64. Rodvold KA (1999) Clinical pharmacokinetics of clarithromycin. Clin Pharmacokinet 37(5):385–398

65. Wilkinson GR (1996) Cytochrome P4503A (CYP3A) metabolism: prediction of in vivo activity in humans. J Pharmacokinet Biopharm 24(5):475–490
66. Prueksaritanont T, Chu X, Gibson C, Cui D, Yee KL, Ballard J et al (2013) Drug-drug interaction studies: regulatory guidance and an industry perspective. AAPS J 15(3):629–645
67. Franke RM, Gardner ER, Sparreboom A (2010) Pharmacogenetics of drug transporters. Curr Pharm Des 16(2):220–230
68. Morton MR, Cooper JW (1989) Erythromycin-induced digoxin toxicity. DICP 23(9):668–670
69. Saha JR, Butler VP Jr, Neu HC, Lindenbaum J (1983) Digoxin-inactivating bacteria: identification in human gut flora. Science 220(4594):325–327
70. Donley TG, Smith RF, Roy B (1990) Reduced oral contraceptive effectiveness with concurrent antibiotic use: a protocol for prescribing antibiotics to women of childbearing age. Compendium 11(6):392–396
71. Meyer B, Muller F, Wessels P, Maree JA (1990) model to detect interactions between roxithromycin and oral contraceptives. Clin Pharmacol Ther 47(6):671–674
72. Toh S, Mitchell AA, Anderka M, de Jong-van den Berg LT, Hernandez-Diaz S, National Birth Defects Prevention S (2011) Antibiotics and oral contraceptive failure – a case-crossover study. Contraception 83(5):418–425
73. Weaver K, Glasier A (1999) Interaction between broad-spectrum antibiotics and the combined oral contraceptive pill. A literature review. Contraception 59(2):71–78
74. Korpela K, Salonen A, Virta LJ, Kekkonen RA, Forslund K, Bork P et al (2016) Intestinal microbiome is related to lifetime antibiotic use in Finnish pre-school children. Nat Commun 7:10410
75. Durek C, Rusch-Gerdes S, Jocham D, Bohle A (2000) Sensitivity of BCG to modern antibiotics. Eur Urol 37(Suppl 1):21–25
76. Goldschmidt N, Azaz-Livshits T, Gotsman, Nir-Paz R, Ben-Yehuda A, Muszkat M (2001) Compound cardiac toxicity of oral erythromycin and verapamil. Ann Pharmacother 35(11):1396–1399
77. Yamreudeewong W, Scavone JM, Paone RP, Lewis GP (1989) Effect of antacid coadministration on the bioavailability of erythromycin stearate. Clin Pharm 8(5):352–354
78. Foulds G, Hilligoss DM, Henry EB, Gerber N (1991) The effects of an antacid or cimetidine on the serum concentrations of azithromycin. J Clin Pharmacol 31(2):164–167
79. Rengelshausen J, Goggelmann C, Burhenne J, Riedel KD, Ludwig J, Weiss J et al (2003) Contribution of increased oral bioavailability and reduced nonglomerular renal clearance of digoxin to the digoxin-clarithromycin interaction. Br J Clin Pharmacol 56(1):32–38
80. Moj D, Hanke N, Britz H, Frechen S, Kanacher T, Wendl T et al (2017) Clarithromycin, midazolam, and digoxin: application of pbpk modeling to gain new insights into drug-drug interactions and co-medication regimens. AAPS J 19(1):298–312
81. Delavenne X, Ollier E, Basset T, Bertoletti L, Accassat S, Garcin A et al (2013) A semi-mechanistic absorption model to evaluate drug-drug interaction with dabigatran: application with clarithromycin. Br J Clin Pharmacol 76(1):107–113
82. Gouin-Thibault I, Delavenne X, Blanchard A, Siguret V, Salem JE, Narjoz C et al (2017) Interindividual variability in dabigatran and rivaroxaban exposure: contribution of ABCB1 genetic polymorphisms and interaction with clarithromycin. J Thromb Haemost 15(2):273–283
83. Gomes T, Mamdani MM, Juurlink DN (2009) Macrolide-induced digoxin toxicity: a population-based study. Clin Pharmacol Ther 86(4):383–386
84. Kalliokoski A, Niemi M (2009) Impact of OATP transporters on pharmacokinetics. Br J Pharmacol 158(3):693–705
85. Seithel A, Eberl S, Singer K, Auge D, Heinkele G, Wolf NB et al (2007) The influence of macrolide antibiotics on the uptake of organic anions and drugs mediated by OATP1B1 and OATP1B3. Drug Metab Dispos 35(5):779–786
86. Hirano M, Maeda K, Shitara Y, Sugiyama Y (2006) Drug-drug interaction between pitavastatin and various drugs via OATP1B1. Drug Metab Dispos 34(7):1229–1236

87. Prueksaritanont T, Tatosian DA, Chu X, Railkar R, Evers R, Chavez-Eng C et al (2017) Validation of a microdose probe drug cocktail for clinical drug interaction assessments for drug transporters and CYP3A. Clin Pharmacol Ther 101(4):519–530

88. Jacobson TA (2004) Comparative pharmacokinetic interaction profiles of pravastatin, simvastatin, and atorvastatin when coadministered with cytochrome P450 inhibitors. Am J Cardiol 94(9):1140–1146

89. Markert C, Hellwig R, Burhenne J, Hoffmann MM, Weiss J, Mikus G et al (2013) Interaction of ambrisentan with clarithromycin and its modulation by polymorphic SLCO1B1. Eur J Clin Pharmacol 69(10):1785–1793

90. Amin ML (2013) P-glycoprotein inhibition for optimal drug delivery. Drug Target Insights 7:27–34

91. Bauer M, Karch R, Tournier N, Cisternino S, Wadsak W, Hacker M et al (2017) Assessment of P-glycoprotein transport activity at the human blood-retinal barrier with (R)-11C-verapamil PET. J Nucl Med 58(4):678–681

92. Chapy H, Saubamea B, Tournier N, Bourasset F, Behar-Cohen F, Decleves X et al (2016) Blood-brain and retinal barriers show dissimilar ABC transporter impacts and concealed effect of P-glycoprotein on a novel verapamil influx carrier. Br J Pharmacol 173(3):497–510

93. Arakawa R, Ito H, Okumura M, Morimoto T, Seki C, Takahashi H et al (2010) No inhibitory effect on P-glycoprotein function at blood-brain barrier by clinical dose of clarithromycin: a human PET study with [(1)(1)C]verapamil. Ann Nucl Med 24(2):83–87

94. Orlando R, De Martin S, Pegoraro P, Quintieri L, Palatini P (2009) Irreversible CYP3A inhibition accompanied by plasma protein-binding displacement: a comparative analysis in subjects with normal and impaired liver function. Clin Pharmacol Ther 85(3):319–326

95. Hisaka A, Kusama M, Ohno Y, Sugiyama Y, Suzuki HA (2009) proposal for a pharmacokinetic interaction significance classification system (PISCS) based on predicted drug exposure changes and its potential application to alert classifications in product labelling. Clin Pharmacokinet 48(10):653–666

96. Vermeer LM, Isringhausen CD, Ogilvie BW, Buckley DB (2016) Evaluation of ketoconazole and its alternative clinical CYP3A4/5 inhibitors as inhibitors of drug transporters: the in vitro effects of ketoconazole, ritonavir, clarithromycin, and itraconazole on 13 clinically-relevant drug transporters. Drug Metab Dispos 44(3):453–459

97. Greenblatt DJ, Harmatz JS (2015) Ritonavir is the best alternative to ketoconazole as an index inhibitor of cytochrome P450-3A in drug-drug interaction studies. Br J Clin Pharmacol 80(3):342–350

98. Greenblatt HK, Greenblatt DJ (2014) Liver injury associated with ketoconazole: review of the published evidence. J Clin Pharmacol 54(12):1321–1329

99. Ohno Y, Hisaka A, Suzuki H (2007) General framework for the quantitative prediction of CYP3A4-mediated oral drug interactions based on the AUC increase by coadministration of standard drugs. Clin Pharmacokinet 46(8):681–696

100. Teo YL, Ho HK, Chan A (2015) Metabolism-related pharmacokinetic drug-drug interactions with tyrosine kinase inhibitors: current understanding, challenges and recommendations. Br J Clin Pharmacol 79(2):241–253

101. Qiu XH, Shao JJ, Mei JG, Li HQ, Cao HQ (2016) Clarithromycin synergistically enhances thalidomide cytotoxicity in myeloma cells. Acta Haematol 135(2):103–109

102. Van Nuffel AM, Sukhatme V, Pantziarka P, Meheus L, Sukhatme VP, Bouche G (2015) Repurposing Drugs in Oncology (ReDO)-clarithromycin as an anti-cancer agent. Ecancermedicalscience 9:513

103. Awan S, Crosby V, Potter V, Hennig I, Baldwin D, Ndlovu M et al (2017) Is clarithromycin a potential treatment for cachexia in people with lung cancer? A feasibility study. Lung Cancer 104:75–78

104. Yano R, Tani D, Watanabe K, Tsukamoto H, Igarashi T, Nakamura T et al (2009) Evaluation of potential interaction between vinorelbine and clarithromycin. Ann Pharmacother 43(3):453–458

105. Tsuji D, Kamezato M, Daimon T, Taku K, Hatori M, Ikeda M et al (2013) Retrospective analysis of severe neutropenia in patients receiving concomitant administration of docetaxel and clarithromycin. Chemotherapy 59(6):407–413

106. Levin PA (2016) Practical combination therapy based on pathophysiology of type 2 diabetes. Diabetes Metab Syndr Obes 9:355–369

107. Fleishaker JC, Phillips JP (1991) Evaluation of a potential interaction between erythromycin and glyburide in diabetic volunteers. J Clin Pharmacol 31(3):259–262

108. Greupink R, Schreurs M, Benne MS, Huisman MT, Russel FG (2013) Semi-mechanistic physiologically-based pharmacokinetic modeling of clinical glibenclamide pharmacokinetics and drug-drug-interactions. Eur J Pharm Sci 49(5):819–828

109. Bussing R, Gende A (2002) Severe hypoglycemia from clarithromycin-sulfonylurea drug interaction. Diabetes Care 25(9):1659–1661

110. Schelleman H, Bilker WB, Brensinger CM, Wan F, Hennessy S (2010) Anti-infectives and the risk of severe hypoglycemia in users of glipizide or glyburide. Clin Pharmacol Ther 88(2):214–222

111. Parekh TM, Raji M, Lin YL, Tan A, Kuo YF, Goodwin JS (2014) Hypoglycemia after antimicrobial drug prescription for older patients using sulfonylureas. JAMA Intern Med 174(10):1605–1612

112. Khamaisi M, Leitersdorf E (2008) Severe hypoglycemia from clarithromycin-repaglinide drug interaction. Pharmacotherapy 28(5):682–684

113. Takanohashi T, Koizumi T, Mihara R, Okudaira K (2007) Prediction of the metabolic interaction of nateglinide with other drugs based on in vitro studies. Drug Metab Pharmacokinet 22(6):409–418

114. Patel CG, Li L, Girgis S, Kornhauser DM, Frevert EU, Boulton DW (2011) Two-way pharmacokinetic interaction studies between saxagliptin and cytochrome P450 substrates or inhibitors: simvastatin, diltiazem extended-release, and ketoconazole. Clin Pharmacol 3:13–25

115. Eadie MJ (2001) Clinically significant drug interactions with agents specific for migraine attacks. CNS Drugs 15(2):105–118

116. Delaforge M, Riviere R, Sartori E, Doignon JL, Grognet JM (1989) Metabolism of dihydroergotamine by a cytochrome P-450 similar to that involved in the metabolism of macrolide antibiotics. Xenobiotica 19(11):1285–1295

117. Francis H, Tyndall A, Webb J (1984) Severe vascular spasm due to erythromycin-ergotamine interaction. Clin Rheumatol 3(2):243–246

118. Horowitz RS, Dart RC, Gomez HF (1996) Clinical ergotism with lingual ischemia induced by clarithromycin-ergotamine interaction. Arch Intern Med 156(4):456–458

119. Bird PA, Sturgess AD (2000) Clinical ergotism with severe bilateral upper limb ischaemia precipitated by an erythromycin – ergotamine drug interaction. Aust N Z J Med 30(5):635–636

120. Ausband SC, Goodman PE (2001) An unusual case of clarithromycin associated ergotism. J Emerg Med 21(4):411–413

121. de Mey C, Althaus M, Ezan E, Retzow A (2001) Erythromycin increases plasma concentrations of alpha-dihydroergocryptine in humans. Clin Pharmacol Ther 70(2):142–148

122. Loder E (2010) Triptan therapy in migraine. N Engl J Med 363(1):63–70

123. Fleishaker JC, Herman BD, Carel BJ, Azie NE (2003) Interaction between ketoconazole and almotriptan in healthy volunteers. J Clin Pharmacol 43(4):423–427

124. Sternieri E, Coccia CP, Pinetti D, Ferrari A (2006) Pharmacokinetics and interactions of headache medications, part I: introduction, pharmacokinetics, metabolism and acute treatments. Expert Opin Drug Metab Toxicol 2(6):961–979

125. Mandrioli R, Mercolini L, Raggi MA (2008) Benzodiazepine metabolism: an analytical perspective. Curr Drug Metab 9(8):827–844

126. Olivera M, Martinez C, Gervasini G, Carrillo JA, Ramos S, Benitez J et al (2007) Effect of common NAT2 variant alleles in the acetylation of the major clonazepam metabolite, 7-aminoclonazepam. Drug Metab Lett 1(1):3–5

127. Mijderwijk H, Klimek M, van Beek S, van Schaik RH, Duivenvoorden HJ, Stolker RJ (2016) Implication of UGT2B15 genotype polymorphism on postoperative anxiety levels in patients receiving lorazepam premedication. Anesth Analg 123(5):1109–1115

128. Senthilkumaran S, Subramanian PT (2011) Prolonged sedation related to erythromycin and midazolam interaction: a word of caution. Indian Pediatr 48(11):909
129. Olkkola KT, Aranko K, Luurila H, Hiller A, Saarnivaara L, Himberg JJ et al (1993) A potentially hazardous interaction between erythromycin and midazolam. Clin Pharmacol Ther 53(3):298–305
130. Yeates RA, Laufen H, Zimmermann T, Schumacher T (1997) Pharmacokinetic and pharmacodynamic interaction study between midazolam and the macrolide antibiotics, erythromycin, clarithromycin, and the azalide azithromycin. Int J Clin Pharmacol Ther 35(12):577–579
131. Greenblatt DJ, Zammit GK (2012) Pharmacokinetic evaluation of eszopiclone: clinical and therapeutic implications. Expert Opin Drug Metab Toxicol 8(12):1609–1618
132. Hesse LM, von Moltke LL, Greenblatt DJ (2003) Clinically important drug interactions with zopiclone, zolpidem and zaleplon. CNS Drugs 17(7):513–532
133. Knops N, Levtchenko E, van den Heuvel B, Kuypers D (2013) From gut to kidney: transporting and metabolizing calcineurin-inhibitors in solid organ transplantation. Int J Pharm 452(1-2):14–35
134. Campana C, Regazzi MB, Buggia I, Molinaro M (1996) Clinically significant drug interactions with cyclosporin. An update. Clin Pharmacokinet 30(2):141–179
135. Page RL 2nd, Ruscin JM, Fish D, Lapointe M (2001) Possible interaction between intravenous azithromycin and oral cyclosporine. Pharmacotherapy 21(11):1436–1443
136. Mignat C (1997) Clinically significant drug interactions with new immunosuppressive agents. Drug Saf 16(4):267–278
137. Mori T, Aisa Y, Nakazato T, Yamazaki R, Ikeda Y, Okamoto S (2005) Tacrolimus-azithromycin interaction in a recipient of allogeneic bone marrow transplantation. Transpl Int 18(6):757–758
138. Kovarik JM, Beyer D, Schmouder RL (2006) Everolimus drug interactions: application of a classification system for clinical decision making. Biopharm Drug Dispos 27(9):421–426
139. Pea F, Cojutti P, Tursi V, Livi U, Baraldo M (2015) Everolimus overexposure in a heart transplant patient receiving clarithromycin for the treatment of pneumonia. Transpl Infect Dis 17(6):926–928
140. Capone D, Palmiero G, Gentile A, Basile V, Federico S, Sabbatini M et al (2007) A pharmacokinetic interaction between clarithromycin and sirolimus in kidney transplant recipient. Curr Drug Metab 8(4):379–381
141. Hill FJ, McCloskey SJ, Sheerin N (2015) From a fish tank injury to hospital haemodialysis: the serious consequences of drug interactions. BMJ Case Rep 2015. pii: bcr2015209961
142. Li DQ, Kim R, McArthur E, Fleet JL, Bailey DG, Juurlink D et al (2015) Risk of adverse events among older adults following co-prescription of clarithromycin and statins not metabolized by cytochrome P450 3A4. CMAJ 187(3):174–180
143. Patel AM, Shariff S, Bailey DG, Juurlink DN, Gandhi S, Mamdani M et al (2013) Statin toxicity from macrolide antibiotic coprescription: a population-based cohort study. Ann Intern Med 158(12):869–876
144. Gu Q, Paulose-Ram R, Burt VL, Kit BK (2014) Prescription cholesterol-lowering medication use in adults aged 40 and over: United States, 2003-2012 NCHS Data Brief (177):1–8.
145. Chatzizisis YS, Koskinas KC, Misirli G, Vaklavas C, Hatzitolios A, Giannoglou GD (2010) Risk factors and drug interactions predisposing to statin-induced myopathy: implications for risk assessment, prevention and treatment. Drug Saf 33(3):171–187
146. Knauer MJ, Urquhart BL, Meyer zu Schwabedissen HE, Schwarz UI, Lemke CJ, Leake BF et al (2010) Human skeletal muscle drug transporters determine local exposure and toxicity of statins. Circ Res 106(2):297–306
147. Law M, Rudnicka AR (2006) Statin safety: a systematic review. Am J Cardiol 97(8A):52C–60C
148. Kantola T, Kivisto KT, Neuvonen PJ (1998) Erythromycin and verapamil considerably increase serum simvastatin and simvastatin acid concentrations. Clin Pharmacol Ther 64(2):177–182
149. Gur S, Kadowitz PJ, Gokce A, Sikka SC, Lokman U, Hellstrom WJ (2013) Update on drug interactions with phosphodiesterase-5 inhibitors prescribed as first-line therapy for patients with erectile dysfunction or pulmonary hypertension. Curr Drug Metab 14(2):265–269

150. Hedaya MA, El-Afify DR, El-Maghraby GM (2006) The effect of ciprofloxacin and clarithromycin on sildenafil oral bioavailability in human volunteers. Biopharm Drug Dispos 27(2):103–110
151. Kivisto KT, Lamberg TS, Kantola T, Neuvonen PJ (1997) Plasma buspirone concentrations are greatly increased by erythromycin and itraconazole. Clin Pharmacol Ther 62(3):348–354
152. Taylor D (1997) Pharmacokinetic interactions involving clozapine. Br J Psychiatry 171:109–112
153. Liukas A, Hagelberg NM, Kuusniemi K, Neuvonen PJ, Olkkola KT (2011) Inhibition of cytochrome P450 3A by clarithromycin uniformly affects the pharmacokinetics and pharmacodynamics of oxycodone in young and elderly volunteers. J Clin Psychopharmacol 31(3):302–308
154. Labroo RB, Paine MF, Thummel KE, Kharasch ED (1997) Fentanyl metabolism by human hepatic and intestinal cytochrome P450 3A4: implications for interindividual variability in disposition, efficacy, and drug interactions. Drug Metab Dispos 25(9):1072–1080
155. Cronnolly B, Pegrum H (2012) Fentanyl-clarithromycin interaction. BMJ Case Rep 2012. pii: bcr0220125936
156. Horton R, Barber C (2009) Opioid-induced respiratory depression resulting from transdermal fentanyl-clarithromycin drug interaction in a patient with advanced COPD. J Pain Symptom Manage 37(6):e2–e5
157. Bolhuis MS, van Altena R, van Soolingen D, de Lange WC, Uges DR, van der Werf TS et al (2013) Clarithromycin increases linezolid exposure in multidrug-resistant tuberculosis patients. Eur Respir J 42(6):1614–1621
158. Varma MV, Kimoto E, Scialis R, Bi Y, Lin J, Eng H et al (2017) Transporter-Mediated Hepatic Uptake Plays an Important Role in the Pharmacokinetics and Drug-Drug Interactions of Montelukast. Clin Pharmacol Ther 101(3):406–415
159. Hegazy SK, Mabrouk MM, Elsisi AE, Mansour NO (2012) Effect of clarithromycin and fluconazole on the pharmacokinetics of montelukast in human volunteers. Eur J Clin Pharmacol 68(9):1275–1280
160. Bouquie R, Deslandes G, Renaud C, Dailly E, Haloun A, Jolliet P (2011) Colchicine-induced rhabdomyolysis in a heart/lung transplant patient with concurrent use of cyclosporin, pravastatin, and azithromycin. J Clin Rheumatol 17(1):28–30
161. Terkeltaub RA, Furst DE, Digiacinto JL, Kook KA, Davis MW (2011) Novel evidence-based colchicine dose-reduction algorithm to predict and prevent colchicine toxicity in the presence of cytochrome P450 3A4/P-glycoprotein inhibitors. Arthritis Rheum 63(8):2226–2237
162. Lee CY, Marcotte F, Giraldeau G, Koren G, Juneau M, Tardif JC (2011) Digoxin toxicity precipitated by clarithromycin use: case presentation and concise review of the literature. Can J Cardiol 27(6):870.e15–870.e16
163. Cohen O, Locketz G, Hershko AY, Gorshtein A, Levy Y (2015) Colchicine-clarithromycin-induced rhabdomyolysis in Familial Mediterranean Fever patients under treatment for Helicobacter pylori. Rheumatol Int 35(11):1937–1941
164. van der Velden W, Huussen J, Ter Laak H, de Sevaux R (2008) Colchicine-induced neuromyopathy in a patient with chronic renal failure: the role of clarithromycin. Neth J Med 66(5):204–206
165. Cheng VC, Ho PL, Yuen KY (2005) Two probable cases of serious drug interaction between clarithromycin and colchicine. South Med J 98(8):811–813
166. Ma TK, Chow KM, Choy AS, Kwan BC, Szeto CC, Li PK (2014) Clinical manifestation of macrolide antibiotic toxicity in CKD and dialysis patients. Clin Kidney J 7(6):507–512
167. Celebi ZK, Akturk S, Oktay EI, Duman N, Keven K (2013) Colchicine-induced rhabdomyolysis following a concomitant use of clarithromycin in a haemodialysis patient with familial Mediterranean fever. Clin Kidney J 6(6):665–666
168. Chan AL, Wang MT, CY S, Tsai FH (2009) Risk of digoxin intoxication caused by clarithromycin-digoxin interactions in heart failure patients: a population-based study. Eur J Clin Pharmacol 65(12):1237–1243

169. Dorani H, Schutzer KM, Sarich TC, Wall U, Logren U, Ohlsson L et al (2007) Pharmacokinetics and pharmacodynamics of the oral direct thrombin inhibitor ximelagatran co-administered with different classes of antibiotics in healthy volunteers. Eur J Clin Pharmacol 63(6):571–581

170. Tobe SW, Siu LL, Jamal SA, Skorecki KL, Murphy GF, Warner E (1995) Vinblastine and erythromycin: an unrecognized serious drug interaction. Cancer Chemother Pharmacol 35(3):188–190

171. Sanguinetti MC, Tristani-Firouzi M (2006) hERG potassium channels and cardiac arrhythmia. Nature 440(7083):463–469

172. Schoenenberger RA, Haefeli WE, Weiss P, Ritz RF (1990) Association of intravenous erythromycin and potentially fatal ventricular tachycardia with Q-T prolongation (torsades de pointes). BMJ 300(6736):1375–1376

173. Tschida SJ, Guay DR, Straka RJ, Hoey LL, Johanning R, Vance-Bryan K (1996) QTc-interval prolongation associated with slow intravenous erythromycin lactobionate infusions in critically ill patients: a prospective evaluation and review of the literature. Pharmacotherapy 16(4):663–674

174. Haefeli WE, Schoenenberger RA, Weiss P, Ritz R (1992) Possible risk for cardiac arrhythmia related to intravenous erythromycin. Intensive Care Med 18(8):469–473

175. Oberg KC, Bauman JLQT (1995) interval prolongation and torsades de pointes due to erythromycin lactobionate. Pharmacotherapy 15(6):687–692

176. Lin YL, Hsiao CL, YC W, Kung MF (2011) Electrophysiologic, pharmacokinetic, and pharmacodynamic values indicating a higher risk of torsades de pointes. J Clin Pharmacol 51(6):819–829

177. Amer S, Hassan S, Shariffi M, Chueca L (2013) QT Interval prolongation associated with azithromycin/methadone combination. West Indian Med J 62(8):864–865

178. Cocco G, Jerie P (2015) Torsades de pointes induced by the concomitant use of ivabradine and azithromycin: an unexpected dangerous interaction. Cardiovasc Toxicol 15(1):104–106

179. Shaffer D, Singer S, Korvick J, Honig P (2002) Concomitant risk factors in reports of torsades de pointes associated with macrolide use: review of the United States Food and Drug Administration Adverse Event Reporting System. Clin Infect Dis 35(2):197–200

180. Samarendra P, Kumari S, Evans SJ, Sacchi TJ, Navarro VQT (2001) prolongation associated with azithromycin/amiodarone combination. Pacing Clin Electrophysiol 24(10):1572–1574

181. Mosholder AD, Mathew J, Alexander JJ, Smith H, Nambiar S (2013) Cardiovascular risks with azithromycin and other antibacterial drugs. N Engl J Med 368(18):1665–1668

182. Svanstrom H, Pasternak B, Hviid A (2013) Use of azithromycin and death from cardiovascular causes. N Engl J Med 368(18):1704–1712

183. Katoh T, Saitoh H, Ohno N, Tateno M, Nakamura T, Dendo I et al (2003) Drug interaction between mosapride and erythromycin without electrocardiographic changes. Jpn Heart J 44(2):225–234

184. Rashid A (2005) The efficacy and safety of PDE5 inhibitors. Clin Cornerstone 7(1):47–56

185. Flockhart DA, Drici MD, Kerbusch T, Soukhova N, Richard E, Pearle PL et al (2000) Studies on the mechanism of a fatal clarithromycin-pimozide interaction in a patient with Tourette syndrome. J Clin Psychopharmacol 20(3):317–324

186. Piquette RK (1999) Torsade de pointes induced by cisapride/clarithromycin interaction. Ann Pharmacother 33(1):22–26

187. Reed M, Wall GC, Shah NP, Heun JM, Hicklin GA (2005) Verapamil toxicity resulting from a probable interaction with telithromycin. Ann Pharmacother 39(2):357–360

188. Darpo B, Sager PT, Fernandes P, Jamieson BD, Keedy K, Zhou M et al (2017) Solithromycin, a novel macrolide, does not prolong cardiac repolarization: a randomized, three-way crossover study in healthy subjects. J Antimicrob Chemother 72(2):515–521

Chapter 3
Quinolones

David R. P. Guay

3.1 Introduction

Drug-drug interactions can be categorized into those originating from pharmacokinetic mechanisms and those originating from pharmacodynamic mechanisms. Pharmacokinetic interactions are those that result in alterations of drug absorption, distribution, metabolism, and elimination; pharmacodynamic interactions occur when one drug affects the actions of another drug. This chapter deals only with the pharmacokinetic and pharmacodynamic interactions of fluoroquinolones (hereafter referred to as quinolones) with non-antimicrobial agents. Additive, synergistic, or antagonistic antimicrobial activity interactions between quinolones and other antimicrobials are not discussed.

Some drug interactions can be predicted from the chemical structure of the agent, its pharmacologic activity, its toxicologic profile, and other characteristics determined in its premarketing evaluation (Fig. 3.1). Other interactions cannot be prospectively predicted and can only be detected through intense, large-scale clinical studies or postmarketing surveillance. The quinolones exhibit drug-drug interactions of both types.

There are a number of problems in the prospective clinical evaluation of drug-drug interactions in humans. First, there may be ethical concerns when administering interacting drug combinations to patients or volunteers, depending on the potential consequences of the interaction. Second, because there are an endless number of drug combinations, doses, and timings of administration that could be

D. R. P. Guay (✉)
Department of Experimental & Clinical Pharmacology, College of Pharmacy,
University of Minnesota, Minneapolis, MN, USA
e-mail: guayx001@umn.edu

© Springer International Publishing AG, part of Springer Nature 2018
M. P. Pai et al. (eds.), *Drug Interactions in Infectious Diseases: Antimicrobial Drug Interactions*, Infectious Disease,
https://doi.org/10.1007/978-3-319-72416-4_3

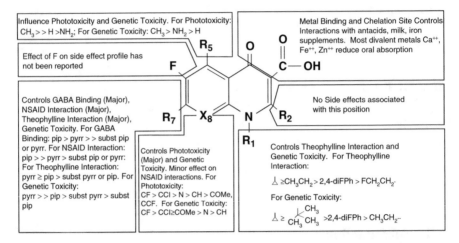

Fig. 3.1 Fluoroquinolone nucleus with the accepted numbering scheme for positions on the molecule. The R indicates possible sites for structural modifications, and the X is a carbon or nitrogen atom, depending on whether it is a quinolone or naphthyridine molecule. Summary of structure-adverse event relationships is also represented. Abbreviations: *GABA* gamma-aminiobutyric acid, *NSAID* nonsteroidal antiinflammatory drug, *pip* piperazine, *pyrr* pyrrolidine, *subst* substituted

investigated, it is economically impossible to fund the study of all possibilities. Third, the prospective evaluation of an interaction in a manageable number of patients is unlikely to uncover a rare interaction. Finally, studies that are carried out in normal volunteers and demonstrate a pharmacokinetic interaction, such as slightly decreased absorption of a drug, may be of uncertain clinical relevance.

Despite these obstacles, delineating the frequencies and types of pharmacokinetic interactions of the quinolones with other drugs is important for several reasons. Since quinolones are often administered orally, absorptive interactions may compromise the efficacy of antimicrobial therapy. Due to their breadth of activity, agents of this class find substantial use in the critically ill and elderly, many of whom receive potentially interacting medications [1–4]. Because elderly individuals have an increased sensitivity to drug-induced toxicity and experience more adverse drug reactions, they may also exhibit an increased incidence and severity of drug-drug interactions. In fact, a retrospective analysis of 505 patients at least 60 years of age hospitalized in a second-level care hospital in Mexico City demonstrated that 62.8% had at least one potential drug-drug interaction and that 14.9% of serious drug-drug interactions were related to fluoroquinolone-hypoglycemic agent interactions [5]. Finally, the quinolones are such a structurally diverse group that the extrapolation of drug-drug interactions from one to another of these agents may not be appropriate.

In this edition, based upon their worldwide regulatory status, the following quinolones will not be covered due to space considerations: clinafloxacin, enoxacin, fleroxacin, gatifloxacin, grepafloxacin, pazufloxacin, rufloxacin, sparfloxacin, temafloxacin, tosufloxacin, and trovafloxacin/alatrofloxacin. Previous editions of this text should be consulted for further information on these agents.

3.2 Pharmacokinetic Interactions

3.2.1 Absorption Interactions

The deleterious effect of multivalent cations on the oral bioavailability of quino-lones was first reported in 1985 [6]. Since this pivotal report, numerous investiga-tions have duplicated and extended this observation; these are detailed in Table 3.1 [6–57].

The concomitant oral administration of magnesium- or aluminum-containing ant-acids has been found to result in six- to ten-fold decreases in the absorption of oral quinolones. Even when dose administrations of the agents were separated by two or more hours, substantial reductions in quinolone absorption persisted [3, 10, 11, 15–17, 20, 25–27, 35, 36, 41, 46, 47, 58, 59]. Studies of the oral coadministration of calcium-containing antacids with oral quinolones have produced conflicting results, with some reporting no significant effect [7, 10, 17–19, 24, 25, 27] and others report-ing significant reductions in absorption [9, 11, 15, 16, 19–21, 23, 26, 46]. Studies have also documented significant reductions in quinolone bioavailability during coadministration with calcium-fortified orange juice and calcium polycarbophil, cal-cium acetate, lanthanum carbonate, and sevelamer hydrochloride [28–31, 40].

Studies have documented substantial reductions in quinolone bioavailability when coadministered with sucralfate. Again, this interaction persisted even when dose administrations of the agents were spaced two or more hours apart [38, 46, 49, 50, 53, 56]. Further studies have documented substantial reductions in quinolone bioavailability when coadministered with iron preparations or multiple vitamins with minerals such as zinc, magnesium, copper, and manganese [35, 36], although one study did not find a significant interaction with iron [10]. A pharmacokinetic-pharmacodynamic model has been created, incorporating the pharmacokinetic data for ciprofloxacin and norfloxacin after metal cation administration at various time intervals before and after quinolone administration and the pharmacodynamic data of complex formulation. This model predicted, in the cases of usual doses of cipro-floxacin with magnesium and aluminum hydroxides (Maalox®) and norfloxacin with sucralfate, that the quinolone should be administered at least 4.5 and 3.5 h after or at least 1 and 0.5 h before the administration of metal cations, respectively, to ensure a relative bioavailability of at least 90% versus control [60].

It is hypothesized that the reduction in quinolone absorption is caused by the formation of insoluble and hence unabsorbable drug-cation complexes or chelates in the gastrointestinal tract [61–65]. This has been confirmed in binding experi-ments utilizing fluoroscopic, UV spectroscopic, and nuclear magnetic resonance spectroscopic techniques [35, 44, 66, 67]. It appears that the complexation or chelation involves the 4-keto and 3-carboxyl groups of the quinolones. In vitro work with the ciprofloxacin-magnesium complex validated the marked stability of these complexed products [68]. The stoichiometry (i.e., the ratio of divalent/trivalent metal cation to quinolone in the stable complexes) varies as a function of the quinolone involved. For example, norfloxacin, nalidixic acid, and ciprofloxa-

Table 3.1 Effects of multivalent cations on quinolone absorption

Quinolone	Cation/preparation/schedule	Mean % change in C_{max}	Mean % change in AUC	References
Levoflox	Al OH/simultaneous with quinolone	−65[a]	−44[a]	[7]
Norflox	Al OH/simultaneous with quinolone	−	−86[a,b]	[8]
Norflox	Al OH/simultaneous with quinolone	−28[c]	−29[c]	[9]
Oflox	Al OH/simultaneous with quinolone	−29[a]	−19	[10]
Pruli	Al OH/1 h prequinolone	−93[a,d]	−85[a,d]	[11]
Pruli	Al OH/3 h prequinolone	−40[a,d]	−35[a,d]	[11]
	2 h prequinolone	−36[d]	−46[d]	
	1 h postquinolone	+6[d]	−13[d]	
	2 h postquinolone	−10[d]	−18[d]	
Oflox	Al phos/simultaneous with quinolone	−10	−3	[12]
Oflox	Al phos/simultaneous with quinolone	−	−7	[13]
Norflox	Bi subsalicylate/simultaneous with quinolone	−	−10[b]	[8]
Cipro	Bi subsalicylate/simultaneous with quinolone	−13	−13	[14]
Cipro	Ca carb/simultaneous with quinolone	−38[a]	−41[a]	[15]
Cipro	Ca carb antacid/simultaneous with quinolone	−47[a]	−42[a]	[16]
Cipro	Ca carb antacid/with meals (PO$_4$ binder)	+13	−	[17]
Cipro	Ca carb/2 h prequinolone	+22[a]	0	[18]
Cipro	Ca carb/simultaneous with quinolone	−	−29[a,b]	[19]
Cipro	Ca carb/(tid × 11 doses) 2 h after dose 10	−24[a]	−14	[20]
Gemi	Ca carb/simultaneous with quinolone	−21[a]	−17[a]	[21]
	2 h prequinolone	−11	−10	
	2 h postquinolone	0	−7	
Levoflox	Ca carb/simultaneous with quinolone	−23	−3	[7]
Levoflox	Ca carb/2 h prequinolone	−9	+16	[22]
	2 h postquinolone	−19[a,e]	−3[e]	
Lomeflox	Ca carb/simultaneous with quinolone	−14[a]	−2	[23]

(continued)

Table 3.1 (continued)

Quinolone	Cation/preparation/schedule	Mean % change in C_{max}	Mean % change in AUC	References
Moxi	Ca carb/simultaneous with quinolone and 12 and 24 h postquinolone	-15^a	-2	[24]
Nemo	Ca carb/simultaneous with quinolone	-14	-18	[25]
Norflox	Ca carb/simultaneous with quinolone	-28^c	$-47^{a,c}$	[9]
Norflox	Ca carb antacid/simultaneous with quinolone	-66^a	-63^a	[26]
Oflox	Ca carb/simultaneous with quinolone	$-$	0^b	[19]
Oflox	Ca carb/simultaneous with quinolone	-18	$+10$	[10]
Oflox	Ca carb antacid/			[27]
	2 h prequinolone	$+3$	-4	
	24 h prequinolone	$+9$	-4	
	2 h postquinolone	$+3$	-3	
Pruli	Ca carb/1 h prequinolone	$-60^{a,d}$	$-55^{a,d}$	[11]
Cipro	Ca acetate/simultaneous with quinolone	-50^a	-51^a	[28]
Cipro	Ca polycarbophil 1200 mg (5.0 mmol Ca)/simultaneous with quinolone	-64^a	-52^a	[29]
Cipro	Ca-fortified orange juice/ simultaneous with quinolone	-41^a	-38^a	[30]
Levo	Ca-fortified orange juice/	-23^a	-14^a	[31]
	Ca-fortified orange juice + milk/	-24^a	-16^a	
	(both simultaneous with ready-to-eat cereal and quinolone)			
Moxi	Ca lact – gluc + carb/			[24]
	immed. before and 12 + 24 h after quinolone	-15^a	-2	
Cipro	Didanosine (+ cations)/3 doses			[32]
	(dose 3 simultaneous with quinolone)	-93^a	-98^a	
Cipro	Didanosine (+ cations)/6 doses			[33]
	(quinolone 2 h pre-didanosine)	-16	-26^a	
Cipro	Didanosine (−cations)/ simultaneous with quinolone	-8	-9	[34]
Cipro	FeSO$_4$/300 mg simultaneous with quinolone	-33^a	-42^a	[35]

(continued)

Table 3.1 (continued)

Quinolone	Cation/preparation/schedule	Mean % change in C_{max}	Mean % change in AUC	References
Cipro	FeSO$_4$/325 mg tid × 7 days	−75[a]	−63[a]	[36]
Cipro	FeSO$_4$/simultaneous with quinolone	−54[a]	−57[a]	[37]
Gemi	FeSO$_4$/			[38]
	3 h prequinolone	−20[a]	−11	
	2 h postquinolone	−4	−10	
Levoflox	FeSO$_4$/simultaneous with quinolone	−45[a]	−19[a]	[7]
Lomeflox	FeSO$_4$/simultaneous with quinolone	−28[a]	−14	[23]
Moxi	FeSO$_4$/simultaneous with quinolone	−59[a]	−39[a]	[39]
Nemo	FeSO$_4$/simultaneous with quinolone	−57[a]	−64[a]	[25]
Norflox	FeSO$_4$/simultaneous with quinolone	−75[a]	−73[a]	[37]
Norflox	FeSO$_4$/simultaneous with quinolone	−97[a,c]	−97[a,c]	[9]
Norflox	FeSO$_4$/simultaneous with quinolone	−	−55[a,b]	[8]
Oflox	FeSO$_4$/simultaneous with quinolone	−	−10[a,b]	[13]
Oflox	FeSO$_4$/simultaneous with quinolone	−36[a]	−25[a]	[37]
Oflox	FeSO$_4$/simultaneous with quinolone	+9	+35	[10]
Pruli	FeSO$_4$/1 h prequinolone	−85[a,d]	−75[a,d]	[11]
Cipro	Lanthanum carbonate/quinolone taken immediately postdose 1 on day 2 of 3-day regimen of 1 g tid with meals	−56[a]	−54[a]	[40]
Norflox	Mg OH/simultaneous with quinolone	−	−90[a,b]	[8]
Levoflox	Mg O/simultaneous with quinolone	−38[a]	−22[a]	[7]
Pruli	Mg O/1 h prequinolone	−61[a,d]	−57[a,d]	[11]
Norflox	Mg trisilicate/simultaneous with quinolone	−72[a,c]	−81[a,c]	[9]
Oflox	Mg trisilicate/simultaneous with quinolone	−2	+19	[10]
Cipro	Mg/Al antacid/simultaneous with quinolone	−81[a]	−84[a]	[16]
Cipro	Mg/Al antacid/			[41]

(continued)

Table 3.1 (continued)

Quinolone	Cation/preparation/schedule	Mean % change in C_{max}	Mean % change in AUC	References
	5–10 min prequinolone	−80[a]	−85[a]	
	2 h prequinolone	−74[a]	−77[a]	
	4 h prequinolone	−13	−30[a]	
	6 h prequinolone	0	+9	
	2 h postquinolone	+32[a]	+7	
Cipro	Mg/Al antacid/10 doses over 24 h prequinolone	−93[a]	−91[a]	[42]
Cipro	Mg/Al antacid/with meals (PO$_4$ binder)	−65	–	[17]
Cipro	Mg/Al antacid/24 h prequinolone	−94[a]	–	[6]
Gemi	Mg/Al antacid/			[43]
	3 h prequinolone	−17[a]	−15	
	10 min postquinolone	−87[a]	−85[a]	
	2 h postquinolone	+10	+3	
Lomeflox	Mg/Al antacid/simultaneous with quinolone	−46[a]	−41[a]	[44]
Moxi	Mg/Al antacid/simultaneous with quinolone	−61[a]	−59[a]	[45]
	2 h postquinolone	−7	−26[a]	
	4 h prequinolone	−1	−23[a]	
Nemo	Mg/Al antacid/		25	
	2 h prequinolone	+3	−9	
	Simultaneous with quinolone	−78[a]	−81[a]	
	4 h prequinolone	−53[a]	−58[a]	
Norflox	Mg/Al antacid/			[26]
	Simultaneous with quinolone	−95[a]	–	
	2 h postquinolone	−24[a]	−20	
Norflox	Mg/Al antacid/			[46]
	Simultaneous with quinolone	−95	−98	
	2 h postquinolone	−24	−22	
Oflox	Mg/Al antacid/			[42]
	10 doses over 24 h prequinolone	−73[a]	−69[a]	
Oflox	Mg/Al antacid/			[27]
	2 h prequinolone	−30[a]	−21[a]	
	24 h prequinolone	−5	−5	
	2 h postquinolone	+3	+5	
Oflox	Mg/Al antacid/simultaneous with quinolone	−24	–	[47]
Peflox	Mg/Al antacid/13 doses (quinolone 1 h after dose 10)	−61[a]	−54[a]	[48]
Cipro	Multivit with Zn/once daily × 7 days	−32[a]	−22[a]	[36]

(continued)

Table 3.1 (continued)

Quinolone	Cation/preparation/schedule	Mean % change in C_{max}	Mean % change in AUC	References
Cipro	Multivit with Fe/Zn/			[35]
	Once simultaneous with quinolone	-53^a	-52^a	
Norflox	NaHCO$_3$/simultaneous with quinolone	$+5^c$	$+5^c$	[9]
Cipro	Sevelamer hydrochloride/seven 403 mg caps simultaneous with quinolone	-34^a	-48^a	[28]
Cipro	Sucralfate/1 gm 6 and 2 h prequinolone	-30^a	-30^a	[49]
Cipro	Sucralfate/1 gm QID × 1 day then simultaneous with quinolone	-90^a	-88^a	[50]
Cipro	Sucralfate/2 gm BID × 5 doses			[51]
	Quinolone simultaneous with dose 5	-95^a	-96^a	
	Quinolone 2 h before dose 5	$+5$	-20	
	Quinolone 6 h before dose 5	0	-7	
Gemi	Sucralfate/			[38]
	2 g 3 h prequinolone	-69^a	-53^a	
	2 g 2 h postquinolone	-2	-8	
Levoflox	Sucralfate/1 gm 2 h postquinolone	$+14$	-5	[52]
Lomeflox	Sucralfate/1 gm 2 h prequinolone	-30^a	-25^a	[53]
Lomeflox	Sucralfate/1 gm simultaneous with quinolone	-65^a	-51^a	[23]
Moxi	Sucralfate/1 gm simultaneous with quinolone and 5, 10, 15, 24 h postquinolone	-71^a	-60^a	[54]
Norflox	Sucralfate/			[46]
	Simultaneous with quinolone	-90	-98	
	2 h prequinolone	-28	-42	
Norflox	Sucralfate/			[55]
	1 gm simultaneous with quinolone	-92^a	-91^a	
	1 gm 2 h postquinolone	$+9$	-5	
Norflox	Sucralfate/			[56]
	1 gm simultaneous with quinolone	-90^a	-98^a	
	1 gm 2 h prequinolone	-28	-43	
Oflox	Sucralfate/			[55]
	1 gm simultaneous with quinolone	-70^a	-61^a	
	1 gm 2 h postquinolone	$+7$	-5	

(continued)

Table 3.1 (continued)

Quinolone	Cation/preparation/schedule	Mean % change in C_{max}	Mean % change in AUC	References
Oflox	Sucralfate/			[57]
	Fasting +1 gm simultaneous with quinolone	−70[a]	−61[a]	
	Nonfasting +1 gm simultaneous with quinolone	−39[a]	−31[a]	
Norflox	Tripotassium citrate/			[9]
	Simultaneous with quinolone	−48[a,b]	−40[a,c]	
Norflox	$ZnSO_4$/simultaneous with quinolone	−	−56[a,b]	[8]

% change change from baseline or placebo control, C_{max} peak serum or plasma concentration, *AUC* area under the plasma or serum concentration-time curve, *oflox* ofloxacin, *cipro* ciprofloxacin, *norflox* norfloxacin, *carb* carbonate, *lomeflox* lomefloxacin, *levoflox* levofloxacin, *tid* 3 times daily, *qid* 4 times daily, *bid* twice daily, *gluc* gluconate, *moxi* moxifloxacin, *gemi* gemifloxacin, *lact* lactate, *multivit* multivitamin, *pruli* prulifloxacin, *nemo* nemonoxacin
[a]Statistically significant change from baseline or placebo control
[b]Based on urinary excretion data
[c]Based on salivary AUC data
[d]Based on ulifloxacin (active metabolite) data
[e]Adults with cystic fibrosis

cin exhibited ratios of 1:1 or 2:1, 1:1–3:1, and 3:1, respectively [60]. In two in vitro studies, calcium, zinc, and iron exposures resulted in respective precipitates of the hydrated ciprofloxacin base, $Zn(SO4)_2(Cl)_2(ciprofloxacin)_2$ x nH_2O, where n can range up to 12, and $Fe(SO4)_2(Cl)_2(ciprofloxacin)_2$ x nH_2O, where n can range up to 12 [69, 70]. In addition, it appears that the presence of these ions results in impaired dissolution of the quinolones (tablet formulations and the agents themselves), at least in vitro [69–76]. However, in vitro permeability studies using human colon carcinoma (Caco-2) cell lines in monolayers suggest that chelate formation and adsorption to cations can only partially explain the interactions between fluoroquinolones and polyvalent metal cations [77]. Chelation has even been reported between ciprofloxacin, ofloxacin, and levofloxacin and the oral magnetic resonance imaging contrast medium manganese chloride tetrahydrate [78]. It is thus recommended to not use magnesium-, aluminum-, or calcium-containing antacids, sucralfate, or iron/vitamin-mineral preparations concomitantly with quinolones. Histamine H_2-receptor antagonists such as ranitidine, cimetidine, and famotidine have not been shown, in general, to alter quinolone absorption. However, these agents do result in significantly decreased absorption of prulifloxacin [7, 11, 42, 45, 52, 53, 79–85]. Omeprazole has also not been shown to alter the pharmacokinetics of quinolones to a clinically significant degree [86–89]. Thus, these agents can be recommended as alternative noninteracting antiulcer and antiesophagitis therapy.

Agents that alter gastric motility may affect quinolone absorption. Pirenzepine, a gastrointestinal tract-specific anticholinergic not available in the USA, delayed gastric emptying and absorption of ciprofloxacin, thus delaying the time to achievement of maximal serum concentration (T_{max}). However, the extent of absorption was not altered [42, 79]. N-butylscopolamine, another anticholinergic, interacted with oral ciprofloxacin in an identical manner [80]. In contrast, absorption of ciprofloxacin was accelerated by the gastrointestinal motility stimulant metoclopramide; again, the extent of absorption was unaltered [80]. These quinolone-drug interactions are thought to be of no clinical importance during usual multiple-dose regimens. In addition, an extended-release mesalamine product (MMX^R mesalamine) had no significant effect on ciprofloxacin pharmacokinetics [90].

The absorption of ciprofloxacin is not significantly altered in the presence of Osmolite® enteral feedings [91]. However, other studies have found significant interaction potential between the quinolones and enteral feedings. Concurrent administration of Osmolite® and Pulmocare® enteral feedings significantly reduced single-dose ciprofloxacin bioavailability as assessed using C_{max} (mean − 26 and −31%, respectively) and area under the serum concentration-vs-time curve (AUC; mean − 33 and −42%, respectively) data [92]. Concurrent administration of Sustacal® enteral feeding significantly reduced single-dose ciprofloxacin bioavailability as assessed using C_{max} (mean − 43%) and AUC (mean − 27%) data. In the same study, continuous administration of Jevity® enteral feeding via gastrostomy and jejunostomy tubes significantly reduced single-dose ciprofloxacin bioavailability as assessed using C_{max} (mean − 37 and −59%, respectively) and AUC (mean − 53 and −67%, respectively) data [93]. Concurrent administration of Ensure® enteral feeding significantly reduced single-dose ciprofloxacin and ofloxacin bioavailability as assessed using C_{max} (mean − 47 and −36%, respectively) and AUC (mean − 27 and −10%, respectively) data. However, the extent of the interaction was significantly greater for ciprofloxacin than for ofloxacin [94].

The interaction potential between quinolones and dairy products appears to be quinolone-specific. Studies have demonstrated no significant interaction between lomefloxacin, moxifloxacin, and ofloxacin and milk (200, 240, or 300 mL) or yogurt (250–300 mL) [23, 95–98]. In contrast, ciprofloxacin, prulifloxacin, and norfloxacin bioavailability is substantially reduced (by 28–58%) by concurrent administration with milk or yogurt [97, 99–101]. Recent in vitro dissolution studies have demonstrated that the major mechanism whereby concurrent milk administration reduces the bioavailability of some quinolones is adsorption on the surface of milk proteins, especially casein. Complexation with calcium is much less important in this regard [102].

3.2.1.1 Therapeutic Implications of Absorption Interactions

Five cases of therapeutic failure due to the interaction of oral quinolones with metal cations have been published [103]. The combination therapies included ciprofloxacin with calcium carbonate, magnesium oxide, and multiple vitamins with minerals and iron; levofloxacin with calcium carbonate and aluminum hydroxide; levofloxacin with calcium carbonate and magnesium oxide; and levofloxacin with calcium carbonate and sucralfate. In all cases, spacing of the administration times of the quinolones and the metal cations, temporary discontinuation of metal cations until the end of quinolone therapy, and/or substitution with noninteracting agents (e.g., histamine H_2-receptor antagonist for sucralfate) allowed subsequent successful oral quinolone therapy.

The potential extent of quinolone-metal cation interactions was explored in a case-control study conducted in a 625-bed tertiary-care medical center in the USA. Data from all patients receiving oral levofloxacin from July 1, 1999, through June 30, 2001, were included. Coadministration was defined as any divalent or tri-valent cation-containing agent being administered within 2 h of levofloxacin admin-istration. Complete coadministration was defined as coadministration complicating every dose of an entire course of levofloxacin. Overall, 1904 (77%) of 2470 doses (427 courses of therapy) were complicated by coadministration. Also, 386/427 courses (90%) had at least 1 dose complicated by coadministration. In 238 courses (56%), complete coadministration occurred. Only three factors were significantly associated with complete coadministration upon multivariate analysis. A higher number of prescribed medications on the first day of levofloxacin therapy was a risk factor (per increase of one drug, odds ratio [OR], 1.05; 95% CI, 1.01–1.10; $p = 0.036$). Two factors were protective (i.e., decreased the risk): location in an intensive care unit (OR, 0.51; 95% CI, 0.30–0.87; $p = 0.013$) and longer duration of levofloxacin therapy (OR, 0.92; 95% CI, 0.88–0.97; $p = 0.001$). Extrapolating these results to all oral levofloxacin recipients at the institution, one in every three doses would be complicated by deleterious coadministration with at least one multivalent cation-containing agent [4].

Therapeutic failure due to quinolone-metal cation interactions may occur not only through the production of subtherapeutic drug concentrations due to malab-sorption. These subtherapeutic drug concentrations may also lead to the emergence of bacterial resistance to the quinolone class. A case-control study of 46 inpatients receiving oral levofloxacin and divalent/trivalent cations was conducted. Of the 46 individuals, 32 (70%) had levofloxacin-resistant pathogens, while 14 (30%) had levofloxacin-susceptible pathogens. Patients with levofloxacin-resistant isolates had previously been exposed to nearly twice as many days of coadministration (as defined previously) compared with those having susceptible isolates (median 5 vs. 3 days, respectively; $p = 0.04$). Upon multivariate analysis, the relationship between the number of days of coadministration and the presence of resistant isolates was no longer statistically significant but nevertheless did show a statistical trend (OR, 1.26; 95% CI, 0.98–1.63; $p = 0.07$). Last, the percentage of subjects with quinolone-resistant isolates varied directly with the number of days of coadministration. Zero–2, 3– 4, 5–7, and 8–31 days of coadministration were linked to 42%, 78%, 68%, and 90% frequencies of quinolone resistance, respectively [104].

A second case-control trial was conducted using data from all inpatients at a tertiary-care hospital and an urban hospital who received oral levofloxacin between January 1, 2001, and December 31, 2005. Coadministration was defined as receipt of levofloxacin and multivalent cation-containing agents during the same day (regardless of the times of administration). A total of 3134 patients had courses of oral levofloxacin at least 3 days in duration. Of the 3134 patients, 895 (29%) had

100% coadministration (i.e., on all days of the course of therapy), while 606 (19%) had partial coadministration (i.e., on ≥ 1 but not all days of the course of therapy). The remainder (52%) had no coadministration at all. Levofloxacin-resistant isolates were found in 198 patients (6%). Coadministration was significantly associated with subsequent identification of a quinolone-resistant isolate (≥ 48 h after initiation of levofloxacin). On univariate analysis, the ORs for resistant pathogens with 100%, $\geq 75\%$, $\geq 50\%$, and $\geq 25\%$ coadministration were 1.49 ($p = 0.005$), 1.48 ($p = 0.005$), 1.61 ($p < 0.001$), and 1.44 ($p = 0.007$), respectively. On multivariate analysis, the ORs for resistant pathogens with 100% (vs. < 100%), $\geq 50\%$ (vs. < 50%), and 100% (vs. 0) coadministration were significant (OR, 1.43; $p = 0.03$; OR, 1.52; $p = 0.006$; and OR, 1.36; $p = 0.05$; respectively). For $\geq 75\%$ (vs. <75%) and $\geq 25\%$ (vs. <25%) coadministration, statistical significance was not achieved [105].

Another important issue is the antimicrobial activity of quinolone-metal cation complexes. Even though these complexes are not bioavailable to the systemic circulation, they theoretically could still be active against bacterial pathogens in the gastrointestinal (GI) tract. Two evaluations of the in vitro activity of drug-cation complexes compared with drug alone have been published [68, 73]. Minimum inhibitory concentrations (MICs) were determined, with a rise in MIC demonstrating reduced susceptibility and a fall in MIC demonstrating increased susceptibility. With ciprofloxacin and magnesium, the MICs for *Escherichia coli, Staphylococcus aureus*, and *Pseudomonas aeruginosa* were significantly higher for the drug-cation complex compared with the drug alone (all $p < 0.05$) [68]. With lomefloxacin and 11 cations, the MICs did not change with *Salmonella typhi, Streptococcus pneumoniae* (except for increases with exposure to nickel and cadmium), and *P. aeruginosa* (except for increases with exposure to copper and zinc). For *Bacteroides fragilis*, MICs increased in the presence of all cations except nickel (for latter, MIC was the same as control). For *S. aureus*, MICs were unaltered in the presence of magnesium, manganese, cobalt, zinc, and cadmium. They were increased upon exposure to calcium, chromium, iron, nickel, and copper [73]. Overall, because of the drug-, cation-, and organism-specific effects on quinolone-cation complex bioactivity, it would be prudent to not consider quinolone-cation complexes to be potentially useful in the therapy of bacterial infections of the GI tract.

3.2.2   Distribution Interactions

The quinolones are plasma protein bound to the extent of only 20–30%. Ciprofloxacin does not displace bilirubin from albumin, which suggests that interactions involving displacement of other drugs from their carrier proteins are unlikely to occur during coadministration of quinolones [106]. The absence of such an interaction with the quinolones may be of particular importance to elderly debilitated patients with hypoalbuminemia who receive multiple drugs.

3.2.3 Metabolism Interactions

The effect of quinolones on the metabolism of antipyrine, a probe drug for hepatic drug metabolism, has been evaluated. Ofloxacin given as 200 mg twice daily for 7 days did not influence antipyrine metabolism significantly [107]. Similarly, 125 mg ciprofloxacin twice daily for 7 or 8 days did not influence antipyrine metabolism significantly [108]. In contrast, a regimen of 500 mg ciprofloxacin twice daily for 8–10 days (a clinically relevant dosing regimen) was associated with a significant mean 39% reduction in antipyrine clearance and mean 58% increase in terminal disposition half-life (t ½) [109].

A number of case reports have documented a clinically significant drug-drug interaction between ciprofloxacin and theophylline, in some cases leading to death [110–116]. A number of the quinolones have been found to reduce the hepatic metabolism of coadministered drugs such as the xanthines theophylline [117–138] and caffeine [139–146] (Table 3.2). In contrast to the absorption interactions with multivalent cations, which appear to be generalizable to the entire quinolone drug class, differences do exist between individual quinolones in their propensity to inhibit hepatic xanthine metabolism. A meta-analysis of quinolone-theophylline interaction studies revealed that ciprofloxacin, antofloxacin (based on Ref. [138]), prulifloxacin (based on Ref. [137]), and norfloxacin (in descending order) are clinically significant inhibitors of theophylline metabolism; ofloxacin, lomefloxacin, and (based on Refs. [85, 131, 134–136]) levofloxacin, moxifloxacin, nemonoxacin, and gemifloxacin are clinically insignificant inhibitors [147]. Using a simple pharmacokinetic model that allowed cross-comparison between quinolone-caffeine interaction studies, Barnett and colleagues developed a relative potency index of quinolone interaction as follows: using enoxacin [100] as a benchmark, ciprofloxacin [11], norfloxacin [9], and ofloxacin (0) [148]. The inhibition of xanthine metabolism may be dose-dependent for those FQs covered herein. The dose dependency of inhibition with enoxacin has been well-demonstrated (see previous editions).

A population-based, nested case-control study of Ontario residents at least 66 years old receiving theophylline therapy between April 1992 and March 2009 evaluated whether there was any relationship between ciprofloxacin use (within 14 days of the event) and hospitalization for theophylline intoxication. Among 77,251 patients, there were 180 case patients who were matched to 9000 control patients. After multivariate adjustment, the adjusted odds ratio (AOR) for the relationship with ciprofloxacin was 1.86 (95% CI, 1.18–2.93). There was no increased risk of hospitalization for theophylline intoxication with comparator antimicrobials (levofloxacin, trimethoprim-sulfamethoxazole, or cefuroxime [AOR, 0.78; 95% CI, 0.38–1.62]) [149].

Few other substrates have been examined. Levofloxacin, norfloxacin, moxifloxacin, gemifloxacin, nemonoxacin, and ciprofloxacin have been shown not to potentiate the anticoagulant effect of warfarin in healthy subjects and patients requiring long-term anticoagulation [85, 150–157]. Based on rat studies, ciprofloxacin and pefloxacin significantly increased prothrombin time (PT)/international normalized

Table 3.2 Effect of quinolones on methylxanthine pharmacokinetics

Quinolone		Mean % change in			References
		C_{ss}	CL	t½	
Theophylline					
Norflox	400 bid	–	–8[a]	+9[a]	[117]
Norflox	400 bid	–	+10	+26	[118]
Cipro	750 bid	–	–31[a]	–	[119]
Oflox	400 bid	+9[a]	–15[a]	–	[120]
Norflox	400 bid	–	–15[a]	+13[a]	[121]
Lomeflox	400 qd	–	–7	+4[a]	[122]
Cipro	500 bid	+66[a]	–30[a]	+42[a]	[123]
Oflox	400 bid	+2	–5	+2	[123]
Oflox	200 tid	–	0	+6	[124]
Norflox	200 tid	–	–7	+15	[124]
Oflox	200 tid	–	0	+6	[125]
Norflox	200 tid	–	–7	+15	[125]
Cipro	500 bid	–	–27[a]	–	[125]
Lomeflox	400 bid	–	–2	–	[126]
Lomeflox	400 x 1 dose	+1	–2	+1	[127]
	400 bid	+8	–7	+7	
Lomeflox	400 bid	–	+7	+3	[128]
Norflox	200 tid	–	–4	–	[129]
Oflox	200 tid	–	–11	–	[129]
Cipro	200 tid	–	–22[a]	–	[129]
Cipro	750 bid	+87[a]	–	–	[130]
Levoflox	500 bid	–	+3	–1	[131]
Oflox	200 bid	–	–5	+5	[132]
Cipro	500 bid	–	–20[a]	+25[a]	[133]
Moxi	200 bid	–	–4	+4	[134]
Moxi	200 bid	–	+5	+3	[135]
Gemi	320 qd	–	–1	–	[136]
Pruli	600 qd	–	–15[a,b]	+14[a,b]	[137]
Antoflox	200 qd	+32[a,c]	–23[a]	–	[138]
Nemonoxacin	500 qd	+15[c]	–17	–	[85]
Caffeine					
Peflox	400 bid	–	–47[a]	+96[a]	[139]
Norflox	800 bid	–	–35[a]	+23	[140]
Cipro	750 bid	–	–45[a]	+58[a]	[141]
Cipro	750 bid	+877[a]	–145[a]	+116[a]	[142]
Oflox	200 bid	–	+2	–3	[142]
Norflox	400 bid	–	–16	+16[a]	[143]
Cipro	100 bid	–	–17	+6	[143]
	250 bid	–	–57[a]	+15[a]	

(continued)

Table 3.2 (continued)

Quinolone		Mean % change in			References
		C_{ss}	CL	t½	
	500 bid	–	-58^a	+26	
Oflox	200 bid	–	+4	–3	[144]
Cipro	250 bid	–	-33^a	$+15^a$	[144]
Lomeflox	400 qd	–6	–3	0	[145]

% change change from baseline or placebo control, C_{ss} steady-state concentration, *CL* total body clearance, *t1/2* elimination half-life, *norflox* norfloxacin, *cipro* ciprofloxacin, *lomeflox* lomefloxacin, *levoflox* levofloxacin, *oflox* ofloxacin, *peflox* pefloxacin, *moxi* moxifloxacin, *gemi* gemifloxacin, *qd* once daily, *bid* twice daily, *tid* three times a day, *pruli* prulifloxacin
[a]Statistically significant change from baseline or placebo control
[b]Ulifloxacin (active metabolite)
[c]Peak concentration

ratio (INR) results after acenocoumarin administration [158]. These results need to be validated in humans. In these same studies, acenocoumarin coadministration significantly enhanced the serum concentrations of both quinolones and their penetration into the mandibular bone but not the femur. The mechanism of this effect is unknown [158]. However, case reports have documented quinolone-associated increases in PT/INR in patients receiving warfarin concurrently with ciprofloxacin, ofloxacin, norfloxacin, levofloxacin, and moxifloxacin [159–176].

Several epidemiological studies have examined the association of antimicrobial use with laboratory and clinical outcomes in warfarin recipients. A nested case-control and case-crossover study using the Medicaid database assessed the risk of hospitalization for gastrointestinal (GI) bleeding in warfarin users who also received oral sulfonamides, azole antifungals, and two quinolones (ciprofloxacin and levofloxacin). When adjusted for all confounders and using cephalexin as a control, neither of the quinolones was statistically associated with the target outcome [177]. A retrospective case-cohort study was designed to measure INR changes occurring in warfarin recipients after initiation of oral azithromycin, levofloxacin, trimethoprim-sulfamethoxazole (TMP-SMX), and terazosin (control) between January 1998 and December 2002. Subjects were outpatients in a university-affiliated VA Medical Center. The mean changes in INR were +0.51, +0.85, +1.76, and −0.15, respectively (all antimicrobial-terazosin pairs, $p < 0.05$). The frequencies of supratherapeutic INRs (i.e., above the upper limit of the desired range) were 31%, 33%, 69%, and 5%, respectively (all antimicrobial-terazosin pairs, $p < 0.05$). The frequencies of INRs exceeding 4.0 were 16%, 19%, 44%, and 0%, respectively (only TMP-SMX vs. terazosin, $p < 0.05$) [178]. A retrospective case-cohort study was conducted in an outpatient oral anticoagulation clinic for patients on long-term warfarin therapy (between January 1, 1998, and March 31, 2003). Forty-three patients received warfarin: 21 were prescribed felodipine (as control), while 22 were prescribed oral levofloxacin (16 on 500 mg/day, 6 on 250 mg/day). The differences in pre- versus post-drug INR values and the proportions of patients requiring a change in dose due to the post-drug INR values were not significantly different between groups. Eight levofloxacin and eight felodipine recipients had INR differences of >0.5, while four levofloxacin and one felodipine recipient had INR differences of >1.0. For the levofloxacin and felodipine groups, 7/9 (78%) and 3/7 (43%) of dose changes were reductions due to supratherapeutic INR values [179]. A retrospective review was conducted of all hospitalized patients in a university hospital in Spain (from 2000 to 2005) who had received levofloxacin and warfarin concurrently. A total of 21 patients were identified and evaluable (mean age 75 years old [range, 49–92]; 57.1% were women). Concurrent therapy lasted for 7 ± 4.4 days (mean ± SD). The routes of administration of levofloxacin were intravenous ($N = 4$), oral ($N = 8$), and intravenous to oral ($N = 9$). Three subjects had bleeding due to INR elevations (to 3.43, 4.8, and 6.32). The mean INR values before, during, and following concurrent therapy were 1.85 (range 1.01–4.08), 2.64 (1–6.32), and 2.32 (1.11–4.15) (before vs. during, $p = 0.001$; before vs. following, $p = $ not significant) [180].

A nested case-control analysis of multiple linked healthcare databases in Ontario, Canada, was conducted between April 1, 1998, and March 31, 2002. Subjects were a minimum of 65 years old. Cases were those on continuous warfarin therapy admitted to the hospital with any type of bleeding. The cohort was a population of elders on continuous warfarin therapy, wherein observation began with the first warfarin prescription following the 66th birthday and ended with the occurrence of one of the following events: first recurrence of hospital admission for bleeding, death, warfarin discontinuation, or the end of the study period. Study medications included oral levofloxacin, ocular antimicrobials, and cefuroxime (as control). A total of 158,510 elders met the inclusion criteria (mean age at start was 79 years old, 48% were women). Cases ($N = 4269$) were matched to 17,048 controls. For 14 days of exposure, the odds ratio of only cefuroxime was significant (1.62; 95% CI, 1.28–2.20). For 28 days of exposure, only the odds ratio of cefuroxime was again significant (1.63; 95% CI, 1.23–1.89). Both times, cefuroxime was associated with an enhanced risk of bleeding. Ocular antimicrobials were weakly associated with a decreased risk of bleeding with 28 days of exposure (OR, 0.94; 95% CI, 0.76–0.98). Levofloxacin was not associated with bleeding in these elderly warfarin recipients [181].

The effects of various antimicrobials, being used to treat UTIs in warfarin recipients, on upper gastrointestinal (UGI) tract hemorrhage rates were evaluated. This trial utilized a population-based, nested case-control trial design using healthcare databases in Ontario, Canada, over the period of April 1, 1997, through March 31, 2007. Cases were Ontario residents 66 years old and older with UGI tract hemorrhage who were being continuously treated with warfarin. Up to ten age- and sex-matched controls were selected for each case. Adjusted odds ratios for antimicrobial exposure within 14 days before UGI tract hemorrhage were calculated for six commonly used antimicrobials for UTI, including ciprofloxacin and norfloxacin. Only trimethoprim-sulfamethoxazole and ciprofloxacin use were significantly associated with UGI tract hemorrhage (AORs of 3.84 and 1.94, respectively). All other agents, including norfloxacin (AOR of 0.38), were not associated with this outcome [182].

Last, the efficacy of preemptively reducing warfarin doses by 10–20% when starting oral TMP-SMX or levofloxacin therapy in warfarin recipients was compared with no preemptive dose reduction. Of 40 patients, 18 were dose-reduced (in 8 TMP-SMX patients, the mean dose reduction was 16.3%, while in 10 levofloxacin patients, the mean dose reduction was 16.2%), and in 22, there was no dose alteration. In the dose-reduced TMP-SMX group, the mean difference in pre- versus on-therapy INR values was not significant, but 25% still developed INR values >4.0, and none had subtherapeutic INR values. In the control TMP-SMX group, there was a significant rise in INR values on-therapy versus pre-therapy ($p < 0.02$), and 89% had INR values >4.0 (25% vs. 89%, $p < 0.02$). In the dose-reduced levofloxacin group, the mean difference in pre- versus on-therapy INR values was not significant, 40% developed subtherapeutic INR values, and none had INR values >4.0. In the control levofloxacin group, there was a significant rise in INR values on-therapy versus pre-therapy ($p < 0.02$), 38.5% developed INR values >4.0, and none had subtherapeutic INR values (differences between levofloxacin groups in proportions with subtherapeutic or supratherapeutic INR values were significant: $p < 0.03$ and $p < 0.02$, respectively). Thus, more pronounced effects were seen with TMP-SMX compared with levofloxacin. For the pooled dose-reduction group, 11% of subjects needed temporary interruption of warfarin therapy due to supratherapeutic INR values. In the pooled control group, 55% of subjects required such interruption ($p = 0.007$) [183].

Pending additional information, patients who are receiving long-term warfarin therapy in whom a quinolone is to be used should be monitored for changes in PT/INR.

Case reports have suggested that the quinolones may reduce the metabolism of cyclosporine and hence potentiate the nephrotoxicity of this agent [184–187]. In addition, results of one study conducted in pediatric renal transplant recipients suggested that norfloxacin may interfere with cyclosporine disposition, as evidenced by the difference in mean daily dose of cyclosporine required to maintain trough blood cyclosporine concentrations of 150–400 ng/mL (4.5 mg/kg/day in norfloxacin recipients versus 7.4 mg/kg/day in non-recipients) [188]. A study was conducted in renal transplant recipients requiring therapy for urinary tract infections wherein the effect of high-dose oral levofloxacin (1 g daily) on cyclosporine pharmacokinetics was evaluated. Levofloxacin therapy resulted in significant increases in cyclosporine C_{max} (mean 23%, $p = 0.0049$), AUC (mean 26%, $p = 0.005$), C_{min} (mean 36%, $p = 0.0013$), and C_{avg} (mean 26%, $p = 0.0005$). A slight fall in polyclonal assay C_{max} (mean 5%, $p = 0.014$), which measures parent compound plus metabolites, was also seen [189]. However, numerous formal in vitro and other pharmacokinetic studies have not found a significant interaction between cyclosporine and ciprofloxacin, pefloxacin, levofloxacin, and moxifloxacin [190–199]. This suggests that these agents with the possible exception of high-dose levofloxacin may be used together with routine monitoring. In addition, high-dose levofloxacin (1 g/day) significantly increased tacrolimus systemic exposure (means 24–28%), and combination therapy would appear to warrant enhanced monitoring [189, 200].

Studies have documented nonsignificant interactions of moxifloxacin, gemifloxacin, and levofloxacin with digoxin [201–203]. However, one case report has suggested a ciprofloxacin-digoxin interaction. The mechanism may be identical to that described with erythromycin and the tetracyclines (i.e., suppression of *Eubacterium lentum* in the gut) as well as competition for P-glycoprotein (reducing digoxin secretion into the gut and its renal clearance) [204]. Findings of reduced quinolone bioavailability have been noted with coadministration of oral ciprofloxacin and intramuscular papaveretum [205]. In contrast, oral oxycodone had no significant effect on oral levofloxacin pharmacokinetics [206]. Ciprofloxacin significantly reduced the total body clearance, renal clearance, and nonrenal clearance and increased $t \frac{1}{2}$ and urinary excretion of R(−) and S(+) mexiletine in both smokers and nonsmokers. However, these changes were modest in degree (≤20%) and suggested the absence of a clinically relevant drug interaction between the two agents [207].

Ciprofloxacin may impair the elimination of diazepam [208], although this is controversial [209].

Multiple-dose delafloxacin (450 mg twice daily) had no significant effect on the pharmacokinetics of midazolam or its 1-hydroxy metabolite [210]. Waite and coworkers demonstrated that elderly subjects are not more sensitive than younger subjects to the inhibitory effect of ciprofloxacin on hepatic metabolism of antipyrine [211]. Similarly, Loi and coworkers demonstrated that elderly subjects are not more sensitive to the inhibitory effect of ciprofloxacin on hepatic metabolism of theophylline [212].

Chandler and colleagues showed that rifampin does not induce the metabolism of ciprofloxacin, suggesting that the two agents may be used concomitantly in standard clinical dosing regimens [213]. A study conducted in rats suggested that levofloxacin pharmacokinetics were also not altered by concurrent rifampin administration [214]. The results of this study must be validated in humans.

Examining the rifampin component of the combination, single-dose ciprofloxacin coadministration significantly increased $t \frac{1}{2}$ and reduced the C_{max} but had no effect on the AUC, volume of distribution, or urinary excretion of single-dose rifampin [215, 216]. Single-dose pefloxacin coadministration significantly increased $t \frac{1}{2}$, C_{max}, AUC (from 0 to 24 h and 0 to ∞), volume of distribution, absorption $t \frac{1}{2}$, and urinary excretion of single-dose rifampin [217, 218]. Single-dose ofloxacin coadministration significantly reduced the C_{max}, absorption $t \frac{1}{2}$, salivary C_{max}, and renal clearance and increased the salivary bioavailability and urinary excretion of rifampin, while single-dose norfloxacin significantly reduced the AUC to infinity, salivary C_{max} and bioavailability, and renal clearance and increased urinary excretion of rifampin [219].

In a multiple-dose healthy volunteer trial evaluating the steady-state interaction of rifampin with moxifloxacin, moxifloxacin C_{max} and AUC were reduced by means of 6% and 27%, respectively ($p \leq 0.047$ and $p < 0.0001$, respectively). Rifampin coadministration also increased C_{max} and AUC of the M-1 metabolite by means of 255% and 116%, respectively, and decreased $t \ 1/2$ of this metabolite by a mean of 68% (no statistical results were presented) [220]. In a study conducted in patients with tuberculosis receiving thrice weekly therapy with rifampin and isoniazid, steady-state moxifloxacin serum concentrations fell in 18/19 patients (95%). Significant ($p < 0.05$) reductions were also found for steady-state moxifloxacin

AUC (mean 31%), C_{max} (mean 32%), C_{min} (mean 62%), and t 1/2 (mean 28%). In addition, significant ($p < 0.05$) increases were found for steady-state T_{max} (median rose from 1.00 to 2.51 h) and total body clearance (mean 45%). The correlations of moxifloxacin AUC (in the absence of rifampin) and rifampin AUC with the change in moxifloxacin AUC in the presence of rifampin were both nonsignificant [221]. It would be of benefit to repeat this study using daily dosing of the same antitubercular agents. In another study conducted in patients with tuberculosis, full moxifloxacin pharmacokinetic characterization was available in ten patients on concurrent rifampin therapy and six patients who were not. Nonsignificant effects were noted on moxifloxacin AUC_{0-24} (mean values of 36.8 vs. 21.3 mg.L/h, respectively) and total body clearance (22.6 vs. 15.5 L/h, respectively), while a significant effect was noted for volume of distribution/F (3.46 vs. 2.90 L/kg, respectively) [222]. The effect of rifampin on moxifloxacin pharmacokinetics is still not established. Two studies have found that ciprofloxacin does not interact pharmacokinetically with isoniazid [223, 224].

The interaction between concomitant multiple-dose ciprofloxacin (500 mg twice daily × 7 days) and multiple-dose itraconazole (200 mg twice daily × 7 days) was evaluated in a three-period crossover study in healthy volunteers. Itraconazole had no significant effect on ciprofloxacin pharmacokinetics. However, ciprofloxacin produced the following significant mean effects on itraconazole pharmacokinetics: increased C_{max} (by 53%), C_{min} (by 86%), AUC_{0-96} (by 49%), and $AUC_{0-\infty}$ (by 82%) and reduced volume of distribution/F (35%) [225].

Single-dose ciprofloxacin coadministration reduced single-dose acetaminophen C_{max} by mean 30% and increased T_{max} and t 1/2 by means of 86% and 29%, respectively (all $p < 0.05$) [226]. The effects of single-dose acetaminophen coadministration on single-dose ciprofloxacin pharmacokinetics were nonsignificant [227]. Similar results were obtained with single-dose acetaminophen and pefloxacin coadministration [228]. Single-dose chloroquine coadministration produced mean reductions in single-dose ciprofloxacin C_{max} and AUC of 18% and 43%, respectively, and a mean increase in cumulative urinary excretion as a percentage of the dose of 1000% (all $p < 0.05$) [229, 230].

The effect of phenazopyridine, a urinary tract analgesic, on the pharmacokinetics of ciprofloxacin was evaluated using a commercially available combination tablet (containing ciprofloxacin 500 mg + phenazopyridine 200 mg). The only significant alterations noted were mean 29% and 30% increases in ciprofloxacin AUC and mean residence time, respectively (both parameters did not fulfill at least one of the lower [80%] or upper [125%] limits of bioequivalence). The mechanism of this effect is not known [231]. A single-dose trial evaluating the effect of ciprofloxacin on sildenafil pharmacokinetics found that ciprofloxacin coadministration resulted in significant increases in t 1/2 (mean 38%), AUC (mean 112%), and C_{max} (mean 117%) (all $p < 0.05$). The 90% CIs for AUC (119–159%) and for C_{max} (127–152%) document a potential drug-drug interaction of considerable magnitude [232].

Two cases have been described of an interaction between ciprofloxacin and levo-thyroxine (T_4), wherein coadministration produced a substantial loss of T_4 pharmacological effect manifested by increases in thyroid-stimulating hormone concentrations to 19 and 44 IU/mL and reductions in free T_4 concentrations to 13 and 4 pmol/L, respectively. In one case, spacing the administration times of the two agents by 6 h led to rapid normalization of thyroid function test results [233]. In a double-blind, randomized, crossover study in healthy volunteers, a single dose of ciprofloxacin 750 mg reduced the baseline-corrected 6-h $AUC_{TOT\ T4}$ by 39% after a single dose of 1000 mcg T_4 ($p = 0.035$). However, no effect was noted on T_3 AUC. The hypothesized mechanism was inhibition of intestinal T_4 uptake transporters (e.g., MCT8, MCT10, LAT1/2) by ciprofloxacin [234].

One case of an interaction between ciprofloxacin and tizanidine has been described. Upon initiation of ciprofloxacin therapy for a urinary tract infection, signs of tizanidine toxicity (bradycardia, hypotension, hypothermia) began almost immediately in this 45-year old with multiple sclerosis. Drowsiness and continuing hypotension led to discontinuation of ciprofloxacin with subsequent improvement and then disappearance of the signs of tizanidine intoxication [235]. The authors then surveyed the medical records of 1165 patients, looking for the combined use of tizanidine with ciprofloxacin. Eight cases were found. Examining these eight cases and comparing them to 11 cases of combined use of tizanidine and fluvoxamine found in the literature (fluvoxamine being another CYP1A2 inhibitor), both combinations were characterized by similar patterns of systolic and/or diastolic hypotension and hypothermia. This suggested that inhibition of CYP1A2-mediated metabolism of tizanidine was the mechanism involved [235]. This mechanism was confirmed by a drug interaction study conducted in ten healthy volunteers. Steady-state ciprofloxacin coadministration led to significant increases in single-dose tizanidine C_{max} (mean 564%), $t\frac{1}{2}$ (mean 23%), and AUC (mean 876%) (all $p \leq 0.007$). Significant decreases were also seen in systolic and diastolic blood pressures (post-tizanidine vs. baseline differences during ciprofloxacin vs. no coadministration of -17 and -11 mm Hg, respectively; both $p < 0.001$). Visual analog scales for drowsiness and drug effect and the Digit Symbol Substitution Test results demonstrated significant negative effects of the combination compared with tizanidine alone ($p = 0.009$–0.02). Correlation analyses with caffeine/paraxanthine concentration ratios (a marker for CYP1A2 activity) supported CYP1A2 inhibition as the mechanism of the interaction [236].

Ciprofloxacin has been used to augment pentoxifylline plasma concentrations in cancer patients undergoing interleukin therapy [237]. Pentoxifylline inhibits interleukin-induced capillary leak syndrome in these patients. In addition, the plasma concentrations of (R)-metabolite-1, an even more potent inhibitor of this syndrome than the parent pentoxifylline, were evaluated during concurrent therapy with ciprofloxacin [237, 238]. In vitro and in vivo (murine) studies revealed that pentoxifylline and metabolite-1 C_{max} and AUC were doubled by coadministration of ciprofloxacin. These two moieties were interconvertible in vivo. The underlying mechanism was inhibition of CYP1A2 (thus increasing pentoxifylline) and induction of CYP2E1 (thus increasing generation of the R-enantiomer of metabolite-1) [239, 240].

Eight case reports have suggested inhibition of clozapine ($N = 6$), olanzapine ($N = 1$), and methadone ($N = 1$) metabolism by ciprofloxacin [241–247]. In a study conducted in seven patients with schizophrenia, ciprofloxacin 250 mg twice daily caused significant elevations in serum clozapine and N-desmethylclozapine concentrations (mean 29 and 31%, respectively) after 1 week of concurrent therapy [248]. One case report has suggested a similar interaction of ciprofloxacin with asenapine, a neuroleptic dependent upon CYP1A2 and UGT1A4 for its metabolism and excretion [249].

Four cases of severe methotrexate toxicity due to concomitant use of ciprofloxacin in oncology patients have been reported. In all cases, elimination of methotrexate after high-dose therapy for cancer was substantially delayed with resultant dermatologic, bone marrow, hepatic, and renal toxicity. The mechanism is unclear but may involve alterations of methotrexate plasma protein binding, reduction in renal function (thus enhancing drug retention), inhibition of hepatic aldolase (thus reducing drug metabolism), or inhibition of renal tubular secretion (again, enhancing retention). Another issue with combination quinolone-high-dose methotrexate therapy is the effect of urinary alkalinization (required for safe high-dose methotrexate use) on the crystalluria risk of the quinolones [250–254]. Case reports have also appeared of severe methotrexate toxicity using the much lower doses used in management of psoriasis and rheumatic diseases, again documenting delayed methotrexate elimination [255].

A case report of lithium toxicity caused by concurrent levofloxacin use has also been reported. It appears that an acute deterioration in renal function occurred, causing retention of lithium. Whether the deterioration in renal function was due to the quinolone or the combination of the two drugs is not known [256].

Ciprofloxacin and moxifloxacin do not interact pharmacokinetically or pharmacodynamically with low-dose oral contraceptives containing 30 μg of ethinyl estradiol and 150 μg desogestrel per tablet [201, 257]. Levofloxacin does not alter the pharmacokinetics of zidovudine, efavirenz, or nelfinavir, and ciprofloxacin does not alter the pharmacokinetics of didanosine [32, 258, 259]. Note that there is an absorption interaction between ciprofloxacin and didanosine with cations as discussed previously [32, 33]. Ciprofloxacin does not interact pharmacokinetically with metronidazole [83].

The effect of quinolones on the pharmacokinetics and pharmacodynamics of ethanol are uncertain. One study using healthy volunteers found no pharmacokinetic or pharmacodynamic interaction with ciprofloxacin [260]. However, another study, again using healthy volunteers, found that ciprofloxacin 750 mg twice daily significantly reduced the ethanol elimination rate (by mean 9%, range 5–18%) and increased the AUC (mean 12%) and time to zero blood ethanol concentration (mean 10%). This pharmacokinetic interaction was felt to be caused by the effect of ciprofloxacin on the ethanol-metabolizing intestinal flora and not its hepatic effects (on enzymes and blood flow) [261]. Perhaps the discrepencies between results of these two studies are caused by differences in subject numbers (statistical power), drug doses, or study design (randomized, parallel group vs. crossover).

The effects of multiple-dose oral ciprofloxacin of the single-dose pharmacokinetics of intravenous ropivacaine have been evaluated in nine healthy volunteers. The clearance of ropivacaine was significantly reduced (mean 31%) during concomitant therapy, with considerable intersubject variability (range 52% reduction to 39%

enhancement). The CYP1A2-mediated formation of 3-OH-ropivacaine was significantly retarded; the AUC and 24-h urinary excretion of this metabolite fell 38% and 27%, respectively. In contrast, the CYP3A4-mediated formation of (S)-2',6'-pipecoloxylidide (PPX) was significantly enhanced, as manifested by mean increases in AUC and 24-h urinary excretion of 71% and 97%, respectively [262].

Pharmacokinetics of single-dose lidocaine and its monoethylglycinexylidide (MEGX) and 3-hydroxylidocaine metabolites were evaluated after multiple-dose oral ciprofloxacin administration. Ciprofloxacin at steady state produced significant increases in lidocaine C_{max} (mean 12%, $p < 0.05$), AUC (mean 26%, $p < 0.01$), and $t\frac{1}{2}$ (mean 7%, $p < 0.01$) and a decrease in total body clearance (mean 22%, $p < 0.01$). Alterations seen with the MEGX metabolite included significant reductions in C_{max} (mean 40%, $p < 0.01$), AUC (mean 21%, $p < 0.01$), and the ratio of MEGX/lidocaine AUCs (mean 40%, $p < 0.001$), while an increase was noted in $t\frac{1}{2}$ (mean 34%, $p < 0.05$). Alterations with the 3-hydroxy metabolite included significant reductions in C_{max} (mean 23%, $p < 0.05$), AUC (mean 14%, $p < 0.01$), and the ratio of 3-hydroxy metabolite to lidocaine AUCs (mean 35%, $p < 0.001$) [263].

A number of case reports have documented substantial reductions in serum phenytoin concentrations when ciprofloxacin therapy was initiated, an unexpected finding for a drug usually associated with enzyme inhibition and reduced drug clearance [264–270]. Indeed, results of a small study revealed that ciprofloxacin cotherapy was associated with nonsignificant reductions in mean steady-state phenytoin C_{max} (4%) and AUC (6%) [271]. The mechanism underlying this interaction may involve CYP2E1 induction by ciprofloxacin [240]. This effect has even been described in a well-documented case report after the use of ciprofloxacin eye drops [270]. Caution is warranted when coadministering phenytoin and quinolones on the basis of this kinetic interaction as well as the epileptogenic potential of the quinolones (when certain quinolones and certain nonsteroidal anti-inflammatory drugs are coadministered see below).

The interaction of single-dose ciprofloxacin (500 mg) with single-dose carbamazepine (200 mg) has been evaluated in healthy volunteers. Ciprofloxacin coadministration produced significant (median) increases in C_{max} (59%), $t\frac{1}{2}$ (31%) and AUC (54%) and reductions in volume of distribution/F (25%), central volume of distribution/F (45%) and total body clearance (42%) [272].

The interaction of multiple-dose ciprofloxacin (500 mg twice daily) and multiple-dose alvimopan (6 mg twice daily) has been evaluated in healthy volunteers. Ciprofloxacin exposure reduced the mean C_{max}, AUC, and C_{trough} of alvimopan by 24%, 12%, and 6%, respectively, and virtually eliminated generation of the active and equipotent metabolite of alvimopan. Thus, concurrent therapy could result in a loss of therapeutic efficacy of alvimopan [273].

The interaction of single-dose zolpidem with multiple-dose ciprofloxacin (500 mg daily × 5 days) pretreatment has been evaluated in healthy volunteers. Ciprofloxacin produced significant (mean) increases in zolpidem C_{max} (6%), AUC_{0-t} (37%), $AUC_{0-\infty}$ (46%), $t\frac{1}{2}$ (40%), and mean residence time (42%) [274].

The effect of combinations of enzyme inhibitors such as ciprofloxacin plus clarithromycin and ciprofloxacin plus cimetidine has been evaluated [212, 275–277].

Interestingly, clarithromycin (1000 mg twice daily) did not significantly augment the effect of ciprofloxacin (500 mg twice daily) on steady-state theophylline pharmacokinetics [276]. In contrast, coadministration of cimetidine (400 mg twice daily or 600 mg four times a daily) plus ciprofloxacin (500 mg twice daily) exerted a greater inhibitory effect on theophylline elimination than each agent alone, although the combined effect was less than the additive sum produced by the individual drugs [212, 275, 277].

Virtually no data are available regarding the interaction potential of quinolones with herbal products. Turmeric (*Curcuma longa*) is a medicinal plant extensively used in Ayurveda, Unani, and Siddha medicine as a home remedy for various disorders. Curcumin is the active moiety of this plant. It is known to inhibit CYP3A4 in the liver and induce P-glycoprotein. After oral curcumin pretreatment for 3 days in rabbits, single oral dose norfloxacin pharmacokinetics were significantly modified compared with those in the non-pretreated group. Mean absorption half-life was decreased (by 23%), while increases were seen in the absorption rate constant (by 41%), $t\frac{1}{2}$ (by 19%), AUC (by 52%), AUC (first moment) (by 69%), mean residence time (by 12%), and Vd (area) (by 31%) (all $p < 0.05$) [278].

The mechanism of these metabolic interactions is largely unexplored. It has been suggested that inhibition of metabolism may be related to the 4-oxo-metabolites of the quinolones, but more recent data suggests that the sequence $N* - C = N - C - N - C$ (where $N*$ = nitrogen on the piperazine ring) is the entity responsible for metabolic inhibition [143, 279].

The structure-activity relationships for in vitro inhibition of human CYP1A2 have been investigated by Fuhr and coworkers. 3'-oxo derivatives had similar or reduced activity, and M1 metabolites (cleavage of piperazinyl substituent) had greater inhibitory activity compared with the parent molecule. Alkylation of the 7-piperazinyl substituent resulted in reduced inhibitory potency. Naphthyridines with an unsubstituted piperazinyl group in position 7 displayed greater inhibitory potency than did corresponding quinolone derivatives. Molecular modeling studies revealed that the keto group, carboxylate group, and core nitrogen at position 1 are likely to be the most important groups for binding to the active site of CYP1A2. These investigators also developed an equation to estimate a priori, using quantitative structure-activity relationship analysis, the potency of a given quinolone to inhibit CYP1A2 [280]. These investigators as well as Sarkar and coworkers have also developed in vitro human liver microsome models that may be useful in qualitatively predicting relevant drug interactions between quinolones and methylxanthines [281, 282].

Antofloxacin and caderofloxacin, new quinolones being developed in China, are derivatives of levofloxacin and ciprofloxacin, respectively. Neither agent appears to inhibit the activities of CYP1A2 or CYP2C9 in human microsomes [283]. However, no data have been published addressing their effects on other CYP isoenzymes in humans, except the drug-drug interaction between antofloxacin and theophylline [138].

Clinically, caution is advised when using any quinolone in combination with a xanthine compound such as theophylline. Close monitoring of serum theophylline concentrations is recommended in any patient receiving these drugs. The clinical significance of inhibited metabolism of other drugs remains largely unclear at present. Until further data become available, clinicians should be aware of the possibility of reduced drug metabolism resulting in adverse effects whenever the quinolones are coadministered with drugs that depend on hepatic metabolism for their elimination.

3.2.4 Excretion Interactions

The quinolone antimicrobials are generally excreted into the urine at a rate higher than creatinine clearance, implying that tubular secretion is a prominent excretory pathway. Indeed, the administration of probenecid, a blocker of the anionic renal tubular secretory pathway, substantially reduces the renal elimination of norfloxacin, levofloxacin, gatifloxacin, ulifloxacin (active metabolite of prulifloxacin), nemonoxacin, and ciprofloxacin, reflecting competitive blockade of quinolone tubular secretion [85, 284–289]. In contrast, probenecid coadministration does not affect the pharmacokinetics of moxifloxacin [290]. In addition, furosemide and ranitidine reduce the renal tubular secretion of lomefloxacin, again reflecting competitive blockade [291, 292]. There is thus a possibility that other drugs or endogenous compounds may interact with the quinolones at this site to competitively impair their mutual renal elimination, thus elevating blood concentrations and perhaps enhancing therapeutic and/or toxic effects. In general, levofloxacin, lomefloxacin, and moxifloxacin exhibit little/no net renal tubular transport, while ofloxacin, ciprofloxacin, and gemifloxacin exhibit net renal tubular secretion [293].

For example, in vitro, the quinolones DX-619 and levofloxacin significantly and dose-dependently inhibited the uptake of creatinine in HEK cells expressing the renal organic cation transporter (hOCT$_2$). At the highest quinolone concentrations tested, creatinine transport fell by 88% with both drugs. Whether these agents can interfere with creatinine clearance estimation in vivo to a clinically relevant degree is unknown [294]. In another study, 13 quinolones were evaluated for their abilities to inhibit OCT$_1$, OCT$_2$, and OCT$_3$. All except enoxacin inhibited OCT$_1$ uptake, while none inhibited OCT$_2$. Moxifloxacin inhibited OCT$_3$ but only weakly (30% at 1000-fold excess) [295].

No data are available regarding the effects of antofloxacin and caderofloxacin on transporters such as P-gp, MRP-2, or organic anion and cation transporters in humans. However, in rats, inhibitors of multiple transporters (P-gp, MRP-2, organic anion and cation transporters, and breast cancer resistance protein) significantly reduced the biliary clearance of antofloxacin (all $p < 0.05$). The effects of antofloxacin on these transporters were not assessed [296].

Another example has been noted in a study of the interaction between ofloxacin and procainamide in healthy volunteers. Ofloxacin coadministration was associated with 22% and 30% falls in procainamide oral total body and renal clearances, respectively. However, neither the pharmacokinetics of N-acetylprocainamide nor the pharmacodynamics of the antiarrhythmic, as assessed by standard 12-lead and signal-averaged electrocardiograms, were affected by ofloxacin coadministration

[297]. A more recent trial compared oral levofloxacin to oral ciprofloxacin with regard to the interaction potential with procainamide and NAPA in ten healthy volunteers. The only significant effects of ciprofloxacin (500 mg twice daily for 5 days) were a mean 15% reduction in procainamide renal clearance and a mean 10% reduction in the ratio of NAPA renal clearance to creatinine clearance (both $p < 0.05$). In levofloxacin (500 mg once daily for 5 days) recipients, the following significant ($p < 0.05$) effects were noted: mean reductions in procainamide total body and renal clearances, fraction excreted in urine in unchanged form, and ratio of procainamide renal clearance to creatinine clearance of 17%, 26%, 11%, and 29%, respectively; mean increase in procainamide $t \frac{1}{2}$ of 19%; and mean reductions in NAPA renal clearance, fraction eliminated in urine in unchanged form, and the ratio of NAPA renal clearance to creatinine clearance of 21%, 20%, and 28%, respectively. This interaction was thus potentially more clinically problematic with levofloxacin than with ciprofloxacin. In fact, of the ten volunteers, only one had a reduction in procainamide total body clearance exceeding 25% with ciprofloxacin, while four had reductions in total body clearance of 30% or greater, and three of the four had reductions in NAPA renal clearance of 30% or greater with levofloxacin [298].

3.3   Pharmacodynamic Interactions

3.3.1   Quinolones and NSAIDs

Central nervous system (CNS) toxicity, including tremulousness and seizures, is rare with quinolones [299–311]. In some cases, concurrent use of nonsteroidal anti-inflammatory drugs (NSAIDs) has been noted [302, 303, 305, 312]. It was the report of multiple cases of seizures associated with the concurrent use of enoxacin and fenbufen (the latter being an NSAID not available in the USA) to Japanese regulatory authorities that led to a plethora of investigations into the possible interaction between quinolones and NSAIDs [305, 312].

Some rat studies have suggested that NSAIDs such as fenbufen may enhance CNS uptake of quinolones such as ciprofloxacin, norfloxacin, and ofloxacin [313, 314]. However, other studies conducted in the same species have documented an absence of a pharmacokinetic interaction between fenbufen and ciprofloxacin and ofloxacin [315, 316]. In addition, human studies have the documented absence of a pharmacokinetic interaction between ciprofloxacin and fenbufen and between pefloxacin or ofloxacin and ketoprofen [317–319]. However, with single-dose diclofenac coadministration, ciprofloxacin pharmacokinetics were modestly affected. Mean ciprofloxacin C_{max}, AUC, and apparent oral clearance increased 58% and 46% and decreased 28%, respectively [320]. Overall, any interaction that occurs between quinolones and NSAIDs is thus probably purely pharmacodynamic in nature. Previous editions of this book should be consulted for mechanistic details of this pharmacodynamic interaction.

Although of theoretical interest, the pharmacodynamic interaction between quinolones and NSAIDs is probably of little clinical relevance so long as fenbufen is not concurrently used with prulifloxacin.

3.3.2 Quinolones and Electrophysiology

Levofloxacin, ciprofloxacin, norfloxacin, nemonoxacin, and moxifloxacin have been associated with prolongation of the QTc interval on the electrocardiogram, which in a few cases has been associated with the development of polymorphous ventricular tachycardia (torsades des pointes), which in turn can degenerate into ventricular fibrillation [85, 321–340]. One case of levofloxacin-associated torsades des pointes in the absence of QTc interval prolongation has also been reported [341]. Grepafloxacin was removed from the market by its manufacturer in October 1999 because of its electrophysiologic adverse event profile.

Extensive but conflicting data are available regarding the epidemiology of electropathophysiology in quinolone recipients. One group utilized the Varese province of Italy as the study database, performing a case-control study of subjects with ventricular arrhythmias or cardiac arrest between July 1998 and December 2003. A total of 1275 cases and 9189 controls formed the study population. The adjusted odds ratio for recent (within 4 weeks) exposure to quinolones was 3.58 (95% CI, 2.51–5.12) [342]. Another group evaluated drug-induced torsades des pointes in patients at least 80 years old. In 24 reports on 25 patients 80–95 years old, the most prevalent risk factors were nonmodifiable (88% were female, 76% had structural heart disease, and 64% were female with structural heart disease). Among potentially modifiable risk factors, 44% received QTc interval-prolonging drugs despite preexisting prolonged QTc intervals, and 36% received two or more concurrent QT interval-prolonging drugs. The most prevalent QTc interval-prolonging drugs were quinolone ($N = 3$) and macrolide ($N = 7$) antimicrobials in 36%. All but three individuals had at least one modifiable risk factor [343].

A nationwide case-control study conducted in Taiwan compared the risks of ventricular arrhythmias (VA) and cardiovascular (CV) death in outpatient recipients of moxifloxacin, ciprofloxacin, levofloxacin, and amoxicillin-clavulanate (A-C) as control (as well as azithromycin and clarithromycin). The primary end points were evaluated within 7 days of antimicrobial initiation. Moxifloxacin therapy was associated with significantly increased risks of VA (odds ratio [OR], 3.30; 95% CI, 2.07–5.25) and CV death (OR, 2.31; 95% CI, 1.39–3.84) compared with A-C. Ciprofloxacin therapy was also associated with an enhanced risk of CV death compared with A-C (OR, 1.77; 95% CI, 1.22–2.59). However, the risks of VA and CV death with moxifloxacin were significantly higher than those with ciprofloxacin. The absolute risks per 1000 patients for VA and CV death for A-C, ciprofloxacin, levofloxacin, and moxifloxacin were 0.12, 0.15, 0.26, and 0.57 and 0.13, 0.12, 0.39, and 0.46, respectively [344].

Two hundred consecutive patients admitted to the intensive care unit who were prescribed one or more of a variety of drugs known to prolong the QTc interval were evaluated for the combined end point of QTc interval prolongation to over 500 milliseconds or QTc interval prolongation exceeding 60 milliseconds over baseline. In 62 recipients of ciprofloxacin, the mean ± SD increase in QTc interval from baseline was 18 ± 33 milliseconds, and prevalence of the combined end point was 40%. For 82 recipients of moxifloxacin, corresponding values were 20 ± 33 milliseconds and 55% [345].

A Canadian case-cohort study conducted in the province of Quebec from January 1990 through December 2005 (with follow-up to March 2007) evaluated the risk of serious arrhythmias (defined as VA or sudden/unattended death as hospital discharge diagnoses) in quinolone recipients. In the cohort of 605,127 patients, 1838 cases were identified (incidence rate = 4.7/10,000 person-years). Current quinolone use was significantly associated (relative risk [RR], 1.76; 95% CI, 1.19–2.59), especially new current use (RR, 2.23; 95% CI, 1.31–3.80). Moxifloxacin and ciprofloxacin were significantly associated as well (RR, 3.30; 95% CI, 1.47–7.37, and RR, 2.15; 95% CI, 1.34–3.46, respectively [as was gatifloxacin]) [346].

A retrospective chart review of hospitalized hematology patients was conducted from September 2008 to January 2010 to evaluate the effect of combined quinolone-azole use on QTc interval. Of 94 patients identified, 88 received levofloxacin, 53 received voriconazole, and 40 received fluconazole. The overall change in QTc interval from baseline was a mean of 6.1 milliseconds (95% CI, 0.2–11.9). Twenty-one patients (22.3%) had a clinically significant change in QTc interval on combination therapy. Major clinical significance was defined as a greater than 60-millisecond prolongation or an absolute value exceeding 500 milliseconds, while moderate clinical significance was defined as a 30–59-millisecond prolongation or an absolute value exceeding 470 milliseconds (males) or 480 milliseconds (females). Significant risk factors in these patients included hypokalemia and a left ventricular ejection fraction below 55%. The relative contributions of the quinolone and azole components could not be discerned [347].

A retrospective case-cohort analysis, conducted within the Veterans Affairs system, evaluated the risk of serious cardiac arrhythmias and all-cause mortality among outpatient recipients of amoxicillin ($N = 979,380$) and levofloxacin ($N = 201,798$) between September 1999 and April 2012 (mean age, 56.8 years). Outcomes were assessed during days 1–5 and 6–10 of antimicrobial therapy. Levofloxacin therapy significantly increased the risks of all-cause mortality (hazard ratio [HR], 2.49; 95% CI, 1.70–3.64) and serious cardiac arrhythmias (HR, 2.43; 95% CI, 1.56–3.79) during days 1–5. Corresponding data for days 6–10 were HR, 1.95; 95% CI, 1.32–2.88, and HR, 1.75; 95% CI, 1.09–2.82 (both significant). Absolute numbers of deaths per million prescriptions were amoxicillin, $N = 154$ and levofloxacin, $N = 384$ at the end of day 5 and amoxicillin, $N = 324$ and levofloxacin, $N = 714$ at the end of day 10 [348].

A retrospective chart review was conducted at the Mayo Clinic (Rochester, MN) from October 9, 2009, to June 12, 2012, to evaluate the epidemiology of ventricular tachycardia (VT)/ventricular fibrillation (VF) in patients receiving levofloxacin who had a prolonged QTc at baseline (defined as a QTc exceeding 450 milliseconds). Over this period, 1004 consecutive patients fulfilled the QTc study criteria. Levofloxacin was administered both orally and intravenously and was dose-adjusted for the level of renal function. The primary outcome was sustained VT on electrocardiogram. The median time from initiation of levofloxacin therapy to hospital discharge was 4 days (range, 1–94 days). Only two patients (0.2%, 95% CI, 0.0–0.7) experienced the primary outcome [349].

A binational (Danish/Swedish) case-cohort study was conducted using Danish data from 1997 to 2011 and Swedish data from 2006 to 2013 to evaluate the CV risks of quinolones in adults 40–79 years old. There were 909,656 quinolone courses (ciprofloxacin, 82.6%; norfloxacin, 12.1%; ofloxacin, 3.2%; moxifloxacin, 1.2%; others, 0.9%) and 909,656 courses of penicillin V (matched on propensity scores). The risk period of interest was days 0–7 of antimicrobial therapy. Subgroup analyses were performed by country, sex, age, underlying CV disease, concomitant use of other drugs increasing the risk of polymorphous VT, quinolone type, and level of arrhythmia risk score. One hundred and forty-four cases of serious arrhythmias occurred during follow-up, 66 among quinolone recipients (3.4/1000 patient-years) and 78 among penicillin recipients (4.0/1000 patient-years), yielding a rate ratio of 0.85 (95% CI, 0.61–1.18). The absolute risk difference was -13 (95% CI, -35–16) per one million courses of quinolones. No statistically significant differences were noted in subgroup analyses. Due to the small numbers of non-ciprofloxacin quinolones used, intra-quinolone class differences could not be ruled out [350].

In a retrospective analysis utilizing the FDA adverse event reporting database from January 1, 1996, through May 2, 2001, the rates of torsades des pointes with ciprofloxacin and levofloxacin were 0.3 and 5.4 cases per 10 million prescriptions, respectively ($p < 0.05$) [323]. However, the numerous potential problems with study design preclude generalizability of these results [323, 324].

Numerous in vitro models have been utilized to elucidate the mechanism underlying the arrhythmogenic effects of these agents: HERG (human ether-a-go-go-related gene) potassium channels, mouse atrial tumor cells, guinea pig myocardium, and canine Purkinje fibers [351–355]. The potency of quinolones in inhibiting HERG-mediated outward potassium currents was moxifloxacin > levofloxacin = ciprofloxacin > ofloxacin in one study; for the other, it was moxifloxacin >> > ciprofloxacin [351, 355]. Similar findings were noted for mouse atrial tumor cell potassium channels [352]. In guinea pig ventricular myocardium, prolongation of action potential duration was 25% for moxifloxacin, and the prolongation with levofloxacin and ciprofloxacin was essentially zero [353]. Similar findings were noted with canine cardiac Purkinje fibers (moxifloxacin > ciprofloxacin) [354]. The maximum degree of blockade of HERG current in transfected HEK293 cells was only 12.3 ± 3.3% at the highest tested concentration (335 µM) of ulifloxacin, the active metabolite of prulifloxacin [356].

In rabbits, levofloxacin and ulifloxacin were equipotent in prolonging the maximum QT interval [357]. In dogs with complete atrioventricular block and dogs under halothane anesthesia, oral and intravenous levofloxacin produced essentially no adverse electrophysiologic and hemodynamic effects [358]. In conscious dogs dosed with oral prulifloxacin 150 mg/kg/day for 5 days and followed by telemetry, no significant effect was noted on the QTc interval at any time [356].

In summary, the in vitro and in vivo (animal) studies revealed that quinolones cause a drug-specific, dose-dependent prolongation in QTc interval by inhibiting outward potassium currents in myocytes. In turn, this prolongation in action potential duration leads to a drug-specific risk of ventricular tachycardia and torsades des pointes. However, the lack of full agreement of the results of evaluations of potassium channel inhibition and QTc interval prolongation, in terms of relative drug potencies, suggests that more than potassium channel inhibition may be involved [352].

In a retrospective review of 23 patients receiving 500 mg levofloxacin once daily in whom pre- and intratherapy electrocardiograms were available, the QTc prolongation exceeded 30 milliseconds in four patients (17%) and 60 milliseconds in two patients (9%), with an absolute QTc prolongation to more than 500 milliseconds in four patients (17%) [359]. Single oral doses of moxifloxacin 400 and 800 mg caused $4.0 \pm 5.1\%$ (mean $\pm$ SD) and $4.5 \pm 3.8\%$ prolongation of the QTc interval at rest (both $p < 0.05$) in healthy volunteers. Significant QTc interval prolongation occurred at all heart rates and across the entire RR interval range (400–1000 milliseconds). The effect was similar in males and females and did not show dose dependence. No significant reverse rate dependence was seen. Statistically significant but weak correlations existed between moxifloxacin plasma concentrations versus QTc interval ($r = 0.35$) and change in QTc interval with placebo ($r = 0.72$) [360].

In another healthy volunteer study, periodic and continuous ECGs were recorded before and after administration of single doses of intravenous levofloxacin 500, 1000, and 1500 mg. Using periodic ECG data, the only significant differences noted were the mean QTc intervals at 1.5 h after administration of 1500 mg (Bazett formula, 415.33 vs. 399.48 milliseconds with placebo; Fredericia correction, 409.67 vs. 400.46 milliseconds with placebo) and 2.0 h after administration of 1500 mg (corresponding values of 414.10 vs. 398.92 and 409.58 vs. 400.10 milliseconds) (all $p < 0.05$). Using continuous ECG data, significant QTc interval prolongation occurred after administration of 1000 mg (Bazett correction, in 3/4 baseline correction methods, mean change ranged from 2.8 to 3.9 milliseconds [$p \leq 0.05$]; Fredericia correction, in 1/4 baseline correction methods, the mean change was 2.8 milliseconds [$p \leq 0.05$]) and 1500 mg (Bazett correction, in 4/4 baseline correction methods, mean change ranged from 6.4 to 7.7 milliseconds [$p \leq 0.001$]; Fredericia correction, in 4/4 baseline correction methods, mean change ranged from 4.9 to 6.9 milliseconds [$p \leq 0.001$]) [361].

Two comparative studies of the effect of quinolones on QTc interval in humans have been published. Single oral doses of 1000 mg levofloxacin, 1500 mg ciprofloxacin, and 800 mg moxifloxacin were compared in healthy volunteers. Mean QT and QTc interval prolongation was significantly greater for moxifloxacin compared to placebo for all end points, but it was generally not so for levofloxacin and cipro-floxacin (the exception was that the postdose QTc and QTc at 1.5, 2, and 2.5 h postdose, using the Bazett method, were significantly increased for levofloxacin vs. placebo). The proportion of subjects with prolongation in QTc interval of 30 milli-seconds or greater was higher with moxifloxacin (72–81%) compared to levofloxa-cin (33–38%) and ciprofloxacin (34–40%) [362].

The second study compared single-dose oral levofloxacin 1000 and 1500 mg with oral moxifloxacin 400 mg in a four-way, placebo-controlled, crossover study in healthy volunteers. The largest difference in QTc for all three agents occurred at 3.5 h postdosing, and mean baseline-corrected QTc prolongations were 4.42, 7.44, and 13.19 milliseconds, respectively (no statistical comparison performed) [363].

In contrast, in thorough QTc studies conducted with delafloxacin, prulifloxacin, and zabofloxacin, there appeared to be clinically significant QTc interval prolonga-tion [364–366]. In addition, the electrophysiologic effects of quinolones do not appear to differ by race/ethnic group, at least for Caucasians, Japanese, Chinese, and Koreans [367–370].

Taking the data above as a whole, it appears that the excess in adverse cardiac events over base rates with ciprofloxacin, levofloxacin, and moxifloxacin is roughly 20–40 cases per million courses of each drug. The number of prescriptions needed to cause 1 additional serious cardiac event is roughly 25,000–50,000. In view of the relative rarity of these adverse events and inconsistencies between studies in risk estimates, the choice of quinolone should not be made primarily on the basis of concern for adverse cardiac events, except in patients at the highest risk for such events [371, 372]. Caution is warranted with the use of these agents in patients receiving other drugs with similar electrophysiologic effects (Table 3.3) [373–375]. In addition, caution is warranted in using these agents in patients with an abnormal pretreatment QTc interval, pretreatment electrolyte abnormalities (e.g., hypokale-mia, hypomagnesemia, rarely hypocalcemia), starvation/liquid-protein fast diets, and a prior or current history of coronary heart disease, bradyarrhythmias, or atrial fibrillation [373–375].

Table 3.3 Drugs prolonging the QTc interval that may potentially interact pharmacodynamically with selected quinolone antimicrobials

Cisapride	
Trimethoprim/sulfamethoxazole	Macrolides (erythromycin, clarithromycin, spiramycin)
Pentamidine	Chloroquine
Halofantrine	Phenothiazines
Quinidine	Tricyclic and tetracyclic antidepressants
Procainamide	Disopyramide
Ibutilide	Lidocaine, mexiletine (rare)
β-blockers (rare)	Amiodarone (rare)
Bepridil	Lidoflazine
Sotalol	Dofetilide
Flecainide	Encainide

Source: From Ref. [373]

3.3.3 Quinolones and Glucose Homeostasis

Case reports have documented pharmacodynamic interactions between quinolones and hypoglycemic agents in patients with type 2 diabetes mellitus, leading to symptomatic, prolonged hypoglycemia. Gatifloxacin was, by far, the most implicated in this and was removed from the market for this reason. Other implicated quinolones have included levofloxacin, ciprofloxacin, and moxifloxacin [376–382].

Glucose homeostasis abnormalities (GHAs) reported to the FDA in the MedWatch® program from November 1997 to September 2003, inclusive, have been reviewed for concurrent use of ciprofloxacin, levofloxacin, and moxifloxacin. These events were identified under 14 unique coding items. Rates were calculated using US retail pharmacy prescriptions as the denominator. The spontaneous reporting rates for the three quinolones combined for both total GHA reports (8/10 million prescriptions) and fatal GHA reports (0.6/10 million prescriptions) were exceedingly low. GHA reports constituted only 1.4% of the combined quinolone adverse event reports [383].

In another population-based analysis, two nested case-control studies were conducted, using data from 1.4 million elderly ($\geq$66 years old) residents of Ontario, Canada. In study 1, case patients were treated in the hospital setting for hypoglycemia after outpatient treatment with macrolide, second-generation cephalosporin, or respiratory quinolone (levofloxacin, moxifloxacin, or ciprofloxacin) agents. In study 2, case patients were those who received hospital care for hyperglycemia. For each case patient, up to five controls were identified, matched by age, sex, presence/absence of diabetes, and timing of antimicrobial therapy [384].

For study 1, all patients treated for hypoglycemia within 30 days of completion of antimicrobial therapy were identified (April 2002–March 2004). As compared with macrolides (e.g., erythromycin, azithromycin, and clarithromycin), levofloxacin recipients were at significantly greater risk for the development of hypoglycemia (adjusted odds ratio [AOR] of 1.5 [95% CI, 1.2–2.0]) in the overall population. In contrast, there was no relationship of the development of hypoglycemia with the use of moxifloxacin, ciprofloxacin, or second-generation cephalosporins (e.g., cefaclor and cefuroxime axetil). Similar findings held true when the analysis was repeated in patients with/without diabetes with one exception. The relationship of hypoglycemia to levofloxacin use in patients without diabetes was not significant (AOR, 2.1; 95% CI, 0.7–6.0) [384].

For study 2, all patients treated for hyperglycemia within 30 days of completion of antimicrobial therapy were identified. Compared with macrolides, none of the three quinolones were significantly associated with the development of hyperglycemia. Similar findings were noted in patients with or without diabetes, with the additional finding of a borderline increase in risk in levofloxacin recipients in the population of patients with diabetes (AOR, 1.5; 95% CI, 1.2–2.0) [384].

Last, all patients $\geq$66 years old treated with antimicrobials during the study period were identified. Those hospitalized within 90 days before receiving an antimicrobial prescription as well as those who received another antimicrobial prescription within 30 days were excluded. Hospital visits for dysglycemia during the 30-day period after the start of antimicrobial therapy were identified for each patient.

In the context of multiple antimicrobial prescriptions, each treatment course was considered separately. The rates of hospitalization for dysglycemia for ciprofloxacin (0.3%), levofloxacin (0.3%), moxifloxacin (0.2%), second-generation cephalosporins (0.2%), and macrolides (0.1%) were comparable [384].

A retrospective case-control study of the risk of severe hypo- or hyperglycemia in recipients of levofloxacin, ciprofloxacin, and azithromycin (control) was conducted using the Veterans Affairs outpatient database (from October 1, 2000, to September 30, 2005). In the final study cohort, 874,682 and 402,566 patients received quinolones and azithromycin, respectively. The event rates for hypoglycemia and hyperglycemia/1000 patients (levofloxacin/ciprofloxacin/azithromycin) were 0.19/0.10/0.07 and 0.18/0.12/0.10, respectively (both p = NS). In patients without diabetes, no quinolone was statistically associated with either hypo- or hyperglycemia, but the statistical power of this analysis was low due to the small number of events (total N = 51). In patients with diabetes, a significant odds ratio (OR) for severe hypoglycemia (versus azithromycin) of 2.1 (95% CI, 1.4–3.3) for levofloxacin was found. A corresponding significant OR for severe hyperglycemia of 1.8 (95% CI, 1.2–2.7) for levofloxacin was also found. Ciprofloxacin was not statistically associated with either glucose perturbation [385].

A retrospective cohort study was conducted using Texas Medicare claims from 2006 to 2009 for patients 66 years old and older who had received prescriptions for glipizide or glyburide and who also filled prescriptions for 1 of 16 antimicrobials most frequently used in this population. The primary outcome evaluated was hospitalization or emergency department visit for hypoglycemia within 14 days of antimicrobial exposure. Five antimicrobials were significantly associated with increased rates of hypoglycemia, including two quinolones: levofloxacin (odds ratio, 2.60; 95% CI, 2.18–3.10) and ciprofloxacin (odds ratio, 1.62; 95% CI, 1.33–1.97). Moxifloxacin had no association (odds ratio, 1.13; 95% CI, 0.65–1.98) [386].

In a population-based cohort study of Taiwanese patients with types 1 and 2 diabetes (N = 78,433, January 2006–November 2007), outpatient new users of levofloxacin, ciprofloxacin, moxifloxacin, cephalosporins, and macrolides were identified. Study outcomes were defined as emergency department visit or hospitalization for dysglycemia within 30 days of the start of antimicrobial therapy. Adjusted odds ratios (AORs) for hyperglycemia compared with macrolides were 1.75 (95% CI, 1.12–2.73) for levofloxacin, 1.87 (95% CI, 1.20–2.93) for ciprofloxacin, and 2.48 (95% CI, 1.50–4.12) for moxifloxacin. Comparable AORs for hypoglycemia compared with macrolides were 1.79 (95% CI, 1.33–2.42) for levofloxacin, 1.46 (95% CI, 1.07–2.00) for ciprofloxacin, and 2.13 (95% CI, 1.44–3.14) for moxifloxacin. Adjusted ORs for moxifloxacin plus insulin, moxifloxacin plus sulfonylureas, and levofloxacin plus sulfonylureas for hypoglycemia were 2.28 (95% CI, 1.22–4.24), 1.93 (95% CI, 1.09–3.42), and 2.03 (95% CI, 1.36–3.03), respectively. Adjusted ORs for hyperglycemia for glitinides, metformin, or thiazolidinediones plus ciprofloxacin, moxifloxacin, and levofloxacin were 3.19 (95% CI, 1.09–9.31), 4.84 (95% CI, 1.39–16.84), and 3.80 (95% CI, 1.34–10.75), respectively [387].

Studies have been conducted to evaluate the mechanism of this interaction. Altered pharmacokinetics of oral hypoglycemics do not appear to be the explanation as ciprofloxacin and moxifloxacin do not significantly alter glyburide pharmacokinetics [201, 388].

In patients with type 2 diabetes mellitus stabilized on diet and exercise therapy, multiple oral dose ciprofloxacin (500 mg twice daily for 10 days) produced no significant effects on the dynamics of the oral glucose tolerance test, fasting serum insulin and glucose profiles over 6 h after dosing on study days 1 and 10 and predose fasting insulin, glucose, and C-peptide concentrations on study days 2, 4, 6, 8, and 11 compared with placebo. The only significant drug-associated effect was a significant increase in the 0- to 6-h postdose fasting serum insulin concentrations on study day 10 [389].

Moxifloxacin has been reported not to alter serum insulin dynamics in patients with type 2 diabetes mellitus stabilized on glyburide therapy. The increases reported in serum glucose 0- to 6-h postdose AUC (mean 7%) and C_{max} (mean 6%), although statistically significant, were felt to be clinically insignificant [201].

3.4 Physicochemical Interactions

Physicochemical interactions involve physical incompatibilities between injectable quinolones and intravenous fluids and admixed medications. Studies of these types of interactions involve combinations of visual inspection (for precipitation), assessment of pH changes, and quantitation of drug and breakdown products. Table 3.4 illustrates the known incompatibilities of the injectable quinolones [390]. It should

Table 3.4 Intravenous fluid and admixed drug incompatibilities with injectable quinolones

Quinolone	Incompatibilities	
	LVP IV fluid	Admixed drugs
Ciprofloxacin	Sodium bicarbonate,[a] sodium phosphate[a]	Amoxicillin, amphotericin B, amoxicillin/clavulanate, clindamycin, floxacillin, furosemide[b], cefepime[b], ceftazidime, cefuroxime, heparin[c], metronidazole, propofol[b], hydrocortisone[b], potassium phosphates, mezlocillin[c], ampicillin/sulbactam[c], piperacillin, ticarcillin, aminophylline[c], teicoplanin, magnesium[b], dexamethasone[b], phenytoin[b], warfarin[b], methylprednisolone[b], TPN[b], pantoprazole[d], azithromycin[c], drotrecogin alfa (activated)[b], lansoprazole[b], pemetrexed[b], sodium phosphates[b], fluorouracil
Levofloxacin	Mannitol, sodium bicarbonate	Acyclovir[b], alprostadil[b], furosemide[b], heparin[b], indomethacin[b], insulin[b], nitroglycerin[b], nitroprusside[b], propofol[b], azithromycin[b], drotrecogin alfa (activated) [b], lansoprazole[b], telavancin[b]
Moxifloxacin	None reported	Not reported

Source: From Ref. [390]

LVP large-volume parenteral, *TPN* total parenteral nutition

[a]Incompatible on simulated Y-site administration as well as when used as an LVP intravenous fluid
[b]Incompatible (evaluated only on simulated Y-site administration)
[c]Incompatible on simulated Y-site administration as well as when admixed into an LVP intravenous fluid
[d]Incompatible (evaluated only in syringe)

be noted that the interaction between ciprofloxacin and hydrocortisone sodium succinate occurs even at low concentrations of the latter administered via Y-site injection [391]. A case report of an interaction between indomethacin and ciprofloxacin, both administered as eye drops following phototherapeutic keratectomy, has been published. The interaction appeared to be physicochemical in nature, as a precipitate containing both drugs was deposited in the cornea [392].

3.5   Summary

The quinolone antimicrobials have proven to be important additions to our therapeutic armamentarium based on their broad spectra of activity, favorable pharmacologic properties, and ease and cost-efficiency of administration. However, with their widespread use comes the realization that drug-drug interactions will occur with these agents. It is important that the clinician be aware of clinically significant interactions with these agents and pay attention to other potential interactions with drugs exhibiting narrow therapeutic/toxic dose ratios.

Note Added in Proof
Delafloxacin was approved by the FDA in June 2017 in oral and injectable formulations [393]. The compatibility of the injectable formulation has only been established in normal saline and 5% dextrose in water large-volume parenteral solutions. The compatibility of other intravenous admixtures (drugs, electrolytes, etc.) with delafloxacin has not been established, and coadministration is not recommended, especially with multivalent cations. Formal bioavailability interaction studies of the oral formulation with orally administered multivalent cations have not been conducted, but similar results to those with other quinolones should be expected.

Table 3.5 Clinically significant pharmacokinetic quinolone-drug interactions

Interacting drug	Results	Comments
Ca, Mg, Al-containing antacids; Ca supplements; iron or mineral preparations; sucralfate; didanosine; lanthanum; sevelamer tripotassium citrate	Reduced quinolone absorption	Avoid quinolone therapy if possible; otherwise space administrations as far apart as possible
Theophylline	Reduced theophylline metabolism	Follow levels if on antofloxacin, ciprofloxacin, nemonoxacin, norfloxacin, or prulifloxacin; watch clinical status if on other quinolones
Caffeine	Reduced caffeine metabolism	Reduce consumption of caffeinated foods/beverages, follow clinical status (see theophylline above)
Warfarin	(?) reduced warfarin metabolism	Follow INR intra- and post-quinolone therapy and adjust warfarin dose accordingly

Delafloxacin should be administered at least 2 h before or 6 or more hours after the administration of any oral multivalent cation preparation. In vitro studies did not find delafloxacin to be a substrate, inducer, or inhibitor of hepatic CYP enzymes at therapeutically relevant drug concentrations. As mentioned previously, delafloxacin did not affect the pharmacokinetics of midazolam, a CYP3A substrate, or its 1-hydroxy metabolite in healthy volunteers [210]. In vitro studies did not find delafloxacin to be a substrate, inducer, or inhibitor of hepatic or renal transporters at therapeutically relevant drug concentrations with the exceptions that it was a substrate of P-gp and BCRP (the clinical relevance of these effects is unknown). As mentioned previously, in a positive-controlled, crossover, thorough QTc study conducted in 51 healthy volunteers, 300 and 900 mg intravenous doses of delafloxacin did not significantly affect the QTc interval [364].

Acknowledgment The author gratefully acknowledges the administrative assistance of Dede Johnston.

References

1. Yuk JH, Williams TW Jr (1991) Drug interaction with quinolone antibiotics in intensive care unit patients [letter]. Arch Intern Med 151:619
2. Lomaestro BM, Lesar TS (1989) Concurrent administration of ciprofloxacin and potentially interacting drugs [letter]. Am J Hosp Pharm 46:1770
3. Bowes J, Graffunder EM, Lomaestro B, Venezia RA (2002) Concomitant administration of drugs known to decrease the systemic availability of gatifloxacin. Pharmacotherapy 22:800–801
4. Barton TD, Fishman NO, Weiner MG et al (2005) High rate of coadministration of di- or tri-valent cation-containing compounds with oral fluoroquinolones: risk factors and potential implications. Infect Cont Hosp Epidemiol 26:93–99
5. Rosas-Carrasco O, Garcia-Pena C, Sanchez-Garcia S et al (2011) The relationship between potential drug-drug interactions and mortality rate of elderly hospitalized patients. Rev Investig Clin 63:564–573
6. Hoffken G, Borner K, Glatzel PD et al (1985) Reduced enteral absorption of ciprofloxacin in the presence of antacids [letter]. Eur J Clin Microbiol 4:345
7. Shiba K, Sakai O, Shimada J, Okazaki O, Aoki H, Hakusui H (1992) Effects of antacids, ferrous sulfate, and ranitidine on absorption of DR-3355 in humans. Antimicrob Agents Chemother 36:2270–2274
8. Campbell NRC, Kara M, Hasinoff B, Haddara WM, McKay DW (1992) Norfloxacin interaction with antacids and minerals. Br J Clin Pharmacol 33:115–116
9. Okhamafe AO, Akerele JO, Chukuka CS (1991) Pharmacokinetic interactions of norfloxacin with some metallic medicinal agents. Int J Pharm 68:11–18
10. Akerele JO, Akhamafe AO (1991) Influence of co-administered metallic drugs on ofloxacin pharmacokinetics. J Antimicrob Chemother 28:87–94
11. Shiba K, Yoshida M, Sakai O et al (1996) Pharmacokinetics and clinical studies on NM441. Jap J Chemother 44(suppl 1):263–278
12. Sanchez Navarro A, Martinez Cabarga M, Dominguez-Gil Hurle A (1994) Oral absorption of ofloxacin administered together with aluminum. Antimicrob Agents Chemother 38:2510–2512

13. Martinez Cabarga M, Sanchez Navarro A, Colino Gandarillas CI, Dominguez-Gil A (1991) Effects of two cations on gastrointestinal absorption of ofloxacin. Antimicrob Agents Chemother 35:2102–2105
14. Rambout L, Sahai J, Gallicano K, Oliveras L, Garber G (1994) Effect of bismuth subsalicylate on ciprofloxacin bioavailability. Antimicrob Agents Chemother 38:2187–2190
15. Sahai J, Healy D, Stotka J, Polk R (1993) The influence of chronic administration of calcium carbonate on the bioavailability of oral ciprofloxacin. Br J Clin Pharmacol 35:302–304
16. Frost DW, Lasseter KC, Noe AJ, Shamblen EC, Lettieri J (1992) Effect of aluminum hydroxide and calcium carbonate antacids on the bioavailability of ciprofloxacin. Antimicrob Agents Chemother 36:830–832
17. Fleming LW, Moreland TA, Stewart WK, Scott AC (1986) Ciprofloxacin and antacids [letter]. Lancet 2:294
18. Lomaestro BM, Baillie GR (1991) Effect of staggered dose of calcium on the bioavailability of ciprofloxacin. Antimicrob Agents Chemother 35:1004–1007
19. Sanchez Navarro A, Martinez Cabarga M, Dominguez-Gil Hurle A (1994) Comparative study of the influence of Ca2+ on absorption parameters of ciprofloxacin and ofloxacin. J Antimicrob Chemother 34:119–125
20. Lomaestro BM, Baillie GR (1993) Effect of multiple staggered doses of calcium on the bioavailability of ciprofloxacin. Ann Pharmacother 27:1325–1328
21. Pletz MW, Petzold P, Allen A, Burkhardt O, Lode H (2003) Effect of calcium carbonate on bioavailability of orally administered gemifloxacin. Antimicrob Agents Chemother 47:2158–2160
22. Pai MP, Allen SE, Amsden GW (2006) Altered steady state pharmacokinetics of levofloxacin in adult cystic fibrosis patients receiving calcium carbonate. J Cystic Fibrosis 5:153–157
23. Lehto P, Kivisto KT (1994) Different effects of products containing metal ions on the absorption of lomefloxacin. Clin Pharmacol Ther 56:477–482
24. Stass H, Wandel C, Delesen H, Moller JG (2001) Effect of calcium supplements on the oral bioavailability of moxifloxacin in healthy male volunteers. Clin Pharmacokinet 40(suppl 1):27–32
25. Zhang YF, Dai XJ, Wang T et al (2014) Effects of an Al(3+)- and Mg(2+)-containing antacid, ferrous sulfate, and calcium carbonate on the absorption of nemonoxacin (TG-873870) in healthy Chinese volunteers. Acta Pharmacol Sin 35:1586–1592
26. Nix DE, Wilton JH, Ronald B et al (1990) Inhibition of norfloxacin absorption by antacids. Antimicrob Agents Chemother 34:432–435
27. Flor S, Guay DRP, Opsahl JA et al (1990) Effects of magnesium-aluminum hydroxide and calcium carbonate antacids on bioavailability of ofloxacin. Antimicrob Agents Chemother 34:2436–2438
28. Kays MB, Overholser BR, Mueller BA, Moe SM, Sowinski KM (2003) Effects of sevelamer hydrochloride and calcium acetate on the oral bioavailability of ciprofloxacin. Am J Kid Dis 42:1253–1259
29. Kato R, Ueno K, Imano H et al (2002) Impairment of ciprofloxacin absorption by calcium polycarbophil. J Clin Pharmacol 42:806–811
30. Neuhofel AL, Wilton JH, Victory JM, Hejmanowsky LG, Amsden GW (2002) Lack of bioequivalence of ciprofloxacin when administered with calcium-fortified orange juice: a new twist on an old interaction. J Clin Pharmacol 42:461–466
31. Amsden GW, Whitaker A-M, Johnson PW (2003) Lack of bioequivalence of levofloxacin when coadministered with a mineral-fortified breakfast of juice and cereal. J Clin Pharmacol 43:990–995
32. Sahai J, Gallicano K, Oliveros L, Khaliq S, Hawley-Foss N, Garber G (1993) Cations in the didanosine tablet reduce ciprofloxacin bioavailability. Clin Pharmacol Ther 53:292–297
33. Knupp CA, Barbhaiya RH (1997) Multiple-dose pharmacokinetic interaction study between didanosine (Videx®) and ciprofloxacin (Cipro®) in male subjects seropositive for HIV but asymptomatic. Biopharm Drug Dispos 18:65–77

34. Damle BD, Mummaneni V, Kaul S, Knupp C (2002) Lack of effect of simultaneously administered didanosine encapsulated enteric bead formulation (Videx EC) on oral absorption of indinavir, ketoconazole, or ciprofloxacin. Antimicrob Agents Chemother 46:385–391

35. Kara M, Hasinoff BB, McKay D, Campbell NRC (1991) Clinical and chemical interactions between iron preparations and ciprofloxacin. Br J Clin Pharmacol 31:257–261

36. Polk RE, Healy DP, Sahai J et al (1989) Effect of ferrous sulfate and multivitamins with zinc on absorption of ciprofloxacin in normal volunteers. Antimicrob Agents Chemother 33:1841–1844

37. Lehto P, Kivisto KT, Neuvonen PJ (1994) The effect of ferrous sulphate on the absorption of norfloxacin, ciprofloxacin and ofloxacin. Br J Clin Pharmacol 37:82–85

38. Allen A, Bygate E, Faessel H, Isaac L, Lewis A (2000) The effect of ferrous sulphate and sucralfate on the bioavailability of oral gemifloxacin in healthy volunteers. Int J Antimicrob Agents 15:283–289

39. Stass H, Kubitza D (2001) Effect of iron supplements on the oral bioavailability of moxifloxacin, a novel 8-methoxyfluoroquinolone, in humans. Clin Pharmacokinet 40(suppl 1):57–62

40. How PP, Fischer JH, Arruda JA et al (2007) Effects of lanthanum carbonate on the absorption and oral bioavailability of ciprofloxacin. Clin J Am Soc Nephrol 2:1235–1240

41. Nix DE, Watson WA, Lener ME et al (1989) Effects of aluminum and magnesium antacids and ranitidine on the absorption of ciprofloxacin. Clin Pharmacol Ther 46:700–705

42. Hoffken G, Lode H, Wiley R et al (1988) Pharmacokinetics and bioavailability of ciprofloxacin and ofloxacin: effect of food and antacid intake. Rev Infect Dis 10(suppl 1):S138–S139

43. Allen A, Vousden M, Porter A, Lewis A (1999) Effect of Maalox^R on the bioavailability of oral gemifloxacin in healthy volunteers. Chemotherapy 45:504–511

44. Shimada J, Shiba K, Oguma T et al (1992) Effect of antacid on absorption of the quinolone lomefloxacin. Antimicrob Agents Chemother 36:1219–1224

45. Stass H, Bottcher MF, Ochmann K (2001) Evaluation of the influence of antacids and H2 antagonists on the absorption of moxifloxacin after oral administration of a 400 mg dose to healthy volunteers. Clin Pharmacokinet 40(suppl 1):39–48

46. Nix DE, Wilton JH, Ronald B et al (1989) Inhibition of norfloxacin absorption by antacids and sucralfate. Rev Infect Dis 11(supp 5):S1096

47. Maesen FPV, Davies BI, Geraedts WH, Sumajow CA (1987) Ofloxacin and antacids [letter]. J Antimicrob Chemother 19:848–849

48. Jaehde U, Sorgel F, Stephan U, Schunack W (1994) Effect of an antacid containing magnesium and aluminum on absorption, metabolism, and mechanism of renal elimination of pefloxacin in humans. Antimicrob Agents Chemother 38:1129–1133

49. Nix DE, Watson WA, Handy L et al (1989) The effect of sucralfate pretreatment on the pharmacokinetics of ciprofloxacin. Pharmacotherapy 9:377–380

50. Garrelts JC, Godley PJ, Peterie JD et al (1990) Sucralfate significantly reduces ciprofloxacin concentrations in serum. Antimicrob Agents Chemother 34:931–933

51. Van Slooten AD, Nix DE, Wilton JH, Love JH, Spivey JM, Goldstein HR (1991) Combined use of ciprofloxacin and sucralfate. DICP Ann Pharmacother 25:578–582

52. Lee L-J, Hafkin B, Lee I-D, Hoh J, Dix R (1997) Effects of food and sucralfate on a single oral dose of 500 mg of levofloxacin in healthy subjects. Antimicrob Agents Chemother 41:2196–2200

53. Nix D, Schentag J (1989) Lomefloxacin (L) absorption kinetics when administered with ranitidine (R) and sucralfate (S). Proceedings of the 29th Interscience Conference on Antimicrobial Agents and Chemotherapy, Houston, TX, September 1989. Abstract 1276

54. Stass H, Schuhly U, Moller JG, Delesen H (2001) Effects of sucralfate on the oral bioavailability of moxifloxacin, a novel 8-methoxyfluoroquinolone. Clin Pharmacokinet 40(suppl 1):49–55

55. Lehto P, Kivisto KT (1994) Effect of sucralfate on absorption of norfloxacin and ofloxacin. Antimicrob Agents Chemother 38:248–251

56. Parpia SH, Nix DE, Hejmanowski LG et al (1989) Sucralfate reduces the gastrointestinal absorption of norfloxacin. Antimicrob Agents Chemother 33:99–102
57. Kawakami J, Matsuse T, Kotaki H et al (1994) The effect of food on the interaction of ofloxacin with sucralfate in healthy volunteers. Eur J Clin Pharmacol 47:67–69
58. Golper T, Hartstein AI, Morthland VH, Christensen JM (1987) Effects of antacids and dialysate dwell times on multiple dose pharmacokinetics of oral ciprofloxacin in patients on continuous ambulatory peritoneal dialysis. Antimicrob Agents Chemother 31:1787–1790
59. Preheim LC, Cuevas TA, Roccaforte JS, Mellencamp MA, Bittner MJ (1986) Ciprofloxacin and antacids [letter]. Lancet 2:48
60. Miyata K, Ohtani H, Tsujimoto M et al (2007) Antacid interaction with new quinolones: dose regimen recommendations based on pharmacokinetic modeling of clinical data for ciprofloxacin, gatifloxacin and norfloxacin and metal cations. Int J Clin Pharmacol Ther 45:63–70
61. Tuncel T, Bergisadi N (1992) In vitro adsorption of ciprofloxacin hydrochloride on various antacids. Pharmazie 47:304–305
62. Wallis SC, Charles BG, Gahan LR, Filippich LJ, Bredhauer MG, Duckworth PA (1996) Interaction of norfloxacin with divalent and trivalent pharmaceutical cations. In vitro complexation and in vivo pharmacokinetic studies in the dog. J Pharm Sci 85:803–809
63. Ross DL, Elkington SK, Knaub SR, Riley CM (1993) Physicochemical properties of the fluoroquinolone antimicrobials VI. Effect of metal-ion complexation on octan-1-ol-water partitioning. Int J Pharmaceutics 93:131–138
64. Ross DL, Riley CM (1993) Physicochemical properties of the fluoroquinolone antimicrobials V. Effect of fluoroquinolone structure and pH on the complexation of various fluoroquinolones with magnesium and calcium. Int J Pharmaceutics 93:121–129
65. Ross DL, Riley CM (1992) Physicochemical properties of the fluoroquinolone antimicrobials III. Complexation of lomefloxacin with various metal ions and the effect of metal ion complexation on aqueous solubility. Int J Pharmaceutics 87:203–213
66. Helena M, Teixeira SF, Vilas-Boas LS, Gil VMS, Teixeira F (1995) Complexes of ciprofloxacin with metal ions contained in antacid drugs. J Chemother 7:126–132
67. Seedher N, Agarwal P (2010) Effect of metal ions on some pharmacologically relevant interactions involving fluoroquinolone antibiotics. Drug Metabol Drug Interact 25:17–24
68. Adepoju-Bello AA, Coker HA, Eboka CJ et al (2008) The physiochemical and antibacterial properties of ciprofloxacin-Mg2+ complex. Nigerian Quart J Hosp Med 18:133–136
69. Stojkovic A, Tajber L, Paluch KJ et al (2014) Biopharmaceutical characterization of ciprofloxacin- metallic ion interactions: comparative study into the effect of aluminium, calcium, zinc, and iron on drug solubility and dissolution. Acta Pharma 64:77–88
70. Parojcic J, Stojkovic A, Tajber L et al (2011) Biopharmaceutical characterization of ciprofloxacin HCl-ferrous sulfate interaction. J Pharm Sci 100:5174–5184
71. Sonia Rodriguez Cruz M, Gonzalez Alonso I, Sanchez-Navarro A, Luisa Sayalero Marinero M (1999) In vitro study of the interaction between quinolones and polyvalent cations. Pharm Acta Helv 73:237–245
72. Sultana N, Arayne MS, Yasmeen N (2007) In vitro availability of ofloxacin in presence of metals essential to human body. Pak J Pharm Sci 20:42–47
73. Sultana N, Arayne MS, Furqan H (2005) In vitro availability of lomefloxacin hydrochloride in presence of essential and trace elements. Pak J Pharm Sci 18:59–65
74. Sultana N, Arayne MS, Yasmeen N (2007) In vitro availability of ofloxacin in presence of metals essential to human body. Pak J Pharm Sci 20:36–42
75. Arayne MS, Sultana N, Hussain F (2005) Interactions between ciprofloxacin and antacids—dissolution and adsorption studies. Drug Metabol Drug Interact 21:117–129
76. Muruganathan G, Nair DK, Bharathi N, Ravi TK (2011) Interaction study of moxifloxacin and lomefloxacin with co-administered drugs. Pak J Pharm Sci 24:339–343
77. Imaoka A, Hattori M, Akiyoshi T, Ohtani H (2014) Decrease in ciprofloxacin absorption by polyvalent metal cations is not fully attributable to chelation or adsorption. Drug Metab Pharmacokinet 29:414–418

78. Hosono M, Yokoyama H, Takayanagi R, Yamada Y (2013) Interactions between new quinolone antibacterials and diagnostic drug containing magnesium. Eur J Drug Metab Pharmacokinet 38:255–259

79. Hoffken G, Lode H, Wiley PD, et al (1986) Pharmacokinetics and interaction in the bioavailability of new quinolones. In: Proceedings of the International Symposium of the New Quinolones, Geneva, July 1986. Abstract 141

80. Wingender W, Foerster D, Beermann D, et al (1985) Effect of gastric emptying time on rate and extent of the systemic availability of ciprofloxacin. In: Proceedings of the 14th International Congress of Chemotherapy, Kyoto, June 1985. Abstract P-37-91

81. Sorgel F, Mahr G, Uwe Koch H, Stephan U, Wiesemann HG, Malter U (1988) Effects of cimetidine on the pharmacokinetics of pefloxacin in healthy volunteers. Rev Infect Dis 10(suppl 1):S137

82. Levofloxacin. Data on file (protocol HR 355/1/GB/101). Raritan, NJ: RW Johnson Pharmaceutical Research Institute

83. Ludwig E, Graber H, Szekely E, Csiba A (1990) Metabolic interactions of ciprofloxacin. Diagn Microbiol Infect Dis 13:135–141

84. Stass HH, Ochmann K (1997) Study to evaluate the interaction between BAY 12–8039 and ranitidine. In: Proceedings of the 20th International Congress of Chemotherapy, Sydney, June–July 1997. Abstract 3357

85. Anonymous (2016) Nemonoxacin product information. Taipei, Taiwan, TaiGen Biotechnology

86. Stuht H, Lode H, Koeppe P, Rost KL, Schaberg T (1995) Interaction study of lomefloxacin and ciprofloxacin with omeprazole and comparative pharmacokinetics. Antimicrob Agents Chemother 39:1045–1049

87. Allen A, Vousden M, Lewis A (1999) Effect of omeprazole on the pharmacokinetics of oral gemifloxacin in healthy volunteers. Chemotherapy 45:496–503

88. Washington C, Hou E, Hughes N et al (2006) Effect of omeprazole on bioavailability of an oral extended-release formulation of ciprofloxacin. Am J Health-Sys Pharm 63:653–656

89. Faruquee CF, Ullah A, Azad MA et al (2010) Interaction study between levofloxacin and omeprazole using urinary pharmacokinetic data. Pak J Pharm Sci 23:143–148

90. Pierce D, Corcoran M, Martin P et al (2014) Effect of MMX mesalamine coadministration on the pharmacokinetics of amoxicillin, ciprofloxacin XR, metronidazole, and sulfamethoxazole: results from four randomized clinical trials. Drug Design Develop Ther 8:529–543

91. Yuk JH, Nightingale CH, Sweeney KR et al (1989) Relative bioavailability in healthy volunteers of ciprofloxacin administered through a nasogastric tube with and without enteral feeding. Antimicrob Agents Chemother 33:1118–1120

92. Noer BL, Angaran DM (1990) The effect of enteral feedings on ciprofloxacin pharmacokinetics. Pharmacotherapy 10:254. Abstract 154

93. Healy DP, Brodbeck MC, Clendenning CE (1996) Ciprofloxacin absorption is impaired in patients given enteral feedings orally and via gastrostomy and jejunostomy tubes. Antimicrob Agents Chemother 40:6–10

94. Mueller BA, Brierton DG, Abel SR, Bowman L (1994) Effect of enteral feeding with Ensure on oral bioavailabilities of ofloxacin and ciprofloxacin. Antimicrob Agents Chemother 38:2101–2105

95. Dudley MN, Marchbanks CR, Flor SC, Beals S (1991) The effect of food or milk on the absorption kinetics of ofloxacin. Eur J Clin Pharmacol 41:569–571

96. Neuvonen PJ, Kivisto KT (1992) Milk and yoghurt do not impair the absorption of ofloxacin. Br J Clin Pharmacol 33:346–348

97. Hoogkamer JFW, Kleinbloesem CH (1995) The effect of milk consumption on the pharmacokinetics of fleroxacin and ciprofloxacin in healthy volunteers. Drugs 49(suppl 2):346–348

98. Stass H, Kubitza D (2001) Effects of dairy products on the oral bioavailability of moxifloxacin, a novel 8-methoxyfluoroquinolone, in healthy volunteers. Clin Pharmacokinet 40(suppl 1):33–38

99. Neuvonen PJ, Kivisto KT, Lehto P (1991) Interference of dairy products with the absorption of ciprofloxacin. Clin Pharmacol Ther 50:498–502
100. Kivisto KT, Ojala-Karlsson P, Neuvonen PJ (1992) Inhibition of norfloxacin absorption by dairy products. Antimicrob Agents Chemother 36:489–491
101. Saito A, Tarao F (1996) Influence of milk on absorption of NM441, a prodrug type of quinolone antibiotic. Jap J Chemother 44(suppl 1):221–228
102. Papai K, Budai M, Ludanyi K et al (2010) In vitro food-drug interaction study: which milk component has a decreasing effect on the bioavailability of ciprofloxacin? J Pharm Biomed Anal 52:37–42
103. Suda KJ, Garey KW, Danziger LH (2005) Treatment failures secondary to drug interactions with divalent cations and fluoroquinolone. Pharm World Sci 27:81–82
104. Quain RD, Barton TD, Fishman NO et al (2005) Coadministration of oral levofloxacin with agents that impair its absorption: potential impact on emergence of resistance. Int J Antimicrob Agents 26:327–330
105. Cohen KA, Lautenbach E, Weiner MG et al (2008) Coadministration of oral levofloxacin with agents that impair absorption: impact on antibiotic resistance. Infect Control Hosp Epidemiol 29:975–977
106. Stutman HR, Parker KM, Marks MI (1985) Potential of moxalactam and other new antimicrobial agents of bilirubin-albumin displacement in neonates. Pediatrics 75:294–298
107. Graber H, Ludwig E, Magyar T, Csiba A, Szekely E (1989) Ofloxacin does not influence antipyrine metabolism. Rev Infect Dis 11(suppl 5):S1093–S1094
108. Ludwig E, Graber H, Szekely E, Csiba A (1989) Effect of ciprofloxacin on antipyrine metabolism. Rev Infect Dis 11(suppl 5):S1100–S1101
109. Ludwig E, Szekely E, Csiba A, Graber H (1988) The effect of ciprofloxacin on antipyrine metabolism. J Antimicrob Chemother 22:61–67
110. Rybak MJ, Bowles SK, Chandrasekar PH, Edwards DJ (1987) Increased theophylline concentrations secondary to ciprofloxacin. Drug Intell Clin Pharm 21:879–881
111. Duraski RM (1988) Ciprofloxacin-induced theophylline toxicity [letter]. South Med J 81:1206
112. Paidipaty B, Erickson S (1990) Ciprofloxacin-theophylline drug interaction [letter]. Crit Care Med 18:685–686
113. Holden R (1988) Probable fatal interaction between ciprofloxacin and theophylline [letter]. Br Med J 297:1339
114. Bem JL, Mann RD (1988) Danger of interaction between ciprofloxacin and theophylline [letter]. Br Med J 296:1131
115. Spivey JM, Laughlin PH, Goss TF, Nix DE (1991) Theophylline toxicity secondary to ciprofloxacin administration. Ann Emerg Med 20:1131–1134
116. Thomson AH, Thomson GD, Hepburn M, Whiting BA (1987) A clinically significant interaction between ciprofloxacin and theophylline. Eur J Clin Pharmacol 33:435–436
117. Davis RL, Kelly HW, Quenzer RW et al (1989) Effect of norfloxacin on theophylline metabolism. Antimicrob Agents Chemother 33:212–214
118. Bowles SK, Popovski Z, Rybak MJ et al (1988) Effect of norfloxacin on theophylline pharmacokinetics at steady state. Antimicrob Agents Chemother 32:510–512
119. Schwartz J, Jauregui L, Lettieri J, Bachmann K (1988) Impact of ciprofloxacin on theophylline clearance and steady-state concentrations in serum. Antimicrob Agents Chemother 32:75–77
120. Gregoire SL, Grasela TH Jr, Freer JP et al (1987) Inhibition of theophylline clearance by coadministered ofloxacin without alteration of theophylline effects. Antimicrob Agents Chemother 31:375–378
121. Ho G, Tierney MG, Dales RE (1988) Evaluation of the effect of norfloxacin on the pharmacokinetics of theophylline. Clin Pharmacol Ther 44:35–38
122. Nix DE, Norman A, Schentag JJ (1989) Effect of lomefloxacin on theophylline pharmacokinetics. Antimicrob Agents Chemother 33:1006–1008

123. Wijnands WJA, Vree TB, van Herwaarden CLA (1986) The influence of quinolone derivatives on theophylline clearance. Br J Clin Pharmacol 22:677–683
124. Sano M, Yamamoto I, Ueda J et al (1987) Comparative pharmacokinetics of theophylline following two fluoroquinolones co-administration. Eur J Clin Pharmacol 32:431–432
125. Sano M, Kawakatsu K, Ohkita C et al (1988) Effects of enoxacin, ofloxacin and norfloxacin on theophylline disposition in humans. Eur J Clin Pharmacol 35:161–165
126. Robson RA, Begg EJ, Atkinson HC, Saunders CA, Frampton CM (1990) Comparative effects of ciprofloxacin and lomefloxacin on the oxidative metabolism of theophylline. Br J Clin Pharmacol 29:491–493
127. LeBel M, Vallee F, St. Laurent M (1990) Influence of lomefloxacin on the pharmacokinetics of theophylline. Antimicrob Agents Chemother 34:1254–1256
128. Wijnands WJA, Cornel JH, Martea M, Vree TB (1990) The effect of multiple dose oral lomefloxacin on theophylline metabolism in man. Chest 98:1440–1444
129. Niki Y, Soejima R, Kawane H et al (1987) New synthetic quinolone antibacterial agents and serum concentration of theophylline. Chest 92:663–669
130. Raoof S, Wollschlager C, Khan FA (1987) Ciprofloxacin increases serum levels of theophylline. Am J Med 82(suppl 4A):115–118
131. Gisclon LG, Curtin CR, Fowler CL, Nayak RK (1995) Absence of pharmacokinetic interaction between intravenous theophylline and orally administered levofloxacin. In: Proceedings of the 35th Interscience Conference on Antimicrobial Agents and Chemotherapy, San Francisco, CA, September 1995. Abstract A39
132. Fourtillan JB, Granier J, Saint-Salvi B et al (1986) Pharmacokinetics of ofloxacin and theophylline alone and in combination. Infection 14(suppl 1):S67–S69
133. Batty KT, Davis TME, Ilett KF, Dusci LJ, Langton SR (1995) The effect of ciprofloxacin on theophylline pharmacokinetics in healthy subjects. Br J Clin Pharmacol 39:305–311
134. Stass HH, Kubitza D, Schweitert H, Wemer R (1997) BAY 12–8039 does not interact with theophylline. In: Proceedings of the 20th International Congress of Chemotherapy, Sydney, Australia, June–July 1997. Abstract 3356
135. Stass H, Kubitza D (2001) Lack of pharmacokinetic interaction between moxifloxacin, a novel 8-methoxyfluoroquinolone, and theophylline. Clin Pharmacokinet 40(suppl 1):63–70
136. Davy M, Allen A, Bird N, Rost KL, Fuder H (1999) Lack of effect of gemifloxacin on the steady-state pharmacokinetics of theophylline in healthy volunteers. Chemotherapy 45:478–484
137. Fattore C, Cipolla G, Gatti G et al (1998) Pharmacokinetic interactions between theophylline and prulifloxacin in healthy volunteers. Clin Drug Invest 16:387–392
138. Liu L, Pan X, Liu HY et al (2011) Modulation of pharmacokinetics of theophylline by antofloxacin, a novel 8-amino-fluoroquinolone, in humans. Acta Pharmacol Sin 32:1285–1293
139. Kinzig-Schippers M, Fuhr U, Zaigler M et al (1999) Interaction of pefloxacin and enoxacin with the human cytochrome P450 enzyme CYP 1A2. Clin Pharmacol Ther 65:262–274
140. Carbo M, Segura J, de la Torre R et al (1989) Effect of quinolones on caffeine disposition. Clin Pharmacol Ther 45:234–240
141. Healy DP, Polk RE, Kanawati L et al (1989) Interaction between oral ciprofloxacin and caffeine in normal volunteers. Antimicrob Agents Chemother 33:474–478
142. Mahr G, Sorgel F, Granneman R et al (1992) Effects of temafloxacin and ciprofloxacin on the pharmacokinetics of caffeine. Clin Pharmacokinet 22(suppl 1):90–97
143. Harder S, Staib AH, Beer C et al (1988) 4-quinolones inhibit biotransformation of caffeine. Eur J Clin Pharmacol 35:651–656
144. Stille W, Harder S, Mieke S et al (1987) Decrease of caffeine elimination in man during coadministration of 4-quinolones. J Antimicrob Chemother 20:729–734
145. Healy DP, Schoenle JR, Stotka J, Polk RE (1991) Lack of interaction between lomefloxacin and caffeine in normal volunteers. Antimicrob Agents Chemother 35:660–664
146. Agrawal B, Chandra P, Goyal RN, Shim YB (2013) Detection of norfloxacin and monitoring its effect on caffeine catabolism in urine samples. Biosens Bioelectron 47:307–312

147. Parent M, LeBel M (1991) Meta-analysis of quinolone-theophylline interactions. DICP Ann Pharmacother 25:191–194
148. Barnett G, Segura J, de la Torre R, Carbo M (1990) Pharmacokinetic determination of relative potency of quinolone inhibition of caffeine disposition. Eur J Clin Pharmcol 39:63–69
149. Antoniou T, Gomes T, Mamdani MM, Juurlink DN (2011) Ciprofloxacin-induced theophylline toxicity: a population-based study. Eur J Clin Pharmacol 67:521–526
150. Rindone JP, Keuey CL, Jones WN, Garewal HS (1991) Hypoprothrombinemic effect of warfarin is not influenced by ciprofloxacin. Clin Pharm 10:136–138
151. Rocci ML Jr, Vlasses PH, Dislerath LM et al (1990) Norfloxacin does not alter warfarin's disposition or anticoagulant effect. J Clin Pharmacol 30:728–732
152. Bianco TM, Bussey HI, Farnett LE, Linn WD, Roush MK, Wong YWJ (1992) Potential warfarin–ciprofloxacin interaction in patients receiving long-term anticoagulation. Pharmacotherapy 12:435–439
153. Israel DS, Stotka JL, Rock W et al (1996) Effect of ciprofloxacin on the pharmacokinetics and pharmacodynamics of warfarin. Clin Infect Dis 22:251–256
154. Levofloxacin. Data on file (protocol LOFBO-PH10–098). Raritan, NJ: RW Johnson Pharmaceutical Research Institute
155. Muller FO, Hundt HKL, Muir AR, et al (1998) Study to investigate the influence of 400 mg BAY 12–8039 (M) given once daily to healthy volunteers on PK and PD of warfarin (W). In: Proceedings of the 38th Interscience Conference on Antimicrobial Agents and Chemotherapy, San Diego, CA, September 1998. Abstract A-13
156. Davy M, Bird N, Rost KL, Fuder H (1999) Lack of effect of gemifloxacin on the steady-state pharmacodynamics of warfarin in healthy volunteers. Chemotherapy 45:491–495
157. Washington C, Hou SY, Hughes NC et al (2007) Ciprofloxacin prolonged-release tablets do not affect warfarin pharmacokinetics and pharmacodynamics. J Clin Pharmacol 47:1320–1326
158. Kotsiou A, Diamanti E, Potamianou A et al (2008) Anticoagulant-induced changes on antibiotic concentrations in the serum and bones. Eur J Drug Metab Pharmacokinet 33:173–179
159. Jolson HM, Tanner LA, Green L, Grasela TH Jr (1991) Adverse reaction reporting of interaction between warfarin and fluoroquinolones. Arch Intern Med 151:1003–1004
160. Kamada A (1990) Possible interaction between ciprofloxacin and warfarin. DICP Ann Pharmacother 24:27–28
161. Leor J, Matetzki S (1988) Ofloxacin and warfarin [letter]. Ann Intern Med 109:761
162. Linville D, Emory C, Graves L (1991) Ciprofloxacin and warfarin interaction [letter]. Am J Med 90:765
163. Linnville T, Matanin D (1989) Norfloxacin and warfarin [letter]. Ann Intern Med 110:751–752
164. Mott FE, Murphy S, Hunt V (1989) Ciprofloxacin and warfarin [letter]. Ann Intern Med 111:542–543
165. Dugoni-Kramer BM (1991) Ciprofloxacin–warfarin interaction [letter]. DICP Ann Pharmacother 25:1397
166. Renzi R, Finkbeiner S (1991) Ciprofloxacin interaction with sodium warfarin. Am J Emerg Med 9:551–552
167. Ravnan SL, Locke C (2001) Levofloxacin and warfarin interaction. Pharmacotherapy 21:884–885
168. Ellis RJ, Mayo MS, Bodensteiner DM (2000) Ciprofloxacin–warfarin coagulopathy: a case series. Am J Hematol 63:28–31
169. Jones CB, Fugate SE (2002) Levofloxacin and warfarin interaction. Ann Pharmacother 36:1554–1557
170. Byrd DC, Gaskins SE, Parrish AM, Freeman LB (1999) Warfarin and ciprofloxacin interaction: case report and controversy. J Am Board Fam Pract 12:486–488
171. Anonymous (2004) Fluoroquinolones and warfarin: suspected interactions. Can Fam Physician 50:1417
172. Vadlamudi RS, Smalligan RD, Ismail HM (2007) Interaction between warfarin and levofloxacin: case series. South Med J 100:720–724

173. Chao CM, Lin SH, Lai CC (2013) Abdominal wall hematoma and hemoperitoneum in an individual with concomitant use of warfarin and moxifloxacin. J Am Geriatr Soc 61:1432–1433
174. Nemoto C, Ikegami Y, Shimada J et al (2012) Acute renal failure caused by severe coagulopathy induced by the interaction between warfarin potassium and levofloxacin: a case report. J Anesth 26:943–944
175. Lee R, Wen A, Berube C (2011) Moxifloxacin-acetaminophen-warfarin interaction during bacille Calmette-Guerin treatment for bladder cancer. Am J Health-Sys Pharm 68:814–817
176. Yew KL, Lee WC (2012) Moxifloxacin-warfarin interaction. Med J Malaysia 67:420–421
177. Schelleman H, Bilker WB, Brensinger CM et al (2008) Warfarin with fluoroquinolones, sulfonamides, or azole antifungals: interactions and the risk of hospitalization for gastrointestinal bleeding. Clin Pharmacol Ther 84:581–588
178. Glasheen JJ, Fugit RV, Prochazka AV (2005) The risk of overanticoagulation with antibiotic use in outpatients on stable warfarin regimens. J Gen Intern Med 20:653–656
179. McCall KL, Scott JC, Anderson HG (2005) Retrospective evaluation of a possible interaction between warfarin and levofloxacin. Pharmacotherapy 25:67–73
180. Mercadal Orfila G, Gracia Garcia B, Leiva Badosa E et al (2009) Retrospective assessment of potential interaction between levofloxacin and warfarin. Pharm World Sci 31:224–229
181. Stroud LF, Mamdami MM, Kopp A et al (2005) The safety of levofloxacin in elderly patients on warfarin. Am J Med 118:1417
182. Fischer HD, Juurlink DN, Mamdani MM et al (2010) Hemorrhage during warfarin therapy associated with cotrimoxazole and other urinary tract anti-infective agents. A population-based study. Arch Intern Med 170:617–621
183. Ahmed A, Stephens JC, Kauc CA et al (2008) Impact of preemptive warfarin dose reduction on anticoagulation after initiation of trimethoprim-sulfamethoxazole or levofloxacin. J Thromb Thrombol 26:44–48
184. Avent CK, Krinsky D, Kirklin JK et al (1988) Synergistic nephrotoxicity due to ciprofloxacin and cyclosporine. Am J Med 85:452–453
185. Elston RA, Taylor J (1988) Possible interaction of ciprofloxacin with cyclosporin A [letter]. J Antimicrob Chemother 21:679–680
186. Thomson DJ, Menkis AH, McKenzie FN (1988) Norfloxacin–cyclosporine interaction. Transplantation 46:312–313
187. Nasir M, Rotellar C, Hand M, Kulczycki L, Alijani MR, Winchester JF (1991) Interaction between ciclosporin and ciprofloxacin [letter]. Nephron 57:245–246
188. McLellan R, Drobitch RK, McLellan H, Acott PD, Crocker JFS, Renton KW (1995) Norfloxacin interferes with cyclosporin disposition in pediatric patients undergoing renal transplantation. Clin Pharmacol Ther 58:322–327
189. Federico S, Carrano R, Capone D et al (2006) Pharmacokinetic interaction between levofloxacin and ciclosporin or tacrolimus in kidney transplant recipients: ciclosporin, tacrolimus and levofloxacin in renal transplantation. Clin Pharmacokinet 45:169–175
190. Kruger HU, Schuler U, Proksch B et al (1990) Investigations of potential interaction between ciprofloxacin and cyclosporin A in patients with renal transplants. Antimicrob Agents Chemother 34:1048–1052
191. Lang J, de Villaine FJ, Guemi A et al (1989) Absence of pharmacokinetic interaction between pefloxacin and cyclosporin A in patients with renal transplants. Rev Infect Dis 11(suppl 5): S1094
192. Lang J, de Villaine FJ, Garraffo R, Touraine J-L (1989) Cyclosporine (cyclosporin A) pharmacokinetics in renal transplant patients receiving ciprofloxacin. Am J Med 87(suppl 5A): 82S–85S
193. Pichard L, Fabre I, Fabre G et al (1990) Screening for inducers and inhibitors of cytochrome P-450 (cyclosporin A oxidase) in primary cultures of human hepatocytes and in liver microsomes. Drug Metab Dispos 18:595–606
194. Robinson JA, Venezio FR, Costanzo-Nordin MR et al (1990) Patients receiving quinolones and cyclosporin after heart transplantation. J Heart Transplant 9:30–31

195. Tan KKC, Trull AK, Shawket S (1989) Co-administration of ciprofloxacin and cyclosporin: lack of evidence for a pharmacokinetic interaction. Br J Clin Pharmacol 28:185–187

196. Hooper TL, Gould FK, Swinburn CR et al (1988) Ciprofloxacin: a preferred treatment for legionella infections in patients receiving cyclosporin A [letter]. J Antimicrob Chemother 22:952–953

197. Van Buren DH, Koestner J, Adedoyin A et al (1990) Effect of ciprofloxacin on cyclosporine pharmacokinetics. Transplantation 50:888–889

198. Levofloxacin. Data on file (protocol N93–059). Raritan, NJ: RW Johnson Pharmaceutical Research Institute

199. Stass H, Delesen H, Kubitza D et al (2010) Moxifloxacin does not alter ciclosporin pharmacokinetics in transplant patients: a multiple-dose, uncontrolled, single-centre study. Clin Drug Invest 30:279–287

200. Capone D, Carrano R, Gentile A et al (2001) Pharmacokinetic interaction between tacrolimus and levofloxacin in kidney transplant recipients [abstract]. Nephrol Dial Transpl 16:A207

201. Stass H, Kubitza D (2001) Profile of moxifloxacin drug interactions. Clin Infect Dis 32(suppl 1):S47–S50

202. Vouden M, Allen A, Lewis A, Ehren N (1999) Lack of pharmacokinetic interaction between gemifloxacin and digoxin in healthy elderly volunteers. Chemotherapy 45:485–490

203. Chien S-C, Rogge MC, Williams RR, Natarajan J, Wong F, Chow AT (2002) Absence of a pharmacokinetic interaction between digoxin and levofloxacin. J Clin Pharm Ther 27:7–12

204. Moffett BS, Valdes SO, Kim JJ (2013) Possible digoxin toxicity associated with concomitant ciprofloxacin therapy. Int J Clin Pharm 35:673–676

205. Morran C, McArdle C, Pettitt L et al (1989) Pharmacokinetics of orally administered ciprofloxacin in abdominal surgery. Am J Med 87(suppl 5A):86S–88S

206. Grant EM, Zhong MK, Fitzgerald JF, Nicolau DP, Nightingale C, Quintiliani R (2001) Lack of interaction between levofloxacin and oxycodone: pharmacokinetics and drug disposition. J Clin Pharmacol 41:206–209

207. Labbe L, Robitaille NM, Lefez C et al (2004) Effects of ciprofloxacin on the stereoselective disposition of mexiletine in man. Ther Drug Monit 26:492–498

208. Kamali F, Thomas SHL, Edwards C (1993) The influence of steady-state ciprofloxacin on the pharmacokinetics and pharmacodynamics of a single dose of diazepam in healthy volunteers. Eur J Clin Pharmacol 44:365–367

209. Wijnands WJA, Trooster JFG, Teunissen PC et al (1990) Ciprofloxacin does not impair the elimination of diazepam in humans. Drug Metab Dispos 18:954–957

210. Paulson SK, Wood-Horrall RN, Hoover R et al (2017) The pharmacokinetics of the CYP3A substrate midazolam after steady-state dosing of delafloxacin. Clin Ther 39:1182–1190

211. Waite NM, Rybak MJ, Krakovsky DJ et al (1991) Influence of subject age on the inhibition of oxidative metabolism by ciprofloxacin. Antimicrob Agents Chemother 35:130–134

212. Loi C-M, Parker BM, Cusack BJ, Vestal RE (1997) Aging and drug interactions. III. Individual and combined effects of cimetidine and ciprofloxacin on theophylline metabolism in healthy male and female nonsmokers. J Pharmacol Exp Ther 280:627–637

213. Chandler MHH, Toler SM, Rapp RP et al (1990) Multiple-dose pharmacokinetics of concurrent oral ciprofloxacin and rifampin therapy in elderly patients. Antimicrob Agents Chemother 34:442–447

214. Murillo O, Pachón ME, Euba G et al (2008) Antagonistic effect of rifampin on the efficacy of high-dose levofloxacin in staphylococcal experimental foreign-body infection. Antimicrob Agents Chemother 52:3681–3686

215. Orisakwe OE, Agbasi PU, Afonne OJ, Ofeofule SI, Obi E, Orish CN (2001) Rifampicin pharmacokinetics with and without ciprofloxacin. Am J Ther 8:151–153

216. Orisakwe OE, Afonne OJ, Agbasi PU, Ofoefule SI (2004) Urinary excretion of rifampicin in the presence of ciprofloxacin. Am J Ther 11:171–174

217. Orisakwe OE, Akunyili DN, Agbasi PU, Ezejiofor NA (2004) Some plasma and saliva pharmacokinetics parameters of rifampicin in the presence of pefloxacin. Am J Ther 11:283–287

218. Orisakwe OE, Agbasi PU, Ofoefule SI et al (2004) Effect of pefloxacin on the urinary excretion of rifampicin. Am J Ther 11:13–16
219. Ezejiofor NA, Brown S, Barikpoar E, Orisakwe OE (2015) Effect of ofloxacin and norfloxacin on rifampicin pharmacokinetics in man. Am J Ther 22:29–36
220. Weiner M, Burman W, Luo C-C et al (2007) Effects of rifampin and multidrug resistance gene polymorphism on concentrations of moxifloxacin. Antimicrob Agents Chemother 51:2861–2866
221. Nijland HMJ, Ruslami R, Juwon Suroto A et al (2007) Rifampin reduces plasma concentrations of moxifloxacin in patients with tuberculosis. Clin Infect Dis 45:1001–1007
222. Pranger AD, van Altena R, Aarnoutsee RE et al (2011) Evaluation of moxifloxacin for the treatment of tuberculosis: 3 years of experience. Eur Resp J 38:888–894
223. Ofoefule SI, Obodo CE, Orisakwe OE et al (2001) Some plasma pharmacokinetic parameters of isoniazid in the presence of a fluoroquinolone antibacterial agent. Am J Ther 8:243–246
224. Ofoefule SI, Obodo CE, Orisakwe OE et al (2002) Salivary and urinary excretion and plasma-saliva concentration ratios of isoniazid in the presence of co-administered ciprofloxacin. Am J Ther 9:15–18
225. Sriwiriyajan S, Samaeng M, Ridtitid W, Mahatthanatrakul W, Wongnawa M (2011) Pharmacokinetic interactions between ciprofloxacin and itraconazole in healthy male volunteers. Biopharm Drug Dispos 32:168–174
226. Issa MM, Nejem RN, El-Abadia NS (2006) Oral ciprofloxacin affects the pharmacokinetics of paracetamol in saliva. Clin Drug Invest 26:223–226
227. Issa MM, Nejem RM, El-Abadia NS et al (2007) Effects of paracetamol on the pharmacokinetics of ciprofloxacin in plasma using a microbiological assay. Clin Drug Invest 27:463–467
228. Gauhar S, Ali SA, Naqvi SB, Shoaib MH (2014) Report: pharmacokinetic and drug interaction studies of pefloxacin with paracetamol (NNAID) in healthy volunteers in Pakistan. Pak J Pharm Sci 27:389–395
229. Ilo CE, Ezejiofor NA, Agbakoba N et al (2008) Effect of chloroquine on the urinary excretion of ciprofloxacin. Am J Ther 15:419–422
230. Ilo CE, Ilondu NA, Okwoli N et al (2006) Effect of chloroquine on the bioavailability of ciprofloxacin in humans. Am J Ther 13:432–435
231. Marcelin-Jimenez G, Angeles AP, Martinez-Rossier L et al (2006) Ciprofloxacin bioavailability is enhanced by oral co-administation with phenazopyridine: a pharmacokinetic study in a Mexican population. Clin Drug Invest 26:323–328
232. Hedaya MA, El-Afify DR, El-Maghraby GM (2006) The effect of ciprofloxacin and clarithromycin on sildenafil oral bioavailability in human volunteers. Biopharm Drug Dispos 27:103–110
233. Cooper JG, Harboe K, Frost SK et al (2005) Ciprofloxacin interacts with thyroid replacement therapy. BMJ 330:1002
234. Goldberg AS, Tirona RG, Asher LJ, Kim RB, Van Uum SH (2013) Ciprofloxacin and rifampin have opposite effects on levothyroxine absorption. Thyroid 23:1374–1378
235. Momo K, Homma M, Kohda Y et al (2006) Drug interaction of tizanidine and ciprofloxacin: case report. Clin Pharmacol Ther 80:717–719. (letter)
236. Granfors MT, Backman JT, Neuvonen M et al (2004) Ciprofloxacin greatly increases concentrations and hypotensive effects of tizanidine by inhibiting its cytochrome P450 1A2-mediated presystemic metabolism. Clin Pharmacol Ther 76:598–606
237. Thompson JA, Bianco JA, Benyunes MC et al (1994) Phase Ib trial of pentoxyfylline and ciprofloxacin in patients treated with interleukin-2 and lymphokine-activated killer cell therapy for metastatic renal cell carcinoma. Cancer Res 54:3436–3441
238. Cleary JD (1992) Ciprofloxacin (CIPRO) and pentoxifylline (PTF): a clinically significant drug interaction. Pharmacotherapy 12:259. (abstract 106)
239. Raoul JM, Peterson MR, Peterson TCA (2007) Novel drug interaction between the quinolone antibiotic ciprofloxacin and a chiral metabolite of pentoxifylline. Biochem Pharmacol 74:639–646

240. Peterson TC, Peterson MR, Wornell PA et al (2004) Role of CYP1A2 and CYP2E1 in the pentoxifylline ciprofloxacin drug interaction. Biochem Pharmacol 68:395–402
241. Markowitz JS, Gill HS, DeVane CL, Mintzer JE (1997) Fluoroquinolone inhibition of clozapine metabolism [letter]. Am J Psychiatry 153:881
242. Markowitz JS, DeVane CL (1999) Suspected ciprofloxacin inhibition of olanzapine resulting in increased plasma concentration [letter]. J Clin Psychopharmacol 19:289–291
243. Herrlin K, Segerdahl M, Gustafsson LL, Kalso E (2000) Methadone, ciprofloxacin, and adverse drug reactions [letter]. Lancet 356:2069–2070
244. Brownlowe K, Sola C (2007) Clozapine toxicity in smoking cessation and with ciprofloxacin [letter]. Psychosomatics 48:170–175
245. Sambhi RS, Puri R, Jones G (2007) Interaction of clozapine and ciprofloxacin: a case report [letter]. Eur J Clin Pharmacol 63:895–896
246. Brouwers EE, Sohne M, Kuipers S et al (2009) Ciprofloxacin strongly inhibits clozapine metabolism: two case reports. Clin Drug Invest 29:59–63
247. van Zuilekorn S, Gijesman HJ (2013) A patient on clozapine in the general hospital: the need for discussion between specialists from different disciplines and for close monitoring of the patient's plasma level. Tidschrift voor Psychiatrie 55:955–959
248. Raaska K, Neuvonen PJ (2000) Ciprofloxacin increases serum clozapine and N-desmethylclozapine: a study in patients with schizophrenia. Eur J Clin Pharmacol 56:585–589
249. Ridout KK, Ridout SJ, Pirnie LF, Puttichanda SP (2015) Sudden-onset dystonia in a patient taking asenapine: interaction between ciprofloxacin and asenapine metabolism. Am J Psychiatr 172:1162–1163
250. Dalle J-H, Auvrignon A, Vassal G, Leverger G (2002) Interaction between methotrexate and ciprofloxacin. J Pediatr Hematol Oncol 24:321–322
251. Aouinti I, Gaies E, Trabelsi S et al (2013) Delayed elimination of methotrexate in a patient receiving ciprofloxacin. Therapie 68:175–177
252. Jarfaut A, Santucci R, Leveque D, Herbrecht R (2013) Severe methotrexate toxicity due to a concomitant administration of ciprofloxacin. Med Malad Infect 43:39–41
253. Liu Y, He Q, Wu M (2015) Levofloxacin-induced crystal nephropathy. Nephrology 20:437–438
254. Kammoun K, Jarraya F, Makni S et al (2014) Ciprofloxacin-induced crystal nephropathy. Iran J Kid Dis 8:240–242
255. Kamangar F, Berger TG, Fazel N, Koo JY (2013) Methotrexate toxicity induced by ciprofloxacin leading to psoriatic plaque ulceration: a case report. Cutis 92:148–150
256. Takahashi H, Higuchi H, Shimizu T (2000) Severe lithium toxicity induced by combined levofloxacin administration [letter]. J Clin Psychiatry 61:949–950
257. Scholten PC, Droppert RM, Zwinkels MGJ, Moesker HL, Nauta JJP, Hoepelman IM (1998) No interaction between ciprofloxacin and an oral contraceptive. Antimicrob Agents Chemother 42:3266–3268
258. Chien SC, Chow AT, Rogge MC, Williams RR, Hendrix CW (1997) Pharmacokinetics and safety of oral levofloxacin in human immunodeficiency virus-infected individuals receiving concomitant zidovudine. Antimicrob Agents Chemother 41:1765–1769
259. Villani P, Viale P, Signorini L et al (2001) Pharmacokinetic evaluation of oral levofloxacin in human immmunodeficiency virus-infected subjects receiving concomitant antiretroviral therapy. Antimicrob Agents Chemother 45:2160–2162
260. Kamali F (1994) No influence of ciprofloxacin on ethanol disposition. Eur J Clin Pharmacol 47:71–74
261. Tillonen J, Homann N, Rautio M, Jousimies-Somer H, Salaspuro M (1999) Ciprofloxacin decreases the rate of ethanol elimination in humans. Gut 44:347–352
262. Jokinen MJ, Olkkola KT, Ahonen J, Neuvonen PJ (2003) Effect of ciprofloxacin on the pharmacokinetics of ropivacaine. Eur J Clin Pharmacol 58:653–657
263. Isohanni MH, Ahonen J, Neuvonen PJ et al (2005) Effect of ciprofloxacin on the pharmacokinetics of intravenous lidocaine. Eur J Anaesthesiol 22:795–799

264. Anonymous (1998) Risk of serious seizures from concomitant use of ciprofloxacin and phenytoin in patients with epilepsy. CMAJ 158:104–105
265. Hull RL (1993) Possible phenytoin–ciprofloxacin interaction [letter]. Ann Pharmacother 27:1283
266. Pollak PT, Slayter KL (1997) Hazards of doubling phenytoin dose in the face of an unrecognized interaction with ciprofloxacin. Ann Pharmacother 31:61–64
267. Dillard ML, Fink RM, Parkerson R (1992) Ciprofloxacin–phenytoin interaction [letter]. Ann Pharmacother 26:263
268. Brouwers PJ, DeBoer LE, Guchelaar H-J (1997) Ciprofloxacin–phenytoin interaction [letter]. Ann Pharmacother 31:498
269. Otero M-J, Moran D, Valverde M-P, Dominguez-Gil A (1999) Interaction between phenytoin and ciprofloxacin [letter]. Ann Pharmacother 33:251–252
270. Malladi SS, Liew EK, Ng XT, Tan RK (2014) Ciprofloxacin eye drops-induced subtherapeutic serum phenytoin levels resulting in breakthrough seizures. Sing Med J 55:e114–e115
271. Job ML, Arn SK, Strom JG, Jacobs NF, D'Souza MJ (1994) Effect of ciprofloxacin on the pharmacokinetics of multiple-dose phenytoin serum concentrations. Ther Drug Monitor 16:427–431
272. Shahzadi A, Javed I, Aslam B et al (2011) Therapeutic effects of ciprofloxacin on the pharmacokinetics of carbamazepine in healthy adult male volunteers. Pak J Pharm Sci 24:63–68
273. Schmith VD, Johnson BM, Vasist LS et al (2010) The effects of a short course of antibiotics on alvimopan and metabolite pharmacokinetics. J Clin Pharmacol 50:338–349
274. Vlase L, Popa A, Neag M, Muntean D, Leucuta SE (2011) Pharmacokinetic interaction between zolpidem and ciprofloxacin in healthy volunteers. Eur J Drug Metab Pharmacokinet 35:83–87
275. Davis RL, Quenzer RW, Kelly HW, Powell JR (1992) Effect of the addition of ciprofloxacin on theophylline pharmacokinetics in subjects inhibited by cimetidine. Ann Pharmcother 26:11–13
276. Gillum JG, Israel DS, Scott RB, Climo MW, Polk RE (1996) Effect of combination therapy with ciprofloxacin and clarithromycin on theophylline pharmacokinetics in healthy volunteers. Antimicrob Agents Chemother 40:1715–1716
277. Loi C-M, Parker BM, Cusack BJ, Vestal RE (1993) Individual and combined effects of cimetidine and ciprofloxacin on theophylline metabolism in male nonsmokers. Br J Clin Pharmacol 36:195–200
278. Pavithra BH, Prakash N, Jayakumar K (2009) Modification of pharmacokinetics of norfloxacin following oral administration of curcumin in rabbits. J Vet Sci 10:293–297
279. Hasegawa T, Nadai M, Kuzuya T et al (1990) The possible mechanism of interaction between xanthines and quinolone. J Pharm Pharmacol 42:767–772
280. Fuhr U, Strobl G, Manaut F et al (1993) Quinolone antibacterial agents: relationship between structure and in vitro inhibition of the human cytochrome P-450 isoform CYP1A2. Mol Pharmacol 43:191–199
281. Fuhr U, Anders E-M, Mahr G, Sorgel F, Staib AH (1992) Inhibitory potency of quinolone antibacterial agents against cytochrome P-450 1A2 activity in vivo and in vitro. Antimicrob Agents Chemother 36:942–948
282. Sarkar M, Polk RE, Guzelian PS, Hunt C, Karnes HT (1990) In vitro effect of fluoroquinolones on theophylline metabolism in human liver microsomes. Antimicrob Agents Chemother 34:594–599
283. Zhang H, Wei M-J, Zhao C-Y et al (2008) Determination of the inhibiting potential of 6 fluoroquinolones on CYP1A2 and CYP2C9 in human liver microsones. Acta Pharmacol Sin 29:1507–1514
284. Wingender W, Beerman D, Foerster D et al (1985) Mechanism of renal excretion of ciprofloxacin, a new quinolone carboxylic acid derivative in humans. Chemioterapia 4(suppl 2):403–404
285. Levofloxacin. Data on file (protocol HR355/1/GB/101). Raritan, NJ: RW Johnson Pharmaceutical Research Institute

286. Shimada J, Yamaji T, Ueda Y, Uchida H, Kusajima H, Irikura T (1983) Mechanism of renal excretion of AM-715, a new quinolone carboxylic acid derivative, in rabbits, dogs, and humans. Antimicrob Agents Chemother 23:1–7

287. Jaehde U, Sorgel F, Reiter A, Sigl G, Naber KG, Schunack W (1995) Effect of probenecid on the distribution and elimination of ciprofloxacin in humans. Clin Pharmacol Ther 58:532–541

288. Totsuka K, Kikuchi K, Shimizu K (1996) Pharmacokinetics in concomitant administration with probenecid and clinical study of NM441. Jap J Chemother 44(suppl 1):279–288

289. Landersdorfer CB, Kirkpatrick CM, Kinzig M et al (2010) Competitive inhibition of renal tubular secretion of ciprofloxacin and metabolite by probenecid. Br J Clin Pharmacol 69:167–178

290. Stass H, Sachse R (2001) Effect of probenecid on the kinetics of a single oral 400 mg dose of moxifloxacin in healthy volunteers. Clin Pharmacokinet 40(suppl 1):71–76

291. Sudoh T, Fujimura A, Shiga T et al (1994) Renal clearance of lomefloxacin is decreased by furosemide. Eur J Clin Pharmacol 46:267–269

292. Sudoh T, Fujimura A, Harada K, Sunaga K, Ohmori M, Sakamoto K (1996) Effect of ranitidine on renal clearance of lomefloxacin. Eur J Clin Pharmacol 51:95–98

293. Mulgaonkar A, Venitz J, Sweet DH (2012) Fluoroquinolone disposition: identification of the contribution of renal secretory and reabsorptive drug transporters. Expert Opin Drug Metab Toxicol 8:553–569

294. Okuda M, Kimura N, Inui K (2006) Interactions of fluoroquinolone antibacterials, DX-619 and levofloxacin, with creatinine transport by renal organic cation transporter hOCT2. Drug Metab Pharmacokinet 21:432–436

295. Mulgaonkar A, Venitz J, Grundemann D, Sweet DH (2013) Human organic cation transporters 1 (SLC22A1), 2 (SLC22A2), and 3 (SLC22A3) as disposition pathways for fluoroquinolone antimicrobials. Antimicrob Agents Chemother 57:2705–2711

296. Hu JH, Liu XD, Xie L et al (2007) Possible multiple transporters were involved in hepatobiliary excretion of antofloxacin in rats. Xenobiotica 37:579–591

297. Martin DE, Shen J, Griener J, Raasch R, Patterson JH, Cascio W (1996) Effects of ofloxacin on the pharmacokinetics and pharmacodynamics of procainamide. J Clin Pharmacol 36:85–91

298. Bauer LA, Black DJ, Lill JS et al (2005) Levofloxacin and ciprofloxacin decrease procainamide and N-acetylprocainamide renal clearances. Antimicrob Agents Chemother 49:1649–1651

299. Christ W (1990) Central nervous system toxicity of quinolones: human and animal findings. J Antimicrob Chemother 26(suppl B):219–225

300. Anastasio GD, Menscer D, Little JM Jr (1988) Norfloxacin and seizures [letter]. Ann Intern Med 109:169–170

301. Lucet J-C, Tilly H, Lerebours G, Gres J-J, Piguet H (1988) Neurological toxicity related to pefloxacin [letter]. J Antimicrob Chemother 21:811–812

302. Rollof J, Vinge E (1993) Neurologic adverse effects during concomitant treatment with ciprofloxacin, NSAIDs, and chloroquine: possible drug interaction. Ann Pharmacother 27:1058–1059

303. Slavich IL, Gleffe RF, Haas EJ (1989) Grand mal epileptic seizures during ciprofloxacin therapy [letter]. JAMA 261:558–559

304. Rumsey S, Wilkinson TJ, Scott SD (1995) Ciprofloxacin-induced seizures—the need for increased vigilance. Aust J Hosp Pharm 25:145–147

305. Yamamoto K, Naitoh Y, Inoue Y et al (1988) Seizure discharges induced by the combination of new quinolone carboxylic acid drugs and non-steroidal anti-inflammatory drugs. Chemotherapy 36(suppl 2):300–324

306. Bellon A, Perez-Garcia G, Coverdale JH et al (2009) Seizures associated with levofloxacin: case presentation and literature review. Eur J Clin Pharmacol 65:959–962

307. Agbaht K, Bitik B, Piskinpasa S et al (2009) Ciprofloxacin-associated seizures in a patient with underlying thyrotoxicosis: case report and literature review. Int J Clin Pharmacol Ther 47:303–310

308. Cone C, Horowitz B (2015) Convulsions associated with moxifloxacin. Am J Health-Syst Pharm 72:910, 912
309. Famularo G, Pizzicannella M, Gasbarrone L (2014) Levofloxacin and seizures: what risk for elderly adults? J Am Geriatr Soc 62:2018–2019
310. Gervasoni C, Cattaneo D, Falvella FS et al (2013) Levofloxacin-induced seizures in a patient without predisposing risk factors: the impact of pharmacogenetics. Eur J Clin Pharmacol 69:1611–1613
311. Mazzei D, Accardo J, Ferrari A, Primavera A (2012) Levofloxacin neurotoxicity and non-convulsive status epilepticus (NCSE): a case report. Clin Neurol Neurosurg 114:1371–1373
312. Matsuno K, Kunihiro E, Yamatoya O et al (1995) Surveillance of adverse reactions due to ciprofloxacin in Japan. Drugs 49(suppl 2):495–496
313. Naora K, Katagiri Y, Ichikawa N, Hayashibara M, Iwamoto K (1991) Enhanced entry of ciprofloxacin into the rat central nervous system induced by fenbufen. J Pharmacol Exp Ther 258:1033–1037
314. Ichikawa N, Naora K, Hayashibara M, Katagiri Y, Iwamoto K (1992) Effect of fenbufen on the entry of new quinolones, norfloxacin and ofloxacin, into the central nervous system in rats. J Pharm Pharmacol 44:915–920
315. Naora K, Katagiri Y, Ichikawa N, Hayashibara M, Iwamoto KA (1990) A possible reduction in the renal clearance of ciprofloxacin by fenbufen in rats. J Pharm Pharmacol 42:704–707
316. Katagiri Y, Naora K, Ichikawa N, Hayashibara M, Iwamoto K (1989) Absence of pharmaco-kinetic interaction between ofloxacin and fenbufen in rats. J Pharm Pharmacol 41:717–719
317. Kamali F (1994) Lack of a pharmacokinetic interaction between ciprofloxacin and fenbufen. J Clin Pharm Ther 19:257–259
318. Fillastre JP, Leroy A, Borsa-Lebas F, Etienne I, Gy C, Humbert G (1993) Lack of effect of ketoprofen on the pharmacokinetics of pefloxacin and ofloxacin [letter]. J Antimicrob Chemother 31:805–806
319. Fillastre JP, Leroy A, Borsa-Lebas F, Etienne I, Gy C, Humbert G (1992) Effects of keto-profen (NSAID) on the pharmacokinetics of pefloxacin and ofloxacin in healthy volunteers. Drugs Exp Clin Res 18:487–492
320. Iqbal Z, Khan A, Naz A et al (2009) Pharmacokinetic interaction of ciprofloxacin with diclof-enac: a single-dose, two-period crossover study in healthy adult volunteers. Clin Drug Invest 29:275–281
321. Lipsky BA, Baker CA (1999) Fluoroquinolone toxicity profiles: a review focusing on newer agents. Clin Infect Dis 28:352–364
322. Springsklee M, Reiter C, Meyer JM (1999) Safety and tolerability profile of moxifloxa-cin. Proceedings of the 13th European Congress of Clinical Microbiology and Infectious Diseases, Berlin, March 1999. Abstract P-0208
323. Frothingham R (2001) Rates of torsades de pointes associated with ciprofloxacin, ofloxacin, levofloxacin, gatifloxacin, and moxifloxacin. Pharmacotherapy 21:1468–1472
324. Samaha FF (1999) QTc interval prolongation and polymorphic ventricular tachycardia in association with levofloxacin [letter]. Am J Med 107:528–529
325. Owens RC Jr, Ambrose PG (2002) Torsades de pointes associated with fluoroquinolones. Pharmacotherapy 22:663–672
326. Amankwa K, Krishnan SC, Tisdale JE (2004) Torsades de pointes associated with fluoroqui-nolones: importance of concomitant risk factors. Clin Pharmacol Ther 75:242–247
327. Nair MK, Patel K, Starer PJ (2008) Ciprofloxacin-induced torsades de pointes in a methadone-dependent patient. Addiction 103:2062–2064
328. Keivanidou A, Arnaoutoglou C, Krommydas A et al (2009) Ciprofloxacin induced acquired long QT syndrome in a patient under class III antiarrhythmic therapy. Cardiol J 16:172–174
329. Knorr JP, Moshfeghi M, Sokoloski MC (2008) Ciprofloxacin-induced Q-T interval prolonga-tion. Am J Health-Syst Pharm 65:547–551
330. Altin T, Ozcan O, Turhan S et al (2007) Torsade de pointes associated with moxifloxacin: a rare but potentially fatal adverse event. Can J Cardiol 23:907–908

331. Kazmierczak J, Peregud-Pogorzelska M, Rzeuski R (2007) QT interval prolongation and torsades de pointes due to a coadministration of ciprofloxacin and azimilide in a patient with implantable cardioverter-defibrillator. Pacing Clin Electrophysiol 30:1043–1046

332. Nykamp DL, Blackmon CL, Schmidt PE et al (2005) QTc prolongation associated with combination therapy of levofloxacin, imipramine, and fluoxetine. Ann Pharmacother 39:543–546

333. Letsas KP, Sideris A, Kounas SP et al (2006) Drug-induced QT interval prolongation after ciprofloxacin administration in a patient receiving olanzapine [letter]. Int J Cardiol 109:273–274

334. Slovacek L, Priester P, Petera J, Slanska I, Kopecky J (2011) Tamoxifen/norfloxacin interaction leading to QT interval prolongation in a female patient with extracranial meningioma. Bratislav Lekar List 112:353–354

335. Haring B, Bauer W (2012) Ciprofloxacin and the risk for cardiac arrhythmias: culprit delicti or watching bystander? Acta Cardiol 67:351–354

336. Tsai LH, Weng YM, Lin CC, Kuo CW, Chen JC (2014) Risk screening for long QT prior to prescribing levofloxacin. Am J Emerg Med 32:1153.e1–1153.e3

337. Ibrahim M, Omar B (2012) Ciprofloxacin-induced torsade de pointes. Am J Emerg Med 30:252.e5–252.e9

338. Patel PD, Afshar H, Birnbaum Y (2010) Levofloxacin-induced torsades de pointes. Texas Heart Inst J 37:216–217

339. Abo-Salem E, Nugent K, Chance W (2011) Antibiotic-induced cardiac arrhythmia in elderly patients. J Am Geriatr Soc 59:1747–1749

340. Zeineh NSA (2010) A toxic combination. Am J Med 123:707–708

341. Paltoo B, O'Donoghue S, Mousavi MS (2001) Levofloxacin induced polymorphic ventricular tachycardia with normal QT interval. Pacing Clin Electrophysiol 24:895–897

342. Zambon A, Polo Friz H, Contiero P et al (2009) Effect of macrolide and fluoroquinolone antibacterials on the risk of ventricular arrhythmia and cardiac arrest: an observational study in Italy using case-control, case-crossover and case-time-control designs. Drug Saf 32:159–167

343. Paran Y, Mashav N, Henis O et al (2008) Drug-induced torsades de pointes in patients aged 80 years or more. Anagolu Kardiyol Derg 8:260–265

344. Chou HW, Wang JL, Chang CH et al (2015) Risks of cardiac arrhythmia and mortality among patients using new-generation macrolides, fluoroquinolones, and beta-lactam/beta-lactamase inhibitors: a Taiwanese nationwide study. Clin Infect Dis 60:566–577

345. Ng TM, Olsen KM, McCartan MA et al (2010) Drug-induced QTc-interval prolongation in the intensive care unit: incidence and predictors. J Pharm Prac 23:19–24

346. Lapi F, Wilchesky M, Kezouh A et al (2012) Fluoroquinolones and the risk of serious arrhythmia: a population-based study. Clin Infect Dis 55:1457–1465

347. Zeuli JD, Wilson JW, Estes LL (2013) Effect of combined fluoroquinolone and azole use on QT prolongation in hematology patients. Antimicrob Agents Chemother 57:1121–1127

348. Rao GA, Mann JR, Shoaibi A et al (2014) Azithromycin and levofloxacin use and increased risk of cardiac arrhythmia and death. Ann Fam Med 12:121–127

349. Stancampiano FF, Palmer WC, Getz TW et al (2015) Rare incidence of ventricular tachycardia and torsades de pointes in hospitalized patients with prolonged QT who later received levofloxacin: a retrospective series. Mayo Clin Proc 90:606–612

350. Inghammar M, Svanstrom H, Melbye M, Pasternak B, Hvid A (2016) Oral fluoroquinolone use and serious arrhythmia: a bi-national cohort study. BMJ 352:i843

351. Kang J, Wang L, Chen XL, Triggle DJ, Rampe D (2001) Interactions of a series of fluoroquinolone antibacterial drugs with the human cardiac K+ channel HERG. Mol Pharmacol 59:122–126

352. Anderson ME, Mazur A, Yang T, Roden DM (2001) Potassium current antagonist properties and proarrhythmic consequences of quinolone antibiotics. J Pharmacol Exp Ther 296:806–810

353. Hagiwara T, Satoh S, Kasai Y, Takasuna KA (2001) Comparative study of the fluoroquinolone antibacterial agents on the action potential duration in guinea pig ventricular myocardia. Jap J Pharmacol 87:231–234

354. Patmore L, Fraser S, Mair D, Templeton A (2000) Effects of sparfloxacin, grepafloxacin, moxifloxacin, and ciprofloxacin on cardiac action potential duration. Eur J Pharmacol 406:449–452

355. Bischoff U, Schmidt C, Netzer R, Pongs O (2000) Effects of fluoroquinolones on HERG currents. Eur J Pharmacol 406:341–343

356. Lacroix P, Crumb WJ, Durando L et al (2003) Prulifloxacin: in vitro (HERG current) and in vivo (conscious dog) assessment of cardiac risk. Eur J Pharmacol 477:69–72

357. Akita M, Shibazaki Y, Izumi M et al (2004) Comparative assessment of prulifloxacin, sparfloxacin, gatifloxacin, and levofloxacin in the rabbit model of proarrhythmia. J Toxicol Sci 29:63–71

358. Chiba K, Sugiyama A, Satoh Y, Shiina H, Hashimoto K (2000) Proarrhythmic effects of fluoroquinolone antibacterial agents: in vivo effects as physiologic substrate for torsades. Toxicol Appl Pharmacol 169:8–16

359. Iannini PB, Doddamani S, Byazrova E, Curciumara I, Kramer H (2001) Risk of torsades de pointes with non-cardiac drugs. Prolongation of QT interval is probably a class effect of fluoroquinolones [letter]. BMJ 322:46–47

360. Demolis JL, Kubitza D, Tenneze L, Funck-Brentano C (2000) Effect of a single oral dose of moxifloxacin (400 mg and 800 mg) on ventricular repolarization in healthy subjects. Clin Pharmacol Ther 68:658–666

361. Noel GJ, Goodman DB, Chien S, Solanki B, Padmanabhan M, Natarajan J (2004) Measuring the effects of supratherapeutic doses of levofloxacin on healthy volunteers using four methods of QT correction and periodic and continuous ECG recordings. J Clin Pharmacol 44:464–473

362. Noel GJ, Natarajan J, Chien S, Hunt TL, Goodman DB, Abels R (2003) Effects of three fluoroquinolones on QT interval in healthy adults after single doses. Clin Pharmacol Ther 73:292–303

363. Taubel J, Naseem A, Harada T et al (2010) Levofloxacin can be used effectively as a positive control in thorough QT/QTc studies in healthy volunteers. Br J Clin Pharmacol 69:391–400

364. Litwin JS, Benedict MS, Thorn MD et al (2015) A thorough QT study to evaluate the effects of therapeutic and supratherapeutic doses of delafloxacin on cardiac repolarization. Antimicrob Agents Chemother 59:3469–3473

365. Rosignoli MT, Di Loreto G, Dionisio P (2010) Effects of prulifloxacin on cardiac repolarization in healthy subjects: a randomized, crossover, double-blind versus placebo, moxifloxacin-controlled study. Clin Drug Invest 30:5–14

366. Anonymous (2016) Zabofloxacin product information. Seoul, Korea, DongWha Pharmaceuticals

367. Taubel J, Ferber G, Lorch U et al (2014) Thorough QT study of the effect of oral moxifloxacin on QTc interval in the fed and fasted state in healthy Japanese and Caucasian subjects. Br J Clin Pharmacol 77:170–179

368. Morganroth J, Wang Y, Thorn M et al (2015) Moxifloxacin-induced QTc interval prolongations in healthy male Japanese and Caucasian volunteers: a direct comparison in a thorough QT study. Br J Clin Pharmacol 80:446–459

369. Chen Q, Liu YM, Liu Y et al (2015) Orally administered moxifloxacin prolongs QTc in healthy Chinese volunteers: a randomized, single-blind, crossover study. Acta Pharmacol Sin 36:448–453

370. Moon SJ, Lee J, An H et al (2014) The effects of moxifloxacin on QTc interval in healthy Korean male subjects. Drugs R&D 14:63–71

371. Abo-Salem E, Fowler JC, Attari M et al (2014) Antibiotic-induced cardiac arrhythmias. Cardiovasc Ther 32:19–25

372. Mehrzad R, Barza M (2015) Weighing the adverse cardiac effects of fluoroquinolones: a risk perspective. J Clin Pharmacol 55:1198–1206

373. Doig JC (1997) Drug-induced cardiac arrhythmias: incidence, prevention and management. Drug Saf 17:265–275

374. Roden DM (1997) A practical approach to torsades de pointes. Clin Cardiol 20:285–290

375. Janeira LF (1995) Torsades de pointes and long QT syndromes. Clin Fam Phys 52:1447–1453

376. Garbel SM, Pound MW, Miller SM (2009) Hypoglycemia associated with the use of levofloxacin. Am J Health-Sys Pharm 66:1014–1019
377. Roberge RJ, Kaplan R, Frank R, Fore C (2000) Glyburide–ciprofloxacin interaction with resistant hypoglycemia. Ann Emerg Med 36:160–163
378. Lin G, Hays DP, Spillane L (2004) Refractory hypoglycemia from ciprofloxacin and glyburide interaction. J Toxicol Clin Toxicol 42:295–297
379. Fusco S, Reitano F, Gambadoro N et al (2013) Severe hypoglycemia associated with levofloxacin in a healthy older woman. J Am Geriatr Soc 61:1637–1638
380. Kapoor R, Blum D, Batra A et al (2012) Life-threatening hypoglycemia with moxifloxacin in a dialysis patient. J Clin Pharmacol 52:269–271
381. Parra-Riffo H, Lemus-Penaloza J (2012) Severe levofloxacin-induced hypoglycaemia: a case report and literature review. Nefrologia 32:546–547
382. Kelesidis T, Canseco E (2010) Quinolone-induced hypoglycemia: a life-threatening but potentially reversible side effect. Am J Med 123:e5–e6
383. Frothingham R (2005) Glucose homeostasis abnormalities associated with the use of gatifloxacin. Clin Infect Dis 41:1269–1276
384. Park-Wyllie LY, Juurlink DN, Kopp A et al (2006) Outpatient gatifloxacin therapy and dysglycemia in older adults. NEJM 354:1352–1361
385. Aspinall SL, Good CB, Jiang R et al (2009) Severe dysglycemia with fluoroquinolones: a class effect? Clin Infect Dis 49:402–408
386. Parekh TM, Raji M, Lin YL et al (2014) Hypoglycemia after antimicrobial drug prescription for older patients using sulfonylureas. JAMA Intern Med 174:1605–1612
387. Chou HW, Wang JL, Chang CH et al (2013) Risk of severe dysglycemia among diabetic patients receiving levofloxacin, ciprofloxacin, or moxifloxacin in Taiwan. Clin Infect Dis 57:971–980
388. Zheng HX, Huang Y, Frassetto LA et al (2009) Elucidating rifampin's inducing and inhibiting effects on glyburide pharmacokinetics and blood glucose in healthy volunteers: unmasking the differential effects of enzyme induction and transporter inhibition for a drug and its primary metabolite. Clin Pharmacol Ther 85:78–85
389. Gajjar DA, LaCreta FP, Kollia GD et al (2000) Effect of multiple-dose gatifloxacin or ciprofloxacin on glucose homeostasis and insulin production in patients with noninsulin dependent diabetes mellitus maintained with diet and exercise. Pharmacotherapy 20(6 pt. 2):76S–86S
390. Trissel LA (2013) Handbook on Injectable Drugs, 17th edn. Bethesda, MD, American Society of Health-System Pharmacists
391. Semark AJ, Venkatesh K, McWhinney BC et al (2013) The compatibility of a low concentration of hydrocortisone sodium succinate with selected drugs during simulated Y-site administration. Crit Care Resusc 15:63–66
392. Szentmary N, Kraszni M, Nagy ZZ (2004) Interaction of indomethacin and ciprofloxacin in the cornea following phototherapeutic keratectomy. Graefes Arch Clin Exp Ophthalmol 242:614–616
393. Anonymous (2017) Delafloxacin product information. Lincolnshire, IL, Melinta Therapeutics

Chapter 4
Glycopeptides, Lipopeptides, and Lipoglycopeptides

Mary A. Ullman and John C. Rotschafer

4.1 Introduction

While glycopeptide antibiotics have been available in practice for over 50 years, lipopeptides and lipoglycopeptide antibiotics are relatively new. Vancomycin is the only glycopeptide antibiotic currently available in the USA. Teicoplanin is a glycopeptide only available in Europe and will not be discussed in this chapter. However, it does not have any significant drug-drug interactions. Daptomycin is the only available lipopeptide. Three agents in the lipoglycopeptide class – telavancin, oritavancin, and dalbavancin – are now available for use in the USA Both dalbavancin and oritavancin have a serum half-life profile quite different than currently available products (150–300 h vs. 6–12 h), and a typical course of therapy with these new agents is one dose (oritavancin) or one to two doses (dalbavancin) of drug [1, 5, 6]. Possible adverse reactions with having a half-life of this magnitude may prove difficult to manage.

While these three antibiotic classes are chemically different, there are many similarities among these compounds. The drugs tend to be large molecules which limits or delays antibiotic penetration to various sites in the body [2–6]. This large molecular weight (Table 4.1) also contributes to a low bioavailability when these compounds are administered orally. All of these compounds with the exception of vancomycin are extensively protein bound (Table 4.1) leaving but a small free fraction of antibiotic that can cross biological barriers and interact with bacteria [2–6]. High levels of protein binding can be theoretically associated with drug-drug binding

M. A. Ullman
Regions Hospital, Department of Pharmacy, St. Paul, MN, USA
e-mail: rotsc001@umn.edu

J. C. Rotschafer (✉)
University of Minnesota College of Pharmacy, Department of Experimental and Clinical Pharmacology, Minneapolis, MN, USA
e-mail: rotsc001@umn.edu

© Springer International Publishing AG, part of Springer Nature 2018
M. P. Pai et al. (eds.), *Drug Interactions in Infectious Diseases: Antimicrobial Drug Interactions*, Infectious Disease,
https://doi.org/10.1007/978-3-319-72416-4_4

139

Table 4.1 Pharmacokinetic properties

	Vancomycin	Daptomyci	Telavancin	Oritavancin	Dalbavancin
Class	Glycopeptide	Lipopeptide	Lipoglycopeptide	Lipoglycopeptide	Lipoglycopeptide
Molecular weight	1485.71	1620.67	1792.1	1989.09	1816.7
Oral bioavailability	–	–	–	–	–
% renal elimination	95	78	72	5	33
Serum half-life (hours)	6–12	8	7.5	245	346 (1000 mg dose)
% protein binding	10–50	>90	>90	85	93
Pregnancy class	C	B	C	C	C

displacement interactions. However, clinically significant protein binding displacement has not been reported with these compounds. The kidneys are primarily responsible for the elimination of vancomycin, daptomycin, and telavancin, thus warranting dose adjustments in renal dysfunction (Table 4.1) [2–6]. Dalbavancin does need a dose reduction in patients with an estimated creatinine clearance less than 30 mL/min but does not require a dose adjustment in hemodialysis patients [5]. While oritavancin has not been studied in severe renal dysfunction, no dose adjustments are needed in mild or moderate renal dysfunction [6]. Many of the older agents in these classes have been associated with nephrotoxicity directly or when used in conjunction with other nephrotoxic agents such as aminoglycosides, nonsteroidal anti-inflammatory agents (NSAIDs), ACE inhibitors, loop diuretics, etc. (Table 4.2). Because these agents are primarily renally eliminated, there is generally modest concerns with the need for dose adjustment in liver failure, liver enzyme induction, or drug-drug interactions associated with CYP liver enzyme metabolism. Another common concern among glyco/lipo/lipoglyco class members is the possibility of infusion reactions, the so-called red man or red neck syndrome (Table 4.2).

As with all class compounds, differences do exist with respect to individual pharmacokinetic parameters particularly in terms of the length of half-life and the size of distribution volume (Table 4.1); these three antibiotic classes are relatively free of typical CYP drug-drug interactions. We will review each member of these three antibiotic classes differentiating each drug in terms of their pharmacokinetic parameters and likely drug interaction potential.

4.2 Vancomycin (Glycopeptide)

Vancomycin has been and remains to date the gold standard antibiotic for the management of methicillin–/oxacillin-resistant *S. aureus* (MRSA/ORSA) infections [7]. While this agent has been available clinically for over 60 years, the clear majority of clinical experience has been at substantially lower doses than what are currently recommended. The IDSA/ASHP/SIDP published a position paper on the therapeutic monitoring of vancomycin in January 2009 [7]. In this consensus opinion and in accordance with previously published treatment guidelines recommending clinicians obtain trough concentrations two to four times the previous standard (5–10 mg/L), a loading dose of 25–30 mg/Kg (actual body weight) and maintenance doses of 15–20 mg/Kg (actual body weight) are to be used when treating patients for serious gram-positive infections. Because of the substantially larger doses currently being used, clinicians should monitor patients carefully for infusion reactions and nephrotoxicity.

4.2.1 Absorption

Because of the large molecular size of vancomycin, very little absorption occurs after oral administration [8, 9]. In patients with normal renal function who received vancomycin 500 mg orally every 6 h, vancomycin serum concentrations were

Table 4.2 Potential drug-drug interactions

	Vancomycin	Daptomycin	Telavancin	Oritavancin	Dalbavancin
Liver CYP				Weak inhibitor 2C9/2C19	
Enzyme effects	–	–	–	Weak inhibitor 2C9/2C19	–
Drug-drug CYP					
Interaction with Metabolism	–	–	–	–	–
Other drug-drug interaction	Nephrotoxic agents[a]	Nephrotoxic agents[a] Statins	Nephrotoxic agents[a] Formulated with hydroxypropyl-beta-cyclodextrin[b]	Heparin Warfarin Live cholera vaccine	Live cholera vaccine
Protein binding					
Displacement	–	–	–	–	–
Antibiotic antagonism	–	–	–	–	–
QT$_c$ prolongation	–	–	5 ms	–	–
Laboratory test	–	–	Anticoagulation tests[d]	Anticoagulation tests[e]	
Interference			Protein dipstick	D-dimer assay	
Infusion reactions and red man and red neck syndrome	Definite	Possible	Possible	Possible	Possible

[a]Concomitant use of loop diuretics, ACE inhibitors, aminoglycosides, amphotericin B, NSAIDs, and polymyxins could contribute to the development of nephrotoxicity; reduced drug elimination of adefovir, cisplatin, cyclosporine, methotrexate, tacrolimus, telbivudine, and tenofovir with concomitant vancomycin is cautioned in individual drug package inserts [64]

[b]Can accumulate in renal dysfunction. Use caution when in combination with other agents using this solubilizer

[c]In itself not likely an issue but used with other agents capable of increasing the QTc interval or in a patient with a QTc interval > 500 ms concomitant use could be a problem

[d]Can artificially alter PT, INR, aPTT, activated clotting time, and factor Xa-based test results – draw blood for these tests as close to next telavancin dose as possible

[e]Can artificially alter PT, INR, aPTT, ACT, silica clotting time, and dilute Russell's viper venom time-based test results – consider monitoring coagulation with chromogenic factor Xa assays or thrombin time

2.4–3.0 mcg/mL [10]. Therefore, to treat systemic infections, vancomycin must be administered intravenously. However, in the management of *Clostridium difficile* infections, oral administration is the preferred route as the drug remains in the intestinal lumen at the site of infection [10, 11]. In the rare cases of a patient with both pseudomembranous colitis and severe renal failure, therapeutic concentrations were achieved in the serum secondary to increased oral absorption due to decreased integrity of the intestinal lumen [11, 12].

4.2.2 Distribution

The pharmacokinetics of vancomycin distribution has been characterized using one-, two-, three-, and non-compartment models. Protein binding is generally estimated to be approximately 50%. Vancomycin binds to albumin and appears to have a low affinity for alpha-1-acid-glycoprotein [4].

4.2.3 Metabolism

Vancomycin does not undergo significant hepatic metabolism and thus is not a source for CYP drug interactions either by induction or competitive metabolism [4].

4.2.4 Elimination

Vancomycin is primarily eliminated renally (95%), and drug clearance correlates well with creatinine clearance [4]. Conventional hemodialysis methods do not extensively remove vancomycin from the serum. However, high flux dialysis methods have been reported to clear vancomycin much more effectively [13]. Additionally, clinicians should be aware concomitant use of drugs that affect a patient's hemodynamics (e.g., dopamine, dobutamine, furosemide) might also result in higher clearance of vancomycin. Patients who continue on vancomycin after discontinuation of these agents may require adjustments due to decreased clearance.

4.2.5 Considerations for Clinical Use

Vancomycin has been associated with nephrotoxicity, although the incidence decreased since the drug's initial introduction as the purity of the drug formulation has improved. However, with current clinical practice recommending much larger loading and maintenance doses of vancomycin plus the almost exclusive use of generic products with potentially higher concentrations of impurities, there may be a greater risk of nephrotoxicity than reported in years past where lower doses were used and the product was branded. In cases of renal dysfunction, vancomycin does accumulate. Monitoring of serum trough concentrations is recommended to prevent nephrotoxicity in patients at high risk of toxicity either due to pre-existing renal impairment, use of aggressive dosing strategies to achieve troughs of 15–20 mg/L, or prolonged courses of vancomycin of 5 days or greater [7]. Potential drug interactions involve the use of vancomycin in conjunction with other nephrotoxic agents, most commonly aminoglycosides. Vancomycin degradation products (VDPs) have been reported to accumulate in patients with renal dysfunction. VDPs have also been reported to result in the reporting of falsely high concentrations of vancomycin as with some assays as VDPs are falsely interpreted as vancomycin (factor B vancomycin) [14]. Additionally, vancomycin can rarely cause and/or contribute to neutropenia; neutrophil counts should be monitored when used with other agents that may cause neutropenia.

4.3 Daptomycin (Lipopeptide)

4.3.1 Pharmacology

Most clinical experience to date with daptomycin has been at daily doses of 4 mg/kg (skin and soft tissue infection) and 6 mg/kg (right-sided endocarditis and bacteremia) [2]. However, as more clinicians consider using higher doses for difficult to treat gram-positive infections, additional adverse events could emerge at the upper limit of daily dosing of daptomycin and are explored [15–22].

While theoretically the potential for a variety of drug-drug interactions exists, daptomycin has remained relatively free of such problems. Like vancomycin and other lipoglycopeptides, daptomycin is primarily cleared renally avoiding induction of CYP liver enzymes or competing with other drugs for metabolism [18]. Even in terms of the drug's antibacterial action combining daptomycin with other antibiotics generally usually results in synergy, an additive effect or indifference [19]. Only rarely is any type of antagonism demonstrated although use with tobramycin has been reported to reduce area under the serum concentration-time curve (AUC) by 15% [2].

Concomitant use of daptomycin with other known nephrotoxic agents may increase the risk of nephrotoxicity, and the use of "statins" in conjunction with daptomycin may increase the risk of muscle enzyme (creatine phosphokinase, CPK) elevation [2, 22–27]. While infusion reactions should be a concern with any of the glycopeptide, lipopeptide, and lipoglycopeptide classes, there are published studies using 2-min intravenous push doses of daptomycin compared to traditional 30-min intravenous infusions demonstrating that the bolus dosing method is as well tolerated by patients as the 30-min infusion [28].

4.3.2   Absorption

Because of the large molecular size of daptomycin, oral absorption would not likely result in therapeutic serum concentrations.

4.3.3   Distribution

The volume of distribution is small for daptomycin, and daptomycin is highly protein bound (90–95%) in a concentration-independent manner. Hepatic dysfunction did not alter the rate of protein binding. Studies indicate a slow distribution of daptomycin into the tissues from the serum [2].

4.3.4   Metabolism

The exact metabolism of daptomycin is not completely understood. Induction or inhibition of cytochrome P450 isoforms has not been demonstrated with daptomycin. While metabolites are detected in the urine, no metabolites are detected in serum, suggesting the possibility of renal metabolism of daptomycin [2].

4.3.5   Elimination

Daptomycin is eliminated renally, with approximately 50–60% excreted as unchanged. Dose adjustments in renal dysfunction are recommended by the manufacturer [2].

4.3.6   Considerations for Clinical Use

Theoretically, because daptomycin is a highly protein-bound drug, concomitant use of other highly protein-bound drugs could result in displacement of drug. While alpha-1 glycoprotein has been identified as one of the proteins that bind daptomycin, other protein targets have not been identified. Alternatively, in critically ill patients who may have reduced levels of serum proteins, higher free drug concentrations of daptomycin may be present.

4.4   Telavancin (Lipoglycopeptide)

4.4.1   Pharmacology

Telavancin attacks the cell wall by inhibiting polymerization of the bacterial cell wall. The drug also interferes with transpeptidation by binding to the d-ala-d-ala terminal sequence. Lastly like daptomycin, telavancin causes depolarization of the outer membrane of the gram-positive cell wall. Telavancin is primarily a gram-positive antibiotic active against staphylococci, streptococci, and enterococci. The drug has no gram-negative activity and has limited activity against anaerobes [1, 3, 29].

4.4.2   Absorption

Like vancomycin, telavancin and other lipopeptides and lipoglycopeptides are not systemically absorbed following oral administration.

4.4.3   Distribution

Like daptomycin, telavancin is highly protein bound (approximately 90%). Albumin is the main protein responsible for binding telavancin; hepatic or renal impairment does not affect the rate of binding. The volume of distribution is small at 0.1 L/kg [3]. In difficult to treat infections such as meningitis, limited animal studies have demonstrated the superior performance of telavancin vs. vancomycin in clearing bacteria causing meningitis and sterilizing CSF [30]. Comparable performance has been demonstrated in animal of osteomyelitis comparing telavancin and vancomycin. Telavancin also appears to penetrate and demonstrate biologic activity in bacterial biofilms [31, 32].

4.4.4   Metabolism

Telavancin has not demonstrated CYP450 3A4 activity following a midazolam probe [33]. Approximately 3–6% of the telavancin dose is converted to a 7-hydroxy metabolite which is excreted in the urine. The mechanism of telavancin's metabolism is unknown at this time. Mild to moderate hepatic impairment does not affect the pharmacokinetics of telavancin, and no dosage adjustment is recommended [3].

4.4.5   Elimination

Two-thirds to three-quarters of telavancin is eliminated renally unchanged. Dose adjustments are recommended in patients with a creatinine clearance of less than 50 mL/min [3]. Clinical outcome in patients with reduced renal function as well as elderly patients (>65 years) who may have age-related renal impairment has not done as well in terms of clinical outcome compared to patients with normal renal function [3]. Current dosage adjustment for renal failure possibly may not be providing the required amount of telavancin to overcome clinical infection keeping in mind that the drug is a concentration-dependent killer. Additionally, one of the excipients in the intravenous solution is hydroxypropyl-beta-cyclodextrin, known to accumulate in patients with renal dysfunction [3]. Hemodialysis has been found to remove ~6% of a single dose of 7.5 mg/kg in patients with end-stage renal disease undergoing hemodialysis. Continuous venovenous hemofiltration is much more efficient at removing telavancin from the bloodstream. The amount of telavancin removal is dependent on the rate of ultrafiltration [3].

4.4.6   Considerations for Clinical Use

While not extensively studied, there does not appear to be any antibiotic-antibiotic antagonistic combinations reported to date with telavancin.

Because of the highly protein-bound nature of telavancin, the potential exists for drug interaction in patients also receiving other drugs that are highly bound to proteins, especially albumin. Although no reports of clinically significant protein binding displacement events have been reported to date, clinicians should be aware of the potential interaction and monitor patients accordingly. A higher incidence of side effects has been reported with telavancin in clinical trials as compared to vancomycin [3]. The most common adverse events with this new agent include nausea, emesis, foamy urine, and a metallic aftertaste following parenteral administration.

Care should be used in patients with renal dysfunction to monitor for drug accumulation but also because the drug is formulated with hydroxypropyl-beta-cyclodextrin; this agent may also accumulate in patients with renal failure. One other intravenous antimicrobial that carries warning about cyclodextrin use in renal impairment is voriconazole [34]. In clinical trials telavancin has been shown to cause more nephrotoxicity than the comparator, vancomycin (15% vs 7%) [35–37]. However, vancomycin dosages were 2 g per day in patients with normal renal function. These differences may not be present at higher, more contemporary dosing of vancomycin. Concern should also be directed at concomitant use of telavancin and other drugs known to contribute to nephrotoxicity.

Telavancin has also been shown to increase QTc intervals [38]. On average, the increase is relatively small, i.e., ~5 ms, but if used in conjunction with other agents or in the background of conditions known to increase QTc, there could be an adverse clinical outcome. Telavancin is also not recommended in pregnancy and is a category C drug [3]. Current dosage recommendations for telavancin call for a daily dose of 10 mg/Kg to be determined based on actual body weight. Possible red man or red neck infusion-related reactions are clearly a risk with the use of telavancin.

Because telavancin binds to phospholipids, there are some significant drug-laboratory test interactions [3]. Telavancin artificially interferes with the determination of prothrombin time, internal normalization ratio, activated partial thromboplastin time, activated clotting time, and coagulation studies based on factor Xa test. A clear distinction needs to be made that the drug interferes with the test result of various coagulation studies but does not actually alter the coagulation state. As telavancin is administered every 24 h, the easiest maneuver to avoid this interaction is to obtain blood for these studies in the terminal 6 h of the dosing interval (i.e., 6 h or less before the next telavancin dose). Telavancin can also interfere with qualitative dipstick protein assay methods.

4.5 Dalbavancin

4.5.1 Pharmacology

Similar to telavancin, dalbavancin interferes with synthesis of the cell wall by binding to the d-ala-d-ala terminal sequence of peptidoglycan [5]. Dalbavancin has demonstrated activity against a large number of staphylococci and streptococci, including resistant isolates. The drug has also demonstrated activity against gram-positive anaerobic bacteria including *Clostridium* species, *Peptostreptococcus* species, and *Actinomyces* species. Dalbavancin does not possess activity against gram-negative organisms or enterococci that possess the van A gene, conferring vancomycin resistance [5].

4.5.2 Absorption

Due to the large molecular size, oral administration of dalbavancin would not result any measurable serum concentration.

4.5.3 Distribution

Dalbavancin is approximately 93% protein bound, with albumin being the primary plasma protein. This binding is not altered by drug concentration or by renal or hepatic impairment. Dalbavancin has demonstrated excellent tissue penetration in blister fluid, with a ratio of fluid concentration to plasma concentration of 0.83:1.1.

The concentration of dalbavancin in blister fluid has been demonstrated to be greater than 30 mg/L for up to 7 days after dose administration [5]. The volume of distribution ranges from 9.75 to 15.7 L [5].

4.5.4 Metabolism

Dalbavancin is not metabolized through the hepatic system. A minor metabolite, hydroxyl-dalbavancin, has been detected in urine samples of healthy volunteers but has not been detected in human plasma [5].

4.5.5 Elimination

Approximately a third of a dose of dalbavancin is eliminated in the urine unchanged, and an additional 12% of the dose is eliminated as the minor metabolite based on elimination studies lasting up to 42 days. Fecal elimination accounted for 20% of the administered dose [5].

4.5.6 Considerations for Clinical Use

Dalbavancin has been FDA approved for skin and soft tissue infections, first as a two-dose strategy 7 days apart and then as a higher single-dose strategy [5, 66]. A two-dose strategy of dalbavancin has also been compared to vancomycin for the treatment of catheter-related bacteremia and demonstrated a higher success rate than vancomycin. The clinical efficacy to use dalbavancin for other infections commonly treated with vancomycin is unknown at this time.

Due to the high protein binding, the same potential issues that exist with the other highly protein lipopeptides and lipoglycopeptides could potential result in drug displacement and/or increased concentrations of dalbavancin. No such studies have demonstrated this effect in a clinical scenario.

No significant drug-drug or drug-lab interactions have been demonstrated with dalbavancin. However, as use increases, additional drug interactions may be discovered; clinicians should consider referring to recently published drug information sources for new potential interactions. In comparison with vancomycin and linezolid for skin and soft tissue infections, dalbavancin was well tolerated and demonstrated similar efficacies.

While a long half-life provides less frequent dosing and the potential for better compliance, in patients that demonstrate a significant reaction to dalbavancin, removal of the offending agent may prove difficult. Data on the ability of hemodialysis or plasmapheresis to remove significant amounts of dalbavancin are not available.

4.6 Oritavancin

4.6.1 Pharmacology

Oritavancin is similar to telavancin in that the mechanism of action is multifaceted: (1) inhibition of peptidoglycan synthesis by binding to peptidoglycan precursors, (2) inhibition of the peptidoglycan cross-linking, and (3) depolarization of the outer membrane [6]. Oritavancin has demonstrated activity against both methicillin-sensitive and methicillin-resistant *Staphylococcus aureus*, *Streptococcus agalactiae*, *Streptococcus dysgalactiae*, *Streptococcus pyogenes*, and *Streptococcus anginosus* group, and vancomycin-susceptible isolates of *Enterococcus faecalis*. Oritavancin does not have any activity against gram-negative or anaerobic organisms [6].

4.6.2 Absorption

As similar to all other agents in this chapter, the large size of oritavancin would limit clinically significant oral absorption.

4.6.3 Distribution

Like other agents in lipoglycopeptide class, oritavancin demonstrates a high percent bound to proteins, 85%. Drug concentrations in blister fluid were 20% of plasma concentration in healthy volunteers. Based on population pharmacokinetics, the average total volume of distribution is 87.6 L [6].

4.6.4 Metabolism

Oritavancin does not appear to be metabolized based on in vitro human liver microsome studies. It has demonstrated weak inhibition and induction of CYP450 enzymes, but the clinical relevance of this activity is unknown at this time [6].

4.6.5 Elimination

Oritavancin, similar to dalbavancin, has a significantly long half-life of 245 h. Oritavancin is slowly excreted unchanged. After 2 weeks of collection, oritavancin was found in feces and urine with concentrations of <1% and 5% of the drug, respectively [6].

4.6.6 Considerations for Clinical Use

FDA approval is currently only for infection of skin and/or subcutaneous tissue. In comparison studies with vancomycin, oritavancin was non-inferior to vancomycin for this indication. However, osteomyelitis was more common in the oritavancin group than the vancomycin group in one study; alternative agents should be considered in patients who have or may develop osteomyelitis [39].

Similar to previous discussions for other agents that demonstrate high protein binding, a theoretical interaction may exist in critically ill patients who demonstrate lower amounts of plasma proteins. The clinical implications of this interaction are unknown at this time. As use increases, additional drug interactions may be discovered; clinicians should consider referring to recently published drug information sources for new potential interactions.

4.7 Important Drug Interactions

4.7.1 Vancomycin

4.7.1.1 Aminoglycosides

Mechanism: Both aminoglycosides and vancomycin are known to cause damage to similar sites in the renal tubules. The use of these two agents concomitantly can result in additive toxicity.

Literature: Several studies have demonstrated the toxicities encountered when using the combination vancomycin and aminoglycosides. In a prospective study with 34 patients and 39 courses of vancomycin performed by Mellor and colleagues, 6 courses of vancomycin alone were compared with 27 courses of vancomycin plus aminoglycoside, either concurrently or within 2 weeks of the first dose of vancomycin [35]. The mean total dose and duration of vancomycin therapy was 28.3 ± 18.1 g and 15.3 ± 9.3 days, respectively, approximately 1.8 g per day. Nephrotoxicity was defined as a rise of ≥0.5 mg/dL if initial creatinine was 3 g/dL or less; if initial creatinine was greater than 3 g/dL, a rise of ≥1.0 mg/dL was indicative of nephrotoxicity. Three patients developed acute nephrotoxicity (7.1%); 9.1% of patients developed nephrotoxicity within the 2 weeks following vancomycin therapy. All cases of nephrotoxicity demonstrated an abrupt rise in creatinine following septicemia or gastrointestinal hemorrhage. Additionally, two patients reported tinnitus and dizziness. One patient was diagnosed with acute hearing loss, but the authors concluded this was not due to drug toxicity.

In an retrospective evaluation of 229 courses of antibiotic therapy, Ciminio and colleagues examined the relationship of serum vancomycin and aminoglycoside concentrations to nephrotoxicity [40]. Antibiotic courses were divided into three groups: aminoglycoside alone (148 cases), vancomycin alone (41 cases), and vancomycin concurrently with aminoglycoside [44]. Nephrotoxicity was defined as a rise of >0.5 mg/dL in serum creatinine, when compared to baseline. Normal values for vancomycin and aminoglycosides were peak 20–40 mg/L and trough <10 mg/L and peak 4–9 mg/L and trough <2 mg/L, respectively. Overall, a 17% incidence of nephrotoxicity was noted. When broken down in the three groups, 18% of patients developed nephrotoxicity with aminoglycosides alone, 15% with vancomycin alone, and 15% with concurrent vancomycin and aminoglycosides. While higher serum creatinine concentrations were associated with increased nephrotoxicity in patients with aminoglycosides alone and higher daily doses of vancomycin (36.3 ± 4.8 mg/kg/day versus 24.0 ± 1.1 mg/kg/day) were associated with nephrotoxicity in patients with vancomycin alone, the only significant relationship of concurrent vancomycin and aminoglycoside therapy to higher incidence of nephrotoxicity was found to be serum drug concentrations which exceeded the normal values.

Rybak et al. studied the nephrotoxicity of vancomycin alone and in combination with an aminoglycoside in 231 courses of antibiotic therapy in 224 patients (168 vancomycin alone, 63 vancomycin and aminoglycoside, 103 aminoglycoside alone) [41]. Nephrotoxicity was defined as an increase of 0.5 mg/dL of serum creatinine or 50% increase above baseline, whichever was greater. Targeted vancomycin peak and trough concentrations were 30–40 mg/L and <15 mg/L, respectively. Targeted aminoglycoside peak and trough concentrations were 4–10 mg/L and <2 mg/L, respectively. The incidence of nephrotoxicity was 5% in the vancomycin alone group, 11% in aminoglycoside alone group, and 22% in patients receiving both vancomycin and an aminoglycoside; these differences were all found to be statistically different. Following a multivariate analysis, increased incidence of nephrotoxicity was found to be associated with concurrent vancomycin and aminoglycoside therapy, treatment with vancomycin of greater than 10 days, and vancomycin serum trough greater than 10 mg/L.

A recent clinical trial compared daptomycin with standard therapy, either antistaphylococcal penicillin or vancomycin in combination with an aminoglycoside, for the treatment of bacteremia and endocarditis [42]. Cosgrove and colleagues utilized patient information collected during the trial to specifically analyze the incidence of nephrotoxicity in relation to use of low-dose gentamicin (mean daily dose of 3.1 mg/kg). A total of 236 patients were analyzed for adverse event data related to kidney function: 120 received daptomycin; 53 received vancomycin, of which 49 also received low-dose gentamicin; and 63 received antistaphylococcal penicillins, of which 59 also received low-dose gentamicin. A clinically significant decrease in creatinine clearance (CrCl) was defined as a decrease in CrCl to <50 mL/min if baseline CrCl was ≥50 mL/min. In patients with a baseline CrCl <50 mL/min, a significant decrease in CrCl occurred if CrCl decrease to <10 mL/min. A sustained decease in CrCl was defined if ≥2 sequential decreased CrCl measurements. CrCl was calculated by Cockcroft-Gault equation. After data evaluation, 8% in the daptomycin arm, 22% in the vancomycin arm, and 25% in the antistaphylococcal penicillin arm were found to have a decreased creatinine clearance. The median length of aminoglycoside therapy with vancomycin and antistaphylococcal penicillins was only 5 and 4 days, respectively. Twenty-two percent of patients who received low-dose gentamicin experienced decreased creatinine clearance versus only 8% of patients who did not, a statistically significant difference. After multivariate analysis, only age greater than 65 and receipt of any gentamicin were individual risk factors for nephrotoxicity. No differences were noted in the incidence of nephrotoxicity when comparing antistaphylococcal penicillins and vancomycin other than nephrotoxicity with antistaphylococcal penicillins that occurred earlier in therapy when compared to vancomycin. The authors suggest this finding may indicate that nephrotoxicity seen with vancomycin and gentamicin is more sustained.

Clinical importance: Guidelines for the treatment of endocarditis recommend the use of low-dose aminoglycosides with vancomycin for treatment with cases due to penicillin-resistant gram-positive strains, particularly MRSA [43]. Some clinicians may also expand the synergistic use of aminoglycosides with vancomycin in other types of infections involving MRSA, based on recommendations for endocarditis. For most of these infections, vancomycin troughs of 15–20 mg/L will be targeted, based on new guidelines released [7].

Management: If possible, patients should be evaluated as to whether the use of an aminoglycoside is essential with vancomycin therapy. Patients receiving concurrent therapy of vancomycin and aminoglycosides should be carefully monitored for development of nephrotoxicity, especially in cases where vancomycin troughs of 15–20 mg/L are targeted and in patients with possibly impaired baseline renal function. If nephrotoxicity develops, switching to alternative agents should be considered.

4.7.1.2 Indomethacin

Mechanism: In infants, indomethacin has been shown to lead to decreased renal elimination of vancomycin.

Literature: In a study of 11 neonates with patent ductus arteriosus (PDA) who received vancomycin, 6 infants received indomethacin, while the 5 others did not and served as controlled [44]. The pharmacokinetics of vancomycin in these two groups was compared. In the neonates who received both indomethacin and vancomycin, the volume of distribution was 0.71 L/kg (control 0.48 L/kg), the half-life was 24.6 h (control 7.0 L/kg), and the serum clearance was 23 mg/L/kg/h (control 54 mL/kg/h).

Clinical importance: The authors of the study recommended, based on these results, neonates with PDA receiving indomethacin should be adequately treated with once-daily doses of vancomycin. Additionally, maintenance dosing of vancomycin should be approximately half of the dose used in neonates not receiving indomethacin.

Management: While these findings have not been confirmed on a larger scale in the neonatal population or in other patient populations, care should be used when administering vancomycin with other renally eliminated drugs, especially in patients demonstrating impaired renal function. Appropriate dose adjustments for all renally eliminated drugs, including vancomycin, should be employed when renal impairment is present.

4.7.1.3 Vecuronium

Mechanism: Vancomycin can depress neuromuscular function as well as skeletal muscle function.

Literature: A case report of a 34 kg patient received vancomycin as surgical prophylaxis during an exploratory laparoscopy [45]. Prior to the administration of the vancomycin, tracheal intubation was performed using vecuronium, and the patient's muscular response was appropriately monitored using the train-of-four (electrical stimulation of the ulnar nerve). T1 function had returned to 35%, and T4 was barely perceptible 20 min after induction of anesthesia. When the infusion of 1 g vancomycin was started, a rapid decrease of T1 to less than 10% and absence of a T4 response was noted. Within 3 min of the completion of the vancomycin infusion, the T1 and T4 response recovered steadily. Neuromuscular function continued to increase after reversal of the vecuronium with atropine and edrophonium after completion of surgery. Five minutes after administration of the edrophonium, responses decreased to levels prior to administration. While the patient was awake and able to control breathing, she was not able to sustain headlift. Twenty minutes after the injection of edrophonium, the patient regained adequate muscle tone response. No other side effects were noted. The serum concentration of the vancomycin 25 min after the start of infusion was found to be 70 mg/L.

Clinical importance: Given the patient's smaller weight in addition to administration of 1 g of vancomycin over 35 min, the peak concentration is larger than what has typically been encountered clinically. However, given new recommendations of aggressive vancomycin dosing, larger serum concentrations are likely to be seen with vancomycin. The authors of the case report also note several papers that also supported evidence of vancomycin influencing neuromuscular function occurring at typical vancomycin peak concentrations of 40–50 mg/L.

Management: Clinicians should be aware of the potential interaction of vancomycin with neuromuscular blockers. Strategies to prevent large peaks of vancomycin such as infusing over at least an hour or more (dependent on the dose) and carefully evaluating the appropriate dose based on patient's actual body weight and renal functions will help to reduce the possibility of vancomycin-related neuromuscular blockade. In patients receiving surgical neuromuscular blockade use or in patients in intensive care units receiving neuromuscular blockers, neuromuscular function should be appropriately monitored when used in conjunction with vancomycin.

4.7.1.4 Heparin

Mechanism: Heparin and vancomycin are incompatible in admixtures or y-sites due the concentration-dependent acid-base reaction that leads to precipitation and inactivation of vancomycin.

Literature: Barg and colleagues present a case of persistent staphylococcal bacteremia in an intravenous drug abuser [46]. The patient presented with fever, shaking chills, and diaphoresis; he admitted to intravenous heroin use for the past 3 years, with the femoral veins as a frequent place of injection. Upon presentation, swelling and tenderness at the injection site in the right groin were noted. A subclavian venous line was placed due to lack of peripheral venous access. Antibiotic treatment was initiated with vancomycin and tobramycin, with the tobramycin being soon discontinued after *Staphylococcus aureus* was identified as the causative pathogen. Additionally, continuous heparin anticoagulation was initiated after the discovery of deep vein thrombosis, confirmed by venogram. Both heparin and vancomycin were infused in the same line. Fevers and positive blood cultures persisted for 7 days despite antibiotic treatment. On the seventh day, another intravenous line was placed. With the availability of two intravenous lines, heparin and vancomycin were administered in separate lines. Within 24 h of the second line placement, fevers dispersed and blood cultures became (and remained) negative. Further investigations performed by the authors of the case examined the effects of co-administering vancomycin and heparin in vitro. At higher vancomycin concentrations of 1–5 mg and heparin concentrations of 1–1000 units/mL (concentrations similar to what would be seen if administered through the same line), a white precipitate was immediately formed. Concentrations of vancomycin similar to those seen in the serum (5, 50, and 100 μg/mL) when combined with less than 1 unit/mL of heparin did not demonstrate the formation of any precipitate.

Clinical importance: Because of the high incidence of both staphylococcal infections and deep vein thrombosis in intravenous drug users, the likelihood of the co-administration of both drugs is high.

Management: Clinicians should be appropriately educated that vancomycin and heparin should not be administered through the same intravenous line. While using sodium chloride for admixtures (instead of dextrose) may reduce the likelihood of precipitation, these two drugs should be administered via separate lines whenever possible.

4.7.1.5   Piperacillin/Tazobactam

Mechanism: Piperacillin may enhance the nephrotoxic effects of vancomycin.

Literature: Many retrospective, observational cohort studies have individually assessed the risk of nephrotoxicity with the combination of vancomycin and piperacillin/tazobactam. A recent meta-analysis of 15 studies compared the risk of nephrotoxicity of the combination of vancomycin and piperacillin/tazobactam to vancomycin alone as well as vancomycin and other beta-lactam combinations [47]. When all eligible studies were included in the meta-analysis, an odds ratio of 3.649 (95% CI 2.157–6.174) was determined for the patients receiving the combination of vancomycin, and piperacillin/tazobactam compared to vancomycin and other beta-lactam therapy had a greater than three times likelihood of developing nephrotoxicity (odds ratio [OR] 3.649, 95% confidence interval [CI] 2.157–6.174). Further analysis also demonstrated an increased risk with high-quality studies and comparing the combination of vancomycin and piperacillin/tazobactam to vancomycin alone [47].

Clinical importance: Broad-spectrum antimicrobials are a cornerstone of sepsis and septic shock management. Vancomycin and piperacillin/tazobactam are commonly used as empiric therapy in septic patients until the source of infection is identified when cultures return. Alternatives to piperacillin/tazobactam often require two separate drugs, and administration may be hampered by the number of intravenous access ports versus the number of IV fluids, pressors, and other drugs used in the management of septic shock. The addition of further prospective clinical studies of this interaction may provide more information for clinical management.

Management: As with any empiric antimicrobial choice, patients should be critically assessed using antimicrobial stewardship principles. Considering the patient's infection risk factors, likely source of infection, and risk of having a multidrug-resistant organism may allow the healthcare provider to choose a less broad agent than piperacillin/tazobactam. Assessing the patient's risk for MRSA with the help of rapid diagnostics may allow a shorter duration of unnecessary vancomycin therapy. In those patients that do require broad-spectrum antimicrobials like the combination of piperacillin/tazobactam and vancomycin, assessment of the patient's risk for renal dysfunction (e.g., age, previous history, concomitant medication, severity of illness, site of infection, hydration status) may prompt use of an alternative agent to piperacillin/tazobactam. Narrowing antimicrobial therapy once a causative organism is identified is also essential.

4.7.1.6 Bile Acid Sequestrants

Mechanism: Agents such as cholestyramine and colestipol are used to treat hyperlipidemia by utilizing their ability to bile acids in the intestines. These agents are also able to bind other materials in the intestines such as co-administered drugs and cytotoxins.

Literature: Taylor and Bartlett studied the binding of *Clostridium difficile* cytotoxins and vancomycin by cholestyramine and colestipol using an in vitro and hamster model of *C. difficile* colitis [48]. The use of cholestyramine and colestipol alone displayed extreme reductions in toxin to below assay sensitivity in vitro. When vancomycin was combined with either of the agents, less than 25% of the vancomycin concentration was detectable. Colestipol bound a greater amount of vancomycin than cholestyramine, but vancomycin was more strongly bound to cholestyramine. In the hamster model, the use of cholestyramine alone, vancomycin alone, and cholestyramine plus vancomycin resulted in a smaller percentage of mortality during the treatment period of 5 days. When following these animals for an additional 11 days, all three treatment arms prevented death longer than controls alone. Vancomycin prevented 100% cumulative mortality for a longer time than use of cholestyramine alone or cholestyramine plus vancomycin. The authors speculate that the 100% mortality reported after the additional 11 days was likely due to reacquisition of *C. difficile* from the environment rather than from the ability of the drug(s) to adequately treat the infection.

Clinical importance: Cholestyramine and colestipol are not recommended as alternative treatments for the treatment of persistent *C. difficile* colitis [7], primarily due to the likelihood that these agents will bind the two medications recommended for treatment, i.e., vancomycin and metronidazole.

Management: Because of the dosing schedule of vancomycin (usually every 6–8 h initially), attempts to avoid interactions by creative scheduling to separate administration times as far apart as possible (at least 1–2 h before administration or 4–6 h after administration) are likely to fail. Use of oral vancomycin for treatment of *C. difficile* should not be in combination with oral binding agents. In cases where patients have been prescribed cholestyramine for the treatment of hyperlipidemia who also need to use vancomycin to treat *C. difficile* infections, alternative agents for hyperlipidemia should be used during the duration of vancomycin treatment.

4.7.2 Daptomycin

4.7.2.1 HMG-CoA Reductase Inhibitors

Mechanism: Both daptomycin and HMG-CoA reductase inhibitors are known to cause increased levels of creatine phosphokinase (CPK), a marker of muscle injury.

Literature: Literature has been published on each agent's individual ability to increase CPK and possible cause rhabdomyolysis and has been included in each product's labeling [2, 49]. Odero and colleagues published a case report of rhabdo-myolysis and acute renal failure associated with the administration of daptomycin, simvastatin, niacin, and esomeprazole [22]. Daptomycin was initially started at 7.2 mg/kg q24h; 4 days after initiation, the dose was changed to 7.2 mg/kg q48h, after noting a serum creatinine of 1.5 mg/dL. After 16 days of treatment with dap-tomycin, patient complaints included muscle weakness and pains in the proximal thighs and arms. Daptomycin was discontinued and linezolid initiated; simvastatin and niacin were continued. A maximal serum CPK level of 8995 units/L and serum creatinine of 3.4 mg/dL were noted. Creatinine levels returned to baseline levels 6 days after daptomycin discontinuation; CPK concentrations were noted to be 125 units/L on the seventh day after daptomycin discontinuation. While the authors note that simvastatin and niacin both have the potential to cause rhabdomyolysis, they also note that the patient had tolerated simvastatin and niacin previously, with no complaints.

Clinical importance: Given the high prevalence of HMG-CoA reductase inhibi-tor use among patients and the increasing incidence of methicillin-resistant *Staphylococcus aureus*, these two agents will be likely used together more often. Clinicians should be aware of the potential interaction and be prepared to monitor these patients frequently who require co-administration of these two agents, espe-cially when daptomycin is used for longer periods of time and/or maximal doses of the HMG-CoA reductase inhibitor are used.

Management: Weekly measurements of CPK levels are recommended in the use of daptomycin alone. More frequent CPK monitoring (two to three times a week) during the concomitant use of these two agents is recommended, particularly in patients who also have renal impairment when receiving both medications [2]. Additionally, patients should be evaluated for unexplained muscle pain and/or weakness, especially in the distal extremities.

When used alone, in patients who have a CPK elevation of greater than 1000 units/L with signs and symptoms or greater than 2000 units/L without signs or symptoms, daptomycin should be discontinued, per manufacturer's recommendations. Furthermore, other drugs associated with rhabdomyolysis, like HMG-CoA reductase inhibitors, should be discontinued temporarily until CPK levels return to baseline.

4.7.2.2 Aminoglycosides

Mechanism: Not fully known.

Literature: While the mechanism is unknown, studies of co-administration of daptomycin and tobramycin showed that daptomycin helped protect against tobramycin-induced nephrotoxicity in rats [50, 51]. In this same experiment, the use of vancomycin and tobramycin demonstrated greater damage to the proximal tubular cells. A further investigation into this phenomenon alluded to the possibility of daptomycin directly interacting with the tobramycin molecule to prevent tobramycin-induced changes in the proximal tubular cells.

Clinical importance: Daptomycin has been studied in vitro and in vivo in animals in combination with aminoglycosides; large-scale trials of the combination of these two drugs in humans have not been performed. The antimicrobial and clinical benefits gained from the addition of an aminoglycoside to a daptomycin regimen have not been fully elucidated. The small number of patients that did receive daptomycin with gentamicin in a large-scale trial comparing daptomycin to vancomycin and antistaphylococcal penicillins demonstrated a smaller cumulative percentage of patients who experienced a significant decrease in creatinine clearance when compared to daptomycin or vancomycin/antistaphylococcal penicillins with or without aminoglycosides.

Management: Administration of both of these drugs should not cause significant negative effects in terms of renal function. Because the antimicrobial benefits (such as synergy) of this combination of drugs have not been fully studied in human trials, the use of both of these agents together should be evaluated on an individual basis [52].

4.7.2.3 PT/INR Laboratory Results

Mechanism: Daptomycin may interact with recombinant thromboplastin reagents, leading to a prolongation of PT and increase in INR due to laboratory artifacts [2, 53]. This interaction is believed to involve reagents that contain phospholipids, which are acted upon by daptomycin during the laboratory procedure [54].

Literature: During clinical trials, a small number (<1%) were noted to have prolonged prothrombin times or elevated INRs. Post-marketing surveillance also demonstrated this trend. One report strictly demonstrated a concentration-related fall in INR following daptomycin administration. Despite elevated INRs, no bleeding complications were demonstrated [55]. Further evaluation of commercially available reagents demonstrated that elevation of INR is reagent specific and may not be applicable to all reagents [55]. Further evaluation of the prothrombin time prolongation was demonstrated to be concentration dependent [56].

Clinical importance: Because of the likelihood of interaction, in patients who are started on daptomycin and need PT and/or INR laboratory results, clinicians should note any unexpected changes in PT and/or INR results. Daptomycin has not been found to interact with warfarin; therefore, the clinician should suppose the interaction to be due to a lab assay interaction and not drug-drug interaction. One case report did demonstrate thrombocytopenia accompanied with INR prolongation, but these two findings were thought to be independent of each other. Thrombocytopenia did resolve after discontinuation of the drug [57].

Management: In cases where an abnormal PT and/or INR measurement is documented, the clinicians should schedule another assessment of PT and INR just prior to the next dose of daptomycin. The clinician may also confer with the laboratory staff which commercial reagents are used for prothrombin and INR testing and compare these to published literature [55]. Additionally, other sources of interactions should be evaluated as relevant.

4.7.3 Telavancin

4.7.3.1 QT$_c$-Prolonging Drugs

Mechanism: Because telavancin has been shown to prolong the QT$_c$ interval during clinical trials, the possibility exists that co-concomitant use with other drugs known to prolong the QT$_c$ interval would have additive effects (Table 4.3). Prolonged QT$_c$ intervals can lead to torsades de points, ventricular arrhythmias, and sudden cardiac death.

Literature: A large body of literature exists on the potential for arrhythmias following administration of one or more drugs known to prolong the QT$_c$ intervals [58]. The website crediblemeds.org provides an up-to-date, well-maintained database of QT$_C$ interactions. For telavancin, a randomized, multidose clinical study conducted in healthy subjects found that mean changes in QT$_c$ intervals were 4.1 and 4.5 ms following administration of 7.5 mg/kg and 15 mg/kg doses, respectively [3, 58]. None of the study subjects demonstrated any significant ECG abnormalities or clinical symptoms beyond the interval changes. Change in QT$_c$ intervals was not found to correlate to concentrations of telavancin.

Table 4.3 QTc-prolonging agents [52, 65]

Select agents with risk of QT$_c$ prolongation	Select agents with possible risk of QT$_c$ prolongation
Amiodarone	Alfuzosin
Astemizole[a]	Amantadine
Bepridil[b]	Atazanavir
Chlorpromazine	Azithromycin
Cisapride[b,c]	Clozapine
Clarithromycin	Dronedarone[b]
Dofetilide	Flecainide
Droperidol	Foscarnet
Erythromycin	Fosphenytoin
Haloperidol	Gatifloxacin
Levomethadyl[b]	Gemifloxacin
Mesoridazine[b]	Levofloxacin
Methadone	Lithium
Pentamidine	Moxifloxacin
Pimozide[b]	Ondansetron
Procainamide	Paliperidone
Quinidine	Risperidone
Sotalol	Tacrolimus
Sparfloxacin[a,b]	Tamoxifen
Terfenadine[a]	Telithromycin
Thioridazine[b]	Venlafaxine
	Voriconazole
	Ziprasidone

[a]No longer available in the USA
[b]Concomitant use with telavancin is contraindicated
[c]Only available through a restricted access program

Clinical importance: Given that telavancin has only recently been approved, no clinical reports of substantial effects of QT$_c$ interval prolongation following the use of telavancin for treatment of infections have been published as of now.

Management: Interactions of this type are best managed by using alternative treatments that do not carry the QT$_c$ prolongation risk, if possible. If no alternatives exist, clinicians should consider cardiac monitoring for patients who require treatment with one or more drugs known to prolong the QT$_c$ interval. Clinicians should also use caution with telavancin in patients who already exhibit QT$_c$ prolongation prior to treatment.

4.7.3.2 Coagulation Panels

Mechanism: Because of the nature of telavancin to bind artificial phospholipid surfaces, telavancin will bind these types of surfaces which are commonly used in anticoagulation tests [3].

Literature: The effects of telavancin on coagulation panels have been studied in in vitro studies with varying commercial reagents and varying concentrations of telavancin [59–61]. A dose-dependent effect of telavancin has been shown with both INR and aPTT; however, the degree of this effect is noted to be dependent on reagents used.

Clinical importance: The degree of binding is dependent upon the commercial assay used, as reagents differ among these assays. False elevations of PT, INR, aPTT, and ACT have been noted, and, in patients where monitoring of these levels is used to dose antithrombotic agents, clinicians should be aware of this possibility and should attempt to schedule telavancin doses and lab draws at times so as not to interfere [62].

Management: The likelihood of the interaction decreases as the plasma concentrations decrease; therefore, in cases of patients who need to daily monitor these lab values, the ideal time for these lab draws to take place is just prior to the administration of the next telavancin dose. In cases where multiple lab draws are needed per day, as in the case of adjustment of heparin by aPTT values, use of another antimicrobial agent might be preferred.

4.7.4 Oritavancin

4.7.4.1 Heparin

Mechanism: Oritavancin falsely elevates aPTT test results, likely by binding to and preventing the action of phospholipid reagents that activate coagulation. Oritavancin does not have effects on coagulation in vivo [6].

Literature: Due to similar agents demonstrating effects on coagulation panels, investigators evaluated the effects of oritavancin on coagulation tests. Dose-dependent effects were noted for aPTT [63].

Clinical importance: Use of heparin within 5 days of administration of oritavancin is contraindicated.

Management: Given the importance of heparin in the management of acute cardiac syndromes, the need to avoid use of heparin in patients who have received oritavancin may prove problematic. Other anticoagulants may need to be considered, such as those used in patients who have heparin allergies. Alternatively, chromogenic factor Xa monitoring is not influenced by oritavancin and may be considered. Xa monitoring of heparin is currently used at some institutions instead of aPTT monitoring of heparin. However, in institutions that do not normally utilize Xa assays for monitoring, the laboratory may not be equipped or appropriately calibrated to provide accurate Xa levels that can be used to guide heparin therapy. Additionally, nursing may be unfamiliar with dose adjustments based on Xa instead of aPTT, and the risk for errors increases with an already high-risk medication.

4.7.4.2 Warfarin

Mechanism: Oritavancin causes false elevations in INR up to 12 h after administration [6].

Literature: Due to similar agents demonstrating effects on coagulation panels, investigators evaluated the effects of oritavancin on coagulation tests. Dose-dependent effects were noted for prothrombin time and INR [63].

Clinical importance: Due to false elevations in INR, test results may be unreliable to adjust warfarin dosing. After screening, drug-drug interaction study indicated that oritavancin may be a nonspecific, weak inhibitor of 2C9 and 2C19, and an additional study was done to assess the effect of a 1200 mg dose on warfarin pharmacokinetics. No effect was demonstrated.

Management: The package insert states to avoid the use of oritavancin in drugs with a narrow therapeutic window, like warfarin, due to the potential of CYP450 inhibition. If warfarin is used in a patient that has also received oritavancin, patients should be monitored for signs and symptoms of bleeding. Defer INR measurements until at least 12 h after the administration of oritavancin. If INR measurements are needed sooner, consider alternative methods to monitor warfarin activity, such as chromogenic factor Xa assays.

4.7.4.3 Anticoagulation Tests

Mechanism: Oritavancin binds and prevents the action of phospholipid reagents in commonly used laboratory coagulation tests, including prothrombin time (PT), INR, aPTT, activated clotting time (ACT), silica clotting time (SCT), dilute Russell's viper venom time (DRVVT), and D-dimer.

Literature: Due to similar agents demonstrating effects on coagulation panels, investigators evaluated the effects of oritavancin on coagulation tests. Dose-dependent effects were noted for prothrombin time, INR, aPTT, and dilute Russell's viper venom time [63].

Clinical importance: The length of coagulation interferences depends on the tests. Oritavancin will only have up to a 12-h effect on PT and INR, an 18-h effect on SCT, a 24-h effect on ACT, and a 72-h effect on DRVVT and D-dimer [63]. The effect on aPTT is significantly longer as discussed under the heparin interaction.

Management: Chromogenic factor Xa and thrombin time are unaffected by oritavancin and may be utilized if necessary [6]. Alternatively, these test measures could be performed after the appropriate amount of time has passed when oritavancin will no longer influence that test result.

4.8 Summary and Conclusions

Use of these agents is likely to increase as the incidence of methicillin-resistant *S. aureus* infections increases. In general, few drug interactions exist for these classes of drugs. The key drug interaction to consider with vancomycin is the concomitant use of other nephrotoxic agents, especially in those patients who may already exhibit decreased renal function. The use of daptomycin with HMG-CoA reductase inhibitors should be limited and monitored closely due to the possibility of rhabdomyolysis. More information regarding possible drug interactions with lipoglycopeptides will likely be available as the drug becomes more widely used and additional FDA indications are investigated. Caution should be used when administering telavancin with other QT_c-prolonging drugs or in patients who present with a QT_c prolongation. In patients requiring anticoagulation with warfarin or heparin, alternative agents to oritavancin should be considered to avoid false elevations in anticoagulation monitoring.

References

1. Guskey MT, Tsuji BT (2010) A comparative review of the lipoglycopeptides: oritavancin, dalbavancin, and telavancin. Pharmacotherapy 30:80–94
2. (2017) Cubicin (daptomycin) product information. Cubist Pharmaceuticals, Inc, Lexington
3. (2016) Vibativ (telavancin) product information. Theravance, Inc., South San Francisco
4. (2017) Vancomycin product information. Hospira, Inc., Lake Forest
5. (2016) Dalvance (dalbavancin) product information. Durata Therapeutics US Ltd, Parsippany
6. (2016) Orbactiv (oritavancin) product information. The Medicines Company, Parsippany
7. Rybak M, Lomaestro B, Rotschafer JC et al (2009) Therapeutic monitoring of vancomycin in adult patients: a consensus review of the American Society of Health-System Pharmacists, the Infectious Diseases Society of America, and the Society of Infectious Diseases Pharmacists. Am J Health Syst Pharm 66:82–98
8. (2011) Vancocin (vancomycin) product information. Viro Pharma Incorporated, Exton
9. Cohen SH, Gerding DN, Johnson S et al (2010) Clinical practice guidelines for Clostridium difficile infection in adults: 2010 update by the society for healthcare epidemiology of America (SHEA) and the infectious diseases society of America (IDSA). Infect Control Hosp Epidemiol 31:431–455
10. Gerding DN, Muto CA, Owens RC Jr (2008) Treatment of Clostridium difficile infection. Clin Infect Dis 46(Suppl 1):S32–S42
11. Dudley MN, Quintiliani R, Nightingale CH, Gontarz N (1984) Absorption of vancomycin. Ann Intern Med 101:144
12. Tedesco F, Markham R, Gurwith M, Christie D, Bartlett JG (1978) Oral vancomycin for antibiotic-associated pseudomembranous colitis. Lancet 2:226–228
13. Boereboom FT, Ververs FF, Blankestijn PJ, Savelkoul TJ, van Dijk A (1999) Vancomycin clearance during continuous venovenous haemofiltration in critically ill patients. Intensive Care Med 25:1100–1104
14. Somerville AL, Wright DH, Rotschafer JC (1999) Implications of vancomycin degradation products on therapeutic drug monitoring in patients with end-stage renal disease. Pharmacotherapy 19:702–707

15. Cunha BA, Eisenstein LE, Hamid NS (2006) Pacemaker-induced Staphylococcus aureus mitral valve acute bacterial endocarditis complicated by persistent bacteremia from a coronary stent: cure with prolonged/high-dose daptomycin without toxicity. Heart Lung 35:207–211
16. Cunha BA, Krol V, Kodali V (2008) Methicillin-resistant Staphylococcus aureus (MRSA) mitral valve acute bacterial endocarditis (ABE) in a patient with Job's syndrome (hyperimmunoglobulin E syndrome) successfully treated with linezolid and high-dose daptomycin. Heart Lung 37:72–75
17. Cunha BA, Mickail N, Eisenstein L (2007) E. faecalis vancomycin-sensitive enterococcal bacteremia unresponsive to a vancomycin tolerant strain successfully treated with high-dose daptomycin. Heart Lung 36:456–461
18. Figueroa DA, Mangini E, Amodio-Groton M et al (2009) Safety of high-dose intravenous daptomycin treatment: three-year cumulative experience in a clinical program. Clin Infect Dis 49:177–180
19. Katz DE, Lindfield KC, Steenbergen JN et al (2008) A pilot study of high-dose short duration daptomycin for the treatment of patients with complicated skin and skin structure infections caused by gram-positive bacteria. Int J Clin Pract 62:1455–1464
20. Lichterfeld M, Ferraro MJ, Davis BT (2010) High-dose daptomycin for the treatment of endocarditis caused by Staphylococcus aureus with intermediate susceptibility to glycopeptides. Int J Antimicrob Agents 35:96
21. Moise PA, Hershberger E, Amodio-Groton MI, Lamp KC (2009) Safety and clinical outcomes when utilizing high-dose (> or =8 mg/kg) daptomycin therapy. Ann Pharmacother 43:1211–1219
22. Lubbert C, Rodloff AC, Hamed K (2015) Real-world treatment of enterococcal infections with daptomycin: insights from a large european registry (EU-CORE). Infect Dise Ther 4:259–271
23. Oleson FB, Berman CL, Li AP (2004) An evaluation of the P450 inhibition and induction potential of daptomycin in primary human hepatocytes. Chem Biol Interact 150:137–147
24. Steenbergen JN, Mohr JF, Thorne GM (2009) Effects of daptomycin in combination with other antimicrobial agents: a review of in vitro and animal model studies. J Antimicrob Chemother 64:1130–1138
25. Odero RO, Cleveland KO, Gelfand MS (2009) Rhabdomyolysis and acute renal failure associated with the co-administration of daptomycin and an HMG-CoA reductase inhibitor. J Antimicrob Chemother 63:1299–1300
26. Schriever CA, Fernandez C, Rodvold KA, Danziger LH (2005) Daptomycin: a novel cyclic lipopeptide antimicrobial. Am J Health Syst Pharm 62:1145–1158
27. Stein GE (2005) Safety of newer parenteral antibiotics. Clin Infect Dis 41(Suppl 5):S293–S302
28. Chakraborty A, Roy S, Loeffler J, Chaves RL (2009) Comparison of the pharmacokinetics, safety and tolerability of daptomycin in healthy adult volunteers following intravenous administration by 30 min infusion or 2 min injection. J Antimicrob Chemother 64:151–158
29. Saravolatz LD, Stein GE, Johnson LB (2009) Telavancin: a novel lipoglycopeptide. Clin Infect Dis 49:1908–1914
30. Stucki A, Gerber P, Acosta F, Cottagnoud M, Cottagnoud P (2006) Efficacy of telavancin against penicillin-resistant pneumococci and Staphylococcus aureus in a rabbit meningitis model and determination of kinetic parameters. Antimicrob Agents Chemother 50:770–773
31. Gander S, Kinnaird A, Finch R (2005) Telavancin: in vitro activity against staphylococci in a biofilm model. J Antimicrob Chemother 56:337–343
32. LaPlante KL, Mermel LA (2009) In vitro activities of telavancin and vancomycin against biofilm-producing Staphylococcus aureus, S. epidermidis, and Enterococcus faecalis strains. Antimicrob Agents Chemother 53:3166–3169
33. Wong SL, Goldberg MR, Ballow CH, Kitt MM, Barriere SL (2010) Effect of Telavancin on the pharmacokinetics of the cytochrome P450 3A probe substrate midazolam: a randomized, double-blind, crossover study in healthy subjects. Pharmacotherapy 30:136–143
34. (2009) Vfend (voriconazole) product information. Pfizer, New York

35. Stryjewski ME, Chu VH, O'Riordan WD et al (2006) Telavancin versus standard therapy for treatment of complicated skin and skin structure infections caused by gram-positive bacteria: FAST 2 study. Antimicrob Agents Chemother 50:862–867
36. Stryjewski ME, Graham DR, Wilson SE et al (2008) Telavancin versus vancomycin for the treatment of complicated skin and skin-structure infections caused by gram-positive organisms. Clin Infect Dis 46:1683–1693
37. Stryjewski ME, O'Riordan WD, Lau WK et al (2005) Telavancin versus standard therapy for treatment of complicated skin and soft-tissue infections due to gram-positive bacteria. Clin Infect Dis 40:1601–1607
38. Barriere S, Genter F, Spencer E et al (2004) Effects of a new antibacterial, telavancin, on cardiac repolarization (QTc interval duration) in healthy subjects. J Clin Pharmacol 44:689–695
39. Corey GR, Kabler H, Mehra P et al (2014) Single-dose oritavancin in the treatment of acute bacterial skin infections. N Engl J Med 370:2180–2190
40. Cimino MA, Rotstein C, Slaughter RL, Emrich LJ (1987) Relationship of serum antibiotic concentrations to nephrotoxicity in cancer patients receiving concurrent aminoglycoside and vancomycin therapy. Am J Med 83:1091–1097
41. Rybak MJ, Albrecht LM, Boike SC, Chandrasekar PH (1990) Nephrotoxicity of vancomycin, alone and with an aminoglycoside. J Antimicrob Chemother 25:679–687
42. Cosgrove SE, Vigliani GA, Fowler VG Jr et al (2009) Initial low-dose gentamicin for Staphylococcus aureus bacteremia and endocarditis is nephrotoxic. Clin Infect Dis 48:713–721
43. Nishimura RA, Carabello BA, Faxon DP et al (2008) ACC/AHA 2008 Guideline update on valvular heart disease: focused update on infective endocarditis: a report of the American College of Cardiology/American Heart Association Task Force on Practice Guidelines endorsed by the Society of Cardiovascular Anesthesiologists, Society for Cardiovascular Angiography and Interventions, and Society of Thoracic Surgeons. J Am Coll Cardiol 52:676–685
44. Spivey JM, Gal P (1986) Vancomycin pharmacokinetics in neonates. Am J Dis Child 140:859
45. Huang KC, Heise A, Shrader AK, Tsueda K (1990) Vancomycin enhances the neuromuscular blockade of vecuronium. Anesth Analg 71:194–196
46. Barg NL, Supena RB, Fekety R (1986) Persistent staphylococcal bacteremia in an intravenous drug abuser. Antimicrob Agents Chemother 29:209–211
47. Giuliano CA, Patel CR, Kale-Pradhan PB (2016) Is the combination of piperacillin-tazobactam and vancomycin associated with development of acute kidney injury? A meta-analysis. Pharmacotyerapy 36:1271–1228
48. Taylor NS, Bartlett JG (1980) Binding of Clostridium difficile cytotoxin and vancomycin by anion-exchange resins. J Infect Dis 141:92–97
49. Chatzizisis YS, Koskinas KC, Misirli G et al (2010) Risk factors and drug interactions predisposing to statin-induced myopathy: implications for risk assessment, prevention and treatment. Drug Saf 33:171–187
50. Beauchamp D, Pellerin M, Gourde P, Pettigrew M, Bergeron MG (1990) Effects of daptomycin and vancomycin on tobramycin nephrotoxicity in rats. Antimicrob Agents Chemother 34:139–147
51. Couture M, Simard M, Gourde P et al (1994) Daptomycin may attenuate experimental tobramycin nephrotoxicity by electrostatic complexation to tobramycin. Antimicrob Agents Chemother 38:742–749
52. Miro JM, Garcia-de-la-Maria C, Armero Y et al (2009) Addition of gentamicin or rifampin does not enhance the effectiveness of daptomycin in treatment of experimental endocarditis due to methicillin-resistant Staphylococcus aureus. Antimicrob Agents Chemother 53:4172–4177
53. van den Besselaar AMHP, Tripodi A (2007) Effect of daptomycin on prothrombin time and the requirement for outlier exclusion in international sensitivity index calibration of thromboplastin. J Thromb Haemost 5:1975–1976
54. van den Besselar AMHP, Breukink E, Koorengevel MC (2010) Phosphatidylglycerol and daptomycin synergistically inhibit tissue factor-induced coagulation in the prothrombin time test. J Thromb Haemost 8:1429–1430

55. Webster PS, Oleson FB, Paterson DL et al (2008) Interaction of daptomycin with two recombinant thromboplastin reagents leads to falsely prolonged patient prothrombin time/internationalized normalized ratio results. Blood Coagul Fibrinolysis 19:32–38

56. Yamada T, Kato R, Oda K et al (2016) False prolongation of prothrombin time in the presence of a high blood concentration of daptomycin. Basic Clin Pharmacol Toxicol 119:353–359

57. Hartmann B, Maus S, Keller F et al (2011) Thrombocytopenia, INR prolongation, and fall in fibrinogen under daptomycin. J Chemother 23:183–184

58. Cubeddu LX (2003) QT prolongation and fatal arrhythmias: a review of clinical implications and effects of drugs. Am J Ther 10:452–457

59. Barriere SL, Goldberg MR, Janc JW et al (2011) Effects of telavancin on coagulation test results. Int J Clin Pract 65:784–789

60. Gosselin R, Dager W, Roberts A (2011) Effect of telavancin (Vibativ) on routine coagulation test results. Am J Clin Pathol 136:848–854

61. Ero MP, Harvey NR, Harbert JL (2014) Impact of telavancin on prothrombin time and activated partial thromboplastin time as determined using point-of-care coagulometers. J Thromb Thrombolysis 38:235–240

62. Amanatullah DF, Lopez MJ, Gosselin RC et al (2013) Artificial elevation of prothrombin time by telavancin. Clin Orthop Relat Res 471:332–335

63. Belley A, Robson R, Francis JL et al (2017) Effects of oritavancin on coagulation tests in the clinical laboratory. AACA 61:e01968–e01916

64. Baxter K (2010) Stockley's drug interactions [online]. Pharmaceutical Press, London. Available at: www.medicinescomplete.com. Date accessed: 28 July 2010

65. Arizona CERT (2017) Available at: www.qtdrugs.org. Date accessed: 24 Jan 2017

66. Micromedex (2016). Available at: www.thomsonhc.com/hcs/librarian. Date accessed: 28 Dec 2016

Chapter 5
Miscellaneous Antibiotics

Gregory M. Susla

5.1 Introduction

This chapter discusses the interactions of antibiotics that may be the only available agents from a class of antibiotics that is used clinically today. Chloramphenicol and tetracycline are older agents that are less frequently prescribed; so many clinicians may not be familiar with their interactions with other medications. Many of the interacting agents also are less frequently prescribed, such as first-generation oral hypoglycemic agents. Since many of the interactions in this chapter are based on single case reports, it is often difficult to determine the mechanism of the interaction and if a true interaction exists. The existence of some interactions may be questioned because of other potential causes that may have been present when the interaction was discovered.

The interactions described in this chapter are summarized in Table 5.1.

5.2 Chloramphenicol

Chloramphenicol is a broad-spectrum antibiotic that has been shown to interact with a number of medications, including analgesics-antipyretics, other antibiotics, oral hypoglycemic agents, anticoagulants, and anticonvulsants. Most of these interactions are limited to case reports with small numbers of patients. The mechanism of the interaction for several of the interactions is unknown or is limited to speculation. Five to fifteen percent of chloramphenicol is excreted as free chloramphenicol in the urine; the remainder of a dose is metabolized in the liver to inactive metabolites, principally the glucuronide metabolite.

G. M. Susla (✉)
Medical Information, MedImmune, Gaithersburg, MD, USA

© Springer International Publishing AG, part of Springer Nature 2018
M. P. Pai et al. (eds.), *Drug Interactions in Infectious Diseases: Antimicrobial Drug Interactions*, Infectious Disease,
https://doi.org/10.1007/978-3-319-72416-4_5

Table 5.1 Antibiotic interactions

Primary drug	Interacting drug	Mechanism	Effects	Comments/management
Chloramphenicol	Acetaminophen	Increased chloramphenicol clearance	Reduced chloramphenicol concentrations Potential for therapeutic failure	Monitor chloramphenicol concentrations, and adjust dose as needed Use alternative agent for antipyresis or analgesia
	Anticonvulsants	Increased chloramphenicol clearance	Reduced chloramphenicol concentrations Potential for therapeutic failure	Monitor chloramphenicol concentrations, and adjust dose as needed patients should be monitored for clinical and microbiologic response to therapy
	Anticonvulsants	Decreased metabolism of phenytoin and phenobarbital	Increased serum concentrations of these anticonvulsants with increased CNS toxicity	Monitor phenytoin and phenobarbital concentrations, and adjust dose as needed
	Oral hypoglycemic agents	Decreased metabolism of tolbutamide and excretion of chlorpropamide	Increased half-life of tolbutamide and chlorpropamide with increased risk of hypoglycemia	Monitor blood glucose, and adjust dose of oral hypoglycemic agents as needed Monitor for clinical signs and symptoms of hypoglycemia
	Penicillins	Antagonism of bactericidal agents	Potential risk of therapeutic failure when both agents are administered concurrently	Monitor clinical and microbiologic response to therapy Monitor MIC and MBC of antibiotic combination and each antibiotic alone. Use alternative class of antibiotic
	Rifampin	Increased chloramphenicol clearance	Reduced chloramphenicol concentrations Potential for therapeutic failure	Monitor chloramphenicol concentrations, and adjust dose as needed patients should be monitored for clinical and microbiologic response to therapy

	Oral anticoagulants	Enhanced metabolism of warfarin Decreased gut production of vitamin K Altered production of prothrombin by hepatic cell	Increased risk of major and minor bleeding	Monitor PT/INR when beginning or discontinuing chloramphenicol therapy Monitor for clinical signs of bleeding
	Immunosuppressive agents	Decreased cyclosporine and tacrolimus clearance	Increased cyclosporine and tacrolimus concentrations Potential for cyclosporine and tacrolimus toxicity	Monitor cyclosporine and tacrolimus concentrations, and adjust dose as needed
	Voriconazole	Decreased voriconazole clearance	Increased voriconazole concentrations	Use alternative class of antibiotic Monitor voriconazole concentrations as appropriate
Clindamycin	Nondepolarizing neuromuscular blocking agents	Local anesthetic effect on myelinated muscle Stimulates nerve terminal and blocks postsynaptic cholinergic receptor Direct depressant action on the muscle	Prolonged duration of neuromuscular blockade	Patients receiving this combination of medications should have their neuromuscular function monitored with peripheral nerve stimulation to access the degree of paralysis induced by these agents Patients should be monitored for the potential development of respiratory failure
	Aminoglycosides	No clear evidence to support the hypothesis that clindamycin leads to an increased risk of nephrotoxicity when prescribed concurrently with aminoglycoside antibiotics		

(continued)

Table 5.1 (continued)

Primary drug	Interacting drug	Mechanism	Effects	Comments/management
	Paclitaxel	No clear evidence to support that clindamycin has any clinically significant effect on paclitaxel pharmacokinetic parameters when prescribed concurrently		
	Rifampin	Increased clindamycin clearance due to induction of CYP 3A4 and 2C8/9 isoenzymes	Reduced clindamycin concentrations Potential for therapeutic failure	Patients should be monitored for clinical and microbiologic response to therapy
Sulfonamides	Oral anticoagulants	Some sulfonamides appear to impair the hepatic metabolism of oral anticoagulants Competition for plasma protein-binding sites may play an additional role	An enhanced hypoprothrombinemic response to warfarin with an increased risk of minor and major bleeding	Monitor PT/INR when beginning or discontinuing sulfonamide therapy Monitor for clinical signs of bleeding
	Oral hypoglycemic agents (sulfonylureas)	Decreased metabolism by CYP2CP	Impair metabolism resulting in severe hypoglycemia	Prescribe alternative antibiotics if possible Monitor serum glucose levels during concomitant therapy
	ACE inhibitors and angiotensin receptor blockers	Decreased potassium renal excretion	Increase incidence of hyperkalemia	Use alternative class of antibiotic
	Spironolactone	Decreased potassium renal excretion	Increase incidence of hyperkalemia	Use alternative class of antibiotic
	MMX mesalamine	Decreased clearance	Nonsignificant increase in serum SMX concentrations	TMP/SMX appears to be safe when coadministered with MMX mesalamine

	Methotrexate	TMP/SMX and MTX are folic acid antagonists. TMP/MTX also inhibits MTX renal excretion	Increased risk for neutropenia, anemia, thrombocytopenia, and mucositis	Monitor for the development of signs and symptoms of bone marrow suppression during concomitant therapy
	Mefloquine	Carboxymefloquine may be responsible for increasing SMX metabolism	Reduced SMX concentrations Potential for therapeutic failure	Patients should be monitored for clinical and microbiologic response to therapy Use alternative class of antibiotic
Tetracyclines	Heavy metals, trivalent cations	Chelate tetracycline products in the gastrointestinal tract	Impair their absorption and decrease bioavailability Potential for therapeutic failure	Tetracycline products should be administered 2 h before or 6 h after an antacid H2-receptor antagonists and proton pump inhibitors may be prescribed in place of antacids Use alternative class of antibiotic Patients should be monitored for clinical and microbiologic response to therapy
	Colestipol	Bind tetracycline products in the gastrointestinal tract	Impair their absorption and decrease bioavailability Potential for therapeutic failure	Tetracycline products should be administered 2 h before or 3 h after colestipol Use alternative class of antibiotic Patients should be monitored for clinical and microbiologic response to therapy
	Digoxin	Tetracycline can suppress the gut flora responsible for metabolizing digoxin in the GI tract	Increased digoxin absorption and bioavailability may result in toxicity	Serum digoxin concentrations should be monitored and the dose adjusted with initiating or discontinuing antibiotic therapy

(continued)

Table 5.1 (continued)

Primary drug	Interacting drug	Mechanism	Effects	Comments/management
	Anticonvulsants	Anticonvulsants increase the hepatic metabolism of doxycycline reducing its serum concentration	Increased potential for therapeutic failure	Patients should be monitored for clinical and microbiologic response to therapy Renally eliminated tetracycline or other classes of antibiotics should be prescribed to avoid this interaction Doxycycline should be administered twice a day in patients on chronic anticonvulsant therapy
	Warfarin	Doxycycline enhances the anticoagulation response to oral anticoagulants	An enhanced hypoprothrombinemic response to warfarin with an increased risk of minor and major bleeding	Patients should be monitored for clinical signs and symptoms of bleeding when these drugs are used concurrently PT and/or INR should be monitored when these drugs are used concurrently Use alternative class of antibiotic
	Lithium	It is unclear if there is a direct interaction between lithium and tetracycline	Potential for increased serum lithium concentrations and lithium toxicity	Patients should be monitored for signs and symptoms of lithium toxicity when receiving lithium and tetracycline concurrently Monitor serum lithium concentrations when patients are receiving lithium and tetracycline
	Psychotropic agents	It is unclear as to the exact mechanism of the interaction	Possible potential for acute psychotic behavior	Monitor for signs and symptoms of acute psychotic behavior Use alternative class of antibiotic
	Theophylline	A reduction in theophylline metabolism	The reduction in clearance appears to be quite variable so that it may be difficult to predict how much of the theophylline concentration will increase following the addition of tetracycline to the medication regimen	Patients should be monitored clinically for signs and symptoms of theophylline toxicity Serum theophylline concentration should be closely monitored in patients at high risk for developing theophylline toxicity

	Oral contraceptives	Prospective trials have failed to document a consistent effect	Unexpected pregnancies	It is not known if noncompliance played a role in some of these unplanned pregnancies. Women should be counseled to use other methods of birth control during tetracycline therapy
	Methotrexate	Decreased methotrexate clearance	Increased methotrexate concentration. Potential for methotrexate toxicity	Monitor methotrexate concentrations. Maintain leucovorin rescue until methotrexate concentrations are below the desired range
	Rifampin	Increased doxycycline clearance	Increased potential for therapeutic failures in patients with *Brucellosis* infections	Monitor clinical and microbiologic response to therapy. Use alternative class of antibiotic
Tigecycline	Warfarin	Decreased warfarin clearance and increased warfarin AUC and half-life	Potential for prolongation of INR and risk of bleeding	Monitor INR, and adjust warfarin dose as needed. Use alternative antibiotic class
	Cyclosporine	Interaction potentially due to inhibition of CYP3A4 pathway	Reduced cyclosporine concentrations. Potential for organ rejection	Monitor cyclosporine concentrations, and adjust dose as needed patients should be monitored for clinical signs of organ rejection
	Tacrolimus	Interaction potentially due to inhibition of CYP3A4 pathway	Reduced tacrolimus concentrations. Potential for organ rejection	Monitor tacrolimus concentrations, and adjust dose as needed patients should be monitored for clinical signs of organ rejection
Aminoglycosides	Amphotericin B	Additive direct nephrotoxicity effects on the kidney	The concurrent administration of aminoglycoside antibiotics and amphotericin B may increase the risk of developing renal failure	Aminoglycoside concentrations should be monitored and the dosage regimen adjusted to maintain serum concentrations within the desired therapeutic range. Attempts should be made to avoid other conditions that increase the risk for developing nephrotoxicity (i.e., hypotension, IV contrast media). Avoid prescribing other agents that cause nephrotoxicity

(continued)

Table 5.1 (continued)

Primary drug	Interacting drug	Mechanism	Effects	Comments/management
	Neuromuscular blocking agents	Aminoglycosides have been shown to interfere with acetylcholine release and exert a postsynaptic curare-like action These agents have membrane-stabilizing properties and exert their effect on acetylcholine release by interfering with calcium ion fluxes at the nerve terminal, an action similar to magnesium ions Aminoglycosides also possess a smaller but significant decrease in postjunctional receptor sensitivity and spontaneous release	These drugs may cause postoperative respiratory depression when administered before or during operations and may also cause a transient deterioration in patents with myasthenia gravis	Patient should be monitored for prolonged postoperative paralysis if they received neuromuscular blocking agents and aminoglycoside antibiotics during the perioperative or immediate postoperative period.
	Indomethacin	Nonsteroidal anti-inflammatory agents may cause renal failure	Increased concentrations of renally eliminated medications	Serum concentrations of medications should be monitored when possible and dosage regimens adjusted to maintain serum concentrations within the accepted therapeutic ranges

Cyclosporine	Additive direct nephrotoxicity effects on the kidney	Concurrent administration of aminoglycoside antibiotics and cyclosporine may increase the risk of developing renal failure	Aminoglycoside and cyclosporine concentrations should be monitored and the dosage regimen adjusted to maintain serum concentrations within the desired therapeutic range Attempts should be made to avoid other conditions that increase the risk for developing nephrotoxicity (i.e., hypotension, IV contrast media) Avoid prescribing other agents that cause nephrotoxicity
Cisplatin	Additive direct nephrotoxicity effects on the kidney	Concurrent administration of aminoglycoside antibiotics and cisplatin-based chemotherapy regimens may increase the risk of developing renal failure	Aminoglycoside concentrations should be monitored and the dosage regimen adjusted to maintain serum concentrations within the desired therapeutic range Attempts should be made to avoid other conditions that increase the risk for developing nephrotoxicity (i.e., hypotension, IV contrast media) Avoid prescribing other agents that cause nephrotoxicity

(continued)

Table 5.1 (continued)

Primary drug	Interacting drug	Mechanism	Effects	Comments/management
	Loop diuretics	Ethacrynic acid may cause direct additive ototoxic effects on the ear	When ethacrynic acid is used alone or in combination with aminoglycosides, it should be used in low doses and titrated to maintain adequate urine output or fluid balance It is unclear whether furosemide directly increases the nephrotoxicity and ototoxicity of aminoglycosides	Aminoglycoside concentrations should be monitored and the dosage regimens adjusted to maintain concentrations within the therapeutic range Furosemide should be used with caution in patients receiving aminoglycoside antibiotics; careful attention should be paid to the patients' weight, urine output, fluid balance, and indices of renal function
	Vancomycin	Unclear if vancomycin increases the nephrotoxicity of aminoglycosides	The development of nephrotoxicity	Aminoglycoside and vancomycin concentrations should be monitored and the dosage regimen adjusted to maintain serum concentrations within the desired therapeutic range Attempts should be made to avoid other conditions that increase the risk for developing nephrotoxicity (i.e., hypotension, IV contrast media) Avoid prescribing other agents that cause nephrotoxicity

| Antipseudomonal penicillins | Penicillins combine with aminoglycoside antibiotics in equal molar concentrations at a rate dependent on the concentration, temperature, and medium composition. The greater the concentration of the penicillin, the greater the inactivation of the aminoglycoside. The inactivation is thought to occur by way of a nucleophilic opening of the beta-lactam ring, which then combines with an amino group of the aminoglycoside, leading to the formation of a microbiologically inactive amide | Unexpected low serum aminoglycoside concentrations for a given dose | Blood samples for aminoglycosides concentrations should be sent to the laboratory within 1–2 h so that the sample can be spun down and frozen if not assayed immediately. The two antibiotics should never be given at the same time; schedule administration time of the antibiotic so that the administration of the aminoglycoside occurs toward the end of the penicillin dosing interval. If a patient is receiving this antibiotic combination and unusually low aminoglycoside concentrations occur, the above factors should be checked |

(continued)

Table 5.1 (continued)

Primary drug	Interacting drug	Mechanism	Effects	Comments/management
Linezolid	Selective serotonin reuptake inhibitors	Decreased serotonin metabolism by inhibition of monoamine oxidase	Development of the serotonin syndrome	Review patient profile before prescribing linezolid Use alternative class of antibiotic If necessary, treat serotonin syndrome with serotonin antagonist cyproheptadine
	Systemic decongestants	Decreased metabolism by inhibition of monoamine oxidase	Increased blood pressure	Review patient profile before prescribing linezolid Use alternative class of antibiotic Consider using topical nasal decongestants
	Meperidine	Decreased metabolism by inhibition of monoamine oxidase	Development of serotonin syndrome	Review patient profile before prescribing meperidine Use alternative class of analgesic If necessary, treat serotonin syndrome with serotonin antagonist cyproheptadine
	Rifampin	Increased linezolid clearance	Unexpected low serum linezolid concentrations Development of therapeutic failures	Use alternative class of antibiotic
	Warfarin	Potential reduction in vitamin K levels	Increased INR and risk of bleeding	Monitor INR, and adjust warfarin dose as needed Use alternative antibiotic class
	Clarithromycin	Increased linezolid serum concentrations	Increased risk of linezolid toxicity	Adjust linezolid dose as needed Use alternative antibiotic class
Quinupristin-dalfopristin	Medications metabolized by cytochrome P450 3A4 enzyme	Decreased metabolism of medications by cytochrome P450 3A4 enzyme	Prolonged therapeutic effects or increased adverse reactions	Review patient profile before prescribing quinupristin-dalfopristin Use alternative class of antibiotic Monitor patients closely for signs of adverse effects

5.2.1 Acetaminophen

Chloramphenicol has been reported to increase, decrease, and have no effect on the half-life of acetaminophen. Spika and colleagues evaluated the effect of multiple doses of acetaminophen on chloramphenicol metabolism in patients with bacterial meningitis [1]. Significant differences in chloramphenicol peak serum concentration, volume of distribution, half-life, and clearance occurred between samples obtained before and during treatment with acetaminophen. Peak serum concentrations fell; volume of distribution and clearance increased, and half-life became shorter. The greatest change was in clearance, which increased by more than 300% from baseline values. During treatment with acetaminophen, the percent of chloramphenicol excreted unchanged in the urine decreased; its succinate metabolite remained unchanged, while the glucuronide metabolite increased by approximately 300%. Kearns also evaluated the effect of acetaminophen in acutely ill pediatric patients [2]. Chloramphenicol pharmacokinetic parameters were compared between a group of patients receiving acetaminophen and a group not receiving acetaminophen. There was no statistical difference in the chloramphenicol pharmacokinetic parameters between the two groups. However, there was a clinically significant increase in chloramphenicol clearance and decrease in half-life between the initial dose and final dose in the patients receiving acetaminophen. Following acetaminophen therapy, the chloramphenicol half-life decreased by approximately 33%, from 3.4 h to 2.2 h, while its clearance increased by more than 50%, from 5.5 to 8.9 mL/min/kg. The peak chloramphenicol serum concentrations were lower after the final dose than at steady state, 15.7 versus 22.7 mg/L, respectively. Stein was unable to document any effect of acetaminophen on chloramphenicol metabolism in hospitalized adult patients [3]. In a randomized crossover design, patients received either chloramphenicol or chloramphenicol with acetaminophen for 48 h. There was no significant difference in peak and trough chloramphenicol concentrations, half-life, or area under the concentration-time curve between the two treatment periods.

Although the mechanism of this interaction is unclear, it appears to be an alteration in clearance. This interaction may take several days to manifest its full effect, and in some studies, patients may not have been studied for a long enough period of time to evaluate fully the effects of acetaminophen on chloramphenicol pharmacokinetic parameters. Although Spika suggested that the increase in chloramphenicol clearance was due to an increased in glucuronidation, other investigators have not confirmed this.

This interaction may be important in patients receiving chloramphenicol for the treatment of central nervous system infections or infections due to organisms that are resistant to more traditional antibiotics. Reduced peak concentrations or increases in clearance without appropriate adjustments in dosage regimens to account for these changes may result in therapeutic failures. Patients receiving chloramphenicol and acetaminophen should have chloramphenicol serum concentrations monitored every 2–3 days during a course of therapy, especially during the later part of therapy when it appears that chloramphenicol concentrations may begin to decline. Dosage regimens should be adjusted to maintain chloramphenicol concentrations within the desired therapeutic range. Other agents such as aspirin or ibuprofen may be used as alternatives to acetaminophen for antipyresis and analgesia.

5.2.2 Anticonvulsants

Anticonvulsants have been shown to increase the metabolism of chloramphenicol by increasing its hepatic metabolism. Phenobarbital has been shown to stimulate the metabolism of chloramphenicol in several case reports [4, 5]. In addition, chloramphenicol has been shown to reduce the metabolism of phenytoin and phenobarbital when both agents are administered concurrently [6–10]; the onset of these interactions appears to be rapid and may persist for several days after chloramphenicol is discontinued.

The reduction in phenytoin and phenobarbital metabolism is mostly likely due to a competition for metabolic enzymes. The clinical significance of the interaction is the potential for patients to develop phenytoin and/or phenobarbital toxicity after beginning chloramphenicol therapy. Patients may show signs of lethargy, excessive sedation, nystagmus, hallucinations, or other mental status changes. Because phenytoin undergoes nonlinear metabolism, toxic serum concentrations may not occur for several days after starting chloramphenicol. After the maximum rate of phenytoin metabolism is exceeded, serum concentrations will rise rapidly and may remain elevated for a period of time after the chloramphenicol is discontinued. Due to phenobarbital's long half-life, its sedative effects can be expected to resolve slowly as the serum concentration falls.

Patients receiving chloramphenicol with either phenytoin or phenobarbital must have their anticonvulsant serum concentrations monitored frequently, preferably every 3–5 days if possible, to detect increases in the concentrations. Patients also should be monitored clinically for the development of signs and symptoms of phenytoin and/or phenobarbital toxicity.

Phenobarbital has been shown to increase the metabolism of chloramphenicol, resulting in a reduction in its peak serum concentrations. Bloxham reported two patients who received chloramphenicol and phenobarbital for the treatment of meningitis [4]. In one patient, peak chloramphenicol serum concentrations fell from 31 mg/L on day 2 and day 3 to less than 5 mg/L on day 5. Patients receiving concurrent therapy with chloramphenicol and phenobarbital should have chloramphenicol concentrations monitored daily to monitor for reductions in the serum concentration. The chloramphenicol dosage regimen needs to be adjusted to maintain therapeutic concentrations and prevent therapeutic failures.

5.2.3 Oral Hypoglycemic Agents

Several investigators have documented chloramphenicol's ability to decrease the hepatic metabolism of tolbutamide, resulting in increases in its half-life and serum concentrations [10, 11]. Patients receiving tolbutamide and chloramphenicol concurrently may experience greater reductions in their serum glucose values and hypoglycemia with its associated complications. However, frank hypoglycemia has not been reported when this combination has been given together.

Petitpierre and Fabre reported the ability of chloramphenicol to inhibit the renal excretion of chlorpropamide [12]. They reported that five patients taking these agents together experienced an increase in their chlorpropamide half-lives from 30 to 36 h up to 40 to 146 h. Hypoglycemia was not documented in these patients.

Patients taking oral hypoglycemic agents should monitor their blood glucose frequently when taking chloramphenicol. The oral hypoglycemic dosage regimen may need to be adjusted to maintain the blood glucose within a desirable range. Patients should also be instructed to monitor for signs of hypoglycemia and to carry glucose-containing products to reverse any episodes of hypoglycemia that may develop. If possible, alternative antibiotics should be selected to avoid this interaction. Since a patient's blood glucose may be controlled on a stable oral hypoglycemic dose, switching oral hypoglycemic agents to avoid this interaction is not recommended.

5.2.4 Antibiotics

5.2.4.1 Penicillins

Chloramphenicol has been reported to antagonize the effect of beta-lactam antibiotics. A number of reports have been published suggesting that bacteriostatic and bactericidal antibiotics may antagonize each other in vitro [13, 14] and in vivo [15, 16]. Despite this information, many authorities do not believe that this is a clinically significant interaction and have used this combination of antibiotics as a standard of practice for many years for the treatment of bacterial meningitis.

French and colleagues described a case in which chloramphenicol and ceftazidime were used together to treat an infant with *Salmonella* meningitis [16]. The combination failed to eradicate the infection, but subsequent treatment with ceftazidime alone was successful. In vitro tests of serum and cerebrospinal fluid taken at that time showed that the serum could inhibit the growth of an inoculum of the salmonella at a dilution of 1:2, and the cerebrospinal fluid at a dilution of 1:16, but that neither fluid could kill the organism at any dilution. A specimen of cerebrospinal fluid taken during treatment with ceftazidime alone inhibited and killed the standard inoculum of salmonella in vitro at a dilution of 1:32.

Minor degrees of antagonism have been demonstrated in occasional laboratory experiments between almost any pairs of drugs, but generally the most consistent interfering drugs are bacteriostatic agents such as chloramphenicol, tetracyclines, and macrolides [14]. All these agents appear to act predominantly as inhibitors of protein synthesis in microorganisms. They actively antagonize agents such as the penicillins, which primarily block the synthesis of cell-wall mucopeptides. It is believed that protein synthesis must proceed actively in order to permit active muco-peptide synthesis; therefore, inhibitors of protein synthesis can antagonize inhibitors of cell-wall synthesis.

5.2.4.2 Rifampin

Prober [17] and Kelly [18] each reported two cases in which the coadministration of rifampin and chloramphenicol resulted in significantly lower chloramphenicol serum concentrations. Two patients were treated with chloramphenicol for *H. influenzae*. During the last 4 days of treatment, the patients received 20 mg/kg/d of rifampin. After 12 doses of chloramphenicol, the peak serum concentrations of chloramphenicol in these two patients were 21.5 and 38.5 mg/L, respectively, and trough concentrations were 13.7 and 28.8 mg/L. After the administration of rifampin, peak chloramphenicol concentrations progressively declined. By day 3 of rifampin coadministration, the peak concentration of chloramphenicol was reduced by 85.5%, to 3.1 mg/L in one patient, and by 63.8%, to 8 mg/L in the second patient. Serum concentrations increased back into the therapeutic range after the daily dose of chloramphenicol was increased to 125 mg/kg/day. The reduction in serum concentrations was most likely due to rifampin stimulating the hepatic metabolism of chloramphenicol, increasing its clearance and decreasing its serum concentrations.

Patients should have chloramphenicol concentrations monitored daily, while they are receiving rifampin. The chloramphenicol dosage regimen may need to be adjusted to maintain concentrations within the therapeutic range, since subtherapeutic concentrations may result in therapeutic failure. Patients also should be monitored clinically for their response to therapy.

5.2.5 Anticoagulants

Chloramphenicol may enhance the hypoprothrombinemic response to oral anticoagulants. Christensen documented a two- to fourfold increase in dicumarol half-life when coadministered with chloramphenicol [10].

Several potential mechanisms may be responsible for this interaction. Chloramphenicol has been shown to inhibit the metabolism of dicumarol, probably by inhibiting hepatic microsomal enzymes [10]. Some investigators have proposed that chloramphenicol decreases vitamin K production by gastrointestinal bacteria [19, 20]; however, bacterial production of vitamin K appears to be less important than dietary intake. Moreover, chloramphenicol does not usually have much effect on bowel flora [21]. Vitamin K depletion by chloramphenicol may affect the production of vitamin K-dependent clotting factors in the hepatocyte [22].

The clinical consequences of an increased prothrombin time (PT) or international normalized ratio (INR) would be increased risk of bleeding. This includes minor bleeding such as nosebleeds and bleeding from the gums but also major bleeding into the gastrointestinal tract, central nervous system, or retroperitoneal space. The PT/INR should be monitored daily when chloramphenicol is started or discontinued in patients taking oral anticoagulants. There may be an increase in clot formation and thromboembolic complications if the warfarin dose is not increased after the chloramphenicol is stopped.

5.2.6 Immunosuppressive Agents

5.2.6.1 Cyclosporine and Tacrolimus

Several reports have appeared in the literature describing an interaction between chloramphenicol and immunosuppressive agents, specifically cyclosporine and tacrolimus. Bui and Huang reported the interaction in a renal transplant patient receiving cyclosporine [23]. The patient is required cyclosporine 50–75 mg twice daily to maintain trough concentrations in the 100–150 µg/L prior to hospital admission. The patient's cyclosporine dose required increasing to 300 mg twice daily during her hospital admission to maintain similar trough concentrations because of rifampin therapy for the treatment of line sepsis. Ten days after the rifampin was stopped, chloramphenicol 875 mg 6 h was started for the treatment of an *Enterococcus* sinusitis. The trough cyclosporine concentration on the following day increased to 280 µg/L. Despite stepwise lowering of the cyclosporine dose to 50–100 mg/day, the concentrations continued to rise for the next 2 weeks, reaching a plateau of 600 µg/L. After stopping the chloramphenicol, the cyclosporine concentration stabilized between 100 and 150 µg/L on a dose of 50 mg twice daily. Steinfort and McConachy reported a similar experience in a heart transplant patient receiving chloramphenicol and cyclosporine [24]. Mathis reported a 41.3% increase in trough cyclosporine concentration in three renal transplant patients following the addition of chloramphenicol to their medication regimens [25]. Mean cyclosporine doses were reduced by 44–49% in order to maintain therapeutic cyclosporine concentrations.

Several reports have documented a similar interaction between chloramphenicol and tacrolimus in transplant patients [26–28]. Schulman and colleagues reported a 7.5-fold increase in tacrolimus dose-adjusted AUC, 22.7 vs 171 µg•h/L, and an increased in tacrolimus half-life from 9.1 to 14.7 h following the addition of chloramphenicol to a stable tacrolimus regimen [26]. Taber and colleagues documented the chloramphenicol-tacrolimus interaction in a liver transplant patient. The patient was stabilized on an outpatient tacrolimus dose of 5 mg twice daily with trough concentrations ranging between 9 and 11 µg/L. The tacrolimus 12-h trough concentration increased to more than 60 µg/L after 3 days of chloramphenicol 1850 mg every 6 h. The patient complained of lethargy, fatigue, headaches, and tremors. The tacrolimus concentration decreased to 8.2 µg/L 7 days after the chloramphenicol was stopped. The tacrolimus regimen was restarted at 5 mg twice daily resulting in stable trough concentrations between 6.7 and 11.0 µg/L [27]. Bakri reported an approximately fourfold increase in the tacrolimus blood concentration of a renal transplant patient after the initiation of chloramphenicol 750 mg four times daily [28]. The tacrolimus concentration ranged between 5 and 11 µg/L on stable regimen of 4 mg twice daily but increased to >30 µg/L within 3 days after starting chloramphenicol. The patient also experienced a slight rise in serum creatinine and a significant increase in his serum potassium level during this time. The tacrolimus dose was reduced to 1.5 mg twice daily, and the blood concentration fell to 18–25 µg/L. Chloramphenicol was stopped after 15 days of therapy, and the patient's tacrolimus blood concentration stabilized between 8 and15 µg/L on a regimen of 3 mg twice daily. Mathis also reported up to a 207% increase in trough tacrolimus concentration in another three renal transplant patients following the addition of chloramphenicol to their medication regimens [25]. Mean tacrolimus doses were reduce by 25–34% in order to maintain therapeutic cyclosporine concentrations.

5.2.7 Antifungal Agents

5.2.7.1 Voriconazole

Chloramphenicol was shown to inhibit the metabolism of voriconazole in a pediatric patient with fungal ventriculitis [29]. A voriconazole dose of approximately 4 mg/kg twice daily resulted in plasma voriconazole trough concentrations of 2.2 and 3.5 mg/L, while the patient was also receiving chloramphenicol. The voriconazole dose had to be increased to 9 mg/kg twice in order to maintain concentrations within the range to treat *Aspergillus* infections.

The mechanism of the interaction with the immunosuppressants is most likely due to chloramphenicol's inhibition of the cytochrome P450 (CYP) 3A4 enzyme that is responsible for the metabolism of cyclosporine and tacrolimus. If chloramphenicol has to be used in a patient receiving cyclosporine or tacrolimus, a prospective decrease in dose may be warranted. Cyclosporine and tacrolimus concentrations should be closely monitored with appropriate dose adjustments, while patients are receiving chloramphenicol. Cyclosporine and tacrolimus administration should be stopped in patients with elevated trough concentrations, especially in patients slowing signs of cyclosporine or tacrolimus toxicity until the concentrations returned to the normal therapeutic range. The agents may be restarted at appropriately adjusted doses to maintain the trough concentrations within the therapeutic range. Chloramphenicol inhibits voriconazole metabolism by inhibiting P450 3A4 and possibly 2C19 isoenzymes. Voriconazole doses will need to be adjusted with the initiation and discontinuation of chloramphenicol therapy. Monitoring voriconazole concentrations may be warranted in order to maintain concentrations within the range needed to effectively treat serious fungal infections.

5.3   Clindamycin

5.3.1   Nondepolarizing Neuromuscular Blocking Agents

Clindamycin has been shown to interact with nondepolarizing neuromuscular blocking agents and aminoglycoside antibiotics. Becker and Miller investigated the neuromuscular blockade induced by clindamycin alone and when mixed with d-tubocurarine or pancuronium in an in vitro guinea pig lumbrical nerve-muscle preparation [30]. Clindamycin initially increased twitch tension, but with higher concentrations, twitch tensions subsequently decreased. With 15–20% twitch depression induced by clindamycin, neostigmine or calcium slightly but not completely antagonized the blockade. Clindamycin at a dose that did not depress twitch tension potentiated d-tubocurarine- and pancuronium-induced neuromuscular blockade.

Several clinical reports document clindamycin's ability to prolong neuromuscular blockade following depolarizing and nondepolarizing neuromuscular blocking agents [31–33]. Best and colleagues reported on a patient who received clindamycin 300 mg IV 30 min before surgery to repair a nasal fracture [31]. Succinylcholine 120 mg was administered to facilitate intubation with no additional nondepolarizing neuromuscular blocking agents administered during the surgery. Approximately 5 h after surgery and 20 min after receiving clindamycin 600 mg intravenously, the patient complained of profound overall body weakness and was noted to have bilateral ptosis, difficulty in speaking, and rapid shallow respirations. After several minutes, her weakness rapidly became more profound, with 1/5 muscle strength noted in all extremities. Nerve stimulation showed marked neuromuscular blockade with the train-of-four (TOF) stimulation noted to be 0/4. The patient was treated with neostigmine 4 mg IV and glycopyrrolate 0.8 mg IV enabling the patient to move all extremities and develop a more normal respiratory pattern. Follow-up nerve stimulation showed a TOF of 4/4, and within 20 min of the reversal agent, the patient returned to baseline muscle strength (5/5) in all extremities.

Clindamycin-induced neuromuscular blockade is difficult to reverse. No reversal could be obtained by using either calcium or neostigmine [34]. The mode of action of clindamycin on neuromuscular function is complex. Although it has a local anesthetic effect on myelinated nerves, it also stimulates the nerve terminal and simultaneously blocks the postsynaptic cholinergic receptor. It appears that its major neuromuscular blocking effect is a direct depressant action on the muscle by the un-ionized form of clindamycin [35]. Clindamycin also has been shown to decrease the quantal content of acetylcholine released with presynaptic stimulation in vitro [36], possibly the result of effects on presynaptic voltage-gated Ca^{+2} channels [37].

This interaction may be of clinical significance in patients receiving clindamycin and depolarizing or nondepolarizing neuromuscular blocking agent during the perioperative period or in an intensive care unit. This interaction may result in a prolonged period of neuromuscular blockade, resulting in recurarization with respiratory failure and an extended period of mechanical ventilation.

Patients receiving this combination of agents should be monitored clinically with peripheral nerve stimulation using train-of-four or other mode of nerve stimulation to assess neuromuscular function and degree of neuromuscular blockade.

5.3.2   Aminoglycosides

One report suggests that clindamycin may increase the risk of nephrotoxicity when administered concurrently with aminoglycoside antibiotics. Butkus and colleagues reported three patients who developed acute renal failure when gentamicin and clindamycin were administered concurrently [38]. The evidence for combined nephrotoxicity consisted of the temporal relationship between administration of the antibiotics and the development of acute renal failure with rapid recovery after the antibiotics were stopped.

5.3.3   Paclitaxel

The pharmacokinetics of paclitaxel 175 mg/m^2 was studied in 16 patients with ovarian cancer [39]. Paclitaxel was administered alone and with clindamycin doses of 600 and 1200 mg per dose. There was a slight reduction in paclitaxel C_{max} and AUC with increasing doses of clindamycin. The baseline paclitaxel C_{max} and AUC were 3.25 ± 1.22 mg/L and 8.40 ± 2.88 μg•h/ml, respectively, but fell progressively with the coadministration of clindamycin 600 mg/dose (3.02 ± 0.81 mg/L and 7.49 ± 1.94 μg•h/ml) and 1200 mg/dose (2.87 ± 0.89 mg/L and 7.45 ± 2.24 μg•h/ml).

This interaction between clindamycin and gentamicin is supported by circumstantial evidence. Although both agents were administered concurrently, none of the patients had gentamicin concentrations monitored during therapy. The reversible renal failure is consistent with that seen with aminoglycosides. It occurs during the course of therapy and resolves rapidly once the aminoglycoside antibiotic is stopped. There is no evidence to suggest that the administration of clindamycin in the setting of appropriately dosed aminoglycoside antibiotics leads to an increased risk of nephrotoxicity. The changes in paclitaxel concentrations following the coadministration of clindamycin are minimal and probably not clinically relevant. No alterations in the paclitaxel dose are recommended when it is coadministered with clindamycin.

5.3.4 Rifampin

Several studies reported on the effect of rifampin on clindamycin serum concentrations in patients taking the combination of clindamycin and rifampin for the prolonged treatment of bone and joint infections [40–42]. In an observational study of 61 patients with bone and joint infections, the median clindamycin daily dose was 1800 mg (range 600 mg– 700 mg/day).The median clindamycin C_{min} was 1.39 mg/L. However, the C_{min} values were 1.52 mg/L in patients being treated with clindamycin alone compared to 0.46 mg/L ($p = 0.034$) in patients taking the antibiotic combination [40]. In a retrospective study of 70 patients being treated for bone and joint infections, the median duration of treatment was 40 days with a median daily clindamycin dose of 2400 mg (range 1200–3600 mg/day). The median serum clindamycin concentrations on days 3 to 14 and days 8 to 28 were 5 mg/L and 6.8 mg/L, respectively. In patients treated with rifampin, their median clindamycin serum concentrations were significantly lower (5.3 mg/L) compared to patients taking clindamycin alone (8.9 mg/L) ($p < 0.02$) [41]. Bernard evaluated 34 patients who randomly received either oral clindamycin-rifampin or clindamycin-levofloxacin for the treatment of staphylococcal osteoarticular infections [42]. Clindamycin trough and peak concentrations were measured at days 1, 15, and 30 during treatment. Mean trough and peak concentrations were lower in the clindamycin-rifampin-treated patients compared to the clindamycin-levofloxacin-treated patients: 0.79 ± 0.3 mg/L (range 0.2–2 mg/L) and 3.48 ± 1.1 mg/L (range 0–8.3 mg/L) versus 4.7 ± 1.2 mg/L (range 0–9.2 mg/L) and 10.2 ± 1.8 mg/L (range 1.1–17.4 mg/L), respectively.

Rifampin is a known potent inducer of CYP isoenzymes 3A4 and 2C8/9 [43]. Clindamycin is >90% metabolized in the liver to active and inactive metabolites. Patients requiring prolonged treatment with clindamycin in combination with other antibiotics should probably avoid rifampin if other agents have activity against the infecting pathogen. High doses of clindamycin may be required in those patients requiring the addition of rifampin. Patients on the combination should be closely monitored for clinical and microbiologic response.

5.4 Sulfonamides

5.4.1 Warfarin

Several reports have described an enhanced hypoprothrombinemic response to warfarin when sulfamethoxazole, usually in combination with trimethoprim, was added to a patient's therapy [44–47]. Two pharmacokinetic studies in healthy adults confirmed that sulfamethoxazole enhances the hypoprothrombinemic response to warfarin in most people [47, 48]. Although the sulfamethoxazole seems more likely to have been responsible than the trimethoprim, a trimethoprim effect cannot be ruled out.

O'Reilly conducted two studies evaluating the stereoselective interaction between trimethoprim-sulfamethoxazole (TMP-SMX) and warfarin. In one study, patients received 1.5 mg/kg of racemic warfarin with and without 320 mg trimethoprim-1600 mg sulfamethoxazole beginning 7 days before warfarin and continuing daily throughout the period of hypoprothrombinemia [48]. There was a significant increase in the areas of the one-stage prothrombin time, from 53 to 83 units, during the administration of TMP-SMX. In a follow-up study, O'Reilly studied the effects of TMP-SMX on each of the warfarin enantiomers [49]. Subjects received each enantiomer alone and in combination with 80 mg trimethoprim-400 mg sulfamethoxazole. TMP-SMX had no effect on the R-isomer. The areas of the one-stage prothrombin time increased by approximately 70%, from 40 to 67 units, when the S-isomer and TMP-SMX were given together. Additional case reports describe the prolongation in PT following the addition of TMP-SMX to medication regimens containing warfarin [44, 47]. Penning-van Beest and colleagues analyzed a retrospective group of approximately 60,000 patients taking coumarin anticoagulants identified in the PHARMO Record Linkage System in the Netherlands [50]. The relative risk of bleeding was calculated for a variety of antibiotics coadministered with the coumarin anticoagulants with the relative risk of bleeding being three to five for TMP-SMX.

A retrospective analysis was conducted in a veteran's population taking warfarin >30 days along with antibiotics considered to be high risk for interaction with warfarin [51]. The high-risk antibiotics included in the analysis were trimethoprim/sulfamethoxazole (TMP-SMX), ciprofloxacin, levofloxacin, metronidazole, fluconazole, azithromycin, and clarithromycin. A total of 22,272 patients were included in the analysis with 14,078 patients taking high-risk antibiotics. One hundred twenty-nine bleeding events were identified during the study. TMP-SMX was associated with an increased risk of serious bleeding when compared to low-risk antibiotics such as clindamycin and cephalexin. 14 of 129 (11%) bleeding events occurred in patients taking TMP-SMX (HR 2.09, 95% CI 1.45–3.02).

Some sulfonamides appear to impair the hepatic metabolism of oral anticoagulants. Competition for plasma protein-binding sites may play an additional role. Although sulfonamides reportedly decrease vitamin K production by the gastrointestinal bacteria, evidence for such an effect is lacking.

Patients should be monitored closely for an increase in PT/INR when sulfamethoxazole-containing products are coadministered with warfarin. Two reports suggest that a preemptive warfarin dose reduction of approximately 10–20% when initiating TMP-SMX therapy is effective in maintaining INR in the therapeutic range [52, 53]. Patients should be monitored clinically for signs of bleeding with initiating TMP-SMX and decreased effects upon discontinuing TMP-SMX or when preemptively reducing the warfarin dose. Other antibiotics may be prescribed to avoid this interaction, or other forms of anticoagulation may be used as alternatives to warfarin.

5.4.2 Oral Hypoglycemic Agents

A retrospective cohort study of Texas Medicare claims from 2006 to 2009 for patients $\geq$66 years was performed to determine the risk of hypoglycemia in patients who were prescribed glyburide or glipizide and an antibiotic [54]. The addition of TMP-SMX to the oral hypoglycemic agent regimen was significantly associated with hypoglycemia. The overall adjusted odds ratio (OR) for an emergency department visit or hospitalization for hypoglycemia within 14 days of exposure to an antimicrobial drug was 2.56 (95% CI 2.12–3.10). The OR ranged between 2.78 and 3.58 in patients co-prescribed with TMP-SMX and glipizide compared to the reference antibiotics azithromycin, amoxicillin, and cephalexin. The OR ranged between 1.63 and 2.37 in patients co-prescribed with TMP-SMX and glyburide compared to the reference antibiotics azithromycin, amoxicillin, and cephalexin. In a second Medicare claims database analysis between 2008 and 2010, 34,239 patients taking glyburide or glipizide were tracked for TMP-SMX prescriptions and subsequent emergency department visits for hypoglycemia [55]. Patients prescribed with the combination of glyburide or glipizide and TMP-SMX had a significantly higher risk of an emergency department visit for hypoglycemia compared with patients prescribed with amoxicillin, a non-interacting reference antibiotic. The OR was 3.89 (95% CI 2.29–6.60) in patients taking glipizide and 3.78 (95% CI 1.81–7.90) in patients taking glyburide. The study also showed that TMP-SMX was prescribed to 16.9% patients taking these oral hypoglycemic agents. Patients with polypharmacy and multiple prescribers were more likely to be prescribed with TMP-SMX. Patients with a documented primary care provider had a 20% lower chance of receiving a TMP-SMX prescription.

These findings are consistent with similar studies documenting a two- to three-fold higher incidence of severe hypoglycemia in patients taking TMP-SMX and glipizide [56] and a two- to fivefold higher incidence of severe hypoglycemia in patients taking TMP-SMX and glyburide [57].

The mechanism of the interaction is TMP-SMX's ability to block CYP2C9 activity interfering with sulfonylurea metabolism. Other antibiotics such as azithromycin, amoxicillin, and cephalexin should be considered when appropriate for patients taking these older-generation oral hypoglycemic agents.

5.4.3 Medications Affecting Potassium Homeostasis

Numerous reports have appeared in the literature over the years documenting TMP-SMX's ability to increase serum potassium concentrations [58–61]. Risk factors for developing hyperkalemia include dose [60, 61], age [59–61], a variety of concomitant medications [61], underlying renal function [58, 59], and diabetes [58, 59]. The mechanism responsible for hyperkalemia is trimethoprim's ability to block apical membrane potentials in the distal nephron similar to amiloride, reducing transepithelial voltage and inhibiting potassium secretion [62].

5.4.3.1 Renin-Angiotensin System Inhibitors

A population-based nested case-control study of patients ≥66 years in Ontario, Canada, was conducted to determine whether a prescription of TMP-SMX along with an angiotensin-converting enzyme inhibitor (ACEI) or angiotensin receptor blocker (ARB) was associated with sudden death [63]. TMP-SMX was associated with an increased risk of sudden death; adjusted OR was 1.38 (95% CI 1.09–1.76) compared to amoxicillin. The risk was slightly higher at 14 days; adjusted OR was 1.54 (95% CI 1.29–1.84). This corresponds to approximately three sudden deaths within 14 days per 1000 SMX-TMP-treated patients. A retrospective medical record review of outpatients receiving high-dose TMP-SMX (at least four double-strength tablets per day) showed that concurrent administration of an ACEI was associated with hyperkalemia [60]. A single case report documented a case of intraoperative hyperkalemia following the administration of TMP-SMX in a patient taking chronic ARB therapy [64].

These reports document the potential interaction between TMP-SMX and medications that affect potassium homeostasis resulting in hyperkalemia and potentially sudden cardiac death. Patients especially at risk include those who are elderly, have reduced renal function, and are receiving higher doses of TMP-SMX. Other antibiotics should be considered in patients taking these chronic cardiac medications.

5.4.3.2 Spironolactone

A similar population-based nested case-control study of patients ≥66 years in Ontario, Canada, was conducted to determine whether the prescription of TMP-SMX along with spironolactone was associated with sudden death [65]. TMP-SMX was associated with an increased risk of sudden death; adjusted OR was 2.46 (95% CI 1.55–3.90) compared to amoxicillin.

5.4.4 MMX Mesalamine

The pharmacokinetic interaction between MMX mesalamine and TMP-SMX was studied in healthy adults [66]. Subjects received MMX mesalamine 4.8 g once daily on days 1–4 and then either placebo or TMP-SMX twice daily on days 1–3 and once daily on day 4. There was a non-statistically significant 12% increase in SMX pharmacokinetic parameters when coadministered with MMX mesalamine compared with placebo. AUC_{ss} increased from 786 to 909 μg•h/ml, $AUC_{0–24}$ increased from 1176 to 1430 μg•h/ml, C_{max} increased from 89.1 to 100 mg/L, and C_{min} increased from 45.1 to 55 mg/L. Half-life was unchanged between the two groups at 2 h.

TMP-SMX appears to be safe when coadministered with MMX mesalamine. There is no need for therapeutic change during therapy.

5.4.5 Methotrexate

Methotrexate (MTX) is an agent widely used for the treatment of malignancies and many autoimmune disorders such as rheumatoid arthritis, Crohn's disease, and psoriasis. Although MTX is usually well tolerated at low doses, it is known to cause myelosuppression, mucositis, hepatotoxicity, and renal injury. Cudmore reported a patient with Crohn's disease who was treated with MTX 25 mg/week for 13 years. The patient was treated with corticosteroids for what appeared to be an exacerbation of her Crohn's disease. TMP-SMX 160 mg/800 mg three times weekly was started concurrently with the steroids as prophylaxis against *P. jirovecii* pneumonia while taking the steroids. Three weeks after starting TMP-SMX, the patient developed signs of leukopenia, thrombocytopenia, anemia, and mucositis. The MTX and TMP-SMX were stopped, and the patient was treated with folinic acid 15 mg for 5 days. Five days after starting folinic acid, the patient's white blood count, hemoglobin, and platelet count returned to normal [67].

Data from the National Ambulatory Care Survey documented that TMP-SMX and MTX were co-prescribed in 22,000 physician visits between 1993 and 2010 [68]. Caution should be used when prescribing these agents together. Both agents are folic acid antagonists and can cause bone marrow suppression. TMP-SMX also inhibits the renal excretion of MTX. Patients should be monitor for the development of signs and symptoms of bone marrow suppression during concomitant therapy.

5.4.6 Mefloquine

A randomized, double-blind, placebo-controlled study was conducted in 124 HIV-positive pregnant women receiving standard doses of TMP-SMX [69]. Seventy-two women received mefloquine 15 mg/kg at monthly intervals. TMP-SMX did not have an impact on mefloquine pharmacokinetics. There was no significant effect on TMP pharmacokinetics. However, there was a 53% decrease in SMX concentrations after mefloquine administration relative to the placebo group. SMX concentrations returned to baseline 28 days after stopping mefloquine. The authors reported a SMX C_{min} of 23.8 mg/L, approximately two-thirds to one-half the concentration reported in previous studies. Although the mechanism of the interaction is unknown, the authors suggested that the presence of carboxymefloquine, the primary metabolite, may be responsible in stimulating the metabolism of SMX and reducing its concentrations.

While the coadministration of both products has no effect on mefloquine or TMP concentrations, the reduction in SMX concentrations may result in an increase in bacterial infections.

5.5   Tetracyclines

Tetracyclines have been documented to interact with a number of medications. The most common interaction is with heavy metals that chelate tetracyclines and impair their absorption from the gastrointestinal tract. Although somewhat controversial, interactions may occur with oral contraceptives, where tetracycline may reduce their effectiveness and increase the risk of pregnancy.

5.5.1   Heavy Metals

Numerous studies have documented the ability of heavy metals to chelate tetracycline products and impair their absorption [70–72]. These products contain divalent and trivalent cations such as aluminum, magnesium, and calcium. Antacids also may impair the dissolution of tetracyclines. Bismuth subsalicylate, a common ingredient in antidiarrheal medications, also has been shown to impair the absorption of tetracyclines through a similar chelation mechanism [73, 74].

This is a pharmacokinetic interaction because it impairs absorption and reduces oral bioavailability. The clinical consequences of this interaction could be the potential of a therapeutic failure because of inadequate tetracycline serum and tissue concentrations.

Oral tetracycline products should be taken 2 h before or 6 h after antacids. This may not completely avoid the interaction but should minimize it. Since this interaction is not based on an alteration in pH, H_2-receptor antagonists and proton pump inhibitors may be alternative medications. Additionally, other antibiotics may be prescribed to avoid the interaction.

Bismuth can reduce the bioavailability of tetracycline, similar to heavy metals. Ericsson and colleagues evaluated the influence of a 60 mL dose of bismuth subsalicylate on the absorption of doxycycline [73]. Doxycycline bioavailability was reduced by 37% and 51% when given simultaneously and as a multiple-dose regimen before doxycycline. Peak serum concentrations of doxycycline were significantly decreased when bismuth subsalicylate was given 2 h before doxycycline but not when given 2 h after doxycycline. Albert and co-workers documented a 34% reduction in doxycycline bioavailability when the two products were administered simultaneously [74]. A further discussion on the effect of various food containing divalent cations is given in Chap. 4.

5.5.2   Colestipol

Colestipol reduces the bioavailability of tetracycline by impairing its absorption in the gastrointestinal tract. Friedman et al. showed that when colestipol and tetracycline were given together, there was a 50% reduction in tetracycline bioavailability

[75]. In a single-dose, three-way, crossover study, subjects ingested 500 mg tetracycline with 180 mL of water, 180 mL of water and 30 g colestipol, and 180 mL of orange juice and 30 g colestipol. There were significant differences in the 48-h urinary excretion of tetracycline. More than 50% of the dose was recovered in the urine when the tetracycline was administered with water. Only 23–24% was recovered in the urine when it was administered with colestipol. There was no significant difference among the three groups in the mean value excretion half-life.

This interaction impairs absorption and reduces oral bioavailability as a result of tetracycline adsorbing onto colestipol-binding sites. The clinical consequences of this interaction could be the potential of a therapeutic failure because of inadequate tetracycline serum and tissue concentrations.

Oral tetracycline should be taken 2 h before or at least 3 h after a dose of colestipol. Additionally, other antibiotics may be prescribed to avoid the interaction.

5.5.3 Digoxin

Tetracycline can reduce the gastrointestinal bacterial flora responsible for metabolizing digoxin in the gastrointestinal tract and increase digoxin absorption and bioavailability in some patients. Lindenbaum and colleagues administered digoxin to healthy volunteers for 22–29 days. After 10 days, 500 mg tetracycline q6h for 5 days was started [76]. During the period of antibiotic administration, digoxin reduction products fell, urine digoxin output rose, and digoxin steady-state serum concentrations increased by as much as twofold in some subjects. Preantibiotic serum digoxin concentrations ranged between 0.37 and 0.76 µg/L and increased to 0.8–1.33 µg/L following antibiotic therapy. It also was noted that these effects persisted for several months after the antibiotics were stopped. There were no reports of digoxin toxicity in the patients who experienced an increase in their digoxin concentrations.

The mechanism of this interaction appears to be the inhibition of digoxin metabolism by suppressing gut bacteria. The clinical implications of this interaction are the possibility that therapy with antibiotics in subjects producing large amounts of digoxin reduction products may precipitate toxicity. Unrecognized changes in gut flora might result in variability in digoxin response, in the direction of either drug toxicity or therapeutic failure.

5.5.4 Anticonvulsants

Phenobarbital and phenytoin have been shown to reduce the serum concentrations of doxycycline [77–79]. Penttilla and Neuvonen conducted three trials to evaluate the effect of anticonvulsants on doxycycline metabolism [77]. In one study they compared the half-life of doxycycline in patients taking long-term phenytoin and/or carbamazepine therapy to a control group of patients not receiving anticonvulsants.

The doxycycline half-life in the patients receiving chronic anticonvulsants ranged between 7 h and 7.5 h compared to 15 h in the control subjects. In a second cross-over trial, they determined the half-life of doxycycline in five patients after 10 days of phenobarbital therapy and in another five patients taking phenobarbital chronically [78]. The half-life of doxycycline was 15 h in the control patients before phenobarbital therapy was begun. After 10 days of therapy, the half-life was reduced to 11 h. The doxycycline half-life was 7 h in the patients taking phenobarbital chronically. In a third trial, they evaluated the effect of chronic anticonvulsant therapy on a variety of tetracycline products and compared this to results in control patients [79]. The doxycycline half-life averaged 7 h, and the peak concentrations were lower in the patients on chronic anticonvulsant therapy compared to the control group. There was no difference in the half-lives of oxytetracycline, methacycline, chlortetracycline, and demethylchlortetracycline between the patients on anticonvulsants and control patients.

Although doxycycline is primarily eliminated in the feces, the enhanced hepatic metabolism of doxycycline appears to be the mechanism of this interaction. The clinical consequences of this interaction could be a reduction in serum doxycycline concentrations and the potential for therapeutic failure. An alternative class of antibiotics should be selected for these patients because they may be receiving anticonvulsants for the control of a seizure disorder, and it would not be wise to switch anticonvulsants to avoid this interaction.

5.5.5   Warfarin

Tetracyclines may be associated with an increased hypoprothrombinemic response in patients taking oral anticoagulants. Several case reports describe patients stabilized on chronic warfarin therapy who experienced increases in PT after the addition of doxycycline to their medication regimens [80, 81]. Westfall described a patient maintained on warfarin therapy with stable PT values approximately two times the control value [80]. After the initiation of 100 mg of doxycycline twice a day, the patient's PT increased to 51 seconds and was associated with an unusually heavy menstrual flow. Upon medical evaluation her hemoglobin and hematocrit had dropped to 5.7 g/dL and 18.9%, respectively.

Caraco and Rubinow described two patients taking chronic oral anticoagulation who presented with severe hemorrhage and disturbed anticoagulation tests after the addition of doxycycline to their medication regimens [81]. In the first patient, the PT ratio increased from 1.49 to 3.82 following the addition of 100 mg of doxycycline daily. In the second patient, the PT ratio increased from between 1.5 and 2.5 to 4.09 following the addition of 100 mg of doxycycline twice daily.

Penning-van Beest estimated the relative risk of bleeding in patients taking coumarin anticoagulants and a tetracycline in the PHARMO Record Linkage System in the Netherlands [50]. The relative risk of bleeding was calculated to be three to five for doxycycline and nine for tetracycline.

Dowd conducted a trial to determine if a 10–20% empiric reduction in warfarin dose before starting doxycycline would prevent nontherapeutic INRs following doxycycline-warfarin coadministration compared to reactive warfarin dose adjustments [82]. The average empiric dose reduction was 11%. Twelve percent of patients in the reactive dosing group experienced an INR above the upper INR goal range compared to none in the empiric dosing group ($p = 0.02$). A higher percentage of patients in the empiric dosing group had a subtherapeutic INR compared to the reactive group (35% vs 6%, $p < 0.05$).

The mechanism of this interaction is unclear but may involve a reduction in the plasma prothrombin activity by impairing prothrombin utilization or decreasing vitamin K production by the gastrointestinal tract.

The clinical significance of this interaction is the increased anticoagulant effect, which may result in an increased risk of bleeding. Patients should be closely monitored for clinical signs of bleeding such as nosebleeds or bleeding from the gums, the PT monitored, and warfarin dose adjusted to maintain the PT/INR in the therapeutic range. Empiric warfarin dose reductions to avoid potential supratherapeutic INR are not warranted and may result in subtherapeutic INR. Other antibiotics may be prescribed to avoid this interaction, or other forms of anticoagulation may be used as alternatives to warfarin.

5.5.6 Lithium

One case report described the increase in lithium concentrations following a course of tetracycline [83]. However, a prospective trial documented small decreases in the serum lithium concentration when both agents were administered concurrently [84].

McGennis reported a patient taking lithium chronically for a history of manic depression [83]. Two days after starting tetracycline, it was noted that her serum lithium concentration increased from 0.81 to1.7 mmol/L. The patient exhibited slight drowsiness, slurred speech, and a fine tremor of both hands consistent with lithium toxicity. At the time lithium and tetracycline were stopped, the serum lithium concentration was 2.74 mmol/L. The concentration declined to within the therapeutic range 5 days after stopping both agents.

Fankhauser evaluated the effect of tetracycline on steady-state serum lithium concentrations in healthy volunteers and compared the frequency and severity of adverse effects in the lithium and lithium-tetracycline treatment phases [84]. There was a significant decrease in the serum lithium concentration between the control and treatment phases (0.51 versus 0.47 mEq/L, $p = 0.01$). It is unclear whether this is a clinically significant decrease in the serum lithium concentration. There was no difference in adverse effects between the control and treatment phases of the trial.

The mechanism of this interaction is not known. One possibility may be that tetracycline-induced renal failure may reduce urinary lithium excretion. Although it is unlikely that a significant interaction exists, patients should be monitored for signs of lithium toxicity when this combination is prescribed. Renal function should also be monitored to prevent increases in the serum lithium concentrations secondary to reductions in renal function. Another class of antibiotics should be prescribed to avoid this interaction.

5.5.7 Psychotropic Agents

Steele and Couturier reported the possible interaction between tetracycline and risperidone and/or sertraline in a 15-year-old male with Asperger's disorder, Tourette's disorder, and obsessive-compulsive disorder [85]. Tetracycline was added to a risperidone-sertraline treatment regimen resulting in an acute exacerbation of motor and vocal tics. The authors postulated that the increase in tics may have resulted from either a tetracycline-risperidone interaction leading to a reduction in risperidone concentrations or a tetracycline-sertraline interaction leading to increased concentrations of sertraline or the natural course of Tourette's disorder. The sertraline dose was increased with no concomitant increase in tics, and subsequent discontinuation of tetracycline resulted in an improvement in tics, which suggests the possibility of an interaction between tetracycline and risperidone. The mechanism of this potential interaction is unknown, but the author recommended that the addition of antibiotics to psychotropic medications requires close monitoring due to the potential for the interaction.

5.5.8 Theophylline

Several case reports describe increases in theophylline serum concentrations during a course of tetracycline administration [86, 87]. However, prospective trials have failed to document a consistent effect [88–91].

Four prospective studies have evaluated the interaction between theophylline and tetracycline. Pfeifer gave nine patients tetracycline for 48 h and did not observe a statistically significant interaction [88]. However, six subjects had a decrease in theophylline clearance during the combined tetracycline-theophylline period, and in four of the subjects, the decrease was greater than 15%. Mathis studied eight healthy volunteers by giving them a single intravenous injection of aminophylline before and after 7 days of tetracycline [89]. Theophylline clearance decreased by an average of 9%, but four patients had greater than 15% decrease in clearance, and one patient had a 32% decrease in clearance. Gotz and Ryerson evaluated the interaction between tetracycline and theophylline in five patients with chronic obstructive airway disease [90]. Theophylline clearance decreased by an average of 11% following the 5-day course of tetracycline. Jonkman evaluated the effects of doxycycline on theophylline pharmacokinetic parameters in healthy volunteers during a 9-day course of theophylline alone and with the coadministration of doxycycline [91]. There was no influence of doxycycline on absorption, elimination, and volume of distribution of theophylline. Mean steady-state plasma concentrations were not significantly different between the two treatment periods.

The mechanism for the interaction is unknown but appears to be a reduction in the hepatic metabolism of theophylline. The reduction in metabolism appears to be quite variable. It may take several days for the interaction to occur, so increases in serum theophylline may not be clinically significant after short courses of tetracycline. Patients taking longer courses of tetracycline may be at risk for developing theophylline toxicity.

Patients should be closely monitored when tetracycline is added to a medication regimen containing theophylline. Although short courses may not result in clinically significant increases in the serum theophylline concentration, patients maintained in the upper end of the therapeutic range may be at risk of developing theophylline toxicity even with modest increases in the serum theophylline concentration. Also, the reduction in clearance appears to be quite variable, so it may be difficult to predict how much of the theophylline will increase following the addition of tetracycline to the medication regimen. All patients should be monitored clinically for signs and symptoms of theophylline toxicity. Serum theophylline concentration should be monitored every 2–3 days in patients at high risk for developing theophylline toxicity.

5.5.9 Oral Contraceptives

Several case reports suggest that tetracycline can reduce the effectiveness of oral contraceptives [92, 93]. One retrospective study showed that the oral contraceptive failure rate was within the expected range associated with the typical pattern of use [94]. However, prospective trials have failed to document a consistent effect [95, 96]. These case reports of unintended pregnancies have occurred following the concurrent administration of tetracycline and other antibiotics with oral contraceptives. Two small controlled studies evaluated the effect of tetracycline on the serum concentrations of ingredients contained in commonly prescribed oral contraceptives. Neely et al. compared the serum concentrations of ethinyl estradiol, norethindrone, and endogenous progesterone during a control period and after a 7-day course of doxycycline starting on day 14 of their cycle [95]. There were no statistically significant differences in serum concentrations of ethinyl estradiol, norethindrone, and endogenous progesterone between the control and treatment phases. Murphy et al. studied the effect of tetracycline on ethinyl estradiol and norethindrone after 24 h and 5–10 days of therapy with tetracycline [96]. There was no significant decrease in ethinyl estradiol and norethindrone concentrations after 24 h or after 5–10 days of therapy. A pharmacokinetic study was performed to investigate whether there was any interaction between etonogestrel and ethinyl estradiol released from the combined contraceptive vaginal ring NuvaRing and concomitant treatment with orally administered doxycycline. Healthy women were randomized to receive either NuvaRing alone for 21 days or NuvaRing for 21 days doxycycline. The doxycycline study measured AUC values over the initial 24 h on days 1 and 10 and the whole of days 1–11 and 1–22. There were no differences in the etonogestrel or ethinyl estradiol serum concentrations between subjects using NuvaRing alone versus those receiving the ring plus doxycycline. Calculation of etonogestrel and ethinyl estradiol interaction/control ratios confirmed the absence of an interaction between these medications [97].

The mechanism for the interaction is unknown but may be due to interference with the enterohepatic circulation of estrogens in the intestines, making this a pharmacokinetic interaction. Other antibiotics have also been reported to reduce the effectiveness of oral contraceptives when administered concurrently. It is not known if nonadherence played a role in some of these unplanned pregnancies. Other more extensive reviews on the interaction between tetracyclines and oral contraceptives have concluded that this interaction is not supported by pharmacokinetic data [98].

Although the evidence of the interaction between tetracycline and oral contraceptives is limited to case reports, women should be counseled to use other methods of birth control during tetracycline therapy.

5.5.10   Methotrexate

Tortajada-Ituren and colleagues reported an interaction between doxycycline and high-dose methotrexate (MTX) [99]. A 17-year-old female was receiving high-dose MTX as part of a chemotherapy regimen. The patient had undergone ten cycles of the regimen without complications. Her mean MTX pharmacokinetic parameters following the ten cycles were a MTX clearance of 2.95 L/h; half-life, 2.96 h; mean residence time, 4.27 h; and volume of distribution, 12.53 liters. On admission to the hospital for the 11th cycle of chemotherapy, the patient was noted to have a palpebral abscess in her left eye which was treated with doxycycline 100 mg twice daily. The high-dose MTX, 18 g, was administered according to her usual protocol. During the first 24 h after the MTX infusion, the patient developed facial erythema, malaise, and vomiting that had not occurred during the first ten cycles. The doxycycline was stopped 48 h after chemotherapy. The pharmacokinetic monitoring was prolonged for 168 h revealing a significant decrease in MTX clearance (1.29 L/h) and significant increase in half-life (6.26 h) and mean residence time (9.03 h) compared to the values obtained during the first ten cycles. Her hospital stay was prolonged to 11 days compared to an average of 7.7 days during the first ten cycles.

Although the mechanism of the interaction is unknown, one proposed theory suggests that tetracyclines may displace MTX from plasma protein-binding sites [100]. In an attempt to validate this mechanism in their patient, the authors determined the degree of MTX plasma protein binding in two plasma samples with similar MTX concentrations from the 7th and 11th cycles. The unbound MTX concentrations were determined with an ultrafiltration process. The unbound MTX fractions during the 7th and 11th cycles were 53% and 41%, respectively.

Although case reports of a tetracycline-MTX interaction are limited, tetracyclines should be avoided in patients receiving high-dose MTX therapy. If therapy with a tetracycline is required, pharmacokinetic monitoring should be continued until the MTX concentrations are below the desired range, and the leucovorin rescue should be continued, if necessary, until all signs and symptoms of MTX toxicity disappear.

5.5.11 Rifampin

Colmenero and colleagues studied the possible interaction between rifampin and doxycycline in 20 patients with brucellosis [101]. Patients were treated with either doxycycline and streptomycin or doxycycline and rifampin. The doxycycline concentrations in the patients treated with rifampin were significantly lower than those patients treated with doxycycline and streptomycin. The doxycycline clearance in patients treated with rifampin was significantly higher than in the patients treated with doxycycline and streptomycin 3.59 L/hours and 1.55 L/hours, respectively. The elimination half-life (4.32 h vs 10.59 h) and area under the concentration-time curve were significantly lower in the rifampin-treated patients (30.4 vs 72.6 µg•h/ml). Additionally, there were lower doxycycline concentrations in the rifampin treatment group who were rapid acetylaters. There were no treatment failures in the patients receiving doxycycline and streptomycin, while there were two treatment failures in the doxycycline-rifampin group.

Rifampin is a potent inducer of hepatic microsomal enzymes. Although doxycycline is only partially metabolized, the effect of rifampin may be significant enough to lower doxycycline concentrations to subtherapeutic levels. Caution should be used when treating patients with combined rifampin and doxycycline therapy. If possible, alternative antibiotic should be prescribed to avoid potential treatment failures.

5.6 Tigecycline

5.6.1 Warfarin

The interaction between tigecycline and warfarin was studied in 19 healthy males [102]. On day 1, the subjects received a single warfarin 25 mg dose. On day 8, they received a 100 mg loading dose of intravenous tigecycline followed by 50 mg every 12 h for eight additional doses. On day 12, they received another warfarin 25 mg dose with their last tigecycline dose. After eight doses of tigecycline, R- and S-warfarin AUCs were increased by 68% and 29%, respectively, and clearance decreased by 40% and 23%, respectively. There was an approximately 50% increase in the *R*-warfarin half-life from 42.4 h to 68.7 h but less than a 20% increase in the S-warfarin half-life from 32 h to 37 h. There was no significant effect on INR.

The reduction in clearance and prolongation in half-life suggest that this effect was due to an increase in warfarin protein binding. Although there was no effect on the INR after single doses in healthy volunteers, all patients on chronic warfarin therapy receiving broad-spectrum antibiotics should have their INR closely monitored and doses adjusted as needed.

5.6.2   Organ Transplant Immunosuppressive Agents

5.6.2.1   Cyclosporine

Stumpf reported a patient who had been maintained on cyclosporine for 5 years after a renal transplant [103] and experienced a urinary tract infection that was ultimately treated with tigecycline. During combined therapy, cyclosporine concentrations rose necessitating a 50% reduction in the daily cyclosporine dose. The cyclosporine dose had to be increased back to the initial dose of 120 mg daily after tigecycline was discontinued. The mechanism of the interaction is unknown, but the authors and Srinivas speculate that it may be due to the inhibition of P-glycoprotein-related efflux transport activity [103, 104].

5.6.2.2   Tacrolimus

Pavan and colleagues report on a similar interaction between tacrolimus and tigecycline [105]. A patient who was stable on tacrolimus for 5 years following renal transplant developed septic shock with *Escherichia coli* sensitive only to tigecycline. During tigecycline therapy tacrolimus serum concentrations rose requiring a reduction in the daily dose of tacrolimus. No other medications known to interact with tacrolimus were administered during this time. The tacrolimus serum concentrations became subtherapeutic after tigecycline was discontinued necessitating an increase in the tacrolimus dose. The authors speculated that the interaction may be due to tigecycline's ability to inhibit CYP3A4.

The mechanism by which tigecycline inhibits the metabolism of cyclosporine and tacrolimus is unclear. Caution should be used when tigecycline is administered to organ transplant patients on stable doses of the antirejection medications. Alternative antibiotics should be prescribed if possible. Serum concentrations of cyclosporine and tacrolimus should be monitored and the doses adjusted, while patients are receiving tigecycline and after it is discontinued.

5.7   Aminoglycosides

Aminoglycoside antibiotics are involved in a number of drug interactions, many of which result in an increased risk of nephrotoxicity.

5.7.1   Amphotericin B

The concurrent use of aminoglycoside antibiotics may lead to an increased risk of developing nephrotoxicity. Churchill and Seely reported four patients that developed nephrotoxicity when both agents were administered together [106]. All of the patients received amphotericin B at an approximate dose of 0.5 mg/kg/day. Two of the four patients had documented gentamicin trough concentrations of 5 mg/L. All patients developed progressive renal failure during the first several days of combined therapy. In the patients who survived, renal function returned to baseline values after both agents were discontinued.

The mechanism of this is the potential of additive nephrotoxicity from both agents. Amphotericin B is associated with a predictable rise in creatinine within the first several days of therapy. Aminoglycoside antibiotics are associated with acute tubular necrosis, especially in the setting of elevated serum concentrations. In the case report, three patients had documented gentamicin concentrations significantly higher than the desired 2 mg/L. This most likely contributed to the development of nephrotoxicity in these patients.

Patients receiving aminoglycoside antibiotics and amphotericin B should be closely monitored for the development of renal failure. The aminoglycoside serum concentrations should be monitored every 2–3 days and the dosage regimen adjusted to maintain peak and trough concentrations within the desired therapeutic range. Every attempt should be made to avoid other conditions that might increase the risk of developing renal failure (i.e., hypotension) and/or administering other medications that might increase the risk of developing renal failure (i.e., IV contrast media, loop diuretics).

5.7.2   Neuromuscular Blocking Agents

Aminoglycoside agents are known to potentiate paralysis from neuromuscular blocking agents [107–110]. Often this has occurred in the setting of the instillation of aminoglycoside-containing irrigation solutions into the intra-abdominal cavity during surgery. Dupuis et al. evaluated prospectively the interaction between aminoglycosides and atracurium and vecuronium in 44 patients [111]. Twenty-two patients had therapeutic concentrations of gentamicin or tobramycin, and 22 patients served as controls. Onset time, clinical duration, and time to spontaneous recovery of a T_1/T_4 ratio of 0.7 after atracurium or vecuronium injection were measured. Although no statistically significant differences were found in onset time, clinical duration was longer in patients receiving tobramycin or gentamicin and paralyzed with vecuronium than in controls. The neuromuscular blockade produced by atracurium was not significantly influenced by the presence of therapeutic serum concentrations of tobramycin or gentamicin. The clinical duration of patients receiving atracurium alone or in the presence of an aminoglycoside was approximately 40 min in each group, and the time to recovery of a T_1/T_4 ratio > 0.7 was approximately 60–70 min. The clinical duration was significantly longer in the vecuronium patients receiving aminoglycosides than in the vecuronium control patients, 30 versus 55 min, respectively. The time to recovery of a T_1/T_4 ratio > 0.7 in the patients receiving vecuronium with aminoglycosides also was longer in the patients receiving an aminoglycoside, 55 versus 105 min, respectively.

Aminoglycosides have been shown to interfere with acetylcholine release and exert a postsynaptic curare-like action [112]. These agents have membrane-stabilizing properties and exert their effect on acetylcholine release by interfering with calcium ion fluxes at the nerve terminal, an action similar to magnesium ions. Aminoglycosides also possess a smaller but significant decrease in postjunctional receptor sensitivity and spontaneous release.

These drugs may cause postoperative respiratory depression when administered before or during operations and may also cause a transient deterioration in patients with myasthenia gravis. Patients should be monitored for prolonged postoperative paralysis if they received neuromuscular blocking agents and aminoglycoside antibiotics during the perioperative or immediate postoperative period.

5.7.3   Indomethacin

Zarfin et al. evaluated the effect of indomethacin on gentamicin and amikacin serum concentration in 22 neonates with patent ductus arteriosus treated with indomethacin and aminoglycosides [113]. The aminoglycoside doses were held stable before the initiation of indomethacin therapy. After the addition of indomethacin, there was a significant rise in aminoglycoside trough and peak concentrations, a reduction in urine output, and a significant rise in serum creatinine. This may have been due to the ability of nonsteroidal anti-inflammatory agents to cause reversible renal failure. In this setting the elimination of all renally eliminated medications would be expected to be reduced with elevation in serum concentrations.

Renal function should be closely monitored in patients receiving nonsteroidal anti-inflammatory agents. If renal failure develops, the doses of all renally eliminated medications should be adjusted to the level of remaining renal function. Serum concentrations of medications should be monitored when possible and dosage regimens adjusted to maintain serum concentrations within the accepted therapeutic ranges.

5.7.4 Cyclosporine

Cyclosporine and aminoglycosides are both nephrotoxic and produce additive renal damage when administered together. Termeer et al. reported that the combined use of gentamicin and cyclosporine in renal transplant patients increased the incidence of acute tubular necrosis to 67%, compared with 5–10% when gentamicin was used alone or when cyclosporine was used with other, non-nephrotoxic antibiotics [114]. Animal studies have also documented the additive nephrotoxicity of aminoglycosides when administered with cyclosporine.

The mechanism appears to be additive injury to the renal tubule. Aminoglycosides induce renal failure by inhibiting the intracellular phospholipases in lysosomes of tubular cells in the proximal tubule. Cyclosporine-induced acute renal failure is related primarily to its effects on the renal blood vessels. Cyclosporine acutely reduces renal blood flow, with a corresponding increase in renal vascular resistance and a reduction in glomerular filtration rate.

Patients receiving aminoglycoside antibiotics and cyclosporine should be closely monitored for the development of renal failure. The aminoglycoside and cyclosporine serum concentrations should be monitored every 2–3 days and the dosage regimen adjusted to maintain peak and trough concentrations within the desired therapeutic range. Every attempt should be made to avoid other conditions that might increase the risk of developing renal failure (i.e., hypotension) and avoid administering other medications that might increase the risk of developing renal failure (i.e., IV contrast media, loop diuretics).

5.7.5 Chemotherapeutic Agents

Numerous reports have documented the additive nephrotoxicity when aminoglycosides are administered to patients receiving cisplatin-type chemotherapeutic agents [115–121]. Cisplatin-type chemotherapeutic agents have been shown to be associated with a reduction in renal function. Patients who received aminoglycoside antibiotics during or after a course of cisplatin-based chemotherapy regimens have demonstrated additional reductions in renal function.

The mechanism appears to be direct injury to the renal tubule. Aminoglycosides induce renal failure by inhibiting the intracellular phospholipases in lysosomes of tubular cells in the proximal tubule. Cisplatin-induced renal failure is mediated by a toxic effect on the renal tubular cells, resulting in acute tubular necrosis.

Prior administration of cisplatin is not an absolute contraindication to the use of aminoglycoside antibiotics. When clinically indicated, patients who have previously received cisplatin and have apparently normal renal function should be treated cautiously with standard doses of aminoglycoside antibiotics, and pharmacokinetic monitoring should be routinely performed, with the dosage regimens adjusted to maintain serum concentrations within the desired therapeutic range.

5.7.6   Loop Diuretics

Several reports describe the increased risk of nephro- and ototoxicity when amino-glycosides and loop diuretics are administered together [122, 123]. Although some case reports suggest there is increased ototoxicity when ethacrynic acid is given in combination with aminoglycosides [125]. The data supporting the association between furosemide and aminoglycosides are controversial [124].

5.7.6.1   Ethacrynic Acid

High doses of ethacrynic acid given alone have been shown to produce hearing loss in patients with renal failure [125, 126]. Hearing loss can range between partial and full deafness and is usually irreversible. When patients receiving ethacrynic acid have been given an aminoglycoside such as kanamycin or streptomycin, hearing loss has been reported to occur within 15 min after an injection of the diuretic and lasting for several hours. Some patients had reduced hearing loss, while others remained deaf [125].

The mechanism of this pharmacodynamic interaction is not known. Ethacrynic acid is thought to produce hearing loss by an alteration in the formation of peri-lymph in the cochlea. This may be disputed because not all patients experience vertigo or nausea. Other possible causes of deafness may be the cysteine adduct of ethacrynic acid, a substance known to be ototoxic, or a direct toxicity to the auditory nerves by ethacrynic acid. Aminoglycosides produce ototoxicity by destroying the sensory hair cells in the cochlea and vestibular labyrinth.

Ethacrynic acid and the older-generation aminoglycosides are rarely used in clinical practice. However, some patients may be unable to take loop diuretics such as furosemide or bumetanide, so ethacrynic acid may be their only available option. When ethacrynic acid is used alone or in combination with aminoglycosides, it should be used in the lowest dose that maintains adequate urine output or fluid balance. Aminoglycoside concentrations should be monitored and the dosage regimens adjusted to maintain concentrations within the therapeutic range. Patients should be monitored with audiograms if therapy is to be continued for an extended duration, and audiograms should be performed in patients who complain of hearing loss.

5.7.6.2   Furosemide

Kaka et al. reported a suspected case of furosemide increasing the peak and trough concentrations of tobramycin in a 72-year-old woman [122]. The patient received intermittent doses of furosemide for the management of congestive heart failure. The patient developed a Gram-negative aspiration pneumonia. Tobramycin was started, with serum concentrations drawn after the loading dose followed by a main-tenance dose of 180 mg IV q8h. Twelve hours after an intravenous dose of 120 mg of furosemide, the tobramycin trough and peak concentrations around the fourth dose were 5.3 and 16.2 mg/L, respectively. The authors concluded that moderate doses of furosemide could increase tobramycin concentrations, thus increasing the risk of ototoxicity and nephrotoxicity in some patients.

It is unclear whether furosemide was the cause of the increased tobramycin concentrations in this patient. Although furosemide has been reported to both increase and decrease the clearance of gentamicin, there are other possible explanations for the elevated tobramycin concentrations in these patients. The authors determined the patient's tobramycin pharmacokinetic parameters after the initial dose and used these parameters to determine the patient's maintenance dosage regimen. The maintenance regimen may have been overly aggressive for the patient's age, weight, and underlying renal function. There was extreme variability in the tobramycin pharmacokinetic parameters between the first and fourth doses, suggesting errors in drug administration or sampling technique rather than changes in the patient's clinical status or the administration of furosemide.

Smith and Liftman analyzed the data from three prospective, controlled, randomized, double-blind clinical trials to determine whether furosemide increased the nephrotoxicity and ototoxicity of aminoglycosides. There was no difference in the incidence of nephrotoxicity or ototoxicity between the groups receiving aminoglycosides alone and the group receiving aminoglycosides and furosemide [124].

It is unclear whether furosemide directly increases the nephrotoxicity and ototoxicity of aminoglycosides. Furosemide may increase the risk of developing nephrotoxicity by causing excessive diuresis, hypovolemia, and a reduction in renal blood flow. Furosemide should be used with caution in patients receiving aminoglycoside antibiotics. Careful attention should be paid to the patient's weight, urine output, fluid balance, and indices of renal function. Aminoglycoside concentrations should be monitored and the dosage regimen adjusted to maintain concentrations within the therapeutic range.

5.7.7 Vancomycin

Several reports have been published evaluating the potential of vancomycin to increase the nephrotoxicity of aminoglycoside antibiotics. Two studies were retrospective reviews and two studies were prospective evaluations. Cimino retrospectively evaluated 229 courses of therapy in 229 oncology patients [127]. Forty patients received vancomycin alone, 148 patients received aminoglycosides alone, and 40 patients received vancomycin and an aminoglycoside antibiotic. The incidence of nephrotoxicity in patients administered with an aminoglycoside was 18%; vancomycin, 15%; and an aminoglycoside and vancomycin, 15%. They could not show that the concurrent administration of vancomycin had an additive effect on the incidence of nephrotoxicity. Pauly et al. retrospectively evaluated the incidence of nephrotoxicity in 105 patients who received at least 5 days of combined therapy [128]. Twenty-eight (27%) patients developed nephrotoxicity during combined vancomycin-aminoglycoside therapy. However, 22 patients had other insults such as amphotericin B, sepsis, or liver disease that could account for the increase in nephrotoxicity. There were no control groups of patients receiving vancomycin or aminoglycosides alone to provide a comparative incidence of nephrotoxicity between these groups. The results of these two studies are limited by their retrospective design, the small number of patients who received vancomycin and an aminoglycoside, and the patients who had other potential causes for developing nephrotoxicity.

Mellor et al. prospectively evaluated 39 courses of vancomycin therapy in 34 patients [129]. Twenty-seven courses were associated with aminoglycoside administration either concurrently or within 2 weeks of the first dose of vancomycin. A reduction in renal function was seen during (7%) and after (9%) vancomycin therapy. There was no evidence of synergistic toxicity between vancomycin and aminoglycosides. One feature of the patients with renal dysfunction was the severity of their underlying disease. Each case of nephrotoxicity occurred in association with either sepsis or gastrointestinal hemorrhage.

Ryback and colleagues prospectively evaluated the incidence of nephrotoxicity in patients receiving vancomycin alone or in combination with an aminoglycoside, following 224 patients receiving 231 courses of therapy [130]. 168 patients received vancomycin alone, 63 patients received vancomycin with an aminoglycoside, and 103 patients received an aminoglycoside alone. 8 patients (5%) receiving vancomycin alone, 14 patients (22%) receiving vancomycin with an aminoglycoside, and 11 patients (11%) receiving an aminoglycoside alone were found to have nephrotoxicity. Factors thought to be associated with an increased risk of nephrotoxicity in patients receiving vancomycin were concurrent therapy with an aminoglycoside, length of treatment with vancomycin (>21 days), and vancomycin trough concentrations (>10 mg/L).

Both of these studies are small prospective studies. Although they had control groups, it is unclear how well matched the control groups were to the group of patients receiving vancomycin and an aminoglycoside for underlying disease states and renal function. The increased risk of nephrotoxicity when vancomycin is administered with an aminoglycoside antibiotic is controversial. The clinical studies published to date do not show a clear association between the combination use of these agents and an increased risk of nephrotoxicity. Patients receiving vancomycin and aminoglycoside antibiotics should be closely monitored for the development of renal failure. The aminoglycoside and vancomycin serum concentrations should be monitored and the dosage regimen adjusted to maintain peak and trough concentrations within the desired therapeutic range. Every attempt should be made to avoid other conditions that might increase the risk of developing renal failure (i.e., hypotension) and to avoid administering other medications that might increase the risk of developing renal failure (i.e., IV contrast media, loop diuretics).

5.8 Linezolid

Linezolid is a synthetic oxazolidinone antibiotic that selectively inhibits bacterial protein synthesis. As a class, oxazolidinones are known to inhibit monoamine oxidase (MAO). Two forms of MAO exit in humans: Type A and Type B. MAO-A preferentially deaminates noradrenaline, adrenaline, and serotonin, while Type B deaminates dopamine. Linezolid has been shown to be a weak, competitive inhibitor of MAO-A.

5.8.1 Selective Serotonin Reuptake Inhibitors (SSRIs)

Numerous reports have documented the development of the serotonin syndrome following the coadministration of linezolid and SSRIs, and this interaction has been extensively reviewed in the literature [131–133]. SSRIs that have been documented to have been associated with the development of the serotonin syndrome following the coadministration with linezolid include paroxetine [131, 134], sertraline [135–138], mirtazapine [139, 140], venlafaxine [141–145], fluoxetine [146], citalopram [140, 144, 147], escitalopram [147], and buspirone [149]. A wide range of complications associated with the serotonin syndrome has been reported involving the central nervous system (altered mental status, paranoia, hallucinations, myoclonus, seizures, dizziness, confusion), delirium, hostility, anger, fatigue, ataxia, and tremors), cardiovascular system (hypertension, tachycardia, palpitations, syncope, cardiac arrest), and gastrointestinal tract (diarrhea). Death has also been associated with the serotonin syndrome. Symptoms can develop anywhere from 1 h to several days after the addition of an SSRI [131].

5.8.2 Meperidine

The serotonin syndrome was reported in a leukemia patient following the coadministration of linezolid and meperidine [148]. The patient had been receiving meperidine as a pretreatment to prevent amphotericin-associated rigors. The patient received meperidine 90 min after his third dose of linezolid and 30 min later developed tremulousness with myoclonus, paranoid ideation with visual hallucinations. The meperidine was stopped, and the patient was treated with methotrimeprazine 4 mg resulting in the resolution of neuropsychiatric symptoms within 2 h.

5.8.3 Rifampin

Two reports have described an interaction between linezolid and rifampin resulting in decreased linezolid serum concentration [150, 151]. Gebhart reported a patient who received rifampin and linezolid for 19 days. Ten days after rifampin was discontinued, the patient's trough linezolid concentration was reported as a trace, and the linezolid dose was increased to 600 mg every 8 h. Rifampin was restarted 11 days after it was initially discontinued and administered for an additional 8 days. Six days after rifampin was discontinued for the second time, linezolid peak and trough concentrations were reported as 7.29 mg/L and 2.04 mg/L, respectively. Follow-up peak and trough concentrations obtained 2 days later 12.46 mg/L and 5.03 mg/L, respectively [150]. Egle administered a single linezolid 600 mg IV dose to eight healthy males [151]. The following day he administered linezolid 600 mg IV with rifampin 600 mg IV. The pooled serum linezolid concentrations were lower after coadministration with rifampin compared to when linezolid was administered alone.

Gervasoni reported a patient who had been maintained on rifampin 600 mg daily and minocycline 100 mg twice daily for chronic osteomyelitis [152]. The minocycline-rifampin regimen was switched to linezolid 600 mg twice daily which was added due to poor response to therapy. The linezolid trough concentration 9 days after stopping rifampin was 0.6 mg/L. Eight and fifteen days later, the linezolid trough concentrations were 1.1 mg/L and 1.4 mg/L, respectively. These trough concentrations were similar to the trough concentration measured in the patients 2 years previously (1.4 mg/L–1.5 mg/L) when the patient was initially treated with linezolid 600 mg twice daily. The authors concluded that the enzyme-inducing ability of rifampin may impact linezolid serum concentrations up to 2–3 weeks after stopping therapy.

Serotonin is removed from the nerve synapse by reuptake into the nerve terminal or degradation by MAO. Linezolid's ability to inhibit MAO degradation of serotonin results in increased serotonin concentrations and the development of the serotonin syndrome. Patient medication profiles should be reviewed for medications that are metabolized by MAO before linezolid is prescribed. When possible, alternative antibiotics should be prescribed to avoid the risk of the development of the serotonin syndrome in susceptible individuals. Due to the long half-lives of some of the SSRIs, the serotonin syndrome may develop in patients whose SSRI was discontinued several days before initiating linezolid therapy. Alternative analgesics such as morphine or hydromorphone should be prescribed in place of meperidine. Management of the serotonin syndrome is primarily supportive with removal of the offending agent with symptoms typically resolve within 24–48 h but may last up to 7–10 days if the agents has a long half-life or active metabolites. If necessary, cyproheptadine appears to be an effective antiserotonin agent. It usually relieves symptoms after the first dose but may be administered every 1–4 h until a therapeutic response is obtained. The mechanism of the interaction between rifampin and linezolid is not known. Linezolid is not metabolized through CYP pathways. Egle has suggested that rifampin may stimulate the induction of P-glycoprotein expression leading to increased linezolid clearance by upregulation of linezolid intestinal secretion [151]. Careful consideration should be used when selecting antibiotics to treat resistant Gram-positive infections. In the event that rifampin and linezolid should be used together, the monitoring of linezolid serum concentrations should be considered, and monitoring may need to be continued up to 2–3 weeks after stopping rifampin.

5.8.4 Cough and Cold Preparations

Many over-the-counter (OTC) cough and cold preparations contain ingredients that are metabolized by MAO or are selective serotonin reuptake inhibitors. Decongestants such as pseudoephedrine and phenylpropanolamine are metabolized by MAO. The cough suppressant, dextromethorphan, has been shown to block serotonin reuptake and has been implicated in precipitating the serotonin syndrome when co-ingested with MAO inhibitors. Hendershot and colleagues reviewed the data from three linezolid clinical trials to evaluate the pharmacokinetic and pharmacodynamic responses to the coadministration of linezolid with pseudoephedrine, phenylpropanolamine, and dextromethorphan [153]. Significant increases in systolic blood pressure (SBP) were observed following the coadministration of linezolid with either pseudoephedrine or phenylpropanolamine. The mean maximum increase from baseline in SBP was 32 mm Hg and 38 mm Hg with the coadministration of pseudoephedrine and phenylpropanolamine, respectively. Treatment-emergent SBP greater than 160 mm Hg was observed following the coadministration of linezolid with pseudoephedrine in five subjects and in two patients in the linezolid-phenylpropanolamine-treated group. Dizziness was the most frequent adverse event when linezolid and pseudoephedrine were given concomitantly, and headache was the most frequent adverse event when linezolid and phenylpropanolamine were given together. There were no statistically or clinically significant effects on heart rate in either treatment group.

There were no statistically or clinically significant changes in blood pressure, heart rate, or temperature and no abnormal neurological examination results in the dextromethorphan-linezolid treatment group.

Linezolid's ability to inhibit the MAO degradation of pseudoephedrine and phenylpropanolamine resulted in the significant increases in blood pressure that was seen when linezolid was coadministered with the decongestants. Patients should be counseled to consult with their pharmacist or physician before taking systemic decongestants while taking linezolid. Topical nasal decongestants such as sodium chloride or oxymetazoline may be alternative agents for patients requiring decongestants while receiving linezolid.

5.8.5   Warfarin

A retrospective study was conducted to evaluate the potential interaction between warfarin and linezolid [154]. Patients had baseline PT/INR obtained before treatment and then at days 4 or 5 and 10, at completion of treatment, and at 1 week after stopping linezolid. The PT/INR increased from 1.62 ± 0.32 at baseline to 3.00 ± 0.83 at day 4 or 5 of concomitant therapy ($p < 0.01$) and declined to 1.65 ± 0.45 at the discontinuation of warfarin and 1.26 ± 0.1 at 1 week after stopping linezolid. Although the mechanism of the interaction is unclear, the authors speculated that the rise in PT/INR with concomitant therapy may be due to linezolid lowering vitamin K levels.

Patients stable on warfarin should have PT/INR monitored closely when adding linezolid to their medication regimen. The patients should be monitored clinically for signs and symptoms of bleeding. Other antibiotics may be prescribed to avoid this interaction, or other forms of anticoagulation may be used as alternatives to warfarin.

5.8.6   Clarithromycin

Bolhuis and colleagues initially reported on a patient who experienced a significant increase in linezolid serum concentrations when linezolid was coadministered with clarithromycin [155]. The patient's baseline 24-h AUC on linezolid 300 mg twice daily was 29 mg•h/L. The 24-h AUC increased to 108 mg•h/L following the addition of clarithromycin 1000 mg once daily. It was also noted that the C_{max} in the absorption phase was delayed. The linezolid dose was decreased to 150 mg twice daily for the remainder of his treatment. Bolhuis and colleagues then followed up with a formal study to investigate the interaction between clarithromycin and linezolid [156]. Five healthy adults were initially started on linezolid 300 mg twice daily for 1 week. After 1 week, clarithromycin 250 mg daily was added for 2 weeks followed by an additional 2 weeks of dosing at 500 mg/day. Linezolid serum concentrations were obtained at baseline, at the 2-week end of dosing clarithromycin 250 mg and 500 mg, and at 1 week after stopping clarithromycin. The linezolid 12-h AUC increased from 36.0 mg•h/L at baseline to 61.0 mg•h/L and 67.2 mg•h/L following the coadministration of clarithromycin 250 mg and 500 mg, respectively. The C_{max} (6 mg/L, 8 mg/L, 9.4 mg/L) and C_{min} (1.2 mg/L, 2.1 mg/L, 2.6 mg/L) values increased accordingly. Linezolid clearance declined with the increasing dose of clarithromycin: 7.0 l/h, 4.0 l/h, 3.5 l/h with a resultant prolongation of linezolid half-life and 4.1 h, 4.9 h, and 5.4 h. The authors speculated that the mechanism could be due to clarithromycin's ability to inhibit CYP3A4 and P-glycoprotein.

Patients receiving linezolid and clarithromycin should be monitored closely for signs of linezolid toxicity. Linezolid serum concentrations should be monitored when possible and the dose adjusted accordingly to maintain serum concentrations above the pathogen-specific minimum inhibitory concentration.

5.9   Quinupristin-Dalfopristin

5.9.1   CYP3A4 Metabolized Drugs

In vitro drug interaction studies have demonstrated that quinupristin-dalfopristin significantly inhibits CYP3A4-mediated metabolism. There are no published drug interaction studies in normal volunteers and only limited reports of interactions in patients receiving quinupristin-dalfopristin for therapeutic indications. The manufacturer's package insert indicates that it is reasonable to expect that the concomitant administration of quinupristin-dalfopristin and other drugs primarily metabolized by the CYP3A4 may likely result in increased plasma concentrations of these drugs that could increase or prolong their therapeutic effect and/or increase adverse reactions [157].

In healthy volunteers, the coadministration of quinupristin-dalfopristin with midazolam increased midazolam C_{max} and AUC by 14% and 33%, respectively. Also in healthy volunteers, the C_{max} and AUC of nifedipine were increased by 18% and 44% when the two agents were coadministered. Additional studies in transplant patients indicate that quinupristin-dalfopristin can inhibit the metabolism of cyclosporine and tacrolimus. Stamatakis and Richard reported an interaction between cyclosporine and quinupristin-dalfopristin in a renal transplant patient [158]. The patient's baseline cyclosporine concentrations ranged from 80 to 105 ng/mL. Two and three days after the initiation of quinupristin-dalfopristin therapy, trough cyclosporine concentrations increased to 261 and 291 ng/mL, respectively. Following the discontinuation of quinupristin-dalfopristin, the cyclosporine blood concentrations decreased, and the dosage was increased to the previous regimen.

Medications known to be metabolized by CYP3A4, especially those with a narrow therapeutic index, should be administered with caution and closely monitored for adverse effects.

5.10   Antipseudomonal Penicillins

Aminoglycosides and penicillins are often administered in combination for their additive or synergistic effects in the treatment of serious Gram-negative infections. Numerous reports have been published documenting the ability of commonly used antipseudomonal penicillins to inactivate aminoglycoside antibiotics in vivo [159–166] and in vitro [167–173]. These have usually documented unusually low aminoglycoside concentrations in patients receiving this combination, despite high doses of aminoglycosides. Carbenicillin inactivates all aminoglycosides at faster rates and to a greater extent than ticarcillin, mezlocillin, and piperacillin. Tobramycin is the least stable and amikacin is the most stable aminoglycoside. Gentamicin has intermediate stability. Pickering and Gearhart evaluated the effect of time on the in vitro interaction between mixtures of four aminoglycosides at two concentrations with carbenicillin, piperacillin, mezlocillin, azlocillin, and mecillinam at three concentrations [169]. The inactivation of the aminoglycoside was shown to be directly proportional to the concentration of the penicillin. Aminoglycoside inactivation was greater at 72 h of incubation with the penicillins than after 24 h of incubation. Inactivation by each penicillin was greater for tobramycin and gentamicin than for netilmicin and amikacin, especially at higher penicillin concentrations. At concentrations of 500 mg/L, significantly less inactivation of amikacin occurred compared to netilmicin. No significant change in aminoglycoside activity occurred when the aminoglycosides were stored with the penicillins at -70 °C for 30 days.

There are several reports of in vivo inactivation of aminoglycosides by ticarcillin and carbenicillin. These have occurred in the patients with renal failure, where the penicillin concentrations would be expected to be high. Thompson and colleagues studied the inactivation of gentamicin by piperacillin and carbenicillin in patients with end-stage renal disease [165]. Patients received a single dose of gentamicin, 4 g piperacillin every 12 h for four doses, or 2 g carbenicillin every 8 h for six doses, and gentamicin plus piperacillin or carbenicillin. Subjects were studied on off-dialysis days. Gentamicin was inactivated to a greater extent by carbenicillin than by piperacillin. In the subjects in the carbenicillin group, the terminal elimination half-life of gentamicin was 61.6 h when gentamicin was administered alone and 19.4 h when gentamicin was administered with carbenicillin. In the subjects in the piperacillin group, the mean gentamicin half-life when gentamicin was given alone was 53.9 h, and it was 37.7 h when it was administered with piperacillin. Control samples verified that no in vitro inactivation occurred.

Penicillins combine with aminoglycoside antibiotics in equal molar concentrations at a rate dependent on the concentration, temperature, and medium composition. The greater the concentration of the penicillin, the greater is the inactivation of the aminoglycoside. The inactivation is thought to occur by way of a nucleophilic opening of the beta-lactam ring, which then combines with an amino group of the aminoglycoside, leading to the formation of a microbiologically inactive amide. The inactivation occurs less in pooled human sera than in other media, including whole blood. Spinning down whole blood can help slow the inactivation. Significant serum inactivation occurs at room temperature and under refrigeration. Only when the blood sample is centrifuged and frozen is the inactivation arrested.

Rich reviewed the procedure for handling aminoglycoside concentrations in patients receiving this combination of antibiotics [174]. Blood samples for aminoglycoside concentrations drawn from patients receiving the combination should be sent on ice to the laboratory within 1–2 h so that the sample can be spun down and frozen to arrest any inactivation. Samples left exposed at room temperature will decay 10% in 1 h. The two antibiotics should not be given at the same time. The administration times should be scheduled so that the administration of the aminoglycoside occurs at the end of the penicillin dosing interval, when its concentrations are the lowest. If a patient is receiving this antibiotic combination and unusually low aminoglycoside concentrations occur, the above factors should be checked. Inactivation with beta-lactam antibiotics is further described in Chap. 7.

References

1. Spika JS, Davis DJ, Martin SR et al (1986) Interaction between chloramphenicol and acetaminophen. Arch Dis Child 61:1211–1124
2. Kearns GL, Bocchini JA, Brown RD et al (1985) Absence of a pharmacokinetic interaction between chloramphenicol and acetaminophen in children. J Pediatr 107:134–139
3. Stein CM, Thornhill DP, Neill P et al (1989) Lack of effect of paracetamol on the pharmacokinetics of chloramphenicol. Br J Clin Pharmacol 27:262–264
4. Bloxham RA, Durbin GM, Johnson T et al (1979) Chloramphenicol and phenobarbitone Na drug interaction. Arch Dis Child 54:76–77
5. Powell DA, Nahata MC, Durrell DC et al (1981) Interactions among chloramphenicol, phenytoin, and phenobarbital in a pediatric patient. J Pediatr 98:1001–1003
6. Koup JR, Gibaldi M, McNamara P et al (1987) Interaction of chloramphenicol with phenytoin and phenobarbital. Clin Pharmacol Ther 24:571–575
7. Ballek RE, Reidenberg MM, Orr L (1973) Inhibition of diphenylhydantoin metabolism by chloramphenicol. Lancet 1:150
8. Greenlaw CW (1979) Chloramphenicol-phenytoin interaction. Drug Intell Clin Pharm 13:609–610
9. Saltiel M, Stephens NM (1980) Phenytoin-chloramphenicol interaction. Drug Intell Clin Pharm 14:221
10. Christensen LK, Skovsted L (1969) Inhibition of metabolism by chloramphenicol. Lancet 2:1397–1399
11. Brunova E, Slabochova Z, Platilova H et al (1977) Interaction of tolbutamide and chloramphenicol in diabetic patients. Int J Clin Pharmacol 15:7–12
12. Petitpierre B, Fabre J (1970) Chlorpropamide and chloramphenicol. Lancet 1:789
13. Wallace JF, Smith RH, Garcia M et al (1967) Studies on the pathogenesis of meningitis. VI. Antagonism between penicillin and chloramphenicol in experimental pneumococcal meningitis. J Lab Clin Med 70:408–418
14. Jawetz E (1968) The use of combinations of antimicrobial drugs. Annu Rev Pharmacol 8:151–170
15. Deritis F, Giammanco G, Manzillo G (1972) Chloramphenicol combined with ampicillin in treatment of typhoid. Br Med J 4:17–18
16. French GL, Ling TKW, Davies DP et al (1985) Antagonism of ceftazidime by chloramphenicol in vitro and in vivo during treatment of gram negative meningitis. Br Med J 291:636–637
17. Prober CG (1985) Effect of rifampin on chloramphenicol levels. N Engl J Med 312:788–789
18. Kelly HW, Couch RC, Davis RL et al (1988) Interaction of chloramphenicol and rifampin. J Pediatr 112:817–820
19. Koch-Weser J, Sellars EM (1971) Drug interactions with coumarin anticoagulants (first of two parts). N Engl J Med 285:487–498
20. Koch-Weser J, Sellars EM (1971) Drug interactions with coumarin anticoagulants (second of two parts). N Engl J Med 285:547–558
21. Finegold SM (1970) Interaction of antimicrobial therapy and intestinal flora. Am J Clin Nutr 23:1466–1471
22. Klippel AP, Pitsinger B (1968) Hypoprothrombinemia secondary to antibiotic therapy and manifested by massive gastrointestinal hemorrhage. Arch Surg 96:266–268
23. Bui LL, Huang DD (1999) Possible interaction between cyclosporine and chloramphenicol. Ann Pharmacother 33:252.253
24. Steinfort CL, McConachy KA (1994) Cyclosporin-chloramphenicol drug interaction in a heart-lung transplant patient. Med J Aust 161:455
25. Mathis AS, Shah N, Knipp GT et al (2002) Interaction of chloramphenicol and the calcineurin inhibitors in renal transplant patients. Transplant Infect Dis 4:169–174
26. Schulman SL, Shaw LM, Jabs K et al (1998) Interaction between tacrolimus and chloramphenicol in a renal transplant recipient. Transplantation 65:1397–1398

27. Taber DJ, Dupuis RE, Hollar KD et al (2000) Drug-drug interaction between chloramphenicol and tacrolimus in a liver transplant recipient. Transplant Proc 32:660–662
28. Bakri R, Breen C, Maclean D et al (2003) Serious interaction between tacrolimus FK506 and chloramphenicol in a kidney-pancreas transplant recipient. Transpl Int 16:441–443
29. Hafner V, Albermann N, Haefeli WE et al (2008) Inhibition of voriconazole metabolism by chloramphenicol in an adolescent with central nervous system aspergillosis. Antimicrob Agents Chemother 52:4172–4174
30. Becker LD, Miller RD (1976) Clindamycin enhances a nondepolarizing neuromuscular blockade. Anesthesiology 45:84–87
31. Best JA, Marashi AH, Pollan LD (1999) Neuromuscular blockade after clindamycin administration: a case report. J Oral Maxillofac Surg 57:600–603
32. al Ahdal O, Bevan DR (1995) Clindamycin-induced neuromuscular blockade. Can J Anaesth 42:614–617
33. Sloan PA, Rasul M (2002) Prolongation of rapacuronium neuromuscular blockade by clindamycin and magnesium. Anesth Analg 94:123–124
34. Rubbo JT, Sokoll MD, Gergis SD (1977) Comparative neuromuscular effects of lincomycin and clindamycin. Anesth Analg 56:329–332
35. Wright JM, Collier B (1976) Characterization of the neuromuscular block produced by clindamycin and lincomycin. Can J Physiol Pharmacol 54:937–944
36. Fiekers J, Henderson F, Marshall I et al (1983) Comparative effects of clindamycin and lincomycin on end-plate currents and quantal content at the neuromuscular junction. J Pharmacol Exp Ther 227:308–315
37. Atchinson W, Adgate L, Beaman C (1988) Effects of antibiotics on the uptake of calcium into isolated nerve terminals. J Pharmacol Exp Ther 245:394–401
38. Butkus DE, de Torrente A, Terman DS (1976) Renal failure following gentamicin in combination with clindamycin. Nephron 17:307–313
39. Fruscio R, Lissoni AA, Frapolli R et al (2006) Clindamycin-paclitaxel pharmacokinetic interaction in ovarian cancer patients. Cancer Chemother Pharmacol 58:319–325
40. Curis E, Pestre V, Jullien V et al (2015) Pharmacokinetic variability of clindamycin and influence of rifampicin on clindamycin concentrations in patients with bone and joint infections. Infection 43:473–481
41. Zeller V, Dzeing-Ella A, Kitzis MD et al (2010) Continuous clindamycin infusion, an innovative approach to treating bone and joint infections. Antimicrob Agents Chemother 54:88–92
42. Bernard A, Kermarrec G, Parize P et al (2015) Dramatic reduction of clindamycin serum concentration in staphylococcal osteoarticular infection patients treated with oral clindamycin-rifampicin combination. J Infect 71:200–206
43. Wynalda MA, Hutzler JM, Koets MD et al (2003) In vitro metabolism of clindamycin in human liver and intestinal microsomes. Drug Metab Dispos 31:878–887
44. Tilstone WJ, Gray JM, Nimmo-Smith RH et al (1977) Interaction between warfarin and sulphamethoxazole. Postgrad Med J 53:388–390
45. Kaufman JM, Fauver HE (1980) Potentiation of warfarin by trimethoprim-sulfamethoxazole. Urology 16:601–603
46. Greenlaw CW (1979) Drug interaction between co-trimoxazole and warfarin. Am J Hosp Pharm 36:1155
47. Errick JK, Keys PW (1978) Co-trimoxazole and warfarin: case report of an interaction. Am J Hosp Pharm 35:1399–1401
48. O'Reilly RA, Motley CH (1979) Racemic warfarin and trimethoprim-sulfamethoxazole interaction in humans. Ann Intern Med 91:34–36
49. O'Reilly RA (1980) Stereoselective interaction of trimethoprim-sulfamethoxazole with the separated enantiomorphism of racemic warfarin in man. N Engl J Med 302:33–35
50. Beest P-v, Koerselman J, Herings RMC (2008) Risk of major bleeding during concomitant use of antibiotic drugs and coumarin anticoagulants. J Thromb Haemost 6:284–290
51. Lane MA, Zeringue A, McDonald JR (2014) Serious bleeding events due to warfarin and antibiotic co-prescription in a cohort of veterans. Am J Med 127:657–663

52. Ahmed A, Stephens JC, Kaus CA et al (2008) Impact of preemptive warfarin dose reduction on anticoagulation after initiation of trimethoprim-sulfamethoxazole or levofloxacin. J Throm Thrombolysis 26:44–48
53. Schalekamp T, van Geest-Daalderop JHH, MHH K et al (2007) Coumarin anticoagulants and co-trimoxazole: avoid the combination rather than manage the interaction. Eur J Clin Pharmacol 63:335–343
54. Parekh TM, Raji M, Lin YL et al (2014) Hypoglycemia after antimicrobial drug prescription for older patients using sulfonylureas. JAMA Int Med 174:1605–1612
55. Tan A, Holmes HM, Kuo YF et al (2015) Co-administration of co-trimoxazole with sulfonylureas: hypoglycemia events and pattern of use. J Gerontol A Biol Sci Med Sci 70:247–254
56. Schelleman H, Bilker WB, Brensinger CM et al (2010) Anti-infective and the risk of severe hypoglycemia in users of glipizide or glyburide. Clin Pharmacol Ther 88:214–222
57. Juurlink DN, Mamdani M, Kopp A et al (2003) Drug-drug interactions among elderly patients hospitalized for drug toxicity. JAMA 289:1652–1658
58. Alappan R, Perazella MA, Buller GK (1996) Hyperkalemia in hospitalized patients treated with trimethoprim-sulfamethoxazole. Ann Int Med 124:316–320
59. Alappan R, Buller GK, Perazella MA (1999) Trimethoprim-sulfamethoxazole therapy in outpatients: is hyperkalemia a significant problem? Am J Nephrol 19:389–394
60. Nguyen AT, Gentry CA, Furrh RZ (2013) A comparison of adverse drug reactions between high- and standard-dose trimethoprim-sulfamethoxazole in the ambulatory setting. Curr Drug Saf 8:114–119
61. Gentry CA, Nguyen AT (2013) An evaluation of hyperkalemia and serum creatinine elevations associated with different dosage levels of outpatient trimethoprim-sulfamethoxazole with and without concomitant medications. Ann Pharmacother 47:1618–1626
62. Velazquez H, Perazella MA, Wright FS et al (1993) Renal mechanism of trimethoprim-induced hyperkalemia. Ann Intern Med 119:296–301
63. Fralick M, Macdonald EM, Gomes T et al (2014) Co-trimoxazole and sudden cardiac death in patients receiving inhibitors of renin-angiotensin system: population based study. BMJ 349:g6196. https://doi.org/10.1136/bmj.g6196
64. Lee SW, Park SW, Kang JM (2014) Intraoperative hyperkalemia induced by administration of TMP-SMX in a patient receiving angiotensin receptor blockers. J Clin Anesth 26:427–428
65. Antoniou T, Hollands S, Macdonald EM et al (2015) Trimethoprim-sulfamethoxazole and risk of sudden death among patients taking spironolactone. CMAJ 187:E138–E142
66. Pierce D, Corcoran M, Martin P et al (2014) Effect of mmx mesalamine coadministration on the pharmacokinetics of amoxicillin, ciprofloxacin xr, metronidazole, and sulfamethoxazole: results from four randomized clinical trials. Drug Des Devel Ther 8:529–543
67. Cudmore J, Seftel M, Sisler J et al (2014) Methotrexate and trimethoprim-sulfamethoxazole. Toxcity from this combination continues to occur. Can Fam Physician 60:53–56
68. Davis SA, Krowchuk DP, Feldman SR (2014) Prescriptions for a toxic combination: use of methotrexate plus trimethoprim-sulfamethoxazole in the United States. South Med J 107:292–293
69. Green M, Otieno K, Katana A et al (2016) Pharmacokinetics of mefloquine and its effect on sulfamethoxazole and trimethoprim steady-state blood levels in intermittent preventive treatment (IPTp) of pregnant HIV-infected women in Kenya. Malar J 17:7. https://doi.org/10.1186/s12936-015-1049-9
70. Jaffe JM, Colonize JL, Pouts RI et al (1973) Effect of altered urinary pH on tetracycline and doxycycline excretion in humans. J Pharmacokinet Bipolar 1:267–282
71. Jaffe JM, Pouts RL, Fled SL et al (1974) Influence of repetitive dosing and altered pH on doxycycline excretion in humans. J Pharm Sci 63:1256–1260
72. Chin TF, Latch JL (1975) Drug diffusion and bioavailability: tetracycline metallic chelation. Am J Hosp Pharm 32:625–529
73. Ericsson CD, Feldman S, Pickering LK et al (1982) Influence of subsalicylate bismuth on absorption of doxycycline. JAMA 247:2266–2267

74. Albert KS, Welch RD, Descanted KA et al (1979) Decreased tetracycline bioavailability caused by a bismuth subsalicylate antidiarrheal mixture. J Pharm Sci 68:586–588

75. Friedman H, Greenbelt DJ, Leduc BW (1989) Impaired absorption of tetracycline by colestipol is not reversed by orange juice. J Clin Pharmacol 29:748–751

76. Lindenbaum J, Round DG, Butler VP et al (1981) Inactivation of digoxin by the gut flora: reversal by antibiotic therapy. N Engl J Med 305:789–794

77. Penttilla O, Neuvonen PJ, Ahoy K et al (1974) Interaction between doxycycline and some anti- epileptic drugs. Br Med J 2:470–472

78. Neuvonen PJ, Penttilla O, Lehtovaara R et al (1975) Effect of antiepileptic drugs on the elimination of various tetracycline derivatives. Eur J Clin Pharmacol 9:147–154

79. Neuvonen PJ, Penttila O (1974) Interaction between doxycycline and barbiturates. Br Med J 1:535–536

80. Westfall LK, Mintzer DL, Wiser TH (1980) Potentiation of warfarin by tetracycline. Am J Hosp Pharm 37:1620–1625

81. Caraco Y, Rubinow A (1992) Enhanced anticoagulant effect of coumarin derivatives induced by doxycycline coadministration. Ann Pharmacother 26:1084–1086

82. McGennis AJ (1978) Lithium carbonate and tetracycline interaction. Br Med J 1:1183

83. Dowd MB, Kippes KA, Witt DM et al (2012) A randomized controlled trial of empiric warfarin dose reduction with the initiation of doxycycline therapy. Thromb Res 130:152–156

84. Fankhauser MP, Lindon JL, Connolly B et al (1988) Evaluation of lithium-tetracycline interaction. Clin Pharm 7:314–317

85. Steele M, Couturier JA (1999) possible tetracycline-risperidone-sertraline interaction in an adolescent. Can J Clin Pharmacol 6:15–17

86. McCormack JP, Reid SE, Lawson LM (1990) Theophylline toxicity induced by tetracycline. Clin Pharm 9:546–549

87. Kawai M, Honda A, Yoshida H et al (1992) Possible theophylline-minocycline interaction. Ann Pharmacother 26:1300–1301

88. Pfeifer HJ, Greenblatt DJ, Friedman P (1979) Effects of three antibiotics on theophylline kinetics. Clin Pharmacol Ther 26:36–40

89. Mathis JW, Prince RA, Weinberger MM et al (1982) Effect of tetracycline hydrochloride on theophylline kinetics. Clin Pharm 1:446–448

90. Gotz VP, Ryerson GG (1986) Evaluation of tetracycline on theophylline disposition in patients with chronic obstructive airways disease. Drug Intell Clin Pharm 20:694–697

91. Jonkman JHG, van der Boon WJV, Schoenmaker R et al (1985) No influence of doxycycline on theophylline pharmacokinetics. Ther Drug Monit 7:92–94

92. Bacon JF, Shenfield GM (1980) Pregnancy attributable to interaction between tetracycline and oral contraceptives. Br Med J 280:293

93. DeSano EA, Hurley SC (1982) Possible interactions of antihistamines and antibiotics with oral contraceptive effectiveness. Fertil Steril 37:853–854

94. Helms SE, Bredle DL, Zajic J et al (1997) Oral contraceptive failure rates and oral antibiotics. J Am Acad Dermatol 36:705–710

95. Neely JL, Abate M, Swinkler M et al (1991) The effect of doxycycline on serum levels of ethinyl estradiol, norethindrone, and endogenous progesterone. Obstet Gynecol 77:416–420

96. Murphy AA, Zacur HA, Charache P et al (1991) The effect of tetracycline on levels of oral contraceptives. Am J Obstet Gynecol 164:28–33

97. Dogterom P, van den Heuvel MW, Thomsen T (2005) Absence of pharmacokinetic interactions of the combined contraceptive vaginal ring NuvaRing with oral amoxicillin or doxycycline in two randomized trials. Clin Pharmacokinet 44:429–438

98. Archer JS, Archer DF (2002) Oral contraceptive efficacy and antibiotic interaction; a myth debunked. J Am Acad Dermatol 46:917–923

99. Tortajada-Ituren JJ, Ordovas-Baines JP, Llopis-Salvia P et al (1999) High-dose methotrexate-doxycycline interaction. Ann Pharmacother 33:804–808

100. Turck M (1984) Successful psoriasis treatment then sudden "cytotoxicity.". Hosp Pract 19:175,6

101. Colmenero JD, Fernandez-Gallardo LC, Agundez JAG et al (1994) Possible implications of doxycycline-rifampin interaction for treatment of brucellosis. Antimicrob Agents Chemother 38:2798–2802
102. Zimmerman JJ, Raible DG, Harper DM et al (2008) Evaluation of potential tigecycline-warfarin drug interaction. Pharmacotherapy 28:895–905
103. Stumpf AN, Schmidt C, Hiddemann W et al (2009) High serum concentrations of ciclosporin related to administration of tigecycline. Eur J Clin Pharmacol 65:101–103
104. Srinivas NR (2009) Tigecycline and cyclosporine interaction-an interesting case of biliary –excreted drug enhancing the oral bioavailability of cyclosporine. Eur J Clin Pharmacol 65:543–544
105. Pavan M, Chaudhari AP, Ranganth R et al (2011) Altered bioavailability of tacrolimus following intravenous administration of tigecycline. Am J Kid Dis 57:352
106. Churchill DN, Seely J (1977) Nephrotoxicity associated with combined gentamicin-amphotericin B therapy. Nephron 19:176–181
107. Kroenfeld MA, Thomas SJ, Turndorf H (1986) Recurrence of neuromuscular blockade after reversal of vecuronium in a patient receiving polymyxin/amikacin sternal irrigation. Anesthesiology 65:93–94
108. Warner WA, Sanders E (1971) Neuromuscular blockade associated with gentamicin therapy. JAMA 215:1153–1154
109. Levanen J, Nordman R (1975) Complete respiratory paralysis caused by a large dose of streptomycin and its treatment with calcium chloride. Ann Clin Res 7:47–49
110. Lippmann M, Yang E, Au E et al (1982) Neuromuscular blocking effects of tobramycin, gentamicin, and cefazolin. Anesth Analg 61:767–770
111. Duouis JY, Martin R, Tetrault JP (1989) Atracurium and vecuronium interaction with gentamicin and tobramycin. Can J Anaesth 36:407–411
112. Singh YN, Marshall IG, Harvey AL (1982) Pre- and postjunctional blocking effects of aminoglycoside, polymyxin, tetracycline and lincosamide antibiotics. Br J Anaesth 54:1295–1306
113. Zarfarin Y, Koren G, Maresky D et al (1985) Possible indomethacin-aminoglycoside interaction in preterm infants. J Pediatr 106:511–513
114. Termeer A, Hoitsma AJ, Koene RAP (1986) Severe nephrotoxicity caused by the combined use of gentamicin and cyclosporine in renal allograft recipients. Transplantation 42:220–221
115. Christensen ML, Stewart CF, Crom WR (1989) Evaluation of aminoglycoside disposition in patients previously treated with cisplatin. Ther Drug Monit 11:631–636
116. Gonzalez-Vitale JC, Hayes DM, Cvitkovic E et al (1978) Acute renal failure after cis-dichlorodiammineplatinum (II) and gentamicin-cephalothin therapies. Cancer Treat Rep 62:693–698
117. Salem PA, Jabboury KW, Khalil MF (1982) Severe nephrotoxicity: a probable complication of cis-dichlorodiammineplatinum (II) and cephalothin-gentamicin therapy. Oncology 39:31–32
118. Kohn S, Fradis M, Podoshin L et al (1997) Ototoxicity resulting from combined administration of cisplatin and gentamicin. Laryngoscope 107:407–408
119. Dentino M, Luft FC, Yum MN et al (1978) Long-term effect of cis-diamminedichloride platinum (CDDP) on renal function and structure in man. Cancer 41:1274–1281
120. Lee EJ, Egorin MJ, Van Echo DA et al (1988) Phase I and pharmacokinetic trial of carboplatin in refractory adult leukemia. J Natl Cancer Inst 80:131–135
121. Bregman CL, Williams PD (1986) Comparative nephrotoxicity of carboplatin and cisplatin in combination with tobramycin. Cancer Chemother Pharmacol 18:117–123
122. Kaka JS, Lyman C, Kilarski DJ (1984) Tobramycin-furosemide interaction. Drug Intell Clin Pharm 18:235–238
123. Mathog RH, Klein WJ (1969) Ototoxicity of ethacrynic acid and aminoglycoside antibiotics in uremia. N Engl J Med 280:1223–1224
124. Smith CR, Lietman PS (1983) Effect of furosemide on aminoglycoside-induced nephrotoxicity and auditory toxicity in humans. Antimicrob Agents Chemother 23:133–137

125. Pillay VKG, Schwartz FD, Aimi K et al (1969) Transient and permanent deafness following treatment with ethacrynic acid in renal failure. Lancet 1:77–79
126. Meriweather WD, Mangi RJ, Serpick AA (1971) Deafness following standard intravenous doses of ethacrynic acid. JAMA 216:795–798
127. Cimino MA, Rotstein C, Slaughter RL et al (1987) Relationship of serum antibiotic concentrations to nephrotoxicity in cancer patients receiving concurrent aminoglycoside and vancomycin therapy. Am J Med 83:1091–1097
128. Pauly DJ, Musa DM, Lestico MR et al (1990) Risk of nephrotoxicity with combination vancomycin-aminoglycoside antibiotic therapy. Pharmacotherapy 10:378382
129. Mellor JA, Kingdom J, Cafferkey M et al (1985) Vancomycin toxicity: a prospective study. J Antimicrob Chemother 15:773–780
130. Ryback MJ, Albrecht LM, Boike SC et al (1990) Nephrotoxicity of vancomycin, alone and with an aminoglycoside. J Antimicrob Chemother 25:679–687
131. Ramsey TD, Lau TTY, Ensom MHH (2013) Serotonergic and adrenergic drug interactions associated with linezolid: a critical review and practical management approach. Ann Pharmacother 47:543–560
132. Douros A, Grabowski K, Stahlmann R (2015) Drug-drug interactions and safety of linezolid, tedizolid, and other oxazolidinones. Expert Opin Drug Metab Toxicol 11:1849–1859
133. Narita M, Tsuji BT, Linezolid-associated YV (2007) Peripheral and optic neuropathy, lactic acidosis, and serotonin syndrome. Pharmacotherapy 27:1189–1197
134. Wigen CL, Goetz MB (2002) Serotonin syndrome and linezolid. Clin Infect Dis 34:1651–1652
135. Hachem RY, Hicks K, Huen A et al (2003) Myelosuppression and serotonin syndrome associated with concurrent use of linezolid and selective serotonin reuptake inhibitors in bone marrow transplant recipients. Clin Infect Dis 37:e8–11
136. Lavery S, Ravi H, McDaniel WW et al (2001) Linezolid and serotonin syndrome. Psychosomatics 42:432–434
137. Clark DB, Andrus MR, Byrd DC (2006) Drug interactions between linezolid and selective serotonin reuptake inhibitors: case report involving sertraline and review of the literature. Pharmacotherapy 26:269–276
138. Sola CL, Bostwick JM, Hart DA et al (2006) Anticipating potential linezolid-SSRI interactions in the general hospital setting an MAOI in disguise. Mayo Clin Proc 81:330–334
139. Aga VM, Barklage NE, Jefferson JW (2003) Linezolid, a monoamine oxidase inhibiting antibiotic, and antidepressants. J Clin Psychiatry 64:609–611
140. Debellas RJ, Schaefer OP, Liquori M et al (2005) Linezolid-associated serotonin syndrome after concomitant treatment with citalopram and mirtazapine in a critically ill bone marrow transplant recipient. J Intensive Care Med 20:303–305
141. Hammerness P, Parada H, Abrams A (2002) Linezolid: MAOI activity and potential drug interactions. Psychosomatics 43:248–249
142. Packer S, Berman SA (2007) Serotonin syndrome precipitated by the monoamine oxidase inhibitor linezolid. Am J Psychiatry 164:346–347
143. Mason LW, Randhawa KS, Carpenter EC (2008) Serotonin toxicity as a consequence of linezolid use in revision hip arthroplasty. Orthopedics 31:1140
144. Bergeron L, Boule M, Perreault S (2005) Serotonin toxicity associated with concomitant use of linezolid. Ann Pharmacother 39:956–961
145. Jones SL, Athan E, O'Brien D (2004) Serotonin syndrome due to co-administration of linezolid and venlafaxine. J Antimicrob Chemother 54:289–290
146. Steinberg M, Morin A (2007) Mild serotonin syndrome associated with concurrent linezolid and fluoxetine. Am J Health Sys Pharm 64:5962
147. Lorenz RA, Vandenberg AM, Canepa EA (2008) Serotonergic antidepressants and linezolid: a retrospective chart review and presentation of cases. Int J Psychiatry Med 38:81–90
148. Das PK, Warkentin DI, Hewko R et al (2008) Serotonin syndrome after concomitant treatment with linezolid with meperidine. Clin Infect Dis 46:264–265
149. Morrison EK, Rowe AS (2012) Probable drug-drug interaction leading to serotonin syndrome in a patient treated with concomitant buspirone and linezolid in the setting of therapeutic hypothermia. J Clin Pharmacol Ther 37:610–613

150. Gebhart BC, Barker BC, Markewitz BA (2007) Decreased serum linezolid levels in a critically ill patient receiving concomitant linezolid and rifampin. Pharmacotherapy 27:476–479
151. Egle H, Trittler R, Kummerer K et al (2005) Linezolid and rifampin: drug interaction contrary to expectations? Clin Pharmacol Ther 77:451–452
152. Gervasoni C, Simonetti FR, Resnati C et al (2015) Prolonged inductive effect of rifampicin in linezolid exposure. Eur J Clin Pharmacol 71:643–644
153. Hendershot PE, Antal EJ, Welshman IR (2002) Linezolid: pharmacokinetic and pharmacodynamic evaluation of coadministration with pseudoephedrine HCl, phenylpropanolamine HCL, and dextromethorphan HBr. J Clin Pharmacol 41:563–572
154. Sakai Y, Naito T, Arima C et al (2015) Potential drug interaction between warfarin and linezolid. Intern Med 54:459–464
155. Bolhuis MS, van Altena R, Uges DRA et al (2010) Clarithromycin significantly increases linezolid serum concentrations. Antimicrob Agents Chemother 54:5418–5419
156. Bolhuis MS, Altena v, van Soolingen D et al (2013) Clarithromycin increases linezolid exposure in multidrug-resistant tuberculosis patients. Eur Respir J 42:1614–1621
157. Synercid IV (2003) (quinupristin-dalfopristin) product information. King Pharmaceutical, Bristol
158. Stamatakis MK, Richards JG (1997) Interaction between quinupristin-dalfopristin and cyclosporine. Ann Pharmacother 31:576–578
159. Lampasona V, Crass RE, Reines HD (1983) Decreased serum tobramycin concentrations in patient with renal failure. Clin Pharm 2:6–9
160. Russo M (1980) Penicillin-aminoglycoside inactivation: another possible mechanism of interaction. Am J Hosp Pharm 37:702–704
161. Chow MSS, Quintiliani R, Nightingale CH (1982) In vivo inactivation of tobramycin by ticarcillin. JAMA 247:658–659
162. Kradjan WA, Burger R (1980) In vivo inactivation of gentamicin by carbenicillin and ticarcillin. Arch Intern Med 140:1668–1670
163. Schentag JJ, Simons GW, Schultz RW et al (1984) Complexation versus hemodialysis to reduce elevated aminoglycoside serum concentrations. Pharmacotherapy 4:374–380
164. Uber WE, Brundage RR, White RL et al (1991) In vivo inactivation of tobramycin by piperacillin. Ann Pharmacother 25:357–359
165. Thompson MIB, Russo ME, Saxon BJ et al (1982) Gentamicin inactivation by piperacillin or carbenicillin in patients with end-stage renal disease. Antimicrob Agents Chemother 21:268–273
166. Ervin FR, Bullock WE, Nuttall CE (1976) Inactivation of gentamicin by penicillins in patients with renal failure. Antimicrob Agents Chemother 9:1004–1031
167. Wallace SM, Chan LY (1985) In vitro interaction of aminoglycosides with β-lactam penicillins. Antimicrob Agents Chemother 28:274–281
168. Henderson JL, Polk RE, Kline BJ (1981) In vitro inactivation of gentamicin, tobramycin, and netilmicin by carbenicillin, azlocillin, or mezlocillin. Am J Hosp Pharm 38:1167–1170
169. Pickering LK, Gearhart P (1979) Effect of time and concentration upon interaction between gentamicin, tobramycin, netilmicin, or amikacin and carbenicillin or ticarcillin. Antimicrob Agents Chemother 15:592–596
170. Pickering LK, Rutherford I (1981) Effect of concentration and time upon inactivation of tobramycin, gentamicin, netilmicin, and amikacin by azlocillin, carbenicillin, mecillinam, mezlocillin, and piperacillin. J Pharmacol Exp Ther 217:345–349
171. Hold HA, Broughall JM, McCarthy M et al (1976) Interactions between aminoglycoside antibiotics and carbenicillin or ticarcillin. Infection 4:107–109
172. Davies M, Morgan JR, Anand C (1975) Interaction of carbenicillin and ticarcillin with gentamicin. Antimicrob Agents Chemother 7:431–434
173. Mclaughlin JE, Reeves DS (1971) Clinical and laboratory evidence of inactivation of gentamicin by carbenicillin. Lancet 1:261–264
174. Rich DS (1983) Recent information about inactivation of aminoglycosides by carbenicillin, and ticarcillin: clinical implications. Hosp Pharm 18:41–43

Chapter 6
Drugs for Tuberculosis

Rocsanna Namdar and Charles A. Peloquin

6.1 Introduction

Tuberculosis (TB) remains a leading infectious killer, particularly in the developing world [1]. Given the high rates of coinfection with TB and HIV, drug interactions are frequent occurrences. In particular, rifamycins commonly produce significant drug interactions that can decrease the effectiveness of highly active antiretroviral therapy in patients with HIV. This chapter assesses drug interactions in patients with TB and briefly in patients with nontuberculous mycobacterial (NTM) infections.

6.2 Standard Treatment for Tuberculosis

The published treatment guidelines for TB generally produce successful outcomes, even in HIV-positive patients [1, 2]. These references are recommended to all practitioners dealing with such patients. For suspected drug-susceptible disease, a regimen of isoniazid (INH), rifampin (RIF), pyrazinamide (PZA), and ethambutol (EMB) is used. Rifabutin (RBN) is an alternative to RIF to reduce cytochrome P450 (CYP) enzyme induction in both the liver and the intestine. When full drug susceptibility is confirmed, EMB can be discontinued. PZA can be discontinued in patients

R. Namdar
Raymond G. Murphy Veterans Affairs Medical Center, Albuquerque, NM, USA

C. A. Peloquin (✉)
Infectious Disease Pharmacokinetics Lab, College of Pharmacy, and Emerging Pathogens Institute, University of Florida, Gainesville, FL, USA
e-mail: peloquin@cop.ufl.edu

© Springer International Publishing AG, part of Springer Nature 2018   221
M. P. Pai et al. (eds.), *Drug Interactions in Infectious Diseases: Antimicrobial Drug Interactions*, Infectious Disease,
https://doi.org/10.1007/978-3-319-72416-4_6

who respond normally to treatment after 8 weeks [1, 2]. INH and either RIF or RBN are continued for an additional 4 months or longer if the patient is slow to respond or has extensive pulmonary cavitary, bone, or central nervous system (CNS) disease. Multidrug-resistant tuberculosis (MDR-TB, defined as resistance to at least INH and RIF) and extensively drug-resistant tuberculosis (XDR-TB, defined as MDR-TB plus resistance to a quinolone and an injectable agent) are much more difficult to treat [1, 2]. The drugs used for DR-TB are less effective and more toxic than INH and RIF, and the duration of treatment for DR-TB often is much longer (24 months or more).

6.3   Oral Absorption

6.3.1   Interactions with Food

INH and RIF show marked decreases in the maximum serum concentration (Cmax, 51% and 36%, respectively) and lesser decreases in area under the serum concentration versus time curve (AUC, 9% and 10%, respectively) when given with high-fat meals (Table 6.1) [3]. EMB shows modest decreases in Cmax (17%) but not AUC, while PZA only shows a modest delay in absorption when these drugs are given with high-fat meals [4, 5]. A meta-analysis evaluating the effect of food on first-line antituberculosis drugs also found that food reduces the Cmax of INH, RIF, and EMB in pooled data analysis [6]. High-fat meals do not adversely affect the absorption of ethionamide (ETA) but decrease the Cmax of cycloserine (CS) by 27% (AUC decreases by only 5%) [7, 8]. Orange juice also decreases the Cmax of cycloserine by about 13% (AUC decreases by only 3%), and presumably this would occur in other acidic beverages [8]. In contrast, high-fat meals increase the Cmax of clofazimine (CF) and p-aminosalicylic acid (PAS) granules [9, 10].

6.3.2   Interactions with Antacids

Of the four most frequently used TB drugs, only EMB appears to be significantly affected by coadministration with antacids (Mylanta®, Table 6.1) [3–5]. Our investigation showed no significant effect when Mylanta® was given 9 h before INH, at the time of dosing, and then with lunch and dinner following dosing. Antacids produced insignificant change in the absorption of RIF, CS, ETA, PAS, CF, and PZA (pyrazinamide) [5–11]. The absorption of fluoroquinolones is markedly decreased when coadministered with antacids or medications containing di- or trivalent cations. These drug interactions can be avoided by ingesting antacids at least 2 h apart from fluoroquinolones [1, 12].

Table 6.1 Effects of food and antacids on the absorption of antituberculous drugs

Drug	Effect of food	Effect of antacids	Clinical implications
Aminosalicylic acid (PAS) granules	Acidic beverage or yogurt prevents release in the stomach, thus reducing nausea; food increases absorption	Small decrease in absorption	Give PAS granules with acidic beverage or with food. Avoid antacids if possible
Ciprofloxacin	Delayed T_{max}, but minimal effect on AUC	Large decrease in C_{max} and AUC	Do not coadminister with di- and trivalent cations, including antacids
Cycloserine	Food decreases C_{max} 27%, no effect on AUC	Antacids slightly increase C_{max}	Do not coadminister with food if possible
Ethambutol	Delayed T_{max}, 16% decrease in C_{max}, but minimal effect on AUC	28% decrease in C_{max} and 10% decrease in AUC	May be given with food. Do not coadminister with antacids
Ethionamide	No significant effect	No significant effect	Can be coadministered with food and antacids
Isoniazid	Food, especially carbohydrate-based meals, significantly reduces isoniazid C_{max} and AUC	0–19% decrease in AUC	Do not coadminister with meals; do not coadminister with antacids whenever possible
Levofloxacin	No significant effect	Large decreases in C_{max} and AUC	Do not coadminister with di- and trivalent cations, including antacids
Pyrazinamide	Delayed T_{max}, no effect on AUC	No significant effect	May be given with food or antacids
Rifabutin	No significant effect	Unknown, not affected by didanosine	May be given with food; do not coadminister with antacids until studied
Rifampin	Delayed T_{max}, 15–36% decrease in C_{max} and 4–23% decrease in AUC	No significant change in serum concentrations, 30% decrease in 24-h urinary excretion	Do not coadminister with food; may be given with ranitidine; avoid coadministration with antacids whenever possible

Adapted from Burman et al. [29]

T_{max} time from drug ingestion to peak (maximal) serum concentration, C_{max} peak (maximal) serum concentration, AUC area under the serum concentration-time curve, C_{max} peak (maximal) serum concentration

6.3.3 Interactions with H₂ Antagonists

RIF is not affected by the coadministration of ranitidine [3]. Data are not available for the other TB drugs.

6.3.4 Malabsorption in Selected Patient Populations

Patients with known or suspected gastroenteropathies may have difficulty absorbing the TB drugs. INH, RIF, and EMB appear to be more prone to malabsorption, with lower Cmax and AUC [13–18]. Recent studies suggest that the dose of RBN often is too low, even in the presence of ritonavir. Minimal data exist for rifapentine [19–21]. Studies show that INH and RBN malabsorption may lead to treatment failures and the selection of drug resistance, especially among AIDS patients [22, 23]. PZA generally is well absorbed [5, 13]. Reasons for drug malabsorption may include HIV-related achlorhydria, HIV enteropathy, and opportunistic infections of the gastrointestinal tract, such as cryptosporidiosis [24–28]. Other populations to observe carefully include patients with cystic fibrosis and diabetes mellitus. Therapeutic drug monitoring (TDM) early during treatment may be used to identify problems and to guide dose adjustments [18].

6.4 Drug and Disease Interactions

6.4.1 Isoniazid Interactions

INH is cleared by N-acetyltransferase 2 (NAT2) to the microbiologically inactive metabolite acetyl isoniazid and subsequently to mono- and di-acetyl-hydrazine [29, 30]. INH is not substantially removed by hemodialysis [31]. INH is a relatively potent inhibitor of several CYP isoenzymes and interacts significantly with phenytoin (a CYP2C9 substrate) and carbamazepine (CYP3A4 and either CYP2C8 or CYP2C9), increasing concentrations of both [31–33]. INH also may inhibit the clearance of valproic acid [33], diazepam (CYP3A4 and CYP2C19), primidone, chlorzoxazone (CYP2E1), theophylline (CYP1A2), warfarin (CYP1A2, CYP2C9, CYP2C19, CYP3A4), serotonergic antidepressants [34], and clozapine [35–42]. Desta and colleagues showed that INH inhibits CYP2C19 and CYP3A4 in a concentration-dependent manner [43]. Significant inhibition of CYP2C9 and CYP1A2 in their human liver microsome system was not shown; however, INH was considered a weak noncompetitive inhibitor of CYP2E1 and a competitive inhibitor of CYP2D6 [42]. The inductive effect of RIF on CYP enzymes outweighs the inhibitory effect of INH. Therefore, the overall effect of combined therapy with RIF and INH is a decrease in the concentrations of drugs [44].

INH can also act as a monoamine oxidase inhibitor, with a potential for interaction with antidepressants. Excess catecholamine stimulation resulting in increased blood pressure has been reported with INH and levodopa therapy [45]. INH causes an initial inhibition, followed by induction of CYP2E1 [40]. Therefore, INH can alter the clearance of ethanol. INH may inhibit or promote the conversion of acetaminophen to its putative toxic intermediate metabolite, N-acetyl-p-benzoquinone imine (NAPQI), depending on the timing of the doses [39]. Therefore, high-dose acetaminophen should be avoided with INH [39, 46–49]. The overall effect of combined therapy with RIF and INH is a decrease in the concentrations of coadministered drugs [43].

The absorption of INH can be affected by other drugs. Coadministration of ciprofloxacin and INH results in a delay (but not a reduction in the extent) of INH absorption [50].

6.4.2 Rifamycin Interactions

The available rifamycins include rifampin (RIF), rifabutin (RBN), and rifapentine (RPNT). They share a similar mechanism of action and generally show cross-resistance. They are cleared by esterases to their 25-desacetyl derivatives, which have roughly half of the parent drugs' activity. RBN undergoes some CYP clearance as well. Most of the parent drug and metabolite are cleared through the biliary tract, with small amounts through the urine [3, 30, 46]. For 25-desacetyl-RBN, subsequent metabolism occurs via CYP3A4. RIF is not substantially removed by hemodialysis [31]. Rifamycins are potent inducers of the P450 enzyme system, especially CYP3A4 and 2C8/9 [28, 50, 51]. Further, RIF also induces the activity of the phase II enzymes uridine glucuronosyltransferase and sulfotransferase and the efflux transporter P-glycoprotein (P-gp) [51, 52].

At least with single doses, RIF is an inhibitor of P-gp and MRP2 in vitro and in animals. However, continued doses of RIF appear to induce MRP2. Rifampin, like cyclosporine and gemfibrozil, inhibits OATP1B1, an uptake transporter for many drugs and endogenous substances [51, 52]. At pharmacological concentrations, RIF induces the expression of MRP transporters both at the apical (MRP2) and basolateral (MRP3) membrane of hepatocytes, while, at higher concentrations, it was shown to exert a competitive inhibitory effect on MRP2 in vitro [53]. Caution should be exercised when reading this literature, since the effects may depend on whether the experiment was done in vitro or in vivo and, for the latter, if the experiment was single dose or multidose. For example, RIF can inhibit OATPs, which correlates with the initial rise in serum bilirubin at the start of RIF treatment [54]. However, it is well known that these values return to normal early in treatment. So, it is possible that some of these effects of RIF on OATP are not sustained over time. Further, compensatory effects across different OATB receptors may be seen. For example, data suggest that SLCO1B1 polymorphism does not affect the extent of induction of hepatic CYP3A4 by rifampicin, probably because other uptake transporters, such as OATP1B3, can compensate for reduced uptake of rifampicin by OATP1B1 [55, 56].

RIF intracellular concentrations and CYP3A induction are strongly correlated with P-gp levels, encoded on the multidrug resistance gene (MDR1) [57, 58]. Polymorphic expression of MDR1 may partially explain the wide inter-patient variability in CYP3A induction by RIF. RPNT is at least 85% as potent an inducer as RIF, and when RPNT is dosed daily, RPNT may be more potent than daily RIF [5960]. RBN is about 40% as potent as RIF [60]. The AUC of RIF may be lower in patients with active tuberculosis, those with the solute carrier organic anion transporter 1B1 (SLCO1B1) c. 463CA genotype, and in TB patients from Africa versus North America [61]. The extent of induction by rifamycins may change with dosing frequency (daily versus intermittently) [62]. For 600 mg RIF daily, near-maximum induction occurs in about 7 days, although true steady state may take weeks [51]. Larger doses may shorten the time to, but not the extent of, induction, which lasts for 7–14 days after the rifamycin is stopped [51, 62, 63]. CYP3A4, and to a lesser degree, CYP2C9, CYP2C19, and CYP2D6, are most affected, leading to shorter half-lives and lower plasma concentrations for many coadministered drugs.

RBN induces and is partly metabolized by CYP3A, leading to complex bidirectional interactions [29, 51]. RBN decreases concentrations of other drugs, while CYP3A inhibitors increase the concentrations of RBN and especially 25-O-desacetyl RBN, sometimes leading to toxicity [29, 51]. RBN often is used in place of RIF if there are drug-drug interactions of concern. In contrast, RPNT is very similar to RIF regarding drug interactions. Unlike RBN, RPNT does not offer any advantage in sparing the drug interactions. Like RIF, RPNT is not a substrate for CYP enzymes, and concentrations of RIF and RPNT do not increase with concurrent enzyme inhibitors.

Significant inter-patient variability in the extent of rifamycin drug interactions can be seen [18, 29, 51, 64]. Most data come from small studies of healthy volunteers and focus on bidirectional interactions. Clinically, complex interactions involving three, four, or more drugs cannot be predicted, especially when factoring in erratic drug absorption in some patients [18, 64]. For example, in solid organ transplant recipients, significant interactions can occur between rifamycin-based regimens and calcineurin inhibitors or rapamycin [65]. Therapeutic drug monitoring (TDM) should be used for such patients. TDM should be considered early in treatment for complicated multidrug interactions involving TB drugs, azoles, protease inhibitors, NNRTI, and macrolides [19, 64]. In contrast, guessing at the doses may be harmful to the patient. A review of rifamycin drug interactions with antimicrobials is described below. A summary is provided in Table 6.2 [11, 66, 67].

6.4.2.1 Azoles

RIF reduces itraconazole AUC 64–88% in both healthy volunteers and in AIDS patients, often resulting in undetectable concentrations [68, 69]. Likewise, RIF reduces the AUC of ketoconazole by 82% in healthy volunteers [70]. Based on these

Table 6.2 Clinically significant drug interactions involving the rifamycins

Drug class	Drugs whose concentrations are substantially decreased by rifamycins	Comments
Antiretroviral agents	HIV-1 protease inhibitors (lopinavir/ritonavir, darunavir/ritonavir, atazanavir, atazanavir/ritonavir)	RFB preferred with protease inhibitors. For ritonavir-boosted regimens, give RFB 150 mg daily. Double-dose lopinavir/ritonavir can be used with RIF but toxicity increased. Do not use RIF with other protease inhibitors
	Non-nucleoside reverse-transcriptase inhibitor (NNRTI) Nevirapine Efavirenz Rilpivirine Complera (fixed-dose combination tablet containing emtricitabine, rilpivirine, TDF) Etravirine	RIF decreases exposure to all NNRTIs. If nevirapine is used with RIF, lead-in nevirapine dose of 200 mg daily should be omitted and 400 mg daily nevirapine dosage given. With RIF, many experts advise that efavirenz be given at standard dosage of 600 mg daily, although FDA recommends increasing efavirenz to 800 mg daily in persons >60 kg. In young children double-dose lopinavir/ritonavir given with RIF results in inadequate concentrations – super-boosted lopinavir/ritonavir is advised (if available) by some experts. Rilpivirine and etravirine should not be given with RIF. RFB can be used with nevirapine and etravirine at usual dosing. Efavirenz and RFB use requires dose increase of RFB to 600 mg daily, as such RIF is preferred. Rilpivirine should not be given with RFB
	Integrase strand transfer inhibitor (INSTI) Raltegravir Dolutegravir Elvitegravir (coformulated with cobicistat, tenofovir and emtricitabine as Stribild) Genvoya® (fixed-dose combination tablet containing elvitegravir, cobicistat, emtricitabine, and tenofovir alafenamide)	Increase dose of raltegravir to 800 mg twice daily with RIF, although clinical trial data show similar efficacy using 400 mg twice daily. Dolutegravir dose should be increased to 50 mg every 12 h with RIF. Do not use RIF with elvitegravir. RFB can be used with all INSTIs
	CCR5 inhibitors Maraviroc	RIF should not be used with maraviroc. RFB can be used with maraviroc

(continued)

Table 6.2 (continued)

Drug class	Drugs whose concentrations are substantially decreased by rifamycins	Comments
Anti-infectives	Macrolide antibiotics (azithromycin, clarithromycin, erythromycin)	Azithromycin has no significant interaction with rifamycins. Coadministration of clarithromycin and RFB results in significant bidirectional interactions that can increase RFB to toxic concentrations increasing the risk of uveitis. Erythromycin is a CYP3A4 substrate, and clearance may increase in setting of rifamycin use
	Doxycycline	May require use of a drug other than doxycycline
	Azole antifungal agents (ketoconazole, itraconazole, voriconazole, fluconazole, posaconazole, isavuconazole)	Itraconazole, ketoconazole, and voriconazole concentrations may be subtherapeutic with any of the rifamycins. Fluconazole can be used with rifamycins, but the dose of fluconazole may have to be increased
	Atovaquone	Consider alternate form of *Pneumocystis jirovecii* treatment or prophylaxis
	Chloramphenicol	Consider an alternative antibiotic
	Mefloquine	Consider alternate form of malaria prophylaxis
	Linezolid	Monitor clinical response and microbiologic cure; may require increased dose
Hormone therapy	Ethinylestradiol, norethindrone	Women of reproductive potential on oral contraceptives should be advised to add a barrier method of contraception when on a rifamycin
	Tamoxifen	May require alternate therapy or use of a non-rifamycin- containing regimen
	Levothyroxine	Monitoring of serum TSH recommended; may require increased dose of levothyroxine
Narcotics	Methadone	RIF and RPT use may require methadone dose increase RFB infrequently causes methadone withdrawal
	Oxycodone	Monitor analgesic effect; may require dose increase
Anticoagulant/antiplatelet	Warfarin	Monitor prothrombin time; may require two- to threefold warfarin dose increase
	Clopidogrel, prasugrel	Monitor clinically
	Ticagrelor, dabigatran, rivaroxaban	Avoid and change to alternate agent

Immunosuppressive agents	Cyclosporine, tacrolimus	RFB may allow concomitant use of cyclosporine and a rifamycin; monitoring of cyclosporine and tacrolimus serum concentrations to assist with dosing
	Corticosteroids	Monitor clinically; may require two- to threefold increase in corticosteroid dose
Anticonvulsants	Phenytoin, lamotrigine	TDM recommended; may require anticonvulsant dose increase
Cardiovascular agents	Verapamil, nifedipine, diltiazem (a similar interaction is also predicted for felodipine and nisoldipine)	Clinical monitoring recommended; may require change to an alternate cardiovascular agent
	Propranolol, metoprolol, atenolol, carvedilol	Clinical monitoring recommended; select beta-blockers may require dose increase or change to an alternate cardiovascular drug
	Enalapril, losartan	Monitor clinically; may require a dose increase or use of an alternate cardiovascular drug
	Digoxin (among patients with renal insufficiency) Digitoxin	TDM recommended; may require digoxin or digitoxin dose increase
	Mexiletine, tocainide, propafenone	Clinical monitoring recommended; may require change to an alternate cardiovascular drug
HMG-CoA reductase inhibitors	Simvistatin, atorvastatin, pravastatin	Monitor lipid panel; increased dose will likely be needed for select HMG-CoA reductase inhibitors
Theophylline	Theophylline	TDM recommended; may require theophylline dose increase
Antidiabetic agents	Tolbutamide, chlorpropamide, glyburide, glimepiride, repaglinide, metformin	Monitor blood glucose; may require dose increase or change to an alternate hypoglycemic drug
Psychotropic drugs	Nortriptyline	TDM recommended; may require dose increase or change to alternate psychotropic drug
	Haloperidol, quetiapine, risperidone	Monitor clinically; may require a dose increase or use of an alternate psychotropic drug
	Benzodiazepines (e.g., diazepam, triazolam, zolpidem, buspirone)	Monitor clinically; may require a dose increase or use of an alternate psychotropic drug

Adapted from: Nahid et al. [1]

TSH thyroid-stimulating hormone, *TDM* therapeutic drug monitoring

data, the concurrent use of RIF and itraconazole and ketoconazole should be avoided due to the risk of therapeutic failure. Schwiesow et al. reported the use of voriconazole and RBN together to successfully treat a combined mycobacterial and *Aspergillus* infection [71]. The approach described may be used as a template for similar types of complex interactions. The goal is to get the patient on safe and effective doses of the relevant drugs as quickly as possible.

RIF may reduce fluconazole's AUC by 23–52% and may cause treatment failures [72–74]. Higher doses of fluconazole may be required if used concomitantly with RIF [1, 75]. There is no significant effect of fluconazole on RIF pharmacokinetics [76]. As noted, RBN demonstrates bidirectional interactions. Fluconazole increased RBN's AUC by 76% and further increased the 25-desacetyl metabolite [77]. Caution should be exercised, and TDM employed, during concurrent use of fluconazole with RBN. In cases of coexisting mycobacterial and fungal infections, careful drug selection and therapeutic monitoring of drug concentrations can allow for combined use of these drugs.

6.4.2.2 Chloramphenicol

Several case reports have described low chloramphenicol serum concentrations in patients treated concomitantly with RIF. The dose of chloramphenicol could be increased to maintain serum concentrations; however, this may put the patient at greater risk for aplastic anemia. Alternative therapies should be considered in patients taking RIF [78, 79].

6.4.2.3 Dapsone

RIF and RBN have been associated with a significant increase (50–70%) in the clearance of dapsone [20, 80], resulting in lower dapsone AUC values. Since lower dapsone exposures may increase the risk of *Pneumocystis jirovecii* pneumonia (PCP). Higher dapsone doses may be needed when used with RBN or RIF [80]. Patients should be monitored closely and TDM should be considered.

6.4.2.4 Doxycycline

Treatment failures have been reported in patients with brucellosis being treated with doxycycline and RIF. Patients receiving RIF and doxycycline had decreased doxycycline AUC values (59%) and higher clearances compared to those receiving doxycycline plus streptomycin. Alternative therapies should be considered in patients taking RIF [81].

6.4.2.5 Fluoroquinolones

Limited data exist regarding the interactions between fluoroquinolones and rifamy-cins. Quinolone clearance may be increased by RIF [82, 83]. Weiner et al. evaluated the effects of RIF on concentrations of moxifloxacin. The AUC of moxifloxacin was decreased by 27%, but peak concentrations were unchanged, secondary to rifampin's inductive effect on phase II clearance pathways noted above [84]. Similar results were seen by Nijland et al. [85]. Currently, there are not enough data to support routine dosage adjustments. That said, TDM can detect patients with low moxi-floxacin concentrations, allowing for dose adjustment. Moxifloxacin does show concentration-dependent activity against *M. tuberculosis* [86]. RPT and RFB may also decrease moxifloxacin serum concentrations. Where available, moxifloxacin plasma concentrations can guide therapy.

6.4.2.6 Isoniazid

The oral bioavailability of RIF was reduced by an average of 32% in volunteers who were administered an INH-RIF fixed-dose combination (FDC) product, compared with RIF alone [87]. This appears to be a function of the FDC formulation and not directly due to an interaction between INH and RIF. This is compensated by giving a slightly higher dose of RIF (in milligrams) as the FDC product. This effect may vary with individual FDC formulations.

6.4.2.7 Macrolides

The combination of clarithromycin and RIF resulted in reduced mean peak clar-ithromycin concentrations (87%) when compared to clarithromycin monotherapy [88]. Overall, concentrations of the metabolite of clarithromycin, 14-OH clarithro-mycin, were not affected, although the usual ratio of parent to metabolite was inverted. While 14-OH clarithromycin is active against *H. influenzae*, it is not active against *M. avium* complex (MAC) [89]. Based on current data, RIF can decrease the efficacy of clarithromycin against MAC by reducing serum concentrations. Azithromycin may be preferred when RIF is used.

Macrolide drug interactions with RBN are complex. RBN decreases macrolide concentrations, while the macrolides, CYP3A inhibitors, increase the concentration of RBN and its active metabolite, occasionally leading to RBN toxicity. The phar-macokinetics of clarithromycin plus RBN has been evaluated in healthy volunteers and in HIV-positive patients [77, 88, 90]. Concomitant administration resulted in increased serum concentration (76%) and AUC (99%) of RBN and its partially active metabolite, 25-O-desacetyl rifabutin. Rifabutin reduced clarithromycin's AUC by 44% and increased concentrations of 14-OH clarithromycin. Reports of significant adverse reactions, including neutropenia, fever, myalgia, and uveitis have been associated with the combination of clarithromycin plus RBN, especially in the presence of additional drugs, such as fluconazole [88, 90]. Based on current data, the combination of RBN and clarithromycin should be avoided when feasible and used cautiously when necessary. Despite azithromycin's reduced affinity for CYP, studies evaluating the combination of azithromycin and RBN also resulted in unusually high rates of neutropenia [88]. Azithromycin may be preferred if RBN + macrolide therapy is necessary. Otherwise, RIF may be easier to use than RBN. TDM should be used to achieve the desired concentrations.

6.4.2.8 Metronidazole

Limited data has shown that RIF increases the clearance of metronidazole and decreases the AUC [91].

6.4.2.9 Non-nucleoside Reverse-Transcriptase Inhibitors

Tables 6.3 and 6.4 summarize the effects of RIF and RBN on the AUC of non-nucleoside reverse-transcriptase inhibitors [28, 92]. Although rarely used today, delavirdine should not be combined with any rifamycin [1, 75, 93, 94]. In general, etravirine should not be used together with RIF but can be used with RBN [95]. Dosage adjustments may be necessary. Rilpivirine is a substrate and inducer of CYP3A4. RIF reduced rilpivirine AUC by 80% and trough concentrations by 89%, so the two drugs should not be coadministered [96].

Nevirapine and efavirenz can be used with either RBN or RIF, although efavirenz generally has been used in the United States. Studies with RIF and efavirenz have shown moderate no significant reductions in concentrations of efavirenz. Patients with certain genetic polymorphisms (CYP 2B6 516 G > T) metabolize efavirenz more slowly, and high concentrations of efavirenz are common [97–99]. Dosage adjustments of efavirenz may be considered if given with RIF in patients who weigh over 50 kg [100–104]. When RBN is used in combination with efavirenz, the concentration of RBN is reduced, and RBN dosage should be adjusted to 600 mg per dose [105, 106]. RIF has been shown to reduce the concentration of nevirapine by 20–55% [107–110]. Efavirenz is the preferred NNRTI to be used with RIF; however, if nevirapine is selected in patients already receiving RIF therapy, it should be initiated without the once daily lead-in dosing to minimize subtherapeutic concentrations [111]. Higher doses of nevirapine have been associated with increased rates of hepatotoxicity [112]. RBN may be an option in patients taking nevirapine as interactions are less likely [113]. Serum concentration changes of RBN and etravirine combinations are 17% and 35%, respectively, when given in combination [95]. Standardized dosage adjustments are not recommended, but TDM could guide dosing in individual patients.

Table 6.3 Coadministration of antiretrovirals and rifampin in adults

	Recommended change in dose of antiretroviral drug	Recommended change in dose of rifampin	Comments
Non-nucleoside reverse-transcriptase inhibitors (NNRTIs)			
Nevirapine	Initiate at a dose of 200 mg twice daily rather than 200 mg once daily (use the same maintenance dose of 200 mg twice daily)	No change (600 mg/day)	Efavirenz is preferred, but if nevirapine must be used, lead-in dosing at 200 mg once daily should be avoided, as this may increase risk of virologic failure. Because of this risk, monitoring of adherence and viral load is recommended. If available, consider therapeutic drug monitoring. 20–58% ↓ in nevirapine exposure
Rilpivirine	Rifampin and rilpivirine should not be used together		Rilpivirine AUC ↓ by 80%, Cmin decreased 89%
Etravirine	Etravirine and rifampin should not be used together		Marked decrease in etravirine predicted, based on data on the interaction with rifabutin. Significant decrease in etravirine exposure possible
Single protease inhibitors			
Atazanavir	Rifampin and atazanavir should not be used together		Atazanavir AUC ↓ by >95%. Increasing the dose to 300 mg twice daily or 400 mg twice daily still resulted in subtherapeutic atazanavir concentrations
Ritonavir-boosted protease inhibitors			
Lopinavir/ritonavir (Kaletra™)	Lopinavir 800 mg plus ritonavir 200 mg twice daily (double dose)	No change (600 mg/day)	Use with caution; this combination resulted in hepatotoxicity in all adult healthy volunteers in an initial study. It was better-tolerated among adult patients already taking lopinavir/ritonavir based ART with increase to 600 mg/150 mg after 1 week, then 800 mg/200 mg 1 week later.
"Super-boosted" lopinavir/ ritonavir (Kaletra™)	Lopinavir 400 mg plus ritonavir 400 mg twice daily (super boosting)	No change (600 mg/day)	Use with caution; this combination resulted in hepatotoxicity among adult healthy volunteers. It has not been adequately tested in patients with HIV.
Atazanavir/ritonavir	Rifampin and atazanavir/ritonavir should not be used together.		Atazanavir trough concentration ↓ by >90%. Doubling the dose to 300/100 twice daily resulted in hepatotoxicity in healthy volunteers.

(continued)

Table 6.3 (continued)

	Recommended change in dose of antiretroviral drug	Recommended change in dose of rifampin	Comments
Darunavir/ritonavir	Rifampin and darunavir/ritonavir should not be used together		No drug interaction studies of darunavir and rifampin have been conducted.
Fosamprenavir/ritonavir	Rifampin and fosamprenavir/ritonavir should not be used together		Fosamprenavir Cmax ↓ by 70%, AUC ↓ 82%, trough ↓ 92%
Saquinavir/ritonavir	Rifampin and saquinavir/ritonavir should not be used together.		The combination of saquinavir (1000 mg twice-daily), ritonavir (100 mg twice-daily), and rifampin caused unacceptable rates of hepatotoxicity among healthy volunteers. In tuberculosis patients, 400/400 twice daily caused similar rates of hepatotoxicity
CCR-5 receptor antagonists			
Maraviroc	Increase maraviroc to 600 mg twice-daily	No change (600 mg/day)	The reductions in maraviroc concentrations related to rifampin co-administration may be overcome by increasing the dose, though the 600 mg twice-daily dose has not been formally tested. Use with caution, as there is no reported clinical experience with increased dose of maraviroc with rifampin. 64% ↓ in maraviroc AUC
Integrase inhibitors			
Raltegravir	Increase dose to 800 mg twice daily	No change (600 mg/day)	Raltegravir trough concentrations ↓ by 53% even with increased dose to 800 mg twice daily despite reasonable overall exposures. The clinical significance of this is unknown. Use this dose with caution and employ viral load monitoring, if available. 40% ↓ in raltegravir AUC and 61% ↓ in raltegravir Cmin
Elvitegravir co-formulated with cobicistat, tenofovir, and emtricitabine (Stribild™)	Stribild™ and rifampin should not be used together		Marked decrease in elvitegravir and cobicistat concentrations predicted based on metabolic pathways of these drugs

Adapted from Centers for Disease Control and Prevention [75]

Table 6.4 Coadministration of antiretrovirals and rifabutin in adults

	Recommended change in dose of antiretroviral drug	Recommended change in dose of rifabutin	Comments
Non-nucleoside reverse-transcriptase inhibitors			
Efavirenz	No change	↑ to 600 mg (daily or thrice weekly)	If efavirenz is used, the rifamycin of choice is rifampin. Efavirenz reduces rifabutin concentrations, so if rifabutin is to be used, increasing rifabutin dose to 600 mg may compensate for the inducing effect of efavirenz. Employ caution as this strategy has not been tested among patients taking rifabutin daily or thrice weekly. Efavirenz should not be used during the first trimester of pregnancy. 38% ↓ in rifabutin exposure
Nevirapine	No change	No change (300 mg daily)	Rifabutin and nevirapine AUC not significantly changed. 16% ↓ in nevirapine Cmin, 17% ↑ in rifabutin AUC, and 24% ↑ in metabolite AUC
Rilpivirine	Rifabutin and rilpivirine should not be used together		Rilpivirine AUC ↓ by 46%, and Cmin ↓ by 49%
Etravirine	No change	No change (300 mg daily)	No clinical experience; etravirine Cmin ↓ by 35%, and rifabutin AUC ↓ by 17%; these changes are unlikely to be clinically relevant, so no dose adjustment is necessary. Since ritonavir-boosted darunavir and saquinavir also diminish etravirine concentrations, the combination of these boosted PIs, etravirine, and rifabutin is not recommended. 37% ↓ in etravirine AUC, and 17% ↓ in rifabutin and metabolite AUC
Single protease inhibitors			
Atazanavir	No change	↓ to 150 mg once daily	No published clinical experience
Dual protease inhibitor combinations			
Lopinavir/ritonavir (Kaletra™)	No change	↓ to 150 mg once daily	In patients with HIV taking lopinavir/ritonavir, 150 mg once daily of rifabutin produces favorable rifabutin pharmacokinetics. Clinical safety data are limited. Monitor closely for potential rifabutin toxicity – Uveitis, hepatotoxicity, and neutropenia. 473% ↑ in rifabutin

(continued)

Table 6.4 (continued)

	Recommended change in dose of antiretroviral drug	Recommended change in dose of rifabutin	Comments
Fosamprenavir/ritonavir	No change	↓ to 150 mg once daily	In healthy volunteers, a dose of 150 mg every other day of rifabutin given together with standard dose boosted fosamprenavir resulted in an increase in amprenavir AUC and Cmax by 35% and no change in Cmin. Limited clinical data among patients with HIV. Monitor closely for uveitis, hepatotoxicity, and neutropenia. 64% ↑ in rifabutin and metabolite AUC
Ritonavir (any dose) with saquinavir, indinavir, amprenavir, fosamprenavir, atazanavir, tipranavir or darunavir	No change	↓ to 150 mg once daily	Rifabutin AUC ↑ and 25-O-des-acetyl rifabutin AUC ↑, by varying degrees. Monitor closely for uveitis, hepatotoxicity, and neutropenia Atazanavir ± ritonavir 110% ↑ in rifabutin AUC and 2101% ↑ in metabolite AUC Darunavir + ritonavir 55% ↑ in rifabutin and metabolite AUC
CCR-5 receptor antagonists			
Maraviroc	No change	No change	No clinical experience; a significant interaction is unlikely, but this has not yet been studied decrease in maraviroc AUC possible
Integrase inhibitors			
Raltegravir	No change	No change	When given with standard-dose rifabutin (300 mg daily), raltegravir AUC ↑ 19%, Cmin ↓ 20%, and Cmax ↑ 39%. These changes are unlikely to be clinically significant. 19% ↑ in raltegravir AUC, 39% ↑ in Cmax, and 20% ↓ in Cmin
Elvitegravir coformulated with cobicistat, tenofovir, and emtricitabine (Stribild™)	Stribild™ and rifabutin should not be used together		

Adapted from Centers for Disease Control and Prevention [75]

6.4.2.10 Nucleoside Reverse-Transcriptase Inhibitors

Zidovudine and lamivudine are not metabolized by the CYP450 enzymes. The efficacy of these drugs is correlated with the intracellular concentrations of the active derivative. The coadministration of RIF with zidovudine resulted in a decrease (43%) in Cmax and AUC (47%) of zidovudine. Decreased plasma concentrations have not been shown to reduce the concentration of the intracellular metabolite [63]. Therefore, RIF is expected to have little impact on the clinical effect and antiviral activity of zidovudine [63].

6.4.2.11 Protease Inhibitors

The protease inhibitors are CYP3A substrates and inhibitors, therefore exhibiting a bidirectional interaction. Tables 6.3 and 6.4 summarize the effects of RIF and RBN on the AUC of protease inhibitors [29, 92]. Due to the profound effects of RIF and RPNT on the AUCs of saquinavir, indinavir, nelfinavir, and amprenavir, concomitant administration is discouraged [75, 114–120].

RIF significantly reduces saquinavir concentrations when but when used with ritonavir, RIF has lesser effects. But, RIF also has modest effects on ritonavir concentrations [121].

Tipranavir, like the other protease inhibitors, both inhibits and induces the cytochrome P450 enzyme system; when used in combination with ritonavir, its net effect on CYP3A4 is inhibition. Tipranavir also induces the P-glycoprotein transporter; tipranavir may alter the concentrations of many other drugs metabolized by these pathways and in some cases may be complex and difficult to predict. Rifampin induces tipranavir metabolism and decreases tipranavir concentrations; RIF should not be given concomitantly with tipranavir. Combining tipranavir with RBN should be done with caution, while toxicity and RBN drug concentrations should be monitored [122]. Darunavir is metabolized by CYP3A4. Rifampin induces darunavir metabolism and decreases darunavir concentrations; RIF should not be given concomitantly with darunavir. Coadministration of darunavir with ritonavir results in a 55% increase in the RBN metabolite AUC values [123] and a concomitant decrease in darunavir concentrations.

Based on current data, ritonavir or saquinavir, ritonavir should be used with caution in combination with RIF [1, 75]. The complexity of this interaction may be overcome by TDM which may help to optimize regimens for the coadministration of these agents.

Because RBN has little effect on the serum concentration of ritonavir-boosted protease inhibitors, RBN can be used if combination therapy is necessary [123, 124]. However, due to the bidirectional interaction and the potential for intolerance, RBN and/or protease inhibitor dosage adjustments may be warranted [117, 119, 125]. The combination of daily nelfinavir with RBN twice-weekly results in what may be considered reasonable RBN AUC's [126]. However, a related study by the same research group found 5% failure or relapse with the selection of acquired rifamycin resistance (ARR) with twice-weekly TB treatment, including RBN [127]. Lopinavir-ritonavir used in combination with RBN three times weekly also lead to the selection of ARR. Therefore, RBN 150 mg daily is recommended when given with boosted protease inhibitors in adult patients. These patients should be monitored for RBN toxicities, and further dosage adjustments can be based on plasma concentration monitoring [128–130].

6.4.2.12 CCR-5 Receptor Antagonists

Maraviroc is the first drug to be approved in the class of CCR-5 receptor antagonists. It is primarily and extensively metabolized by CYP3A4. In the presence of CYP3A4 inducers, such as RIF, dosage of maraviroc should be increased due to a 64% decrease in AUC [131]. Current clinical experience with RBN is limited, but a decrease maraviroc AUC is possible.

6.4.2.13 Integrase Inhibitors

In vitro and in vivo studies indicate raltegravir does not have any significant induction or inhibitory effects on the CYP enzymes [132]. Interactions have been observed with RIF; however, efficacy data is required [132]. Raltegravir is metabolized by glucuronidation. The enzyme responsible for the metabolite of raltegravir appears to be the UDP-glucuronosyltransferase (UGT) 1A1 subtype. Rifampin upregulates the synthesis of UGT. A 40% reduction in raltegravir AUC occurred when it was dosed as 400 mg twice daily with RIF 600 mg daily. An increased dose of raltegravir (800 mg twice daily) demonstrated that the negative effect of RIF on raltegravir exposure could be reversed [133]. Therefore, raltegravir 800 mg twice daily should be used in adults taking RIF. Caution is warranted if the combination of raltegravir and RIF is selected, and RBN may be preferred [133, 134].

Drug doses for persons with HIV coinfection who are being treated with highly active antiretroviral therapy (HAART) often must be adjusted when rifamycins are used concurrently. The extent of these interactions typically is drug- and patient-specific, and an effort should be made to obtain expert consultation and the latest available information to guide dosing. Note that plasma concentration changes seen in two-way interaction studies may be very different from those seen clinically when three or more interacting drugs are used concurrently. For example, the combination of elvitegravir, cobicistat, tenofovir, and emtricitabine (Stribild®) should not be given together with RIF as it is expected to reduce the concentrations of elvitegravir and cobicistat.

6.4.2.14 Bedaquiline

Bedaquiline, approved by the FDA in 2012, is an ATP synthase inhibitor. RIF and RBN increase the clearance of bedaquiline. Cmax concentrations when administered with RIF are decreased 21% and with RPNT concentrations are decreased 25%. However, bedaquiline metabolites may accumulate, and the parent and metabolites display very long elimination half-lives. Concomitant administration of rifamycin antibiotics and bedaquiline is not recommended [135].

6.4.2.15 Sulfamethoxazole and Trimethoprim

The effect of RIF on concentrations of sulfamethoxazole/trimethoprim (SMZ/TMP) was evaluated in HIV-positive patients [136]. A decrease (23%) in mean AUC of SMZ/TMP was observed. The clinical significance of this interaction has not been established, but the reduced efficacy of SMZ/TMP may be of concern.

Rifamycins interact with several other classes of drugs beyond those listed above. Additional information regarding rifamycin interactions can be found in the article by Niemi et al. and in several other papers [19, 136, 137].

6.4.3 Pyrazinamide Interactions

Pyrazinamide (PZA) is metabolized to pyrazinoic acid and 5-hydroxypyrazinoic acid, which are subsequently cleared renally [5, 30]. PZA is removed by hemodialysis [30]. It is not associated with a large number of drug interactions. Because PZA can compete with uric acid for excretion, patients will accumulate uric acid while on PZA. In most cases, this is not a clinically significant problem, which may precipitate a flare-up of the disease. Allopurinol inhibits the clearance of PZA's primary metabolite, pyrazinoic acid, thereby exacerbating the metabolite's inhibition of uric acid secretion [138, 139]. Further, probenecid may be significantly less effective as a uricosuric agent in the presence of PZA [140]. Thus, the most effective management of PZA-induced elevations of uric acid and arthralgias may be to hydrate the patient and withhold PZA.

The combination of RIF and PZA in the absence of INH leads to an unexpectedly high incidence of hepatotoxicity in HIV-negative patients [141–143]. It is important to stress this was in the context of 2-month regimens of RIF and PZA for latent TB infection (LTBI) and not during the treatment of active TB disease with INH, RIF, PZA, and EMB. Therefore, this 2-month RIF+ PZA LTBI regimen should no longer be used [142]. PZA combined with ofloxacin or levofloxacin for LTBI due to multidrug-resistant TB also is poorly tolerated [144–146]. It is possible that PZA or its metabolites compete with quinolones for renal tubular secretion, although this has not been proven. This regimen also cannot be recommended at this time. In some cases, patients receiving zidovudine may have diminished concentrations of pyrazinamide.

6.4.4 Ethambutol Interactions

EMB is cleared both hepatically and renally [4, 29, 30]. EMB is not significantly removed by hemodialysis [4, 30]. Anecdotal experience suggests that it is not cleared by peritoneal dialysis, and accumulation can occur. The specific pathways involved in its hepatic clearance are not well established, and EMB has few documented interactions. The effect of concurrent antacids has been described above. Because EMB can cause optic neuritis, patients receiving other potential agents associated with ocular injury (RBN, cidofovir) should be monitored carefully. While RBN and cidofovir are associated with uveitis and not optic neuritis, additive effects may adversely affect vision [37, 130].

6.4.5 Aminoglycoside and Polypeptide Interactions

The aminoglycosides amikacin, kanamycin, and streptomycin, as well as the polypeptides capreomycin and viomycin, are all primarily cleared renally [30, 148, 149]. Aminoglycosides are removed by hemodialysis [150]. However, under clinical conditions, especially in the intensive care setting, traditional hemodialysis removal may be limited. Other methods of renal replacement, however, potentially can remove substantially more drug. Aminoglycosides can adversely affect vestibular, auditory, and renal function. Reported differences in the incidence of these toxicities among the agents reflect, in part, differences in doses and frequencies used. Elevated serum creatinine values due to non-oliguric acute tubular necrosis are usually reversible; renal wasting of cations also may occur [30, 46, 148, 151]. Periodic monitoring (every 2–4 weeks) of the serum blood urea nitrogen, creatinine, calcium, potassium, and magnesium should be considered, especially if other nephrotoxins (such as amphotericin B) are being used [1]. Vestibular changes may be noted on physical exam and may occur independently of, or in conjunction with, tinnitus and auditory changes [149–151]. Auditory changes are best detected by monthly audiograms for those patients requiring prolonged treatment or those receiving concurrent potential ototoxins (clarithromycin, ethacrynic acid, furosemide) [1, 150]. Aminoglycosides and polypeptides can potentiate the neuromuscular blocking agents or may precipitate neuromuscular blockade in patients with myasthenia gravis [153]. Therefore, these drugs should be used cautiously in those settings.

6.4.6 Cycloserine Interactions

There is little known regarding the potential for drug interactions with cycloserine (CS) [152]. This drug is renally cleared, and there are no known metabolites [30, 151]. CS is cleared by hemodialysis [152]. It can cause a variety of central nervous system disturbances, including anxiety, confusion, memory loss, dizziness, lethargy, and depression, including suicidal tendencies [30]. Therefore, other agents associated with any of these effects (INH, ethionamide, and quinolones) may have additive CNS toxicities. CS should be used cautiously in patients with a history of depression or psychosis or those receiving treatment for these conditions. It is not clear if cycloserine can alter the potential for seizures in patients predisposed to these events. Caution is advised, as is TDM, to ensure that concentrations do not exceed the recommended range of 20–35 mcg/ml [18]. Older literature suggests that CS may decrease the clearance of phenytoin, possibly leading to toxicity [37].

6.4.7   Ethionamide Interactions

ETA is a prodrug that must be converted to its active form. It is extensively metabolized to a sulfoxide metabolite that is active against mycobacteria [7, 30]. Flavin-containing monooxygenase (FMO) has been found to play an important role in the pharmacokinetics of ethionamide and its hepatotoxicity. The specific hepatic microsomal enzymes primarily responsible for metabolism of ethionamide's metabolism are CYP2C8, CYP2C19, and to a lesser extent CYP3A4, CYP2B6, and CYP1A2 [153]. The inhibitory effects of ethionamide on CYP2C8 enzymes may cause clinically significant drug interactions [153, 154]. Little unchanged drug is excreted in the urine/feces or cleared by hemodialysis [152, 153]. ETA causes significant gastrointestinal (GI) distress which may improve when taken with food. ETA may cause CNS effects, including headache, drowsiness, depression, psychosis, and visual changes, although a causative role has not been established [153]. Therefore, additive effects with INH, CS, or fluoroquinolones are possible. ETA may cause peripheral neuritis, so caution should be exercised in patients receiving other agents, such as nucleoside reverse-transcriptase inhibitors that share this toxicity. ETA can cause hepatotoxicity, up to 5 months after initiation, and goiter, with or without hypothyroidism. The latter is worsened by the concurrent use of PAS [153–155]. Thyroid-stimulating hormone (TSH) concentrations should be monitored periodically in patients receiving ETA.

6.4.8   Para-aminosalicylic Acid Interactions

PAS is metabolized by N-acetyltransferase 1 (NAT1) to acetyl-PAS that is subsequently cleared renally [9, 152, 155]. Little PAS is cleared by hemodialysis, but acetyl-PAS is cleared by hemodialysis, and post-hemodialysis doses may be necessary [156]. PAS can cause diarrhea, and this can affect the pharmacokinetics of other drugs. Various types of malabsorption with PAS have been described, including steatorrhea, vitamin B_{12}, folate, xylose, and iron. With the possible exception of digoxin, it is not known if PAS can cause specific drugs to be malabsorbed [157].

Hypersensitivity reactions with fever, including hepatitis, can occur, and desensiti-zation to PAS-induced hypersensitivity is *not* recommended [157]. PAS is known to produce goiter, with or without myxedema, and this is more frequent with concomi-tant ETA therapy. This can be prevented or treated with thyroxine. Older tablet forms of PAS that contained bentonite reduced serum RIF concentrations; this should not occur with the granule form [157]. Due to the reported greater risk of crystalluria, the concurrent use of ammonium chloride with PAS is not recom-mended [156, 157].

6.4.9 Clofazimine Interactions

Clofazimine (CF) is a weak anti-TB drug, and has a very unusual pharmacokinetic profile [10, 30]. It is highly tissue-tropic, and as a result, displays a very long elimi-nation half-life. It is primarily excreted in feces, but the precise mechanisms have not been described. CF is negligibly removed by hemodialysis [158]. As noted above, oral absorption is improved when CF is given with a high-fat meal. The most serious adverse reactions associated with CF are dose-related gastrointestinal (GI) toxicities, and these can be additive with other drugs' effects [30, 149]. Skin discol-oration may also occur, and other drugs, including amiodarone and RBN, may make this worse. CF can produce a statistically significant increase in RIF's Tmax, but this interaction is unlikely to be clinically significant. The large accumulation of CF in macrophages may affect the function of these cells, but this has not been well defined. It is at least theoretically possible that such affects contribute to the worse outcome seen in some AIDS patients who received CF as part of their regimen for disseminated MAC infection (DMAC). CF is being studied as a potential adjuvant to cancer chemotherapy, in the hope that it may reduce or reverse acquired multi-drug resistance (MDR) [158].

6.4.10 Bedaquiline

Bedaquiline is an ATP synthase inhibitor, approved for the treatment of pulmonary multidrug-resistant tuberculosis. The hepatic isoenzyme, CYP3A4, is primarily responsible for metabolism of bedaquiline. Bedaquiline has not been observed to induce or inhibit CYP isoenzyme activity in vitro; however given its metabolism by CYP3A4, there is a potential for drug interaction during coadministration with CYP3A4 inducers (rifampin, rifabutin, carbamazepine, efavirenz, etravirine) and inhibitors (ciprofloxacin, ketoconazole, ritonavir) [159, 160].

6.4.11   Delamanid

Delamanid is a new agent that inhibits mycolic acid synthesis and is currently being used under the guidance of the WHO for MDR-TB treatment when other treatments are ineffective. Coadministration of delamanid with strong CYP3A4 inhibitors has been associated with increase in the concentration of metabolites. Clinically significant drug interactions and increased risk of QTc prolongation have been associated with concomitant administration of CYP3A4 inhibitors and with fluoroquinolones. Caution must be exercised in patients at risk for QTc prolongation. Strong CYP 3A4 inducers can reduce the exposure of delamanid by up to 45%. In drug interaction studies in healthy subjects, delamanid did not affect the concentration of rifampicin, isoniazid, or pyrazinamide but increased concentrations of ethambutol by approximately 25% [162].

6.5   Management of Patients Coinfected with HIV and TB

The Centers for Disease Control and Prevention has published guidelines for the management of TB in patients coinfected with HIV [1, 75]. First, clinicians should look for a paradoxical worsening of TB symptoms upon the introduction of highly active antiretroviral therapy (HAART) as a consequence of immune reconstitution inflammatory syndrome (IRIS) [1, 75]. Next, the guidelines generally recommend the use of RBN instead of RIF for patients receiving HAART in an attempt to minimize drug interactions. It is very important to bear in mind that most interaction studies involving RIF or RBN and HAART were performed in small numbers of healthy volunteers. Results seen in HIV-positive patients can be very different. For example, combinations of RIF and ritonavir are very poorly tolerated in healthy subjects, with high rates of hepatic enzyme elevations. This is not nearly as common in patients with HIV, based on available information. Table 6.5 summarizes the current data available on drug interactions between protease inhibitors and antituberculosis drugs other than rifamycins [29, 75]. It is our practice to measure the serum concentrations of the various interacting drugs (antimycobacterial drugs, oral azole antifungals, and anti-HIV protease inhibitors) in order to verify adequate dosing [18, 28]. As noted, low RBN plasma concentrations have been associated with acquired rifamycin resistant (ARR) mycobacteria in patients with HIV [75, 127]. Intermittent dosing of TB drugs in patients with HIV may become a thing of the past. However, the best daily dose of RBN for patients receiving HAART requires further study.

Table 6.5 Predicted potential for drug-drug interaction between HIV protease inhibitors and antituberculosis drugs other than rifamycins

Drug	Metabolism	Effect on CYP3A	Effect of drug X on PI (predicted)[a]	Effect of PI on drug X (predicted)[a]
Isoniazid	Acetylation > CYP	Mild inhibitor	No change in indinavir	None
Pyrazinamide	Deamidase > xanthine oxidase	None known	(none)	(none)
Ethambutol	Renal > CYP	None known	(none)	(none)
Ethionamide	CYP	None known	Unknown	(increase)
PAS	Acetylation	None known	(none)	(none)
Quinolones	Renal > CYP	None known	(none)	(none)
Aminoglycosides	Renal	None known	(none)	(none)

Adapted from Burman et al. [29] with permission
[a]Predicted using existing knowledge regarding metabolic pathways for the two drugs

6.6   Nontuberculous Mycobacterial (NTM) Infections

The NTM comprise a substantial list of infections caused by various slow-growing and rapid-growing mycobacteria. The management of such infections has been summarized elsewhere [161]. Clinicians should be aware that there are differences between HIV-infected and non-infected hosts as far as disease presentation and management. It is important to consider the drug interactions described above for TB, as many of these same drugs are used to treat NTM. Advanced generation macrolides (azithromycin, clarithromycin) are frequently used to treat NTM, such as *M. avium* complex, and clarithromycin has been associated with many CYP3A4 interactions [46, 162]. In particular, bidirectional interactions involving RBN and clarithromycin should be anticipated [160]. RIF causes a more pronounced decline in clarithromycin concentrations than RBN [162–164].

6.7   Future Considerations

Several new TB drugs are under development, including PA-824 and SQ109 [165–168]. SQ109 is metabolized by CYP2D6 and CYP2C19, and up to 58% of parent is metabolized in 10-min incubation with microsomes. Insignificant metabolism is found in the presence of CYP3A4 [168]. The optimal dosing for combined use of these new TB drugs with RIF or other rifamycins in humans has not been studied to date.

6.8 Conclusion

The above discussion highlights the need for the careful introduction of the TB drugs into existing drug regimens. In particular, rifamycins can seriously disrupt ongoing treatment, with potentially serious consequences. While the role of TDM remains to be better defined for these situations, it does offer the potential to untangle multidirectional drug interactions.

References

1. Nahid P, Dorman SE, Alipanah N et al (2016) Executive summary: Official American Thoracic Society/Centers for Disease Control and Prevention/Infectious Diseases Society of America Clinical Practice Guidelines: treatment of drug-susceptible tuberculosis. Clin Infect Dis 63(7):853–867
2. Peloquin CA (2014) Tuberculosis. In: JT DP, Talbert RL, Yee GC, Matzke GR, Wells BG, Posey LM (eds) Pharmacotherapy: a pathophysiologic approach, 9th edn. McGraw Hill, New York, pp 1917–1937
3. Peloquin CA, Namdar R, Singleton MD, Nix DE (1999) Pharmacokinetics of rifampin under fasting conditions, with food, and with antacids. Chest 115:12–18
4. Peloquin CA, Bulpitt AE, Jaresko GS, Jelliffe RW, Childs JM, Nix DE (1999) Pharmacokinetics of ethambutol under fasting conditions, with food, and with antacids. Antimicrob Agents Chemother 43:568–572
5. Peloquin CA, Bulpitt AE, Jaresko GS, Jelliffe RW, James GT, Nix DE (1998) Pharmacokinetics of pyrazinamide under fasting conditions, with food, and with antacids. Pharmacotherapy 18:1205–1211
6. Lin MY, Lin SJ, Chan LC, Lu YC (2010) Impact of food and antacids on the pharmacokinetics of anti-tuberculosis drugs : systematic review and meta-analysis. Int J Tuberc Lung Dis 14:806–818
7. Auclair B, Nix DE, Adam RD, James GT, Peloquin CA (2001) Pharmacokinetics of ethionamide under fasting conditions, with orange juice, food, and antacids. Antimicrob Agents Chemother 45:810–814
8. Zhu M, Nix DE, Adam RD, Childs JM, Peloquin CA (2001) Pharmacokinetics of cycloserine under fasting conditions, with orange juice, food, and antacids. Pharmacotherapy 21:891–897
9. Peloquin CA, Zhu M, Adam RD, Godo PG, Nix DE (2001) Pharmacokinetics of p-aminosalicylate under fasting conditions, with orange juice, food, and antacids. Ann Pharmacother 35:1332–1338
10. Nix DE, Zhu M, Adam RD, Godo PG, Peloquin CA (in press) Pharmacokinetics of clofazimine under fasting conditions, with orange juice, food, and antacids. Tuberculosis
11. Sahai J, Gallicano K, Swick L et al (1997) Reduced plasma concentrations of antiturberculous drugs in patients with HIV infection. Ann Intern Med 127:289–293
12. Peloquin CA, Nitta AT, Burman WJ et al (1996) Low antituberculosis drug concentrations in patients with AIDS. Ann Pharmacother 30:919–925
13. Fish DN, Chow AT (1997) The clinical pharmacokinetics of levofloxacin. Clin Pharmacokinet 32:101–119
14. Berning SE, Huitt GA, Iseman MD, Peloquin CA (1992) Malabsorption of antituberculosis medications by a patient with AIDS. N Engl J Med 327:1817–1818
15. Kumar AK, Chandrasekaran V, Kannan T et al (2017) Anti-tuberculosis drug concentrations in tuberculosis patients with and without diabetes mellitus. Eur J Clin Pharmacol 73(1):65–70

16. Bento J, Duarte R, Brito MC et al (2010) Malabsorption of antimycobacterial drugs as a cause of treatment failure in tuberculosis. BMJ Case Rep. https://doi.org/10.1136/bcr.12.2009.2554

17. Peloquin CA, MacPhee AA, Berning SE (1993) Malabsorption of antimycobacterial medications (letter). N Engl J Med 329:1122–1123

18. Gordon SM, Horsburgh CR Jr, Peloquin CA et al (1993) Low serum levels of oral antimycobacterial agents in patients with disseminated mycobacterium avium complex disease. J Infect Dis 168:1559–1562

19. Alsultan A, Peloquin CA (2013) Therapeutic drug monitoring in the treatment of tuberculosis: an update. Drugs 74:839–854

20. Baciewicz AM, Chrisman AR, Finch CK, Self TH (2013) Update on rifampin, rifabutin, and rifapentine drug interactions. Curr Med Res Opin 29(1):1–12

21. Gatti G, Di Biagio A, De Pascalis CR, Guerra M, Bassetti M, Bassetti D (1999) Pharmacokinetics of rifabutin in HIV-infected patients with or without wasting syndrome. Br J Clin Pharmacol 48:704–711

22. Keung AC, Owens RC Jr, Eller MG, Weir SJ, Nicolau DP, Nightingale CH (1999) Pharmacokinetics of rifapentine in subjects seropositive for the human immunodeficiency virus: a phase I study. Antimicrob Agents Chemother 43:1230–1233

23. Weiner M, Burman W, Vernon A, Benator D, Peloquin CA, Khan A, Weis S, King B, Shah N, Hodge T (2003) The tuberculosis trials consortium. Low isoniazid concentration associated with outcome of tuberculosis treatment with once-weekly isoniazid and rifapentine. Am J Respir Crit Care Med 167:1341–1347

24. Egelund EF, Dupree L, Huesgen E, Peloquin CA (2016) The pharmacologic challenges of treating tuberculosis and HIV coinfections. Expert Rev Clin Pharmacol 28:1–11. [Epub ahead of print]

25. Gurumurthy P, Ramachandran G, Hemanthkumar AK et al (2004) Malabsorption of rifampin and isoniazid in HIV-infected patients with and without tuberculosis. Clin Infect Dis 38(2):280–283

26. Kotler DP, Francisco A, Clayton F, Scholes JV, Orenstein JM (1990) Small intestinal injury and parasitic diseases in AIDS. Ann Intern Med 113:444–449

27. Kotler DP, Giang TT, Thiim M, Nataro JP, Sordillo EM, Orenstein JM (1995) Chronic bacterial enteropathy in patients with AIDS. J Infect Dis 171:552–558

28. Blum RA, D'Andrea DT, Florentino BM et al (1991) Increased gastric pH and the bioavailability of fluconazole and ketoconazole. Ann Intern Med 114:755–757

29. Burman WJ, Gallicano K, Peloquin C (1999) Therapeutic implications of drug interactions in the treatment of HIV-related tuberculosis. Clin Infect Dis 28:419–430

30. Peloquin CA (1991) Antituberculosis drugs: pharmacokinetics. In: Heifets L (ed) Drug susceptibility in the chemotherapy of mycobacterial infections. CRC Press, Boca Raton, pp 59–88

31. Malone RS, Fish DN, Spiegel DM, Childs JM, Peloquin CA (1999) The effect of hemodialysis on isoniazid, rifampin, pyrazinamide, and ethambutol. Am J Respir Crit Care Med 159:1580–1584

32. Murray FJ (1962) Outbreak of unexpected reactions among epileptics taking isoniazid. Am Rev Respir Dis 86:729–732

33. Valsalan VC, Cooper GI (1982) Carbamazepine intoxication caused by interaction with isoniazid. BMJ 285:261–262

34. Dockweiler U (1987) Isoniazid induced valproic acid toxicity, or vice versa. Lancet 2:152

35. Judd FK, Mijch AM, Cockram A, Norman TR (1994) Isoniazid and antidepressants: is there cause for concern? Int Clin Psychopharmacol 9:123–125

36. Angelini MC, MacCormack-Gagnon J, Dizio S (2009) Increase in plasma levels of clozapine after the addition of isoniazid. J Clin Psychopharmacol 29(2):190

37. Bertz RJ, Granneman GR (1997) Use of in vitro and in vivo data to estimate the likelihood of metabolic pharmacokinetic interactions. Clin Pharmacokinet 32:210–258

38. Ochs HR, Greenblatt DJ, Roberts GM, Dengler HJ (1981) Diazepam interaction with antituberculosis drugs. Clin Pharmacol Ther 29:671–678
39. Sutton G, Kupferberg HJ (1975) Isoniazid as an inhibitor of primidone metabolsim. Neurology 25:1179–1181
40. Zand R, Nelson SD, Slattery JT, Thummel KE, Kalhorn TF, Adams SP, Wright JM (1993) Inhibition and induction of cytochrome P4502E1-catalyzed oxidation by isoniazid in humans. Clin Pharmacol Ther 54:142–149
41. Baciewicz AM, Self TH (1985) Isoniazid interactions. South Med J 78:714–718
42. Rosenthal AR, Self TH, Baker ED et al (1977) Interaction of isoniazid and warfarin. JAMA 238:2177
43. Desta Z, Soukhova NV, Flockhart DA (2001) Inhbition of cytochrome P450 (CYP450) isoforms by isoniazid: potent inhibition of CYP2C19 and CYP3A4. Antimicrob Agents Chemother 45:382–392
44. Wen X, Wang JS, Neuvonen PJ et al (2001) Isoniazid is a mechanism based inhibitor of cytochrome P450 1A2, 2A6, 2C19, and 3A4 isoforms in human liver microenzymes. Eur J Clin Pharmacol 57:799–804
45. Wenning GK, O'Connell MT, Patsalos PN et al (1995) A clinical and pharmacokinetic case study of an interaction of levodopa and antituberculous therapy in Parkinson's disease. Mov Disord 10:664–667
46. Nolan CM, Sandblom RE, Thummel KE, Slattery JT, Nelson SD (1994) Hepatotoxicity associated with acetaminophen usage in patients receiving multiple drug therapy for tuberculosis. Chest 105:408–411
47. McEvoy GK (ed) (2010) AHFS drug information. American Society of Health-Systems Pharmacists, Bethesda
48. Murphy R, Swartz R, Watkins PB (1990) Severe acetaminophen toxicity in a patient receiving isonizid. Ann Intern Med 113:799–800
49. Moulding TS, Redeker AG, Kanel GC (1991) Acetaminophen, isoniazid and hepatic toxicity. Ann Intern Med 114:431
50. Ofoefule SI, Obodo CE, Orisakwe OE, Ilondu NA, Afonne OJ, Maduka SO, Anusiem CA, Agbassi PU (2001) Some plasma pharmacokinetic parameters of isoniazid in the presence of a fluoroquinolone antibacterial agent. Am J Ther 8:243–246
51. Burman WJ, Gallicano K, Peloquin C (2001) Comparative pharmacokinetics and pharmacodynamics of the rifamycin antibacterials. Clin Pharmacokinet 40:327–341
52. Oswald S, Giessmann T, Luetjohann D, Wegner D, Rosskopf D, Weitschies W, Siegmund W (2006) Disposition and sterol-lowering effect of ezetimibe are influenced by single-dose coadministration of rifampin, an inhibitor of multidrug transport proteins. Clin Pharmacol Ther 80(5):477–485
53. Payen L, Sparfel L, Courtois A, Vernhet L, Guillouzo A, Fardel O (2002) The drug efflux pump MRP2: regulation of expression in physiopathological situations and by endogenous and exogenous compounds. Cell Biol Toxicol 18(4):221–233
54. Corpechot C, Ping C, Wendum D, Matsuda F, Barbu V, Poupon R (2006) Identification of a novel 974C→G nonsense mutation of the MRP2/ABCC2 gene in a patient with Dubin-Johnson syndrome and analysis of the effects of rifampicin and ursodeoxycholic acid on serum bilirubin and bile acids. Am J Gastroenterol 101:2427–2432
55. Vavricka SR, Van Montfoort J, Ha HR, Meier PJ, Fattinger K (2002) Interactions of rifamycin SV and rifampicin with organic anion uptake systems of human liver. Hepatology 36(1):164–172
56. Niemi M, Kivistö KT, Diczfalusy U, Bodin K, Bertilsson L, Fromm MF, Eichelbaum M (2006) Effect of SLCO1B1 polymorphism on induction of CYP3A4 by rifampicin. Pharmacogenet Genomics 16(8):565–568
57. Williamson B, Dooley KE, Zhang Y, Back DJ, Owen (2013) Induction of influx and efflux transporters and cytochrome P450 3A4 in primary human hepatocytes by rifampin, rifabutin and rifapentine. Antimicrob Agents Chemother 57:6366–6369

58. Schuetz EG, Schinkel AH, Relling MV, Schuetz JD (1996) P-glycoprotein: a major determinant of rifampicin inducible expression of cytochrome P4503A in mice and humans. Proc Natl Acad Sci U S A 93:4001–4005
59. Hoffmeyer S, Burk O, von Richter O (2000) Functional polymorphisms of the human multidurg resistance gene:multiple sequence variations and correlation of one allele with P-glycoprotein expression and activity in vivo. Proc Natl Acad Sci U S A 97:3473–3478
60. Li AP, Reith MK, Rasmussen A, Gorski JC, Hall SD, Xu L, Kaminski DL, Cheng LK (1997) Primary human hepatocytes as a tool for the evaluation of structure-activity relationship in a cytochrome P450 induction potential of xenobiotics: evaluation of rifampin, rifapentine, and rifabutin. Chem Biol Interact 107:17–30
61. Weiner M, Peloquin C, Burman MD et al (2010) Tuberculosis Trials Consortium. The effects of tuberculosis, race and human gene SLCO1B1 polymorphisms on rifampin concentrations. Antimicrob Agents Chemother 54(10):4192–4200
62. Keung AC, Reith K, Eller MG (1999) Enzyme induction observed in healthy volunteers after repeated administration of rifapentine and its lack of effect on steady-state rifapentine pharmacokinetics. Int J Tuber Lung Dis 3:426–436
63. Gallicano KD, Sahai J, Shukla VK, Seguin I, Pakuts A, Kwok D, Foster BC, Cameron DW (1999) Induction of zidovudine glucuronidation and amination pathways by rifampicin in HIV-infected patients. Br J Clin Pharmacol 48:168–179
64. Peloquin CA (2003) What is the right dose of rifampin? Int J Tuberc Lung Dis 7:3–5
65. Aguado JM, Torre-Cisneros J, Fortun J et al (2009) Tuberculosis in solid-organ transplant recipients: consensus statement of the group for the study of infection in transplant recipients (GESITRA) of the Spanish Society of Infectious Diseases and Clinical Microbiology. Clin Infect Dis 48:1276–1284
66. Regazzi M, Carvalho AC, Villani P, Matteelli A (2014) Treatment optimization in patients co-infected with HIV and Mycobacteriium tuberculosis infections: focus on drug-drug interactions with rifamycins. Clin Pharmacokinet 53:489–507
67. Sahasrabudhe V, Zhy T, Vaz A, Tse S (2015) Drug metabolism and drug interactions: potential application to antituberculsosis drugs. JID 211:S107–S114
68. Jaruratanasirikul S, Sriwiriyajan S (1998) Effect of rifampicin on the pharmacokinetics of itraconazole in normal volunteers and AIDS patients. Eur J Clin Pharmacol 54:155–158
69. Drayton J, Dickenson G, Rinaldi MG (1994) Coadministration of rifampin and itraconazole leads to undetectable levels of serum itraconazole. Clin Infect Dis 18:266
70. Doble N, Shaw R, Rowland-Hill C, Lush M, Warnock DW, Keal EE (1998) Pharmacokinetic study of the interaction between rifampicin and ketoconazole. J Antimicrob Chemother 21:633–635
71. Schwiesow JN, Iseman MD, Peloquin CA (2008) Concomitant use of voriconazole and rifabutin in a patient with multiple infections. Pharmacotherapy 28(8):1076–1080
72. Nicolau DP, Crowe HM, Nightingale CH, Quintiliani R (1995) Rifampin-fluconazole interaction in critically ill patients. Ann Pharmacother 29:994–996
73. Apseloff G, Hilligoss DM, Gardner MJ, Henry EB, Inskeep PB, Gerver N (1991) Induction of fluconazole metabolism by rifampin: in vivo study in humans. J Clin Pharmacol 31:358–361
74. Coker RJ, Tomlinson DR, Prakin J, Harris JRW, Pinching AJ (1990) Interaction between fluconazole and rifampicin. BMJ 301:818
75. CDC (2013) Managing drug interactions in the treatment of HIV-related tuberculosis [online]. Available from URL: http://www.cdc.gov/tb/TB_HIV_Drugs/default.htm
76. Jaruratanasirikul S, Kleepkaew A (1996) Lack of effect of fluconazole on the pharmacokinetics of rifampicin in AIDS patients. J Antimicrob Chemother 38:877–880
77. Jordan MK, Polis MA, Kelly G, Narang PK, Masur H, Piscitelli SC (2000) Effects of fluconazole and clarithromycin on rifabutin and 25-O-desacetylrifabutin pharmacokinetics. Antimicrob Agents Chemother 44:2170–2172
78. Kelly HW, Couch RC, Davis RL, Cushing AH, Knott R (1988) Interaction of chloramphenicol and rifampin. J Pediatr 112:817–820

79. Prober CG (1985) Effect of rifampin on chloramphenicol levels. N Engl J Med 312:788
80. Mirochnick M, Cooper E, Capparelli E, McIntosh K, Lindsey J, Xu J, Jacobus D, Mofenson L, Bonagura VR, Nachman S, Yogev R, Sullivan JL, Spector SA (2001) Population pharmocokinetics of dapsone in children with human immunodeficiency virus infection. Clin Pharmacol Ther 70:24–32
81. Colmenero JD, Fernandez-Gallardo LC, Agundez JAG, Sedeno J, Benitez J, Valverde E (1994) Possible implications of doxycycline-rifampin interaction for treatment of brucellosis. Antimicrob Agents Chemother 38:2798–2802
82. Temple ME, Nahata MC (1999) Interaction between ciprofloxacin and rifampin. Ann Pharmacother 33:868–870
83. Chandler MH, Toler SM, Rapp RP, Muder RR, Korvick JA (1990) Multiple dose pharmacokinetics of concurrent oral ciprofloxacin and rifampin therapy in elderly patients. Antimicrob Agents Chemother 34:442–447
84. Weiner M, Burman W, Luo CC et al (2007) Effects of rifampin and multidrug resistance gene polymorphism on concentrations of moxifloxacin. AACA 51:2861–2866
85. Nijland HM, Ruslami R, Suroto AJ et al (2007) Rifampicin reduces plama concentrations of moxifloxacin in patients with tuberculosis. Clin Infect Dis 45(8):1001–1007
86. Drusano GL, Sgambati N, Eichas A, Brown D, Kulawy R, Louie A (2011) Effect of administration of moxifloxacin plus rifampin against mycobacterium tuberculosis for 7 of 7 days versus 5 of 7 days in an in vitro pharmacodynamic system. M Bio 2(4):e00108–e00111
87. Shishoo CJ, Shah SA, Rathod IS, Savale SS, Vora MJ (2001) Impaired bioavailabilty of rifampicin in presence of isoniazid from fixed dose combination (FDC) formulation. Int J Pharm 228:53–67
88. Apseloff G, Foulds G, LaBoy-Goral L, Willavize S, Vincent J (1998) Comparison of azithromycin and clarithromycin in their interactions with rifabutin in healthy volunteers. J Clin Pharmacol 38:830–835
89. Van Ingen J, Simons S, De Zwaan R et al (2010) Comparative study on genotypic and phenotypic second-line drug resistance testing of mycobacterium tuberculosis complex isolates. J Clin Microbiol 48(8):2749–2753
90. Hafner R, Bethel J, Power M, Landry B, Banach M, Mole L, Standiford HC, Follansbee S, Kumar P, Raasch R, Cohn D, Mushatt D, Drusano G (1998) Tolerance and pharmacokinetic interactions of rifabutin and clarithromycin in human immunodeficiency virus-infected volunteers. Antimicrob Agents Chemother 42:631–639
91. Djojosaputro M, Mustofa S, Donatus IA, Santoso B (1990) The effects of doses and pretreatment with rifampicin on the elimination kinetics of metronidazole. Eur J Pharmacol 183:1870
92. McIlleron H, Meintjes G, Burman WJ et al (2007) Complications of antiretroviral therapy in patients with tuberculosis: drug interactions, toxicity, and immune reconstitution inflammatory syndrome. J Infect Dis 196(Suppl 1):S63–S75
93. Borin MT, Chambers JH, Carel BJ, Gagnon S, Freimuth WW (1997) Pharmacokinetic study of the interaction between rifampin and delavirdine mesylate. Clin Pharmacol Ther 61:544–553
94. Borin MT, Chambers JH, Carel BJ, Freimuth WW, Aksentijevich S, Piergies AA (1997) Pharmacokinetic study of the interaction between rifabutin and delavirdine mesylate in HIV1 infected patients. Antivir Res 35:53–63
95. Kakuda TN, Scholler-Gyure M, Hoetelmans RM (2011) Pharmacokinetic interaction between etravirine and non-antiretroviral drugs. Clin Pharmacokinet 50:25–39
96. Edurant package insert. © Janssen Products, LP (2011) Issued June 2013. Available at http://www.edurant.com/sites/default/files/EDURANT-PI.pdf
97. Cohen K, Grant A, Dandara C et al (2009) Effect of rifampicin-based antitubercular therapy and the cytochrome P450 2B6 516G>T polymorphism on efavirenz concentrations in adults in South Africa. Antivir Ther 14:687–695

98. Kwara A, Lartey M, Sagoe KW et al (2011) Paradoxically elevated efavirenz concentrations in HIV/tuberculosis-coinfected patients with CYP2B6 516TT genotype on rifampin-containing antituberculous therapy. AIDS 25:388–390

99. Gengiah TN, Holford NH, Botha JH et al (2011) The influence of tuberculosis treatment on efavirenz clearance in patients co-infected with HIV and tuberculosis. Eur J Clin Pharmacol 68:689–695

100. Boulle A, Van Cutsem G, Cohen K et al (2008) Outcomes of nevirapine- and efavirenz-based antiretroviral therapy when coadministered with rifampicin-based antitubercular therapy. JAMA 300:530–539

101. Manosuthi W, Sungkanuparph S, Tantanathip P et al (2009) A randomized trial comparing plasma drug concentrations and efficacies between 2 nonnucleoside reverse-transcriptase inhibitor-based regimens in HIV-infected patients receiving rifampicin: the N2R study. Clin Infect Dis 48:1752–1759

102. Brennan-Benson P, Lyus R, Harrison T et al (2005) Pharmacokinetic interactions between efavirenz and rifampicin in the treatment of HIV and tuberculosis: one size does not fit all. AIDS 19:1541–1543

103. Manosuthi W, Sungkanuparph S, Tantanathip P et al (2009) Body weight cutoff for daily dosage of efavirenz and 60-week efficacy of efavirenz-based regimen in human immunodeficiency virus and tuberculosis coinfected patients receiving rifampin. Antimicrob Agents Chemother 53:4545–4548

104. Villar J, Sanchez P, Gonzalez A et al (2011) Use of non-nucleoside analogues together with rifampin in HIV patients with tuberculosis. HIV Clin Trials 12:171–174

105. Kwara A, Ramachandran G, Swaminathan S (2010) Dose adjustment of the non-nucleoside reverse transcriptase inhibitors during rifampicin containing tuberculosis therapy: one size does not fit all. Expert Opin Drug Metab Toxicol 6:55–68

106. Weiner M, Benator D, Peloquin CA et al (2005) Evaluation of the drug interaction between rifabutin and efavirenz in patients with HIV infection and tuberculosis. CID 41(1):1343–1349

107. Ribera E, Pou L, Lopez RM et al (2001) Pharmacokinetic interaction between nevirapine and rifampicin in HIV-infected patients with tuberculosis. J Acquir Immune Defic Syndr 28:450–453

108. Ramachandran G, Hemanthkumar AK, Rajasekaran S et al (2006) Increasing nevirapine dose can overcome reduced bioavailability due to rifampicin coadministration. J Acquir Immune Defic Syndr 42:36–41

109. Manosuthi W, Ruxrungtham K, Likanonsakul S et al (2007) Nevirapine levels after discontinuation ofrifampicin therapy and 60-week efficacy of nevirapine-based antiretroviral therapy in HIV-infected patients with tuberculosis. Clin Infect Dis 44:141–144

110. Autar RS, Wit FW, Sankote J et al (2005) Nevirapine plasma concentrations and concomitant use of rifampin in patients coinfected with HIV-1 and tuberculosis. Antivir Ther 10:937–943

111. Lamorde M, Byakika-Kibwika P, Okaba-Kayom V et al (2011) Nevirapine pharmacokinetics when initiated at 200 mg or 400 mg daily in HIV-1 and tuberculosis co-infected Ugandan adults on rifampicin. J Antimicrob Chemother 66:180–183

112. Avihingsanon A, Manosuthi W, Kantipong P et al (2008) Pharmacokinetics and 48-week efficacy of nevirapine: 400 mg versus 600 mg per day in HIV-tuberculosis coinfection receiving rifampicin. Antivir Ther 13:529–536

113. Back D, Gibbons S, Khoo S (2003) Pharmacokinetic drug interactions with nevirapine. JAIDS 34:S8–S14

114. Crauwels HM, Kakuda TN (2010) Drug interactions with new investigational antiretrovirals. Clin Pharm 49(1):67–68

115. Barry M, Mulcahy F, Merry C (1999) Pharmacokinetics and potential interactions amongst antiretroviral agents used to treat patients with HIV infection. Clin Pharmacokinet 36:289–304

116. Cato A, Cavanaugh J, Shi H, Hsu A, Leonard J, Granneman GR (1998) The effect of multiple doses of ritonavir on the pharmacokinetics of rifabutin. Clin Pharmacol Ther 63:414–421

117. Kerr BM, Daniels R, Cledeninn N (1999) Pharmacokinetic interaction of nelfinavir with half-dose rifabutin. Can J Infect Dis 10:21B
118. Moyle J, Buss NE, Goggin T, Snell P, Higgs C, Hawkins DA (2002) Interaction between saquinavir soft- gel and rifabutin in patients infected with HIV. Br J Clin Pharmacol 54:178–182
119. Moreno S, Podzamczer D, Blazquez R, Tribarren JA, Ferror B, Reparez J, Pena JM, Carero E, Usan L (2001) Treatment of tuberculosis in HIV-infected patients: safety and antiretroviral efficacy of theconcomitant use of ritonavir and rifampin. AIDS 15:1185–1187
120. Polk RE, Brophy DF, Israel DS, Patron R, Sadler BM, Chittick GE, Symonds WT, Lou Y, Kristoff D, Stein DS (2001) Pharmacokinetic interaction between amprenavir and rifabutin or rifampin in healthy males. Antimicrob Agents Chemother 45:502–508
121. Ribera E, Azuaje C, Lopez RM (2007) Pharmacokinetic interaction between rifampicin and the once daily combination of saquinavir and low dose ritonavir in HIV infected patients with tuberculosis. JAntimicrob Chemother 59:690–697
122. Tipranavir [package insert] (2006) US prescribing information. Boehringer Ingelheim, Germany
123. Darunavir [package insert] (2006, July) Tibotec Therapeutics. East Bridgewater
124. Narita M, Stambaugh JJ, Hollender ES, Jones D, Pitchenik AE, Ashkin D (2000) Use of rifabutin with protease inhibitors for human immunodeficiency virus-infected patients with tuberculosis. Clin Infect Dis 30:779–783
125. Peloquin CA (2001) Tuberculosis drug serum levels (letter). Clin Infect Dis 33:584–585
126. Hamzeh FM, Benson C, Gerber J, Currier J, McCrea J, Deutsch P, Ruan P, Wu H, Lee J, Flexner C (2003) AIDS clinical trials group 365 study team. Steady-state pharmacokinetic interaction of modified-dose indinavir and rifabutin. Clin Pharmacol Ther 73:159–169
127. Benator D, Weiner M, Burman W, et al. for the Tuberculosis Trials Consortium (2007) Clinical evaluation of the Nelfinavir – Rifabutin interaction in patients with HIV infection and tuberculosis. Pharmacotherapy 27:793–800
128. Weiner M, Benator D, Burman W, et al., for the Tuberculosis Trials Consortium (2005) Association between acquired Rifamycin resistance and the pharmacokinetics of Rifabutin and isoniazid among patients with HIV and tuberculosis. Clin Infect Dis 40:1481–1491
129. Boulaner C, Hollender E, Farrell K et al (2009) Pharmacokinetic evaluation of rifabutin in combination with lopinavir-ritonavir in patients with HIV infection and active tuberculosis. CID 49:1305–1311
130. Tseng AL, Walmsley SL (1995) Rifabutin-associated uveitis. Ann Pharmacother 29:1149–1155
131. Abel S, Back DJ, Vourvahis M (2009) Maraviroc: pharmacokinetics and drug interactions. Antivir Ther 14(5):607–618
132. Iwamoto M, Kassahun K, Troyer MD (2008) Lack of a pharmacokinetic effect of raltegravir on midazolam; in vitro/in vivo correclation. J Clin Pharmacol 48:209–214
133. Wenning LA, Hanley WD, Brainard DM, Petry AS, Ghosh K et al (2009) Effect of rifampin, a potent inducer of drug metabolizing enzymes, on the pharmacokinetics of raltegravir. Antimicrob Agents Chemother 53:2852–2856
134. Weiner M, Egelund EF, Engle M et al (2014) Pharmacokinetic interaction of rifapentine and raltegravir in healthy volunteers. J Antimicrob Chemother 69(4):1079–1085
135. Svensson EM, Murray S, Karlsson MO, Dooley KE (2015) Rifampicin and rifapentine significantly reduce concentrations of bedaquiline, a new anti-TB drug. J Antimicrob Chemother 70:1106–1114
136. Ribera E, Pou L, Fernandez-Sola A, Campos F, Lopez RM, Ocana I, Ruiz I, Pahissa A (2001) Rifampin reduces concentrations of trimethoprim and sulfamethoxazole in serum in human immunodeficiency virus-infected patients. Antimicrob Agents Chemother 45:3238–3241
137. Niemi M, Backman JT, Fromm MF, Neuvonen PJ, Kivisto KT (2003) Pharmacokinetic interactions with rifampin: clinical relevance. Clin Pharm 42:815–850
138. Lin JH, Lu AYH (1998) Inhibition and induction of cytochrome P450 and the clinical implications. ClinPharmacokinet 35:361–390

139. Lacroix C, Guyonnaud C, Chaou M, Duwoos H, Lafont O (1988) Interaction between allopurinol and pyrazinamide. Eur Respir J 1:807–811
140. Urban T, Maquarre E, Housset C, Chouaid C, Devin E, Lebeua B (1995) Allopurinol hypersensitivity. A possible cause of hepatitis and mucocutaneous eruptions in a patient undergoing antitubercular treatment. Rev Mal Respir 12:314–316
141. Louthrenoo W, Hongsongkiat S, Kasitanon N, Wangkawe S, Jatuworkapruk K (2015) Effect of antituberculous drugs on serum uric acid and urine uric acid excretion. J Clin Rheumatol 7:346–348
142. Jasmer RM, Saukkonen JJ, Blumberg HM, Daley CL, Bernardo J, Vittinghoff E, King MD, Kawamura LM, Hopewell PC (2002) Short-course rifampin and pyrazinamide for tuberculosis infection (SCRIPT) study investigators. Short-course rifampin and pyrazinamide compared with isoniazid for latent tuberculosis infection: a multicenter clinical trial. Ann Intern Med 137:640–647
143. Centers for Disease Control and Prevention (CDC); American Thoracic Society (2003) Update: adverse event data and revised American Thoracic Society/CDC recommendations against the use of rifampin and pyrazinamide for treatment of latent tuberculosis infection - United States, 2003. MMWR 52:735–739
144. Kunimoto D, Warman A, Beckon A et al (2003) Severe hepatotoxicity associated with rifampin-pyrazinamide preventative therapy requiring transplantation in an individual at low risk for hepatotoxicity. Clin Infect Dis 36:158–161
145. Ridzon R, Meador J, Maxwell R, Higgins K, Weismuller P, Onorato IM (1997) Asymptomatic hepatitis in persons who received alternative preventive therapy with pyrazinamide and ofloxacin. Clin Infect Dis 24:1264–1265
146. Papastavros T, Dolovich LR, Holbrook A, Whitehead L, Loeb M (2002) Adverse events associated with pyrazinamide and levofloxacin in the treatment of latent multidrug-resistant tuberculosis. CMAJ 167:131–136
147. Lou HX, Shullo MA, McKaveney TP (2002) Limited tolerability of levofloxacin and pyrazinamide for multidrug-resistant tuberculosis prophylaxis in a solid organ transplant population. Pharmacotherapy 22:701–704
148. Tseng AL, Mortimer CB, Salit IE (1999) Iritis associated with intravenous cidofovir. Ann Pharmacother 33:167–171
149. Nicolau DP, Quintiliani R (1998) Aminoglycosides. In: Yu VL, Merigan TC, Barriere S, White NJ (eds) Antimicrobial chemotherapy. Williams and Wilkins, Baltimore, pp 621–637
150. Kucers A, Bennett NMK (eds) (2010) The use of antibiotics, 6th edn. JB Lippencott Co, Philadelphia
151. Peloquin CA, Berning SE, Nitta AT, Simone PM, Goble M, Huitt GA, Iseman MD, Cook JL, Curran-Everett D (2004) Aminoglycoside toxicity: daily versus thrice-weekly dosing for treatment of mycobacterial diseases. Clin Infect Dis 38:1538–1544
152. Berning SE, Peloquin CA (1998) Antimycobacterial agents: cycloserine. In: Yu VL, Merigan TC, Barriere S, White NJ (eds) Antimicrobial chemotherapy. Williams and Wilkins, Baltimore, pp 638–642
153. Malone RS, Fish DN, Spiegel DM, Childs JM, Peloquin CA (1999) The effect of hemodialysis on cycloserine, ethionamide, para-aminosalicylate, and clofazimine. Chest 116(4):984–990
154. Vale N, Gomes P, Santos HA (2013) Metabolism of the antituberculosis drug ethionamide. Curr Drug Metab 14(1):151–158
155. Shimokawa Y, Yoda N, Kondo S, Yamamura Y, Takiguchi Y, Umehara K (2015) Inhibitory potential of twenty five anti-tuberculosis drugs on CYP activities in human liver microsomes. Biol Pharm Bull 38(9):1425–1429
156. Berning SE, Peloquin CA (1998) Antimycobacterial agents: para-aminosalicylic acid. In: Yu VL, Merigan TC, Barriere S, White NJ (eds) Antimicrobial chemotherapy. Williams and Wilkins, Baltimore, pp 663–668

157. Jacobus Pharmaceutical Co (1996) Paser® granules (aminosalicylic acid granules) prescribing information. In: Physicians' desk reference, 57th edn, Medical Economics Company Inc, Montvale; 2003, pp 1770–1771
158. Dey T, Brigden G, Cox H, Shubber Z, Cooke G, Ford N (2012) Outcomes of clofazamine for the treatment of drug resistant tuberculosis: a systematic review and meta- analysis. J Antimicrob Chemother 68:284–293
159. van Heeswijk RPG, Dannemann B, Hoetelmans RMW (2014) Bedaquiline: a review of human pharmacokinetics and drug–drug interactions. J Antimicrob Chemother 69:2310–2318
160. European Medicines Agency (2017) Website: http://www.ema.europa.eu/ema/index.jsp?curl=pages/medicines/human/medicines/002552/human_med_001699.jsp&mid=WC0b01ac058001d124. Accessed 20 Aug 2017
162. Griffith DE, Aksamit T, Brown-Elliott BA et al (2007) An official ATS/IDSA statement: diagnosis, treatment, and prevention of nontuberculous mycobacterial diseases. Am J Respir Crit Care Med 175:367
163. Wallace RJ, Brown BA, Griffith DE (1995) Reduced serum levels of clarithromycin in patients treated with multidrug regimens including rifampin or rifabutin for Mycobacterium avium-intracellulare infection. J Infect Dis 171:747–750
164. Peloquin CA, Berning SE (1996) Evaluation of the drug interaction between clarithromycin and rifampin. J Infect Dis Pharmacother 2:19–35
165. Lalloo UG, Ambaram A (2010) New antituberculous drugs in development. Curr HIV/AIDS Rep 7(3):143–151
166. Burman WJ (2010) Rip van winkle wakes up: development of tuberculosis treatment in the 21st century. Clin Infect Dis 50(Suppl 3):S165–S172
167. Brigden G, Hewison C, Varaine F (2015) New developments in the treatment of drug resistant tuberculosis; clinical utility of bedaquiline and delamanid. Infect Drug Resis 8:367–378
168. Borsari C, Ferrari S, Venturelli A, Costi MP (2016) Target based approaches for the discovery of new antimycobacterial drugs. Drug Discov Today pii: S1359-6446(16)30432-9. https://doi.org/10.1016/j.drudis.2016.11.014. [Epub ahead of print] Review

Chapter 7
Drug Interactions in HIV: Protease and Integrase Inhibitors

Parul Patel and Stan Louie

7.1 Introduction

Treatment of HIV disease has greatly enhanced our understanding of the molecular basis leading to drug-drug interactions. HIV protease inhibitors (PIs) and nonnucleoside reverse transcriptase inhibitors (NNRTIs), critical components of early antiretroviral therapy (ART), are subject to a number of clinically significant drug-drug interactions. Early effective ART consisted of either a PI- or NNRTI-based regimen in combination with two nucleosides. Although effective, these two therapeutic platforms had major drawbacks, which included a large number of pills, multiple daily dosing, intolerance, and significant impact on metabolic clearance. For example, PIs can alter metabolic clearance by both inhibiting a wide variety of cytochrome P450 (CYP) enzymes and inducing the expression of phase II enzymes such as glucuronidation. In contrast, NNRTIs are well-known inducers of CYP450 isoenzymes that increase the clearance of concomitantly administered drugs.

In the past decade, ART regimens have undergone dramatic improvements in terms of tolerability and simplification (given once or twice daily). Two early changes significantly improved ART tolerability and adherence. One was the development of more tolerable agents such as the NNRTI, efavirenz, and PI, atazanavir, with pharmacokinetic (PK) profiles that also permitted once daily administration. The second was a fortuitous discovery that the PI ritonavir (RTV) could prolong

P. Patel (✉)
Clinical Pharmacology, ViiV Healthcare, Research Triangle Park, NC, USA
e-mail: parul.x.patel@viivhealthcare.com

S. Louie
University of Southern California, School of Pharmacy, Los Angeles, CA, USA
e-mail: slouie@usc.edu

© Springer International Publishing AG, part of Springer Nature 2018
M. P. Pai et al. (eds.), *Drug Interactions in Infectious Diseases: Antimicrobial Drug Interactions*, Infectious Disease,
https://doi.org/10.1007/978-3-319-72416-4_7

255

Table 7.1 Fixed-dose combinations (FDCs) of antiretroviral agents

Name of FDC	Components	No. of tablets per day	Daily dose
Atripla	600 mg efavirenz, 300 mg tenofovir DF, 200 mg emtricitabine (FTC)	1	Once daily
Combivir	300 mg zidovudine, 300 mg lamivudine	1	Twice daily
Descovy	10 mg tenofovir alafenamide (TAF) and 200 mg emtricitabine (FTC) – Not available in the United States 25 mg tenofovir alafenamide (TAF) and 200 mg emtricitabine (FTC)	1	Once daily Once daily
Eviplera or Complera	25 mg rilpivirine (RPV), 300 mg tenofovir DF (TDF), 200 mg emtricitabine (FTC)	1	Once daily
Epzicom or Kivexa	600 mg abacavir (ABC), 300 mg lamivudine (3TC)	1	Once daily
Evotaz	300 mg atazanavir (ATV), 150 mg cobicistat (COBI)	1	Once daily
Genvoya	150 mg elvitegravir, 150 mg cobicistat, 10 mg tenofovir alafenamide (TAF), 200 mg emtricitabine (FTC)	1	Once daily
Kaletra	200 mg lopinavir (LPV), 50 mg ritonavir (RTV)	2	Twice daily
Odefsey	25 mg rilpivirine, 25 mg tenofovir alafenamide (TAF), 200 mg emtricitabine (FTC)	1	Once daily
Prezcobix or Rezolsta	800 mg darunavir (DRV), 150 mg cobicistat (COBI)	1	Once daily
Stribild	150 mg elvitegravir (EVG), 150 mg cobicistat (COBI), 300 mg tenofovir DF (TDF), 200 mg emtricitabine (FTC)	1	Once daily
Triumeq	50 mg dolutegravir (DTG), 600 mg abacavir (ABC), 300 mg lamivudine (3TC)	1	Once daily
Trizivir	600 mg abacavir (ABC), 300 mg zidovudine (AZT), 300 mg lamivudine (3TC)	1	Twice daily
Truvada	300 mg tenofovir DF (TDF), 200 mg emtricitabine (FTC)	1	Once daily

systemic PK of a concomitantly administered PI, opening the potential to dose PIs once daily with low doses of RTV. The ability to reduce dosing frequency also improved adherence which ultimately improved clinical outcomes. In addition, the development of new ARVs with increased potency meant lower daily dosages could be used and permitted co-formulation of various individual components into a single pill, more commonly referred to as fixed-dose combination tablets or FDC (Table 7.1).

More recently, a potent new class of ART, integrase inhibitors (INIs), has set a new standard as the most potent and well-tolerated agents available with a high genetic barrier to resistance. Raltegravir (RAL) was the first INI approved which did not require use of RTV as a pharmacoenhancing agent. Subsequently, elvitegravir (EVG) was co-formulated with the first alternative pharmacokinetic booster, cobicistat (COBI), to enable once daily dosing. Elvitegravir is available as a complete, single-tablet regimen (STR) which includes COBI and dual NRTIs, emtricitabine and tenofovir disoproxil fumarate (TDF), in Stribild™ or with another tenofovir prodrug, tenofovir alafenamide (TAF), in Genvoya™. Dolutegravir (DTG) is the first second-generation INI with a PK profile that also supports once daily dosing but without the need for pharmacokinetic boosting agents such as RTV and COBI. Dolutegravir has also been co-formulated as a STR with dual NRTIs, lamivudine (3TC) and abacavir (ABC), in Triumeq™. Among the recently approved INIs, elvitegravir FDC regimens have similar potential for drug-drug interactions as PIs or NNRTIs, as elvitegravir is a CYP3A substrate co-formulated with a potent CYP3A inhibitor. In contrast, raltegravir and dolutegravir have a lower potential for drug-drug interactions because their metabolic pathway involves glucuronidation and thus have lower propensity for being a victim or perpetrator of drug-drug interactions.

This chapter will make extensive use of tables for reference, while the text will provide historical context and discussions about the molecular basis of drug-drug interactions and address key drug-drug interactions issues with PIs and INIs. Internet websites are continually updated with the latest antiretroviral drug-drug interaction information that may be useful to the reader [1–4].

7.2 Pharmacology of ART

7.2.1 Protease Inhibitors

PIs were the first class of antiretrovirals to dramatically improve HIV morbidity and mortality [5]. PIs are potent ARVs that exhibit durable antiretroviral suppression and have a high genetic barrier to resistance. To enhance efficacy, PIs are commonly co-administered with other ARV agents from different therapeutic classes to produce additive to synergistic antiviral activity. The introduction of PIs as part of the antiretroviral regimen led to the term, highly active antiretroviral therapy (HAART), which has since been simplified to antiretroviral therapy (ART). HIV protease inhibitors competitively inhibit HIV protease. PIs are peptidomimetic agents that bind and inhibit viral proteases from liberating the active peptide moieties from a virally produced pro-peptide preventing further viral propagation from infected cells. Numerous PIs have been approved since their introduction in the early 1990s; however, no new agents have been approved in this class since 2006, apart from the new pharmacokinetic boosting agent, cobicistat. Atazanavir, darunavir, and lopinavir are still employed in the treatment of HIV-infected adults and children in combination with a pharmacokinetic enhancer (ritonavir or cobicistat) and other antiretroviral agents, but the use of other HIV PIs has declined.

In general, PIs are large-molecular-weight compounds that often have low water solubility leading to poor oral bioavailability and necessitating large oral doses and concomitant administration of food in many cases to increase bioavailability. Atazanavir has pH-dependent absorption and should be used with caution with gastric-acid modifiers. The majority are highly protein bound and undergo extensive hepatic metabolism via CYP3A and may require multiple daily dosing in the absence of a pharmacoenhancing agent. Hepatic impairment increases PI exposures; however, renal excretion is typically minimal supporting recommendations for no dosage adjustment in renal impairment. PIs are metabolically eliminated by hepatic clearance and are also known to inhibit CYP3A activity or UGT activity [6]. RTV is a potent inhibitor of CYP3A which was a fortuitous discovery that ultimately shifted its use away from a primary ART agent to its use as a low-dose PK-enhancing drug. This boosting effect of RTV was discovered during evaluation of RTV with another PI, saquinavir (SQV), to determine if the combination could enhance antiretroviral activity. Initially, the reduction of circulating HIV was attributed to synergistic antiviral activity but was later found to be the result of RTV's ability to prolong the plasma half-life of SQV and increase its total drug exposure. The added antiviral activities were found to be a drug-drug interaction, whereby RTV potently blocked CYP3A4and reduced the metabolic elimination of the co-administered PI. This strategy was later employed to reduce the pill burden and dosing frequency of other PIs such as indinavir (IDV), lopinavir (LPV), atazanavir (ATV), darunavir (DRV), and tipranavir (TPV) using much lower doses of RTV that still retained potent CYP inhibition but exerted little antiretroviral activity. The majority of DDI studies with PIs have focused on their ability to act as substrates, inhibitors, or inducers of CYP3A4 and P-gp transport; however, PIs are also known to inhibit active transport processes with the most prominent interactions involving OATPs and OCTs. Key drug-drug interactions with protease inhibitors are summarized in Table 7.2.

7.2.2   Integrase Inhibitors

HIV integrase is one of three viral enzymes that are critical for viral replication. Since it is only expressed in virally infected cells, this has been an obvious target to block HIV proliferation and an area of active research for decades. After viral penetration into the host cell and viral DNA replication, the viral DNA integrates into the host genome through a series of DNA cutting and joining reactions. The function of this multipurpose viral enzyme is to remove the 3′-end from the viral DNA to enable the viral DNA strand to be transferred into the host genome. The resultant viral DNA is then joined and inserted into the host genome. After viral DNA insertion, cellular enzymes are activated to repair the single gaps found in the DNA and remove the unpaired nucleotides. HIV integrase inhibitors (INIs) bind to Mg^{+2} within the enzyme catalytic site to effectively disrupt binding to viral DNA, thus blocking the viral strand transfer step [7].

Table 7.2 Drug-drug interactions with HIV protease inhibitors

Concomitant Drugs	PI	Effect	Recommendation
Gastrointestinal agents			
Antacids Al(OH)$_3$, Mg(OH)$_2$	ATV ATV/RTV ATV/COBI	↑ gastric pH will ↓ ATV AUC	Stagger ATV administration for at least 2 h prior or 2 h after antacids
	FPV	↑ gastric pH will ↓ APV AUC 18%	Stagger FPV administration for at least 2 h prior or 1 h after antacids
	TPV/RTV	↑ gastric pH will ↓TPV AUC 27%	Stagger TPV administration for at least 2 h prior or 1 h after antacids
Histamine-2 receptor Antagonists	ATV	↑ gastric pH will ↓ATV AUC	ATV administration must be at least 2 h prior or 10 h after H$_2$ antagonist administration. H$_2$ antagonist dosage should not exceed the equivalence of 20 mg famotidine
	ATV/RTV ATV/COBI	↑ gastric pH will ↓ATV AUC	ATV administration must be at least 2 h prior or 10 h after H2 antagonist administration. H2 antagonist dosage should not exceed the equivalence of 40 mg famotidine twice daily for treatment naïve patients If ARV regimen includes TDF with H2 antagonist, use 400 mg ATV/150 mg COBI or 100 mg RTV
	DRV/COBI DRV/RTV LPV/RTV	↑ gastric pH does not significantly affect these PIs	No adjustment needed
	FPV	↑ gastric pH will ↓APV AUC 30%	FPV administration must be at least 2 h prior to H2 antagonist administration. May consider FPV/RTV

(continued)

Table 7.2 (continued)

Concomitant Drugs	PI	Effect	Recommendation
Proton pump inhibitors	ATV ATV/COBI ATV/RTV	↑ gastric will ↓ ATV AUC	PPI dosage should not exceed the equivalence of 20 mg omeprazole daily. PPI should be administered at least 12 h prior ATV/COBI or ATV/RTV
	DRV/COBI	↑ gastric pH does not significant affect these DRV	No adjustment needed
	DRV/RTV	↑ gastric pH does not significant affect these DRV but ↓ omeprazole AUC 42%	No adjustment needed for DRV Dosage adjustment of omeprazole may be necessary
	FPV FPV/RTV LPV/RTV	↑ gastric pH does not significantly affect these PIs	No adjustment needed
	SQV/RTV	↑ gastric pH will ↑AUC 82%	Monitor for signs and symptoms of toxicities
	TPV/RTV	↓ omeprazole AUC	Dosage adjustment of omeprazole may be necessary
Anticoagulants/ antiplatelets agents			
Apixaban	All PIs	↑apixaban	Avoid concomitant use
Dabigatran	All PIs/RTV	↑dabigatran	No dosage adjustment if CrCl >50 mL/min Avoid concomitant use if CrCl < 50 mL/min
Edoxaban	All PIs	↑rivaroxaban	Avoid concomitant use
Ticagrelor	All PIs	↑ticagrelor	Avoid concomitant use
Vorapaxar	PI/RTV	↑vorapaxar	Avoid concomitant use
Warfarin	PI/RTV	↓warfarin	Monitor INR when either initiating or stopping PI/RTV
Warfarin	ATV/COBI DRV/COBI	No data	Monitor INR when either initiating or stopping PI/COBI.

Anti-seizure medications

Carbamazepine (CBZ)	ATV FPV	↓PI concentrations	Do not combine these PIs with CBZ
	ATV/COBI DRV/COBI	↓COBI concentration ↓PI concentration	Do not combine these PI with CBZ
	ATV/RTV FPV/RTV LPV/RTV SQV/RTV TPV/RTV DRV/RTV	RTV will ↑CBZ concentrations No changes with DRV	Do not combine LPV/RTV or FPV/RTV with CBZ Consider using Keppra as an alternative anticonvulsant
Ethosuximide	All PIs	↑ethosuximide	Monitor for signs and symptoms of ethosuximide toxicities
Lamotrigine	ATV	No effect	No dosage adjustment
	ATV/RTV	↓lamotrigine AUC 32%	↑lamotrigine dosage is necessary or consider alternative anticonvulsant
	LPV/RTV	↓lamotrigine AUC 50% No change in LPV	
	ATV/COBI DRV/COBI	No data	Monitor lamotrigine concentrations
Phenobarbital (PB)	ATV/COBI DRV/COBI	↓COBI ↓PI levels	Do not recommend combining PB with ATV/COBI or DRV/COBI
	PI or PI/RTV	↓PI levels	Do not recommend combining PB with PI

(continued)

Table 7.2 (continued)

Concomitant Drugs	PI	Effect	Recommendation
Phenytoin (PHT)	ATV FPV	↓PI levels	Do not recommend combining PHT with PI Consider ATV/RTV or FPV/RTV Consider alternative anticonvulsants
	ATV/RTV DRV/RTV SQV/RTV TPV/RTV	↓PI levels ↓PHT levels	Do not recommend combining PHT with PI Consider alternative anticonvulsants Assess virologic efficacy
	ATV/COBI DRV/COBI	↓COBI concentration ↓PI concentration	Consider alternative anticonvulsants Assess virologic efficacy
	FPV/RTV	↓PHT AUC 22% ↑APV AUC 20%	Adjust dose according to PHT levels No change in FPV/RTV
	LPV/RTV	↓PHT AUC 31% ↓LPV AUC 33%	Consider alternative anticonvulsant Assess virologic efficacy
Valproic (VPA)	LPV/RTV	↓ or ↔VPA ↑LPV AUC 75%	Monitor VPA concentration and assess virologic efficacy Monitor for LPV toxicities
Antimycobacterial			
Clarithromycin	ATV	↑clarithromycin AUC 94%	Consider alternative macrolide (e.g., azithromycin) ↑risk for QTc prolongation ↑risk for increased liver function test
	PI/RTV ATV/COBI DRV/COBI FPV	↑clarithromycin DRV/RTV ↑Clari AUC 57%	Consider alternative macrolide (e.g., azithromycin) ↑risk for QTc prolongation ↑risk for increased liver function test ↓clarithromycin dose by 50%

Rifabutin	ATV	↑rifabutin	↓Rifabutin to 150 mg daily or 300 mg TIW
	ATV/COBI	↑rifabutin	↓Rifabutin to 150 mg daily or 300 mg TIW
	ATV/RTV	↑rifabutin and ↑rifabutin metabolite AUC 2101%	↓Rifabutin to 150 mg daily or 300 mg TIW Monitor for signs and symptoms of toxicities
	DRV/RTV	Rifabutin ↔ but rifabutin metabolite AUC ↑ 881%	↓Rifabutin to 150 mg daily or 300 mg TIW Monitor for signs and symptoms of toxicities
	FPV	No data	Consider alternative PI or anti-mycobacterial
	FPV/RTV	↑rifabutin and ↑rifabutin metabolite AUC 64%	↓Rifabutin to 150 mg daily or 300 mg TIW Monitor for signs and symptoms of toxicities
	LPV/RTV	↑rifabutin and ↑rifabutin metabolite AUC 473%	↓Rifabutin to 150 mg daily or 300 mg TIW Monitor for signs and symptoms of toxicities
	SQV	↑rifabutin	↓Rifabutin to 150 mg daily or 300 mg TIW Monitor for signs and symptoms of toxicities
	SQV/RTV	Expect ↑rifabutin	↓Rifabutin to 150 mg daily or 300 mg TIW Monitor for signs and symptoms of toxicities
	TPV/RTV	↑rifabutin and ↑rifabutin metabolite AUC 333%	↓Rifabutin to 150 mg daily or 300 mg TIW Monitor for signs and symptoms of toxicities
Rifampin	PI PI/RTV PI/COBI	↓ PI concentration by >75%	Do not co-administer rifampin with PIs. Addition of RTV can overcome drug-interactions but may increase potential for drug-induced hepatotoxicity Addition of COBI is not recommended. Consider substituting rifabutin for rifampin
Rifapentine	PI	↓ PI concentration	Do not co-administer Consider substituting rifabutin for rifapentine

(continued)

Table 7.2 (continued)

Concomitant Drugs	PI	Effect	Recommendation
Steroid contraceptives			
Oral contraceptives	ATV	↑ethinyl estradiol AUC 48%	Recommend contraceptives containing <30 µg ethinyl estradiol Recommend alterative form of contraceptives
	ATV	↑norethindrone AUC 110%	Contraceptives containing ethinyl estradiol or progestins <25 µg have not been studied
	ATV/RTV	↓ethinyl estradiol AUC 19% & C_{min} ↓ 37% Norgestimate ↑ 85% Norethindrone AUC ↑ 51% and C_{min} ↑ 67%	Dosage of oral contraceptives should contain at least 35 µg ethinyl estradiol
	ATV/COBI DRV/COBI	No data	Consider alternatives or change ARV regimen
	FPV	↑ethinyl estradiol ↑norethindrone C_{min} ↓APV C_{min} 20%	Dosage of oral contraceptive should contain <30 µg of ethinyl estradiol
	DRV/RTV FPV/RTV LPV/RTV	↓ethinyl estradiol AUC 37%–48% ↓norethindrone AUC 14%–34% TPV/RTV: Norethindrone AUC↔	Consider alternatives or change ARV regimen
Depo-medroxyprogesterone IM	LPV/RTV	↑MPA AUC 46%; C_{min} ↔	No adjustment needed

Opioid agents			
Buprenorphine	ATV	↑buprenorphine AUC 66% ↑norbuprenorphine AUC 105%	Not recommended
	ATV/RTV	No data	Titrate buprenorphine to effect
	ATV/COBI DRV/COBI	Buprenorphine ↔ ↑norbuprenorphine AUC 46% & C_{min} ↑71%	Titrate buprenorphine to effect
	DRV/RTV	Buprenorphine ↔ ↓norbuprenorphine AUC 15%	Titrate buprenorphine to effect
	FPV/RTV LPV/RTV	Buprenorphine ↔ LPV/RTV alters Norbuprenorphine	Titrate buprenorphine to effect
	TPV/RTV	↓TPV C_{min} 19–40%	Titrate buprenorphine to effect Monitor TPV levels
Fentanyl	PIs	PIs ↔	Monitor patient for respiratory depression
Methadone	ATV	ATV ↔	No adjustment needed
	ATV/COBI DRV/COBI	No data	Titrate methadone to effect
	FPV	APV ↓ R-methadone C_{min} 21%,	Titrate methadone to effect
	PI/RTV	ATV/RTV, DRV/RTV, & FPV/ RTV↓ R-methadone AUC 16–18% LPV: ↓ R-methadone AUC 26–53%	Titrate methadone to effect

Adapted from DHHS Table 19a

The HIV integrase strand transfer inhibitor or integrase inhibitor (INSTI or INI) class of drugs includes raltegravir, elvitegravir, and dolutegravir. Their introduction has revolutionized the management of HIV infection for several reasons. They are characterized as among the most potent and efficacious antiretroviral agents as exemplified by an initial rapid achievement of virologic suppression that is durable over time. They exhibit favorable safety and tolerability profiles, have a high genetic barrier to resistance without cross-resistance to other agents, and have manageable drug-drug interaction profiles (Table 7.3). Most are available in fixed-dose combinations (FDC) with other antiretrovirals to constitute a complete, single-tablet regimen that can be administered once daily (Table 7.1). These features explain why these compounds have transformed how HIV is currently being managed.

7.2.2.1 Raltegravir

Raltegravir (RAL) was the first integrase inhibitor to receive FDA approval for the treatment of HIV. Raltegravir is a small molecule with a beta-diketo acid moiety that selectively inhibits the strand transfer step of viral integration. Its introduction into clinical use was a paradigm changing moment as it was one of the most potent treatment options developed at the time and with a completely new mechanism of action which was especially important for heavily treatment-experienced patients in desperate need for alternative options that were also well tolerated with potential to significantly improve patient adherence.

Raltegravir is approved for twice daily administration (400 mg twice daily) and, more recently, for once daily administration (1200 mg once daily) (Isentress PI). Raltegravir does not require PK boosting with RTV or COBI. It has a low oral bioavailability of 32% with large interindividual variability in exposure. Potential reasons for its high pharmacokinetic variability include whether it is dosed with food, pH-dependent solubility impacting absorption, differences in UGT1A1 expression or polymorphisms, and drug interactions. For example, pharmacokinetic variability increases when RAL is dosed with food, as a high fat meal increased AUC and C_{max} by approximately twofold and increased C12h by 4.1-fold, while T_{max} was delayed for up to 12 h in some healthy volunteers [8]. It has an initial distribution half-life of 1 h, which is followed by a beta terminal half-life that is approximately 9 h. Similar to other INI, raltegravir is eliminated through UGT1A1-mediated metabolism. At clinically achievable concentrations, raltegravir is not a CYP substrate which explains why it has few drug-drug interactions. In addition, raltegravir is not an inhibitor or inducer of CYP expression.

7.2.2.2 Elvitegravir

Elvitegravir (EVG, E) is the second INI that received FDA approval. Similar to raltegravir, EVG is a potent INI with antiviral activity in the subnanomolar range. It is available as a single entity and as part of two complete single-tablet antiretroviral

Table 7.3 Drug-drug interactions with INIs

Concomitant drug	Integrase inhibitor (INI)	Effect	Dosing recommendations and clinical comments
Acid reducers			
Aluminum, magnesium +/– Calcium-containing antacids	DTG	↓DTG AUC (74%) when simultaneously administered ↓DTG AUC (26%) when administered 2 h prior to antacid	Stagger DTG dosing at least 2 hours before or at least 6 hours after antacids containing polyvalent cations
	EVG/COBI EVG plus PI/RTV	↓EVG AUC (40–50%) when simultaneously administered with antacid ↓EVG AUC (15–20%) when administered 2 h before or after antacid	Stagger EVG/COBI/TDF/FTC and antacid administration by more than 2 h No changes when administered 4-h apart
	RAL	Al-mg (OH): ↓RAL C_{min} (54–63%) $CaCO_3$:↓RAL C_{min} 32%	*Do not co-administer RAL and Al-Mg(OH).* Use alternative acid reducing agent No dosing separation necessary when co-administering RAL and $CaCO_3$, antacids
H2-receptor antagonists	EVG/COBI	No significant effect	No dosage adjustment necessary
	EVG plus PI/RTV	↔ EVG	No dosage adjustment necessary for EVG
Proton pump inhibitor	DTG	No significant effect	No dosage adjustment necessary
	EVG/COBI	No significant effect	No dosage adjustment necessary
	EVG plus PI/RTV	No significant effect	No dosage adjustment necessary
	RAL	RAL AUC ↑ 212% and C_{min} ↑ 46%	No dosage adjustment necessary
Anticoagulants and antiplatelets			
Apixaban	EVG/COBI EVG plus PI/RTV	↑apixaban expected	*Avoid concomitant use*

(continued)

Table 7.3 (continued)

Concomitant drug	Integrase inhibitor (INI)	Effect	Dosing recommendations and clinical comments
Dabigatran	EVG/COBI EVG plus PI/RTV	↑dabigatran potential	No dosage adjustment for dabigatran if CrCl >50 mL/min. *Avoid co-administration if CrCl < 50 mL/min*
Edoxaban	EVG/COBI EVG plus PI/RTV	↑edoxaban expected	*Avoid concomitant use*
Rivaroxaban	EVG/COBI EVG plus PI/RTV	↑rivaroxaban expected	*Avoid concomitant use*
Ticagrelor	EVG/COBI EVG plus PI/RTV	↑ticagrelor expected	*Avoid concomitant use*
Vorapaxar	EVG/COBI EVG plus PI/RTV	↑vorapaxar expected	*Avoid concomitant use*
Warfarin	EVG/COBI EVG plus PI/RTV	Warfarin levels may be affected	Monitor INR and adjust warfarin dose accordingly
Anticonvulsants			
Carbamazepine	DTG	↓DTG expected	Consider alternative anticonvulsant
Phenobarbital Phenytoin	EVG/COBI	↑carbamazepine AUC 43% ↓EVG AUC 69% and ↓C_{min} >99% ↓COBI expected	*Contraindicated. Do not co-administer*
	EVG plus PI/RTV	↓EVG	Consider alternative anticonvulsant.
Ethosuximide	EVG/COBI EVG plus PI/RTV	↑ethosuximide potential	Clinically monitor for ethosuximide toxicities

Oxcarbazepine	DTG EVG/COBI EVG plus PI/RTV	↓ INSTI potential	Consider alternative anticonvulsant
Antidepressants/anxiolytics/antipsychotics Also see 'sedative/hypnotics' section below			
Bupropion	EVG/COBI	↑ or ↓ bupropion potential	Titrate bupropion dose based on clinical response
	EVG plus PI/RTV	↓ bupropion potential	Titrate bupropion dose based on clinical response
Buspirone	EVG/COBI EVG plus PI/RTV	↑buspirone potential	Initiate buspirone at a low dose. Dose reduction may be necessary
Fluvoxamine	EVG/COBI EVG plus PI/RTV	↑ or ↓ EVG potential	Consider alternative antidepressant or ARV
Quetiapine	EVG/COBI EVG plus PI/RTV	↑ quetiapine AUC expected	*Initiation of quetiapine in a patient receiving EVG/COBI:* Start quetiapine at the lowest dose and titrate up as needed monitor for quetiapine efficacy and adverse effects *Initiation of EVG/c in a patient receiving a stable dose of quetiapine:* Reduce quetiapine dose to 1/6 of the original dose, and closely monitor for quetiapine efficacy and adverse effects
SSRIs Citalopram Escitalopram	EVG/COBI	↑ SSRI potential	Initiate with lowest dose of SSRI and titrate dose carefully based on antidepressant response
Fluoxetine Paroxetine	EVG plus PI/RTV	↑ or ↓SSRI potential	Titrate SSRI dose based on clinical response
Sertraline	RAL	↔ RAL ↔ citalopram	No dosage adjustment necessary

(continued)

Table 7.3 (continued)

Concomitant drug	Integrase inhibitor (INI)	Effect	Dosing recommendations and clinical comments
TCAs Amitriptyline Desipramine Doxepin Imipramine Nortriptyline	EVG/COBI	↑Desipramine AUC (65%)	Initiate with lowest dose of TCA and titrate dose carefully
	EVG plus PI/RTV	↑ TCA expected	Initiate with lowest dose of TCA and titrate dose carefully based on antidepressant response and/or drug levels
Trazodone	EVG/COBI EVG plus PI/RTV	↑ trazodone potential	Initiate with lowest dose of trazodone and titrate dose carefully
Antifungals			
Isavuconazole	EVG/COBI	↑ isavuconazole expected ↑ EVG and COBI potential	If co-administered, consider monitoring isavuconazole concentrations and assess virologic response
	EVG plus PI/RTV	Changes in isavuconazole and EVG potential	If co-administered, consider monitoring isavuconazole concentrations and assessing virologic response
Itraconazole	EVG/COBI	↑ itraconazole expected ↑ EVG and COBI potential	Consider monitoring itraconazole level to guide dosage adjustments. High itraconazole doses (>200 mg/day) are not recommended unless dose is guided by itraconazole levels.
	EVG plus PI/RTV	↑ EVG potential	Consider monitoring itraconazole level to guide dosage adjustments. Doses >200 mg/day are not recommended with PI/RTV, ATV/COBI, or DRV/COBI unless dosing is guided by itraconazole levels
Posaconazole	EVG/COBI	↑ EVG and COBI potential ↑ posaconazole potential	If co-administered, monitor posaconazole concentrations
	EVG plus PI/RTV	↑ EVG potential	If co-administered, consider monitoring posaconazole concentrations. Monitor for PI adverse effects

Voriconazole	EVG/COBI	↑ voriconazole expected ↑ EVG and COBI potential	Risk/benefit ratio should be assessed to justify use of voriconazole. If administered, consider monitoring voriconazole level. Adjust dose accordingly
	EVG plus PI/RTV	Changes in voriconazole and EVG potential	Do not co-administer voriconazole and RTV or COBI unless benefit outweighs risk. If co-administered, consider monitoring voriconazole concentration and adjust dose accordingly
Antimycobacterials			
Clarithromycin	EVG/COBI	↑clarithromycin potential ↑ COBI potential	CrCl 50–60 mL/min: Reduce clarithromycin dose by 50% CrCl <50 mL/min: EVG/COBI is not recommended
Rifabutin	DTG	Rifabutin (300 mg once daily):↓DTG C_{min} (30%)	No dosage adjustment necessary
	EVG/COBI	Rifabutin 150 mg every other day with EVG/COBI once daily compared to Rifabutin 300 mg once daily alone: ↔ rifabutin AUC 25-o-desacetyl-rifabutin AUC ↑ 625% EVG AUC ↓ 21%, C_{min} ↓ 67%	*Do not co-administer*
	EVG plus PI/RTV	↔ EVG ↔ rifabutin AUC 25-O-desacetyl-rifabutin AUC ↑ 951%	Rifabutin 150 mg once daily or 300 mg three times a week. Monitor for antimycobacterial activity and consider therapeutic drug monitoring PK data reported in this table are results from healthy volunteer studies. Lower rifabutin exposure has been reported in HIV-infected patients than in the healthy study participants
RAL		↑RAL AUC (19%) and ↓C_{min} (20%)	No dosage adjustment necessary

(continued)

Table 7.3 (continued)

Concomitant drug	Integrase inhibitor (INI)	Effect	Dosing recommendations and clinical comments
Rifampin	DTG	*Rifampin:* ↓DTG AUC (54%), ↓C_{min} (72%) *rifampin:* ↑DTG AUC (33%), ↑C_{min} (22%)	DTG 50 mg BID (instead of 50 mg once daily) for patients without suspected or documented INIs mutation *Alternative to rifampin should be used in patients with certain suspected or documented INI-associated resistance substitutions Consider using rifabutin*
	EVG/COBI EVG plus PI/RTV	Significant ↓EVG and COBI expected	*Do not co-administer*
	RAL	*RAL 400 mg:* ↓RAL AUC (40%), ↓C_{min} (61%) *Compared with RAL 400 mg BID alone, rifampin with RAL 800 mg BID:* ↑RAL AUC (27%), ↓C_{min} (53%)	*Dose:* RAL 800 mg BID Monitor closely for virologic response or consider using rifabutin as an alternative rifamycin
Rifapentine	DTG	Significant ↓DTG expected	*Do not co-administer*
	EVG/COBI EVG plus PI/RTV	Significant ↓ EVG and COBI expected	*Do not co-administer*
	RAL	*Rifapentine 600 mg once daily:* ↓RAL C_{min} (41%) *Rifapentine 900 mg once weekly:* RAL AUC ↑ 71%, C_{min} ↓ 12%	*Do not co-administer with once-daily rifapentine* for once-weekly rifapentine, use standard doses

Cardiac medications

Antiarrhythmics Amiodarone Bepridil Digoxi Disopyramide Dronedarone FlecainideSystemic Lidocaine Mexiletine Propafenone Quinidine	EVG/COBI	↑ antiarrhythmics potential digoxin C_{max} ↑ 41% and AUC no significant change	Use antiarrhythmics with caution. Therapeutic drug monitoring, if available, is recommended for antiarrhythmics
	EVG plus PI/RTV	↑ antiarrhythmics potential	Not recommended
Bosentan	EVG/COBI	↑ bosentan potential	*In patients on EVG/COBI ≥ 10 days:* Start bosentan at 62.5 mg once daily or every other day based on individual tolerability *In patients on bosentan who require EVG/COBI:* Stop bosentan ≥36 h before EVG/COBI initiation. At least 10 days after initiation of EVG/COBI, resume bosentan at 62.5 mg once daily or every other day based on individual tolerability
	EVG plus PI/RTV	↑ bosentan potential	In patients on a PI (other than unboosted ATV) >10 days: Start bosentan at 62.5 mg once daily or every other day In patients on Bosentan who require a PI (other than Unboosted ATV): Stop bosentan ≥36 h before PI initiation and 10 days after PI initiation restart bosentan at 62.5 mg once daily or every other day When switching between COBI and RTV: Maintain same bosentan dose
Beta-blockers (e.g., metoprolol, timolol)	EVG/COBI EVG plus PI/RTV	↑ beta-blockers potential	Beta-blocker dose may need to be decreased; adjust dose based on clinical response consider using beta-blockers that are not metabolized by CYP450 enzymes (e.g., atenolol, labetalol, nadolol, sotalol)

(continued)

Table 7.3 (continued)

Concomitant drug	Integrase inhibitor (INI)	Effect	Dosing recommendations and clinical comments
CCBs	EVG/Cobi EVG plus PI/RTV	↑ CCBs potential	Use with caution. Titrate CCB dose and monitor closely. ECG monitoring is recommended when CCB used with ATV and SQV
Dofetilide	DTG	↑ dofetilide expected	*Do not co-administer*
Eplerenone	EVG/COBI EVG plus PI/RTV	↑ eplerenone expected	*Contraindicated. Do not co-administer*
Ivabradine	EVG/COBI EVG plus PI/RTV	↑ ivabradine expected	*Contraindicated. Do not co-administer*
Corticosteroids			
Dexamethasone (systemic)	EVG/COBI EVG plus PI/RTV	↓ EVG and COBI potential ↓ EVG potential	Use systemic dexamethasone with caution. Monitor virologic response to ART. Consider alternative corticosteroid
Fluticasone Inhaled/intranasal	EVG/COBI EVG plus PI/RTV	↑ fluticasone potential	Co-administration may result in adrenal insufficiency and Cushing's syndrome. Consider alternative therapy (e.g., beclomethasone), particularly for long-term use
Methylprednisolone, prednisolone, triamcinolone Local injections, including intra-articular, epidural, intra-orbital	EVG/COBI EVG plus PI/RTV	↑ glucocorticoids expected	Co-administration may result in adrenal insufficiency and Cushing's syndrome. *Do not co-administer*

Hepatitis C direct-acting antivirals

Daclatasvir	DTG	daclatasvir ↔	No dosage adjustment necessary.
	EVG/COBI	↑ daclatasvir	Decrease daclatasvir dose to 30 mg once daily
	EVG plus PI/RTV	↑ daclatasvir expected	Decrease daclatasvir dose to 30 mg once daily, regardless of which PI/RTV is used, except for TPV/RTV. *Do not co-administer EVG plus TPV/r with daclatasvir*
	RAL	No data	No dosage adjustment necessary
Dasabuvir plus Ombitasvir/ paritaprevir/r	DTG	No data	No dosing recommendations at this time
	EFG plus PI/RTV EVG/COBI	No data	*Do not co-administer*
	RAL	RAL AUC ↑ 134%	No dosage adjustment necessary
Elbasvir/grazoprevir	DTG	Elbasvir ↔ Grazoprevir ↔ DTG ↔	No dosage adjustment necessary
	EVG plus PI/RTV		Contraindicated. Do not co-administer. May increase the risk of ALT elevations due to a significant increase in grazoprevir plasma concentrations caused by OATP1B1/3 inhibition
	EVG/COBI	↑ elbasvir, grazoprevir expected	*Co-administration is not recommended*
	RAL	Elbasvir ↔ Grazoprevir ↔ RAL ↔ with elbasvir RAL AUC ↑ 43% with grazoprevir	No dosage adjustment necessary

(continued)

Table 7.3 (continued)

Concomitant drug	Integrase inhibitor (INI)	Effect	Dosing recommendations and clinical comments
Ledipasvir/sofosbuvir	EVG/COBI/TDF/FTC	↑ TDF and ↑ ledipasvir expected	*Do not co-administer*
	EVG/COBI/TAF/FTC	↔ EVG/COBI/TAF/FTC expected	No dosage adjustment necessary
	EVG plus PI/RTV	↔ EVG expected	No dosage adjustment necessary
			Co-administration of ledipasvir/sofosbuvir with TDF and a PI/RTV results in increased exposure to TDF. The safety of the increased TDF exposure has not been established. Consider alternative HCV or ARV drugs to avoid increased TDF toxicities. If co-administration is necessary, monitor for TDF-associated adverse reactions
	DTG RAL	↔ DTG or RAL	No dosage adjustment necessary
Simeprevir	DTG	↔ DTG expected	No dosage adjustment necessary
	EVG/COBI	↑ simeprevir expected	*Co-administration is not recommended*
	EVG plus PI/RTV	↔ EVG expected	*Co-administration is not recommended*
	RAL	No significant effect	No dosage adjustment necessary
Sofosbuvir	All INSTIs	No significant effect expected	No dosage adjustment necessary
Herbal products			
St. John's wort	DTG	↓ DTG potential	*Do not co-administer*
	EVG/COBI EVG plus PI/RTV	↓ EVG and COBI potential	*Do not co-administer*

Hormonal contraceptives

Hormonal contraceptives			
Norgestimate/ Ethinyl estradiol	RAL	No clinically significant effect	No dosage adjustment necessary
	DTG	No significant effect	No dosage adjustment necessary
	EVG/COBI	Norgestimate AUC, C_{max}, and C_{min} ↑ >2-fold Ethinyl estradiol AUC ↓ 25% and C_{min}↓ 44%	The effects of increases in progestin (norgestimate) are not fully known and can include insulin resistance, dyslipidemia, acne, and venous thrombosis. Weigh the risks and benefits of the drug, and consider alternative contraceptive method
	EVG plus PI/RTV	↔EVG	Recommendations when used with PI/RTV Refer to Table 7.2

HMG-CoA reductase inhibitors

Atorvastatin	EVG/COBI	↑ atorvastatin potential	Titrate statin dose slowly and use the lowest dose potential
	EVG plus PI/RTV	↔ EVG expected	Titrate atorvastatin dose carefully and use lowest dose necessary
Lovastatin	EVG/COBI	Significant ↑ lovastatin expected	*Contraindicated. Do not co-administer*
	EVG plus PI/RTV		
Pitavastatin	EVG/COBI	No data	No dosage recommendation
Pravastatin	EVG plus PI/RTV	↔ EVG expected	Use lowest possible starting dose of pravastatin with careful monitoring
Rosuvastatin	EVG/COBI	Rosuvastatin AUC ↑ 38% and C_{max} ↑ 89%	Titrate statin dose slowly and use the lowest dose potential
	EVG plus PI/RTV	↔ EVG expected	Titrate rosuvastatin dose carefully and use the lowest necessary dose while monitoring for toxicities
Simvastatin	EVG/COBI	Significant ↑ simvastatin expected	*Contraindicated. Do not co-administer*
	EVG plus PI/RTV		

(continued)

Table 7.3 (continued)

Concomitant drug	Integrase inhibitor (INI)	Effect	Dosing recommendations and clinical comments
Immunosuppressants			
Cyclosporine Everolimus Sirolimus Tacrolimus	EVG/COBI EVG plus PI/RTV	↑ immunosuppressant potential	Initiate with an adjusted immunosuppressant dose to account for potential increased concentration and monitor for toxicities. Therapeutic drug monitoring of immunosuppressant is recommended. Consult with specialist as necessary
Narcotics/treatment for opioid dependence			
Buprenorphine Sublingual/buccal/implant	EVG/COBI	Buprenorphine AUC ↑ 35%, C_{max} ↑ 12%, and C_{min} ↑ 66% Norbuprenorphine AUC ↑ 42%, C_{max} ↑ 24%, and C_{min} ↑ 57%	No dosage adjustment necessary. Clinical monitoring is recommended When transferring buprenorphine from transmucosal to implantation, monitor to ensure buprenorphine effect is adequate and not excessive
	EVG plus PI/RTV	↔ EVG expected	Re titrate buprenorphine dose using the lowest initial dose. Dose adjustment of buprenorphine may be needed. It may be necessary to remove implant and treat with a formulation that permits dose adjustments. Clinical monitoring is recommended
	RAL	No significant effect observed (sublingual) or expected (implant)	No dosage adjustment necessary
Methadone	DTG	No significant effect	No dosage adjustment necessary
	EVG/COBI	No significant effect	No dosage adjustment necessary
	EVG plus PI/RTV	↓ methadone	Opioid withdrawal unlikely but may occur. Dosage adjustment of methadone is not usually required. Monitor for opioid withdrawal and increase methadone dose as clinically indicated
	RAL	No significant effect	No dosage adjustment necessary
Neuroleptics			

Perphenazine Risperidone Thioridazine	EVG/COBI	↑ neuroleptic potential	Initiate neuroleptic at a low dose. Decrease in neuroleptic dose may be necessary
PDE5 inhibitors			
Avanafil	EVG/COBI EVG plus PI/RTV	No data	*Co-administration is not recommended*
Sildenafil	EVG/COBI	↑ sildenafil expected	*For treatment of erectile dysfunction:* Start with sildenafil 25 mg every 48 h and monitor for adverse effects of sildenafil *For treatment of PAH:* *Contraindicated*
Tadalafil	EVG/COBI EVG plus PI/RTV	↑ tadalafil expected	*For treatment of erectile dysfunction:* Start with tadalafil 5-mg dose and do not exceed a single dose of 10 mg every 72 h. Monitor for adverse effects of tadalafil *For treatment of PAH* *In patients on EVG/COBI > 7 days:* Start with tadalafil 20 mg once daily and increase to 40 mg once daily based on tolerability *In patients on tadalafil who require EVG/COBI:* Stop tadalafil ≥24 h before EVG/COBI initiation. Seven days after EVG/COBI initiation, restart tadalafil at 20 mg once daily, and increase to 40 mg once daily based on tolerability
Vardenafil	EVG/COBI EVG plus PI/RTV	↑ vardenafil expected	Start with vardenafil 2.5 mg every 72 h and monitor for adverse effects of vardenafil
Sedative/hypnotics			

(continued)

Table 7.3 (continued)

Concomitant drug	Integrase inhibitor (INI)	Effect	Dosing recommendations and clinical comments
Clonazepam Clorazepate Diazepam Estazolam Flurazepam	EVG/COBI EVG plus PI/RTV	↑ benzodiazepines potential	Dose reduction of benzodiazepine may be necessary. Initiate with low dose and clinically monitor. Consider alternative benzodiazepines to diazepam, such as lorazepam, oxazepam, or temazepam
Midazolam Triazolam	DTG	*With DTG 25 mg*: Midazolam AUC ↔	No dosage adjustment necessary
	EVG/COBI EVG plus PI/RTV	↑ midazolam expected ↑ triazolam expected	*Do not co-administer triazolam or oral midazolam and EVG/COBI or (EVG plus PI)*. Parenteral midazolam can be used with caution in a closely monitored setting. Consider dose reduction, especially if more than one dose is administered
Suvorexant	EVG/COBI EVG plus PI/RTV	↑ suvorexant expected	*Co-administration is not recommended*
Zolpidem	EVG/COBI EVG plus PI/RTV	↑ zolpidem expected	Initiate zolpidem at a low dose. Dose reduction may be necessary
Miscellaneous drugs			
Colchicine	EVG/COBI EVG plus PI/RTV	↑ colchicine expected	*Do not co-administer in patients with hepatic or renal impairment*. *For treatment of gout flares:* Colchicine 0.6 mg for 1 dose, followed by 0.3 mg 1 h later. Do not repeat dose for at least 3 days. *For prophylaxis of gout flares:* If original dose was colchicine 0.6 mg BID, decrease to colchicine 0.3 mg once daily. If regimen was 0.6 mg once daily, decrease to 0.3 mg every other day. *For treatment of familial Mediterranean fever:* Do not exceed colchicine 0.6 mg once daily or 0.3 mg BID

Flibanserin	EVG/ COBIEVG plus PI/ RTV	↑ flibanserin expected	*Contraindicated. Do not co-administer*
Metformin	DTG	*DTG 50 mg once daily plus metformin 500 mg BID:* Metformin AUC ↑ 79%, C_{max} ↑ 66% *DTG 50 mg BID plus metformin 500 mg BID:* Metformin AUC↑ 2.4 fold, C_{max} ↑2 fold	Limit metformin dose to no more than 1000 mg per day. When starting/stopping DTG in patient on metformin, dose adjustment of metformin may be necessary to maintain optimal glycemic control and/or minimize GI symptoms
Polyvalent cation supplements Mg, Al, Fe, Ca, Zn, including multivitamins with minerals Note: Please refer to the acid reducers section in this table for recommendations on use with Al-, Mg-, and Ca-containing antacids	All INSTIs	↓ INSTI potential DTG ↔ when administered with Ca or Fe supplement simultaneously with food	If co-administration is necessary, give INSTI at least 2 h before or at least 6 h after supplements containing polyvalent cations, including but not limited to the following products: Cation-containing laxatives; Fe, ca, or mg supplements; and sucralfate. Monitor for virologic efficacy. DTG and supplements containing ca or Fe can be taken simultaneously with food. Many oral multivitamins also contain varying amounts of polyvalent cations; the extent and significance of chelation is unknown
Salmeterol	EVG/ COBIEVG plus PI/ RTV	↑ salmeterol potential	*Do not co-administer* due to potential increased risk of salmeterol-associated cardiovascular events

Adapted from DHHS Table 19b

Key to acronyms: Al aluminum, *ART* antiretroviral therapy, *ARV* antiretroviral, *ATV/r* atazanavir/ritonavir, *AUC* area under the curve, *BID* twice daily, *Ca* calcium, *CaCO3* calcium carbonate, *CCB* calcium channel blocker, *Cmax* maximum plasma concentration, *Cmin* minimum plasma concentration, *c or COBI* cobicistat, *CrCl* creatinine clearance, *CYP* cytochrome P, *DTG* dolutegravir, *EVG* elvitegravir, *EVG/COBI* elvitegravir/cobicistat, *Fe* iron, *GI* gastrointestinal, *INR* international normalized ratio, *INSTI* integrase strand transfer inhibitor, *Mg* magnesium, *PAH* pulmonary arterial hypertension, *PI* protease inhibitor, *PI/RTV* ritonavir-boosted protease inhibitor, *PPI* proton pump inhibitor, *RAL* raltegravir, *SQV/r* saquinavir/ritonavir, *SSRI* selective serotonin reuptake inhibitor, *TCA* tricyclic antidepressant, *Zn* zinc

regimens with cobicistat (COBI, C), emtricitabine (FTC, F), and either tenofovir disoproxil fumarate (E/C/F/TDF, Stribild™) or tenofovir alafenamide (E/C/F/TAF, Genvoya™). Since its plasma half-life alone is short, it must be co-administered with a pharmacoenhancer like COBI or RTV to prolong the half-life to approximately 9.5 h to permit once daily dosing [9].

Elvitegravir is metabolized primarily via CYP3A4 with a minor component via UDP-glucuronosyltransferase (UGT) 1A1 and 1A3. As stated earlier, EVG requires pharmacoenhancement by COBI and the presence of food to improve its systemic exposure [10]. Thus, EVG-containing regimens are prone to complex drug interactions via CYP3A similar to those observed with PI and NNRTI classes. EVG is a weak inhibitor of P-gp transport; however, COBI and RTV are both clinically relevant P-gp inhibitors. Additionally, EVG and COBI are substrates and inhibitors of the transporters breast cancer resistant protein (BCRP), organic anionic transporter-B1 (OATP1B1), and OATP1B3 [11, 12]. The use of TAF as an alternative to TDF-based fixed-dose combination tablets of EVG does not generally change the overall drug interaction liability of EVG FDC tablets, rather improved reduction in the incidence of TDF-related renal and bone adverse effects has been reported in some studies.

Elvitegravir has a number of clinically relevant drug-drug interactions. Elvitegravir is contraindicated when used with drugs that are dependent on CYP3A-mediated clearance with a narrow therapeutic index (Table 7.3). In particular, EVG is contraindicated when used together with rifamycins like rifampin and rifapentine. Combining EVG with the potent CYP3A inducers, rifapentine or rifampin, will reduce EVG levels below the threshold required for HIV suppression. When elvitegravir has to be used in combination with a rifamycin, rifabutin may be a good alternative in this scenario since this is a weak CYP3A inducer. In healthy volunteers, rifabutin administered at an adjusted dose of 150 mg every other day significantly reduced elvitegravir C_{min} by 67% with concurrent COBI administration [13]. However, doubling the EVG dose to 300 mg once daily and switching the pharmacokinetic booster to low dose RTV mitigate this interaction and permit co-administration with dose-adjusted rifabutin [14]. However, increased monitoring for rifabutin-associated adverse effects is required as a result of increased rifabutin and desacetyl-rifabutin metabolite concentrations with EVG/r co-administration. Other potent CYP3A enzyme inducers should not be co-administered with EVG, regardless if COBI, within fixed dose tablet regimens, or RTV, with boosted PI regimens, is used for pharmacoenchancement.

In treatment-experienced subjects receiving RTV-boosted PIs with atazanavir or lopinavir, the dose of EVG must be reduced as a result of potent UGT1A1 inhibition by these PIs. In healthy volunteers, EVG exposures increased up to 2.8-fold when co-administered with ATV 300 mg/RTV 100 mg daily and increased by 2.3-fold when co-administered with LPV 400 mg/RTV 100 mg twice daily [15]. Thus, concomitant use with these RTV-boosted protease inhibitors requires a dose reduction in EVG from 150 to 85 mg once daily.

Overall, EVG-containing regimens are prone to complex CYP3A-mediated drug interactions similar to those observed with PI and NNRTI classes but unlike other members of the integrase class. EVG is contraindicated for use with drugs dependent on CYP3A-mediated clearance possessing a narrow therapeutic index and drugs with potent CYP3A induction potential. Healthcare providers are advised to evaluate the potential for drug-drug interactions when prescribing EVG-containing regimens due to potent enzyme and transporter inhibition by RTV and COBI.

7.2.2.3 Dolutegravir

Dolutegravir (DTG) is a potent, second-generation INI with antiviral activity against raltegravir- and elvitegravir-resistant virus. It exhibits low to moderate variability in plasma concentrations and does not require COBI or RTV boosting or food for optimal drug exposures. DTG is available as a single entity and as a component of a STR (Triumeq™) which is combined with two nucleosides abacavir (ABC) and lamivudine (3TC) (Table 7.1).

DTG is primarily metabolized by UGT 1A1 with a minor contribution from CYP3A4 and has a favorable drug interaction profile similar to raltegravir, which also undergoes glucuronidation. Although DTG is a substrate for UGT1A3, UGT1A9, and the efflux transporters P-gp and BCRP, it has a low potential to cause drug-drug interactions. DTG does not induce or inhibit CYP or UGT enzymes. Additionally, it has relatively little impact on major efflux transporters such as P-gp or MRPs [16]. DTG has been shown to potently inhibit renal organic cation transporter, OCT2 and MATE-1, which results in a reduction in creatinine secretion in the renal proximal tubule. This causes a benign increase in serum creatinine concentrations which consequently causes an artificial decline in calculated glomerular filtration rate.

Dolutegravir exposures can be impacted when co-administered with potent CYP3A inducers such as tipranavir, fosamprenavir, efavirenz, and rifampin. When co-administered with moderate to strong UGT1A1 and CYP3A inducers, DTG exposures can be reduced requiring either dose adjustment or employment of mitigating strategies. Rifampin, a potent inducer of CYP and efflux transporters, can significantly decrease dolutegravir plasma concentrations. This interaction can be mitigated by simply increasing the dolutegravir dosage to 50 mg BID. Similar to raltegravir, but unlike elvitegravir, dolutegravir may be administered with rifabutin without dose adjustment [17].

NNRTIs like efavirenz and etravirine are known CYP3A inducers, and both agents are able to reduce DTG plasma concentrations. Efavirenz can significantly reduce DTG plasma C_{min} by 75%; therefore, the dose of DTG must be increased to 50 mg twice daily to overcome this interaction [18]. Surprisingly, etravirine dramatically reduced plasma DTG C_{min} by 88% likely due to combined UGT and CYP3A induction; however, this interaction can be attenuated with the addition of specific RTV-boosted PIs, either DRV/RTV or LPV/RTV [19]. Therefore, no dose adjustment in DTG is necessary when co-administered with etravirine and RTV-boosted PIs, DRV/RTV and LPV/RTV. Similarly, when DTG is combined with enzyme-inducing PIs like tipranavir or fosamprenavir, even the addition of RTV cannot mitigate the enzyme-inducing effect; therefore, DTG requires dose adjustment when co-administered with these agents.

When DTG is combined with atazanavir (ATV), a potent competitive inhibitor of UGT-mediated metabolism, plasma DTG C_{min} increased by 2.8-fold and AUC by 1.9-fold compared with ATV alone. However, when RTV was added to ATV, the increase in DTG exposure was attenuated possibly via RTV-mediated induction of UGT metabolism [19, 20]. Additionally, there were no safety issues as a result of higher DTG concentrations with atazanavir co-administration in clinical trials; therefore, no dose adjustment of DTG is required.

7.2.3 Drug Interactions to Improve Bioavailability and Dosing

7.2.3.1 Pharmacoenhancement of PIs

As previously mentioned, RTV is a potent CYP3A4 inhibitor found in high levels in the gut wall and liver [21]. These inhibitory properties can be exploited to increase plasma concentrations of co-administered PIs with RTV. The resultant effect is increased PI exposure leading to improved antiviral activity while permitting a lower dosage and dosing frequency of the co-administered PI and potentially reducing the incidence of adverse effects.

RTV is ideally suited to boosting, as it acts both on first-pass metabolism and on hepatic clearance. Thus, CYP3A4 and P-gp found in the intestinal walls can be inhibited using RTV and thus overcome poor intrinsic bioavailability. When RTV 200 mg was co-administered with SQV, a three- to eightfold increase SQV exposure was realized when compared with unboosted SQV [22]. For PIs with good bioavailability but a short half-life, RTV-mediated inhibition of CYP3A4 can decrease hepatic clearance and therefore extend the half-life of a co-administered PI.

Ideally, boosting should maintain drug concentrations that are within the therapeutic window. This means that the C_{min} should fall within a zone above the minimum effective concentration for viral inhibition, while C_{max} should fall below the threshold for toxicity. If an agent has a narrow therapeutic window, boosting is likely to increase the frequency or magnitude of adverse effects, even if boosting has a relatively minor effect on C_{max}. For example, IDV has good bioavailability, but its relatively short half-life requires frequent dosing to prevent suboptimal plasma levels, a shortcoming that can be overcome by boosting with RTV [23]. Although the increase in IDV C_{max} from boosting is small, it is apparently sufficient to substantially increase risk of nephrolithiasis [24]. In contrast, addition of RTV to a PI with low bioavailability, like SQV or LPV, results in greatly increased C_{max} without increased toxicity, probably because these drugs have intrinsically lower toxicity [25].

7.2.3.2 Cobicistat

Cobicistat (COBI, Tybost™) is a structural analog of RTV developed as an alternative pharmacokinetic boosting agent. Although RTV co-administration can improve the pharmacokinetic profile of a co-administered PI, low-dose RTV is also associated with many metabolic side effects including hyperlipidemia and hypercholesterolemia. Advantages of COBI include its lack of antiretroviral activity at the dosage employed, its lower potential to alter lipid metabolism compared with RTV, and its availability as FDC with PIs such as darunavir (DRV) and atazanavir (ATV) and the integrase inhibitor, elvitegravir (EVG), to block CYP3A-mediated metabolism of the parent compound to enable once a day dosing. COBI and RTV share some similarities in pharmacokinetics. For example, COBI has a plasma $t_{1/2}$ of approximately 3 h, which is similar to that found for RTV ($t_{1/2}$ 3–5 h), and COBI inactivates CYP3A enzymes in a time- and concentration-dependent manner similar to RTV. COBI and RTV have both been shown to inhibit the activity of multiple transporters. In vitro, COBI is an inhibitor of Pgp, BCRP, organic anion-transporting polypeptide 1B1 and 1B3 (OATP 1B1 and 1B3), and renal tubular cell drug transporters like multidrug and toxin extrusion protein 1 (MATE-1) and organic cation transporter 1 (OCT1) [26].

In general, COBI and RTV are considered to be equipotent in CYP3A inhibition potential and therefore can be used interchangeably to boost systemic exposures of CYP3A substrates. However, there are notable differences in their ability to inhibit or induce other CYP isoenzymes and glucuronyltransferase activity which can make it difficult to predict the magnitude and direction of drug-drug interactions when using RTV or COBI with concomitant drugs that are metabolized by multiple CYPs or that undergo glucuronidation. Moreover, given that RTV is co-formulated with various other PIs and COBI is co-formulated with EVG, the parent PI, EVG, and the choice of PK boosting agent will all contribute to the net interaction observed with a concomitantly administered agent. COBI is distinct from RTV in that it does not inhibit CYP2C8, is considered a weak 2D6 inhibitor, and does not induce CYP isoenzymes 1A2, 2B6, 2C9, and 2C19 or glucuronyltransferase activity [27]. In addition, other drugs may impact COBI exposures. For example, combining COBI 150 mg BID with tipranavir 500 mg BID resulted in a significantly lower COBI C_{max} and a reduction in COBI $t_{1/2}$ to 2.2 h [13, 28]. These results further support the observation that tipranavir can significantly induce metabolic enzymes leading to enhanced COBI clearance. Therefore, when COBI is combined with tipranavir, a dosage increase of COBI to 200 mg BID may be necessary; however, there is no recommended TPV/COBI dosing regimen approved in current product labeling [26].

Regimens containing 150 mg of cobicistat and darunavir, atazanavir, and elvitegravir are considered to provide bioequivalent exposures compared to these same regimens containing 100 mg of RTV. Drug interaction studies comparing COBI vs RTV-containing regimens and third agents are limited and based on extrapolation of interaction data with RTV. Relative bioavailability studies have been conducted to compare pharmacokinetic exposures of concomitantly administered PIs and COBI versus RTV. In healthy volunteers receiving DRV 800 mg/COBI 150 mg, DRV C_{max} and AUC_{24} were comparable to those receiving 800 mg DRV/100 mg RTV, as the geometric mean ratio was 1.03, 0.694, and 1.02 for C_{max}, C_{tau}, and AUC_{tau}, respectively. However, both C_0 and C_{24} were approximately 30% lower when DRV was combined with 150 mg COBI as compared to 100 mg RTV [29]. However, DRV AUC_{0-12}, C_{max}, and C_{tau} were comparable between the two pharmacoenhancers when DRV/COBI 600/150 mg BID was compared to DRV/RTV 600/100 BID, yet the use of DRV/COBI BID is not considered interchangeable and is not recommended per product labeling [26]. When either EVG or etravirine was added to DRV/COBI 600/150 BID, no changes in DRV PK were noted. Similarly, bioequivalent exposures were demonstrated between ATV 300 mg/COBI 150 mg and ATV 300 mg/RTV 100 mg, as geometric mean ratios (GMR) for C_{max}, C_{tau}, and AUC were 0.923, 0.976, and 1.01, respectively, following a meal in healthy volunteers.

7.3 Issues with ARV Drug-Drug Interactions

Specific interactions with PIs and INIs and concomitant medications are summarized in Tables 7.2 and 7.3. For additional information on interactions with PIs and INIs with antifungals, antimalarials, antimycobacterials, and hepatitis C agents, the reader is referred to Chaps. 12, 15, 17, and 18, respectively.

7.3.1 Interactions with pH-Altering Agents and Polyvalent Cations

Drug-drug interactions exist between ARVs that require low gastric pH for absorption and drugs that raise gastric pH such as proton pump inhibitors (PPI), histamine-2 (H_2) receptor blockers, and antacids. In one study involving 200 HIV patients, 88% reported taking a PPI, H_2-blocker, or antacid, individually or in combination [30]. More importantly, 95% used over-the-counter agents alone or in combination with a prescription drug, underscoring the importance of asking patients about non-prescription medications.

HIV PIs are susceptible to drug-drug interactions with agents that alter gastric pH. When PIs are co-administered with antacids, reductions in atazanavir (ATV), tipranavir (TPV), and amprenavir/fosamprenavir exposure have been noted; therefore, it is recommended to stagger PI doses 2 h before or 1–2 h after antacid dosing. Concomitant H_2 antagonists reduced unboosted atazanavir AUC by 41% with famotidine and reduced amprenavir AUC by 30% with ranitidine. Therefore, it is recommended that H_2 antagonists should not exceed 20 mg of famotidine or its equivalent with unboosted ATV or 40 mg of famotidine or its equivalent with boosted ATV and that a 2–10-h dose separation strategy is employed [31]. When proton pump inhibitors are used together with PIs, there is generally no significant interaction, except when using atazanavir; therefore, it is not recommended to use PPIs in combination with atazanavir.

HIV INIs demonstrate varying propensities of interaction with concomitant cation-containing products, such as aluminum, magnesium, and calcium-containing antacids, which may be a result of chelation and/or pH-related changes in drug absorption. Cation interactions arise from the ability of INIs to form a complex with divalent cations in the active site of the integrase enzyme, which can effectively reduce the amount of drug available for absorption. Concurrent or staggered administration of aluminum- and/or magnesium-containing antacids is not recommended with RAL because these agents can reduce RAL exposure by greater than 50%. Raltegravir exhibits pH-dependent solubility which may explain why C_{max} and AUC were not significantly altered with concurrent magnesium antacid administration, as increasing pH improves RAL solubility initially; however, the net impact is an overall reduction in RAL exposures, likely a consequence of large quantities of metal cations available for chelation [32–34]. However, the impact on RAL exposures is product dependent as concurrent calcium carbonate antacid administration only modestly reduced RAL C_{min} by 32%. A 2-h dose separation is required between EVG and concomitant aluminum- or magnesium-containing antacid administration to avoid a 40–50% reduction in EVG AUC with simultaneous administration. Dolutegravir requires a 2- to 6-h dose separation with concurrent polyvalent cation administration as simultaneous administration of magnesium- or aluminum-containing antacids reduced DTG AUC by 74%. Alternatively, products containing iron and calcium-containing products may be co-administered at the same time as DTG if also administered with food [35].

7.3.2 Interactions with Anticoagulants

PI interactions with anticoagulants or antiplatelets have been well-established. The co-administration of thrombin inhibitors such as apixaban, dabigatran, edoxaban, and rivaroxaban with any PI will increase thrombin inhibitor concentrations; thus, it is recommended these agents be avoided with PIs. The only exception is dabigatran with PI, when the creatinine clearance is greater than 50 mL/min where no dosage adjustment is required. Similarly, the thrombin receptor antagonist, vorapaxar, should be avoided with PIs. When PI/RTV is co-administered with warfarin, warfarin levels are expected to decrease, and thus titration of warfarin dosage may be necessary to attain targeted anticoagulation levels.

7.3.3   Interactions with Anticonvulsants

As expected, the co-administration of anticonvulsants and HIV PIs will lead to significant drug-drug interactions. In general, PIs cause an increase in levels of anticonvulsants such as carbamazepine and ethosuximide. In contrast, phenytoin and phenobarbital are potent inducers of CYP, and thus co-administration of these anticonvulsants with PIs will lead to reduced PI levels, and the combination may severely impact efficacy of both classes of drugs. It is recommended that other classes of ART such as integrase inhibitors be substituted or a change to a noninteracting anticonvulsant, such as levetiracetam, be considered. All currently marketed integrase inhibitors are substrates for CYP isoenzymes and/or undergo glucuronidation, and therefore anticonvulsants with potent enzyme induction potential, such as phenytoin, phenobarbital, carbamazepine, and oxcarbazepine, are not recommended with concomitant INIs.

7.3.4   Interactions with Steroids

Steroids are widely used immunosuppressive agents for the treatment of inflammation and allergic reactions. HIV patients receiving ART and steroid therapy should be aware of potential drug-drug interactions, particularly when immunosuppressive steroids such as dexamethasone, fluticasone, and methylprednisolone are utilized. Increased fluticasone systemic concentrations have been shown when inhaled fluticasone is co-administered with ART combinations containing COBI or RTV. Thus, chronic use of steroids in combination with RTV- or COBI-containing regimens may enhance the risk for adrenal insufficiency and Cushing's syndrome and is therefore not recommended.

7.3.5   Interactions with Oral Contraceptives

Hormonal contraceptives are frequently co-administered in HIV-infected populations, and drug interactions with various antiretroviral classes have been demonstrated. Hormonal contraceptives consist of steroids that are eliminated by phase I and II metabolic enzymes. Unboosted PI therapy typically results in slightly increased ethinyl estradiol or norethindrone concentrations partly from phase I enzyme inhibition. For example, ATV is a potent inhibitor of UGT that can increase ethinyl estradiol AUC by 48% and norethindrone AUC by 110%. However, many boosted PIs cause a reduction in ethinyl estradiol exposures, and alternative forms of contraception are recommended. RTV alters the expression of the phase II metabolic enzyme, uridine 5′-diphospho-glucuronosyltransferase (UGT).

PI-induced expression of phase II metabolism may partially explain why it is able to interact with nucleosides, steroidal compounds, methadone, and antipsychotic agents. Nucleosides such as abacavir and zidovudine are metabolized via glucuronidation, and thus co-administration of PIs with nucleosides can alter their plasma concentrations. Fortunately, nucleosides have wide therapeutic indices, so no dosage adjustments are necessary.

When PIs are combined with steroid compounds such as components of birth control pills like ethinyl estradiol (EE) and norethindrone (NE), their systemic concentrations can be modified requiring a dose alteration. When an alternative to estradiols is needed, depo-medroxyprogesterone acetate (DMPA) can be utilized. One study evaluating 65 HIV-infected women receiving ART demonstrated no significant interaction between DMPA and EFV, NVP, or NFV [36]. There was no evidence of ovulation, although the study was limited to assessment of interactions with EFV, NVP, and NFV.

Although PIs are most notable in their interaction with oral contraceptives, integrase inhibitors also utilize this pathway for elimination, and thus potential drug-drug interaction between integrase inhibitors and hormonal contraceptives should always be considered. Although both estrogen- and progestin-containing products are available in addition to progestin-only formulations, the progestin component is generally considered to be most important for contraceptive activity.

No significant interaction was seen with either RAL or DTG when co-administered with hormonal contraceptives. However, co-administration of EVG/COBI/FTC/TDF with a norgestimate (NGNM)- and ethinyl estradiol (EE)-containing oral contraceptive resulted in a significant increase in progestin concentrations and reduced EE concentrations. Mean plasma AUC of NGMN increased by 2.26-fold, while the mean plasma AUC of EE was lowered by 0.75-fold. However, increasing the amount of EE to compensate for lower EE concentrations observed with EVG/COBI/FTC/TDF would likely require a higher strength NGNM/EE tablet, further increasing the potential for NGNM-related adverse events. Therefore, caution should be exercised with concomitant administration of NGMN/EE and EVG/COBI/FTC/TDF, and non-hormonal forms of contraception may be considered. However, when EVG was combined with PI/RTV, no change in EVG concentration was noted.

7.3.6 Interactions with Opioid and Psychotropic Agents

Methadone concentrations can be decreased when co-administered with various antiretroviral agents. While EFV and NVP are prominent for their negative interactions with methadone, several PIs, including NFV and LPV/r, also significantly lower methadone levels. This interaction can lead to opioid withdrawal whereby an inability to adjust methadone doses could prompt patients to interrupt or discontinue ART.

When INI are co-administered with methadone, no alteration in methadone levels was observed. However, when INI is combined with PI/RTV, the presence of RTV can induce glucuronidation resulting in lower methadone levels. In this scenario, it is advised that signs and symptoms of opioid withdrawal be closely monitored. Alternative opioids such as buprenorphine can be considered. EVG in combination with COBI increased buprenorphine AUC by 35% and increased buprenorphine metabolite concentrations by 42%. However, despite these PK changes, no adjustment is recommended.

Trazodone is a CYP3A4 substrate, and interactions with IDV and RTV can raise trazodone concentrations. A study in 10 healthy volunteers demonstrated that boosting doses of RTV (200 mg BID for 2 days) led to a 2.4-fold increase in exposure to trazodone. Study participants experienced nausea, hypotension, and syncope [37]. These observations suggest that caution is warranted whenever PIs are used concurrently with trazodone.

7.3.7 Interactions with HMG-Coenzyme a Reductase Inhibitors (Statins)

Limited data are available characterizing ARV drug interactions with HMG-coenzyme A reductase inhibitors (statins). Statins are frequently prescribed in the treatment of dyslipidemias in the HIV population which can be associated to the use of certain ARVs such as protease inhibitors (PIs). In general, statins are predominately metabolized by CYP isoenzymes and can involve drug transport via OATP, P-gP, BCRP, and OAT3 [38]. The majority of PIs inhibit the metabolism of statins through potent CYP3A enzyme inhibition which can increase statin concentrations leading to a clinically significant risk of myopathy and rhabdomyolysis. All PIs, with the exception of fosamprenavir, inhibit the OATP transporter. The degree of interaction varies by PI and statin; however, the potential for statin inhibition is considered greatest with simvastatin and lovastatin followed by atorvastatin and rosuvastatin. Simvastatin and lovastatin are contraindicated with PIs to avoid risk of myopathy. RTV-boosted PIs (ATV/r, DRV/r, and LPV/r) have been documented to increase rosuvastatin AUC from 50% to 200% and C_{max} from 90% to 600%, resulting in cautious use and titrating with the lowest dose possible [39–41] which may be explained by PI-mediated inhibition of OATP transport as rosuvastatin is not metabolized by CYP3A to any significant extent or to inhibition of breast cancer resistance protein (BCRP). Overall, atorvastatin, rosuvastatin, or pravastatin may be considered in patients receiving PI-based regimens; however, dosing adjustments may still be required depending on the components of the ARV regimen. Pitavastatin and fluvastatin do not exhibit significant interactions with PIs and can generally be used without dose adjustments. No dose adjustment of statins is required when co-administered HIV integrase inhibitors, with the exception of elvitegravir co-administered with potent CYP3A inhibitors containing boosted PIs, RTV, or COBI which is similarly contraindicated with simvastatin and lovastatin, but cautious use with rosuvastatin is permitted [1].

7.3.8 INI Interactions with Renal Transporters

As previously stated, dolutegravir potently inhibits renal organic cation transporter, OCT2, and multidrug and toxin extrusion transporter MATE1. DTG may increase the concentrations of drugs such as metformin and dofetilide or endogenous molecules such as creatinine that are dependent on OCT2-mediated transport for renal excretion. A modest 10–15% increase in serum creatinine observed in DTG clinical trials is attributed to potent OCT2 inhibition by DTG and does not indicate renal toxicity or actual reduction of glomerular filtration [42]. In healthy volunteers, metformin mean AUC increased by 1.8- to 2.5-fold and mean C_{max} by 1.7- to 2.1-fold following metformin 500 mg twice daily administration with either once or twice daily DTG administration; therefore, limiting the total daily dose of metformin and monitoring of blood glucose are recommended [43]. Drugs that have a narrow therapeutic index, such as dofetilide and pilsicainide, which are dependent on OCT2-mediated transport, are contraindicated for use with DTG.

EVG is not known to modulate renal drug transporters; however, it requires pharmacokinetic boosting using either RTV or COBI. RTV demonstrates inhibitory activity for a wide range of transporters, including MATE-1, OAT1, OAT3, MRP2, MRP4, and P-gp, and thus may affect transport of drugs requiring these transporters. COBI is known to inhibit MATE-1 and P-gp drug transporters [44]. Given EVG/COBI is co-formulated with a nucleotide associated with renal toxicity (TDF) or the alternative TFV prodrug, TAF, renal function changes have been reported during use of EVG-containing regimens. COBI has been associated with reports of mild nonprogressive elevations in serum creatinine from inhibition of creatinine active tubular secretion without effecting glomerular filtration or overall renal function in clinical trials [45, 46]. Tenofovir is a substrate for OAT1, OAT3, and MRP4, and therefore inhibition of TFV influx or efflux from renal tubular cells may increase TFV exposure contributing to the observed 10–15% reduction in eGFR among patients receiving TDF and RTV or COBI-boosted PI- or INI-containing regimens [44]. TAF has less nephrotoxic potential when compared to TDF; however, limited long-term data exist on the renal impact of TAF-containing regimens given its recent approval [47–49]. The effect of COBI on creatinine clearance (CrCl) and iohexol-measured GFR demonstrated decreased CrCl but no effect on measured GFR (German) in healthy volunteers, suggesting the observed increase in renal-related adverse events with EVG-containing regimens may be attributed to COBI-mediated inhibition of TDF efflux by P-gp rather than COBI-mediated effects on MATE-1 transport [50, 51]. Overall, the inability to independently evaluate the renal effects of elvitegravir and COBI independently of TAF, TDF, or RTV limits our understanding of the predominant mechanisms that mediated renal function changes in EVG−/COBI-containing regimens.

In summary, although a wide spectrum of co-administered drugs may interact with antiretrovirals, much is presently known about these interactions and how to avoid their effects. A thorough history of prescription and nonprescription drugs, supplements, and herbs should identify key pharmacologic areas where potential interactions may be lurking.

7.4 Summary

Antiretroviral agents with novel mechanisms of action are constantly in development with hopes that these agents may constitute a curative treatment as seen with hepatitis C viral infection. While new classes of ART are expected to provide significant therapeutic benefits, they are also likely to further expand the number of potential drug-drug interactions. Integrase inhibitors have transformed the antiretroviral landscape with their intrinsic high-level potency, rapid attainment of virologic suppression, availability of complete single tablet regimens for once daily dosing, significant improvements in safety and tolerability, and high genetic barriers to resistance affording high efficacy rates. New advances in formulation technology are paving the way for longer-acting medications, including all injectable antiretroviral regimens, as an alternative to daily oral dosing and freedom from the daily reminder of HIV infection. Healthcare providers must remain vigilant in monitoring for drug interactions between antiretrovirals and medications used for the treatment of comorbidities. The vast array of websites and mobile applications are increasingly aiding in rapid identification of potential drug interactions at the bedside; however, the careful management of HIV disease requires a multidisciplinary focus optimized by contributions from experienced practitioners.

References

1. Panel on Antiretroviral Guidelines for Adults and Adolescents (2016) Guidelines for the use of antiretroviral agents in HIV-1-infected adults and adolescents. Department of Health and Human Services. Available at http://aidsinfo.nih.gov/contentfiles/lvguidelines/AdultandAdolescentGL.pdf. Section accessed 12/21/16
2. www.hiv-druginteractions.org
3. http://arv.ucsf.edu/insite?page=ar-00-02&post=7
4. www.medscape.com
5. van Heeswijk RP, Veldkamp A, Mulder JW et al (2001) Combination of protease inhibitors for the treatment of HIV-1-infected patients: a review of pharmacokinetics and clinical experience. Antivir Ther 6:201–229
6. Zhang D, Chando TJ, Everett DW, Patten CJ, Dehal SS, Humphreys WG (2005) Vitro inhibition of UDP glucuronosyltransferases by atazanavir and other HIV protease inhibitors and the relationship of this property to in vivo bilirubin glucuronidation. Drug Metab Dispos 33(11):1729–1739

7. Mouscadet J-F, Tchertanov L (2009) Raltegravir: molecular basis of its mechanism of action. Eur J Med Res 14(Suppl 3):5–16
8. Brainard DM, Friedman EJ, Jin B et al (2011) Effect of low-, moderate-, and high-fat meals on raltegravir pharmacokinetics. J Clin Pharmacol 51:422–427
9. Ramanathan S, Mathias AA, German P et al (2011) Clinical pharmacokinetic and pharmacodynamic profile of the HIV integrase inhibitor elvitegravir. Clin Pharmacokinet 50:229–244
10. Ramanathan S, Wright M, West S, et al. Pharmacokinetics, metabolism and excretion of ritonavir-boosted GS-9137(elvitegravir). 8th International Workshop on clinical pharmacology of HIV therapy. April 16–18, 2007, Budapest, Abstract 30
11. Lepist E-I, Phan TK, Roy A et al (2012) Cobicistat boosts the intestinal absorption of transport substrates, including HIV protease inhibitors and GS-7340, in vitro. Antimicrob Agents Chemother 56:5409–5413
12. Stribild (elvitegravir, cobicistat, emtricitabine, tenofovir disoproxil fumarate) package insert (2016) Gilead Sciences Inc., Foster City
13. Ramanathan S, Wang H, Szwarcberg J, Kearney BP. Safety/tolerability, pharmacokinetics, and boosting of twice-daily cobicistat administered alone or in combination with Darunavir or Tipranavir. 13th HIV clinical pharmacology workshop Apr 16–18 2012, Barcelona
14. German P, West S, Hui J, et al. Pharmacokinetic interaction between elvitegravir/ritonavir and dose-adjusted rifabutin. 9th International Workshop on clinical pharmacology of HIV therapy. April 7–9, 2008, New Orleans, Abstract 19
15. Vitekta (elvitegravir) package insert (2015) Gilead Sciences Inc, Foster City
16. Cottrell ML, Hadzic T, Kashuba AD (2013) Clinical pharmacokinetic, pharmacodynamic and drug-interaction profile of the integrase inhibitor dolutegravir. Clin Pharmacokinet 52:981–994
17. Dooley KE, Sayre P, Borland J et al (2013) Safety, tolerability, and pharmacokinetics of the HIV integrase inhibitor dolutegravir given twice daily with rifampin or once daily with rifabutin: results of a phase 1 study among healthy subjects. J Acquir Immune Defic Syndr 62:21–27
18. Song I, Borland J, Chen S, Guta P, Lou Y, Wilfret D et al (2014) Effects of enzyme inducers efavirenz and tipranavir/ritonavir on the pharmacokinetics of the HIV integrase inhibitor dolutegravir. Eur J Clin Pharmacol 70:1173–1179
19. Song I, Borland J, Min S et al (2011a) Effects of etravirine alone and with ritonavir-boosted protease inhibitors on the pharmacokinetics of dolutegravir. Antimicrob Agents Chemother 55:3517–3521
20. Song I, Borland J, Chen S et al (2011b) Effect of atazanavir and atazanavir/ritonavir on the pharmacokinetics of the next-generation HIV integrase inhibitor, S/GSK1349572. Br J Clin Pharmacol 72:103–108
21. Kirby BJ, Collier AC, Kharasch ED, Whittington D, Thummel KE, Unadkat JD (2011) Complex drug interactions of HIV protease inhibitors 1: inactivation, induction, and inhibition of cytochrome P450 3A by ritonavir or nelfinavir. Drug Metab Dispos 39(6):1070–1078
22. Buss N, Snell P, Bock J, Hsu A, Jorga K (2001) Saquinavir and ritonavir pharmacokinetics following combined ritonavir and saquinavir (soft gelatin capsules) administration. Br J Clin Pharmacol 52(3):255–264
23. Ghosn J, Lamotte C, Ait-Mohand H, Wirden M, Agher R, Schneider L, Bricaire F, Duvivier C, Calvez V, Peytavin G, Katlama C (2003) Efficacy of a twice-daily antiretroviral regimen containing 100 mg ritonavir/400 mg indinavir in HIV-infected patients. AIDS 17(2):209–214
24. Collin F, Chêne G, Retout S, Peytavin G, Salmon D, Bouvet E, Raffi F, Garraffo R, Mentré F, Duval X, ANRS CO8 Aproco-Copilote Study Group (2007) Indinavir trough concentration as a determinant of early nephrolithiasis in HIV-1-infected adults. Ther Drug Monit 29(2):164–170
25. Guiard-Schmid JB, Poirier JM, Meynard JL, Bonnard P, Gbadoe AH, Amiel C, Calligaris F, Abraham B, Pialoux G, Girard PM, Jaillon P, Rozenbaum W (2003) High variability of plasma drug concentrations in dual protease inhibitor regimens. Antimicrob Agents Chemother 47(3):986–990
26. Tybost (cobicistat) package insert (2016) Gilead Sciences Inc, Foster City

27. Marzolini C, Gibbons S, Khoo S, Back D (2016) Cobicistat versus ritonavir boosting and differences in the drug-drug interaction profiles with co-medications. J Antimicrob Chemother 71:1755–1758
28. Ramanathan S, Wang H, Stondell T, et al. Pharmacokinetics and drug interaction profile of cobicistat boosted-EVG with atazanavir, rosuvastatin or rifabutin. 13th International Workshop on clinical pharmacology of HIV therapy. April 16–18, 2012, Barcelona, Abstract O_03
29. Mathias A, Liu HC, Warren D, Sekar V, Kearney BP. Relative bioavailability and pharmacokinetics of darunavir when boosted with the pharmacoenhancer GS-9350 versus ritonavir. 11th International Workshop on clinical pharmacology and HIV therapy. April 7–9, 2010. Sorrento. Abstract 28
30. Luber A, Garg V, Gharakhanian S, et al. Survey of medications used by HIV-infected patients that affect gastrointestinal (GI) acidity and potential for negative drug interactions with HAART. Poster #206. 7th International Congress on Drug Therapy in HIV Infection; 14–18 November 2004
31. Reyataz [package insert] (2004) Bristol-Myers Squibb Company, Princeton
32. Isentress (raltegravir) package insert (2017) Merck & Co., Whitehouse Station
33. Kiser JJ, Bumpass JB, Meditz AL et al (2010) Effect of antacids on the pharmacokinetics of Raltegravir in human immunodeficiency virus-seronegative volunteers. Antimicrob Agents Chemother 54:4999–5003
34. Moss DM, Siccardi M, Murphy M et al (2012) Divalent metals and pH alter raltegravir disposition In Vitro. Antimicrob Agents Chemother 56:3020–3026
35. Tivicay (dolutegravir) package insert (2016) Viiv Healthcare, Research Triangle Park
36. Cohn SE, Park JG, Watts DH et al (2007) Depo-medroxyprogesterone in women on antiretroviral therapy: effective contraception and lack of clinically significant interactions. Clin Pharmacol Ther 81(2):222–227
37. Greenblatt DJ, von Moltke LL, Harmatz JS, Fogelman SM, Chen G, Graf JA, Mertzanis P, Byron S, Culm KE, Granda BW, Daily JP, Shader RI (2003) Short-term exposure to low-dose ritonavir impairs clearance and enhances adverse effects of trazodone. J Clin Pharmacol 43(4):414–422
38. Wiggins BS, Lamprecht DG, Page RL et al (2017) Recommendations for Managing Drug-Drug Interactions with Statins and HIV Medications. Am J Cardiovasc Drugs 17:375
39. Busti AJ, Bain AM, Hall RG et al (2008) Effects of atazanavir/ritonavir or fosamprenavir/ritonavir on the pharmacokinetics of rosuvastatin. J Cardiovasc Pharmacol 51:605–610
40. Fichtembaum C, Samineni D, Moore E, et tal. Darunavir/ritonavir incresaes rosuvastatin concentrations but does not alter lipid-lowering effect in healthy volunteers [poster WePE0101]. 18th international AIDS conference, July 18–23, 2010; Vienna
41. Hoody DW, Kiser JJ, et al. Drug-drug interaction between lopinavir/ritonavir and rosuvastatin [poster #564]. 14th conference on retroviruses and opportunistic infections; 2007 Feb 25–28; Los Angeles
42. Koteff J, Borland J, Chen S et al (2013) A phase 1 study to evaluate the effect of dolutegravir on renal function via measurement of iohexol and para-aminohippurate clearance in healthy subjects. Br J Clin Pharmacol 75(4):990–996
43. Song IH, Zong J, Borland J et al (2016) The effect of Dolutegravir on the pharmacokinetics of metformin in healthy subjects. J Acquir Immune Defic Syndr 72:400–407
44. Yombi JC, Pozniak A, Boffito M et al (2014) Antiretrovirals and the kidney in current clinical practice: renal pharmacokinetics, alterations of renal function and renal toxicity. AIDS 28(5):621–632
45. DeJesus E, Rockstroh JK, Henry K, Molina JM, Gathe J, Ramanathan S et al (2012) Co-formulated elvitegravir, cobicistat, emtricitabine, and tenofovir disoproxil fumarate versus ritonavir-boosted atazanavir plus co-formulated emtricitabine and tenofovir disoproxil fumarate for initial treatment of HIV-1 infection: a randomised, double-blind, phase 3, noninferiority trial. Lancet 379:2429–2243

46. Sax PE, DeJesus E, Mills A, Zolopa A, Cohen C, Wohl D et al (2012) Co-formulated elvitegravir, cobicistat, emtricitabine, and tenofovir versus co-formulated efavirenz, emtricitabine, and tenofovir for initial treatment of HIV-1 infection: a randomised, double-blind, phase 3 trial, analysis of results after 48 weeks. Lancet 379:2439–2448
47. Genvoya (elvitegravir/cobicistat/emtricitabine/tenofovir alafenamide) package insert (2017) Gilead Sciences Inc, Foster City
48. Descovy (emtricitabine/tenofovir alafenamide) package insert (2017) Gilead Sciences Inc, Foster City
49. Novick TK, Choi MJ, Rosenberg AZ, McMahon BA, Fine D, Atta MG (2017) Tenofovir alafenamide nephrotoxicity in an HIV-positive patient: a case report. Medicine (Baltimore) 96(36):e8046
50. German P, Liu HC, Szwarcberg J et al (2012) Effect of cobicistat on glomerular filtration rate in subjects with normal and impaired renal function. J Acquir Immune Defic Syndr 61:32–40
51. Milburn J, Jones R, Levy JB (2017) Renal effects of novel antiretroviral drugs. 2017. Nephrol Dial Transplant 32(3):434–439

Chapter 8
Drug Interactions in HIV: Nucleoside, Nucleotide, and Nonnucleoside Reverse Transcriptase Inhibitors and Entry Inhibitors

Check for updates

Lauren R. Cirrincione and Kimberly K. Scarsi

Abbreviations

ART	Antiretroviral therapy
AUC	Area under the concentration-time curve
BCRP	Breast cancer resistance protein
BMD	Bone mineral density
CCB	Calcium channel blockers
CCR5	C-C motif co-receptor 5
CES1	Hepatic carboxylesterase
Cmax	Maximum concentration
COC	Combined oral contraceptives
CrCl	Creatinine clearance
CSF	Cerebrospinal fluid
CYP	Cytochrome P450
FDC	Fixed dose combination
HMG-CoA	3-Hydroxy-3-methylglutaryl-coenzyme A
INR	International normalized ratio
MAC	*Mycobacterium avium* complex
MRP4	Multidrug resistance protein 4
NNRTI	Nonnucleoside reverse transcriptase inhibitors
NRTI	Nucleoside/nucleotide reverse transcriptase inhibitor
OAT	Organic anion transporter
OATP	Organic anion-transporting polypeptides

L. R. Cirrincione • K. K. Scarsi (✉)
University of Nebraska Medical Center, College of Pharmacy,
Department of Pharmacy Practice, Omaha, NE, USA
e-mail: kim.scarsi@unmc.edu

© Springer International Publishing AG, part of Springer Nature 2018
M. P. Pai et al. (eds.), *Drug Interactions in Infectious Diseases: Antimicrobial Drug Interactions*, Infectious Disease,
https://doi.org/10.1007/978-3-319-72416-4_8

OCT	Organic cation transporters
PBMCs	Peripheral blood mononuclear cells
PDE-5	Phosphodiesterase type-5
P-gp	P-glycoprotein
PrEP	Pre-exposure prophylaxis
SNP	Single-nucleotide polymorphisms
TAF	Tenofovir alafenamide fumarate
TDF	Tenofovir disoproxil fumarate
TFV	Tenofovir
UGT	Uridine-5′-diphosphate glucuronosyltransferase

8.1   Introduction

It is estimated that 36.7 million persons are living with HIV worldwide, with approximately 2.1 million new infections per year [1]. Fortunately, combination antiretroviral therapy (ART) used to treat HIV infection is highly effective at decreasing both HIV-related and HIV-unrelated morbidity and mortality. Therefore, both the World Health Organization and the US Department of Health and Human Services recommend ART for all persons living with HIV [2, 3]. In addition to ART for the treatment of HIV, there is an emerging role for antiretroviral agents as both short- and long-term therapies for HIV prevention in high-risk populations [4].

Drug interactions are a common medication-related problem for persons living with HIV [5], particularly in aging populations who may also require therapy for other chronic and acute conditions. To optimize the selection of antiretrovirals with concomitant medications, understanding common drug interactions, their clinical significance, and management strategies is critical for healthcare providers who care for patients at risk for, or living with, HIV.

8.1.1   Nucleoside and Nucleotide Reverse Transcriptase Inhibitors (NRTIs)

Nucleoside and nucleotide reverse transcriptase inhibitors (NRTIs) are analogs of endogenous purines and pyrimidines. During replication, viral reverse transcriptase enzyme inserts the NRTI into the growing viral DNA in place of thymidine (zidovudine or stavudine), cytosine (lamivudine or emtricitabine), guanosine (abacavir), or adenosine (didanosine or tenofovir), resulting in viral DNA chain termination. NRTIs commonly included in modern ART regimens are abacavir, emtricitabine, lamivudine, tenofovir alafenamide (TAF), and tenofovir disoproxil fumarate (TDF). Didanosine, stavudine, and zidovudine are used less frequently but remain options in select patient populations. Zalcitabine is no longer recommended in the United States and will not be addressed in this chapter. Coformulated tenofovir disoproxil fumarate and emtricitabine are also used for pre-exposure prophylaxis (PrEP) in patients at risk for HIV infection.

All NRTIs are tri-phosphorylated intracellularly to their active form [6], except tenofovir, which is di-phosphorylated intracellularly [7]. Most NRTIs are renally eliminated and hence less commonly associated with drug-drug interactions than other antiretroviral drug classes. Known mechanisms of NRTI interactions include interference with intracellular phosphorylation, alterations in drug transport, and competition for renal elimination.

8.1.2 Nonnucleoside Reverse Transcriptase Inhibitors (NNRTIs)

Unlike NRTIs, NNRTIs are not nucleoside/nucleotide analogs, nor do they undergo intracellular phosphorylation. NNRTIs are noncompetitive inhibitors of the HIV reverse transcriptase enzyme by binding to reverse transcriptase and causing a conformational change in the enzyme, which inhibits the enzyme's polymerase ability and stops transcription of HIV-RNA to HIV-DNA. Rilpivirine is commonly used in the United States and is being evaluated as part of a long-acting intramuscularly administered regimen for HIV therapy [8, 9]. Etravirine is reserved for treatment-experienced patients. Efavirenz and, to a lesser extent, nevirapine remain important components of ART in many countries. Delavirdine is no longer recommended in the United States and will not be addressed in this chapter.

The NNRTIs are each metabolized by one or more cytochrome P450 (CYP) isoenzymes to inactive metabolites, with minimal renal elimination. Simultaneously, most NNRTIs induce, and sometimes inhibit, numerous CYP enzymes. Therefore, most drug-drug interactions within this class are a consequence of alterations in metabolism of the NNRTI or the coadministered medication; however, rilpivirine is also affected by pH-dependent absorption. In contrast to the NRTIs, NNRTI-related drug-drug interactions are not commonly mediated by drug transport proteins [10].

8.1.3 Entry Inhibitors

Entry inhibitors include a fusion inhibitor (enfuvirtide) and a C-C motif co-receptor 5 (CCR5) antagonist (maraviroc). Enfuvirtide is a fusion inhibitor that binds to the viral gp41 subunit, which inhibits conformational changes of HIV-1 structural protein gp41, thereby restricting viral entry into CD4 cells [11]. Maraviroc is a small molecule receptor antagonist, inhibiting the interaction between HIV-1 structural protein gp120 and CCR5. Neither enfuvirtide nor maraviroc is commonly used in modern ART. However, clinical trials involving maraviroc as a PrEP strategy are underway, indicating a potential expansion of maraviroc use. Enfuvirtide is not associated with any drug interactions, whereas maraviroc is highly susceptible to drug-drug interactions involving the CYP enzyme system and p-glycoprotein (P-gp) efflux transporter.

8.2 NRTI Pharmacology and Potential Mechanisms of Drug-Drug Interactions

8.2.1 NRTI Absorption

Unless indicated below, all NRTIs are extensively absorbed and may be administered irrespective of meals [7, 12–16]. NRTIs reach maximum concentration (Cmax) within 2 h after oral administration [7, 12–17]. In contrast, didanosine has <50% bioavailability after oral administration under fasted conditions [17]. It is unstable in acidic conditions, and both oral tablets (no longer manufactured in the United States) and oral solution contain a buffering agent [18], while delayed-release capsules contain enteric-coated beadlets to prevent degradation by stomach acid [17]. Didanosine should be administered under fasted conditions in adults, because the Cmax and area under the concentration-time curve (AUC) are reduced with food. A pediatric pharmacokinetic study found prolonged absorption and a reduced didanosine Cmax under fed conditions, but there was no reduction in overall exposure (AUC) [19]. Given the complexities of administering didanosine on an empty stomach to infants who feed frequently, clinicians often recommend didanosine oral solution without regard to meals to improve adherence [20].

Tenofovir (TFV) is a dianion at physiological pH, leading to low oral bioavailability [21]. Therefore, TFV is formulated as a prodrug—either tenofovir alafenamide fumarate (TAF) or tenofovir disoproxil fumarate (TDF). While both prodrugs are hydrolyzed to TFV, TDF is rapidly hydrolyzed in plasma, and TAF is hydrolyzed primarily within target cells [21, 22], resulting in approximately 90% lower plasma TFV concentrations if the dose is administered as TAF versus TDF [21, 23, 24]. TAF and TDF may be given irrespective of meals [22, 25].

8.2.2 Protein Binding and Distribution of NRTI

Most NRTIs have low plasma protein binding: didanosine, emtricitabine, TFV, and stavudine (all <5%) and lamivudine and zidovudine (both <40%) [7, 13–17]. Abacavir is moderately protein bound (50%) [12], and TAF is approximately 80% bound to plasma protein [25]. Didanosine, emtricitabine, lamivudine, TAF, tenofovir, and zidovudine have wide tissue distribution [7, 14, 17, 26, 27], while abacavir and stavudine distribution approximates total body water [12]. NRTIs have variable penetration into the blood-brain barrier, with cerebrospinal fluid (CSF): plasma concentration ratio of ~5% (tenofovir, lamivudine), 20–30% (abacavir, didanosine), and 60% (stavudine, zidovudine) [13, 14, 28]. No data are available to describe TAF distribution to the CSF.

8.2.3 Metabolism and Elimination of NRTIs

Few NRTIs are metabolized, with most undergoing renal elimination unchanged via active tubular secretion and glomerular filtration [13, 15, 16]. Except abacavir, all NRTIs require dose adjustment in patients with renal impairment. Neither NRTIs nor their metabolites are CYP or uridine-5′-diphosphate glucuronosyltransferase (UGT) enzyme inducers or inhibitors [16, 17, 25, 29].

Abacavir is metabolized by hepatic alcohol dehydrogenase and UGT1A1 [12, 29, 30], didanosine by intestinal and hepatic xanthine oxidase to uric acid [31], and zidovudine via glucuronidation during first-pass metabolism [14]. Subsequently, all three are renally eliminated as inactive metabolites (50–80%) [6, 14, 29]. TDF is rapidly metabolized by esterases in the blood to TFV [32], followed by renal elimination (80%) of TFV [7]. TAF is metabolized intracellularly by both cathepsin A, which is a protease in lymphoid tissue and macrophages, and hepatic carboxylesterase (CES1) [25, 26]. The majority of TAF (80%) is eliminated by metabolism to TFV, with the remaining TAF primarily eliminated in the feces [25].

Phosphorylated NRTI moieties are characterized by long intracellular half-lives, allowing for less frequent dosing than suggested by their comparatively short plasma half-lives. For instance, TFV-diphosphate has an estimated elimination half-life of 87 h in peripheral blood mononuclear cells (PBMCs)—five times the elimination half-life of TFV in plasma [7, 33].

8.2.4 Role of Drug Transporters in NRTI Drug-Drug Interactions

Influx and efflux transporters are a mechanism of NRTI drug-drug interactions by competition between agents that share these transport pathways. Drug transporters are discussed in detail in Chap. 3 of this textbook; Fig. 3.1 and Table 3.3 provide a brief description of relevant drug transporters involved in NRTI transport. NRTIs with transporter-related drug-drug interactions include lamivudine, TFV, and both TAF and TDF prodrugs [7, 23, 34]; therefore, these NRTIs will be the focus of this drug transport section.

While TAF and TDF are substrates of the efflux pump P-gp, TFV is not [34]. TAF, TDF, and TFV are all breast cancer resistance protein (BCRP) efflux transporter substrates [7, 23, 34], resulting in decreased oral bioavailability when coadministered with BCRP inducers [35]. Agents that alter P-gp and BCRP, such as protease inhibitors, are expected to affect the disposition of TDF and TAF (see Sect. 3.4.4) [7, 23, 25].

Renal transporters involved in NRTI tubular secretion include organic anion transporters (OAT) and organic cation transporters (OCT). An example of an interaction by shared tubular secretion between agents is increased lamivudine exposure with trimethoprim-sulfamethoxazole, as trimethoprim is a substrate and inhibitor of

OCT2 [36, 37]. Given lamivudine's favorable safety profile, this interaction is not clinically significant in patients with normal renal function. However, it provides a mechanistic view of potential drug-drug interactions between renally secreted agents. TFV is secreted by OAT1 and OAT3 which may result in competitive inhibition with other renally secreted agents [21, 24, 38, 39]. In vitro, OCT1 and OCT2 are inhibited by lamivudine and TAF [25, 34, 40, 41]. TDF, TAF, and TFV are substrates of multidrug resistance protein 4 (MRP4) [7, 21, 24, 39]; however, competition for active tubular secretion with other MRP4 substrates has not been evaluated in vivo [39].

Organic anion-transporting polypeptides (OATP) mediate uptake of compounds into the liver for metabolism [35, 42]. TAF is a substrate of hepatic uptake transporters OATP1B1 and OATP1B3 [23, 25], and modulation of these transporters is responsible for clinically significant drug-drug interactions with hepatitis C direct-acting antivirals.

8.3 NRTI Drug-Drug Interactions

Given the importance of NRTI intracellular concentrations, the clinical significance of alterations in NRTI plasma concentrations is unclear, yet few drug-drug interaction studies evaluate intracellular concentrations of NRTIs. Therefore, most clinically relevant drug-drug interactions relate to supra-therapeutic plasma concentrations of the NRTI or coadministered agent, leading to exposure-related adverse events. Clinical relevance of interactions that decrease NRTI plasma concentrations is difficult to interpret without associated pharmacodynamic antiviral effectiveness measures. Additionally, the role of coadministered antiretrovirals in the summative drug-drug interaction potential of the ART regimen must be considered.

The magnitude of pharmacokinetic NRTI drug interactions and clinical management recommendations discussed in this section are summarized in Table 8.1.

8.3.1 Anticonvulsants

Carbamazepine is a P-gp inducer that reduces TAF plasma exposure [25]. Oxcarbazepine, phenobarbital, and phenytoin have not been studied with TAF; however, they also induce P-gp efflux activity, and use with TAF should be avoided. Interactions between carbamazepine, oxcarbazepine, phenobarbital, and phenytoin with other NRTIs are not expected [7, 12–17]. Valproic acid increases zidovudine exposure [14], likely due to inhibition of its glucuronidation [61], and routine monitoring of zidovudine-related adverse events (e.g., bone marrow suppression) should be considered. No clinically significant interactions are expected between valproic acid and other NRTIs.

Table 8.1 Drug-drug interactions involving NRTIs

Drug	Effect on NRTI pharmacokinetics (change in NRTI)	Clinical recommendation
Abacavir		
Ethanol [43]	↑ AUC 41%	No dose adjustment required
Methadone [44]	↓ Cmax 34%	No dose adjustment required; ↑CL 23% of methadone also observed; monitor for methadone withdrawal symptoms
Tipranavir/ritonavir [45]	↓ AUC 40%	No dose adjustment required
Didanosine		
Allopurinol [17]	↑ AUC 113%	Avoid coadministration
Atazanavir [46], atazanavir/ritonavir [46], atazanavir/cobicistat [47]	↓ AUC 34% (didanosine EC plus ATV alone or ATV/r, fed conditions)	Atazanavir AUC ↓ 87% with buffered didanosine tablets. Because both atazanavir and didanosine have pH-dependent absorption, separate administration time, and give atazanavir with food and didanosine EC on an empty stomach; separate liquid atazanavir solution 2 h before, or 1 h after, buffered didanosine solution
Cidofovir/probenecid [48, 49]	↑ AUC 60%	No dose adjustment required; monitor for didanosine-related adverse events
Darunavir/ritonavir [50], darunavir/cobicistat [51]	...	Because darunavir must be given with food, separate didanosine EC dose either an hour before, or 2 h after, darunavir administration
Ganciclovir [52]	Concomitant dosing: ↑ AUC 107% Separated dosing: ↑ AUC 115%	Do not separate administration; give agents concomitantly; monitor for didanosine-related adverse events
Methadone [17]	↓ AUC 17% (didanosine EC)	No dose adjustment required
Rilpivirine [53]	↔ AUC Separated dosing (didanosine EC)	Administer rilpivirine with food 4 h before or 2 h after didanosine
Tenofovir disoproxil fumarate [2]	Didanosine ↑ AUC 48%–60%	Avoid coadministration due to inferior clinical outcomes. If coadministration is required, reduce didanosine dose to 250 mg per day in patients > 60 kg

(continued)

Table 8.1 (continued)

Drug	Effect on NRTI pharmacokinetics (change in NRTI)	Clinical recommendation
Tipranavir/ritonavir [45]	↓ AUC 3%–33%	Because tipranavir must be given with food, separate didanosine EC dosing from tipranavir by 2 h
Stavudine		
Methadone [54]	↓ AUC 23%	No dose adjustment required
Tenofovir alafenamide[a]		
Anticonvulsants: Carbamazepine [55] Oxcarbazepine, phenobarbital, phenytoin	With carbamazepine: ↓ AUC 55%: other anticonvulsants: ↓ expected	Avoid coadministration
Antivirals: Acyclovir, cidofovir, ganciclovir, valganciclovir [25]	…	Potential for overlapping nephrotoxicity, monitor renal function routinely
Atazanavir/cobicistat [55]	↑ AUC 75%	No dose adjustment required
Atazanavir/ritonavir [25]	↑ AUC 91%	No dose adjustment required
Darunavir/cobicistat or darunavir/ritonavir [25]	↔ AUC	No change in plasma concentrations, but intracellular TFV-diphosphate concentrations increase significantly; no dose adjustment required
Elvitegravir/cobicistat [56]	↑ AUC 128%	TAF 10 mg plus elvitegravir/cobicistat/ emtricitabine results in equivalent TAF exposure as compared to TAF 25 mg alone
Lopinavir/ritonavir [25]	↑ AUC 47%	No dose adjustment required
Rifabutin, rifampin, rifapentine [25]	↓ expected	Avoid coadministration
St. John's wort (*Hypericum perforatum*) [25]	↓ expected	Avoid coadministration
Tipranavir/ritonavir [25]	↓ expected	Avoid coadministration
Tenofovir disoproxil fumarate[b]		
Antivirals: Acyclovir, cidofovir, ganciclovir, valganciclovir [25]	…	Potential for overlapping nephrotoxicity; monitor renal function routinely

Drug	PK change	Recommendation
Atazanavir/ritonavir [46]	↑ AUC 37%	Tenofovir decreases atazanavir exposure (↓ AUC 25%) with and without ritonavir; give atazanavir 300 mg plus ritonavir 100 mg with TDF as single daily dose
Atazanavir/cobicistat [47]; darunavir/cobicistat [51]	↑ expected	No dosage adjustment required; avoid coadministration if CrCl < 70 ml/min; monitor for tenofovir-associated nephrotoxicity
Darunavir/ritonavir [50]	↑ AUC 22%	No dosage adjustment required; monitor for tenofovir-associated nephrotoxicity
Elvitegravir/cobicistat [7, 57]	↑ AUC 33%	Avoid initiation if CrCl <70 ml/min; monitor for tenofovir-associated nephrotoxicity
Lopinavir/ritonavir [58]	↑ AUC 32%	No dose adjustment required; monitor for tenofovir-associated nephrotoxicity
Raltegravir [59]	↔ AUC	No dose adjustment required
Rifampin [60]	↔ AUC	No dose adjustment required
Tipranavir/ritonavir [2]	↔ AUC	No dose adjustment required
Zidovudine		
Atovaquone [2]	↑ AUC 31%	No dose adjustment required; monitor for zidovudine-associated adverse events
Cidofovir/probenecid [14]	↑ AUC 106%	Decrease zidovudine dose by 50% on days of cidofovir/probenecid infusion; or consider alternative therapy. Interaction is related to probenecid; cidofovir alone does not influence zidovudine
Ganciclovir [52]	↑ AUC 20%	No dose adjustment required; monitor for renal and zidovudine-related adverse events
Methadone [2]	↑ AUC 29%–43%	No dose adjustment required; monitor for zidovudine-related adverse events
Ribavirin [14]	↔ AUC	Avoid coadministration due to exacerbation of zidovudine-related anemia
Tipranavir/ritonavir [2]	↓ AUC 35%	No dose adjustment required
Valproic acid [14]	↑ AUC 80%	No dose adjustment required; monitor for zidovudine-related adverse events

Pharmacokinetic information is summarized as follows: decreased plasma concentration (↓ %), increased plasma concentration (↑ %), or no change (↔, indicating <10% increase or decrease in plasma concentration). For drug combinations with potential for an interaction, but no available evidence, the predicted result of the interaction is described as *expected* or *possible*, which indicates the strength of the prediction based on the pharmacologic characteristics of each drug

Abbreviations: *ATV/r* atazanavir plus ritonavir, *AUC* area under the plasma concentration time curve, *CL* renal clearance, *CrCL* creatinine clearance, *EC* enteric coated, *TAF* tenofovir alafenamide fumarate, *TFV* tenofovir, *TDF* tenofovir disoproxil fumarate

[a]Reported percent changes reflect tenofovir alafenamide (TAF) plasma concentrations

[b]Reported percent changes reflect tenofovir (TFV) plasma concentrations

8.3.2 Antigout

Allopurinol and its metabolite are potent inhibitors of xanthine oxidase [31], and didanosine exposure is substantially increased with allopurinol. Therefore, coadministration is contraindicated due to the risk of didanosine-related toxicity [17]. Because xanthine oxidase is not involved in the metabolism of other NRTIs, allopurinol interactions are not expected with other NRTIs [62].

8.3.3 Antimycobacterials

Drug interactions involving antimycobacterial agents are reviewed in Chap. 12. Briefly, rifamycin antimycobacterial agents (rifampin, rifabutin, and rifapentine) are known P-gp inducers [25], which may interact with some NRTIs. Specifically, based on observed data with carbamazepine, TAF plasma concentrations are expected to be decreased with rifamycins; therefore, rifamycins are not recommended with TAF [55]. In contrast, rifampin does not affect TFV exposure after administration of TDF [60]; therefore, until further evidence is available for TAF, TDF is the preferred TFV formulation with rifamycins. No significant interactions are expected with other NRTIs.

8.3.4 Antiretrovirals

8.3.4.1 Integrase Strand Transfer Inhibitors

Elvitegravir/cobicistat/emtricitabine formulated with TDF 300 mg results in higher TFV exposure compared to TDF alone (AUC 4.4 versus 3.3 mcg*hr/mL, respectively) [7, 57]. Due to increased risk of nephrotoxicity, this fixed dose combination (FDC) ART regimen should be avoided in patients with creatinine clearance (CrCl) <70 ml/min. Furthermore, it should be discontinued if CrCl falls below 50 ml/min during therapy, as dose adjustment of TDF is limited as a component of an FDC [57], and renal function should be monitored (urinalysis and CrCl) throughout use. Elvitegravir/cobicistat/emtricitabine formulated with TAF 10 mg results in over 90% lower concentrations of circulating TFV yet achieves fivefold higher intracellular concentrations in PBMCs as compared to the TDF 300 mg-containing FDC described above [63, 64]. Given the decrease in TFV plasma concentrations, the TAF-containing FDC regimen can be used in patients with CrCl ≥30 ml/min [57]. Similar TAF plasma exposures are found between TAF 10 mg with cobicistat and TAF 25 mg alone (AUC 250 versus 278 ng*hr/mL, respectively), likely related to P-gp inhibition by cobicistat [56].

No change in raltegravir exposure was observed with TDF coadministration [59].

8.3.4.2 Nonnucleoside Reverse Transcriptase Inhibitors

Interactions between NRTIs and NNRTIs are minimal. Didanosine and rilpivirine must be separated due to incongruent food requirements [53]. No other clinically significant interactions are expected.

8.3.4.3 Nucleoside and Nucleotide Reverse Transcriptase Inhibitors

Didanosine and stavudine should never be combined due to a pharmacodynamic drug-drug interaction, including increased adverse effects such as peripheral neuropathy, pancreatitis, and lactic acidosis [17]. Didanosine exposure is significantly higher when combined with TDF, increasing the risk of didanosine-associated adverse events [39]. While decreasing the didanosine dose was a strategy to overcome this pharmacokinetic drug-drug interaction [17], ART combining TDF plus didanosine resulted in significantly lower CD4 cell counts, despite achieving virologic suppression [65]. As both agents are adenosine analogs, purine accumulation inside CD4 cells is suggested as the potential mechanism of this pharmacodynamic interaction. For these reasons, didanosine and TDF should be avoided [2].

Similarly, lamivudine and emtricitabine should not be combined, as both are cytidine analogs, and there is no evidence supporting combined use [2].

8.3.4.4 Protease Inhibitors

Protease inhibitors are typically administered with a pharmacokinetic enhancer, cobicistat or ritonavir. Not only do both agents inhibit CYP3A4-mediated metabolism, they also inhibit P-gp efflux and renal transporters, affecting elimination of NRTIs [41, 66]. When TDF is combined with most pharmacokinetically enhanced protease inhibitors, inhibition of P-gp and renal transporters results in increased TFV plasma exposure (↑ 22–37%) [2, 7, 46, 58].

TAF plasma exposure is also increased with pharmacokinetically enhanced protease inhibitors (↑ 47–91%) [25, 55]. Additionally, TAF plus darunavir/cobicistat/emtricitabine results in intracellular TFV-diphosphate concentrations in PBMCs that are nearly 6.5 times higher than TFV-diphosphate concentrations observed with the identical TDF-containing regimen [67]. Despite shared increases in plasma exposure, circulating TFV remains significantly lower with TAF compared to TDF.

Increased TFV plasma exposure, specifically when administered as TDF, increases the risk of tenofovir-associated adverse effects. In addition to nephrotoxicity (described further in Sect. 3.10), TDF-based ART regimens are associated with declining bone mineral density (BMD) [7, 39, 68]. The mechanism of TDF-associated bone loss is suspected to be related to proximal tubular toxicity, which leads to phosphate wasting and increased bone turnover [7, 68]. Studies have found greater BMD reduction with TDF plus pharmacokinetically enhanced protease inhibitor-based regimens [39]. While a direct relationship between declining BMD and fracture risk has not been established, protease inhibitors plus TDF should be avoided in patients with osteoporosis. Clinical trials to date suggest that TAF results in less BMD decline as compared to TDF and may be preferred in patients with osteoporosis [67].

In contrast to other protease inhibitors, tipranavir is both a substrate and potent inducer of P-gp [45], persisting even when combined with ritonavir. Tipranavir/ritonavir decreases abacavir, enteric-coated didanosine, and zidovudine exposure; however, no dose adjustments are recommended [45]. Similarly, a decrease in TAF exposure is expected, and coadministration is not recommended [25].

In addition to protease inhibitors' effect on tenofovir exposure, TDF decreases atazanavir exposure [46]. Therefore, atazanavir must be pharmacokinetically enhanced when combined with TDF [46]. Atazanavir also requires acidic conditions for optimal absorption [17, 46], which results in a drug-drug interaction with didanosine. Under fasted conditions, atazanavir plus buffered didanosine results in lower atazanavir exposure, which can be overcome by taking atazanavir 1 h following buffered didanosine administration [46]. Studies under fed conditions of enteric-coated didanosine plus atazanavir/ritonavir, and atazanavir alone, identified lower didanosine exposure, but no impact on atazanavir pharmacokinetics. In addition to these specific pH-based interactions, several other protease inhibitors should be taken with food to improve absorption, while enteric-coated didanosine should be administered on an empty stomach (see specific timing of dose separation in Table 8.1).

8.3.5 Antivirals

Ganciclovir and cidofovir/probenecid are primarily eliminated by renal tubular secretion [39, 52], which leads to drug-drug interactions by inhibition of renal tubular transporters. Exposure of didanosine and zidovudine is increased by both ganciclovir and cidofovir/probenecid, and monitoring for NRTI-related toxicities is recommended [14, 17, 48, 49, 52]. In addition, zidovudine dose adjustment is recommended if concomitant administration of cidofovir/probenecid is required (see Table 8.1). Clinically significant interactions between didanosine or zidovudine and other antivirals are not expected [62, 69, 70].

Abacavir, emtricitabine, lamivudine, TDF, TAF, and stavudine have not been evaluated with most antivirals; however, clinically significant interactions are not expected. Famciclovir was studied with emtricitabine, but no significant interaction was found [69]. Likewise, foscarnet does not affect stavudine in vitro, and interactions with other NRTIs are not expected. Overall, overlapping toxicities between

NRTIs and other antivirals require routine monitoring of renal function during coadministration [2].

8.3.6 Hepatitis C Direct-Acting Antiviral Agents

Drug interactions involving direct-acting antiviral agents are reviewed in Chap. 15. Briefly, ribavirin should not be used in combination with didanosine due to an increase in the phosphorylated form of didanosine and an increased risk of didanosine-associated toxicities [17]. Exacerbation of anemia is possible with zidovudine and ribavirin and the combination should be avoided [14]. No clinically significant interactions have been found between ribavirin and TDF [7], and no interaction is expected between ribavirin plus emtricitabine or TAF. Tenofovir concentrations are increased with ledipasvir/sofosbuvir and sofosbuvir/velpatasvir [71, 72]. Whether higher tenofovir concentrations during ledipasvir/sofosbuvir or sofosbuvir/velpatasvir treatment increases the risk of renal toxicity is presently unclear.

8.3.7 Herbal Products

St. John's wort (*Hypericum perforatum*) induces CYP3A4 and P-gp [35, 73]. Because TAF is a P-gp substrate, St. John's wort plus TAF should be avoided, as decreased TAF concentrations are expected [25]. No clinically significant interactions are expected with other NRTIs.

8.3.8 Narcotics or Treatment for Opioid Dependence

Methadone is biotransformed via CYP enzymes, and its metabolites are excreted in urine and bile (described in further detail in Sect. 5.18) [44]. Abacavir absorption is decreased with methadone ($\downarrow$ Cmax 34%), and an increase in methadone clearance ($\uparrow$ 23%) is also observed [44]. Similarly, methadone moderately reduces didanosine and stavudine exposure, likely by decreased intestinal motility and subsequent gastric degradation of these agents [54]. Finally, methadone increases zidovudine exposure, potentially due to inhibition of glucuronidation, and zidovudine-related toxicities are reported with coadministration [74, 75]; therefore, patients should be monitored for zidovudine-related anemia. No dose adjustment is recommended for any of these NRTIs when combined with methadone, and no interaction is expected between methadone and other NRTIs [7, 62].

No drug-drug interactions have been observed, or are expected, between buprenorphine/naloxone and NRTIs [2, 74, 76].

8.3.9   Miscellaneous

Ethanol alters pharmacokinetic parameters of abacavir due to overlapping metabolism via alcohol dehydrogenase. However, this increase is not considered clinically relevant [43].

Atovaquone increases systemic exposure of zidovudine, possibly related to inhibition of glucuronidation [77]. No dose adjustment is recommended, but monitoring for zidovudine-related adverse events is suggested [78].

8.3.10   Pharmacokinetic and Pharmacodynamic Mechanisms of Tenofovir-Associated Nephrotoxicity

Proximal renal tubulopathy, acute kidney injury, or Fanconi syndrome leading to renal impairment or, in rare cases, renal failure are reported with TDF [7, 79]. Concomitant nephrotoxic agents, including aminoglycosides, certain non-ART antivirals, high-dose NSAIDs, amphotericin B, foscarnet, pentamidine, vancomycin, or interleukin-2, increase the occurrence of tenofovir-associated nephrotoxicity [7]. Combining pharmacokinetic enhancers with TDF may also increase the occurrence of nephrotoxicity (see Sect. 3.4.4). To decrease the likelihood of tenofovir-associated nephrotoxicity, renal function monitoring (CrCl and serum phosphate) is advised before initiating TDF. After initiation of TDF therapy, renal function monitoring is recommended after 2 weeks and at least every 6 months thereafter [39]. Compared to TDF, TAF results in lower TFV plasma concentrations, and less nephrotoxicity has been observed in clinical evaluations to date; therefore, TAF may be preferred for patients at risk for nephrotoxicity [21, 67].

8.4   NNRTI Pharmacology and Potential Mechanisms of Drug-Drug Interactions

8.4.1   NNRTI Absorption

Most NNRTIs require specific gastric conditions for optimal absorption. Nevirapine is the exception, with high oral bioavailability (>90%) under both fasted or fed conditions [80]. Conversely, giving efavirenz with high-fat, high-calorie food increases the Cmax achieved ($\uparrow$ 39%) compared to fasted conditions [81]. Despite improved absorption under fed conditions, efavirenz is nearly 50% bioavailable and achieves adequate exposure without food; therefore, efavirenz should be administered on an empty stomach or with a low-calorie meal to decrease the likelihood of dose-related side effects [82]. Absolute bioavailability of etravirine or rilpivirine is unknown; however, both agents should be administered with meals, as absorption is reduced on an empty stomach [53, 83]. Unique in the NNRTI class, rilpivirine absorption is pH-dependent.

All NNRTIs reach Cmax within 5 h after oral administration, with the exception of extended release nevirapine, which reaches Cmax within 24 h following oral administration [53, 80, 81, 84].

8.4.2   Protein Binding and Distribution of NNRTI

All NNRTIs demonstrate a high volume of distribution and are highly protein bound to plasma albumin (>99%), except nevirapine, which has moderate protein binding (62%) [83]. Efavirenz, etravirine, and rilpivirine minimally penetrate the blood-brain barrier (CSF/plasma concentration ratio 0.007 to 0.014) [81, 85–87], whereas nevirapine achieves higher CSF penetration (~50% plasma concentration) [82].

8.4.3   Metabolism and Elimination of NNRTIs

NNRTIs have long elimination half-lives, ranging from 30 to 50 h [53, 80, 81, 84], and undergo extensive CYP-mediated biotransformation. Specific CYP and UGT pathways of NNRTI metabolism are described in Table 8.2.

Pharmacogenomic characteristics of patients are known to affect the disposition of NNRTIs and may influence the extent of some drug interactions. CYP2B6-associated single-nucleotide polymorphisms (SNPs) that reduce CYP2B6 activity correspond with a 62% increase in efavirenz plasma concentrations [85, 98]. Expression of loss-of-function SNPs in the CYP2C19 gene substantially decreases the etravirine hydroxylated metabolites (75–100%). SNPs in the CYP2B6 gene reduce nevirapine elimination, increasing plasma concentrations by nearly 25% [98].

After metabolism, NNRTIs are eliminated via both renal and hepatic routes, mostly as inactive metabolites, except for etravirine. Renal elimination accounts for up to 34% of the efavirenz dose (<1% unchanged), and up to 61% of dose is eliminated in feces unchanged [81]. Etravirine elimination is mainly in feces (94% of dose), mostly as unchanged drug [84], and renal elimination is minimal. Nevirapine metabolites are eliminated via renal elimination (80% of dose) [93], while fecal elimination accounts for 10% of dose (approximately 4% unchanged drug). For rilpivirine, 25% of unchanged drug is found in feces, with trace amounts of drug detected in the urine (<1%) [53].

The role of P-gp efflux in efavirenz and nevirapine disposition is unclear [98]. Although induction of P-gp by both agents is observed in vitro, this occurs at supratherapeutic plasma concentrations [99]. Etravirine and rilpivirine are suggested to inhibit P-gp in vitro; however, both are unlikely to cause clinically significant inhibition at standard doses [10, 100–102].

Table 8.2 NNRTI metabolism, induction, and inhibition of cytochrome P450 (CYP) and uridine-5′-diphosphate glucuronosyltransferase (UGT) isoenzymes

NNRTI	CYP substrate	UGT substrate	CYP inhibition	CYP/UGT induction
Efavirenz [85, 88–91]				
	Primary: CYP2B6 Minor: CYP2A6, CYP3A4	Direct: UGT2B7	CYP3A4, CYP2C8, CYP2C9, CYP2C19	CYP3A4, CYP2B6 UGT1A1, UGT1A4
Etravirine [83, 92]				
	Primary: CYP3A4, CYP2C9, CYP2C19	Metabolites: UGT1A3, UGT1A8	CYP2C9, CYP2C19	CYP3A4
Nevirapine [93–95]				
	Primary: CYP3A4 Minor: CYP2B6, CYP2D6		CYP1A2 (weak)	CYP2B6, CYP3A4 (weak)
Rilpivirine [96, 97]				
	Primary: CYP3A4 Minor: CYP1A1/2, CYP3A5, CYP3A7, CYP1B1, CYP2C8/9/10/18/19	Direct: UGT1A4 Metabolites: UGT1A1		

Abbreviations: *CYP* cytochrome P450, *NNRTI* nonnucleoside reverse transcriptase inhibitor, *UGT* uridine-5′-diphosphate glucuronosyltransferase

8.5 NNRTI Drug-Drug Interactions

A common mechanism of NNRTI-related drug-drug interactions is via drug metabolism. NNRTIs are all CYP and/or UGT isoenzyme substrates, making them susceptible to drug-drug interactions by other medications that affect these enzymes. In addition, most NNRTIs induce or inhibit CYP and UGT enzymes; specific enzymes are described in Table 8.2. In vitro, rilpivirine induces or inhibits a range of transporters, UGT, and CYP enzymes. However, these effects were observed at supratherapeutic concentrations and are unlikely to be observed at therapeutic concentrations [103].

In addition to metabolism-based interactions, rilpivirine has pH-dependent solubility, resulting in decreased bioavailability as gastric pH increases [96]. Gastric acid-reducing agents, including proton pump inhibitors and H2-receptor inhibitors, have variable effects on rilpivirine [104].

Only a few drug-drug interactions result in a recommendation for dose adjustment of the NNRTI component. The clinical impact of a change in NNRTI plasma pharmacokinetic exposure will depend upon the NNRTI's therapeutic range, pharmacologic properties, and the individual patient characteristics. For example, high efavirenz exposure has been associated with central nervous system adverse effects [105]; therefore, drug-drug interactions resulting in increased exposure to efavirenz may increase the risk of adverse effects. In contrast, evidence is emerging that a reduced dose of efavirenz may maintain adequate virologic suppression, despite overall lower pharmacokinetic exposure [106]. The individual NNRTI's exposure-related adverse events and the minimum pharmacokinetic parameter associated with virologic outcomes should be considered when evaluating drug-drug interactions that influence NNRTI exposure.

The magnitude of pharmacokinetic NNRTI drug interactions and clinical management recommendations discussed in this section are summarized in the Tables: efavirenz, etravirine, nevirapine, and rilpivirine are individually summarized in Tables 8.3, 8.4, 8.5 and 8.6.

8.5.1   NNRTI Interactions Involving Additive/Synergistic Electrocardiographic Abnormalities

Individuals living with HIV may require the use of QT interval-prolonging antimicrobial agents (e.g., clarithromycin, trimethoprim-sulfamethoxazole, clindamycin, and bedaquiline) [142–144]. An increase in QTcF through 48 weeks of use has been reported with efavirenz-based ART [145]. Furthermore, a healthy volunteer study found a concentration-response relationship between increased efavirenz exposure and extent of QTcF prolongation, most notable in individuals with impaired efavirenz metabolism (homozygous for the *CYP2B6*6* decreased functional allele) [143].

Rilpivirine may also prolong the QT interval, as healthy volunteers receiving supratherapeutic rilpivirine (75 mg or 300 mg daily) had increased QTcF intervals of 10.7–23.3 ms [53]. Further studies of standard dose rilpivirine (25 mg daily) found a gradual increase in QTcF from baseline until 48 weeks; however, the QTcF stabilized beyond this time point [145–147].

QTcF increased by rilpivirine or efavirenz alone is not considered to be clinically significant [145]. Combination with other QT interval-prolonging agents may increase overall QT prolongation risk, and efavirenz or rilpivirine should be used with caution or alternative agents considered [53, 81]. Similarly, agents that inhibit CYP3A4 may also potentiate QT prolongation by yielding supratherapeutic plasma concentrations of efavirenz or rilpivirine, and the potential risk should be considered.

8.5.2   Acid-Reducing Agents

Acid-reducing agents decrease absorption and systemic exposure of rilpivirine when given simultaneously. Proton-pump inhibitors are contraindicated with rilpivirine [53], as coadministration of omeprazole results in rilpivirine concentrations below the desired therapeutic concentration [104]. H2-receptor antagonists also decrease rilpivirine plasma concentrations when administered simultaneously;

Table 8.3 Pharmacokinetic drug-drug interactions with efavirenz

Drug	Effect on coadministered agent	Effect on efavirenz	Clinical recommendation
Anticoagulants and antiplatelet agents			
Apixaban [107]	↓ possible		Consider alternative therapy
Clopidogrel [81, 107]	Active metabolite: ↓ possible		Avoid coadministration and consider alternative antiplatelet agent
Dabigatran [107]	↔ expected		No interaction expected
Prasugrel [107]	↔ expected		No interaction observed with other potent CYP inducers; therefore, may be combined with efavirenz
Rivaroxaban [107]	↓ expected		Consider alternative therapy
Ticagrelor [108]	Active metabolite: ↓ expected		Both ticagrelor and its metabolite have antiplatelet activity; coadministration may alter effectiveness or safety
Warfarin [81]	↓ or ↑ possible		Conflicting reports of higher and lower warfarin doses required to maintain a therapeutic INR; monitor INR and adjust warfarin dose as necessary
Anticonvulsants			
Carbamazepine [109]	↓ AUC 27%	↓ AUC 36%	Monitor virologic outcomes, seizure control, and plasma concentrations of anticonvulsants or antiretrovirals (when available)
Phenobarbital, phenytoin [81, 110]	↓ expected	↓ expected	
Ethosuximide, lacosamide, lamotrigine, tiagabine, zonisamide	↓ possible	↔ expected	
Gabapentin, levetiracetam, pregabalin, topiramate, valproic acid	↔ expected	↔ expected	No clinically significant interactions expected
Antidepressants [81]			
Bupropion	↓AUC 55% Active metabolite: ↔ AUC	↔ expected	Monitor the antidepressant effect and titrate dose as necessary
Citalopram, escitalopram	↓ possible	↔ expected	
Fluoxetine, fluvoxamine	↔ expected	↔ expected	No clinically significant interaction expected
Nefazodone	↓ expected	↑ possible	Monitor the antidepressant effect and titrate dose as necessary and

Paroxetine	↔ AUC		No clinically significant interaction reported
Sertraline	↓ AUC 39%	↔ AUC	Monitor therapeutic effects of sertraline and titrate dose as necessary
Antifungals			
Fluconazole [81]	↔ AUC	↑ AUC 16%	No dose adjustment recommended
Isavuconazole [2]	↓ expected	↑ possible	Consider alternative antifungal treatment; if coadministered, closely monitor antifungal plasma concentrations and signs and symptoms of fungal infection
Itraconazole [81]	↓ AUC 39% Active metabolite: ↓ AUC 37%	↔ AUC	
Posaconazole [91, 111]	↓ AUC 50%	↔ AUC	
Voriconazole [81]	↓ AUC 77%	↑ AUC 44%	Contraindicated at standard dosing of each agent. If combined, increase the voriconazole maintenance dose to 400 mg every 12 h and decrease efavirenz to 300 mg once daily, using the capsule formulation
Antimalarials			
Artemether/lumefantrine [2, 81, 112, 113]	Artemether: ↓ AUC 51%–79% Lumefantrine: ↓ AUC 21%–70% Dihydroartemisinin: ↓ AUC 46%–75%	↔ AUC	Consider alternative antimalarial; if combination required, monitor for antimalarial efficacy and malaria recurrence
Atovaquone/proguanil [2, 114]	Atovaquone: ↓ AUC 75% Proguanil: ↓ AUC 43%	↔ expected	Consider alternative antimalarial prophylaxis
Doxycycline, mefloquine [115]	↓ possible	↔ expected	If decreased exposure occurs, it is unlikely to be clinically significant
Antimycobacterials (limited to Mycobacterium avium complex therapy)			

(continued)

Table 8.3 (continued)

Drug	Effect on coadministered agent	Effect on efavirenz	Clinical recommendation
Clarithromycin [81]	↓ AUC 39% Active metabolite: ↑ AUC 34%	↔ AUC	Consider alternative macrolide therapy (e.g., azithromycin)
Rifabutin [116]	↓ AUC 38%	↔ AUC	Increase rifabutin to 450–600 mg/day
Azithromycin, aminoglycosides, ethambutol, fluoroquinolones	↔ expected	↔ expected	No clinically significant interaction expected
Antiretrovirals			
Integrase inhibitors [2]			
Dolutegravir	↓ AUC 57% ↓ Cmin 75%	↔ expected	If no suspected or confirmed integrase inhibitor resistance, increase dolutegravir to 50 mg twice daily. If integrase inhibitor resistance (suspected or confirmed), consider alternative therapy
Elvitegravir (plus cobicistat or ritonavir)	↓ expected	↑ possible	Avoid coadministration
Raltegravir	↓ AUC 36%	↔ expected	No dose adjustment recommended
Protease inhibitors			
Atazanavir [46]	↓ AUC 74%	↔ expected	Avoid coadministration
Atazanavir/cobicistat [2, 47]	↓ expected	↔ expected	Avoid coadministration in treatment-experienced patients. If ART-naïve, use atazanavir 400 mg plus cobicistat 150 mg once daily; use standard dosing of efavirenz
Atazanavir/ritonavir [46]	↓ AUC 74% (atazanavir without ritonavir) ↔ AUC (atazanavir 400 mg/ritonavir 100 mg)	↑ or ↓ possible	Only use with ritonavir-boosted atazanavir; avoid coadministration in treatment-experienced patients. If antiretroviral-naïve, administer atazanavir 400 mg plus ritonavir 100 mg once daily with standard dose efavirenz; give atazanavir and ritonavir with food and efavirenz on an empty stomach at bedtime
Darunavir/cobicistat [51, 81]	↓ expected	↔ expected	Avoid coadministration

Darunavir/ritonavir [50]	↓ AUC 13% ↓ Cmin 31%	↑ AUC 21%	No dose adjustment recommended
Fosamprenavir/ritonavir [2]	↓ Cmin 31%	↑ or ↓ possible	Administer fosamprenavir 1400 mg plus ritonavir 300 mg if fosamprenavir/ritonavir is given once daily; fosamprenavir 700 mg plus ritonavir 100 mg if twice daily; use standard dosing of efavirenz
Indinavir/ritonavir [117]	↓ AUC 25% ↓ Cmin 50%	↔ AUC	Dose adjustment not established
Lopinavir/ritonavir [58]	↓ AUC 19% ↓ Cmin 39%	↓ AUC 16%	Dose adjustment required: increase lopinavir/ritonavir to 500/125 mg tablets twice daily (533/133 mg [6.5 mL] oral solution twice daily) in combination with efavirenz; neither lopinavir/ritonavir tablets nor oral solution should be administered once daily in combination with efavirenz; use standard dosing of efavirenz
Saquinavir/ritonavir [81]	↓ AUC 62% (SQV alone)	↓ AUC 12% (SQV alone)	Only use with ritonavir-boosted saquinavir; no dose adjustment recommended; use standard dosing of efavirenz
Tipranavir/ritonavir [2, 118]	↔ AUC	↔ AUC	No dose adjustment necessary
Benzodiazepines [119]			
Alprazolam, chlordiazepoxide, clobazam, clonazepam, clorazepate, diazepam, flurazepam	↓ possible		Monitor therapeutic effect of benzodiazepine
Midazolam, triazolam [2]	↓ or ↑ possible		Avoid coadministration; single-dose parenteral midazolam may be used for procedural sedation with monitoring
Lorazepam [81]	↔ AUC		No clinically significant interaction found
Oxazepam, temazolam	↔ expected		No clinically significant interaction expected
Calcium channel blockers [81]			
Diltiazem	↓ AUC 69%	↑ AUC 11%	Titrate calcium channel blocker based on therapeutic response
Amlodipine, felodipine, nicardipine, nifedipine, verapamil	↓ expected	↔ expected	

(continued)

Table 8.3 (continued)

Drug	Effect on coadministered agent	Effect on efavirenz	Clinical recommendation
Corticosteroids [2]			
Dexamethasone (systemic)	↓ possible	↓ possible	Alternative corticosteroid should be used for long-term therapy; if long-term dexamethasone required, monitor virologic response
Herbal products [81]			
St. John's wort		↓ expected	Avoid coadministration
Hormonal contraceptives			
Depot medroxyprogesterone [120]	↔ AUC		No clinically significant interaction observed
Ethinyl estradiol/norgestimate [2, 81, 121]	Ethinyl estradiol: ↔ AUC Norgestimate active metabolites: ↓ AUC 64% and ↓ AUC 83%		Consider alternative or additional contraceptive methods
Ethinyl estradiol/desogestrel [120]	Desogestrel active metabolite: ↓ Cmin 61%		
Etonogestrel implant [81, 120, 140]	↓ AUC 66–82%		Consider alternative or additional contraceptive methods. Unintended pregnancies have been observed during concomitant efavirenz and implant use
Levonorgestrel subdermal implant [122]	↓ AUC 47%		
Levonorgestrel emergency contraception [2, 120]	↓ AUC 58%		Effectiveness of emergency contraception may be reduced

3-Hydroxy-3-methylglutaryl-coenzyme A (HMG-CoA) reductase inhibitors

Atorvastatin [81]	↓ AUC 43% Active metabolites: ↓ AUC 32%	↔ AUC	Monitor serum lipid levels and adjust statin dose if necessary
Pitavastatin [123]	↓ AUC 11%	↔ expected	
Pravastatin [81]	↓ AUC 44%	↔ AUC	
Simvastatin [81]	↓ AUC 68% Active metabolite: ↓ AUC 60%	↔ AUC	
Lovastatin	↓ expected	↔ expected	
Fluvastatin	↑ expected		
Rosuvastatin	↔ expected		
Immunosuppressants [2]			
Cyclosporine, sirolimus, tacrolimus	↓ expected		Increase in immunosuppressant dose may be necessary; therapeutic drug monitoring of immunosuppressant is recommended
Narcotics or treatment for opioid dependence			
Buprenorphine [124]	↓ AUC 49% Active metabolite: ↓ AUC 71%		Monitor therapeutic response to buprenorphine and adjust dose as necessary
Methadone [81, 125]	↓ AUC 52–57%		Monitor for withdrawal symptoms and increase methadone dose if necessary
Phosphodiesterase-5 inhibitors			
Avanafil, sildenafil, tadalafil, vardenafil	↓ possible		No dose adjustment recommended; monitor for desired effect of PDE-5 inhibition therapy

Pharmacokinetic information is summarized as follows: decreased plasma concentration (↓ %), increased plasma concentration (↑ %), or no change (↔), indicating <10% increase or decrease in plasma concentration). For drug combinations with potential for an interaction, but no available evidence, the predicted result of the interaction is described as *expected* or *possible*, which indicates the strength of the prediction based on the pharmacologic characteristics of each drug

Abbreviations: *AUC* area under the plasma concentration time curve, *INR* international normalized ratio, *Cmax* maximum plasma concentration, *Cmin* minimum plasma concentration, *PDE-5* phosphodiesterase-5, *SQV* saquinavir

Table 8.4 Pharmacokinetic drug-drug interactions with etravirine

Drug	Effect on coadministered agent	Effect on etravirine	Clinical recommendation
Anticoagulants and antiplatelet agents			
Apixaban [107]	↓ possible		Consider alternative therapy
Clopidogrel [84, 126]	Active metabolite: ↓ possible		Avoid coadministration and consider alternative antiplatelet agent
Dabigatran [2]	↑ expected		Monitor for dabigatran toxicity or consider alternative therapy
Prasugrel	↔ expected		No interaction observed with other potent CYP inducers and, therefore, may be combined with etravirine
Rivaroxaban [107]	↓ possible		Consider alternative therapy
Ticagrelor [108]	Active metabolite: ↓ possible		Both ticagrelor and its metabolite have antiplatelet activity; coadministration may alter effectiveness or safety
Warfarin [84]	↓ or ↑ possible		Monitor INR, and adjust warfarin dose as necessary
Anticonvulsants			
Carbamazepine, phenobarbital, phenytoin [84]	↓ possible	↓ possible	Avoid coadministration
Ethosuximide, lacosamide, tiagabine, zonisamide	↓ possible	↔ expected	Monitor virologic outcomes, seizure control, and plasma concentrations of anticonvulsants or antiretrovirals (when available)
Oxcarbazepine [127]	↔ expected	↓ possible	
Valproic acid	↔ expected	↑ possible	No dose adjustment recommended, monitor for etravirine-related adverse events
Gabapentin, lamotrigine, levetiracetam, pregabalin, topiramate	↔ expected	↔ expected	No clinically significant interactions expected
Antidepressants			
Bupropion, fluoxetine, fluvoxamine	↔ expected	↔ expected	No clinically significant interaction expected
Citalopram, escitalopram, sertraline	↓ possible	↔ expected	Monitor the antidepressant effect and titrate dose as necessary
Nefazodone	↓ possible	↑ possible	Monitor the antidepressant effect and titrate dose as necessary and monitor for etravirine related adverse events

	↔ AUC	↔ AUC	
Paroxetine [84]	↔ AUC	↔ AUC	No clinically significant interaction reported
Antifungals			
Fluconazole [84, 126]	↔ AUC	↑ AUC 86%	Use with caution due to increased etravirine exposure; monitor antifungal concentrations and clinical effectiveness
Voriconazole [84]	↓ AUC 14%	↑ AUC 36%	
Isavuconazole, itraconazole	↓ possible	↑ expected	
Posaconazole [84]	↔ expected	↑ expected	
Antimalarials			
Artemether/lumefantrine [128]	Artemether: ↓ AUC 38% Dihydroartemisinin: ↓ AUC 15% Lumefantrine: ↓ AUC 13%	↔ AUC	Monitor for antimalarial efficacy
Atovaquone-proguanil [129]	↑ possible	↑ possible	Increased etravirine reported in one case report; no dose adjustment recommended
Doxycycline, mefloquine	↓ possible	↔ expected	If decreased exposure occurs, it is unlikely to be clinically significant
Antimycobacterials (limited to Mycobacterium avium complex therapy) [84]			
Clarithromycin	↓ AUC 39% Active metabolite: ↑ AUC 21%	↑ AUC 42%	Consider alternative macrolide therapy (e.g., azithromycin)
Rifabutin	↓ AUC 17%	↓ AUC 37%	Use standard dose of rifabutin only if etravirine is not combined with a protease inhibitor; monitor for antiretroviral effectiveness
Azithromycin, aminoglycosides, ethambutol, fluoroquinolones	↔ expected	↔ expected	No clinically significant interaction expected
Antiretrovirals			

(continued)

Table 8.4 (continued)

Drug	Effect on coadministered agent	Effect on etravirine	Clinical recommendation
Integrase inhibitors [2]			
Dolutegravir	↓ AUC 71%	↔ expected	Do not coadminister without PK-enhanced atazanavir, darunavir, or lopinavir. If no suspected or confirmed integrase inhibitor resistance, use 50 mg dolutegravir once daily. If integrase inhibitor resistance (suspected or confirmed), increase dolutegravir to 50 mg twice daily
Elvitegravir/cobicistat	↓ possible	↑ possible	Avoid coadministration
Elvitegravir/ritonavir	↔ expected	↔ expected	Elvitegravir dose will depend on coadministered protease inhibitor
Raltegravir	↓ Cmin 34%	↓ Cmin 17%	No dose adjustment recommended
Protease inhibitors [84]			
Atazanavir	↓ AUC 17% ↓ Cmin 47%	↑ AUC 50% ↑ Cmin 58%	Avoid coadministration
Atazanavir/cobicistat [47]	↓ possible	↑ or ↓ possible	
Atazanavir/ritonavir [84]	↓ AUC 4%–14%	↑ AUC 30%	Only use with ritonavir-enhanced atazanavir; no further dose adjustment recommended
Darunavir/cobicistat [51]	↓ possible	↑ or ↓ possible	Avoid coadministration
Darunavir/ritonavir	↑ AUC 15%	↓ AUC 37% ↓ Cmin 49%	Because the efficacy of this combination was evaluated in clinical trials, decreased etravirine is not expected to be clinically significant; no dose adjustment necessary
Fosamprenavir/ritonavir	↑ AUC 69%	↑ or ↓ possible	Avoid coadministration
Indinavir/ritonavir	↔ expected	↑ or ↓ possible	Administer only with ritonavir-enhanced indinavir

Lopinavir/ritonavir	↓ AUC 13%	↓ AUC 35%	Decreased etravirine not expected to be clinically significant; no dose adjustment necessary	
Saquinavir/ritonavir [130]	↔ AUC	↓ AUC 33%		
Tipranavir/ritonavir	↑ AUC 18%	↓ AUC 76%	Avoid coadministration	
Benzodiazepines				
Clobazam [131]	Active metabolite: ↑ possible		No dosage adjustment recommended	
Diazepam [126]	↑ possible		Monitor therapeutic effect of diazepam, and consider dose reduction	
Alprazolam, chlordiazepoxide, clonazepam, clorazepate, flurazepam, midazolam, triazolam	↓ possible		Monitor therapeutic effect of benzodiazepine and withdrawal symptoms	
Lorazepam, oxazepam, temazepam	↔ expected		No clinically significant interaction expected	
Calcium channel blockers				
Amlodipine, felodipine, nifedipine	↓ possible		Titrate calcium channel blocker based on therapeutic response	
Diltiazem, verapamil, nicardipine	↓ possible	↑ possible		
Corticosteroids [2]				
Dexamethasone (systemic)	↓ possible	↓ possible	Alternative corticosteroid should be used for long-term therapy; if long-term dexamethasone required, monitor virologic response	
Herbal products [84]				
St. John's wort		↓ possible	Avoid coadministration	
Hormonal contraceptives [132]				
Ethinyl estradiol/norethindrone	Ethinyl estradiol: ↑ AUC 22% Norethindrone: ↔ AUC		No dose adjustment recommended	

(continued)

Table 8.4 (continued)

Drug	Effect on coadministered agent	Effect on etravirine	Clinical recommendation
3-Hydroxy-3-methylglutaryl-coenzyme A (HMG-CoA) reductase inhibitors			
Atorvastatin [84]	↓ AUC 37% Active metabolite: ↑ AUC 27%	↔ AUC	Monitor serum lipid levels and adjust statin dose as necessary
Lovastatin, simvastatin	↓ possible	↔ expected	
Fluvastatin, pitavastatin	↑ possible		
Pravastatin, rosuvastatin [2]	↔ expected		
Immunosuppressants [2]			
Cyclosporine, sirolimus, tacrolimus	↓ possible		Increase in immunosuppressant dose may be necessary; therapeutic drug monitoring of immunosuppressant is recommended
Narcotics or treatment for opioid dependence [84]			
Buprenorphine	↓ AUC 25% Active metabolite: ↓ AUC 12%		Monitor therapeutic response to buprenorphine and adjust dose as necessary
Methadone	↔ AUC		Monitor for withdrawal symptoms and increase methadone dose if necessary
Phosphodiesterase-5 inhibitors			
Sildenafil [84]	↓ AUC 57% Active metabolite: ↓ AUC 41%		No dose adjustment recommended; monitor for desired effect of PDE-5 inhibition therapy
Avanafil, tadalafil, vardenafil	↓ expected		

Pharmacokinetic information is summarized as follows: decreased plasma concentration (↓ %), increased plasma concentration (↑ %), or no change (↔, indicating <10% increase or decrease in plasma concentration). For drug combinations with potential for an interaction, but no available evidence, the predicted result of the interaction is described as *expected* or *possible*, which indicates the strength of the prediction based on the pharmacologic characteristics of each drug

Abbreviations: *AUC* area under the plasma concentration time curve, *INR* international normalized ratio, *PK* pharmacokinetic, *PDE-5* phosphodiesterase-5

Table 8.5 Pharmacokinetic drug-drug interactions with nevirapine

Drug	Effect on coadministered agent	Effect on nevirapine	Clinical recommendation
Anticoagulants and antiplatelet agents [107]			
Apixaban	↓ possible		Consider alternative therapy
Clopidogrel	↔ expected		No clinically significant interaction expected
Dabigatran	↔ expected		No interaction expected
Prasugrel	↔ expected		No interaction observed with other potent CYP inducers and, therefore, may be combined with nevirapine
Rivaroxaban	↓ possible		Consider alternative therapy
Ticagrelor [108]	Active metabolite: ↓ possible		Both ticagrelor and its metabolite have antiplatelet activity; coadministration may alter effectiveness or safety
Warfarin [133]	↓ or ↑ possible		Case reports suggest higher doses of warfarin are required to maintain therapeutic INR when combined with nevirapine; monitor INR and adjust warfarin dose as necessary
Anticonvulsants			
Carbamazepine, phenobarbital, phenytoin [80]	↓ possible	↓ possible	Monitor virologic outcomes, seizure control, and plasma concentrations of anticonvulsants or antiretrovirals (when available)
Ethosuximide, lacosamide, tiagabine, zonisamide	↓ possible	↔ expected	
Oxcarbazepine	↔ expected	↓ possible	
Gabapentin, lamotrigine, levetiracetam, pregabalin, topiramate, valproic acid	↔ expected	↔ expected	No clinically significant interactions expected

(continued)

Table 8.5 (continued)

Drug	Effect on coadministered agent	Effect on nevirapine	Clinical recommendation
Antidepressants			
Bupropion	↓ possible	↔ expected	Monitor the antidepressant effect and titrate dose as necessary
Citalopram, escitalopram, sertraline	↓ possible	↔ expected	
Fluoxetine [134]	Parent drug and metabolite: ↓ possible	↔ expected	
Fluvoxamine [134]	↔ expected	↑ expected	Nevirapine clearance decreased 33.7%; use caution and monitor for nevirapine-associated toxicities; no change in fluvoxamine concentrations were observed when combined with nevirapine in one report
Nefazodone	↓ possible	↑ possible	Monitor the antidepressant effect and titrate dose as necessary and monitor for nevirapine-related adverse events
Paroxetine	↔ expected	↔ expected	No clinically significant interactions expected
Antifungals			
Fluconazole [2]	↔ AUC	↑ AUC 110%	Consider alternative therapy; monitor for nevirapine-associated hepatotoxicity
Isavuconazole	↓ expected	↑ expected	Dose adjustments for isavuconazole may be necessary; consider monitoring isavuconazole level and antifungal response
Itraconazole [80]	↓ AUC 61%	↔ expected	Avoid coadministration if possible
Posaconazole [111]	↔ expected	↑ expected	Monitor for nevirapine-associated hepatotoxicity
Voriconazole [135]	↓ expected	↑ expected	Monitor for nevirapine-associated hepatotoxicity; monitor antifungal concentrations and clinical effectiveness

Antimalarials

Artemether/lumefantrine [2, 112, 113, 136, 137]	Artemether: ↓ AUC 67%–72% Dihydroartemisinin: ↔ AUC to ↓ AUC 37% Lumefantrine: ↓ AUC 25%–58% ↑ AUC 56% [137]	↔ AUC [136]; ↓ AUC 46% [112]	Studies report varying results for both lumefantrine and nevirapine exposure during combination therapy; the clinical significance of decreased concentrations is unclear; monitor virologic effect, antimalarial efficacy, and lumefantrine toxicity
Atovaquone-proguanil	↔ expected	↔ expected	No clinically significant interaction expected
Doxycycline, mefloquine	↓ possible	↔ expected	If decreased exposure occurs, it is unlikely to be clinically significant

Antimycobacterials (limited to Mycobacterium avium complex therapy) [80]

Clarithromycin	↓ AUC 31% Active metabolite: ↑ AUC 42%	↑ AUC 26%	Avoid coadministration and administer alternative macrolide therapy (e.g., azithromycin)
Rifabutin	↑ AUC 17% Active metabolite: ↑ AUC 24%	↓ Cmin 16%	No dose adjustment recommended; use with caution
Azithromycin, aminoglycosides, ethambutol, fluoroquinolones	↔ expected	↔ expected	No clinically significant interaction expected

Antiretrovirals

Integrase inhibitors [2]

Dolutegravir	↓ AUC 19% ↓ Cmin 34%	↔ expected	No dose adjustment recommended
Elvitegravir (plus cobicistat or ritonavir)	↓ expected	↑ possible	Avoid coadministration
Raltegravir	↔ expected	↔ expected	No dose adjustment recommended
Protease inhibitors			

(continued)

Table 8.5 (continued)

Drug	Effect on coadministered agent	Effect on nevirapine	Clinical recommendation
Atazanavir	↓ expected	↑ possible	Avoid coadministration
Atazanavir/cobicistat [47]	↓ AUC	↑ possible	
Atazanavir/ritonavir [46]	↓ AUC 42%	↑ AUC 25%	
Darunavir/cobicistat [51]	↓ expected	↑ possible	
Darunavir/ritonavir [50]	↑ AUC 24%	↑ AUC 27%	No dose adjustment recommended
Fosamprenavir/ritonavir [2, 80]	↓ AUC 11%	↑ Cmin 22%	Fosamprenavir without ritonavir should not be combined with nevirapine (fosamprenavir AUC ↓ 33%). Administer fosamprenavir 700 mg plus ritonavir 100 mg twice daily; no adjustment for nevirapine
Indinavir/ritonavir [80]	↓ expected	↑ possible	Dose adjustment not established
Lopinavir/ritonavir [58]	↓ AUC 27% ↓ Cmin 51%	↑ possible	Dose adjustment required: increase lopinavir/ritonavir tablets to 500/125 mg twice daily (or 533/133 mg [6.5 mL] oral solution twice daily) in combination with nevirapine; neither lopinavir/ritonavir tablets nor oral solution should be administered once daily in combination with nevirapine
Saquinavir/ritonavir [2]	↓ AUC 24% (saquinavir 600 mg three times daily without ritonavir)	↔ AUC	Dose adjustment not established
Tipranavir/ritonavir [45, 138]	↔ expected	↔ AUC	No dose adjustment necessary
Benzodiazepines [119]			
Alprazolam, chlordiazepoxide, clobazam, clonazepam, clorazepate, diazepam, flurazepam midazolam, triazolam	↓ possible		Monitor therapeutic effect of benzodiazepine
Lorazepam, oxazepam, temazepam	↔ expected		No clinically significant interaction expected
Calcium channel blockers [2]			
Amlodipine, felodipine, nifedipine	↓ possible		Titrate calcium channel blocker based on therapeutic response

Diltiazem, nicardipine, verapamil	↓ possible		↓ possible	
Corticosteroids [2]				
Dexamethasone (systemic)	↓ possible		↓ possible	Alternative corticosteroid should be used for long-term therapy; if long-term dexamethasone required, monitor virologic response
Herbal products [80]				
St. John's wort			↓ expected	Avoid coadministration
Hormonal contraceptives				
Depot medroxyprogesterone [120]	...			No pharmacokinetic data available, but no clinically significant interaction observed
Ethinyl estradiol/norethindrone [2, 139]	Ethinyl estradiol: ↓ AUC 29% ↓ Cmin 58% Norethindrone: ↓ AUC 19%			Consider alternative or additional contraceptive methods
Ethinyl estradiol/desogestrel [2, 120]	Desogestrel active metabolite: ↓ Cmin 22%			Consider alternative or additional contraceptive methods
Etonogestrel implant [140]	↔ AUC			No dose adjustment recommended
Levonorgestrel implant [122]	↑ AUC 30%			
3-Hydroxy-3-methylglutaryl-coenzyme A (HMG-CoA) reductase inhibitors				
Atorvastatin, lovastatin, simvastatin	↓ possible		↔ expected	Monitor serum lipid levels and adjust statin dose as necessary
Fluvastatin, rosuvastatin, pitavastatin, pravastatin	↔ expected		↔ expected	
Immunosuppressants [2]				
Cyclosporine, sirolimus, tacrolimus	↓ possible			Increase in immunosuppressant dose may be necessary; therapeutic drug monitoring of immunosuppressant is recommended

(continued)

Table 8.5 (continued)

Drug	Effect on coadministered agent	Effect on nevirapine	Clinical recommendation
Narcotics or treatment for opioid dependence			
Buprenorphine [141]	↔ AUC Active metabolite: ↓ AUC 14%		No clinically significant interaction expected
Methadone [80, 94]	↓ AUC 50%		Monitor for withdrawal symptoms and increase methadone dose if necessary
Phosphodiesterase-5 inhibitors			
Avanafil, sildenafil, tadalafil, vardenafil	↓ possible		No dose adjustment recommended; monitor for desired effect of PDE-5 inhibition therapy

Pharmacokinetic information is summarized as follows: decreased plasma concentration (↓ %), increased plasma concentration (↑ %), or no change (↔), indicating <10% increase or decrease in plasma concentration). For drug combinations with potential for an interaction, but no available evidence, the predicted result of the interaction is described as *expected* or *possible*, which indicates the strength of the prediction based on the pharmacologic characteristics of each drug

Abbreviations: *AUC* area under the plasma concentration time curve, *INR* international normalized ratio, *Cmin* minimum plasma concentration, *PDE-5* phosphodiesterase-5

Table 8.6 Pharmacokinetic drug-drug interactions with rilpivirine

Drug	Effect on coadministered agent	Effect on rilpivirine	Clinical recommendation
Acid-reducing agents [53]			
Antacids		↓ expected	Administer 2 h before, or 4 h after, rilpivirine dose
H-2 receptor antagonists		↑ AUC 76% when given simultaneously with famotidine	Separate acid-reducing agent from rilpivirine: administer 12 h before, or 4 h after, rilpivirine dose
Proton-pump inhibitors		↓ AUC 40% with omeprazole	Avoid coadministration
Anticonvulsants [53]			
Carbamazepine, oxcarbazepine, phenobarbital, phenytoin	↔ expected	↓ expected	Avoid coadministration
Antidepressants			
Nefazodone	↔ expected	↑ possible	Monitor for rilpivirine related adverse events
Antifungals			
Fluconazole, isavuconazole, itraconazole, posaconazole, voriconazole	↔ expected	↑ expected	No dose adjustment suggested; monitor for rilpivirine-associated adverse events
Antimycobacterials (limited to Mycobacterium avium complex therapy) [53]			
Clarithromycin	↔ expected	↑ expected	Consider alternative macrolide therapy (e.g., azithromycin)
Rifabutin	↔ expected	↓ AUC 42%	Increase rilpivirine to 50 mg once daily; rilpivirine 50 mg daily + rifabutin results in similar exposure to rilpivirine 25 mg daily alone

(continued)

Table 8.6 (continued)

Drug	Effect on coadministered agent	Effect on rilpivirine	Clinical recommendation
Antiretrovirals			
Integrase inhibitors [2]			
Dolutegravir	↔ AUC	↔ AUC	No dose adjustment recommended for either agent
Elvitegravir/cobicistat	↔ expected	↑ expected	Avoid coadministration
Elvitegravir/ritonavir	↔ expected	↑ expected	Use standard elvitegravir dose based on coadministered protease inhibitor; no rilpivirine dose adjustment recommended
Raltegravir	↑ Cmin 27%	↔ AUC	No dose adjustment recommended
Protease inhibitors [2]			
Atazanavir (alone or PK-enhanced)	↔ expected	↑ expected	Standard dosing recommended
Other PK-enhanced protease inhibitors	↔ expected	↑ expected	
Calcium channel blockers			
Diltiazem, verapamil, nicardipine	↔ expected	↑ possible	No dose adjustment recommended
Corticosteroids [53]			
Dexamethasone	↔ expected	↓ possible	Avoid coadministration with more than a single dose of systemic dexamethasone
Herbal products [53]			
St. John's wort		↓ expected	Avoid coadministration

3-Hydroxy-3-methylglutaryl-coenzyme A (HMG-CoA) reductase inhibitors

Atorvastatin [53]	↔ AUC Active metabolites: ↑ AUC 23%–39%	↔ AUC	Monitor serum lipid levels and adjust statin dose if necessary
Fluvastatin, lovastatin, pitavastatin, pravastatin, rosuvastatin, simvastatin	↔ expected	↔ expected	No clinically significant interactions expected

Narcotics or treatment for opioid dependence

Buprenorphine	↔ expected		No clinically significant interaction expected
Methadone [53]	↓ AUC 16%		Monitor for withdrawal symptoms and increase methadone dose if necessary

Pharmacokinetic information is summarized as follows: decreased plasma concentration (↓ %), increased plasma concentration (↑ %), or no change (↔, indicating <10% increase or decrease in plasma concentration). For drug combinations with potential for an interaction, but no available evidence, the predicted result of the interaction is described as *expected* or *possible*, which indicates the strength of the prediction based on the pharmacologic characteristics of each drug

Abbreviations: *AUC* area under the plasma concentration time curve, *INR* international normalized ratio, *PK* pharmacokinetic

however, famotidine administered 12 h prior to rilpivirine does not reduce rilpivirine exposure [53]. Therefore, H2-receptor antagonists should be given 12 h before, or 4 h after, rilpivirine. Effects of antacids have not been studied [104]; product labeling suggests that antacids may be administered if separated by at least 2 h before, or 4 h after, the rilpivirine dose [53].

8.5.3 Anticoagulants and Antiplatelet Agents

Warfarin is a racemic compound containing R- and S-enantiomers. S-warfarin is the more potent enantiomer, accounting for the majority of anticoagulant effect, and is metabolized by CYP2C9. R-warfarin is metabolized by multiple CYP enzymes (CYP3A4, CYP1A, and CYP2C) [148].

Given the metabolism of R- and S-warfarin, both efavirenz and etravirine may exert variable effects on warfarin's metabolism. One case report describes a patient with a supratherapeutic international normalized ratio (INR) and macrohematuria upon coadministration of warfarin 5 mg daily plus efavirenz [149]. Conversely, a case-control study found that patients receiving efavirenz-based ART required higher doses of warfarin to maintain a therapeutic INR ($n = 8$) [148]. If warfarin is combined with efavirenz, careful INR monitoring and warfarin dose titration are recommended.

A case series reported outcomes of nevirapine-based ART and warfarin coadministration in three adult men [133]. In each patient, higher than expected warfarin dosing was required when combined with nevirapine. Therefore, if warfarin and nevirapine are combined, the INR must be monitored regularly with appropriate warfarin titration [2].

Effectiveness of antiplatelet agents must also be considered when used with NNRTIs. Clopidogrel is biotransformed via CYP2C19 to an active metabolite [107]. Efavirenz may decrease clopidogrel's active metabolite via CYP2C19 inhibition [81]. Similarly, etravirine is expected to reduce clopidogrel metabolism via the same mechanism. When possible, avoid clopidogrel and consider alternative antiplatelet therapy with both efavirenz and etravirine [2, 84, 150]. Neither nevirapine nor rilpivirine is expected to have a clinically significant effect on clopidogrel biotransformation [107]. Ticagrelor is metabolized to an active metabolite primarily by CYP3A4, and both ticagrelor and its major metabolite have antiplatelet activity [151]. CYP3A4 induction by NNRTIs, other than rilpivirine, may alter ticagrelor's pharmacokinetic parameters, and it should be coadministered with caution with these agents [108]. Prasugrel is an oral prodrug that is hydrolyzed by carboxylesterases and subsequently oxidized, primarily by CYP3A4 and CYP2B6. Coadministration of prasugrel and a strong CYP3A4 inducer did not result in a significant change in prasugrel's active metabolite. Therefore, all NNRTIs may be combined with prasugrel [107].

New oral anticoagulants have not been studied in combination with NNRTIs. Rivaroxaban is primarily eliminated by CYP3A4, and efavirenz-mediated CYP3A4 induction is expected to reduce rivaroxaban exposure, decreasing anticoagulation [107]. A case report describes a patient on nevirapine-based ART developed venous thromboembolism after two doses of rivaroxaban following a knee replacement surgery [152]. With the exception of rilpivirine, decreased rivaroxaban concentrations are likely with all NNRTIs, and anticoagulant effect should be monitored, or alternative anticoagulants should be considered [107]. Apixaban is metabolized primarily by CYP3A4, but it is also metabolized to a lesser extent by CYP1A2, CYP2C8, CYP2C9, and CYP2C19. Therefore, the impact of NNRTIs on the effectiveness of apixaban is difficult to anticipate but may be impaired by CYP3A4 induction by efavirenz, etravirine, and nevirapine. Neither dabigatran nor the dabigatran etexilate prodrug are significantly metabolized by CYP isoenzymes, and no interactions are anticipated between dabigatran and NNRTIs via this pathway. While weak P-gp inhibition by etravirine may increase dabigatran, the clinical significance of this potential interaction is unclear. Potential drug interactions should be considered before combining these new anticoagulant agents with NNRTIs.

8.5.4 Anticonvulsants

Early generation anticonvulsants are biotransformed by CYP3A4 (carbamazepine) and CYP2C19 (phenobarbital, phenytoin); in addition, they also induce CYP3A4 [126]. A crossover study found that combining efavirenz with carbamazepine significantly decreased exposure of both agents, but there was no change in carbamazepine's active metabolite, suggesting anticonvulsant efficacy may be maintained [109]. One case report describes efavirenz-based ART in combination with phenytoin [110]. In this individual, efavirenz concentrations became subtherapeutic within 1 week of combined use (0.34 mcg/mL), while phenytoin concentrations increased from 17 to 23.5 mg/L after doubling the phenytoin dose. Based on anticipated interactions and these sparse data, caution should be used when combining efavirenz plus carbamazepine, phenytoin, or phenobarbital, with appropriate therapeutic drug monitoring of the anticonvulsant and virologic response [2]. A small study ($n = 4$) identified a decrease in nevirapine's half-life with carbamazepine or phenytoin, but not with phenobarbital [153]. Decreased plasma concentrations of anticonvulsants and nevirapine are possible; therefore, anticonvulsant plasma concentrations and virologic response should be monitored [80]. Neither etravirine nor rilpivirine should be used with carbamazepine, phenobarbital, or phenytoin, due to an expected decrease in antiretroviral exposure [53, 84].

Valproic acid is primarily metabolized by UGT isoenzymes, and it inhibits CYP2C9 [154]. No significant pharmacokinetic changes in either efavirenz or valproic acid were observed when coadministered [155], and no clinically significant interactions are expected with other NNRTIs. Oxcarbazepine inhibits CY2C19 and induces CYP3A4, though to a lesser extent than carbamazepine [156]. No data are available to describe the potential interactions between NNRTIs and oxcarbazepine; however, rilpivirine exposure is expected to be significantly decreased, so it should not be combined with oxcarbazepine [53]. Other anticonvulsants are metabolized by multiple pathways and have not been evaluated in combination with NNRTIs. Specific CYP3A4 substrates include ethosuximide, lacosamide, tiagabine, and zonisamide [154, 157]. Potential metabolism-mediated interactions between these agents and specific NNRTIs (Table 8.2) should be considered when selecting an anticonvulsant or antiretroviral therapy. Use of renally eliminated anticonvulsants (gabapentin, pregabalin, levetiracetam, and topiramate) with NNRTIs may be preferable [158, 159].

8.5.5 Antidepressants

Second-generation antidepressants undergo extensive multiple CYP enzyme-mediated biotransformation to active and inactive metabolites, except bupropion and nefazodone, which are metabolized by a single pathway: CYP2B6 and CYP3A4, respectively [160]. In addition, many antidepressants inhibit various CYP enzymes, highlighting the potential for bidirectional drug-drug interactions between antidepressants and NNRTIs. Overall, studies investigating interactions between antidepressants and NNRTIs are lacking, with data available only for efavirenz and nevirapine combined with some selective serotonin reuptake inhibitors and bupropion (see Tables 8.3 and 8.5) [84, 117, 134, 161].

Despite scarce evidence regarding NNRTI-antidepressant interactions, cotherapy of HIV and depression is important to the overall care of a patient living with HIV. Prior to combining antidepressants with NNRTI therapy, it is important to consult a reliable drug information resource to evaluate overlapping CYP pathways to determine the most appropriate antidepressant. Generally, the clinical impact of NNRTI-antidepressant interactions may be minimized by selecting antidepressants with a primary route of metabolism other than enzymes known to be induced or inhibited by the NNRTI used as part of ART (Table 8.2). Antidepressants with single CYP-mediated metabolic pathways may be at a higher risk for drug-drug interactions related to efavirenz (CYP2B6 and CYP3A induction), etravirine, or nevirapine (CYP3A induction). Similarly, antidepressants that are strong CYP3A inhibitors, like nefazodone, should be cautiously combined with NNRTIs metabolized primarily by CYP3A4. Finally, combination of antidepressants that are associated with QT prolongation should be considered when NNRTI therapy containing rilpivirine or efavirenz is required.

8.5.6 Azole Antifungals

Detailed information about drug-drug interactions associated with all antifungal agents is provided in Chap. 17. Relevant to NNRTI interactions, azole antifungals are primarily biotransformed by CYP isoenzymes (itraconazole, isavuconazole, voriconazole), except posaconazole (UGT1A4) and fluconazole (renal elimination) [162]. All azoles inhibit CYP3A, and most also inhibit CYP2C19 and CYP2C9, except for itraconazole and posaconazole [162, 163]. These effects on shared CYP metabolic pathways may affect the disposition of both azoles and NNRTIs.

Apart from fluconazole, efavirenz decreases azole antifungal concentrations by 37–77% (Table 8.3). Voriconazole is decreased the most and also significantly increases efavirenz exposure [81, 117]. This combination is contraindicated at standard dosing but can be used if the efavirenz dose is reduced 50% (300 mg daily) and the voriconazole dose is increased 100% (400 mg twice daily). Fluconazole, dose-adjusted voriconazole, or non-azole antifungals should be considered in combination with efavirenz when possible. If isavuconazole, itraconazole, or posaconazole are required clinically with efavirenz, antifungal plasma concentrations and clinical response should be closely monitored [81, 111, 163, 164].

Etravirine exposure is increased with both fluconazole and voriconazole, and a similar increase is expected with all other azoles [2, 84, 126]. No change is expected in fluconazole or posaconazole concentrations, but lower isavuconazole and itraconazole concentrations are expected. Overall, azole antifungal dosing should be adjusted based on desired concentrations and clinical response, with monitoring for etravirine-related adverse events [53, 84].

Nevirapine should be avoided with itraconazole due to suboptimal antifungal concentrations [80, 164]. Isavuconazole and voriconazole may also be reduced when combined with nevirapine, so close clinical and therapeutic drug monitoring is required [135, 163]. In contrast, nevirapine is not expected to influence posaconazole metabolism [111]. Nevirapine exposure increases substantially with fluconazole [80], and similarly, posaconazole and voriconazole are expected to increase nevirapine exposure [111, 135]. If coadministration is required, antifungal dosing should be adjusted based on desired concentrations and clinical response, with monitoring for nevirapine-associated hepatotoxicity.

An increase in rilpivirine exposure is possible with all azole antifungals, but rilpivirine is not expected to influence azole antifungal exposure [53].

8.5.7 Antimalarials

Antimalarials are described in Chap. 18. This chapter will review only artemether-lumefantrine for treatment of malaria and antimalarial prophylaxis for travelers.

Both artemether and lumefantrine are primarily biotransformed by CYP3A4 to active metabolites, with minor metabolism by multiple CYP isoenzymes [112, 136, 165]. Dihydroartemisinin is a potent active metabolite of artemether and is subsequently glucuronidated by UGT isoenzymes [112, 136]. Significant decreases in artemether, dihydroartemisinin, and lumefantrine plasma concentrations are consistently described with efavirenz coadministration [2, 112, 113, 166]. The clinical significance of lower exposure is not established, but some studies suggest that lower antimalarial exposure influences clinical outcomes [167], and alternative antimalarial therapy should be considered. However, if coadministration is required, the patient should be monitored for antimalarial efficacy and malaria recurrence [2].

Clinical pharmacokinetic data examining nevirapine with artemether-lumefantrine are conflicting. Studies consistently describe lower artemether plasma exposure in combination with nevirapine [112, 113, 136, 137, 167]. However, lumefantrine exposure may be either increased [113, 137] or possibly decreased [112, 136, 167]. Likewise, dihydroartemisinin plasma exposure is reported to be either decreased [112, 113] or unchanged [136, 167]. Artemisinins also induce CYP3A and CYP2C19, and one study observed lower nevirapine plasma exposure when combined with artemether-lumefantrine [112]; however, this is not observed in all studies [136]. The combination of nevirapine-based ART and artemether-lumefantrine should be used with caution, with monitoring for antimalarial efficacy [2].

Etravirine decreases artemether, dihydroartemisinin, and lumefantrine exposure [128], and patients should be monitored for antimalarial efficacy [2, 128]. In contrast, rilpivirine is not expected to influence the disposition of artemether-lumefantrine. Artemisinins may decrease the exposure of both etravirine and rilpivirine via CYP induction; and lumefantrine plus rilpivirine should be used with caution due to potential risk of additive QT interval prolongation [53, 168].

Atovaquone-proguanil, doxycycline, and mefloquine are commonly used for malaria prevention. Atovaquone is a UGT1A1 substrate, while proguanil is metabolized mainly by CYP2C19 [114]. Atovaquone and proguanil's single-dose plasma exposure is decreased with efavirenz, and alternative antimalarial prophylaxis may be considered [2]. Clinically significant interactions are not expected between other NNRTIs and atovaquone-proguanil. Mefloquine is a CYP3A4 substrate [115], and the metabolism of doxycycline is not fully understood, but strong CYP3A4 inducers decrease doxycycline's half-life [169]. Despite this, no clinically significant pharmacokinetic interactions are expected between NNRTIs and mefloquine or doxycycline [170]. Given the potential for additive central nervous system adverse events with efavirenz and mefloquine, alternative antimalarial prophylaxis may be considered. Similarly, mefloquine and rilpivirine are both suggested to prolong the QT interval [53, 168].

8.5.8 Antimycobacterials

It is critical to consider interactions between NNRTIs and antimycobacterial agents, particularly rifamycins, prior to coadministration. Notably, significant interactions between rifampin and NNRTIs may result in inadequate exposure to etravirine, nevirapine, and rilpivirine. Therefore, only efavirenz may be combined with rifampin-containing tuberculosis therapy. Refer to Chap. 12 for a complete discussion of antituberculosis regimens.

Azithromycin and clarithromycin are key components of *Mycobacterium avium* complex (MAC) prevention and treatment. Azithromycin is not metabolized, so has minimal drug-drug interaction potential with NNRTIs. In contrast, clarithromycin is a CYP3A4 inhibitor that undergoes CYP3A4-mediated metabolism to an active metabolite that is less effective against MAC than the parent compound [80, 84]. All NNRTIs, except rilpivirine, decrease clarithromycin exposure, while increasing its less-potent metabolite [80, 81, 84, 117]. Etravirine and nevirapine plasma concentrations are increased, and a similar increase is expected for rilpivirine [53]. Azithromycin is the preferred macrolide therapy in combination with all NNRTIs.

The treatment of MAC requires the combination of a macrolide plus ethambutol, and in some cases, a quinolone, aminoglycoside, or rifabutin is required. Ethambutol, quinolones, and aminoglycosides are not expected to have a clinically significant interaction with NNRTIs. Rifabutin may result in significant, bidirectional interactions with NNRTIs. Refer to the "Tuberculosis" chapter for details regarding rifabutin metabolism and discussion of these interactions, as well as Tables 8.3, 8.4, 8.5 and 8.6 for management of these interactions. Use caution when combining rilpivirine and efavirenz with macrolides or quinolones, due to the overlapping risk of QT prolongation.

8.5.9   Antiretrovirals

Drug-drug interactions between NNRTIs and NRTIs are addressed in Sect. 3.4.2 and Table 8.1.

8.5.9.1   Integrase Inhibitors

Integrase inhibitors, except raltegravir, are susceptible to interactions with all NNRTIs via CYP3A4 metabolism. Briefly, dolutegravir can be administered at standard doses with nevirapine or rilpivirine but must be given twice daily when combined with efavirenz or etravirine [2]. Furthermore, dolutegravir plus etravirine should not be used without a concomitant protease inhibitor. Elvitegravir/cobicistat should be avoided with all NNRTIs; however, elvitegravir/ritonavir may be used with etravirine or rilpivirine plus a concomitant protease inhibitor.

8.5.9.2   Protease Inhibitors

Protease inhibitors may result in bidirectional drug-drug interactions with most NNRTIs. Efavirenz may only be combined with dose-adjusted, pharmacokinetically enhanced protease inhibitors; however, efavirenz may be used with standard doses of darunavir/ritonavir, saquinavir/ritonavir, and tipranavir/ritonavir (Table 8.3). Efavirenz may not be combined with darunavir/cobicistat, or with atazanavir/cobicistat in treatment-experienced patients, and should only be used with atazanavir/ritonavir in patients without antiretroviral resistance. Etravirine is only recommended with ritonavir-enhanced atazanavir, darunavir, indinavir, lopinavir, or saquinavir (Table 8.4). Nevirapine should only be combined with standard doses of darunavir/ritonavir or tipranavir/ritonavir and dose-adjusted fosamprenavir/ritonavir and lopinavir/ritonavir (Table 8.5). Rilpivirine does not have clinically significant interactions with protease inhibitors (Table 8.6).

8.5.10 Benzodiazepines

Most benzodiazepines are metabolized to variably active metabolites via CYP pathways, many primarily by CYP3A4, which may result in drug-drug interactions with some NNRTIs (see Tables 8.3, 8.4 and 8.5) [119, 126, 131]. A few agents (e.g., lorazepam, oxazepam, and temazepam) are metabolized by direct UGT isoenzyme-mediated glucuronidation to mostly inactive metabolites and are expected to have less interaction potential with NNRTIs [119, 126]. Overall, the clinical impact of NNRTI-benzodiazepine interactions may be minimized by selecting benzodiazepines with a primary route of metabolism other than enzymes affected by NNRTIs (Table 8.2) and by dose adjustment of the benzodiazepine based on clinical effectiveness and adverse effects.

8.5.11 Calcium Channel Blockers

Dihydropyridine and non-dihydropyridine calcium channel blockers (CCB) are primarily metabolized by CYP3A4. Nicardipine, diltiazem, and verapamil also inhibit CYP3A4 [171]. Data are limited for NNRTI-CCB interactions; however, efavirenz decreased exposure of diltiazem and its two active metabolites [81]. Similarly, decreased exposure of other CCBs is expected when combined with either efavirenz, etravirine, or nevirapine [80, 81, 84]. The clinical response to CCBs should be monitored and the dose titrated to effect.

8.5.12 Corticosteroids

Dexamethasone is a CYP3A4 substrate and inducer [96]. Single-dose systemic dexamethasone may be used with all NNRTIs. Multiple doses of systemic dexamethasone are expected to decrease rilpivirine exposure, and long-term systemic dexamethasone should be avoided with rilpivirine [53, 96]. Other NNRTIs have not been studied with systemic dexamethasone, but decreased NNRTI or dexamethasone exposure is possible [2].

8.5.13 Hepatitis C Direct-Acting Antiviral Agents

Drug interactions involving direct-acting antiviral agents are reviewed in Chap. 15.

8.5.14 Herbal Products

St. John's wort (*Hypericum perforatum*) is an inducer of CYP3A4, and it is expected to substantially reduce NNRTI plasma concentrations. For this reason, coadministration of St. John's wort is contraindicated with any NNRTI [81, 94, 96, 126].

8.5.15 Hormonal Contraceptives

Hormonal contraceptives are metabolized by CYP and UGT enzymes and are susceptible to interactions with medications that induce or inhibit these pathways [120, 132]. Ethinyl estradiol is extensively hydroxylated by primarily CYP3A4 and CYP2C9 during first-pass metabolism and also metabolized by UGT1A1 [121]. In vitro, ethinyl estradiol inhibits CYP2B6, CYP2C19, and CYP3A4 [120, 132], but the clinical significance of this inhibition is unknown. All progestins contained in hormonal contraceptives are substrates of CYP3A4 and are susceptible to CYP3A4-mediated interactions [120]. Inhibition or induction of CYP3A by exogenous progestins has been suggested, primarily by in vitro reports; however, the clinical significance is unclear. Overall, hormonal contraceptives do not appear to significantly influence ART systemic exposure [120].

Most data surrounding NNRTI's influence on hormonal contraceptives involve efavirenz-based ART. With the exception of depot medroxyprogesterone [172, 173], both oral and implantable hormonal contraceptive exposures are significantly reduced with efavirenz-based ART [121, 122, 140, 174–176]. Efavirenz reduces the progestin component of contraceptives the most significantly, ranging from 47 to 83% reduction in exposure. Notably, this reduction in progestin exposure corresponds to a higher risk of unintended pregnancy, particularly for contraceptives with low overall exposure, such as implants [122, 177]. Therefore, careful consideration of the choice of hormonal contraceptive is required for patients receiving efavirenz-based ART.

When combined oral contraceptives (COC) were given with etravirine monotherapy, ethinyl estradiol increased, potentially related to CYP2C9 and CYP2C19 inhibition, while there was no effect on norethindrone exposure [120, 132]. No dose modification of COCs is recommended, and the change in ethinyl estradiol exposure is not expected to influence contraceptive effectiveness, but estrogen-related adverse event should be monitored.

In patients receiving nevirapine-based ART plus COCs, moderately lower ethinyl estradiol, etonogestrel (the active metabolite of desogestrel), and norethindrone are observed, likely through CYP3A4 induction [122, 175, 178]. This decrease is less profound than what is observed with efavirenz, and clinical studies do not demonstrate a higher rate of contraceptive failures in women receiving nevirapine-based ART [177]. With non-oral routes of hormonal contraceptives, nevirapine does not decrease progestin exposure (levonorgestrel, etonogestrel, and medroxyprogesterone) [122, 140, 172]. Overall, nevirapine is not likely to reduce the effectiveness of hormonal contraception.

In one study with COCs, rilpivirine monotherapy did not significantly affect the pharmacokinetic parameters of ethinyl estradiol or norethindrone [179]. Based on what is known about rilpivirine's effect on CYP and UGT, it is not expected to clinically influence hormonal contraceptive pharmacokinetics.

8.5.16   3-Hydroxy-3-Methylglutaryl-Coenzyme A (HMG-CoA) Reductase Inhibitors

Atorvastatin, lovastatin, and simvastatin undergo extensive CYP3A4 metabolism [126, 180], whereas fluvastatin and rosuvastatin are primarily metabolized by CYP2C9, with only minimal metabolism of rosuvastatin [181, 182]. Lovastatin and simvastatin exposures are all expected to significantly decrease, while atorvastatin exposure is expected to moderately decrease, with NNRTIs except rilpivirine [126, 180]. No studies have evaluated NNRTIs with fluvastatin or rosuvastatin, but no clinically significant interactions are expected [2, 84].

Pravastatin and pitavastatin are primarily metabolized via glucuronidation [123], with minimal CYP isoenzyme involvement [126]. Efavirenz decreased pravastatin exposure [123, 180], possibly due to altered glucuronidation; however, clinically significant drug-drug interactions are not expected between NNRTIs and pitavastatin or pravastatin [2, 96, 123]. When coadministered, serum lipid levels should be monitored, with HMG-CoA reductase inhibitor dose adjustments as needed.

8.5.17   Immunosuppressants

Cyclosporine, sirolimus, and tacrolimus are substrates of CYP3A4, and decreased concentrations are expected by CYP3A4 induction with all NNRTIs except rilpivirine [80, 81, 84]. A small case series found increased oral clearance of tacrolimus with efavirenz (↑ 179%) [183], while no significant changes in cyclosporine pharmacokinetic parameters were reported when combined with efavirenz or nevirapine. Overall, dose adjustments should be guided by immunosuppressant concentrations and the desired therapeutic range [2].

8.5.18   Narcotics or Treatment for Opioid Dependence

Buprenorphine is primarily biotransformed via CYP3A4 to norbuprenorphine, its active metabolite [124]. Despite lower buprenorphine and norbuprenorphine exposure observed with efavirenz and etravirine coadministration [2, 124], buprenorphine can be administered at standard doses with either agent if the therapeutic response is monitored [2, 124]. No clinically significant interactions are expected with nevirapine or rilpivirine [141].

Methadone is a racemic compound; it primarily exerts its therapeutic effect through the R-enantiomer, which is primarily metabolized by CYP3A4 and CYP2B6, while the less-potent S-methadone is metabolized by CYP2C19 [184]. Methadone exposure is significantly reduced with efavirenz and nevirapine, accompanied by opioid withdrawal symptoms [94, 125]. Withdrawal symptoms should be monitored with efavirenz or nevirapine, and methadone dose titration may be necessary [2, 125]. Methadone was not significantly affected with etravirine or rilpivirine [96, 126], but monitoring for symptoms of methadone withdrawal is advised.

8.5.19   Phosphodiesterase Type-5 (PDE-5) Inhibitors

PDE-5 inhibitors are primarily biotransformed by CYP3A4 [126]. Drug interaction data are only available for etravirine plus sildenafil, which found significantly lower exposure of both sildenafil and its active metabolite. Efavirenz, etravirine, and nevirapine are expected to decrease plasma exposure of all PDE-5 inhibitors, and therapeutic response should be monitored [126].

8.6   Entry Inhibitors

8.6.1   Entry Inhibitor Absorption

Enfuvirtide is only available as a powder for subcutaneous injectable administration, and it reaches maximum plasma concentrations within 4–8 h [11]. Maraviroc is available as an oral tablet, with an absolute bioavailability of 23% under fasted conditions and peak plasma concentration between 0.5 and 4 h after administration [185]. Administration with high-fat breakfast decreases systemic exposure by 33%, but maraviroc may be administered without regard to food.

8.6.2   Protein Binding and Distribution of Entry Inhibitors

Enfuvirtide is highly bound to plasma proteins (92%), primarily albumin, and does not appreciably cross the blood-brain barrier [11]. Its low volume of distribution reflects its confinement to blood (~5.5 L) and failure to distribute into tissue [186]. In contrast, maraviroc has moderate protein binding (76%) to albumin and alpha-1 acid glycoprotein and achieves high tissue distribution (volume of distribution ~194 L) [185].

8.6.3   Metabolism and Elimination of Entry Inhibitors

As a peptide, enfuvirtide is expected to undergo catabolism to its constituent amino acids [186]. Clinical trials have found no clinically significant alteration of enfuvirtide with ritonavir, saquinavir, or rifampin, describing the low drug-drug interaction potential of enfuvirtide. Enfuvirtide's elimination half-life is markedly lower (~4 h) than maraviroc's half-life (14–18 h) [185, 186]. Maraviroc undergoes metabolism primarily via CYP3A to inactive metabolites; it is also a substrate for P-gp efflux, making it susceptible to drug-drug interactions [185]. The majority of the total dose is recovered in feces, with 25% recovered as unchanged drug, while 8% of the total dose is recovered unchanged in urine.

8.7   Entry Inhibitor Drug-Drug Interactions

Enfuvirtide is not associated with any known drug-drug interactions [186]. Maraviroc is often a victim of drug-drug interactions via CYP3A and P-gp; it is not known to be the perpetrator of drug-drug interactions with other medications [187].

Higher maraviroc exposure is expected when coadministered with CYP3A4 and/ or P-gp inhibitors; therefore, maraviroc dose reduction is required when combined with strong CYP3A4 inhibitors (Table 8.7). Furthermore, in patients with compromised renal function (CrCl <30 ml/min), maraviroc should not be coadministered with strong CYP3A4 inhibitors [185]. Moderate CYP3A4 inhibitors, such as diltiazem, may increase maraviroc plasma concentrations; however, no clinical studies have fully investigated this interaction, and maraviroc dose-reduction is not recommended.

Table 8.7 Drug-drug interactions effecting maraviroc dosing

Drug	Effect on maraviroc	Clinical recommendation
Integrase strand transfer inhibitors		
Elvitegravir/ritonavir	↑ AUC 186%	Maraviroc 150 mg twice daily[a]
Dolutegravir	↔ expected	Maraviroc 300 mg twice daily[b]
Raltegravir	↓ AUC 14%	
Nucleoside/nucleotide reverse transcriptase inhibitors		
Emtricitabine, lamivudine, stavudine, TAF, TDF, zidovudine [25, 187]	↔ expected	Maraviroc 300 mg twice daily[b]
Nonnucleoside reverse transcriptase inhibitors		
Efavirenz	↓ AUC 45%	Maraviroc dose 600 mg twice daily;
Etravirine	↓ AUC 53%	avoid coadministration if CrCl < 30 mL/min[a]
Nevirapine	↔ AUC	Maraviroc 300 mg twice daily[b]
Protease inhibitors		
Atazanavir [125, 184, 185, 187]	↑ AUC 257%	Maraviroc 150 mg twice daily[a]
Atazanavir/ritonavir	↑ AUC 388%	
Darunavir/ritonavir	↑ AUC 305%	
Lopinavir/ritonavir +/− efavirenz	↑ AUC 295% ↑ AUC 153% (with efavirenz)	
Tipranavir/ritonavir	↔ AUC	Maraviroc 300 mg twice daily[b]
Miscellaneous strong CYP3A inducers		
Carbamazepine, phenobarbital, phenytoin	↓ expected	Maraviroc 600 mg twice daily[a]
Rifampin	↓ AUC 63%	
Miscellaneous strong CYP3A inhibitors		
Itraconazole, ketoconazole, clarithromycin	↑ expected	Maraviroc 150 mg twice daily[a]

Pharmacokinetic information is summarized as follows: decreased plasma concentration (↓ %), increased plasma concentration (↑ %), or no change (↔, indicating <10% increase or decrease in plasma concentration). For drug combinations with potential for an interaction, but no available evidence, the predicted result of the interaction is described as *expected* or *possible*, which indicates the strength of the prediction based on the pharmacologic characteristics of each drug

Abbreviations: *AUC* area under the plasma concentration time curve, *CrCL* creatinine clearance, *TAF* tenofovir alafenamide fumarate, *TDF* tenofovir disoproxil fumarate

[a]Avoid coadministration in patients with a CrCl < 30 mL/min or who are receiving hemodialysis

[b]Reduce to 150 mg twice daily if symptoms of postural hypotension occur in patients with CrCl < 30 mL/min or who are receiving regular hemodialysis

In contrast, maraviroc plasma concentrations are expected to decrease when combined with CYP3A4 and/or P-gp inducers, including some anticonvulsants and NNRTIs (such as efavirenz and etravirine) [84, 185]. If maraviroc is coadministered with strong or moderate CYP3A4 inducers, a dose increase is recommended (Table 8.7) [185]. However, if maraviroc is combined with both a CYP3A4 inducer and CYP3A4 inhibitor, maraviroc dosing is based on the coadministered CYP3A4 inhibitor (Table 8.7) [84, 188].

Notably, tipranavir/ritonavir, which is both a CYP3A inhibitor and P-gp inducer, does not influence maraviroc exposure, demonstrating the ability of P-gp induction to overcome the influence of CYP3A inhibition on maraviroc's disposition [185]. However, in general, maraviroc dosing is guided by coadministration of CYP3A inhibitors or inducers.

8.8   Conclusion

This chapter highlights drug-drug interactions associated with three classes of anti-retroviral agents. For the antiretrovirals discussed in this chapter, NNRTIs and the entry inhibitor maraviroc have the greatest potential for clinically significant interactions. Because modern ART consists of multiple antiretrovirals used concurrently, the entire ART combination, rather than individual agents, must be considered when evaluating potential interactions. Overlapping mechanisms of pharmacokinetic interactions are often present in a complete ART regimen, which will impact the extent of any resulting drug-drug interaction. Patients receiving ART should be closely monitored for new or changing co-prescribed medication. Frequent review of up-to-date patient medication lists, along with close collaboration and communication between healthcare providers, will allow potential drug-drug interactions to be identified and managed in a timely manner.

References

1. UNAIDS Global AIDS Update 2016. 31 May 2016. Available at: http://www.unaids.org/en/resources/documents/2016/Global-AIDS-update-2016. Accessed 14 Dec 2016
2. Panel on Antiretroviral Guidelines for Adults and Adolescents Guidelines for the use of antiretroviral agents in HIV-1-infected adults and adolescents. Department of Health and Human Services. Available at: http://aidsinfo.nih.gov/ContentFiles/AdultandAdolescentGL.pdf. Accessed 1 Dec 2016
3. World Health Organization (2015) Guideline on when to start antiretroviral therapy and on pre-exposure prophylaxis for HIV. World Health Organization, Geneva
4. Centers for Disease Control and Prevention, U.S. Public Health Service (2014) Pre-exposure prophylaxis for the prevention of HIV infection in the United States—2014: a clinical practice guideline. Available at https://www.cdc.gov/hiv/pdf/prepguidelines2014.pdf. Accessed 14 Dec 2016

5. Evans-Jones JG, Cottle LE, Back DJ, Gibbons S, Beeching NJ, Carey PB et al (2010) Recognition of risk for clinically significant drug interactions among HIV-infected patients receiving antiretroviral therapy. Clin Infect Dis: Off Publ Infect Dis Soc Am 50(10):1419–1421. https://doi.org/10.1086/652149

6. Barry M, Mulcahy F, Merry C, Gibbons S, Back D (1999) Pharmacokinetics and potential interactions amongst antiretroviral agents used to treat patients with HIV infection. Clin Pharmacokinet 36(4):289–304. https://doi.org/10.2165/00003088-199936040-00004

7. Viread(R) (tenofovir disproxil fumarate) [package insert] (2016) Gilead Sciences, Inc., Foster City

8. Margolis DA, Brinson CC, Smith GH, de Vente J, Hagins DP, Eron JJ et al (2015) Cabotegravir plus rilpivirine, once a day, after induction with cabotegravir plus nucleoside reverse transcriptase inhibitors in antiretroviral-naive adults with HIV-1 infection (LATTE): a randomised, phase 2b, dose-ranging trial. Lancet Infect Dis 15(10):1145–1155. https://doi.org/10.1016/s1473-3099(15)00152-8

9. Margolis DA, Podzamczer D, Stellbrink HJ, Lutz T, Angel JB, Richmond G, Clotet B, Gutierrez F, Sloan L, Griffith SK, St Clair M, Dorey D, Ford S, Mrus J, Crauwels H, Smith KY, Williams PE, Spreen WR (2016) Cabotegravir + rilpivirine as long-acting maintenance therapy: LATTE-2 week 48 results. AIDS 2016, 18–22 July 2016, Durban. Oral late breaker abstract THAB0206LB

10. Moss DM, Liptrott NJ, Curley P, Siccardi M, Back DJ, Owen A (2013) Rilpivirine inhibits drug transporters ABCB1, SLC22A1, and SLC22A2 in vitro. Antimicrob Agents Chemother 57(11):5612–5618. https://doi.org/10.1128/AAC.01421-13

11. FUZEON® (enfuvirtide) [package insert] (2015) Hoffmann-La Roche Inc., South San Francisco

12. Ziagen(R) (abacavir) [package insert] (2013) GlaxoSmithKline, Research Triangle Park

13. Zerit(R) (stavudine) [package insert] (2012) Bristol-Myers Squibb Company, Princeton

14. Retrovir(R) (zidovudine) [package insert] (2008) GlaxoSmithKline, Research Triangle Park

15. Epivir(R) (lamivudine) [package insert] (2011) GlaxoSmithKline, Research Triangle Park

16. Emtriva(R) (emtricitabine) [package insert] (2012) Gilead Sciences, Inc.,Foster City

17. Videx(R) EC (didanosine EC) [package insert] (2015) Bristol-Myers Squibb Company, Princeton

18. Videx(R) (didanosine) [package insert] (1999) Bristol-Myers Squibb Company, Princeton

19. Stevens RC, Rodman JH, Yong FH, Carey V, Knupp CA, Frenkel LM (2000) Effect of food and pharmacokinetic variability on didanosine systemic exposure in HIV-infected children. Pediatric AIDS Clinical Trials Group Protocol 144 Study Team. AIDS Res Hum Retrovir 16(5):415–421. https://doi.org/10.1089/088922200309070

20. Panel on Antiretroviral Therapy and Medical Management of HIV-Infected Children Guidelines for the use of antiretroviral agents in pediatric HIV infection. Department of Health and Human Services. Available at: http://aidsinfo.nih.gov/guidelines/html/2/pediatric-treatment-guidelines/0/. Accessed 16 Dec 2016

21. Ray AS, Fordyce MW, Hitchcock MJ (2016) Tenofovir alafenamide: a novel prodrug of tenofovir for the treatment of human immunodeficiency virus. Antivir Res 125:63–70. https://doi.org/10.1016/j.antiviral.2015.11.009

22. Kearney BP, Flaherty JF, Shah J (2004) Tenofovir disoproxil fumarate: clinical pharmacology and pharmacokinetics. Clin Pharmacokinet 43(9):595–612. https://doi.org/10.2165/00003088-200443090-00003

23. Gibson AK, Shah BM, Nambiar PH, Schafer JJ (2016) Tenofovir Alafenamide: a review of its use in the treatment of HIV-1 infection. Ann Pharmacother. https://doi.org/10.1177/1060028016660812

24. Bam RA, Yant SR, Cihlar T (2014) Tenofovir alafenamide is not a substrate for renal organic anion transporters (OATs) and does not exhibit OAT-dependent cytotoxicity. Antivir Ther 19(7):687–692. https://doi.org/10.3851/IMP2770

25. Descovy(R) (emtricitabine/tenofovir alafenamide) [package insert] (2016) Gilead Sciences, Inc., Foster City

26. Markowitz M, Zolopa A, Squires K, Ruane P, Coakley D, Kearney B et al (2014) Phase I/II study of the pharmacokinetics, safety and antiretroviral activity of tenofovir alafenamide, a new prodrug of the HIV reverse transcriptase inhibitor tenofovir, in HIV-infected adults. J Antimicrob Chemother 69(5):1362–1369. https://doi.org/10.1093/jac/dkt532

27. Emtriva(R) (emtricitabine) Summary of product characteristics. Gilead Sciences International Ltd., September 2008. Available at: http://www.ema.europa.eu/docs/en_GB/document_library/EPAR_-_Product_Information/human/000533/WC500055586.pdf. Accessed 12 Oct 2016

28. Calcagno A, Bonora S, Simiele M, Rostagno R, Tettoni MC, Bonasso M et al (2011) Tenofovir and emtricitabine cerebrospinal fluid-to-plasma ratios correlate to the extent of blood-brainbarrier damage. AIDS 25(11):1437–1439. https://doi.org/10.1097/QAD.0b013e3283489cb1

29. Yuen GJ, Weller S, Pakes GE (2008) A review of the pharmacokinetics of abacavir. Clin Pharmacokinet 47(6):351–371. https://doi.org/10.2165/00003088-200847060-00001

30. Michaud V, Bar-Magen T, Turgeon J, Flockhart D, Desta Z, Wainberg MA (2012) The dual role of pharmacogenetics in HIV treatment: mutations and polymorphisms regulating antiretroviral drug resistance and disposition. Pharmacol Rev 64(3):803–833. https://doi.org/10.1124/pr.111.005553

31. Ray AS, Olson L, Fridland A (2004) Role of purine nucleoside phosphorylase in interactions between 2′,3′-dideoxyinosine and allopurinol, ganciclovir, or tenofovir. Antimicrob Agents Chemother 48(4):1089–1095

32. Moss DM, Neary M, Owen A (2014) The role of drug transporters in the kidney: lessons from tenofovir. Front Pharmacol 5(248). https://doi.org/10.3389/fphar.2014.00248

33. Baheti G, Kiser JJ, Havens PL, Fletcher CV (2011) Plasma and intracellular population pharmacokinetic analysis of tenofovir in HIV-1-infected patients. Antimicrob Agents Chemother 55(11):5294–5299. https://doi.org/10.1128/aac.05317-11

34. Ray AS, Cihlar T, Robinson KL, Tong L, Vela JE, Fuller MD et al (2006) Mechanism of active renal tubular efflux of tenofovir. Antimicrob Agents Chemother 50(10):3297–3304. https://doi.org/10.1128/AAC.00251-06

35. Giacomini KM, Huang S-M, Tweedie DJ, Benet LZ, Brouwer KLR, Chu X et al (2010) Membrane transporters in drug development. Nat Rev Drug Discov 9(3):215–236. https://doi.org/10.1038/nrd3028

36. Moore KH, Yuen GJ, Raasch RH, Eron JJ, Martin D, Mydlow PK et al (1996) Pharmacokinetics of lamivudine administered alone and with trimethoprim-sulfamethoxazole. Clin Pharmacol Ther 59(5):550–558. https://doi.org/10.1016/s0009-9236(96)90183-6

37. Bactrim(R) (trimethoprim-sulfamethoxazole) [package insert] (2013) Mutual Pharmaceutical Company, Inc., Philadelphia

38. Odefsey(R) (rilpivirine/emtricitabine/tenofovir alafenamide) Assessment Report. Committee for Medicinal Products for Human Use (CHMP): European Medicines Agency., April 2016

39. Viread(R) (tenofovir) Summary of product characteristics. Gilead Sciences Ltd., December 2011. Available at: http://www.ema.europa.eu/docs/en_GB/document_library/EPAR_-_Product_Information/human/000419/WC500051737.pdf. Accessed 1 Oct 2016

40. Jung N, Lehmann C, Rubbert A, Knispel M, Hartmann P, van Lunzen J et al (2008) Relevance of the organic cation transporters 1 and 2 for antiretroviral drug therapy in human immunodeficiency virus infection. Drug Metab Dispos 36(8):1616–1623. https://doi.org/10.1124/dmd.108.020826

41. Gutierrez F, Fulladosa X, Barril G, Domingo P (2014) Renal tubular transporter-mediated interactions of HIV drugs: implications for patient management. AIDS Rev 16(4):199–212

42. Shitara Y (2011) Clinical importance of OATP1B1 and OATP1B3 in drug-drug interactions. Drug Metab Pharmacokinet 26(3):220–227. https://doi.org/10.2133/dmpk.DMPK-10-RV-094

43. McDowell JA, Chittick GE, Stevens CP, Edwards KD, Stein DS (2000) Pharmacokinetic interaction of abacavir (1592U89) and ethanol in human immunodeficiency virus-infected adults. Antimicrob Agents Chemother 44(6):1686–1690

44. Gourevitch MN, Friedland GH (2000) Interactions between methadone and medications used to treat HIV infection: a review. Mt Sinai J Med NY 67(5–6):429–436
45. Aptivus(R) (tipranavir) [package insert] (2016) Boehringer Ingelheim Pharmaceuticals, Inc., Ridgefield
46. Reyataz(R) (atazanavir) [package insert] (2016) Bristol-Myers Squibb Company, Princeton
47. Evotaz(TM) (atazanavir/cobicistat) [package insert} (2016) Bristol-Myers Squibb Company, Princeton
48. Vistide(R) (cidofovir) [package insert] (2010) Gilead Sciences, Inc., Foster City
49. Luber A, L J, Rooney J, Jaffe H, Flaherty J Drug-drug interaction study with intravenous cidofovir (CDV) and either trimethoprim-sulfamethoxazole (TMP/SMX), didanosine (DDI) or fluconazole (FLU) in HIV-infected individuals. 42nd Interscience Conference on Antimicrobial Agents and Chemotherapy; September 27–30; Chicago, Illinois 2002
50. Prezista(R) (darunavir) [package insert] (2016) Janssen Therapeutics, Titusville
51. Prezcobix(R) (darunavir/cobicistat) [package insert] (2016) Janssen Therapeutics, Division of Janssen Products, LP, Titusville
52. Cimoch PJ, Lavelle J, Pollard R, Griffy KG, Wong R, Tarnowski TL et al (1998) Pharmacokinetics of oral ganciclovir alone and in combination with zidovudine, didanosine, and probenecid in HIV-infected subjects. J Acquir Immune Defic Syndr Hum Retrovirol: Off Publ Int Retrovirol Assoc 17(3):227–234
53. Edurant® (rilpivirine) [package insert] (2014) Janssen Therapeutics, Titusville
54. Rainey PM, Friedland G, McCance-Katz EF, Andrews L, Mitchell SM, Charles C et al (2000) Interaction of methadone with didanosine and stavudine. J Acquir Immune Defic Syndr 24(3):241–248
55. Descovy(R) (emtricitabine/TAF) Summary of Product Characteristics. Gilead Sciences International Ltd., April 2016. Available at: http://www.ema.europa.eu/docs/en_GB/document_library/EPAR_-_Product_Information/human/004094/WC500207650.pdf. Accessed 1 Dec 2016
56. Ramanathan S, Wei X, Custodio J, Wang H, Dave A, Cheng A, Kearney B Pharmacokinetics of a novel EVG/COBI/FTC/GS-7340 single tablet regimen. Paper presented at: 13th international workshop on clinical pharmacology of HIV therapy, Barcelona 2012
57. STRIBILD® (elvitegravir, cobicistat, emtricitabine, tenofovir disoproxil fumarate) [package insert] (2016) Gilead Sciences, Inc., Foster City
58. Kaletra(R) (lopinavir/ritonavir) [package insert] (2016) AbbVie Inc., North Chicago
59. Wenning LA, Friedman EJ, Kost JT, Breidinger SA, Stek JE, Lasseter KC et al (2008) Lack of a significant drug interaction between raltegravir and tenofovir. Antimicrob Agents Chemother 52(9):3253–3258. https://doi.org/10.1128/aac.00005-08
60. Droste JA, Verweij-van Wissen CP, Kearney BP, Buffels R, Vanhorssen PJ, Hekster YA et al (2005) Pharmacokinetic study of tenofovir disoproxil fumarate combined with rifampin in healthy volunteers. Antimicrob Agents Chemother 49(2):680–684. https://doi.org/10.1128/aac.49.2.680-684.2005
61. Lertora JJ, Rege AB, Greenspan DL, Akula S, George WJ, Hyslop NE Jr et al (1994) Pharmacokinetic interaction between zidovudine and valproic acid in patients infected with human immunodeficiency virus. Clin Pharmacol Ther 56(3):272–278
62. University of Liverpool. HIV drug interactions. Available at: http://www.hiv-druginteractions.org/. Accessed 13 Oct 2016
63. Sax PE, Zolopa A, Brar I, Elion R, Ortiz R, Post F et al (2014) Tenofovir alafenamide vs. tenofovir disoproxil fumarate in single tablet regimens for initial HIV-1 therapy: a randomized phase 2 study. J Acquir Immune Defic Syndr 67(1):52–58. https://doi.org/10.1097/qai.0000000000000225
64. Sax PE, Wohl D, Yin MT, Post F, DeJesus E, Saag M et al (2015) Tenofovir alafenamide versus tenofovir disoproxil fumarate, coformulated with elvitegravir, cobicistat, and emtricitabine, for initial treatment of HIV-1 infection: two randomised, double-blind, phase 3, non-inferiority trials. Lancet (London, England) 385(9987):2606–2615. https://doi.org/10.1016/s0140-6736(15)60616-x

65. Barrios A, Rendon A, Negredo E, Barreiro P, Garcia-Benayas T, Labarga P et al (2005) Paradoxical CD4+ T-cell decline in HIV-infected patients with complete virus suppression taking tenofovir and didanosine. AIDS 19(6):569–575
66. Yombi JC, Pozniak A, Boffito M, Jones R, Khoo S, Levy J et al (2014) Antiretrovirals and the kidney in current clinical practice: renal pharmacokinetics, alterations of renal function and renal toxicity. AIDS 28(5):621–632. https://doi.org/10.1097/qad.0000000000000103
67. Mills A, Crofoot G Jr, McDonald C, Shalit P, Flamm JA, Gathe J Jr et al (2015) Tenofovir alafenamide versus tenofovir disoproxil fumarate in the first protease inhibitor-based single-tablet regimen for initial HIV-1 therapy: a randomized phase 2 study. J Acquir Immune Defic Syndr 69(4):439–445. https://doi.org/10.1097/qai.0000000000000618
68. McComsey GA, Kitch D, Daar ES, Tierney C, Jahed NC, Tebas P et al (2011) Bone mineral density and fractures in antiretroviral-naive persons randomized to receive abacavir-lamivudine or tenofovir disoproxil fumarate-emtricitabine along with efavirenz or atazanavir-ritonavir: Aids Clinical Trials Group A5224s, a substudy of ACTG A5202. J Infect Dis 203(12):1791–1801. https://doi.org/10.1093/infdis/jir188
69. Famvir(R) (famciclovir) [package insert] (2016) Novartis Pharmaceuticals Corporation, East Hanover
70. Aweeka FT, Gambertoglio JG, van der Horst C, Raasch R, Jacobson MA (1992) Pharmacokinetics of concomitantly administered foscarnet and zidovudine for treatment of human immunodeficiency virus infection (AIDS Clinical Trials Group protocol 053). Antimicrob Agents Chemother 36(8):1773–1778
71. Harvoni (R) (ledipasvir and sofosbuvir) [package insert] (2017) Gilead Sciences, Inc., Foster City
72. Epclusa(R) (sofosbuvir and velpatasvir) [package insert] (2017) Gilead Sciences, Inc., Foster City
73. Rahimi R, Abdollahi M (2012) An update on the ability of St. John's wort to affect the metabolism of other drugs. Expert Opin Drug Metab Toxicol 8(6):691–708. https://doi.org/10.151 7/17425255.2012.680886
74. McCance-Katz EF, Rainey PM, Friedland G, Kosten TR, Jatlow P (2001) Effect of opioid dependence pharmacotherapies on zidovudine disposition. Am J Addict 10(4):296–307
75. McCance-Katz EF, Rainey PM, Jatlow P, Friedland G (1998) Methadone effects on zidovudine disposition (AIDS Clinical Trials Group 262). JAIDS J Acquir Immune Defic Syndr 18(5):435–443
76. Baker J, Rainey PM, Moody DE, Morse GD, Ma Q, McCance-Katz EF (2010) Interactions between buprenorphine and antiretrovirals: nucleos(t)ide reverse transcriptase inhibitors (NRTI) didanosine, lamivudine, and tenofovir. Am J Addict 19(1):17–29. https://doi.org/10.1111/j.1521-0391.2009.00004.x
77. Lee BL, Tauber MG, Sadler B, Goldstein D, Chambers HF (1996) Atovaquone inhibits the glucuronidation and increases the plasma concentrations of zidovudine. Clin Pharmacol Ther 59(1):14–21. https://doi.org/10.1016/s0009-9236(96)90019-3
78. Khoo SH GS, Seden K, Black D Systematic review: Drug-drug Interactions between Antiretrovirals and medications used to treat TB, Malaria, Hepatitis B&C and opioid dependence. 2009. Available at: http://www.who.int/hiv/topics/treatment/drug_drug_interactions_review.pdf. Accessed 1 Dec 2016
79. Hall AM, Hendry BM, Nitsch D, Connolly JO (2011) Tenofovir-associated kidney toxicity in HIV-infected patients: a review of the evidence. Am J kidney Dis: Off J Natl Kidney Found 57(5):773–780. https://doi.org/10.1053/j.ajkd.2011.01.022
80. Viramune(R) (nevirapine) [package insert] (2014) Boehringer Ingelheim Pharmaceuticals, Inc., Ridgefield
81. Sustiva® (efavirenz) [package insert] (2014) Princeton, Bristol-Myers Squibb Company
82. Usach I, Melis V, Peris JE (2013) Non-nucleoside reverse transcriptase inhibitors: a review on pharmacokinetics, pharmacodynamics, safety and tolerability. J Int AIDS Soc 16:1–14. https://doi.org/10.7448/IAS.16.1.18567

83. Dickinson L, Khoo S, Back D (2010) Pharmacokinetic evaluation of etravirine. Expert Opin Drug Metab Toxicol 6(12):1575–1585. https://doi.org/10.1517/17425255.2010.535811

84. Intelence(R) (etravirine) [package insert] (2013) Janssen Therapeutics, Titusville

85. Aouri M, Barcelo C, Ternon B, Cavassini M, Anagnostopoulos A, Yerly S et al (2016) Vivo profiling and distribution of known and novel phase I and phase II metabolites of efavirenz in plasma, urine, and cerebrospinal fluid. Drug Metab Dispos 44(1):151–161. https://doi.org/10.1124/dmd.115.065839.

86. Tiraboschi JM, Niubo J, Vila A, Perez-Pujol S, Podzamczer D (2012) Etravirine concentrations in CSF in HIV-infected patients. J Antimicrob Chemother 67(6):1446–1448. https://doi.org/10.1093/jac/dks048

87. Mora-Peris B, Watson V, Vera JH, Weston R, Waldman AD, Kaye S et al (2014) Rilpivirine exposure in plasma and sanctuary site compartments after switching from nevirapine-containing combined antiretroviral therapy. J Antimicrob Chemother 69(6):1642–1647. https://doi.org/10.1093/jac/dku018

88. Cho DY, Ogburn ET, Jones D, Desta Z (2011) Contribution of N-glucuronidation to efavirenz elimination in vivo in the basal and rifampin-induced metabolism of efavirenz. Antimicrob Agents Chemother 55(4):1504–1509. https://doi.org/10.1128/AAC.00883-10

89. Lee LS, Pham P, Flexner C (2012) Unexpected drug-drug interactions in human immunodeficiency virus (HIV) therapy: induction of UGT1A1 and bile efflux transporters by Efavirenz. Ann Acad Med Singap 41(12):559–562

90. McDonagh EM, Lau JL, Alvarellos ML, Altman RB, Klein TE (2015) PharmGKB summary: Efavirenz pathway, pharmacokinetics. Pharmacogenet Genomics 25(7):363–376. https://doi.org/10.1097/FPC.0000000000000145

91. Krishna G, Moton A, Ma L, Martinho M, Seiberling M, McLeod J (2009) Effects of oral posaconazole on the pharmacokinetics of atazanavir alone and with ritonavir or with efavirenz in healthy adult volunteers. J Acquir Immune Defic Syndr 51(4):437–444

92. Yanakakis LJ, Bumpus NN (2012) Biotransformation of the antiretroviral drug etravirine: metabolite identification, reaction phenotyping, and characterization of autoinduction of cytochrome P450-dependent metabolism. Drug Metab Dispos 40(4):803–814. https://doi.org/10.1124/dmd.111.044404

93. Riska P, Lamson M, MacGregor T, Sabo J, Hattox S, Pav J et al (1999) Disposition and biotransformation of the antiretroviral drug nevirapine in humans. Drug Metab Dispos 27(8):895–901

94. Back DGS, Khoo S (2003) Pharmacokinetic drug interactions with nevirapine. J Acquir Immune Defic Syndr 34:8–14

95. von Moltke LL, Greenblatt DJ, Granda BW, Giancarlo GM, Duan SX, Daily JP et al (2001) Inhibition of human cytochrome P450 isoforms by nonnucleoside reverse transcriptase inhibitors. J Clin Pharmacol 41(1):85–91. https://doi.org/10.1177/00912700122009728

96. Crauwels H, van Heeswijk RP, Stevens M, Buelens A, Vanveggel S, Boven K et al (2013) Clinical perspective on drug-drug interactions with the non-nucleoside reverse transcriptase inhibitor rilpivirine. AIDS Rev 15(2):87–101

97. Lade JM, Avery LB, Bumpus NN (2013) Human biotransformation of the nonnucleoside reverse transcriptase inhibitor rilpivirine and a cross-species metabolism comparison. Antimicrob Agents Chemother 57(10):5067–5079. https://doi.org/10.1128/AAC.01401-13

98. Heil SG, van der Ende ME, Schenk PW, van der Heiden I, Lindemans J, Burger D et al (2012) Associations between ABCB1, CYP2A6, CYP2B6, CYP2D6, and CYP3A5 alleles in relation to efavirenz and nevirapine pharmacokinetics in HIV-infected individuals. Ther Drug Monit 34(2):153–159. https://doi.org/10.1097/FTD.0b013e31824868f3

99. Störmer E, von Moltke LL, Perloff MD, Greenblatt DJ (2002) Differential modulation of P-glycoprotein expression and activity by non-nucleoside HIV-1 reverse transcriptase inhibitors in cell culture. Pharm Res 19(7):1038–1045. https://doi.org/10.1023/a:1016430825740

100. Kakuda TN, Van Solingen-Ristea RM, Onkelinx J, Stevens T, Aharchi F, De Smedt G et al (2014) The effect of single- and multiple-dose etravirine on a drug cocktail of representative cytochrome P450 probes and digoxin in healthy subjects. J Clin Pharmacol 54(4):422–431. https://doi.org/10.1002/jcph.214

101. Schöller-Gyüre M, Kakuda TN, Raoof A, De Smedt G, Hoetelmans RM (2009) Clinical pharmacokinetics and pharmacodynamics of etravirine. Clin Pharmacokinet 48. https://doi.org/10.2165/10895940-000000000-00000

102. Zembruski NCL, Haefeli WE, Weiss J (2011) Interaction potential of etravirine with drug transporters assessed in vitro. Antimicrob Agents Chemother 55(3):1282–1284. https://doi.org/10.1128/AAC.01527-10

103. Weiss J, Haefeli WE (2013) Potential of the novel antiretroviral drug rilpivirine to modulate the expression and function of drug transporters and drug-metabolising enzymes in vitro. Int J Antimicrob Agents 41(5):484–487. https://doi.org/10.1016/j.ijantimicag.2013.01.004

104. Lewis JM, Stott KE, Monnery D, Seden K, Beeching NJ, Chaponda M et al (2016) Managing potential drug-drug interactions between gastric acid-reducing agents and antiretroviral therapy: experience from a large HIV-positive cohort. Int J STD AIDS 27(2):105–109. https://doi.org/10.1177/0956462415574632

105. Gutiérrez F, Navarro A, Padilla S, Antón R, Masiá M, Borrás J et al (2005) Prediction of neuropsychiatric adverse events associated with long-term Efavirenz therapy, using plasma drug level monitoring. Clin Infect Dis 41(11):1648–1653. https://doi.org/10.1086/497835

106. Carey D, Puls R, Amin J, Losso M, Phanupak P, Foulkes S et al (2015) Efficacy and safety of efavirenz 400 mg daily versus 600 mg daily: 96-week data from the randomised, double-blind, placebo-controlled, non-inferiority ENCORE1 study. Lancet Infect Dis 15(7):793–802. https://doi.org/10.1016/s1473-3099(15)70060-5

107. Egan G, Hughes CA, Ackman ML (2014) Drug interactions between antiplatelet or novel oral anticoagulant medications and antiretroviral medications. Ann Pharmacother 48(6):734–740. https://doi.org/10.1177/1060028014523115

108. Brilinta(R) (ticagrelor) [package insert] (2016) AstraZeneca Pharmaceuticals LP, Wilmington

109. Ji P, Damle B, Xie J, Unger SE, Grasela DM, Kaul S (2008) Pharmacokinetic interaction between efavirenz and carbamazepine after multiple-dose administration in healthy subjects. J Clin Pharmacol 48(8):948–956. https://doi.org/10.1177/0091270008319792

110. Robertson SM, Penzak SR, Lane J, Pau AK, Mican JMA (2005) Potentially significant interaction between efavirenz and phenytoin: a case report and review of the literature. Clin Infect Dis: Off Publ Infect Dis Soc Am 41(2):e15–e18. https://doi.org/10.1086/431208

111. Noxafil(R) (posaconazole) [package insert] (2016) Merck & Co., Inc., Whitehouse Station

112. Byakika-Kibwika P, Lamorde M, Mayito J, Nabukeera L, Namakula R, Mayanja-Kizza H et al (2012) Significant pharmacokinetic interactions between artemether/lumefantrine and efavirenz or nevirapine in HIV-infected Ugandan adults. J Antimicrob Chemother 67(9):2213–2221. https://doi.org/10.1093/jac/dks207

113. Hoglund RM, Byakika-Kibwika P, Lamorde M, Merry C, Ashton M, Hanpithakpong W et al (2015) Artemether-lumefantrine co-administration with antiretrovirals: population pharmacokinetics and dosing implications. Br J Clin Pharmacol 79(4):636–649. https://doi.org/10.1111/bcp.12529

114. van Luin M, Van der Ende ME, Richter C, Visser M, Faraj D, Van der Ven A et al (2010) Lower atovaquone/proguanil concentrations in patients taking efavirenz, lopinavir/ritonavir or atazanavir/ritonavir. AIDS 24(8):1223–1226. https://doi.org/10.1097/QAD.0b013e3283389129

115. Dooley KE, Flexner C, Adriana AS (2008) Drug interactions involving combination antiretroviral therapy and other anti-infective agents: repercussions for resource-limited countries. J Infect Dis 198(7):948–961. https://doi.org/10.1086/591459

116. Panel on Antiretroviral Guidelines for Adults and Adolescents. Guidelines for prevention and treatment of opportunistic infections in HIV-infected adults and adolescents. Department of Health and Human Services. Available at: https://aidsinfo.nih.gov/contentfiles/lvguidelines/Adult_OI.pdf. Accessed 14 Jan 17

117. Sustiva(R) (efavirenz). Summary of product characteristics. Bristol-Myers Squibb Pharma EEIG., 2014. Available at: http://www.ema.europa.eu/docs/en_GB/document_library/EPAR_-_Product_Information/human/000249/WC500058311.pdf. Accessed 1 Oct 2016

118. la Porte CJ, Sabo JP, Beique L, Cameron DW (2009) Lack of effect of efavirenz on the pharmacokinetics of tipranavir-ritonavir in healthy volunteers. Antimicrob Agents Chemother 53(11):4840–4844. https://doi.org/10.1128/aac.00462-09

119. Bailey L, Ward M, Musa MN (1994) Clinical pharmacokinetics of benzodiazepines. J Clin Pharmacol 34(8):804–811. https://doi.org/10.1002/j.1552-4604.1994.tb02043.x

120. Scarsi KK, Darin KM, Chappell CA, Nitz SM, Lamorde M (2016) Drug-drug interactions, effectiveness, and safety of hormonal contraceptives in women living with HIV. Drug Saf 39(11):1053–1072. https://doi.org/10.1007/s40264-016-0452-7

121. Sevinsky H, Eley T, Persson A, Garner D, Yones C, Nettles R et al (2011) The effect of efavirenz on the pharmacokinetics of an oral contraceptive containing ethinyl estradiol and norgestimate in healthy HIV-negative women. Antivir Ther 16(2):149–156. https://doi.org/10.3851/imp1725

122. Scarsi KK, Darin KM, Nakalema S, Back DJ, Byakika-Kibwika P, Else LJ et al (2016) Unintended pregnancies observed with combined use of the levonorgestrel contraceptive implant and efavirenz-based antiretroviral therapy: a three-arm pharmacokinetic evaluation over 48 weeks. Clin Infect Dis: Off Publ Infect Dis Soc Am 62(6):675–682. https://doi.org/10.1093/cid/civ1001

123. Malvestutto CD, Ma Q, Morse GD, Underberg JA, Aberg JA (2014) Lack of pharmacokinetic interactions between pitavastatin and efavirenz or darunavir/ritonavir. J Acquir Immune Defic Syndr 67(4):390–396. https://doi.org/10.1097/qai.0000000000000333

124. McCance-Katz EF, Moody DE, Morse GD, Friedland G, Pade P, Baker J et al (2006) Interactions between buprenorphine and antiretrovirals. I. The nonnucleoside reverse-transcriptase inhibitors efavirenz and delavirdine. Clin Infect Dis: Off Publ Infect Dis Soc Am 43(Suppl 4):S224–S234. https://doi.org/10.1086/508187

125. Clarke SM, Mulcahy FM, Tjia J, Reynolds HE, Gibbons SE, Barry MG et al (2001) The pharmacokinetics of methadone in HIV-positive patients receiving the non-nucleoside reverse transcriptase inhibitor efavirenz. Br J Clin Pharmacol 51(3):213–217

126. Kakuda TN, Schöller-Gyüre M, Hoetelmans RMW (2011) Pharmacokinetic interactions between etravirine and non-antiretroviral drugs. Clin Pharmacokinet 50(1):25–39. https://doi.org/10.2165/11534740-000000000-00000

127. Perucca E (2006) Clinically relevant drug interactions with antiepileptic drugs. Br J Clin Pharmacol 61(3):246–255. https://doi.org/10.1111/j.1365-2125.2005.02529.x

128. Kakuda TN, DeMasi R, van Delft Y, Mohammed P (2013) Pharmacokinetic interaction between etravirine or darunavir/ritonavir and artemether/lumefantrine in healthy volunteers: a two-panel, two-way, two-period, randomized trial. HIV Med 14(7):421–429. https://doi.org/10.1111/hiv.12019

129. Tommasi C, Bellagamba R, Tempestilli M, D'Avolio A, Gallo AL, Ivanovic J et al (2011) Marked increase in etravirine and saquinavir plasma concentrations during atovaquone/proguanil prophylaxis. Malar J 10(1):141. https://doi.org/10.1186/1475-2875-10-141

130. Intelence(R) (etravirine) Summary of Product Characteristics. Boehringer Ingelheim International GmbH., October 2010. Available at: http://www.ema.europa.eu/docs/en_GB/document_library/EPAR_-_Product_Information/human/000900/WC500034180.pdf. Accessed 29 Mar 2017

131. Naccarato M, Yoong D, Kovacs C, Gough KA (2012) Case of a potential drug interaction between clobazam and etravirine-based antiretroviral therapy. Antivir Ther 17(3):589–592. https://doi.org/10.3851/imp1953

132. Scholler-Gyure M, Kakuda TN, Woodfall B, Aharchi F, Peeters M, Vandermeulen K et al (2009) Effect of steady-state etravirine on the pharmacokinetics and pharmacodynamics of ethinylestradiol and norethindrone. Contraception 80(1):44–52. https://doi.org/10.1016/j.contraception.2009.01.009

133. Dionisio D, Mininni S, Bartolozzi D, Esperti F, Vivarelli A, Leoncini F (2001) Need for increased dose of warfarin in HIV patients taking nevirapine. AIDS 15(2):277–278

134. de Maat MMR, Huitema ADR, Mulder JW, Meenhorst PL, van Gorp ECM, Mairuhu ATA et al (2003) Drug interaction of fluvoxamine and fluoxetine with nevirapine in HIV-1-infected individuals. Clin Drug Investig 23(10):629–637. https://doi.org/10.2165/00044011-200323100-00002

135. Vfend(R) (voriconazole) [package insert] (2015) Pfizer inc., New York

136. Parikh S, Fehintola F, Huang L, Olson A, Adedeji WA, Darin KM et al (2015) Artemether-lumefantrine exposure in HIV-infected Nigerian subjects on nevirapine-containing antiretroviral therapy. Antimicrob Agents Chemother 59(12):7852–7856. https://doi.org/10.1128/aac.01153-15
137. Kredo T, Mauff K, Van der Walt JS, Wiesner L, Maartens G, Cohen K et al (2011) Interaction between artemether-lumefantrine and nevirapine-based antiretroviral therapy in HIV-1-infected patients. Antimicrob Agents Chemother 55(12):5616–5623. https://doi.org/10.1128/aac.05265-11
138. Aptivus(R) (tipranavir) Summary of Product Characteristics. Boehringer Ingelheim International GmbH., October 2010. http://www.ema.europa.eu/docs/en_GB/document_library/EPAR_-_Product_Information/human/000631/WC500025936.pdf. Accessed 1 Oct 2016
139. Mildvan D, Yarrish R, Marshak A, Hutman HW, McDonough M, Lamson M et al (2002) Pharmacokinetic interaction between nevirapine and ethinyl estradiol/norethindrone when administered concurrently to HIV-infected women. J Acquir Immune Defic Syndr 29(5):471–477
140. Chappell CA, Lamorde M, Nakalema S, Chen BA, Mackline H, Riddler SA et al (2017) Efavirenz decreases etonogestrel exposure: a pharmacokinetic evaluation of implantable contraception with antiretroviral therapy. AIDS 31(14):1965-1972. http://doi:10.1097/qad.0000000000001591.
141. McCance-Katz EF, Moody DE, Morse GD, Ma Q, Rainey PM (2010) Lack of clinically significant drug interactions between nevirapine and buprenorphine. Am J Addict 19(1):30–37. https://doi.org/10.1111/j.1521-0391.2009.00006.x
142. Sirturo(R) (bedaquiline) (2016) Summary of product characteristics. Janssen-Cilag International NV
143. Abdelhady AM, Shugg T, Thong N, Lu JB, Kreutz Y, Jaynes HA et al (2016) Efavirenz inhibits the Human Ether-A-Go-Go Related Current (hERG) and induces QT interval prolongation in CYP2B6*6*6 allele carriers. J Cardiovasc Electrophysiol 27(10):1206–1213. https://doi.org/10.1111/jce.13032
144. Sani MU, Okeahialam BN (2005) QTc interval prolongation in patients with HIV and AIDS. J Natl Med Assoc 97(12):1657–1661
145. Cohen CJ, Molina JM, Cahn P, Clotet B, Fourie J, Grinsztejn B et al (2012) Efficacy and safety of rilpivirine (TMC278) versus efavirenz at 48 weeks in treatment-naive HIV-1-infected patients: pooled results from the phase 3 double-blind randomized ECHO and THRIVE Trials. J Acquir Immune Defic Syndr 60(1):33–42. https://doi.org/10.1097/QAI.0b013e31824d006e
146. Cohen CJ, Molina JM, Cassetti I, Chetchotisakd P, Lazzarin A, Orkin C et al (2013) Week 96 efficacy and safety of rilpivirine in treatment-naive, HIV-1 patients in two Phase III randomized trials. AIDS 27(6):939–950. https://doi.org/10.1097/QAD.0b013e32835cee6e
147. Pozniak AL, Morales-Ramirez J, Katabira E, Steyn D, Lupo SH, Santoscoy M et al (2010) Efficacy and safety of TMC278 in antiretroviral-naive HIV-1 patients: week 96 results of a phase IIb randomized trial. AIDS 24(1):55–65. https://doi.org/10.1097/QAD.0b013e32833032ed
148. Esterly JS, Darin KM, Gerzenshtein L, Othman F, Postelnick MJ, Scarsi KK (2013) Clinical implications of antiretroviral drug interactions with warfarin: a case–control study. J Antimicrob Chemother 68(6):1360–1363. https://doi.org/10.1093/jac/dkt043
149. Liedtke MD, Rathbun RC (2010) Drug interactions with antiretrovirals and warfarin. Expert Opin Drug Saf 9(2):215–223. https://doi.org/10.1517/14740330903493458
150. Plavix(R) (clopidogrel) [package insert] (2016) Bristol-Myers Squibb/Sanofi Pharmaceuticals Partnership, Bridgewater
151. Htun WW, Steinhubl SR (2013) Ticagrelor: the first novel reversible P2Y(12) inhibitor. Expert Opin Pharmacother 14(2):237–245. https://doi.org/10.1517/14656566.2013.757303
152. Bates D, Dalton B, Gilmour J, Kapler J (2013) Venous thromboembolism due to suspected interaction between rivaroxaban and nevirapine. Can J Hosp Pharm 66(2):125–129

153. Birbeck GL, French JA, Perucca E, Simpson DM, Fraimow H, George JM et al (2012) Evidence-based guideline: antiepileptic drug selection for people with HIV/AIDS: report of the quality standards subcommittee of the American Academy of Neurology and the Ad Hoc Task Force of the Commission on Therapeutic Strategies of the International League Against Epilepsy. Neurology 78(2):139–145. https://doi.org/10.1212/WNL.0b013e31823efcf8

154. Liedtke MD, Lockhart SM, Rathbun RC (2004) Anticonvulsant and antiretroviral interactions. Ann Pharmacother 38(3):482–489. https://doi.org/10.1345/aph.1D309

155. DiCenzo R, Peterson D, Cruttenden K, Morse G, Riggs G, Gelbard H et al (2004) Effects of valproic acid coadministration on plasma efavirenz and lopinavir concentrations in human immunodeficiency virus-infected adults. Antimicrob Agents Chemother 48(11):4328–4331. https://doi.org/10.1128/aac.48.11.4328-4331.2004

156. Goicoechea M, Best B, Capparelli E, Haubrich R (2006) Concurrent use of efavirenz and oxcarbazepine may not affect efavirenz plasma concentrations. Clin Infect Dis: Off Publ Infect Dis Soc Am 43(1):116–117. https://doi.org/10.1086/504952

157. Vimpat(R) (lacosamide) [package insert] (2015) UCB, Inc., Smyrna

158. Topamax(R) (topiramate) [package insert] (2014) Janssen Pharmaceuticals, Inc., Titusville

159. Kirmani BF, Mungall-Robinson D (2014) Role of anticonvulsants in the management of AIDS related seizures. Front Neurol 5:10. https://doi.org/10.3389/fneur.2014.00010

160. Spina E, Santoro V, D'Arrigo C (2008) Clinically relevant pharmacokinetic drug interactions with second-generation antidepressants: an update. Clin Ther 30(7):1206–1227

161. Robertson SM, Maldarelli F, Natarajan V, Formentini E, Alfaro RM, Penzak SR (2008) Efavirenz induces CYP2B6-mediated hydroxylation of bupropion in healthy subjects. J Acquir Immune Defic Syndr 49(5):513–519. https://doi.org/10.1097/QAI.0b013e318183a425

162. Brüggemann RJM, Alffenaar J-WC, Blijlevens NMA, Billaud EM, Kosterink JGW, Verweij PE et al (2009) Clinical relevance of the pharmacokinetic interactions of azole antifungal drugs with other coadministered agents. Clin Infect Dis 48(10):1441–1458. https://doi.org/10.1086/598327

163. Cresemba (R) (isavuconazonium sulfate) [package insert] (2015) Astellas Pharma US, Inc., Northbrook

164. Sriwiriyajan S, Mahatthanatrakul W, Ridtitid W, Jaruratanasirikul S (2007) Effect of efavirenz on the pharmacokinetics of ketoconazole in HIV-infected patients. Eur J Clin Pharmacol 63(5):479–483. https://doi.org/10.1007/s00228-007-0282-8

165. Huang L, Parikh S, Rosenthal PJ, Lizak P, Marzan F, Dorsey G et al (2012) Concomitant efavirenz reduces pharmacokinetic exposure to the antimalarial drug artemether-lumefantrine in healthy volunteers. J Acquir Immune Defic Syndr 61(3):310–316. https://doi.org/10.1097/QAI.0b013e31826ebb5c

166. Maganda BA, Ngaimisi E, Kamuhabwa AA, Aklillu E, Minzi OM (2015) The influence of nevirapine and efavirenz-based anti-retroviral therapy on the pharmacokinetics of lumefantrine and anti-malarial dose recommendation in HIV-malaria co-treatment. Malar J 14:179. https://doi.org/10.1186/s12936-015-0695-2

167. Parikh S, Kajubi R, Huang L, Ssebuliba J, Kiconco S, Gao Q et al (2016) Antiretroviral choice for HIV impacts antimalarial exposure and treatment outcomes in Ugandan children. Clin Infect Dis: Off Publ Infect Dis Soc Am 63(3):414–422. https://doi.org/10.1093/cid/ciw291

168. Lariam(R) (mefloquine) [package insert] (2008) Roche Laboratories Inc., Nutley

169. Doryx(R) (doxycycline hyclate) [package insert] (2015) Mayne Pharma, Greenville

170. Khoo S, Back D, Winstanley P (2005) The potential for interactions between antimalarial and antiretroviral drugs. AIDS 19(10):995–1005

171. Peyriere H, Eiden C, Macia JC, Reynes J (2012) Antihypertensive drugs in patients treated with antiretrovirals. Ann Pharmacother 46(5):703–709. https://doi.org/10.1345/aph.1Q546

172. Cohn SE, Park JG, Watts DH, Stek A, Hitti J, Clax PA et al (2007) Depo-medroxyprogesterone in women on antiretroviral therapy: effective contraception and lack of clinically significant interactions. Clin Pharmacol Ther 81(2):222–227. https://doi.org/10.1038/sj.clpt.6100040

173. Nanda K, Amaral E, Hays M, Viscola MA, Mehta N, Bahamondes L (2008) Pharmacokinetic interactions between depot medroxyprogesterone acetate and combination antiretroviral therapy. Fertil Steril 90(4):965–971. https://doi.org/10.1016/j.fertnstert.2007.07.1348

174. Carten ML, Kiser JJ, Kwara A, Mawhinney S, Cu-Uvin S (2012) Pharmacokinetic interactions between the hormonal emergency contraception, levonorgestrel (Plan B), and Efavirenz. Infect Dis Obstet Gynecol 2012:137192. https://doi.org/10.1155/2012/137192

175. Landolt NK, Phanuphak N, Ubolyam S, Pinyakorn S, Kerr S, Ahluwalia J et al (2014) Significant decrease of ethinylestradiol with nevirapine, and of etonogestrel with efavirenz in HIV-positive women. J Acquir Immune Defic Syndr 66(2):e50–e52. https://doi.org/10.1097/QAI.0000000000000134

176. Vieira CS, Bahamondes MV, de Souza RM, Brito MB, Rocha Prandini TR, Amaral E et al (2014) Effect of antiretroviral therapy including lopinavir/ritonavir or efavirenz on etonogestrel-releasing implant pharmacokinetics in HIV-positive women. J Acquir Immune Defic Syndr 66(4):378–385. https://doi.org/10.1097/QAI.0000000000000189

177. Patel RC, Onono M, Gandhi M, Blat C, Hagey J, Shade SB et al (2015) Pregnancy rates in HIV-positive women using contraceptives and efavirenz-based or nevirapine-based antiretroviral therapy in Kenya: a retrospective cohort study. Lancet HIV 2(11):e474–e482. https://doi.org/10.1016/S2352-3018(15)00184-8

178. Stuart GS, Moses A, Corbett A, Phiri G, Kumwenda W, Mkandawire N et al (2011) Combined oral contraceptives and antiretroviral PK/PD in Malawian women: pharmacokinetics and pharmacodynamics of a combined oral contraceptive and a generic combined formulation antiretroviral in Malawi. J Acquir Immune Defic Syndr 58(2):e40–e43. https://doi.org/10.1097/QAI.0b013e31822b8bf8

179. Crauwels HM, van Heeswijk RP, Buelens A, Stevens M, Hoetelmans RM (2014) Lack of an effect of rilpivirine on the pharmacokinetics of ethinylestradiol and norethindrone in healthy volunteers. Int J Clin Pharmacol Ther 52(2):118–128. https://doi.org/10.5414/CP201943

180. Gerber JG, Rosenkranz SL, Fichtenbaum CJ, Vega JM, Yang A, Alston BL et al (2005) Effect of efavirenz on the pharmacokinetics of simvastatin, atorvastatin, and pravastatin: results of AIDS Clinical Trials Group 5108 Study. J Acquir Immune Defic Syndr 39(3):307–312

181. Crestor(R) (rosuvastatin) [package insert] (2016) AstraZeneca Pharmaceuticals LP, Wilmington

182. Lescol(R) XL (fluvastatin extended release) [package insert] (2012) Novartis Pharmaceuticals CoEast, Hanover

183. Teicher E, Vincent I, Bonhomme-Faivre L, Abbara C, Barrail A, Boissonnas A et al (2007) Effect of highly active antiretroviral therapy on tacrolimus pharmacokinetics in hepatitis C virus and HIV co-infected liver transplant recipients in the ANRS HC-08 study. Clin Pharmacokinet 46(11):941–952. https://doi.org/10.2165/00003088-200746110-00002

184. Gerber JG, Rhodes RJ, Gal J (2004) Stereoselective metabolism of methadone N-demethylation by cytochrome P4502B6 and 2C19. Chirality 16(1):36–44. https://doi.org/10.1002/chir.10303

185. Selzentry® (maraviroc) [package insert] (2016) ViiV Healthcare group of companies, Research Triangle Park

186. Patel IH, Zhang X, Nieforth K, Salgo M, Buss N (2005) Pharmacokinetics, pharmacodynamics and drug interaction potential of enfuvirtide. Clin Pharmacokinet 44(2):175–186. https://doi.org/10.2165/00003088-200544020-00003

187. Abel S, Back DJ, Vourvahis M (2009) Maraviroc: pharmacokinetics and drug interactions. Antivir Ther 14. https://doi.org/10.3851/imp1297

188. Tybost(R) (cobicistat) [package insert] (2016) Gilead Sciences, Inc., Foster City

Chapter 9
Hepatitis B and Hepatitis C Antiviral Agents

Christine E. MacBrayne and Jennifer J. Kiser

9.1 Introduction

Half a billion people worldwide have chronic hepatitis B (HBV) or hepatitis C virus (HCV) [1, 2]. These individuals are at risk for complications of chronic liver disease including potential cirrhosis, hepatocellular carcinoma, and death [2]. Fortunately, HCV can be cured with a combination of direct-acting antiviral agents (DAAs). However, many individuals are unaware of their HCV infection, are not engaged in care, or are unable to access treatment due to its expense. Current treatments for HBV can suppress viral replication which significantly reduces the development of complications from chronic liver disease. However, current HBV therapies are not completely curative, and only a small proportion achieves clearance of hepatitis B surface antigen and development of hepatitis B surface antibody (known as a "functional cure"). Following the success of DAAs for HCV, many agents are in clinical development for the treatment of HBV [3].

HBV is currently treated with a single nucleoside or nucleotide analog (e.g., entecavir, tenofovir, adefovir, lamivudine) or a finite course of pegylated interferon alfa. These therapies have a low potential for drug interactions but are not devoid of interactions. Nucleos(t)ide analogs typically require long-term treatment, and thus continued vigilance around the potential for drug interactions is necessary.

Combinations of agents including inhibitors of the NS5B polymerase, NS5A, and/or NS3 protease used with or without the purine nucleoside analog ribavirin are used to treat HCV. These therapies are typically administered for 12 weeks (range 8–24 weeks) and achieve cure rates of at least 90% in most patient populations [4]. Despite the short course of treatment, DAAs participate in a number of clinically

C. E. MacBrayne • J. J. Kiser (✉)
Department of Pharmaceutical Sciences, Skaggs School of Pharmacy and Pharmaceutical Sciences, University of Colorado Denver, Aurora, CO, USA
e-mail: christine.macbrayne@ucdenver.edu; jennifer.kiser@ucdenver.edu

© Springer International Publishing AG, part of Springer Nature 2018
M. P. Pai et al. (eds.), *Drug Interactions in Infectious Diseases: Antimicrobial Drug Interactions*, Infectious Disease,
https://doi.org/10.1007/978-3-319-72416-4_9

357

significant drug interactions. A systematic approach to screening, identifying, and managing drug interactions before and during HCV treatment is imperative.

The HBV nucleos(t)ide analogs and DAAs are well tolerated and have wide therapeutic indices. Thus, large increases in exposures would be required to cause toxicities. Interactions that result in a reduced exposure to the HBV nucleos(t)ide analogs or DAAs are of greater concern and could have significant clinical consequences. Suboptimal concentrations of antiviral drugs may lead to therapeutic failure and the development of drug resistance. HBV and HCV therapies may also alter the exposures of other drugs. One study found that HCV-infected Americans take an average of ten medications (excluding HCV therapies), so there is considerable potential for drug-drug interactions with this disease [5].

9.2 Pharmacology and Drug Interaction Potential of HBV Agents

9.2.1 Interferon

9.2.1.1 Pharmacology of Interferon

Interferons are cytokines that regulate the innate immune system [6]. Exogenous interferon, created using recombinant DNA technology, has non-specific antiviral, immunomodulatory, and antiproliferative effects [7, 8]. Interferon alfa-2a was the first biologic agent to receive regulatory approval [6]. This compound is used in the treatment of several cancers and viruses. For decades, interferon was a fundamental component of the treatment of HCV, but it has been replaced with less toxic oral agents that directly target specific steps in the HCV life cycle (also known as DAAs). However, interferon remains a first-line treatment for HBV. To improve the pharmacokinetics and allow once-weekly subcutaneous dosing, a polyethylene glycol (PEG) moiety was attached to interferon. Pegylated interferon alfa-2a consists of an ester derivative of a branched-chain 40,000-Da PEG bonded to interferon alfa-2a. This injection is administered once weekly for 48 weeks for the treatment of HBV. Patients with end-stage renal disease require dose reduction. Interferons are contraindicated in patients with advanced liver disease because they can precipitate clinical deterioration and increase the risk of bacterial infections.

9.2.1.2 Interferon Interactions

Peginterferon alfa-2a does not participate in many interactions, but it may increase the exposure of theophylline (a CYP1A2 substrate) [9]. It should also be used with caution with other myelosuppressive agents (e.g., zidovudine). For a detailed review of interactions with interferon, see Chap. 10, "Drug Interactions for Antiviral Agents."

9.2.2 Nucleoside and Nucleotide Analogs

9.2.2.1 Pharmacology of HBV Nucleos(t)ide Analogs

Nucleoside analogs are prodrugs. These compounds undergo intracellular conversion by host enzymes to the active moiety which is a triphosphate. The triphosphate is an analog of an endogenous base. If the drug triphosphate is incorporated into replicating HBV rather than the endogenous triphosphate, then replication is halted. A nucleotide analog differs from a nucleoside analog in that it already contains a phosphate group and thus requires one less phosphorylation step to become activated. Adefovir and tenofovir are nucleotides, while entecavir and lamivudine are nucleosides. Adefovir and tenofovir are adenosine analogs. Entecavir is a guanosine analog. Lamivudine is a cytidine analog. Telbivudine is a thymidine analog. The production of telbivudine was recently discontinued by its manufacturer due to a limited market for this agent and thus will not be reviewed here.

In addition to their use in the treatment of HBV, tenofovir and lamivudine are also used to treat HIV. Additional information on the pharmacology of tenofovir and lamivudine can be found in Chap. 8, "Drug Interactions in HIV: Nucleoside and Nonnucleoside Reverse Transcriptase Inhibitors and Entry Inhibitors." Nonadherence or abrupt discontinuation of the nucleos(t)ide analogs could lead to liver toxicity as the result of reactivation of HBV replication. Additionally, all individuals initiating chemotherapy or other immunosuppressive treatments should be tested for HBV as these individuals may require a nucleos(t)ide analog to prevent HBV reactivation [10].

Adefovir

Adefovir dipivoxil is a diester prodrug of adefovir. The active form of adefovir, adefovir diphosphate, inhibits the HBV polymerase. The bioavailability of adefovir is ~30–60% [11, 12]. Food does not affect adefovir absorption. Binding to plasma proteins is minimal (<3%) [11]. Adefovir is renally cleared and may cause nephrotoxicity. The adefovir dipivoxil dose should be reduced in patients with creatinine clearance <50 mL/min or ideally avoided in patients with renal impairment. Adefovir pharmacokinetics are not significantly different in individuals with decompensated cirrhosis compared with individuals without hepatic impairment.

Entecavir

Entecavir triphosphate inhibits the HBV polymerase. The bioavailability is estimated at ~70% for this drug [13]. Food decreases the extent of entecavir absorption, so this medication should be taken on an empty stomach. Entecavir is minimally (~13%) bound to serum proteins [14]. Entecavir is primarily eliminated unchanged by the kidneys via a combination of glomerular filtration and active tubular secretion. The dose should be reduced in patients with creatinine clearance <50 mL/min. Entecavir pharmacokinetics are similar in those with decompensated cirrhosis to individuals without hepatic impairment. A higher dose is used in adults with decompensated cirrhosis and in adults with lamivudine or telbivudine resistance.

Lamivudine

Lamivudine triphosphate inhibits HBV polymerase. The bioavailability of lamivudine is 86% [15]. There is no significant difference in lamivudine pharmacokinetics in the fasted vs. fed state. Binding of lamivudine to human plasma proteins is approximately 36% [15]. Lamivudine is renally cleared. The dose should be reduced in patients with creatinine clearance less than 50 mL/min. The pharmacokinetics of lamivudine were not significantly different in patients with hepatic decompensation compared with individuals without hepatic impairment.

Tenofovir

Tenofovir, an acyclic nucleoside phosphonate, is available in two different prodrug formulations, tenofovir disoproxil fumarate (TDF) and tenofovir alafenamide fumarate (TAF).

Tenofovir Disoproxil Fumarate

With TDF, cellular esterases cleave the diester, yielding tenofovir in plasma. Tenofovir then enters cells and is phosphorylated by cellular kinases to the active form, tenofovir diphosphate. Tenofovir diphosphate is a competitive inhibitor of HBV polymerase.

The bioavailability of tenofovir when given as TDF is approximately 25% [16]. Tenofovir exposures are increased by 40% with a high-fat meal [16]. Binding to plasma proteins is negligible (<8%) [16]. Tenofovir is renally cleared by a combination of glomerular filtration and active tubular secretion. Dose adjustments are necessary for creatinine clearance less than 50 mL/min. Tenofovir pharmacokinetics are similar in patients with decompensated cirrhosis to those in individuals without hepatic impairment.

Tenofovir Alafenamide Fumarate

TAF is more stable in plasma than TDF; thus, TAF is taken up directly into cells where it is de-esterified, concentrated, and phosphorylated to tenofovir diphosphate. Plasma concentrations of tenofovir when administered as TAF are 90% less than when administered as TDF [17]. The renal adverse effects of tenofovir are mediated through uptake by human organic anion transporter 1 (OAT1) in the kidney. TAF is not a substrate for OAT1, so less tenofovir is delivered to the kidneys, and there is less renal toxicity with this agent. However, for several other cell types, TAF is preferentially taken up, and cellular concentrations of tenofovir diphosphate are actually higher than those achieved with TDF.

9.2.2.2 Nucleos(t)ide Interactions

Adefovir, entecavir, lamivudine, and tenofovir participate in very few drug interactions. These compounds are not substrates, inhibitors, or inducers of CYP enzymes; however, these agents may participate in drug-drug interactions at the level of renal transporters. The interaction potential of lamivudine, TDF, and TAF is reviewed in Chap. 8, "Drug Interactions in HIV: Nucleoside and Nonnucleoside Reverse Transcriptase Inhibitors and Entry Inhibitors." All four nucleos(t)ide analogs have activity against HIV and therefore could induce nucleoside reverse transcriptase inhibitor resistance if used in individuals with HIV/HBV coinfection without fully suppressive antiretroviral therapy. Adefovir and tenofovir should be used with caution in combination with other nephrotoxic agents.

9.3 Pharmacology and Drug Interaction Potential of HCV Agents

In this section, the pharmacology of various HCV medications is reviewed followed by a summary of interaction potential with CYP and transporter probes and then clinically significant/serious drug interactions. Table 9.1 highlights the pharmacokinetic properties of each HCV agent. Table 9.2 shows the effects of renal and hepatic impairment on the pharmacokinetics of each drug and drug dosing in renal and hepatic impairment. Interactions with HCV agents and concomitant medications are summarized in Table 9.3.

9.3.1 Sofosbuvir

9.3.1.1 Pharmacology of Sofosbuvir

Sofosbuvir (SOF) is a HCV NS5B polymerase inhibitor [36]. SOF is used in combination with other DAAs and in some cases ribavirin for the treatment of HCV. SOF is administered as a phosphoramidate prodrug of the uridine nucleotide analog GS-331007 monophosphate [37]. Once SOF is inside cells, it is hydrolyzed by cathepsin A and/or carboxyesterase 1 to GS-331007 monophosphate [37, 38]. GS-331007 monophosphate is then phosphorylated by uridine monophosphate-cytidine monophosphate kinase to the GS-331007 diphosphate form, which is then

Table 9.1 Pharmacology of drugs used to treat HCV

	Route of metabolism	Transporter substrate	Enzyme inhibition	Enzyme induction	Transporter inhibition
Sofosbuvir	Phosphorylated to the triphosphate, GS-461203, SOF and plasma metabolite, GS-331007 are renally cleared	P-gp and BCRP	None	None	None
Ledipasvir	Slow oxidative metabolism via an unknown mechanism	P-gp	None	None	P-gp, BCRP
Daclatasvir	Primarily hepatic via CYP 3A	P-gp	CYP3A4	None	P-gp, BCRP, OATP1B1/3
Simeprevir	Hepatic via CYP 3A	P-gp, MRP2, BCRP, OATP1B1/3, and OATP2B1	CYP2A6, CYP2C8, CYP2D6, CYP2C19, CYP1A2, and intestinal CYP3A	None	P-gp, OATP1B1, NTCP MRP2, and BSEP
Velpatasvir	Hepatic via CYP 3A4, 2C8, and 2B6	P-gp	None	None	P-gp, BCRP, OATP1B1/1B3, and OATP2B1
Ritonavir-boosted paritaprevir	Hepatic via CYP3A4 and to a lesser extent by CYP3A5	P-gp, OATP1B1, BCRP	CYP2C8, UGT1A1 (ritonavir inhibits CYP3A)	None	P-gp, OATP1B1/3, BCRP
Ombitasvir	Amide hydrolysis followed by oxidative metabolism	P-gp, BCRP	CYP2C8, UGT1A1	None	
Dasabuvir	Hepatic via CYP2C8 and to a lesser extent by CYP3A	P-gp, BCRP	UGT1A1	None	P-gp, BCRP
Grazoprevir	Hepatic via CYP3A4	P-gp and OATP1B1	CYP2C8, UGT1A1	None	UGT1A1 and BCRP
Elbasvir	Hepatic via CYP3A4	P-gp	None	None	BCRP and P-gp
Ribavirin	Phosphorylated to RBV-TP, RBV is renally cleared	ENT1, CNT2, CNT3	None	None	None

Abbreviations: *SOF* sofosbuvir, *CYP* cytochrome P450, *RBV-TP* ribavirin-triphosphate, *RBV* ribavirin, *P-gp* p-glycoprotein, *BCRP* breast cancer resistance protein, *MRP2* multidrug resistance-associated protein 2, *OATP* organic anion-transporting polypeptide, *ENT* equilibrative nucleoside transporter, *CNT* concentrative nucleoside transporter, *NTCP* Na-taurocholate cotransporting polypeptide, *BSEP* bile salt export pump, *UGT* uridine diphosphate-glucuronosyltransferase

Table 9.2 Pharmacokinetics of hepatitis C therapies in hepatic and renal impairment

	Hepatic excretion, %	Renal excretion, %	Hepatic impairment	Renal impairment
Sofosbuvir	14%	80%	SOF AUC is 126% and 143% higher in patients with moderate and severe hepatic impairment (i.e., Child-Pugh B and C), respectively, but GS-331007 AUC is unchanged	SOF AUC is 61%, 107%, and 171% higher in mild (GFR >50 and < 80 mL/min/1.73 m2), moderate (GFR >30 and <50 mL/min/1.73 m2), and severe (GFR <30 mL/min/1.73 m2) renal impairment, respectively, whereas GS-331007 AUC is increased by 55%, 88%, and 451% in mild, moderate, and severe renal impairment, respectively. In patients with ESRD on hemodialysis, sofosbuvir and GS-331007 are 28–60% and 1300–2000% higher, respectively
Ledipasvir	86%	1%	Not significantly altered	Not significantly altered
Daclatasvir	88%	7%	Total exposures of DCV are 43%, 38%, 36% lower in patients with Child-Pugh A, B, and C decompensated cirrhosis, respectively, but unbound concentrations are unchanged	DCV AUC in those with end-stage renal disease (eGFR less than 15 mL/min/1.73 m2 receiving hemodialysis), eGFR 30–59 mL/min/1.73 m2, and eGFR 15–29 mL/min/1.73 m2 was 80%, 60%, and 26% higher, respectively, compared to those with normal renal function, which is more than expected given only that 7% of the drug is renally eliminated
Simeprevir	91%	<1%	SIM exposures are increased by 140% and 420% in patients with Child-Pugh B and C hepatic impairment, respectively	There is a 62% increase in exposure of SIM in HCV uninfected subjects with eGFR < 30 mL/min/1.73 m2
Velpatasvir	77%	<1%	Not significantly altered	Not significantly altered

(continued)

Table 9.2 (continued)

	Hepatic excretion, %	Renal excretion, %	Hepatic impairment	Renal impairment
PrOD/PrO	88% – paritaprevir 90.2% – ombitasvir 94.4% – dasabuvir	8.8% – paritaprevir 1.91% – ombitasvir ~2% – dasabuvir	Paritaprevir exposures increased by 29% and 945% in mild and severe hepatic impairment, respectively. Ombitasvir exposures increased by 8% and 30% in mild and moderate impairment, respectively, and decreased by 54% in severe hepatic impairment Dasabuvir exposure increases by 17%, 16%, and 325% in patients with mild, moderate, and severe hepatic impairment, respectively	Paritaprevir AUC is 19%, 33%, and 45% higher in in mild (GFR >50 and <80 mL/min/1.73 m2), moderate (GFR >30 and <50 mL/min/1.73 m2), and severe (GFR < 30 mL/min/1.73 m2) renal impairment, respectively Ombitasvir is not significantly altered Dasabuvir AUC is 21%, 37%, and 50% higher in mild (GFR >50 and <80 mL/min/1.73 m2), moderate (GFR >30 and <50 mL/min/1.73 m2), and severe (GFR <30 mL/min/1.73 m2) renal impairment, respectively
Grazoprevir/ elbasvir	>90%	< 1%	GZR exposures are increased by 70% in those with mild (Child-Pugh A), 400% in those with moderate (Child-Pugh B), and 1100% in those with severe (Child-Pugh C) hepatic impairment relative to those with no hepatic impairment Total concentrations of EBR are 24% and 14% lower in patients with mild and moderate hepatic insufficiency, respectively	GZR and EBR AUCs are increased by 65% and 86%, respectively, in non-HCV-infected individuals with eGFR <30 mL/min/1.73 m2 not receiving dialysis In patients with ESRD requiring dialysis, GZR/EBR AUCs were increased by 11% and 25%, respectively
Ribavirin	Not reported	5–15%	RBV exposures are ~30% higher in those with Child-Pugh C following a single dose, and Cmax is doubled	RBV is given in alternating daily doses of 200 mg and 400 mg in individuals with a GFR of 30–50 mL/min/1.73 m2 and 200 mg orally once daily for individuals with a GFR <30 mL/min/1.73 m2 or receiving hemodialysis

Abbreviations: *PrOD or PrO* ritonavir-boosted paritaprevir, ombitasvir with or without dasabuvir, *AUC* area under the concentration-time curve, *DCV* daclatasvir, *SIM* simeprevir, *GZR* grazoprevir, *EBR* elbasvir, *SOF* sofosbuvir, *GFR* glomelular filtration rate, *HCV* hepatitis C virus, *ESRD* end-stage renal disease, *RBV* ribavirin

Table 9.3 Drug interactions with hepatitis C therapies and commonly used medications

	Sofosbuvir	Ledipasvir	Daclatasvir	Simeprevir	Velpatasvir	PrOD/PrO	Grazoprevir/elbasvir
Anticonvulsants	X – carbamazepine, oxcarbazepine, phenytoin, and phenobarbital √ – levetiracetam, lamotrigine, and ethosuximide	X – carbamazepine, oxcarbazepine, phenytoin, and phenobarbital √ – levetiracetam, lamotrigine, and ethosuximide	X – carbamazepine, oxcarbazepine, phenytoin, and phenobarbital √ – levetiracetam, lamotrigine, and ethosuximide	X – carbamazepine, oxcarbazepine, phenytoin, and phenobarbital √ - levetiracetam, lamotrigine, and ethosuximide	X – carbamazepine, oxcarbazepine, phenytoin, and phenobarbital √ - levetiracetam, lamotrigine, and ethosuximide	X – carbamazepine, oxcarbazepine, phenytoin, phenobarbital, lamotrigine and ethosuximide √ - levetiracetam	X – carbamazepine, oxcarbazepine, phenytoin, and phenobarbital √ - levetiracetam, lamotrigine, and ethosuximide
Opioid and opioid substitutes	√	√	√ [18]	√ [19]	√	Δ- monitor patients; a dose ↓ with titration to effect may be necessary for certain agents [20]	√ [21]
Amiodarone	Δ – refer to text for management of this interaction	Δ – due to coformulation with SOF, the same restrictions apply	Δ – theoretically this interaction could occur with all DAAs, must monitor	Δ – theoretically this interaction could occur with all DAAs, must monitor	Δ – due to coformulation with SOF, the same restrictions apply	Δ – theoretically this interaction could occur with all DAAs, must monitor	Δ – theoretically this interaction could occur with all DAAs, must monitor
Oral contraceptives	√ [22]	√ - ↑ ethinyl estradiol by 20% but still okay to use concomitantly [22]	√ [23]	√ [24]	√ [25]	Δ- refer to text for management of interaction [26]	√ [27]

(continued)

Table 9.3 (continued)

	Sofosbuvir	Ledipasvir	Daclatasvir	Simeprevir	Velpatasvir	ProD/PrO	Grazoprevir/elbasvir
HMG-CoA reductase inhibitors	Δ – consider a dose reduction	Δ – consider a dose reduction	No data	Δ – ↑ atorvastatin and rosuvastatin AUC by 112% and 181%, respectively. Max of 40 mg atorvastatin and 10 mg rosuvastatin [28]	Δ – ↑ rosuvastatin AUC by 170%, max of 10 mg. Consider dose reduction and lowest possible dose for other statins [25]	X – ↑ pravastatin by 80% and rosuvastatin by 160% [29]	Δ – ↑ atorvastatin by 94% and rosuvastatin by 126%. Max of 20 mg atorvastatin and 10 mg rosuvastatin [30, 31]
Amlodipine	Δ	Δ	Δ	Δ	Δ	Δ – ↑ amlodipine by 160%, monitor	Δ
Gastric acid modifiers	Δ – coformulated with LDV and VEL	Δ – refer to text for management of interaction	√ – ↓DCV AUC by 16% with omeprazole and 18% with famotidine [32]	√	Δ – refer to text for management of interaction	√	√
Calcineurin inhibitors	√	√	√ – cyclosporine ↑ AUC of DCV by 40% and ↑ tacrolimus by 5% [33]	X – cyclosporine ↑ simeprevir AUC by 481%; Δ – monitor; tacrolimus ↑ simeprevir AUC by 85%	√ [34]	Δ – must monitor levels and adjust for both cyclosporine and tacrolimus. Use 1/5 of cyclosporine dose during treatment and 0.5 mg tacrolimus every 7 days [35]	X – cyclosporine not recommended; √ – ↑ tacrolimus by 43% but no dose adjustment necessary

Abbreviations: *SOF* sofosbuvir, *LDV* ledipasvir, *VEL* velpatasvir, *DCV* daclatasvir, *AUC* area under the concentration-time curve, *DAAs* direct-acting antivirals, Δ a change in dose or drug may be required, X contraindicated, √ safe to coadminister

phosphorylated by nucleotide diphosphate kinase to the triphosphate moiety (GS-461203) [37, 38]. GS-461203 is the pharmacologically active nucleoside analog triphosphate metabolite form of SOF which is incorporated into HCV RNA by the NS5B polymerase, allowing for HCV replication to be halted. The major drug-related substance found in plasma is GS-331007, which has no antiviral activity.

The absolute bioavailability of SOF has not been determined in humans but is estimated to be at least 80% based on recovery of SOF and GS-331007, following a radiolabeled dose [38]. If given concomitantly with a high-fat meal, SOF area under the curve (AUC) is increased by 67–91%, but the drug is approved without regard to food [38]. SOF is 61–65% protein bound and is predominantly renally excreted [36]. SOF and GS-331007 exposures are significantly increased in renal impairment, and thus, SOF is not recommended for individuals with CrCl less than 30 mL/min [39]. However, case series are emerging on the safety and efficacy of SOF-based therapy in patients with renal impairment [40–42]. For the metabolism of SOF, refer to Table 9.1 [43], and for the pharmacokinetic alterations in hepatic and renal impairment, refer to Table 9.2 [36, 38, 44].

9.3.1.2 Probe Interactions

SOF is not a substrate, inhibitor, or inducer of CYP enzymes and has a low potential for drug interactions. SOF is used in combination with other DAAs, however, and thus consideration must be given to the potential for interactions with the concomitant DAA. SOF is a substrate for p-glycoprotein (P-gp) and breast cancer resistance protein (BCRP); thus, potent inducers of P-gp or BCRP should not be used with SOF [36].

9.3.1.3 Clinically Significant Drug Interactions

Amiodarone

Serious and life-threatening cases of symptomatic bradycardia and at least one case of fatal cardiac arrest have occurred in patients taking amiodarone with SOF and another DAA [36]. In vitro and animal studies suggest that this interaction is not pharmacokinetic in nature but rather appears to be the result of inhibition of a calcium channel or disrupted intracellular calcium handling [45–47]. There are conflicting data on the contribution of other DAAs to the interaction with sofosbuvir and amiodarone [45, 47]. Based on this interaction, the combination of sofosbuvir and amiodarone is not recommended. However, for patients taking amiodarone who have no other alternative, counseling patients on the risk of serious symptomatic bradycardia is necessary, and inpatient cardiac monitoring is advised for the first 48 h of DAA treatment and then daily heart rate monitoring for 2 weeks thereafter [36].

9.3.2 Daclatasvir

9.3.2.1 Pharmacology of Daclatasvir

Daclatasvir (DCV) is an HCV NS5A inhibitor. The absolute bioavailability of DCV is 67%. A high-fat and high-calorie meal decreases DCV exposures by 23%, but a low-fat meal has no effect. DCV is 99% protein bound and has minimal renal excretion [48]. For the metabolism of DCV, refer to Table 9.1 [43], and for properties of hepatic and renal impairment, refer to Table 9.2 [48]. DCV is the only DAA with different dosing strengths available to accommodate dose adjustments for drug interactions.

9.3.2.2 Probe Interactions

DCV is mainly a victim of drug interactions rather than a perpetrator, though it does increase digoxin AUC (a P-gp substrate) by 27% and rosuvastatin AUC (an OATP1B1 and BCRP substrate) by 58%. Due to DCV being highly reliant on CYP3A for its metabolism, dose adjustments of DCV are necessary in the presence of strong or moderate CYP3A inhibitors and moderate inducers. Ketoconazole (a potent CYP3A4 inhibitor) increases the AUC of DCV by 200%. Cyclosporine (a P-gp inhibitor and weak CYP3A4 inhibitor) increases DCV AUC by 40% [33]. Multi-dose rifampin (a potent CYP3A4 inducer) decreases the AUC of DCV by 79% [48, 49]; thus, strong inducers are contraindicated with DCV.

9.3.2.3 Clinically Significant Drug Interactions

HIV Antiretroviral Agents

The DCV dose should be increased from 60 to 90 mg with efavirenz and etravirine which are inducers. The DCV dose should be decreased with cobicistat-containing regimens and ritonavir-boosted atazanavir from 60 to 30 mg [50, 51]. No dose adjustment of DCV is necessary with ritonavir-boosted lopinavir or darunavir [52].

9.3.3 Ledipasvir

9.3.3.1 Pharmacology of Ledipasvir

Ledipasvir (LDV) is an inhibitor of NS5A. LDV is only available coformulated with SOF. LDV absorption is pH-dependent. The bioavailability of LDV in humans is not known but ranges from 30 to 50% in rats, monkey, and dogs [53]. LDV concentrations are similar when given fasted vs. with a moderate- or high-fat meal. LDV is greater than 99.8% protein bound. LDV is primarily eliminated unchanged in the feces [44]. For the metabolism of LDV, refer to Table 9.1 [43]. For the pharmacokinetics of LDV in hepatic and renal impairment, refer to Table 9.2 [44]. LDV pharmacokinetics are not significantly altered by hepatic or renal impairment, but since the drug is coformulated with SOF, the same limitations on use in those with renal impairment apply [44].

9.3.3.2   Probe Interactions

LDV inhibits P-gp and BCRP, which may increase the concentrations of rosuvastatin (an OATP1B1 and BCRP substrate); thus, this combination is not recommended. Rosuvastatin AUC was increased by 699% with LDV, GS-9451 (an investigational protease inhibitor), and tegobuvir (an investigational non-nucleoside NS5B inhibitor) [54]; thus, it is unknown whether LDV would cause this effect on rosuvastatin in the absence of these other DAAs [55].

9.3.3.3   Clinically Significant Drug Interactions

Gastric Acid-Modifying Agents

LDV is dependent on an acidic environment for optimal absorption; thus, gastric acid modifiers should be used with caution. There are conflicting data on whether the use of proton pump inhibitors (PPI) compromises the likelihood of achieving cure also known as a sustained virologic response (SVR) [56, 57]. If gastric acid modifiers must be used, temporal separation is necessary with antacids (by 4 h). Histamine-2 receptor antagonists (e.g., famotidine, ranitidine) should not exceed the equivalent of 40 mg famotidine twice daily. PPI (omeprazole, lansoprazole, etc.) doses should not exceed the equivalent of 20 mg omeprazole once daily and should be administered simultaneously with LDV/SOF in the fasted state [44, 58].

HIV Antiretroviral Agents

LDV increases tenofovir exposures by 30–60%. This may increase the risk of renal toxicity in HIV-infected individuals taking TDF with a ritonavir-boosted protease inhibitor or cobicistat [59]. Efavirenz reduces LDV concentrations by 30% [60]. This reduction is unlikely to compromise SVR unless coupled with negative prognostic factors such as imperfect adherence, black race, or shortened (8 weeks) treatment duration [61–63].

9.3.4 Simeprevir

9.3.4.1 Pharmacology of Simeprevir

Simeprevir (SMV) is a HCV NS3/4A protease inhibitor. The absolute bioavailability of SMV is 62%. A 533 kcal meal and 928 kcal meal increase SMV AUC by 69% and 61%, respectively; thus, it is recommended to take with food. SMV is 99.9% protein bound, primarily to albumin, and less than 1% of SMV is renally cleared [64]. For the metabolism of SMV, refer to Table 9.1 [43], and for the pharmacokinetics of simeprevir in hepatic and renal impairment, refer to Table 9.2 [64].

9.3.4.2 Probe Interactions

As a perpetrator in drug interactions, SMV increases the AUC of oral midazolam (a CYP3A substrate) by 45%, caffeine (a CYP1A2 substrate) by 26%, omeprazole (a CYP2C19 substrate) by 21%, digoxin (a P-gp substrate) by 39%, and rosuvastatin (an OATP1B1 and BCRP substrate) by 2.8-fold in healthy volunteers [65, 66]. As a victim, SMV is altered by moderate or strong inducers and inhibitors of CYP3A. Multi-dose rifampin reduces SMV AUC by 48%. Cyclosporine had no effect on SMV exposures in healthy volunteers, but SMV exposures were six fold higher in HCV-infected patients' post-liver transplant taking SMV, DCV, and ribavirin with cyclosporine vs. historical values [64, 67]. Thus, cyclosporine should not be used with SMV.

9.3.4.3 Clinically Significant Drug Interactions

HIV Antiretroviral Agents

Due to the effects of potent CYP3A inhibitors and inducers on SMV, ARV options are more limited. Efavirenz reduces SMV AUC by 71% [65]. SMV exposures are increased by 2.6-fold by ritonavir-boosted darunavir, even after an empiric dose reduction of SMV from 150 to 50 mg. Given these interactions, ritonavir- or cobicistat-boosted HIV protease inhibitors and efavirenz are not recommended with SMV.

9.3.5 Velpatasvir

9.3.5.1 Pharmacology of Velpatasvir

Velpatasvir (VEL) is an NS5A inhibitor [25, 68]. Like LDV, VEL absorption is pH-dependent. When administered with moderate-fat and high-fat meals, there is a 34% and 21% increase in VEL exposures, respectively [25, 69]. VEL is >99.5% bound to proteins and predominantly excreted in feces as parent drug (77%), and less than 1% of the dose is excreted in urine. For the metabolism of VEL, refer to Table 9.1 [43], and for the pharmacokinetics of VEL in hepatic and renal impairment, refer to Table 9.2 [34, 69]. LDV pharmacokinetics are not significantly altered by hepatic or renal impairment, but since the drug is coformulated with SOF, the same limitations on use in those with renal impairment apply.

9.3.5.2 Probe Interactions

In terms of VEL's ability to act as a perpetrator in interactions, pravastatin (an OATP1B1 substrate) AUC increased 35%, and rosuvastatin (an OATP1B1 and BCRP substrate) AUC increased approximately 170% when coadministered with VEL in healthy volunteers. Digoxin (a P-gp substrate) AUC increased 34%. In terms of its ability to act as a victim in interactions, single-dose rifampin (an OATP1B1 inhibitor) increased VEL AUC by 47%, while multiple-dose rifampin reduced VEL AUC by 81%. VEL AUC increased 103% with a single dose of cyclosporine (a mixed OATP/P-gp/MRP2 inhibitor). Ketoconazole (a CYP3A inhibitor) increased VEL AUC by 70% [34].

9.3.5.3 Clinically Significant Drug Interactions

Gastric Acid-Modifying Agents

VEL relies heavily on an acidic environment for optimal absorption [25]. If gastric acid modifiers must be used, temporal separation is necessary with antacids (by 4 h). Histamine-2 receptor antagonists (e.g., famotidine, ranitidine) should not exceed the equivalent of 40 mg famotidine twice daily. As with LDV/SOF, PPI doses should not exceed the equivalent of omeprazole 20 mg once daily, but unlike LDV/SOF, SOF/VEL should be administered 4 h before the PPI in the fed state [25].

HIV Antiretroviral Agents

VEL AUC is reduced by 50% with TDF/emtricitabine/efavirenz; thus, this combination should be avoided. As with LDV/SOF, SOF/VEL increased tenofovir exposures by 40–81% [70], and this could increase the risk for nephrotoxicity in individuals receiving TDF with ritonavir or cobicistat.

9.3.6   Grazoprevir/Elbasvir

9.3.6.1   Pharmacology of Grazoprevir/Elbasvir

Grazoprevir and elbasvir (GZR/EBR) are inhibitors of the NS3 and NS5A proteins, respectively. GZR/EBR is available as a coformulated tablet administered once daily. GZR/EBR is a preferred HCV treatment for patients with renal impairment. Ribavirin is given with GZR/EBR in individuals with genotype 1a infection and pre-existing NS5A viral variants to increase the likelihood of achieving SVR.

The bioavailability of EBR is 30%. GZR bioavailability ranges from 10 to 40%. Administration of GZR/EBR with a high-fat meal to healthy subjects results in decreases in EBR AUC and Cmax of approximately 11% and 15%, respectively, and increases in GZR AUC and Cmax of approximately 1.5-fold and 2.8-fold, respectively [30]. GZR is at least 98% protein bound and EBR is more than 99% bound [30]. GZR and EBR are hepatically metabolized and less than 1% of GZR and EBR are renally eliminated. For the metabolism of GZR/EBR, refer to Table 9.1 [43], and for the pharmacokinetics of GZR and EBR in hepatic and renal impairment, refer to Table 9.2 [30, 71].

9.3.6.2   Probe Interactions

As a perpetrator in drug interactions, pravastatin (OATP1B1 substrate) and rosuvastatin (OATP1B1 and BCRP substrate) exposures are increased by 33% and 126%, respectively, with GZR/EBR [30]. Atorvastatin and rosuvastatin doses should not exceed 20 mg and 10 mg daily, respectively, with GZR/EBR. In terms of being victims of drug interactions, OATP1B1 inhibitors significantly raise GZR exposures. Single-dose rifampin (an OATP1B1 inhibitor) raises GZR AUC by eight- to tenfold, and cyclosporine (OATP/P-gp/MRP2 inhibitor) raises the GZR AUC by 15-fold. GZR/EBR should therefore not be used with cyclosporine. GZR is also susceptible to potent CYP3A inhibitors. Ketoconazole (a CYP3A4 inhibitor) increases GZR AUC by threefold [31, 72]. EBR AUC is increased by 80% with ketoconazole [30]. Coadministration of ketoconazole with GZR/EBR is not recommended.

9.3.6.3   Clinically Significant Drug Interactions

HIV Antiretroviral Agents

In healthy volunteers, GZR and EBR are increased by 10.6-fold and 4.76-fold, respectively, by ritonavir-boosted atazanavir; 7.5-fold and 1.66-fold, respectively, by ritonavir-boosted darunavir; and 12.86-fold and 3.71-fold, respectively, by ritonavir-boosted lopinavir. Thus, GZR/EBR should not be used with ARV regimens that include ritonavir or cobicistat. GZR and EBR are reduced by 83% and 54%, respectively, with efavirenz [30, 73]. Thus, inducers like efavirenz and etravirine should not be used with GZR/EBR.

9.3.7 Ritonavir-Boosted Paritaprevir, Ombitasvir with or Without Dasabuvir

9.3.7.1 Pharmacology of Ritonavir-Boosted Paritaprevir, Ombitasvir, and Dasabuvir

Paritaprevir is an NS3 protease inhibitor, ombitasvir is an NS5A inhibitor, and dasabuvir is a non-nucleoside NS5B polymerase inhibitor. Ritonavir-boosted paritaprevir, ombitasvir, and dasabuvir (PrOD) is used for the treatment of HCV genotype 1. In individuals with genotype 1a, ribavirin is used with PrOD to increase the likelihood of SVR. Ritonavir-boosted paritaprevir and ombitasvir (PrO) is used with ribavirin, but without dasabuvir, for 12 weeks for the treatment of individuals with genotype 4 disease [74]. Ritonavir is used in this combination to pharmacokinetically enhance the exposures of paritaprevir via inhibition of CYP3A. Ritonavir has no HCV activity.

The absolute bioavailability of paritaprevir, ombitasvir, and ritonavir is unknown. The absolute bioavailability of dasabuvir is approximately 70%. Moderate- and high-fat meals increase exposures of all four components, and the treatment is approved for administration with a meal. Protein binding is high for all four drugs: paritaprevir 97–98.6%, ombitasvir 99.9%, ritonavir greater than 99%, and dasabuvir greater than 99.5% [75]. For the metabolism of PrOD/PrO, refer to Table 9.1 [43], and for the pharmacokinetics of paritaprevir, ombitasvir, and dasabuvir in hepatic and renal impairment, refer to Table 9.2 [76].

9.3.7.2 Probe Interactions

Paritaprevir and ritonavir are primarily metabolized by CYP3A. Ombitasvir is primarily metabolized via amide hydrolysis. Dasabuvir is primarily metabolized by CYP2C8 enzymes. Ombitasvir, paritaprevir, dasabuvir, and ritonavir are substrates of P-gp. Ombitasvir, paritaprevir, and dasabuvir are substrates of BCRP. Paritaprevir is a substrate of OATP1B1 and OATP1B3.

Dasabuvir, ombitasvir, and paritaprevir are inhibitors of UGT1A1, and ritonavir is an inhibitor of CYP3A4. Paritaprevir is an inhibitor of OATP1B1 and OATP1B3, and dasabuvir, paritaprevir, and ritonavir are inhibitors of BCRP.

Based on the pharmacology of these agents, the use of PrOD/PrO can lead to a number of clinically important drug interactions. Clinicians are encouraged to screen for interactions with PrOD/PrO (and other DAAs) using the University of Liverpool website, www.hep-druginteractions.org. Coadministration with drugs that are highly dependent on CYP3A for clearance, moderate or strong inducers of CYP3A or strong inducers of CYP2C8, and strong inhibitors of CYP2C8 is contraindicated with PrOD.

In terms of acting as a perpetrator, rosuvastatin (substrate for OATP1B1 and BCRP) and pravastatin (substrate for OATP1B1) exposures are increased by 159% and 82%, respectively, by PrOD [26]. PrOD increases ketoconazole (a CYP3A substrate) AUC by twofold. PrOD lowers exposures of some drugs including omeprazole, a CYP2C19 substrate.

In terms of acting as victims of interactions, ketoconazole (a CYP3A inhibitor) increases ombitasvir, paritaprevir, and dasabuvir by 17%, 98%, and 42%, respectively. Gemfibrozil (a potent CYP2C8 inhibitor) increases dasabuvir exposures by 11-fold. Carbamazepine (CYP3A and P-gp inducer) reduces ombitasvir, paritaprevir, and dasabuvir by 31%, 70%, and 70%, respectively.

9.3.7.3 Clinically Significant Drug Interactions

HIV Antiretroviral Agents

With the exception of ritonavir-boosted atazanavir, ritonavir-boosted protease inhibitors and cobicistat-based ARV regimens should not be used with PrOD. In healthy volunteers, PrOD reduced darunavir trough concentrations by 48% with once-daily ritonavir-boosted darunavir and 43% with twice-daily ritonavir-boosted darunavir. Without dasabuvir however (i.e., PrO), the decrease in darunavir troughs is not as prominent. Rilpivirine AUC is increased by 225% with PrOD [75]. This is concerning due to potential QTc prolongation with increased rilpivirine exposures. A study of efavirenz and PrOD in healthy volunteers was prematurely discontinued due to toxicities. Thus, efavirenz and etravirine are not recommended with PrOD or PrO.

Hormonal Contraceptives

PrOD/PrO was studied with progestin-only (norethindrone-only) oral contraception and ethinyl estradiol plus either norgestimate or norethindrone [26]. There was no effect on norethindrone pharmacokinetics. With norgestimate, the norelgestromin and norgestrel metabolites were increased by 2.6-fold, but this increase is not expected to have clinical relevance. Liver function test (LFT) elevations were noted in the studies of the combined ethinyl estradiol and progestin-containing contraceptive. While there was no increase in the pharmacokinetics of ethinyl estradiol with PrOD/PrO, grade 3 or 4 LFT elevations were noted in 5 of 21 women. Thus, estrogen-containing oral contraceptives, the transdermal patch, and vaginal ring should be avoided during PrOD/PrO treatment and can be reinitiated 2 weeks after completing HCV treatment. Progestin-only oral contraceptives, the etonogestrel implant, and levonorgestrel intrauterine device can be safely used with PrOD.

Calcineurin Inhibitors

Cyclosporine AUC and trough concentration are increased by 5.82-fold and 15.8-fold, respectively, with PrOD. Modeling and simulation suggest using one fifth of the cyclosporine dose during PrOD treatment with careful monitoring of cyclosporine levels and dose titration as needed [35]. Tacrolimus AUC and trough concentration are increased by 57-fold and 17-fold, respectively, with PrOD. Modeling and simulation suggest using 0.5 mg of tacrolimus every 7 days during PrOD treatment with careful monitoring of tacrolimus levels and dose titration as needed [35].

Psychotropics

Few formal interaction studies have been performed with PrOD/PrO and psychotropic medications, but theoretically, interactions may occur based on overlapping clinical pharmacology. The psychotropic medications are unlikely to affect the pharmacokinetics of PrOD/PrO; rather, the pharmacokinetics of psychotropic medications may be altered by PrOD/PrO. Clinicians are advised to screen for interactions with PrOD/PrO and psychotropic medications using the University of Liverpool website, www.hep-druginteractions.org. Psychotropics that are contraindicated with PrOD/PrO include orally administered midazolam, St. John's wort, pimozide, and quetiapine. A recent and helpful review is available on interactions with psychotropics and DAAs [77].

9.3.8   Ribavirin

9.3.8.1   Pharmacology of Ribavirin

Ribavirin (RBV) is a purine nucleoside analog. RBV is used in combination with DAAs for the treatment of chronic HCV infection in certain situations. RBV dosing is typically weight based (less than 75 kg receives 1000 mg daily; ≥75 kg receives 1200 mg daily). The dose is often divided and given twice daily.

Host cell enzymes phosphorylate RBV to mono-, di-, and triphosphate derivatives. The exact mechanism(s) of action of RBV and/or its phosphorylated derivatives in vivo are unknown [78, 79].

The absolute bioavailability of RBV is 64%. RBV concentrations are increased with a high-fat meal and decreased with purine-rich foods such as margarine, tuna, ham, or whole milk [80, 81]. RBV is not protein bound, and 61% is recovered in the urine. Dose adjustments are required for patients with creatinine clearance values of <50 mL/min. For the metabolism of RBV, refer to Table 9.1 [43], and for the pharmacokinetics of RBV in hepatic and renal impairment, refer to Table 9.2 [81].

9.3.8.2 Drug Interaction Potential of Ribavirin

RBV has a low potential for drug interactions because it is not a substrate, inhibitor, or inducer of CYP enzymes. RBV is a substrate for concentrative (CNT) and equilibrative nucleoside transporters (ENT), though no clinically significant drug interactions have been attributed to CNT- and ENT-mediated interactions. RBV should not be used with the HIV nucleoside analog, didanosine, due to an increase in the formation of the triphosphorylated form of didanosine which raises the risk of mitochondrial toxicity [82]. Zidovudine and RBV both cause the adverse effect of anemia and thus, this combination should also be avoided.

9.4 Summary

DAAs have revolutionized the treatment of HCV. Current therapies provide cure rates in excess of 90%. With these new treatments, comes an additional consideration in the treatment of this disease, the avoidance of unfavorable drug-drug interactions. The use of DAAs necessitates the identification and management of potential interactions before, during, and after treatment. The treatment of HBV involves the use of pegylated interferon and nucleos(t)ide analogs. These therapies have minimal interactions, but several new agents are in development for the treatment of HBV which will likely necessitate more careful consideration of drug interactions for the treatment of this virus as well.

References

1. World Health Organization (WHO) (2016) Hepatitis B Virus. Updated July 2016. Accessed 17 Aug 2016
2. World Health Organization (WHO) (2016) Hepatitis C Virus. Updated July 2016. Accessed 17 Aug 2016
3. Liang TJ, Block TM, McMahon BJ, Ghany MG, Urban S, Guo JT et al (2015) Present and future therapies of hepatitis B: from discovery to cure. Hepatology 62(6):1893–1908
4. Flisiak R, Pogorzelska J, Flisiak-Jackiewicz M, Hepatitis C (2017) Efficacy and safety in real life. Liver Int. 37(Suppl 1):26–32
5. Lauffenburger JC, Mayer CL, Hawke RL, Brouwer KL, Fried MW, Farley JF (2014) Medication use and medical comorbidity in patients with chronic hepatitis C from a US commercial claims database: high utilization of drugs with interaction potential. Eur J Gastroenterol Hepatol. 26(10):1073–1082
6. Asmuth DM, Utay NS, Pollard RB (2016) Peginterferon alpha-2a for the treatment of HIV infection. Expert Opin Investig Drugs. 25(2):249–257
7. Antonelli G, Scagnolari C, Moschella F, Proietti E (2015) Twenty-five years of type I interferon-based treatment: a critical analysis of its therapeutic use. Cytokine Growth Factor Rev. 26(2):121–131
8. Konerman MA, Lok AS (2016) Interferon Treatment for Hepatitis B. Clin Liver Dis. 20(4):645–665

9. Brennan BJ, ZX X, Grippo JF (2013) Effect of peginterferon alfa-2a (40KD) on cytochrome P450 isoenzyme activity. Br J Clin Pharmacol. 75(2):497–506

10. Gonzalez SA, Perrillo RP, Hepatitis B (2016) Virus reactivation in the setting of cancer chemotherapy and other immunosuppressive drug therapy. Clin Infect Dis. 62(Suppl 4):S306–S313

11. Noble S, Goa KL (1999) Adefovir dipivoxil. Drugs. 58(3):479–487. discussion 88-9

12. Rivkin AM (2004) Adefovir dipivoxil in the treatment of chronic hepatitis B. Ann Pharmacother. 38(4):625–633

13. Baraclude – European medicines agency summary of product characteristics June 26 2011 [Available from: http://www.ema.europa.eu/docs/en_GB/document_library/EPAR_-_Product_Information/human/000623/WC500051984.pdf

14. Matthews SJ (2006) Entecavir for the treatment of chronic hepatitis B virus infection. Clin Ther. 28(2):184–203

15. Johnson MA, Verpooten GA, Daniel MJ, Plumb R, Moss J, Van Caesbroeck D et al (1998) Single dose pharmacokinetics of lamivudine in subjects with impaired renal function and the effect of haemodialysis. Br J Clin Pharmacol. 46(1):21–27

16. Kearney BP, Flaherty JF, Shah J (2004) Tenofovir disoproxil fumarate: clinical pharmacology and pharmacokinetics. Clin Pharmacokinet. 43(9):595–612

17. Ray AS, Fordyce MW, Hitchcock MJ (2016) Tenofovir alafenamide: a novel prodrug of tenofovir for the treatment of Human Immunodeficiency Virus. Antiviral Res. 125:63–70

18. Garimella T, Wang R, Luo WL, Wastall P, Kandoussi H, DeMicco M et al (2015) Assessment of drug-drug interactions between daclatasvir and methadone or buprenorphine-naloxone. Antimicrob Agents Chemother. 59(9):5503–5510

19. Ouwerkerk-Mahadevan S B-MM, De Smedt G, et al (2011) The pharmacokinetic interaction between the investigational NS3/4A HCV protease inhibitor TMC435 and methadone. In: 62nd annual meeting of teh American Association for the Study of Liver Diseases (AASLD), San Francisco, CA, 4–8 Nov 2011

20. Polepally AR, King JR, Ding B, Shuster DL, Dumas EO, Khatri A et al (2016) Drug-drug interactions between the anti-hepatitis C virus 3D Regimen of Ombitasvir, Paritaprevir/Ritonavir, and dasabuvir and eight commonly used medications in healthy volunteers. Clin Pharmacokinet. 55(8):1003–1014

21. Fraser IPYW, Reitmann C, et al (2013) Lack of PK interaction between the HCV protease inhibitor MK-5172 and methadone or buprenorphine/naloxone in subjects on opiate maintenance therapy. [abstract O_16_PK]. In: 8th international workshop on clinical pharmacology of hepatitis therapy, Cambridge, MA, June 2013

22. German P, Moorehead L, Pang P, Vimal M, Mathias A (2014) Lack of a clinically important pharmacokinetic interaction between sofosbuvir or ledipasvir and hormonal oral contraceptives norgestimate/ethinyl estradiol in HCV-uninfected female subjects. J Clin Pharmacol. 54(11):1290–1298

23. Bifano M, Sevinsky H, Hwang C, Kandoussi H, Jiang H, Grasela D et al (2014) Effect of the coadministration of daclatasvir on the pharmacokinetics of a combined oral contraceptive containing ethinyl estradiol and norgestimate. Antivir Ther. 19(5):511–519

24. Ouwerkerk-Mahadevan SSA, Spittaels K, et al (2012) No pharmacokinetic interaction between the investigational HCV protease inhibitor simeprevir (TMC435) and an oral contraceptive containing ethinylestradiol and norethindrone. Poster. In: 63rd annual meeting of teh American Association for the Study of Liver Diseases (AASLD),Boston, MA, 9–13 Nov 2012

25. Epclusa. sofosbuvir and velpatasvir tablets.

26. Menon RM, Badri PS, Wang T, Polepally AR, Zha J, Khatri A et al (2015) Drug-drug interaction profile of the all-oral anti-hepatitis C virus regimen of paritaprevir/ritonavir, ombitasvir, and dasabuvir. J Hepatol. 63(1):20–29

27. Marshall WYW, Caro L, et al (2013) No Pharmacokinetic interaction between the hepatitis c virus non-structural protein 5a inhibitor MK-8742 and ethinyl estradiol and levonorgestrel. [abstract 53]. HEPDART 2013: Frontiers in drug development fo rViral hepatitis, Big Island, 8–12 Dec 2013

28. Ouwerkerk-Mahadevan S, Snoeys J, Peeters M, Beumont-Mauviel M, Simion A (2016) Drug-drug interactions with the NS3/4A protease inhibitor simeprevir. Clin Pharmacokinet. 55(2):197–208

29. Kohli A, Kapoor R, Sims Z, Nelson A, Sidharthan S, Lam B et al (2015) Ledipasvir and sofos-buvir for hepatitis C genotype 4: a proof-of-concept, single-centre, open-label phase 2a cohort study. Lancet Infect Dis. 15(9):1049–1054

30. Zepatier (2016) elbasvir/grazoprevir. prescribing Information. Merck and Co., INC

31. Caro LTJ, Guo Z et al (2013) Pharmacokinetic interaction between the HCV protease inhibitor MK-5172 and midazolam, pitavastatin, and atorvastatin in healthy volunteers. [abstract 477]. Hepatology 58(4 (suppl) AASLD Abstracts):437A

32. Bifano M CS, Hwang C, et al. The effect of co-administration f the proteon-pump inhibitor omeprazole on the pharmacokinetics of daclatasvir in healthy subjects. In: 48th annual meet-ing of the European Association for the Study of the Liver, Amsterdam, 24–28 Apr 2013

33. Bifano M, Adamczyk R, Hwang C, Kandoussi H, Marion A, Bertz RJ (2015) An open-label investigation into drug-drug interactions between multiple doses of daclatasvir and single-dose cyclosporine or tacrolimus in healthy subjects. Clin Drug Investig. 35(5):281–289

34. Mogalian E, German P, Kearney BP, Yang CY, Brainard D, McNally J et al (2015) Use of multiple probes to assess transporter- and Cytochrome P450-mediated drug-drug interaction potential of the Pangenotypic HCV NS5A inhibitor Velpatasvir. Clin Pharmacokinet.

35. Badri P, Dutta S, Coakley E, Cohen D, Ding B, Podsadecki T et al (2015) Pharmacokinetics and dose recommendations for cyclosporine and tacrolimus when coadministered with ABT-450, ombitasvir, and dasabuvir. Am J Transplant. 15(5):1313–1322

36. Sovaldi (2015) Sofosbuvir. Prescribing information

37. Denning J, Cornpropst M, Flach SD, Berrey MM, Symonds WT (2013) Pharmacokinetics, safety, and tolerability of GS-9851, a nucleotide analog polymerase inhibitor for hepatitis C virus, following single ascending doses in healthy subjects. Antimicrob Agents Chemother. 57(3):1201–1208

38. Kirby BJ, Symonds WT, Kearney BP, Mathias AA (2015) Pharmacokinetic, pharmacody-namic, and drug-interaction profile of the hepatitis C virus NS5B polymerase inhibitor sofos-buvir. Clin Pharmacokinet. 54(7):677–690

39. Product Information: Sovaldi, sofosbuvir (2015) Gilead Sciences, Inc, Foster City, CA

40. Desnoyer A, Pospai D, Le MP, Gervais A, Heurgue-Berlot A, Laradi A et al (2016) Pharmacokinetics, safety and efficacy of a full dose sofosbuvir-based regimen given daily in hemodialysis patients with chronic hepatitis C. J Hepatol. 65(1):40–47

41. Saxena V, Koraishy FM, Sise ME, Lim JK, Schmidt M, Chung RT et al (2016) Safety and efficacy of sofosbuvir-containing regimens in hepatitis C-infected patients with impaired renal function. Liver Int. 36(6):807–816

42. Bhamidimarri KR, Czul F, Peyton A, Levy C, Hernandez M, Jeffers L et al (2015) Safety, efficacy and tolerability of half-dose sofosbuvir plus simeprevir in treatment of Hepatitis C in patients with end stage renal disease. J Hepatol. 63(3):763–765

43. MacBrayne CE, Kiser JJ (2016) Pharmacologic Considerations in the Treatment of Hepatitis C Virus in Persons With HIV. Clin Infect Dis. 63(Suppl 1):S12–S23

44. Harvoni (2015) ledipasvir and sofosbuvir. Prescribing information. Gilead Sciences.

45. Liu GTC, Chang C, Rajamani S, Ray AS, Stamm LM, Vick J, Willichinsky M, McHutchinson JG, Brainard DM (2016) Effect of amiodarone and HCV direct-acting antiviral agents on car-diac conduction in nonclinical stuides. In: EASL-The International Liver Congress, Barcelona, 13–17 Apr 2016

46. Millard DC, Strock CJ, Carlson CB, Aoyama N, Juhasz K, Goetze TA et al (2016) Identification of Drug-Drug interactions in vitro: a case study evaluating the effects of sofosbuvir and amio-darone on hiPSC-derived cardiomyocytes. Toxicol Sci. 154(1):174–182

47. Regan CP, Morissette P, Regan HK, Travis JJ, Gerenser P, Wen J et al (2016) Assessment of the clinical cardiac drug-drug interaction associated with the combination of hepatitis C virus nucleotide inhibitors and amiodarone in guinea pigs and rhesus monkeys. Hepatology 64(5):1430–1441

48. Daklinza (2015) daclatasvir (prescribing information)
49. Garimella T, You X, Wang R, Huang SP, Kandoussi H, Bifano M et al (2016) A review of daclatasvir drug-drug interactions. Adv Ther. 33(11):1867–1884
50. Eley TYX, Wang R, et al (2014) Daclatasvir: overview of drug-drug interactions with antiretroviral agents and other common concomitant drugs. HIV DART Miami, FL, 9–12 Dec 2014
51. Bifano M, Hwang C, Oosterhuis B, Hartstra J, Grasela D, Tiessen R et al (2013) Assessment of pharmacokinetic interactions of the HCV NS5A replication complex inhibitor daclatasvir with antiretroviral agents: ritonavir-boosted atazanavir, efavirenz and tenofovir. Antivir Ther. 18(7):931–940
52. Gandhi YAR, Wang R, et al (2015) Assessment of drug-drug interactions between daclatasvir and darunavir/ritonavir or lopinavir/ritonavir. In: 16th international workshop on clinical pharmacology of HIV and hepatitis therapy, Washington, DC, 26–28 May 2015. Abstract 80
53. Center for drug evaluation and research: clinical pharmacology and biopharmaceutics review(s) ledipasvir/sofosbuvir (2014)
54. German P, Fang L, et al (2014) Drug-drug interaction profile of the fixed dose combination tablet ledipasvir/sofosbuvir [abstract 1976]. In: American Association for the study of liver diseases, Boston, MA, 7–11 Nov 2014
55. Kirby B (2014) Transporters: role in clinical development of HCV compounds. Presentation at the 15th international workshop on clinical pharmacology of HIV and hepatitis therapy,Washington, DC.
56. Tapper EB, Bacon BR, Curry MP, Dieterich DT, Flamm SL, Guest LE et al (2016) Evaluation of proton pump inhibitor use on treatment outcomes with ledipasvir and sofosbuvir in a real-world cohort study. Hepatology. 64(6):1893–1899
57. Terrault NA, Zeuzem S, Di Bisceglie AM, Lim JK, Pockros PJ, Frazier LM et al (2016) Effectiveness of ledipasvir-sofosbuvir combination in patients with hepatitis C virus infection and factors associated with sustained virologic response. Gastroenterology. 151(6):1131–1140. e5
58. Mascolini M (2014) Impact of food and antacids on levels of ledipasvir and sofosbuvir. In: 15th international workshop on clinical pharmacology of HIV and hepatitis therapy, Washington, DC, 19–21 May 2014
59. German PGK, Pang PS et al (2015) Drug interactions between the anti-HCV regimen ledipasvir/sofosbuvir and ritonavir boosted protease inhibitors plus Emtricitabine/Tenofovir DF. In: 22nd conference on retroviruses and opportunistic infections seattle Washington, DC, 23–26 Feb 2015
60. German PPP, West S et al (2014) Drug interaction between direct acting anti-HCV antivirals sofosbuvir and ledipasvir and HIV antiretrovirals. [Abstract 06.]. In: 15th international workshop on clinical pharmacology of HIV and hepatitis therapy, Washington, DC, 19–21 May 2014
61. Naggie S, Cooper C, Saag M, Workowski K, Ruane P, Towner WJ et al (2015) Ledipasvir and sofosbuvir for HCV in patients coinfected with HIV-1. N Engl J Med. 373(8):705–713
62. Su F, Green PK, Berry K, Ioannou GN (2017) The association between race/ethnicity and the effectiveness of direct antiviral agents for hepatitis C virus infection. Hepatology. 65(2):426–438
63. O'Brien TR, Kottilil S, Feld JJ, Morgan TR, Pfeiffer RM (2017) Race or genetic makeup for hepatitis C virus treatment decisions? Hepatology. 65(6):2124–2125
64. Olysio. Simeprevir. Prescribing Information (2015)
65. Sivi Ouwerkerk-Mahadevan As, Monika Peeters, et al (2013) Summary of pharmacokinetic drug-drug interactions for simeprevir (TMC435), a hepatitis C virus Ns3/4A protease inhibitor. In: 14th European AIDS conference, Brussels, 16–19 Oct 2013
66. Sekar VVR, Meyvisch P, et al (2010) Evaluation of metabolic interactions for TMC435 via cytochromie P450 (CYP) enzymes in healthy volunteers [poster]. In: 45th annual meeting of the European Association for the Study of the Liver 9EASL), Vienna, AU, 14–18 Apr 2010
67. Tischer S, Fontana RJ (2014) Drug-drug interactions with oral anti-HCV agents and idiosyncratic hepatotoxicity in the liver transplant setting. J Hepatol. 60(4):872–884

68. Pianko S, Flamm SL, Shiffman ML, Kumar S, Strasser SI, Dore GJ et al (2015) Sofosbuvir plus velpatasvir combination therapy for treatment-experienced patients with Genotype 1 or 3 Hepatitis C virus infection: a randomized trial. Ann Intern Med. 163(11):809–817
69. Mathias A (2013) Clinical Pharmacology of DAA's for HCV: What's New and What's in the Pipeline. In: 14th international workshop on clinical pharmacology of HIV therapy, 24 Apr 2013
70. Mogalian E SL, Osinusi A, et al (2015) Drug-drug interaction studies between Hepatitis C virus antivirals sofosbuvir and Velpatasvir (GS-5816) and HIV antiretoviral therapies. In: 66th annual meeting of the American Association for the Study of Liver Diseases Boston, MA, 13–17 2015
71. Jacobson IMPF, Firpi-Morell R, et al (2015) Efficacy and safety of grazoprevir and elbasvir in hepatitis C genotype 1—infected patients with Child-Pugh class B cirrhosis (C-SALT Part A). The 50th Annual Meeting of the European Association for the Study of the Liver, Vienna, AU, 22–26 Apr 2015
72. Yeh WMW, Caro L, et al (2014) Pharmacokinetic interaction of HCV NS5A inhibitor MK-8742 and ketoconazole in healthy subjects {P_27}. In: 15th international workshop on clinical pharmacology of HIV and hepatitis therapy, Washington, DC, 19–21 May 2014
73. Yeh W (2015) Drug-drug interactions with grazoprevir/elbasvir: practical considerations for the care of HIV/HCV co-infected patients. In: international workshop on clinical pharmacology of HIV & hepatitis therapy, 28 May 2015
74. TECHNIVIE (2015) Ombitasvir, paritaprevir, and ritonavir tablets. Prescribing information
75. VIEKIRA PAK (2015) ombitasvir, paritaprevir, and ritonavir tablets; dasabuvir tablets. Prescribing information
76. Pockros PJRK, Mantry PS, et al (2015) Safety of Ombitasvir/Paritaprevir/Ritonavir plus dasabuvir For treating HCV GT1 infection in patients with severe renal impairment or end-stage renal disease: the Ruby-I study in: European Association for the Study of the Liver, Vienna, AU, 22–26 Apr 2015
77. Smolders EJ, de Kanter CT, de Knegt RJ, van der Valk M, Drenth JP, Burger DM (2016) Drug-drug interactions between direct-acting antivirals and psychoactive medications. Clin Pharmacokinet. 55(12):1471–1494
78. Werner JM, Serti E, Chepa-Lotrea X, Stoltzfus J, Ahlenstiel G, Noureddin M et al (2014) Ribavirin improves the IFN-gamma response of natural killer cells to IFN-based therapy of hepatitis C virus infection. Hepatology. 60(4):1160–1169
79. Thomas E, Ghany MG, Liang TJ (2013) The application and mechanism of action of ribavirin in therapy of hepatitis C. Antivir Chem Chemother. 23(1):1–12
80. Li L, Koo SH, Limenta LM, Han L, Hashim KB, Quek HH et al (2009) Effect of dietary purines on the pharmacokinetics of orally administered ribavirin. J Clin Pharmacol. 49(6):661–667
81. Copegus (2015) Ribavirin. Prescribing Information
82. Salmon-Ceron D, Chauvelot-Moachon L, Abad S, Silbermann B, Sogni P (2001) Mitochondrial toxic effects and ribavirin. Lancet. 357(9270):1803–1804

Chapter 10
Drug Interactions of Non-HIV Antiviral Agents

Douglas N. Fish

10.1 Introduction

Viruses continue to be recognized as frequent and important pathogens in many populations. Critically ill and immunocompromised patients, particularly within the transplant and human immunodeficiency virus (HIV)-infected populations, are at particularly high risk for severe viral infections such as those caused by cytomegalovirus (CMV), adenovirus, and disseminated herpes simplex virus (HSV) and varicella zoster virus (VZV). Infections caused by hepatitis B virus (HBV) and hepatitis C virus (HCV) are also quite prevalent, and influenza continues to be a significant public and global health problem. Despite the frequent need for effective management of such infections, the development of new antiviral agents for the prophylaxis and/or treatment of non-HIV viral infections has been relatively sluggish. With the exception of drugs used for HIV and the newer direct-acting antiviral (DAA) agents for HCV, few novel antiviral agents have been introduced to the US marketplace in recent years. Many of the currently available antiviral agents have been in clinical use for many years, and clinically significant drug interactions are reasonably well characterized, while interaction data are often quite limited for newer agents. This chapter summarizes available data regarding pharmacokinetic and toxic interactions with the current antiviral agents, excluding those used specifically for HIV infections as well as the DAAs. While many relevant drug interactions are summarized in Tables 10.1, 10.2, 10.3, and 10.4, there is still a great need for additional studies related to interactions with this clinically important group of drugs.

D. N. Fish (✉)
Department of Clinical Pharmacy, University of Colorado School of Pharmacy, Aurora, CO, USA

University of Colorado Hospital, Aurora, CO, USA
e-mail: Doug.Fish@ucdenver.edu

© Springer International Publishing AG, part of Springer Nature 2018   381
M. P. Pai et al. (eds.), *Drug Interactions in Infectious Diseases: Antimicrobial Drug Interactions*, Infectious Disease,
https://doi.org/10.1007/978-3-319-72416-4_10

Table 10.1 Interactions with drugs used for treatment of herpes simplex and varicella zoster virus infections

Primary drug	Interacting drug	Effects	Mechanisms	Comments/management
Acyclovir	Food	AUC ↓ 18%	Decreased absorption	Not clinically significant, may administer without regard to meals
	Cytarabine	ACV C_{max} ↓ 43%, F ↓ 38%	Decreased absorption caused by mucosal damage	Clinical significance unknown, monitor response to antiviral therapy
	Phenytoin, valproic acid	↓ serum concentrations of AEDs	Decreased absorption, unknown mechanism	Clinical significance unknown, monitor for effectiveness of AED therapy
	Probenecid	ACV AUC ↑ 40%	Decreased renal tubular secretion	May be clinically significant, consider reduction in ACV dose
	Benzylpenicillin	ACV AUC ↑ 30%	Decreased renal tubular secretion	May be clinically significant, but not reported in humans
	Lithium	300% ↑ lithium concentration	Competitive renal tubular secretion	Clinical significance unknown, monitoring of lithium concentrations recommended
	Cyclosporine	No effect		No additional monitoring necessary
	Mycophenolate mofetil	ACV C_{max} ↑ 18%; MMF AUC ↑ 9%	Competitive renal tubular secretion	Not likely to be clinically significant
	Nephrotoxins (aminoglycosides, amphotericin B, cidofovir, foscarnet, intravenous pentamidine, vancomycin, etc.); also reports of nephrotoxicity during ceftriaxone therapy	Additive renal toxicity	Overlapping adverse effects	Avoid combination if possible; close clinical monitoring of renal function required

(continued)

Table 10.1 (continued)

Primary drug	Interacting drug	Effects	Mechanisms	Comments/ management
	Theophylline	THEO CL ↓ 30%, AUC ↑ 45%	Decreased oxidative metabolism	Clinical significance unknown, monitoring of theophylline concentrations recommended
	Zidovudine	Improved survival in HIV-infected patients	Synergistic anti-HIV activity	Not consistently demonstrated, clinical significance unknown; may be less relevant with HAART
Valacyclovir	High-fat meal	No effect		May administer without regard to meals
	Al^{3+}/Mg^{2+} antacids	No effect		Not significant
	Probenecid	ACV C_{max} ↑ 22%, AUC ↑ 49%	Decreased renal tubular secretion	May be clinically significant, consider reduction in VACV dose
	Cimetidine	ACV C_{max} ↑ 8%, AUC ↑ 32%	Decreased renal tubular secretion	May be clinically significant
	Probenecid + cimetidine	ACV C_{max} ↑ 30%, AUC ↑ 78%	Decreased renal tubular secretion	May be clinically significant
	Digoxin	No effect		No additional monitoring necessary
	Mycophenolate mofetil	ACV C_{max} ↑ 40%, AUC ↑ 31%; MMF ↔	Competitive renal tubular secretion	Not likely to be clinically significant but monitor antiviral therapy
	Nephrotoxins (as with ACV)	Additive renal toxicity	Overlapping side effects	Avoid combination if possible; close clinical monitoring of renal function required
	NSAIDs	Increased renal toxicity	Inhibition of ACV tubular secretion	May be clinically significant
	Tipranavir/ritonavir	ACV C_{max} ↓ 5%, AUC ↑ 7%		Not significant

(continued)

Table 10.1 (continued)

Primary drug	Interacting drug	Effects	Mechanisms	Comments/ management
Famciclovir/ penciclovir	Food	PCV C_{max} ↓ 53%, ↑ T_{max}, AUC ↔	Decreased rate of absorption	Not clinically significant, may administer without regard to meals
	Digoxin	No effect on either drug		No additional monitoring necessary

ACV acyclovir, *AED* antiepileptic drug, *AUC* area under the plasma concentration vs. time curve, *CL* total systemic clearance, C_{max} maximum plasma concentration, *F* bioavailability, *HAART* highly active antiretroviral therapy, *MMF* mycophenolate, *NSAIDs* nonsteroidal anti-inflammatory drugs, *PCV* penciclovir, *THEO* theophylline, T_{max} time to maximum plasma concentration, *VACV* valacyclovir

10.2 Drugs for Treatment of HSV and VZV Infections

10.2.1 Acyclovir/Valacyclovir

Potentially important drug interactions with acyclovir and valacyclovir are summarized in Table 10.1. Although the effects of administration with food will differ depending on whether acyclovir is administered as the parent drug or the valacyclovir prodrug, other potential drug interactions involving systemic acyclovir would be assumed to occur after administration of either compound. Studies evaluating potential interactions between acyclovir and valacyclovir with food have shown no significant or clinically relevant changes in the areas under the plasma concentration vs. time curve (AUC) compared to administration of the drugs in the fasted state. Both oral acyclovir and valacyclovir may thus be administered without regard to meals [1, 2]. The concurrent administration of a single dose of aluminum- or magnesium-containing antacid was also shown to have no significant effects on the pharmacokinetics of valacyclovir after oral administration of a 1 g dose [2].

Acyclovir is commonly administered for suppression of HSV disease during remission-induction chemotherapy in patients with acute myelogenous leukemia (AML). Malabsorption of D-xylose as a probable result of damage to the intestinal mucosa has previously been observed after cytarabine therapy in such patients. While the pharmacokinetics of intravenously administered acyclovir were unchanged after chemotherapy, both the peak plasma concentration (C_{max}) and bioavailability of oral acyclovir were substantially decreased after chemotherapy [3]. It is not known whether these changes are clinically important or whether such pharmacokinetic alterations are also seen with valacyclovir; however, effectiveness of antiviral therapy should be carefully monitored when these agents are orally administered in similar clinical scenarios.

Two case reports have described the possibly decreased oral absorption of phenytoin and valproic acid (VPA) after the use of oral acyclovir in children; the

Table 10.2 Interactions with drugs used for treatment of cytomegalovirus infections

Primary drug	Interacting drug	Effects	Mechanisms	Comments/management
Ganciclovir	High-fat meal	GCV C_{max} ↑ 15%, AUC ↑ 20%	Increased absorption	Administer oral ganciclovir with high-fat meal
Valganciclovir	High-fat meal	GCV C_{max} ↑ 14%, AUC ↑ 23–57%	Increased absorption	Administer oral valganciclovir with high-fat meal
Ganciclovir/ valganciclovir	Probenecid	GCV AUC ↑ 53%	Decreased renal tubular secretion	Avoid combination if possible due to increased risk of GCV toxicities; consider empiric reduction of GCV dose, close patient monitoring required
	Zidovudine	GCV AUC ↓ 17%, ZDV AUC ↑ 19%	Unknown, may involve competition for renal secretion	Pharmacokinetic changes not clinically significant, but potential for increased hematologic toxicities requires close patient monitoring
	Zidovudine	Increased myelosuppression	Overlapping adverse effects	Potential for increased hematologic toxicities requires close patient monitoring; avoid combination if possible
	Zidovudine	Decreased survival in AIDS patients	Unknown	Clinical significance unknown; may be less relevant with HAART
	Myelosuppressives (dapsone, flucytosine, antineoplastics, amphotericin B, intravenous pentamidine, TMP/SMX, pyrimethamine, trimetrexate, zidovudine, etc.)	Increased myelosuppression	Overlapping adverse effects	Avoid combination if possible; close patient monitoring required due to potential for increased hematologic toxicities
	Didanosine	GCV AUC ↑ 21%; ddI C_{max} ↑ 36–49%, AUC ↑ 50–115%	Unknown, does not appear to involve tubular secretion	Avoid combination if possible; consider ddI dose reduction, close monitoring of both drugs required due to potential for increased toxicities

(continued)

Table 10.2 (continued)

Primary drug	Interacting drug	Effects	Mechanisms	Comments/management
	Didanosine	Pancytopenia, persistently decreased CD4+ count	Decreased ddI metabolism, ↑ toxicity	Avoid combination if possible; consider ddI dose reduction, close monitoring for increased ddI toxicities
	Cyclosporine	No effect		No additional monitoring required
	Mycophenolate mofetil	No effect on either drug		No additional monitoring required
	Valproic acid	Decreased antiviral activity of GCV, foscarnet, cidofovir against CMV	VPA stimulation of viral replication	Clinical significance unknown
	Imipenem	Seizures	Overlapping adverse effects	
Cidofovir	Probenecid	CDV C_{max} ↑ 35–70%, AUC ↑ 29–45%	Decreased renal tubular secretion	Interaction used therapeutically to increase CDV exposure and improve efficacy; recommended probenecid dose is 2 g orally 3 h prior to CDV and then 1 g orally 2 and 8 h after completion of CDV infusion
	Zidovudine	ZDV AUC ↑ in presence of probenecid; no effect with CDV alone	Decreased renal tubular secretion	Temporarily discontinue ZDV or reduce dose by 50%
	Nephrotoxins (aminoglycosides, amphotericin B, acyclovir, NSAIDs, foscarnet, intravenous pentamidine, vancomycin, etc.)	Additive renal toxicity	Overlapping adverse effects	Close clinical monitoring of renal function required; discontinue other agents prior to starting CDV if possible
	Rifabutin	Additive ocular toxicity, uveitis	Overlapping adverse effects	Avoid combination if possible; close clinical monitoring for ocular toxicities required
Foscarnet	Probenecid	Not well described	Decreased renal tubular secretion	Avoid combination if possible; close clinical monitoring of renal function required

Primary drug	Interacting drug	Effects	Mechanisms	Comments/management
	Nephrotoxins (as with CDV)	Additive renal toxicity	Overlapping adverse effects	Avoid combination if possible; close clinical monitoring of renal function required
	Ciprofloxacin	Seizures	Enhanced ciprofloxacin inhibition of GABA receptors	Clinical significance unknown; avoid combination if possible, close clinical monitoring recommended
	Pentamidine (IV)	Severe hypocalcemia	Additive nephrotoxicity, enhanced renal calcium wasting	Avoid combination if possible; close clinical monitoring is required
	Zidovudine	Synergistic effects against HIV and CMV in vitro	Unknown	Clinical significance unknown
	Ganciclovir	No change in either GCV or foscarnet		No additional monitoring required

AUC area under the plasma concentration *vs.* time curve, *CDV* cidofovir, *C*max maximum plasma concentration, *CMV* cytomegalovirus, *ddI* didanosine, *GABA* gamma-aminobutyric acid, *GCV* ganciclovir, *HAART* highly active antiretroviral therapy, *HIV* human immunodeficiency virus, *IV* intravenous, *NSAIDs* non-steroidal anti-inflammatory drugs, *PO* oral, *TMP/SMX* trimethoprim/sulfamethoxazole, *ZDV* zidovudine

Table 10.3 Interactions with drugs used for prevention and treatment of influenza

Primary drug	Interacting drug	Effects	Mechanisms	Comments/management
Amantadine	Food	No effect		May administer without regard to meals
	Triamterene/hydrochlorothiazide	AMA AUC ↑ 50%	Decreased renal tubular secretion	Clinical monitoring of patients for enhanced CNS toxicities advised
	TMP/SMX	Decreased AMA clearance, increased CNS toxicity	Decreased renal tubular secretion by TMP	Clinical monitoring of patients for enhanced CNS toxicities advised
	Quinine, quinidine	AMA renal clearance ↓ 30%	Decreased renal tubular secretion	Clinical monitoring of patients for enhanced CNS toxicities advised
	Agents with CNS toxicities (antihistamines, psychotropics, anticholinergics, bupropion, phenylpropanolamine)	Additive CNS toxicity	Overlapping adverse effects	Clinical monitoring of patients for enhanced CNS toxicities advised; consider avoiding combination if possible
	Oseltamivir	No effect		No significant interaction
	Intranasal LAIV	Potential for reduced vaccine efficacy	Inhibition of viral replication	LAIV should not be administered within 2 weeks before, or 48 h after, administration of AMA
Rimantadine	Food	No effect		Administer without regard to meals
	Intranasal LAIV	Potential for reduced vaccine efficacy	Inhibition of viral replication	LAIV should not be administered within 2 weeks before, or 48 h after, administration of RIM
	Anti-influenza antivirals (oseltamivir, peramivir)	No effects		No significant interaction
	Cimetidine	No effects		No significant interaction
	Aspirin, acetaminophen	RIM C_{max} and AUC ↓ 10% with ASA, ↓ 11% with APAP		Not clinically significant with either ASA or APAP
Oseltamivir	Food	OTV C_{max} ↓ 16%, AUC ↓ 2%		Not clinically significant, may administer without regard to meals

Primary drug	Interacting drug	Effects	Mechanisms	Comments/management
	Al^{3+}/Mg^{2+} or calcium carbonate antacids	No effect		No significant interaction
	Milk	OTV C_{max} ↓ 31%, AUC ↓ 65%		Clinical significance unknown, may be advisable to avoid combination if possible
	Clopidogrel	Conversion of OTV to active OTV carboxylate ↓ 90%	Competitive inhibition of hepatic carboxylesterase 1 enzyme	Probably not clinically significant due to methodological problems with in vitro studies
	Probenecid	OTV AUC ↑ 100%	Decreased renal tubular secretion	Probably not clinically significant due to favorable side effect profile of OTV; may allow for alternative dosing regimens to reduce daily dosing requirements of OTV
	Methotrexate	Interaction not evaluated but potential for ↑ AUC of MTX	Potential for decreased renal tubular secretion	Caution with use of OTV during MTX therapy; close patient monitoring advised
	Warfarin	No effects on warfarin kinetics or coagulation parameters in well-designed studies, but case reports of significantly elevated INR after starting OTV		Clinical significance of potential interaction unclear; monitoring of patients for increased INR advised
	Anti-influenza antivirals (amantadine, rimantadine, zanamivir, peramivir)	No significant effects on either OTV or other antivirals		No significant interactions

(continued)

Table 10.3 (continued)

Primary drug	Interacting drug	Effects	Mechanisms	Comments/management
	Intranasal LAIV	Potential for reduced vaccine efficacy	Inhibition of viral replication	LAIV should not be administered within 2 weeks before, or 48 h after, administration of OTV
	Sotalol	↑ QTc, TdP	Unknown	Case reports of two patients with multiple risk factors for ↑ QTc, clinical significance unclear
Zanamivir	Oseltamivir	C_{max} of IV zanamivir		
	LAIV intranasal vaccine	Potential for reduced vaccine efficacy	Inhibition of viral replication	LAIV vaccine should not be administered within 2 weeks before, or 48 h after, administration of zanamivir
Peramivir	Rimantadine	No significant effects on either drug		No significant interaction
	Oseltamivir	No significant effects on either drug		No significant interaction
	Probenecid	No significant effect		No significant interaction
	Oral contraceptives containing ethinyl estradiol and levonorgestrel	No significant effect		No significant interaction
	Intranasal LAIV	Potential for reduced vaccine efficacy	Inhibition of viral replication	LAIV should not be administered within 2 weeks before, or 48 h after, administration of peramivir

AMA amantadine, *ASA* aspirin, *APAP* acetaminophen, *AUC* area under the plasma concentration vs. time curve, C_{max} maximum plasma concentration, *CNS* central nervous system, *LAIV* live attenuated influenza vaccine, *MTX* methotrexate, *OTV* oseltamivir, *QTc* corrected Q-T interval, *RIM* rimantadine, *TdP* torsades de pointes, *TMP* trimethoprim, *TMP/SMX* trimethoprim/sulfamethoxazole

Table 10.4 Interactions with ribavirin and the interferons

Primary drug	Interacting drug	Effects	Mechanisms	Comments/management
Ribavirin	Food	RBV C_{max} ↑ 66%, T_{max} ↑ 100%, AUC ↑ 42%	Increased absorption	Administer oral ribavirin with meals
	Interferon-α	Enhanced efficacy in treatment of HCV; no pharmacokinetic alterations in RBV or IFN	Synergistic antiviral activity; primary mechanism unknown but seems to be based on pharmacology rather than immunologic or pharmacokinetic effects	No role for ribavirin monotherapy, combination therapy with pegylated interferon or DAAs required
	Interferon-α	Increased hemolytic anemia	Overlapping adverse effects	Close patient monitoring required
	Al^{3+}/Mg^{2+} plus simethicone antacids	RBV AUC ↓ 14%	Decreased absorption	No significant interaction
	Myelosuppressives (azathioprine, clozapine, dapsone, flucytosine, antineoplastics [including 5-fluorouracil, hydroxyurea, melphalan], thalidomide, amphotericin B, intravenous pentamidine, TMP/SMX, pyrimethamine, trimetrexate, zidovudine, etc.)	Increased myelosuppression	Overlapping adverse effects; with azathioprine, ribavirin causes alteration in metabolic pathways leading to increase in toxic metabolites	Avoid combination if possible; close patient monitoring required due to potential for increased hematologic toxicities
	Zidovudine	Reduced anti-HIV activity of ZDV; may also affect other nucleoside analogues including lamivudine, stavudine, emtricitabine, and tenofovir; no change in pharmacokinetics of RBV or ZDV	Decreased intracellular phosphorylation of nucleoside analogues through competitive inhibition by RBV	Clinical significance unknown; in vitro findings have not been confirmed by clinical data in humans

(continued)

Table 10.4 (continued)

Primary drug	Interacting drug	Effects	Mechanisms	Comments/management
	Didanosine	Increased ddI toxicities (lactic acidosis, pancreatitis); may also affect stavudine and abacavir	RBV inhibition of intracellular enzyme leading to accumulation of dideoxyadenosine-5′-triphosphate, increased mitochondrial toxicity	Avoid combination if possible; close patient monitoring required due to potential for increased mitochondrial toxicities of antiretrovirals
	Atazanavir	Increased hyperbilirubinemia	RBV-induced hemolysis in combination with ATV inhibition of bilirubin conjugation pathways	Discontinuation of ATV, change to different antiretroviral if possible
	Raltegravir	RBV T_{max} ↑ 39%, C_{max} ↓ 21%		
	Abacavir	No changes in RBV C_{min} or antiviral response		No significant interaction
	Warfarin	Decreased warfarin activity	Unknown mechanism	Clinical significance unknown
Interferon-α (including pegylated interferons)	Drugs metabolized by CYP1A (theophylline, caffeine, antipyrine, TCAs, olanzapine, clozapine) and CYP3A (azole antifungals, macrolide antibiotics, many antiretroviral agents, some immunosuppressants, SSRIs, TCAs, statins, antipsychotics, benzodiazepines, barbiturates, calcium channel antagonists)	Theophylline AUC ↑ 25–100%; erythromycin AUC ↑ 15–35%, hexobarbital CL ↓ 7%; antipyrine CL ↓ 16–22%; caffeine less effected	IFN-related downregulation/inhibition of CYP activity	Changes highly variable and may be dose-related; alterations not always clinically significant, but patients should be monitored for enhanced adverse effects of concomitant drugs

Primary drug	Interacting drug	Effects	Mechanisms	Comments/management
	Drugs metabolized by CYP2C8/9 (warfarin, phenytoin, NSAIDs, angiotensin receptor blockers, certain statins, sulfonylureas) or CYP2D6 (β-blockers, lidocaine, flecainide, TCAs, SSRIs, opiate analgesics, antipsychotics)	Enzyme activity increased by 28–66%; effects on warfarin, phenytoin, and flecainide specifically mentioned by drug manufacturers	IFN-related upregulation/ induction of CYP activity	Changes highly variable and may be dose-related; alterations not always clinically significant, but patients should be monitored for altered activity or enhanced adverse effects of concomitant drugs
	Methadone	Methadone AUC ↑ 10–15%, no change in IFN	Inhibition of hepatic CYP3A4 metabolism	Clinical significance unknown, but patients should be closely monitored for enhanced methadone side effects
	Ribavirin	Increased hemolytic anemia, hepatic decompensation	Overlapping adverse effects	Close patient monitoring required
	Myelosuppressives (as with RBV)	Increased myelosuppression	Overlapping adverse effects	Avoid combination if possible; close patient monitoring required due to potential for increased hematologic toxicities
	Ethanol	Decreased clinical response of HCV infection to interferon	Ethanol-mediated ↑ viral replication, ↑ inflammation/ fibrosis; inhibition of intracellular signaling pathways and ↓ expression of interferon-induced antiviral gene products	Avoid ethanol consumption during treatment with interferon

(continued)

Table 10.4 (continued)

Primary drug	Interacting drug	Effects	Mechanisms	Comments/management
	Selected antidepressants: venlafaxine, mirtazapine, amitriptyline	Decreased short-term clinical response of HCV infection to interferon	Modulation of central and/or peripheral inflammatory pathways and circulating cytokines	Clinical significance unknown

ATV atazanavir, *AUC* area under the plasma concentration *vs.* time curve, *CL* total systemic clearance, C_{max} maximum plasma concentration, C_{min} minimum plasma concentration, *CYP* cytochrome P450, *DAA* direct-acting antivirals for HCV, *ddI* didanosine, *HCV* hepatitis C virus, *HIV* human immunodeficiency virus, *IFN* interferon, *NSAID* nonsteroidal anti-inflammatory drug, *RBV* ribavirin, *SSRI* serotonin-specific reuptake inhibitor, *TCA* tricyclic antidepressant, T_{max} time to maximum plasma concentration, *TMP/SMX* trimethoprim/sulfamethoxazole, *ZDV* zidovudine

subsequently decreased serum concentrations of the antiepileptic drugs were associated with increased seizure frequency in previously stable patients [4]. The mechanisms underlying this potential interaction are unknown.

Acyclovir is primarily eliminated by the kidneys through a combination of glomerular filtration and active renal tubular secretion. Potential drug interactions involving drugs which may inhibit the tubular secretion of acyclovir have therefore been evaluated. The AUC of acyclovir was increased by approximately 40% after concomitant administration of probenecid, with renal clearance and urinary excretion of acyclovir also correspondingly decreased [1, 5]. Similar results have been observed with valacyclovir [2, 6]. The C_{max} and AUC of acyclovir were also increased by 8% and 32%, respectively, in subjects receiving valacyclovir orally after concomitant administration of a single 800 mg dose of cimetidine [2, 6]. Additionally, acyclovir C_{max} and AUC were increased to an even greater extent after concomitant administration of valacyclovir along with the combination of both probenecid and cimetidine [2, 6]. Similar results on acyclovir tubular secretion have been described in rat models where the concomitant administration of acyclovir and benzylpenicillin was associated with a 30% increase in acyclovir AUC, presumably through interactions with human organic anion transporters (hOAT) 1 and 3 in the renal tubules [7]. Although both probenecid and cimetidine caused substantial alterations in acyclovir pharmacokinetics, these changes are unlikely to be clinically relevant unless using high doses of acyclovir where accumulation to excessively high concentrations may increase the risk of drug-related adverse effects. Conversely, although acyclovir is not otherwise known to inhibit the renal secretion of other drugs, one case report described a fourfold increase in lithium serum concentrations following the addition of high-dose intravenous acyclovir to a chronic lithium carbonate regimen [8].

No significant pharmacokinetic alterations were observed when valacyclovir was administered together with multiple doses of thiazide diuretics [2]. Likewise, the pharmacokinetics of neither acyclovir nor digoxin were significantly altered when valacyclovir was administered concomitantly [2].

The pharmacokinetics of neither acyclovir nor cyclosporine were significantly affected when the two drugs were administered concomitantly [1]. However, studies assessing the possible interaction of acyclovir and mycophenolate mofetil (MMF) have had conflicting results [9–11]. One study found that the AUC of MMF and its primary glucuronide conjugate metabolite, MPAG, were increased by 9% and 10%, respectively, while the acyclovir C_{max} was also increased by 18% [9]. A second study found that the pharmacokinetics of neither MMF nor acyclovir were significantly affected [10]. In a final study, 15 patients were randomized to receive oral acyclovir alone, valacyclovir alone, MMF alone, acyclovir plus MMF, or valacyclovir plus MMF in a single-dose, crossover study [11]. After co-administration with MMF, acyclovir C_{max}, time to peak plasma concentration (T_{max}), and AUC were significantly increased compared to those parameters after administration of acyclovir alone. The mean renal clearance of acyclovir was also reduced by 19%, possible due to competition with MPAG for renal tubular secretion. Valacyclovir T_{max} was also significantly increased by 0.5 h when co-administered with MMF, and alterations in C_{max}, T_{max}, and AUC were similar to those seen with oral acyclovir. Mycophenolate pharmacokinetics were not significantly altered after co-administration with either acyclovir or valacyclovir, with the exception that the AUC of MPAG was decreased by 12% after concomitant administration of valacyclovir. Overall, interactions between acyclovir, valacyclovir, and MMF were felt by the authors to be clinically insignificant in otherwise healthy subjects with good renal function [11].

Because high-dose acyclovir and valacyclovir have been associated with adverse renal effects including acute tubular necrosis, crystalluria, and acute renal failure, caution should be exercised when using these antivirals in combination with other drugs that also have potential for additive nephrotoxicity, e.g., aminoglycosides, amphotericin B products, cidofovir, foscarnet, intravenous pentamidine, vancomycin, and others [1]. The concomitant use of ceftriaxone has also been reported to increase the nephrotoxicity of acyclovir [12]. A retrospective case-control evaluation utilizing the Food and Drug Administration (FDA) Adverse Event Reporting System database found the combination of valacyclovir and loxoprofen, a nonsteroidal anti-inflammatory drug (NSAID), to be associated with a 26-fold increased rate of acute kidney injury (AKI) in elderly patients ≥ 65 years of age compared to patients of similar age exposed to neither agent (odds ratio 26.0, 95% confidence interval 19.2–35.3) [13]. An earlier case report also described AKI occurring in an elderly patient receiving valacyclovir and an NSAID [14]. The mechanism of potential renal injury is unknown, although NSAIDS are known to inhibit hOAT1 and hOAT3 in the renal tubules which may in turn lead to excessive accumulation of acyclovir [13].

A study conducted in five healthy subjects demonstrated a mean 30% reduction in total systemic theophylline clearance and 45% increase in AUC when theophylline was administered together with acyclovir [15]. The results of this study suggest that acyclovir interferes with the oxidative metabolism of theophylline, although such metabolic interactions with acyclovir have not been previously reported and the clinical significance is unknown.

The combination of acyclovir and zidovudine has been suggested to have synergistic antiviral effects and to be associated with improved survival in HIV-infected patients. However, the benefits of this combination have not been consistently demonstrated, and no pharmacokinetic or in vitro pharmacologic interactions between these two drugs have been demonstrated [16–21]. The clinical significance of this interaction is therefore unknown and may be less relevant with the current use of highly active antiretroviral therapy (HAART) for treatment of HIV infection. The combination of tipranavir plus ritonavir has been shown to have no significant pharmacokinetic interactions with acyclovir [22]; other antiretroviral agents have not been formally evaluated.

10.2.2   Famciclovir/Penciclovir

Famciclovir is the prodrug of penciclovir. Dideacetylation of famciclovir occurs in the blood and possibly in the intestinal wall, followed by 6-oxidation of the intermediary metabolite to form the active antiviral agent penciclovir. Conversion of

6-deoxy-penciclovir to penciclovir is catalyzed by the aldehyde oxidase enzyme. Cimetidine and promethazine are both in vitro inhibitors of aldehyde oxidase, but interaction studies have not shown any relevant effects of these drugs on the formation of penciclovir [23].

The effects of food on the pharmacokinetics of famciclovir were evaluated in two studies [24, 25]; the results of these studies are summarized in Table 10.1. The administration of oral famciclovir with food was associated with a 53% decrease in penciclovir C_{max} and an increase in T_{max} of approximately 2 h compared to the fasting state. However, the penciclovir AUC was not significantly altered, indicating that the rate of famciclovir absorption was altered but the overall bioavailability was not affected. Famciclovir may thus be administered without regard to meals [23–25].

Like acyclovir, probenecid may affect the renal tubular excretion of penciclovir, but this interaction is not considered to be clinically relevant under most circumstances [23]. An in vitro study using human liver microsomes found no inhibition of cytochrome (CYP) 3A4 enzymes by penciclovir [23]. In other studies, no clinically significant effects on penciclovir pharmacokinetics were observed after pretreatment with multiple doses of allopurinol, magnesium/aluminum hydroxide antacids, cimetidine, promethazine, theophylline, thiazides, emtricitabine, or zidovudine (either the parent drug or the zidovudine glucuronide metabolite) followed by administration of single doses of famciclovir [23]. Finally, the steady-state pharmacokinetics of digoxin were also not affected by either single or multiple doses of famciclovir [23].

10.3   Drugs Used for Prophylaxis and Treatment of CMV Infections

10.3.1   Ganciclovir and Valganciclovir

Potentially important drug interactions with ganciclovir and valganciclovir are summarized in Table 10.2. Although the effects of administration with food will differ depending on whether ganciclovir is administered as the parent drug or the valganciclovir prodrug, other potential drug interactions involving systemic ganciclovir would generally occur after administration of either compound. The C_{max} and AUC of oral ganciclovir were increased by 15% and 20%, respectively, when the drug was administered with a high-fat meal [26]. Although the preferred use of the better-absorbed valganciclovir prodrug makes this interaction of little relevance, similar results have also been observed with valganciclovir; C_{max} of ganciclovir was increased by 14%, and the AUC increased by 23–57% when valganciclovir was administered with standard or high-fat meals [27, 28]. Based on these studies, valganciclovir tablets should be administered with food [27–29].

As with acyclovir and penciclovir, ganciclovir is approximately 90% eliminated as unchanged drug through a combination of glomerular filtration and renal tubular secretion. The ganciclovir steady-state AUC was increased by $53 \pm 91\%$ (range -14% to 299%), and the renal clearance was decreased by $22 \pm 20\%$ (range -54% to -4%) when oral ganciclovir was administered concomitantly with probenecid, consistent with a probenecid-induced decrease in renal tubular transport [28]. The potential for acyclovir to competitively inhibit ganciclovir secretion was also assessed, but no interaction was found [28–30].

Because of the high incidence of CMV disease in HIV-infected patients and the overlapping myelosuppressive toxicities (i.e., anemia, leukopenia, thrombocytopenia) of ganciclovir and zidovudine, potential interactions between these two drugs have been assessed [30]. In a study in which oral ganciclovir was administered concomitantly with zidovudine, the steady-state ganciclovir AUC was decreased by $17 \pm 25\%$ (range -52% to 23%), while the steady-state zidovudine AUC was increased by $19 \pm 27\%$ (range -11 to 74%). Neither of these alterations was considered to be clinically significant [30]; no studies have been conducted with intravenous ganciclovir.

Although studies did not indicate a high risk for significant pharmacokinetic interactions, the overlapping myelosuppressive toxicities of ganciclovir and zidovudine are of concern. The combination of ganciclovir with zidovudine was found to be associated with high rates of drug intolerance and hematologic toxicity [31–35]. The combination of zidovudine plus intravenous ganciclovir was associated with hematologic toxicity in 82% and with severe neutropenia in 55% of 40 AIDS patients. These toxicities required dose reduction or drug discontinuation to effectively manage the adverse effects [31]. In a second study, 113 patients with AIDS or AIDS-related complex received zidovudine for a median duration of 152 days. Statistical analysis showed that the concomitant use of ganciclovir was associated with significantly increased risk of anemia and thrombocytopenia [32]. Similar high rates of anemia and neutropenia were reported in other studies as well [36]. Because of the high risk of hematologic toxicities and the current availability of a number of other antiretroviral agents, the use of zidovudine in combination of ganciclovir or valganciclovir should be avoided if possible. If such a combination must be used for clinical reasons, frequent monitoring of complete blood counts is required. Caution should also be exercised in combining ganciclovir or valganciclovir with other drugs with myelosuppressive potential; such drugs include dapsone, flucytosine, various antineoplastic agents, intravenous pentamidine, pyrimethamine, amphotericin B products, trimethoprim/sulfamethoxazole (TMP/SMX), and trimetrexate [30, 37].

The product information for ganciclovir states that ganciclovir pharmacokinetics are not affected by didanosine [30]. However, in a multiple-dose crossover study in 13 HIV-positive patients, a minor pharmacokinetic interaction was reported when oral ganciclovir was administered after didanosine [38]. In this study didanosine was administered either simultaneously with ganciclovir or sequentially, i.e., 2 h before ganciclovir. Significantly increased AUC was reported for didanosine during both simultaneous and sequential administration [115% and 108% increased AUC, respectively ($P < 0.001$)]. In addition, the AUC of ganciclovir was also decreased by 21% when administered 2 h after didanosine ($P = 0.002$) [38]. In a second study, intravenous ganciclovir plus didanosine resulted in significantly increases in didanosine AUC (70 ± 40%, range 3% to 121%) and C_{max} (49 ± 48%, range − 28% to 125%) [30, 39]. In a third study, ganciclovir combined with didanosine resulted in the steady-state didanosine AUC being increased by 50 ± 26% (range 22% to 110%) and C_{max} being increased by 36 ± 36% (range − 27% to 94%) [30, 40]. The mechanism for the apparent two-way pharmacokinetic interaction between ganciclovir and didanosine is unknown, but does not appear to involve competition for renal tubular excretion of either drug [40]. Finally, a case report has described pancytopenia and persistently decreased CD4+ lymphocyte counts in an HIV-infected patient receiving the combination of valganciclovir and didanosine. The proposed mechanism is a ganciclovir-induced inhibition of purine nucleoside phosphorylase, an enzyme responsible for catalyzing the breakdown of didanosine and endogenous purines [41, 42]. Because of the potential for increased toxicities of didanosine in association with significantly increased drug exposure, this interaction should be approached with caution, and the concomitant use of ganciclovir or valganciclovir plus didanosine should be avoided. If this combination is required for clinical reasons, careful monitoring for adverse effects of both drugs is required.

In vitro models have previously suggested that ganciclovir has antagonistic effects on the anti-HIV effects of zidovudine and didanosine, while foscarnet plus zidovudine has synergistic activity [35]. Studies evaluating the use of ganciclovir and foscarnet in the treatment of CMV retinitis in 234 patients with AIDS found that patients receiving foscarnet had a 3-month relative survival advantage compared to patients receiving ganciclovir and that this difference in survival could not be attributed solely to differences in drug exposures or toxicities [43, 44]. Although these clinical data provide some support for potential antagonistic or synergistic anti-HIV effects of ganciclovir and foscarnet, it is difficult to prove that the observed mortality differences were due to pharmacological effects of the drugs alone. The clinical relevance of these findings is unknown.

Serum creatinine elevations to greater than 2.5 mg/dL have been reported in up to 20% of bone marrow transplant and heart transplant patients during ganciclovir therapy [45, 46]. Most of these patients received cyclosporine and, in many cases, amphotericin B as well. Whether ganciclovir played a role in increasing the nephrotoxicity of other drugs through pharmacokinetic interactions or additive toxicities is unknown. However, a retrospective study of 93 liver transplant patients receiving ganciclovir concomitantly with oral cyclosporine found no evidence of effects on cyclosporine whole blood concentrations which may have predisposed to enhanced toxicities [30]. No changes in the pharmacokinetics of either ganciclovir or MMF were observed during concomitant administration of these two agents.

Valproic acid, an inhibitor of histone deacetylase (HDAC), has been shown in vitro to stimulate the replication of CMV and significantly impair the antiviral activity of ganciclovir, cidofovir, and foscarnet through mechanisms probably related to HDAC-related stimulatory effects on CMV itself [47]. Effects were most pronounced in cells that had been pretreated with VPA; when added during or after infection, VPA did not inhibit antiviral actions of the other drugs. The clinical relevance of these findings is unknown. Finally, seizures have been reported in patients receiving ganciclovir together with imipenem [30, 48]. Whether these seizures were related to the combination therapy or solely to imipenem is unclear.

10.3.2 Cidofovir

Potential drug interactions involving cidofovir are summarized in Table 10.2. Cidofovir is approximately 90% renally excreted with a high degree of renal tubular secretion. Probenecid has been shown to significantly reduce cidofovir clearance with a corresponding increase in AUC. This significant interaction serves as the basis for the FDA-approved use of cidofovir in combination with probenecid in the treatment of CMV infection. The recommended dose of probenecid is 2 g orally 3 h prior to the cidofovir dose, followed by 1 g orally at 2 and 8 h after completion of the cidofovir infusion [49]. Although cidofovir is routinely used in combination with probenecid in order to improve the pharmacokinetic profile of cidofovir and enhance clinical efficacy, the use of probenecid also appears to increase the overall incidence of drug-related adverse effects observed during cidofovir therapy [50–52]. Up to half of patients receiving cidofovir plus probenecid may develop constitutional symptoms of fever, chills, nausea, vomiting, fatigue, headache, GI upset, and rash; serious reactions including systemic hypotension may occur in 3% of patients and often result in discontinuation of cidofovir/probenecid therapy. Although difficult to determine whether such adverse effects are primarily due to cidofovir or probenecid, they seem to be most closely related to the administration of probenecid [53, 54]. Probenecid is also known to interact with the renal tubular secretion of many other drugs including acetaminophen, angiotensin-converting enzyme inhibitors, aminosalicylic acid, barbiturates, benzodiazepines, bumetanide, clofibrate, methotrexate, famotidine, furosemide, NSAIDs, theophylline, and zidovudine. Concomitant medications should be carefully evaluated as part of the overall assessment and monitoring of cidofovir/probenecid therapy [49]. Although no consistent change in zidovudine AUC has been observed when combined with cidofovir [55], and although the combination of cidofovir plus zidovudine did not appear to increase the incidence of drug-related myelosuppression [33], the manufacturer of cidofovir recommends that zidovudine should either be temporarily discontinued or the dose decreased by 50% when co-administered with probenecid on the day of cidofovir dosing [49].

The combination of cidofovir with other potentially nephrotoxic agents such as aminoglycosides, acyclovir, amphotericin B products, foscarnet, intravenous pentamidine, vancomycin, and NSAIDs should be avoided whenever possible due to potential for additive nephrotoxicity [49]. The manufacturer of cidofovir recommends that other potential nephrotoxins be discontinued at least 7 days prior to starting therapy with cidofovir [49]. If the use of other potentially nephrotoxic agents cannot be avoided due to clinical considerations, serum creatinine and other markers of renal function should be carefully monitored before and after each dose of cidofovir. Although mechanisms of additive nephrotoxicity have not been described in detail for most drug combinations and are likely multifactorial, animal and tissue culture models indicate that potential interactions between cidofovir and amphotericin B do not involve amphotericin B-related inhibition of renal organic acid transport proteins [56].

A number of case reports have described the occurrence of ocular toxicities during the administration of cidofovir concomitantly with rifabutin [57–60]. Both agents have been associated with uveitis, but whether the combination substantially increases the risk of ocular toxicities is unknown. Nevertheless, caution should be exercised when using these two agents together.

10.3.3   Foscarnet

Important drug interactions involving foscarnet are summarized in Table 10.2. Like many other antiviral agents, foscarnet undergoes significant renal tubular secretion. Although the potential interaction between foscarnet and probenecid has not actually been well described, there is also potential for competition between foscarnet and certain other drugs such as didanosine and zalcitabine which are also renally secreted. However, no pharmacokinetic alterations with didanosine or zalcitabine have been observed [61, 62]. While the combination of foscarnet plus ganciclovir should rarely be required, no alterations in the pharmacokinetics of either drug were noted when the combination was studied in 13 patients [61].

The risk of overlapping and potentially additive toxicity with other drugs is a major consideration during foscarnet therapy. The concomitant use of foscarnet and other potentially nephrotoxic agents such as aminoglycosides, acyclovir, amphotericin B products, cidofovir, cyclosporine, intravenous pentamidine, and vancomycin should be approached with caution and avoided whenever possible [61, 63]. Abnormal renal function has also been noted with combinations of foscarnet plus ritonavir, as well as combined foscarnet, ritonavir, plus saquinavir [61]. The potential mechanisms or significance of observed renal dysfunction during combined therapy with these antiretroviral drugs is unknown.

Additive central nervous system toxicity resulting in seizures in two patients has been reported with the concomitant use of foscarnet and ciprofloxacin [64]. Although both patients in whom these seizures occurred were receiving multiple medications and a direct causal effect is unclear, a study in mice has also reported increased seizure potential with the combination of foscarnet plus ciprofloxacin (but not enoxacin) [65]. Although this interaction appears to involve alteration of gamma-aminobutyric acid (GABA) activity in the central nervous system, it has not been commonly reported with fluoroquinolones and the clinical importance is unknown.

Severe hypocalcemia has been reported during combined therapy of foscarnet and intravenous pentamidine [61]. Post-marketing surveillance by the manufacturer found that four patients in the United Kingdom who were treated with the combination of foscarnet and pentamidine may have developed drug-related hypocalcemia; one of these patients reportedly died of severe hypocalcemia. This potential additive toxicity would be expected to occur only with intravenous pentamidine since the systemic absorption of pentamidine after aerosolized administration is negligible. The combination of foscarnet and intravenous pentamidine should be avoided when possible; close patient monitoring is required if concomitant therapy with these drugs is required.

In vitro models have demonstrated additive or synergistic activity against HIV and CMV when the combination of foscarnet plus zidovudine was studied at clinically relevant concentrations [66, 67]. The mechanisms of such enhanced effects are unknown, and no pharmacokinetic interactions between the two drugs have been observed [68]. Although the clinical significance of these potential interactions is unknown, it is likely of low clinically relevance given the use of multiple-drug antiretroviral combinations during HAART.

10.4 Drugs Used for Prevention and Treatment of Influenza

10.4.1 Amantadine and Rimantadine

Potentially important drug interactions with the adamantane drugs (amantadine and rimantadine) are summarized in Table 10.3. Pharmacokinetic studies of adamantanes have demonstrated that administration of these drugs with food has no significant effects on their bioavailability compared to administration in the fasting state [69, 70]. The adamantanes may thus be administered without regard to meals.

As with many other antiviral agents, the adamantanes undergo renal tubular secretion [71, 72]. A number of drugs have been associated with decreased renal clearance and increased adverse effects of amantadine. Concomitant use of amantadine with the diuretic triamterene/hydrochlorothiazide was associated with a 50% increase in amantadine concentrations and occurrence of central nervous system toxicity. The mechanism of this interaction is assumed to be decreased renal clearance through inhibition of tubular secretion, although it is unknown which component of the diuretic combination was responsible [73]. Use of TMP/SMX was also reported to cause decreased renal clearance and neurologic toxicity when administered with amantadine, presumably due to TMP-induced reduction in the tubular secretion of amantadine [74]. Co-administration of quinine or quinidine also reportedly reduces the renal clearance of amantadine by approximately 30% [71]. Cationic drugs with active tubular secretion could theoretically compete with the renal tubular secretion of rimantadine, but no cases have been reported [75]. No significant interaction was found with cimetidine, nor with aspirin or acetaminophen [72].

Increased central nervous system toxicities, particularly of amantadine, have been reported when the antiviral agents are used concomitantly with a number of other agents with overlapping toxicity profiles. The neurotoxic effects of amantadine may reportedly be increased by antihistamines, psychotropic agents including thioridazine, and drugs with pronounced anticholinergic activity. Reports have been particularly frequent among patients taking anticholinergic agents for treatment of Parkinson's disease [71, 72, 76–78]. Neurotoxicity has also been reported with combinations of amantadine with either phenylpropanolamine [79] or bupropion [80]. Reversible central nervous system toxicity was reported to occur in six of eight patients receiving bupropion within 1 week of beginning amantadine treatment; the mechanism of toxicity is believed to be related to the dopamine-stimulating effects of the two drugs [80]. In general, caution should be used when amantadine is combined with other central nervous system stimulants and patients carefully monitored for evidence of neurologic toxicities [71].

Due to concerns regarding severe pandemics with novel influenza strains such as H5N1, combination therapy with amantadine plus oseltamivir has been suggested in order to increase the potential for both increased antiviral efficacy and decreased resistance. A randomized, crossover trial was conducted in which 17 subjects received amantadine alone or in combination with oseltamivir for 5 days to evaluate any pharmacokinetic interactions between the two drugs [81]. Co-administration with oseltamivir had no significant effects on amantadine AUC or C_{max}. Similarly, amantadine co-administration had no significant effects on the pharmacokinetics of either oseltamivir or oseltamivir carboxylate. No evidence of increased adverse effects of either drug were noted [81]. Similarly, no significant pharmacokinetic interactions were found between rimantadine and either oseltamivir or peramivir [82, 83].

The adamantanes exert their effects against the influenza virus by inhibiting viral replication. There is a theoretical potential for these agents to reduce the efficacy of the intranasal live attenuated influenza vaccine (LAIV) by inhibiting replication of the live virus after vaccine administration. It is therefore recommended that the LAIV should not be administered within 2 weeks before, or 48 h after, administration of amantadine. The use of amantadine or rimantadine should not affect administration of the injectable influenza vaccine (IIV) containing inactivated virus [71, 72].

10.4.2   Oseltamivir

Potential drug interactions involving oseltamivir are shown in Table 10.3. Co-administration of oral oseltamivir with food has no significant effect on either the C_{max} or AUC of oseltamivir carboxylate, the active compound which is rapidly formed after administration of oseltamivir phosphate [84]. In addition, the rate and extent of oseltamivir absorption were not affected by either magnesium hydroxide/ aluminum hydroxide or calcium carbonate antacids after concomitant administration to healthy volunteers [85]. In contrast, concomitant administration with 400 mL of milk was reportedly associated with a 31% reduction in C_{max}, 65% reduction in

AUC, and 22% reduction in the urinary recovery of oseltamivir. The proposed mechanism for this interaction is reduced absorption of oseltamivir through inhibition of proton-coupled oligopeptide transporter 1 (PEPT1) caused by milk peptides [86]. This potential interaction has not been reported elsewhere and the clinical significance is unknown.

Oseltamivir has low potential for drug interactions based on the characteristics of low protein binding, lack of hepatic metabolism, and renal elimination through glomerular filtration and anionic tubular secretion [84, 87]. In vitro studies also suggest that neither oseltamivir nor oseltamivir carboxylate are good substrates for CYP450 mixed-function oxidases or glucuronyl transferases. The conversion of oseltamivir to oseltamivir carboxylate occurs via human carboxylesterase 1 (HCE1), which is located predominantly in the liver. Activity of HCE1 was reportedly inhibited by as much as 90% in the presence of the antiplatelet drug clopidogrel, which is also hydrolyzed by HCE1 but has a greater affinity for the enzyme than does oseltamivir [88]. The inhibition of HCE1 in vitro was dependent on the clopidogrel dose/concentration used [89], and studies describing clopidogrel interactions have been criticized in part because the relative concentrations of oseltamivir and clopidogrel evaluated (50 μM and 2.5–50 μM) were approximately 240-fold and 400–8000-fold higher, respectively, than plasma concentrations achieved with typical oseltamivir and clopidogrel dosing regimens. The true significance of this potential clopidogrel interaction is unknown but is not likely to be clinical important based on methodological problems with the in vitro studies [90]. Drug interactions involving competition for, or inhibition of, these esterases otherwise have not been extensively reported in literature.

Systemic clearance of oseltamivir carboxylate primarily occurs through renal secretion via renal tubular hOAT1. Clinically important drug interactions involving oseltamivir could thus potentially occur with other drugs that inhibit renal tubular secretion through this pathway [87]. Cimetidine is a potent inhibitor of O-carboxylates 1 and 2, two active pathways for transport and secretion of cationic drugs by renal tubular epithelial cells. Not surprisingly because of the difference in transporter systems affected, no interaction was observed when cimetidine was administered concomitantly with oseltamivir [87, 91]. However, probenecid reduced the renal clearance of oseltamivir carboxylate by 50% and increased the AUC by 100% [84, 87]. No interaction was noted between oseltamivir and concomitant administration of amoxicillin (also secreted into urine by hOAT) in healthy volunteers [87, 91]. Even though oseltamivir has only weak inhibitory effects on renal tubular anionic secretory transporters, it is recommended that care be exercised with co-administration of methotrexate because of common secretory pathways and potential for increased methotrexate toxicities [91].

The interaction between oseltamivir and probenecid has been suggested as likely not clinically important because these two drugs are seldom used together and because oseltamivir lacks serious toxicities [87]. However, in response to concerns regarding potential influenza pandemics and limited supplies of oseltamivir [92, 93], at least two studies have evaluated the feasibility of oseltamivir/probenecid combinations as a means of reducing oseltamivir dosing requirements [94, 95]. In the first study, 48 healthy volunteers were randomized to receive oseltamivir 75 mg once daily, oseltamivir 75 mg every 48 h plus probenecid 500 mg four times daily, or oseltamivir 75 mg every 48 h plus probenecid 500 mg twice daily [94]. Oseltamivir and oseltamivir carboxylate C_{max} and T_{max} did not significantly differ between the three groups. However, the steady-state apparent oral clearances of oseltamivir carboxylate were significantly decreased in the probenecid groups compared to oseltamivir alone, confirming inhibitory effects of probenecid on oseltamivir renal excretion. Arithmetic mean concentrations at 48 h were not significantly different between the oseltamivir and oseltamivir plus four-times-daily probenecid group (42 ± 76 ng/mL vs. 81 ± 54 ng/mL, respectively, $P = 0.194$); however, concentrations in the twice-daily probenecid group were significantly decreased compared to oseltamivir alone (23 ± 26 ng/mL vs. 81 ± 54 ng/mL, respectively, $P = 0.012$). The results of this study suggested that co-administration of oseltamivir 75 mg every 48 h plus probenecid 500 mg four times daily was equivalent to everyday dosing of oseltamivir and that this regimen might be a feasible way of allowing for reduction of oseltamivir doses without compromising clinical efficacy [94]. A second study found that reducing oseltamivir doses to 45 mg twice daily plus probenecid maintained oseltamivir exposures which were comparable to the typical 75 mg twice daily regimen without probenecid [95]. Although the daily dose of oseltamivir could potentially be reduced from 150 mg/day to 90 mg/day through combination therapy with probenecid, the authors of the study noted that the potential for increased adverse effects and nonadherence related to probenecid use requires careful consideration prior to routine recommendations for such a dosing strategy [95]. The combination of oseltamivir plus probenecid has also been associated with thrombocytopenia, lending some credence to concerns regarding toxicity [96].

No interactions have been observed between oseltamivir and either single-dose acetaminophen or single-dose aspirin [91, 97, 98]. In addition, no interactions have been observed between oseltamivir and cyclosporine, MMF, or tacrolimus [91, 99, 100]. Similarly, no significant pharmacokinetic interactions were found between oseltamivir and other antiviral agents used for prevention or treatment of influenza including amantadine, rimantadine, intravenous zanamivir, and peramivir [81–83, 101].

In a study evaluating potential interactions between oseltamivir and warfarin, subjects received oseltamivir 75 mg twice daily for a total of nine doses either with or without warfarin with an appropriate wash-out period of 4–8 days between treatment periods. No statistical differences in international normalized ratio (INR), factor VIIa levels, or vitamin K1 concentrations were found when the drugs were administered concomitantly. Also, no effects of oseltamivir on warfarin pharmacokinetics were noted [84, 87, 102]. A large retrospective database analysis likewise found no significantly increased risk of bleeding events within 14 days of beginning oseltamivir therapy in 13,406 predominantly elderly patients (adjusted odds ratio 1.24, 95% confidence interval 0.97–1.57) [103]. Despite data from well-designed pharmacokinetic and outcome studies showing no apparent interactions between oseltamivir and warfarin, clinical case reports suggest that changes in INR may occur during oseltamivir therapy in patients receiving warfarin. A published case of a single pediatric patient hospitalized with hypoplastic left heart syndrome experienced a 250% increase in INR (from approximately 2.0 to 7.46) over a 5-day course of oseltamivir therapy; INR values promptly returned to the original stable values within 2 days of discontinuing oseltamivir [104]. An additional case series of 15 hospitalized Korean patients who were previously stable on warfarin therapy reported an alteration of INR values in 7 patients (46%) after initiation of oseltamivir therapy; the average INR increase in these patients was 150% [105]. While it can be postulated that acute illness and clinical instability may be responsible for the reported INR alterations rather than a true drug interaction with oseltamivir, it is nevertheless advisable to monitor INR more carefully after initiation of oseltamivir therapy in patients previously receiving warfarin.

As with the adamantanes, there is a theoretical potential for oseltamivir to reduce the efficacy of the intranasal LAIV by inhibiting replication of the live virus after vaccination. It is therefore recommended that LAIV not be administered within 2 weeks before, or 48 h after, administration of oseltamivir. The use of oseltamivir should not affect administration of IIV [84].

Published case reports describe two patients previously stable on sotalol therapy who developed corrected QT-interval (QTc) prolongation and torsades de pointes (TdP) after being treated with oseltamivir [106]. In one case the TdP occurred after six doses of oseltamivir; in the second case, the arrhythmia occurred 6 days after completion of a 5-day course of oseltamivir therapy. No potentially feasible mechanisms of oseltamivir-induced QTc prolongation or drug-drug interaction with sotalol were offered by the authors of the report, and it is important to note that multiple other risk factors for QTc prolongation were present including older age (>60 years), female sex, antiarrhythmic drug therapy, borderline or already prolonged QTc intervals at baseline, and hypokalemia and bradycardia in one patient [106]. The true role of oseltamivir in QTc prolongation and subsequent TdP in these case reports is unknown.

10.4.3   Zanamivir

Like oseltamivir, zanamivir has a low potential for significant drug interactions due to its low degree of protein binding, lack of hepatic metabolism, and elimination primarily by glomerular filtration and tubular secretion [107]. Zanamivir does not appear to serve as a substrate or otherwise affect CYP450 enzymes (i.e., 1A1/2, 2A6, 2C9, 2C18, 2D6, 2E1, or 3A4) in human liver microsomes [108]. Few specific pharmacokinetic drug interactions have been evaluated or observed because of the use of aerosolized zanamivir with its resultant minimal systemic exposure. However, the availability of intravenous zanamivir may prompt additional interaction studies in the future. One study has assessed potential pharmacokinetic interactions between intravenous zanamivir and oral oseltamivir [101]. The C_{max} of zanamivir was increased 10% when administered concurrently with oseltamivir; however, this pharmacokinetic alteration was not felt to be clinically significant.

Several drugs have been assessed for their potential effects on the antiviral activity of zanamivir. Aspirin, ibuprofen, acetaminophen, promethazine, oxymetazoline, phenylephrine, and amoxicillin/clavulanate were all shown to have no effect on the antiviral activity of zanamivir against influenza A in vitro [107]. Although codeine and diphenhydramine have been shown to enhance zanamivir's antiviral activity in vitro through direct antiviral effects by some unknown mechanism, the concentrations of codeine and diphenhydramine used in the in vitro studies were many times higher than would be achieved with typical doses of these agents, and the clinical relevance is therefore unlikely to be important [107, 108].

As with oseltamivir, LAIV should not be administered within 2 weeks before or 48 h after zanamivir in order to avoid the theoretical concern regarding decreased vaccine efficacy (Table 10.3) [108].

10.4.4 Peramivir

Like oseltamivir and zanamivir, peramivir has a low potential for significant drug interactions due to its low degree of protein binding (<30%), lack of hepatic metabolism, and elimination primarily by glomerular filtration [109]. Peramivir does not appear to serve as a substrate or otherwise affect CYP450 enzymes, does not affect glucuronidation pathways, and is not a substrate or inhibitor of P-glycoprotein transporters [109]. Limited studies to date have found no significant drug-drug interactions with rimantadine, oseltamivir, or oral contraceptives containing ethinyl estradiol and levonorgestrel. In contrast to many other antiviral agents, peramivir does not undergo significant renal tubular secretion and is therefore not significantly affected by the concomitant administration of oral probenecid [109].

10.5 Miscellaneous Antiviral Agents

10.5.1 Ribavirin

A large number of potential drug interactions involving ribavirin exist and are summarized in Table 10.4. The prevalence of infection with HCV is a growing problem worldwide, particularly among HIV-infected patients in whom rates of coinfection with HIV and HCV may be as high as 90% [110, 111]. Ribavirin monotherapy is ineffective in the chronic treatment of HCV infections with a sustained virological response rate (SVR) of close to 0% in clinical studies, while the SVR of interferon-α alone is approximately 20% [111–114]. However, the combination of ribavirin plus interferon-α is associated with SVR rates of approximately 40% [111, 112, 114]. The synergistic SVR associated with ribavirin plus interferon-α combination therapy occurs through mechanisms which are not completely understood, but which may involve drug-induced stimulation of an anti-HCV immune response and/or direct antiviral effects of the drugs. In vitro models indicate that direct, synergistic antiviral effects of the drugs occur at physiologically relevant concentrations [115]. Whether interferon-α stimulation of infected cells renders them more susceptible to the effects of ribavirin or vice versa is not clear. However, regardless of potential mechanisms, the combination of ribavirin with pegylated interferon-α was a standard of care in the treatment of HCV until supplanted by the direct-acting antiviral (DAA) agents, and this regimen is still used in patients who are intolerant of, or do not have access to, DAAs [115, 116]. No pharmacokinetic interactions have been noted with combined administration of ribavirin and interferon alfa-2b or peginterferon alfa-2b [117, 118]. However, hemolytic anemia with hemoglobin values of less than 10 g/dL was reported in approximately 10% of patients receiving combination therapy with ribavirin and interferon alfa-2b, usually occurring within 1–2 weeks of initiating ribavirin therapy [119].

No studies have specifically evaluated the potential for interactions involving inhaled ribavirin; however, the manufacturer recommends that inhaled ribavirin not be administered together with other drugs given by the inhaled route [120]. The remainder of this section deals only with drug interactions involving orally or intravenously administered ribavirin.

When single oral doses of ribavirin were co-administered with a high-fat meal, T_{max} of ribavirin was doubled, the AUC was increased by 42%, and the C_{max} was increased by 66% compared to the fasting state. Ribavirin should thus be routinely administered with food in order to maximize oral absorption [117, 118]. Concomitant administration of magnesium, aluminum, and simethicone antacids reduced the AUC of oral ribavirin by 14% [117, 118]. This reduction in bioavailability may be related to either increases in intestinal transit time or a change in gastrointestinal pH; however, it is not considered to be clinically relevant [117, 118].

Ribavirin elimination is accomplished through a mixture of hepatic and renal processes including reversible phosphorylation, degradation by deribosylation and amide hydrolysis, and elimination of unchanged drug through the kidneys with evidence of both glomerular filtration and renal tubular secretion [121]. Ribavirin does not appear to be influenced by, nor to be a substrate of, CYP450 enzyme systems based on in vitro human and rat microsomal liver preparations; there is furthermore no evidence for induction or inhibition of 2C9, 2C19, 2D6, or 3A4 enzymes [117, 118]. A single case report describes two patients in whom the addition of the DAAs boceprevir and telaprevir to an existing ribavirin plus pegylated interferon-α regimen was associated with the new onset of seizures [122]. Although the DAAs are well known to interact with the CYP450 system, this does not seem to be a likely explanation for this potential interaction, and the clinical significance of this report is unclear.

Due to ribavirin's potential for hematologic toxicities, concomitant therapy with other myelosuppressive agents should be used with caution and carefully monitored [117, 118, 123]. Several case reports describe severe myelosuppression with the combination of ribavirin plus azathioprine in the treatment of inflammatory bowel diseases and coexistent HCV infection. The proposed mechanism is interference with the normal clearance of azathioprine intermediate metabolites through ribavirin-induced inhibition of inosine monophosphate dehydrogenase (IMPDH). Inhibition of IMPDH leads to increased levels of methylated azathioprine metabolites, e.g., 6-methylthioinosine monophosphate (6-MTIMP), which have been associated with myelotoxicity [124–127]. In addition, patients with HIV/HCV coinfection who were administered zidovudine in combination with pegylated interferon-α plus ribavirin developed severe neutropenia and anemia more frequently than did patients not receiving zidovudine. This increased incidence of hematologic toxicity with ribavirin plus zidovudine is apparently due to overlapping toxicities (rather than pharmacokinetic alterations) and is usually able to be clinically managed through dose reduction or drug discontinuation [117, 118, 128, 129].

Ribavirin is a guanosine analogue and may compete with zidovudine, lamivudine, stavudine, emtricitabine, and other nucleosides for intracellular phosphorylation [117, 118, 129]. In vitro studies indicate that ribavirin induces an increase in deoxythymidine triphosphate which results in feedback inhibition of thymidine kinase and decreased intracellular formation of phosphorylated zidovudine [130, 131]. These effects of ribavirin may potentially increase zidovudine toxicities while also reducing clinical efficacy of the drug in HIV-infected patients [132].

Although myelosuppression with the combination of ribavirin and zidovudine may occur, several published studies of HIV/HCV coinfection showed no evidence of adverse clinical outcomes related to antiviral drug failure with combination therapy [133–135]. Another study of 14 subjects receiving zidovudine also found no significant impact on zidovudine triphosphate AUC, plasma zidovudine AUC, or the ratio of zidovudine triphosphate to zidovudine AUC after the addition of ribavirin [136]. Since the clinical pharmacology of zidovudine does not appear to be altered despite in vitro findings, dosage adjustment does not appear to be needed with concomitant ribavirin therapy. Likewise, although in vitro studies indicate that the anti-HIV activity of tenofovir may be antagonized by ribavirin [137], no interaction between oral ribavirin and tenofovir was observed in a multiple-dose interaction study in 23 healthy subjects [138]. Other studies have also failed to find evidence of adverse clinical outcomes in patients receiving ribavirin for HCV treatment along with HAART regimens; the clinical significance of in vitro studies remains unknown but does not appear to be highly relevant [128, 139, 140].

Ribavirin-induced inhibition of IMPDH in patients receiving didanosine for treatment of HIV infection promotes formation of dideoxyadenosine-5′-triphosphate, elevated levels of which are in turn associated with the mitochondrial toxicity of didanosine. There are a number of reports describing lactic acidosis and pancreatitis in patients receiving concomitant ribavirin and didanosine therapy [118, 128, 141–143]. In one study, the incidence of symptomatic lactic acidosis was 33/1000 patient years in those patients treated for HCV with ribavirin and receiving HAART versus 13.5/1000 patient years in those receiving HAART only. In this study, both didanosine and stavudine were significantly associated with increased risk of symptomatic lactic acidosis ($P < 0.01$ and $P = 0.04$) [134]. Since no pharmacokinetic interactions have been observed between ribavirin and didanosine [144], toxicities are presumed to be caused by ribavirin-induced mitochondrial toxicity

[111, 145–148]. It has been stated that didanosine-related lactic acidosis and pancreatitis occur more rapidly in the presence of ribavirin than with didanosine alone, usually within the first 3 months of therapy [117, 118]. Finally, this toxic interaction may persist for up to 1–2 months based on the very long half-life of ribavirin (approximately 120–170 h) [117, 118]. Extreme caution should therefore be used when combining ribavirin with didanosine, and concomitant use of the two drugs should be avoided if possible. Combination therapy with stavudine should also be approached with caution [117, 118, 128]. There is no indication of risk with other antiretroviral drugs such as non-nucleoside reverse transcriptase inhibitors or protease inhibitors [117, 118].

The addition of ribavirin plus pegylated interferon for HCV treatment in HIV-infected patients receiving the protease inhibitor atazanavir has been associated with a significantly increased incidence of hyperbilirubinemia [149]. A total of 72 patients with HIV/HCV coinfection were evaluated following the addition of HCV therapy to existing antiretroviral drug regimens. By 4 weeks, patients also receiving atazanavir had a significantly greater increase in total bilirubin levels ($P = 0.003$). The proportion of patients experiencing increases of more than 1 mg/dL was also significantly greater in patients receiving atazanavir (45% vs. 3%, $P = 0.001$). The proposed mechanism of toxicity is ribavirin-induced hemolysis of red blood cells and increased production of bilirubin, followed by an inhibitory competition by atazanavir of uridine glucuronosyltransferase (UGT) 1A1, an enzyme which is normally responsible for bilirubin conjugation. This combination of effects thus leads to increased serum bilirubin levels and jaundice [149].

Potential interactions between oral ribavirin and raltegravir, an HIV-1 integrase inhibitor, were investigated in 14 healthy volunteers [150]. Ribavirin T_{max} was increased by 39% and the C_{max} reduced by 21% when the drugs were co-administered; other pharmacokinetic parameters including half-life and AUC were not significantly different. No significant changes in raltegravir pharmacokinetic parameters were observed. Although the mechanism underlying the apparently altered absorption of ribavirin is unknown, this interaction was felt to be clinically insignificant [150].

A prospective study in 124 HIV−/HCV-coinfected patients compared the pharmacokinetics of ribavirin and HCV virologic responses in patients receiving pegylated interferon plus ribavirin, with or without the addition of abacavir [151]. Median ribavirin C_{min} was not different between patients who received abacavir compared to those who did not, and no statistically significant differences in rapid virological response (RVR) measured at 4 weeks after end of treatment, early virological response (EVR) measured at 12 weeks, or SVR measured at 24 weeks were observed between groups. The study concluded that there are no pharmacokinetic or virologic interactions between ribavirin and abacavir, and abacavir-containing regimens were also found to be safe with no observed increase in adverse events [151].

Finally, the use of ribavirin in patients receiving chronic warfarin therapy has reportedly caused a decrease in prothrombin time. Although the mechanism involved is unknown, the interaction was clinically significant in the reported case [152].

10.5.2   The Interferons

Very few formal studies of potential drug interactions have been conducted for most interferon drugs [153–158]. Despite this, a considerable amount of information (although often conflicting) is available regarding potential interactions with the interferons, particularly related to interferon-induced alterations in CYP450-related drug metabolism; these potential interactions are summarized in Table 10.4. The effects of various interferons on CYP450 activity are highly variable and probably depend on the use of specific interferons, specific CYP enzyme families studied, and interferon doses. The following summaries will therefore focus on peginterferon alfa-2a (Pegasys®) and peginterferon alfa-2b (PegIntron®), two interferon products specifically FDA-indicated for use in the treatment of chronic HCV infection.

The interferons as a class have long been associated with reduced CYP450 activity after it was determined that CYP downregulation during acute viral infections was primarily mediated by interferons [159–162]. Interferon effects on CYP450 metabolism occur through unclear mechanisms and may be attributed to either increased degradation, suppressed synthesis, or direct inhibition of the enzymes [153, 154, 158]. With few exceptions, the interferons have consistently demonstrated decreased clearance of various drugs metabolized by CYP1A and CYP3A subfamilies. Studies which failed to show significant changes in hepatic drug metabolism were using low doses of interferon-α (e.g., three million units three times per week) [163–165]. Chronic administration of low-dose interferon-α was associated with a moderate decrease in theophylline metabolism, minimal effect on antipyrine clearance, and minimal effect on hexobarbital metabolism [163–166]. However, larger doses of interferons have been associated with more pronounced reductions in theophylline and antipyrine metabolism, suggesting that the effects of interferons on CYP1A2 drug metabolism are likely to be dose-dependent [153, 167]. Once-weekly administration of interferon-α for 4 weeks in healthy subjects resulted in inhibition of CYP450 1A2 and a 25% increase in theophylline AUC [153], while other studies have reported 100% increases in theophylline concentrations after interferon treatment [168, 169]. Interferon-α has also been shown to inhibit CYP3A4 metabolism using the ^{14}C–erythromycin breath test as a marker of CYP3A4 activity [170]. Despite data indicating alterations in CYP1A2 and CYP3A4 activity, these changes are not always consistently reported [153] and/or may not always be considered clinically significant with required changes in drug dosing [154]. In light of potentially variable and dose-related effects, patients receiving drugs metabolized by CYP1A2 (e.g., theophylline, caffeine, antipyrine, tricyclic antidepressants [TCAs], olanzapine, clozapine) or CYP3A (e.g., azole antifungals, macrolide antibiotics, many antiretroviral agents, some immunosuppressants, some serotonin-specific reuptake inhibitors [SSRIs], TCAs, certain statins, opiate analgesics, benzodiazepines, antipsychotics, barbiturates, calcium channel antagonists) should be carefully monitored during interferon therapy.

Although it has been stated that interferon-α has no effect on the pharmacokinetics of drugs metabolized by CYP2C9, CYP2C11, CYP2C19, or CYP2D6 [153, 171], it has elsewhere been reported that the activities of CYP2C8/9 and CYP2D6 were actually induced by 28%–66% in 22 patients with chronic HCV who received interferon-α for 4 weeks [154]. These effects were highly variable, however, with 40% of patients exhibiting either inhibition of CYP activity or no change rather than increased activity [154]. Close monitoring is therefore recommended during interferon therapy with the concomitant use of drugs metabolized by CYP2C8/9 (e.g., warfarin, phenytoin, NSAIDs, angiotensin receptor blockers, certain statins, sulfonylureas) or CYP2D6 (e.g., β-blockers, lidocaine, flecainide, TCA, SSRIs, opiate analgesics, antipsychotics) as the therapeutic effects of these drugs may be either decreased, increased, or unchanged [154]. As a case in point, increased effects of warfarin during interferon therapy have been previously described in two case reports [172, 173].

The pharmacokinetics of methadone were assessed in 18 patients with chronic HCV who received concomitant administration of interferon-α2b [154]. All patients were stable on chronic methadone treatment at the time of interferon initiation. A mean 16% increase in methadone AUC was observed after 4 weeks of interferon therapy, but the AUC was increased by 100% in two patients. This interaction is probably related to inhibition of CYP3A4 metabolism. The clinical significance of this interaction is unknown and likely highly variable; cautious monitoring of sedative and respiratory effects of methadone is warranted during the first few weeks of combined therapy [154].

The combination of ribavirin plus pegylated interferon-α has the potential for increased incidence and/or severity of myelosuppression due to overlapping potentials for hematologic toxicity [117, 118, 153, 154]. As previously described, hemolytic anemia occurred in 10% of patients receiving combination therapy with ribavirin and interferon alfa-2b within 1–2 weeks of initiating ribavirin therapy [154]. Such interactions between ribavirin and the interferons have not been not consistently reported [117, 118]; however, close monitoring for hematologic toxicities is required during therapy with these agents. A published case report describes a patient receiving pegylated interferon-α plus ribavirin for chronic HCV who developed severe anemia after addition of oseltamivir for treatment of influenza [174]. Whether this case describes a new drug-drug interaction as postulated by the authors of the report, or merely reflects the potential toxicities with ribavirin-interferon combinations previously discussed, is unknown.

Caution should also be exercised during combined use of interferons with potentially myelosuppressive drugs. As previously described, the administration of zidovudine in combination with pegylated interferon-α plus ribavirin was associated with increased rates of severe neutropenia and anemia [128]. Interferon was associated with significantly decreased zidovudine clearance and increased AUC in eight patients with AIDS who were started on interferon-β after 8 weeks of zidovudine monotherapy [175]. Interferon has also been associated with significant changes in zidovudine metabolic rates, plasma elimination rates, and decreased ratio of parent drug to glucuronide metabolite after initiation of interferon; such metabolic alterations may contribute to the increased risk of myelosuppression with the combination of interferon and zidovudine. In contrast to the significant effects on zidovudine metabolism, no interaction was found between interferon-α and didanosine [176]. Severe and irreversible granulocytopenia has also been reported in several patients during concomitant use of interferon alfa-2a and angiotensin-converting enzyme (ACE) inhibitors including both captopril and enalapril [177, 178]. Potential drug interactions resulting in increased drug toxicities have been reported during concomitant use of interferon-α and antineoplastic agents including 5-fluorouracil (myelosuppression) [179], hydroxyurea (myelosuppression, vasculitis) [180], and melphalan (myelosuppression) [181, 182]. Studies have not consistently shown alterations in pharmacokinetic parameters such as C_{max} or AUC, and the mechanisms behind these potential interactions with certain antineoplastic agents are unknown [183–188]. However, the potential for severe toxicities necessitates careful patient monitoring during combined use of these agents. Suspected additive myelosuppression during combined peginterferon and thalidomide therapy has also been reported [189]. Finally, hematologic toxicity has been reported in a patient receiving combined interferon-α and clozapine; although the specific mechanisms for this interaction were not defined, clozapine is known to be a CYP1A2 substrate [190].

A number of studies have demonstrated a decreased response to interferon-α therapy in the treatment of HCV infection among patients who consume alcohol [191–195]; SVR rates were directly related to the level of ethanol consumption. In one study nondrinkers had a 53% response rate to interferon, while responses were 43% among light drinkers (<70 g of ethanol/day, or approximately 2.5 ounces) and 0% among heavy drinkers (>70 g ethanol/day); the difference between nondrinkers and heavy drinkers was statistically significant ($P < 0.01$) [191]. In another study, only 11% of patients failing to respond to interferon therapy were nondrinkers compared to 63% of patients with any level of alcohol consumption; furthermore, overall non-response rates directly increased according to the level of alcohol consumption [194]. While the association between ethanol consumption and response of HCV infection to treatment with interferon-α is clear, the actual cause is unclear. Alcohol has been shown to accelerate the course of HCV disease through increased HCV replication, enhanced oxidative stress, increased inflammatory and fibrotic effects, and modulation of the immune response to HCV infection, therefore indicating that the effects of alcohol on interferon response are more attributable to effects on the underlying infectious process [196]. However, in vitro data also suggest that alcohol may directly inhibit the actions of interferon-α through effects on intracellular signaling pathways which are activated after binding of interferon to cellular receptors. Specifically, alcohol has been shown in vitro to inhibit phosphorylation and activation of specific cytoplasmic transcription factors (signal transducers and activators of transcription, or STATs); the inhibition of these STATs then results in downstream decreases in expression of antiviral interferon-stimulated genes (ISGs) which are responsible for the efficacy of interferon against

HCV [196]. The effects of alcohol on interferon response rates are thus multifactorial and likely involve both direct inhibition of interferon's pharmacologic activity and indirect disease state-mediated effects on HCV infection. Patients infected with HCV should thus abstain from any level of alcohol consumption while receiving therapy with interferon-α.

Finally, a prospective study evaluated whether exposure to an antidepressant medication during treatment of HCV with interferon-α and ribavirin influenced treatment response [197]. Although antidepressant exposure was associated with a statistically significant reduction in end-of-treatment response (ETR) at the conclusion of 24 or 48 weeks of interferon-α therapy ($P = 0.016$), multivariate logistic regression found that this reduced ETR was specifically associated with exposure to antidepressant drugs which enhance norepinephrine activity in the brain (e.g., venlafaxine, mirtazapine, amitriptyline) (odds ratio 0.15, 95% CI 0.04–0.60; $P = 0.008$). However, this association of certain antidepressants with treatment response only affected the ETR; the more important study endpoint of SVR at 6 months post-therapy was not significantly affected (odds ratio 0.39, 95% CI 0.11–1.34; $P = 0.136$). Antidepressant use as a whole was also not significantly associated with reduced SVR ($P = 0.316$). Of note, patient numbers in this study were quite small; only 47 patients received antidepressants and only 12 of these received norepinephrine-enhancing agents [197]. The overall significance of these findings as well as the mechanistic basis for any potential interaction is difficult to determine, and additional study is clearly needed.

10.5.3 Direct-Acting Antiviral (DAA) Agents Used for Treatment of HCV

Nearly a dozen DAAs have been approved for treatment of HCV, and their use is considered the current standard of care for most patients with chronic HCV infection [116]. A review of the many known or potential drug-drug interactions associated with DAAs is beyond the scope of this chapter. However, it is worth noting that up to this point in time no significant drug-drug interactions have been documented between the DAAs and any of the many antiviral agents discussed in this chapter [198–203]. The various antiviral agents used in the management of infections due to HSV, VZV, CMV, and influenza are not likely to be problematic in terms of drug-drug interactions with DAAs because of primarily renal excretion, lack of significant CYP450 interactions, and other properties discussed elsewhere in this chapter. However, new interactions may potentially become known as the use of DAAs becomes more widespread and clinicians should be vigilant for any new information in this regard.

References

1. Zovirax (acyclovir) product information. GlaxoSmithKline, Research Triangle Park (2016)
2. Valtrex (valacyclovir) product information. GlaxoSmithKline, Research Triangle Park (2015)
3. Sitar D, Aoki FY, Bow EJ (2008) Acyclovir bioavailability in patients with acute myelogenous leukemia treated with daunorubicin and cytarabine. J Clin Pharmacol 48:995–998
4. Parmeggiani A, Riva R, Posar A, Rossi PG (1995) Possible interaction between acyclovir and antiepileptic treatment. Ther Drug Monit 17:312–315
5. Laskin OL, de Miranda P, Kinh DH et al (1982) Effects of probenecid on the pharmacokinetics and elimination of acyclovir in humans. Antimicrob Agents Chemother 21:691–715
6. De Bony F, Tod M, Bidault R, On NT, Posner J, Rolan P (2002) Multiple interactions with cimetidine and probenecid with valaciclovir and its metabolite acyclovir. Antimicrob Agents Chemother 46:458–463
7. Ye J, Liu Q, Wang C et al (2013) Benzylpenicillin inhibits the renal excretion of acyclovir by OAT1 and OAT3. Pharmacol Rep 65:505–512
8. Sylvester RK, Leitch J, Granum C (1996) Does acyclovir increase serum lithium levels? Pharmacotherapy 16:466–468
9. Bullingham R, Nicholls A, Kamm B (1998) Clinical pharmacokinetics of mycophenolate mofetil. Clin Pharmacokinet 34:429–455
10. Shah J, Juan D, Bullingham R (1994) A single dose drug interaction study of mycophenolate mofetil and acyclovir in normal subjects [abstract]. J Clin Pharmacol 34:1029–1033
11. Gimenez F, Foeillet E, Bourdon O et al (2004) Evaluation of pharmacokinetic interactions after oral administration of mycophenolate mofetil and valacyclovir or acyclovir to healthy subjects. Clin Pharmacokinet 43:685–692
12. Vomiero G, Carpenter B, Robb I, Filler G (2002) Combination of ceftriaxone and acyclovir – an underestimated nephrotoxic potential? Pediatr Nephrol 17:633–637
13. Yue Z, Shi J, Jiang P, Sun H (2014) Acute kidney injury during concomitant use of valacyclovir and loxoprofen: detecting drug-drug interactions in a spontaneous reporting system. Pharmacoepidemiol. Drug Saf 23:1154–1159
14. Sugimoto T, Yasuda M, Sakaguchi M et al (2008) Oliguric acute renal failure following oral valacyclovir therapy. QJM 101:164–166
15. Maeda Y, Konishi T, Omoda K et al (1996) Inhibition of theophylline metabolism by aciclovir. Biol Pharm Bull 19:1591–1595
16. Tartaglione TA, Collier AC, Opheim K, Gianola FG, Benedetti J, Corey L (1991) Pharmacokinetic evaluations of low-and high-dose zidovudine plus high-dose acyclovir in patients with symptomatic human immunodeficiency virus infection. Antimicrob Agents Chemother 35:2225–2231
17. Cooper DA, Pedersen C, Aiuti F et al (1991) The efficacy and safety of zidovudine with or without acyclovir in the treatment of patients with AIDS-related complex. The European-Australian Collaborative Group. AIDS 5:933–943
18. Pedersen C, Cooper DA, Brun-Vezinet F et al (1992) The effect of treatment with zidovudine with or without acyclovir on HIV p24 antigenaemia in patient with AIDS or AIDS-related complex. AIDS 6:821–825
19. Cooper DA, Pehrson PO, Pedersen C et al (1993) The efficacy and safety of zidovudine alone or as cotherapy with acyclovir for the treatment of patients with AIDS and AIDS-related complex: a double-blind randomized trial. European-Australian Collaborative Group. AIDS 7:197–207
20. Stein DS, Graham NM, Park LP et al (1994) The effect of the interaction of acyclovir with zidovudine on progression to AIDS and survival. Analysis of data in the Multicenter AIDS Cohort Study. Ann Intern Med 121:100–108
21. Gallant JE, Moore RD, Keruly J, Richman DD, Chaisson RE (1995) Lack of association between acyclovir use and survival in patients with advanced acquired immunodeficiency

virus disease treated with zidovudine. Zidovudine Epidemiology Study Group. J Infect Dis 172:346–352

22. Sabo JP, Cong X, Kraft M-F et al (2011) Lack of a pharmacokinetic interaction between steady-state tipranavir/ritonavir and single-dose valacyclovir in healthy volunteers. Eur J Clin Pharmacol 67:277–281

23. Famvir (famciclovir) product information. Novartis Pharmaceuticals Corporation, East Hanover (2016)

24. Fowles SE, Pierce MC, Prince WT et al (1990) Effect of food on the bioavailability and pharmacokinetics of penciclovir, a novel antiherpes agent, following oral administration of the prodrug, famciclovir. Br J Clin Pharmacol 29:620P–621P

25. Fowles SE, Fairless AJ, Pierce DM et al (1991) A further study of the effect of food on the bioavailability and pharmacokinetics of penciclovir after oral administration of famciclovir. Br J Clin Pharmacol 32:657P

26. Lavella J, Follansbee S, Trapnell CB et al (1996) Effect of food on the relative bioavailability of oral ganciclovir. J Clin Pharmacol 36:238–241

27. Brown F, Banken L, Saywell K et al (1999) Pharmacokinetics of valganciclovir and ganciclovir following multiple oral dosages of valganciclovir in HIV- and CMV-seropositive volunteers. Clin Pharmacokinet 37:167–176

28. Markham A, Faulds D (1994) Ganciclovir: an update of its therapeutic use in cytomegalovirus infection. Drugs 48:455–484

29. Valcyte (valganciclovir) product information. Genentech USA, Inc, South San Francisco (2015)

30. Cytovene (ganciclovir) product information. Genentech USA, Inc, South San Francisco (2016)

31. Hochster H, Dieterich D, Bozzette S et al (1990) Toxicity of combined ganciclovir and zidovudine for cytomegalovirus disease associated with AIDS: an AIDS clinical trials group study. Ann Intern Med 113:111–117

32. Pinching AJ, Helbert M, Peddle B, Robinson D, Janes K, Gor D (1989) Clinical experience with zidovudine for patients with acquired immune deficiency syndrome and acquired immune deficiency syndrome-related complex. J Infect 18(Suppl1):33–40

33. Snoeck R, Lagneaux L, Delforge A et al (1990) Inhibitory effects of potent inhibitors of human immunodeficiency virus and cytomegalovirus on the growth of human granulocyte-macrophage progenitor cells in vitro. Eur J Clin Micro Infect Dis 9:615–618

34. Causey D (1991) Concomitant ganciclovir and zidovudine treatment for cytomegalovirus retinitis in patients with HIV infection: an approach to treatment. J Acquir Immune Defic Synd 4(Suppl 1):S16–S21

35. Burger DM, Meenhorst PL, Koks CHW, Beijnen JH (1993) Drug interactions with zidovudine. AIDS 7:445–460

36. Millar AB, Miller RF, Patou G, Mindel A, Marsh R, Semple SJG (1990) Treatment of cytomegalovirus retinitis with zidovudine and ganciclovir in patients with AIDS: outcome and toxicity. Genitourin Med 66:156–158

37. Freitas VR, Fraser-Smith EB, Matthews TR (1993) Efficacy of ganciclovir in combination with other antimicrobial agents against cytomegalovirus in vitro and in vivo. Antivir Res 20:1–12

38. Cimoch PJ, Lavelle J, Pollard R et al (1998) Pharmacokinetics of oral ganciclovir alone and in combination with zidovudine, didanosine, and probenecid in HIV-infected subjects. J Acquir Immune Defic Syndr Hum Retrovirol 17:227–234

39. Frascino RJ, Gaines Griffy K, Jung D, Yu S (1995) Multiple dose crossover study of IV ganciclovir induction dose (5 mg/kg IV q12h) and didanosine (200 mg po q12h) in HIV-infected persons. In: Abstracts of the 35th interscience conference on antimicrobial agents and chemotherapy, San Francisco, CA, 17–20 September, Abstract A-27

40. Jung D, Griffy K, Dorr A et al (1998) Effect of high-dose oral ganciclovir on didanosine disposition in human immunodeficiency virus (HIV)-positive patients. J Clin Pharmacol 38:1057–1062

41. Tseng AL, Salit IE (2007) CD4+ cell count decline despite HIV suppression: a probable didanosine-valganciclovir interaction. Ann Pharmacother 41:512–517

42. Ray AS, Olson L, Fridland A (2004) Role of purine nucleoside phosphorylase in interactions between 2′,3′-dideoxyinosine and allopurinol, ganciclovir, or tenofovir. Antimicrob Agents Chemother 48:1089–1095

43. Studies of Ocular Complications of AIDS Research Group in collaboration with the AIDS Clinical Trials Group (1992) Mortality in patients with the acquired immune deficiency syndrome treated with either foscarnet or ganciclovir for cytomegalovirus retinitis. N Engl J Med 326:213–220

44. Studies of Ocular Complications of AIDS Research Group in collaboration with the AIDS Clinical Trials Group (1995) Morbidity and toxic effects associated with ganciclovir or foscarnet therapy in a randomized cytomegalovirus retinitis trial. Arch Intern Med 155:65–74

45. Merigan TC, Renlund DG, Keay S et al (1992) A controlled trial of ganciclovir to prevent cytomegalovirus disease after heart transplantation. N Engl J Med 326:1182–1186

46. Schmidt GM, Horak DA, Niland JC et al (1991) A randomized, controlled trial of prophylactic ganciclovir for cytomegalovirus pulmonary infection in recipients of allogeneic bone marrow transplants. N Engl J Med 324:1005–1011

47. Michaelis M, Ha TAT, Doerr HW, Cinati J Jr (2008) Valproic acid interferes with antiviral treatment in human cytomegalovirus-infected endothelial cells. Cardiovasc Res 77:544–550

48. Faulds D, Heel RC (1990) Ganciclovir – a review of its antiviral activity, pharmacokinetic properties and therapeutic efficacy in cytomegalovirus infections. Drugs 39:597–638

49. Vistide (cidofovir) product information. Gilead Sciences, Inc., Foster City (2010)

50. Lalezari J, Jaffe HS, Schaker T et al (1997) A randomized, double-blind placebo controlled trial of cidofovir gel for the treatment of acyclovir-unresponsive mucocutaneous herpes simplex virus infection in patients with AIDS. J Infect Dis 176:892–898

51. Lalezari JP, Stagg RJ, Kuppermann BD et al (1997) Intravenous cidofovir for peripheral cytomegalovirus retinitis in patients with AIDS. Ann Intern Med 126:257–263

52. Studies of Ocular Complications of AIDS Research Group in Collaboration with the AIDS Clinical Trials Group (1997) Parenteral cidofovir for cytomegalovirus retinitis in patients with AIDS: the HPMPC peripheral cytomegalovirus retinitis trial. Ann Intern Med 126:264–274

53. Lalezari JP, Drew WL, Glutzer E et al (1995) (S)-1-[3-hydroxy-2-(phosphonylmethoxy)propyl]cytosine (cidofovir): results of a phase I/II study of a novel antiviral nucleoside analogue. J Infect Dis 171:788–796

54. Polis MA, Spooner KM, Baird BF et al (1995) Anticytomegaloviral activity and safety of cidofovir in patients with human immunodeficiency virus infection and cytomegalovirus viruria. Antimicrob Agents Chemother 39:882–886

55. Wachsman M, Petty BG, Cundy KC et al (1996) Pharmacokinetics, safety and bioavailability of HPMPC (cidofovir) in human immunodeficiency virus-infected subjects. Antivir Res 29:153–161

56. Trejtnar F, Mandlikova J, Kocincova M, Volkova M (2014) Renal handling of amphotericin B and amphotericin B-deoxycholate and potential renal drug-drug interactions with selected antivirals. Antimicrob Agents Chemother 58:5650–5657

57. Tseng AL, Walmsley SL (1995) Rifabutin-associated uveitis. Ann Pharmacother 29:1149–1155

58. Davis JL, Taskintuna I, Freeman WR, Weinberg DV, Feuer WR, Leonard RE (1997) Iritis and hypotony after treatment with intravenous cidofovir for cytomegalovirus retinitis. Arch Ophthalmol 115:733–737

59. Palau LA, Tufty GT, Pankey GA (1997) Recurrent iritis after intravenous administration of cidofovir. Clin Infect Dis 25:337–338

60. Tseng AL, Mortimer CB, Salit IE (1999) Iritis associated with intravenous cidofovir. Ann Pharmacother 33:167–171

61. Foscavir (foscarnet sodium) product information. Clinigen Group Plc, Yardley (2011)

62. Aweeka FT, Brody SR, Jacobson M, Botwin K, Martin-Munley S (1998) Is there a pharma-cokinetic interaction between foscarnet and zalcitabine during concomitant administration? Clin Ther 20:232–243

63. Morales JM, Muñoz MA, Fernandez Zatarain G et al (1995) Reversible acute renal failure caused by the combined use of foscarnet and cyclosporin in organ transplanted patients. Nephrol Dial Transplant 10:882–883

64. Fan-Havard P, Sanchorawala V, Oh J, Moser EM, Smith SP (1994) Concurrent use of foscarnet and ciprofloxacin may increase the propensity for seizures. Ann Pharmacother 28:869–872

65. Matsuo H, Ryu M, Nagata A et al (1998) Neurotoxicodynamics of the interaction between ciprofloxacin and foscarnet in mice. Antimicrob Agents Chemother 42:691–694

66. Koshida R, Vrang L, Gilljam G, Harmenberg J, Oberg B, Wahren B (1989) Inhibition of human immunodeficiency virus in vitro by combinations of 3′-azido-3′-deoxythymidine and foscarnet. Antimicrob Agents Chemother 33:778–780

67. Eriksson BF, Schinazi RF (1989) Combinations of 3′-azido-3′-deoxythymidine (zidovudine) and phosphonoformate (foscarnet) against human immunodeficiency virus type 1 and cyto-megalovirus replication in vitro. Antimicrob Agents Chemother 33:778–780

68. Aweeka FT, Gambertoglio JG, van der Horst C, Raasch R, Jacobson MA (1992) Pharmacokinetics of concomitantly administered foscarnet and zidovudine for treatment of human immunodeficiency virus infection (AIDS Clinical Trials Group protocol 053). Antimicrob Agents Chemother 36:1773–1778

69. Aoki FY, Sitar DS (1988) Clinical pharmacokinetics of amantadine hydrochloride. Clin Pharmacokinet 14:35–51

70. Wills RJ, Rodriguez LC, Choma N, Oakes M (1987) Influence of meal on the bioavailability of rimantadine HCL. J Clin Pharmacol 27:821–823

71. Amantadine hydrochloride product information. Upsher-Smith Laboratories, Inc., Maple Grove (2015)

72. Rimantadine hydrochloride product information. Impax Laboratories, Inc., Hayward (2016)

73. Wilson TW, Rajput AH (1983) Amantadine-dyazide interaction. Can Med Assoc J 12:974–975

74. Speeg KV, Leighton JA, Maldonado AL (1989) Toxic delirium in a patient taking amantadine and trimethoprim-sulfamethoxazole. Am J Med Sci 29:410–412

75. Wills RJ (1989) Update on rimantadine's clinical pharmacokinetics. J Resp Dis 10:S20–S25

76. Millet VM, Dreisbach M, Bryson YJ (1982) Double-blind controlled study of central nervous system side effects of amantadine, rimantadine, and chlorpheniramine. Antimicrob Agents Chemother 211:1–4

77. Harper RW, Knothe BU (1973) Coloured Lilliputian hallucinations with amantadine. Med J Aust 19:444–445

78. Postma JU, Tilburg WV (1975) Visual hallucinations and delirium during treatment with amantadine (Symmetrel). J Am Geriatr Soc 23:212

79. Stroe AE, Hall J, Amin F (1995) Psychotic episode related to phenylpropanolamine and amantadine in a healthy female. Gen Hosp Psychiatry 17:457–458

80. Trappler B, Miyashiro AM (2000) Bupropion-amantadine-associated neurotoxicity. J Clin Psychiatry 61:61–62

81. Morrison D, Roy S, Rayner C et al (2007) A randomized, crossover study to evaluate the pharmacokinetics of amantadine and oseltamivir administered alone and in combination. PLoS One 2:e1305

82. Cirrincione-Dall G, Brennan B, Ballester-Sanchis RM, Navarro MT, Davies BE (2012) Pharmacokinetics and safety of coadministered oseltamivir and rimantadine in healthy volunteers: an open-label, multiple-dose, randomized crossover study. J Clin Pharmacol 52:1255–1264

83. Atiee G, Lasseter K, Baughmann S et al (2012) Absence of pharmacokinetic interaction between intravenous peramivir and oral oseltamivir or rimantadine in humans. J Clin Pharmacol 52:1410–1419

84. Tamiflu (oseltamivir) product information. Genentech USA, Inc., South San Francisco (2016)

85. Snell P, Oo C, Dorr A, Barrett J (2002) Lack of pharmacokinetic interaction between the oral anti-influenza neuraminidase inhibitor prodrug oseltamivir and antacids. Br J Clin Pharmacol 544:372–377
86. Morimoto K, Kishimura K, Nagami T et al (2011) Effect of milk on the pharmacokinetics of oseltamivir in healthy volunteers. J Pharm Sci 100:3854–3861
87. Hill G, Cihlar T, Oo C et al (2002) The anti-influenza drug oseltamivir exhibits low potential to induce pharmacokinetic drug interactions via renal secretion – correlation of in vivo and in vitro studies. Drug Metab Dispos 301:13–19
88. Tang M, Mukundan M, Yang J et al (2006) Antiplatelet agents aspirin and clopidogrel are hydrolyzed by distinct carboxylesterases and the hydrolyses are markedly altered with certain polymorphistic variants. J Pharmacol Exp Ther 319:1467–1476
89. Shi D, Yang J, Yang D et al (2006) Anti-influenza prodrug oseltamivir is activated by carbo-xylesterase human carboxylesterase 1, and the activation is inhibited by antiplatelet agent clopidogrel. J Pharmacol Exp Ther 319:1477–1484
90. Fowler S, Lennon SM, Hoffmann G, Rayner C (2006) Comments on "anti-influenza prodrug oseltamivir is activated by carboxylesterase human carboxylesterase 1, and the activation is inhibited by antiplatelet agent clopidogrel" [letter]. J Pharmacol Exp Ther 322:422–423
91. Davies BE (2010) Pharmacokinetics of oseltamivir: an oral antiviral for the treatment and prophylaxis of influenza in diverse populations. J Antimicrob Chemother 65(Suppl 2):ii5–i10
92. Butler D (2005) Wartime tactic doubles power of scarce bird-flu drug. Nature 438:6
93. Howton JC (2006) Probenecid with oseltamivir for human influenza A (H5N1) virus infection? N Engl J Med 354:879–880
94. Holodniy M, Penzak SR, Stright TM et al (2008) Pharmacokinetics and tolerability of oseltamivir combined with probenecid. Antimicrob Agents Chemother 52:3013–3021
95. Rayner CR, Chanu P, Gieschke R, Boak LM, Jonsson EN (2008) Population pharmacokinetics of oseltamivir when coadministered with probenecid. J Clin Pharmacol 48:935–947
96. Raisch DW, Straight TM, Holodniy M (2009) Thrombocytopenia from combination treatment with oseltamivir and probenecid: case report, MedWatch data summary, and review of the literature. Pharmacotherapy 29:988–992
97. He G, Massarella J, Aitken M et al (1999) The safety and pharmacokinetics of the neuraminidase inhibitor RO 64–0796 when administered concurrently with paracetamol. Clin Microbiol Infect 5:150. [abstract P.247]
98. Oo C, Barrett J, Dorr A, Liu B, Ward P (2002) Lack of pharmacokinetic interaction between the oral anti-influenza prodrug oseltamivir and aspirin. Antimicrob Agents Chemother 466:1993–1995
99. Lam H, Jeffery J, Sitar DS, Aoki FY (2011) Oseltamivir, an influenza neuraminidase inhibitor drug, does not affect the steady-state pharmacokinetic characteristics of cyclosporine, mycophenolate, or tacrolimus in adult renal transplant patients. Ther Drug Monit 33:699–704
100. Kute V, Goplani KR, Godara SM, Shah PR, Vanikar AV, Trivedi HL (2011) Post-exposure prophylaxis for H1N1 with oseltamivir in renal allograft recipient – safe and effective without any immunosuppressive drug interaction. J Assoc Physicians India 59:49–51
101. Pukrittayakamee S, Jittamala P, Stepniewska K et al (2011) An open-label crossover study to evaluate potential pharmacokinetic interactions between oral oseltamivir and intravenous zanamivir in healthy Thai adults. Antimicrob Agents Chemother 55:4050–4057
102. Davies BE, Baldo PA, Lennon-Chrimes S, Brewster M (2010) Effect of oseltamivir treatment on anticoagulation: a cross-over study in warfarinized patients. Br J Clin Pharmacol 70:834–843
103. Mosholder AD, Racoosin JA, Young S et al (2013) Bleeding events following concurrent use of warfarin and oseltamivir by Medicare beneficiaries. Ann Pharmacother 47:1420–1428
104. Wagner J, Abdel-Rahman SM (2015) Oseltamivir-warfarin interaction in hypoplastic left heart syndrome: case report and review. Pediatrics 135:e1333–e1336
105. Lee S-H, Kang H-R, Jung J-W et al (2012) Effect of oseltamivir on bleeding risk associated with warfarin therapy. Clin Drug Investig 32:131–137

106. Wells Q, Hardin B, Raj SR, Darbar D (2010) Sotalol-induced torsades de pointes precipitated during treatment with oseltamivir for H1N1 influenza. Heart Rhythm 7:1454–1457

107. Daniel MJ, Barnett JM, Pearson BA (1999) The low potential for drug interactions with zanamivir. Clin Pharmacokinet 36:41–50

108. Relenza (zanamivir) product information. GlaxoSmithKline, Research Triangle Park (2016)

109. Rapivab (peramivir) product information. BioCryst Pharmaceuticals, Inc., Durham (2016)

110. Rockstroh JK (2003) Management of hepatitis B and C in HIV co-infected patients. J Acquir Immune Defic Syndr 34:S59–S65

111. Ghany MG, Strader DB, Thomas DL, Seeff LB (2009) Diagnosis, management, and treatment of hepatitis C: an update. Hepatology 49:1337–1374

112. Gutfreund KS, Bain VC (2000) Chronic viral hepatitis C: management update. Can Med Assoc J 162:827–833

113. Bodenheimer HC Jr, Lindsay KL, Davis GL et al (1997) Tolerance and efficacy of oral ribavirin treatment of chronic hepatitis C: a multicenter trial. Hepatology 26:473–477

114. Buckwold VE (2004) Implications of finding synergistic *in vitro* drug-drug interactions between of interferon-α and ribavirin for the treatment of chronic hepatitis C. J Antimicrob Chemother 53:413–414

115. Buckwold VE, Wei J, Wenzel-Mathers M et al (2003) Synergistic in vitro interactions between alpha interferon and ribavirin against bovine viral diarrhea virus and yellow fever virus as surrogate models of hepatitis C viral replication. Antimicrob Agents Chemother 47:2293–2298

116. American Association for the Study of Liver Diseases and the Infectious Diseases Society of America. HCV Guidance: Recommendations for Testing, Managing, and Treating Hepatitis C. Available at http://hcvguidelines.org/full-report-view. Last accessed 2/7/2017

117. Rebetol (ribavirin) product information. Merck & Co., Inc., Whitehouse Station (2016)

118. Copegus (ribavirin) product information. Genentech USA, South San Francisco (2015)

119. Intron A (interferon alfa-2b) product information. Merck & Co., Inc., Whitehouse Station (2016)

120. Virazole (ribavirin for inhalation) product information. Valeant Pharmaceuticals North America, Costa Mesa (2016)

121. Paroni R, Del Puppo M, Borghi C, Sirtori CR, Galli Kienle M (1989) Pharmacokinetics of ribavirin and urinary excretion of the major metabolite 1,2,4-triazole-3-carboxamide in normal volunteers. Int J Clin Pharmacol Ther Toxicol 27:302–307

122. Milazzo L, Falvella FS, Magni C et al (2013) Seizures in patients with chronic hepatitis C treated with NS3/4A protease inhibitors: does a pharmacological interaction play a role? Pharmacology 92:235–237

123. Huggins JW (1989) Prospects for treatment of viral hemorrhagic fevers with ribavirin, a broad-spectrum antiviral drug. Rev Infect Dis 2:S750–S761

124. Chaparro M, Trapero-Marugan M, Moreno-Otero R, Gisbert JP (2009) Azathioprine plus ribavirine treatment and pancytopenia [letter]. Aliment Pharmacol Ther 30:955–963

125. Peyrin-Biroulet L, Cadranel JF, Nousbaum JB et al (2008) Interaction of ribavirin with azathioprine metabolism potentially induces myelosuppression. Aliment Pharmacol Ther 28:984–993

126. Hindorf U, Lindqvist M, Peterson C et al (2006) Pharmacogenetics during standardized initiation of thiopurine treatment in inflammatory bowel disease. Gut 55:1423–1431

127. Gilissen LP, Derijks LJ, Verhoeven HM et al (2007) Pancytopenia due to high 6-methylmercaptopurine levels in a 6-mercaptopurine treated patient with Crohn's disease. Dig Liver Dis 39:182–186

128. Perronne C (2006) Antiviral hepatitis and antiretroviral drug interactions. J Hepatol 44:119–125

129. Retrovir [package insert]. GlaxoWellcome, Research Triangle Park (2016)

130. Vogt MW, Hartshorn KL, Furman PA et al (1987) Ribavirin antagonizes the effect of azidothymidine on HIV replication. Science 235:1376–1379

131. Sim SM, Hoggard PG, Sales SD, Phiboonbanakit D, Hard CA, Back DJ (1998) Effect of ribavirin on zidovudine efficacy and toxicity in vitro: a concentration-dependent interaction. AIDS Res Hum Retrovir 14:1661–1667

132. Tornevik Y, Ullman B, Balzarini J, Wahren B, Eriksson S (1995) Cytotoxicity of 3′-azido-3′deoxythymidine correlates with 3′-azidothymidine-5′-monophosphate (AZTMP) levels, whereas anti-human immunodeficiency virus (HIV) activity correlates with 3′-azidothymidine-5′-triphosphate (AZTTTP) levels in cultured CEM T-lymphoblastoid cells. Biochem Pharmacol 49:829–837

133. Chung RT, Andersen J, Volberding P et al (2004) Peginterferon Alfa-2a plus ribavirin versus interferon alfa-2a plus ribavirin for chronic hepatitis C in HIV-coinfected persons. N Engl J Med 351:451–459

134. Torriani FJ, Rodriguez-Torres M, Rockstroh JK et al (2004) Peginterferon Alfa-2a plus ribavirin for chronic hepatitis C virus infection in HIV-infected patients. N Engl J Med 351:438–450

135. Gries J-M, Torriani FJ, Rodriguez-Torres M et al (2005) Effect of ribavirin on intracellular and plasma pharmacokinetics of nucleoside reverse transcriptase inhibitors in patients with human immunodeficiency virus-hepatitis C coinfection: results of a randomized clinical study. Antimicrob Agents Chemother 49:3997–4008

136. Aweeka FT, Kang M, Yu J-Y et al (2007) Pharmacokinetic evaluation of the effects of ribavirin on zidovudine triphosphate formation: ACTG 5092s Study Team. HIV Med 8:288–294

137. Margot NA, Miller MD (2006) In vitro combination studies of tenofovir and other nucleoside analogues with ribavirin against HIV-1. Antivir Ther 10:343–348

138. Ramanathan S, Cheng A, Mittan A, Ebrahimi R, Kearney BP (2006) Absence of clinically relevant pharmacokinetic interaction between ribavirin and tenofovir in healthy subjects. J Clin Pharmacol 46:559–566

139. Morsica G, De Bona A, Foppa CU, Sitia G, Finazzi R, Lazzarin A (2000) Ribaviring therapy for chronic hepatitis C does not modify HIV viral load in HIV-1 positive patients under antiretroviral treatment. AIDS 154:1656–1658

140. Landau A, Batisse D, Piketty C, Jian R, Kazatchkine MD (2000) Lack of interference between ribavirin and nucleoside analogues in HIV/HCV co-infected individuals undergoing concomitant antiretroviral and anti-HCV combination therapy. AIDS 14:1875–1858

141. Balzarini J, Naesens L, Robins MJ, DeClercq E (1990) Potentiating effects of ribavirin on the in vitro and in vivo antiretrovirus activities of 2′,3′-dideoxyinosine and 2′,3′-dideoxy-2,6-diaminiopurine riboside. J AIDS 3:1140–1147

142. Videx (didanosine) product information. Bristol-Myers Squibb Company, Princeton (2015)

143. Butt AA (2003) Fatal lactic acidosis and pancreatitis associated with ribavirin and didanosine therapy. AIDS Read 13:344–348

144. Japour AJ, Lertora JJ, Meehan PM et al (1996) A phase-1 study of the safety, pharmacokinetics, and antiviral activity of combination didanosine and ribavirin in patients with HIV-1 disease. J AIDS 13:235–246

145. Lafeuillade A, Hittinger G, Chadapaud S (2001) Increased mitochondrial toxicity with ribavirin in HIV/HCV coinfection. Lancet 357:280–281

146. Salmon-Ceron D, Chauvelot-Moachon L, Abad S, Silbermann B, Sogni P (2001) Mitochondrial toxic effects and ribavirin [letter]. Lancet 357:1803–1804

147. Kakuda TN, Brinkman K (2001) Mitochondrial toxic effects and ribavirin [letter]. Lancet 357:1802–1803

148. Soriano V, Sulkowski M, Bergin C et al (2002) Care of patients with chronic hepatitis C and HIV co-infection: recommendations from the HIV-HCV International Panel. AIDS 16:813–828

149. Rodriguez-Novoa S, Morello J, Gonzalez M et al (2008) Increase in serum bilirubin in HIV/hepatitis-C virus co-infected patients on atazanavir therapy following initiation of pegylated-interferon and ribavirin [letter]. AIDS 22:2535–2548

150. Ashby J, Garvey L, Erlwein OW et al (2011) Pharmacokinetic and safety profile of raltegravir and ribavirin, when dosed separately and together, in healthy volunteers. J Antimicrob Chemother 66:1340–1345

151. Solas C, Pambrun E, Winnock M et al (2012) Ribavirin and abacavir drug interaction in HIV-HCV coinfected patients: fact or fiction? AIDS 26:2193–2199

152. Schulman S (2002) Inhibition of warfarin activity by ribavirin. Ann Pharmacother 36:72–74

153. Pegasys (peginterferon alfa-2a) product information. Genentech USA, South San Francisco (2015)

154. PegIntron (peginterferon alfa-2b) product information. Merck & Co., Inc., Whitehouse Station (2015)

155. Alferon N (interferon alfa-n3, human leukocyte derived). Hemispherx Biopharma, Inc., Philadelphia (2010)

156. Rebif (interferon beta-1a) product information. EMD Serono, Inc., Rockland (2015)

157. Avonex (interferon beta-1a) product information. Biogen, Inc., Cambridge, MA (2015)

158. Actimmune (interferon gamma-1b) product information. HZNP USA, Inc., Lake Forest (2016)

159. Bleau AM, Levitchi MC, Maurice H, du Souich P (2000) Cytochrome P450 inactivation by serum from humans with a viral infection and serum from rabbits with a turpentine-induced inflammation: the role of cytokines. Br J Pharmacol 130:1777–1784

160. Renton KW, Mannering GJ (1976) Depression of hepatic cytochrome P-450-dependent monooxygenase systems with administered interferon inducing agents. Biochem Biophys Res Commun 2:343–348

161. Singh G, Renton KW (1981) Interferon-mediated depression of cytochrome P450-dependent drug biotransformation. Mol Pharmacol 20:681–684

162. Okuno H, Kitao Y, Takasu M et al (1990) Depression of drug-metabolizing activity in the human liver by interferon-α. Eur J Clin Pharmacol 39:365–367

163. Israel BC, Blouin RA, McIntyre W, Shedlofsky SI (1993) Effects of interferon-α monotherapy on hepatic drug metabolism in cancer patients. Br J Clin Pharmacol 36:229–235

164. Pageaux GP, LeBricquir Y, Berthou F et al (1998) Effects of interferon-α on cytochrome P-450 isoforms 1A2 and 3A activities in patients with chronic hepatitis C. Eur J Gastroenterol Hepatol 10:491–495

165. Carlson TJ, Billings RE (1996) Role of nitric oxide in the cytokine-mediated regulation of cytochrome P-450. Mol Pharmacol 49:796–801

166. Brockmeyer NH, Barthel B, Mertins L, Goos M (1998) Changes in antipyrine pharmacokinetics during influenza and after administration of interferon-alpha and –beta. Int J Clin Pharmacol Ther 36:309–311

167. Williams SJ, Farrell GC (1986) Inhibition of antipyrine metabolism by interferon. Br J Clin Pharmacol 22:610–612

168. Williams SJ, Baird-Lambert JA, Farrell GC (1987) Inhibition of theophylline metabolism by interferon. Lancet 2:939–941

169. Okuno H, Takasu M, Kano H et al (1993) Depression of drug-metabolizing activity in the human liver by interferon-β. Hepatology 17:65–69

170. Craig PI, Tapner M, Farrell GC (1993) Interferon suppresses erythromycin metabolism in rats and human subjects. Hepatology 17(2):230–235

171. Sewer MB, Morgan ET (1997) Nitric oxide-independent suppression of P450 2CII expression by interleukin-1β and endotoxin in primary rat hepatocytes. Biochem Pharmacol 54:729–737

172. Adachi Y, Yokoyama Y, Nanno T, Yamamoto T (1995) Potentiation of warfarin by interferon. BMJ 311:292

173. Serratrice J, Durand JM, Morange S (1998) Interferon-alpha 2b interaction with acenocoumarol. Am J Hematol 57:89

174. Simon-Talero M, Buti M, Esteban R (2012) Severe anemia related to oseltamivir during treatment of chronic hepatitis C: a new drug interaction? J Viral Hepat 19(Suppl 1):14–17

175. Nokta M, Loh JP, Douidar SM, Ahmed AE, Pollard RE (1991) Metabolic interaction of recombinant interferon-β and zidovudine in AIDS patients. J Interf Res 11:159–164

176. Piscitelli SC, Amantea MA, Vogel S et al (1996) Effects of cytokines on antiviral pharmacokinetics: an alternative approach to assessment of drug interactions using bioequivalence guidelines. Antimicrob Agents Chemother 40:161–165

177. Casato M, Pucillo LP, Leoni M et al (1995) Granulocytopenia alter combined therapy with interferon and angiotensin-converting enzyme inhibitors: evidence for a synergistic hematologic toxicity. Am J Med 99:386–391

178. Jacquot C, Caudwell V, Belenfant X (1996) Granulocytopenia after combined therapy with interferon and angiotensin-converting enzyme inhibitors: evidence for a synergistic hematologic toxicity. Am J Med 101:235–236

179. Raderer M, Scheithauer W (1995) Treatment of advanced colorectal cancer with 5-fluorouracil and interferon-alpha: an overview of clinical trials. Eur J Cancer 31:1002–1008

180. Al-Zahrani H, Gupta V, Minden MD, Messner HA, Lipton JH (2003) Vascular events associated with alpha interferon therapy. Leuk Lymphoma 44:471–475

181. Österborg A, Björkholm M, Björeman M et al (1993) Natural interferon-α in combination with melphalan/prednisone versus melphalan/prednisone in the treatment of multiple myeloma stages II and III: a randomized study from the Myeloma Group of Central Sweden. Blood 81:1428–1434

182. Cooper MR, Dear K, McIntyre OR et al (1993) A randomized clinical trial comparing melphalan/prednisone with or without interferon alfa-2b in newly diagnosed patients with multiple myeloma: a Cancer and Leukemia Group B study. J Clin Oncol 11:155–160

183. Pittman K, Perren T, Ward U et al (1993) Pharmacokinetics of 5-fluorouracil in colorectal cancer patients receiving interferon. Ann Oncol 4:515–516

184. Seymour MT, Patel N, Johnston A, Joel SP, Slevin ML (1994) Lack of effect of interferon α2a upon fluorouracil pharmacokinetics. Br J Cancer 70:724–728

185. Schueller J, Czejka MJ, Schernthaner G, Fogl U, Jaeger W, Micksche M (1992) Influence of interferon alfa-2b with or without folinic acid on pharmacokinetics of fluorouracil. Semin Oncol 19(Suppl 2):93–97

186. Danhauser LL, Freimann JH, Gilchrist TL et al (1993) Phase I and plasma pharmacokinetic study of infusional fluorouracil combined with recombinant interferon alfa-2b in patients with advanced cancer. J Clin Oncol 11:751–761

187. Schueller J, Czejka M (1995) Pharmacokinetic interaction of 5-fluorouracil and interferon alpha-2b with or without folinic acid. Med Oncol 12:47–53

188. Ehrsson H, Eksborg S, Wallin I, Österborg A, Mellstedt H (1990) Oral melphalan pharmacokinetics: influence of interferon-induced fever. Clin Pharmacol Ther 47:86–90

189. Gomez-Rangel JD, Ruiz-Delgado GJ, Ruiz-Arguelles GJ (2003) Pegylated-interferon induced severe bone marrow hypoplasia in a patient with multiple myeloma receiving thalidomide. Am J Hematol 74:290–291

190. Hoffmann RM, Ott S, Parhofer KG, Batrl R, Pape GR (1998) Interferon-a induced agranulocytosis in a patient on long-term clozapine therapy. J Hepatol 29:170–175

191. Okazaki K, Yoshihara H, Suzuki K et al (1994) Efficacy of interferon therapy in patients with chronic hepatitis C. Scand J Gastroenterol 29:1039–1043

192. Mochida S, Ohnishi K, Matsuo S, Kakihara K, Fujiwara K (1996) Effect of alcohol intake on the efficacy of interferon therapy in patients with chronic hepatitis C as evaluated by multivariate logistic regression analysis. Alcohol Clin Exp Res 20:A371–A377

193. Ohnishi K, Matsuo S, Matsutani K et al (1996) Interferon therapy for chronic hepatitis C in habitual drinkers: comparison with chronic hepatitis C in infrequent drinkers. Am J Gastroenterol 91:1374–1379

194. Loguercio C, Di Pierro M, Di Marino MP et al (2000) Drinking habits of subjects with hepatitis C virus-related chronic liver disease: prevalence and effect on clinical, virological and pathological aspects. Alcohol Alcohol 35:296–301

195. Safdar K, Schiff ER (2004) Alcohol and hepatitis C. Semin Liver Dis 24:305–315

196. McCartney EM, Beard MR (2010) Impact of alcohol on hepatitis C virus replication and interferon signaling. World J Gastroenterol 16:1337–1343
197. Fialho R, Burridge A, Pereira M et al (2016) Norepinephrine-enhancing antidepressant exposure associated with reduced antiviral effect of interferon alpha on hepatitis C. Psychopharmacology 233:1689–1694
198. Daklinza (daclatasvir) product information. Bristol-Myers Squibb Company, Princeton (2016)
199. Epclusa (sofosbuvir and velpatasvir) product information. Gilead Sciences, Inc., Foster City (2016)
200. Harvoni (ledipasvir and sofosbuvir) product information. Gilead Sciences, Inc., Foster City (2016)
201. Olysio (simeprevir) product information. Janssen Products, LP, Titusville (2016)
202. Viekira XR (dasabuvir, ombitasvir, paritaprevir, and ritonavir) product information. AbbVie, Inc., North Chicago (2016)
203. Zepatier (elbasvir and grazoprevir) product information. Merck & Co., Inc., Whitehouse Station (2017)

Chapter 11
Antifungal Agents

Jarrett R. Amsden and Paul O. Gubbins

11.1　Introduction

Clinicians now have multiple antifungal therapy options when managing systemic mycoses. Clear differences in the spectrum of activity, toxicity, and drug interaction potential exist between and, in some cases, within the antifungal therapeutic class. These differences can be exploited to tailor therapy against a specific pathogen. When choosing systemic antifungal therapy, clinicians consider available susceptibility data, the drug's spectrum of activity, and potential toxicities. The significant potential for a systemic antifungal agent to interact with other medicines is often difficult to overlook, but if not considered, drug-drug interactions involving systemic antifungal agents may lead to enhanced toxicity of the concomitant medication(s) or ineffective antifungal treatment. Therefore, clinicians must understand the drug interaction profile of antifungal agents.

11.1.1　Amphotericin B Pharmacology

Amphotericin B binds to ergosterol, a key component of the fungal cell membrane, which disrupts the fungal cell membrane integrity allowing cellular components to leak out causing cell death. Amphotericin B produces infusion-related toxicities,

J. R. Amsden (✉)
Department of Pharmacy Practice, Butler University College of Pharmacy & Health Sciences, Indianapolis, IN, USA
e-mail: jamsden@butler.edu

P. O. Gubbins
Division of Pharmacy Practice and Administration, UMKC School of Pharmacy at MSU, Springfield, MO, USA

© Springer International Publishing AG, part of Springer Nature 2018　　　425
M. P. Pai et al. (eds.), *Drug Interactions in Infectious Diseases: Antimicrobial Drug Interactions*, Infectious Disease,
https://doi.org/10.1007/978-3-319-72416-4_11

including hypotension, fever, rigors, and chills, and dose-dependent adverse effects, such as nephrotoxicity, azotemia, renal tubular acidosis, electrolyte imbalance, cardiac arrhythmias, and anemia [1]. Infusion-related adverse effects rarely limit the use of amphotericin B or other agents, but dose-dependent adverse effects often do.

11.1.1.1 Distribution

Amphotericin B deoxycholate is protein bound (>95%), primarily to albumin and α_1-acid glycoprotein [2]. Amphotericin B deoxycholate apparently distributes extensively in tissue (apparent volume of distribution (Vd) $\approx$ 2–4 L/kg) [2, 3]. Formulating amphotericin B in a lipid vehicle alters its distribution by increasing reticuloendothelial system drug uptake. This alteration reduces renal distribution and thereby decreases the propensity for acute kidney injury with lipid amphotericin B formulations compared to amphotericin B deoxycholate [2–4].

11.1.1.2 Elimination

Amphotericin B deoxycholate is cleared slowly from its distribution sites [3]. More than 90% of a dose can be recovered up to 1 week after administration. Amphotericin B deoxycholate is mostly excreted as unchanged drug in the urine (20.6%) and feces (42.5%) [3]. The formulation of amphotericin B with lipids significantly alters its elimination [3].

11.1.2 5-Fluorocytosine (5-FC) Pharmacology

5-Fluorocytosine (5-FC) is used only in combination with amphotericin B in the treatment of cryptococcal meningitis. The oral absorption of 5-FC is rapid and complete. In the fasting state, 5-FC bioavailability is approximately 90% [5]. 5-FC is minimally bound to plasma proteins and its Vd approximates total body water [5]. Renal clearance (CL_R) of 5-FC correlates highly with creatinine clearance (CL_{CR}), and its half-life ($t_{1/2}$) increases as CL_{CR} declines, because nearly 90% of a dose is renally excreted as unchanged drug [5]. When administered orally, 5-FC is deaminated by intestinal microflora, resulting in elevated 5-fluorouracil (5-FU) serum concentrations, which can cause myelosuppression and gastrointestinal mucosa toxicity [5]. The incidence of this toxicity is approximately 20–40% [6, 7].

11.1.3 Azole Pharmacology

Commonly prescribed systemic azoles (itraconazole, fluconazole, voriconazole, posaconazole, isavuconazole) inhibit fungal cytochrome P450 (CYP)-dependent C-14 α-demethylase, which converts lanosterol to ergosterol. This inhibition depletes ergosterol, the essential sterol of the fungal cell membrane, and compromises cell membrane integrity. In general, the azoles are weak bases, and most, with the exception of fluconazole, are lipophilic with poor water solubility [8]. Due to their lipophilicity, itraconazole, voriconazole, and posaconazole cannot be formulated as an intravenous (i.v.) dosage form without the use of a solubilizing agent [9–12]. Although the active form of isavuconazole is also lipophilic and poorly soluble in water, it is administered as a highly water-soluble prodrug isavuconazonium sulfate, which consists of triazolium salt bound to isavuconazole by an ester moiety and does not require a solubilizing agent [13, 14]. Phase I (CYP), phase II (conjugative enzymes), and transport proteins all have a role in the disposition and many drug interactions involving azole antifungal agents. The systemic azoles are CYP and conjugative enzyme substrates and inhibitors to varying degrees. Certain azoles are also substrates and inhibitors of transport proteins in the ATP-binding cassette (ABC) transporter protein family (i.e., P-glycoprotein (P-gp), various multidrug resistance-associated proteins (MRP), breast cancer resistance protein (BCRP), and bile salt export pump (BSEP)) and other transporter proteins [15–18]. As discussed in Chap. 3, some of these transport proteins share substrate specificity or are co-localized with CYP3A in the intestine, liver, and kidney [17–19].

11.1.3.1 Oral Absorption of the Systemic Azoles

In the USA, itraconazole is only marketed as a capsule and as 40% hydroxypropyl-β-cyclodextrin (HP-βCD) solution for oral use. Absorption from the capsule is slow, variable, and optimal under acidic gastric conditions or in the fed state [20]. Due to insufficient and variable intestinal concentrations, itraconazole capsules are subject to extensive intestinal and hepatic CYP3A4 ("first-pass") metabolism before reaching the systemic circulation [21, 22]. Unlike the capsule, itraconazole in oral solution requires no dissolution, so its absorption is rapid and unaffected by alterations in gastric pH but is optimal in the fasting state [21, 23]. The oral solution delivers high itraconazole concentrations to the intestinal epithelium that may transiently saturate intestinal CYP3A4 and minimize first-pass metabolism [21, 22]. The formulations are clinically bioequivalent, but the absolute bioavailability of the oral solution is higher than that of the capsule [9, 20].

Oral fluconazole is rapidly and nearly completely absorbed, and its absorption is independent of gastric acidity or the presence of food [24]. Fluconazole is more hydrophilic than the other systemic azoles; thus, no solubilizing agent is needed to formulate it as an i.v. solution. Moreover, fluconazole undergoes less hepatic metabolism than the other azoles.

Voriconazole is available in both i.v. and oral formulations. Oral voriconazole absorption is rapid and nearly complete, and its dissolution is unaffected by changes in gastric pH [11, 25]. However, fatty foods and enteral feedings decrease voriconazole bioavailability up to 22% and its C_{max} by 34% [11, 25]. Intravenous voriconazole is formulated with the solubilizing agent sulfobutyl ether β-cyclodextrin (SBECD).

Originally, posaconazole was marketed solely as an oral suspension. This original formulation exhibits saturable absorption at doses exceeding 800 mg daily; it is also influenced by gastric pH, food, and divided dosing [26–33]. Recently, posaconazole has been formulated as a delayed release tablet and i.v. solution. As with other lipophilic azoles, the i.v. solution of posaconazole is formulated in SBECD; however, the daily dose SBECD is lower compared to voriconazole [34, 35]. The delayed release tablets are formulated as pH-sensitive polymer hypromellose acetate succinate via hot-melt extrusion technology [36]. This delivery system improves solubility and ultimate bioavailability of posaconazole. This drug delivery system prevents its dissolution in low pH environments of the stomach and allows release in the higher pH intestines [36, 37]. In contrast to the oral suspension, posaconazole delayed release tablets are unaffected by food, pH, or GI motility agents [37–39].

Isavuconazonium sulfate is an orally or intravenously administered prodrug that is rapidly and nearly completely (99%) converted by gut or plasma esterases (i.e., plasma butyrylcholinesterase), respectively, to the active moiety isavuconazole, leaving negligible concentrations of the prodrug or cleavage product [13]. The apparent oral bioavailability of isavuconazole is approximately 98%, which is achieved in 1–3 h [14]. Absorption of isavuconazole after oral administration of isavuconazonium sulfate is not significantly altered by food or changes in pH [40]. Because isavuconazonium sulfate is a water-soluble prodrug, a cyclodextrin carrier molecule is not required to solubilize the drug for i.v. administration [13, 14].

11.1.3.2 Protein Binding and Distribution of the Systemic Azoles

Itraconazole, posaconazole, and isavuconazole are highly protein bound (95–99%) [13, 20, 23, 41]. Itraconazole, posaconazole, voriconazole, and isavuconazole extensively distribute throughout the body [13, 20, 23, 41]. Binding to plasma proteins is moderate (58%) for voriconazole and minimal for fluconazole (11%) [11, 24, 25]. Fluconazole distributes into a variety of body fluids and hepatic and renal tissues [24]. Unlike itraconazole, fluconazole and voriconazole adequately distribute into the cerebrospinal fluid (CSF) and central nervous system (CNS) tissues [23, 24, 42]. Based upon limited data, posaconazole minimally distributes to the CSF and CNS tissue [43]. The amount of isavuconazole distribution into human CSF and CNS tissues is unknown; however, it was effective in an experimental cryptococcal meningitis animal model and a single case report of disseminated mucormycosis involving the brain [44, 45].

11.1.3.3 Metabolism/Elimination of the Systemic Azoles

Several CYPs including CYP3A4, 2C19, and 2C9 catalyze azole biotransformation to varying extents [8]. CYP3A5, another member of the CYP3A subfamily, is 85% homologous to CYP3A4, and although it often shares substrate specificity with CYP3A4, it catalyzes the biotransformation of the isavuconazole, but not other azoles [8, 46].

Itraconazole

Very little (≈ 2%) of an itraconazole dose is excreted unchanged in the urine [10, 20]. Itraconazole formulations are comprised of four stereoisomers, and they exhibit dose-dependent elimination. Itraconazole undergoes extensive stereoselective sequential CYP3A4 metabolism of only a pair of its stereoisomers to produce three metabolites (hydroxyitraconazole, keto-itraconazole, and N-desalkyl-itraconazole) [47–49].

Fluconazole

Approximately 91% of oral fluconazole is excreted in the urine, most of which (80%) is excreted as unchanged drug. Two metabolites including a glucuronide conjugate (fluconazole β-D glucuronide) and fluconazole N-oxide account for the majority of metabolites recovered in the urine [50]. The N-oxide metabolite is formed from the heme-coordinating triazole moiety, and the reaction is likely catalyzed by CYP2C9 and CYP3A4 [51]. The formation of fluconazole β-D glucuronide is catalyzed by the uridine diphosphate glucuronosyltransferase (UGT) isoform UGT2B7 [52].

Voriconazole

Less than 2% of a voriconazole dose is excreted unchanged in the urine [25, 53]. Voriconazole is extensively metabolized to eight metabolites by hepatic CYP enzymes [54]. The CYP metabolism of voriconazole involves CYP2C19, CYP3A4, and CYP2C9, but not CYP3A5 [25, 53]. CYP2C19 and CYP2C9 also exhibit genetic polymorphisms. CYP2C19, the primary CYP enzyme involved in voriconazole metabolism, has eight variant alleles which, if expressed, manifest as a poor metabolizing (PM) phenotype. The PM phenotype is expressed among Pacific Islanders and less frequently among Asians, Caucasian, and African American populations [53, 55]. Drug exposure is increased nearly a fourfold in the CYP2C19 homozygous PM phenotype compared to the homozygous efficient metabolizing (EM) phenotype. Furthermore, drug exposure is nearly double in the CYP2C19 heterozygous EM phenotype compared to the homozygous EM phenotype [25].

CYP2C9 metabolism has 32 known variant alleles, of which 18 are associated with no or reduced enzyme activity [56]. The frequencies of the CYP2C9 variant alleles differ between racial/ethnic groups [55, 57, 58]. CYP3A4 expression varies widely and likely contributes to interindividual variability in voriconazole pharmacokinetics. In addition to CYP, enzymes of the flavin-containing monooxygenase (FMO) family (primarily FMO1 and FMO3) and UGT1A4 also catalyze the formation of voriconazole metabolites [8, 58, 59].

Posaconazole

Posaconazole is primarily eliminated unchanged in the feces (77%) and urine (13%) [60]. Only 17% of a dose undergoes biotransformation, of which little (2%) is metabolized by CYP3A4 [60, 61]. Most metabolites are glucuronide conjugates formed by UGT pathways [61]. The primary posaconazole metabolite is formed by UGT1A4 [61].

Isavuconazole

Very little isavuconazole (<1%) is excreted unchanged in the urine. Active isavuconazole is extensively metabolized by CYP3A4/5 and undergoes further modification by enzymes in the UGT pathway prior to being excreted in the feces (46%) and bile (46%) [46, 62, 63]. Preclinical in vitro studies indicate that isavuconazole is not a substrate of CYP2B6 and CYP2C9 [46, 64]. However, isavuconazole displays mild CYP2B6 induction effects [64].

11.1.4 Echinocandin Pharmacology

The echinocandins, caspofungin, micafungin, and anidulafungin, disrupt cell wall synthesis by inhibiting 1, 3,-β-D-glucan synthase. These compounds are large molecular weight semisynthetic lipopeptides that are administered intravenously [65].

11.1.4.1 Distribution of the Echinocandins

Caspofungin binds extensively to albumin and has multiphasic distribution. The drug first distributes to plasma and extracellular fluid [66]. Caspofungin rapidly and reversibly binds to the surface of hepatocytes. Then, slowly via active transport by organic anion transport protein 1B1 (OATP1B1), it distributes into the liver and, to a lesser extent, other tissues [66, 67]. This slow active transport influences the caspofungin $t_{1/2}$ [67]. Micafungin binds extensively (>99%) to albumin and, to a lesser extent, α1-acid glycoprotein. In vitro data suggest that like caspofungin, the hepatic uptake of micafungin involves transport proteins, specifically, Na+ -taurocholate co-transporting polypeptide (NTCP) [68]. In addition, hepatocyte uptake may also involve a transporter from the OATP family [68]. Micafungin is not a P-gp substrate. Anidulafungin distribution in humans is not fully understood. Compared to other echinocandins, anidulafungin is less protein bound, has a larger volume of distribution, and achieves lower peak (C_{max}) serum concentrations [69].

11.1.4.2 Metabolism/Elimination of the Echinocandins

The echinocandins exhibit linear pharmacokinetics. However, echinocandins differ in how they are metabolized or degraded but are not appreciably metabolized by CYP. Less than 2% of a caspofungin dose is excreted unchanged in the urine [70]. Caspofungin is slowly degraded in the liver via N-acetylation and peptide hydrolysis to several metabolites, which are excreted in the bile and feces [71]. Less than 1% of a micafungin dose is eliminated unchanged in the urine, with the majority (90%) undergoing biliary excretion [72]. Following micafungin hepatic uptake via NTCP, micafungin is metabolized to several metabolites that are formed by arylsulfatase, catechol-O-methyltransferase, and, to a minor extent, ω-1 hydroxylation via CYP, which undergo biliary elimination with the parent compound [69, 72, 73]. Data suggest the canalicular membrane efflux transporter, BSEP, is an important mediator of micafungin biliary excretion [68]. Less than 10% of an anidulafungin dose is excreted in the feces or urine as unchanged drug [74, 75]. Anidulafungin is not hepatically metabolized. Rather, in the plasma it undergoes slow nonenzymatic chemical degradation to a peptide breakdown product [74, 75].

11.2 Drug Interaction Potential of Antifungal Agents

The potential for drug interactions involving amphotericin B formulations is related to its associated nephrotoxicity and whether the concomitant medication is eliminated renally or shares other toxicities. This potential is high when amphotericin B formulations are administered with other nephrotoxic or renally eliminated medications. Because 5-FC is renally eliminated and used with amphotericin B, its potential for an interaction is high. The azoles cause drug interactions at various sites (intestine, liver, blood brain barrier, kidneys, etc.) via several mechanisms (alterations in pH, interference with transport proteins, and oxidative or conjugative enzymatic drug metabolism processes). Many of the azole-drug interactions occur class-wide; thus, the potential for azoles to cause an interaction is high. The echinocandins are relatively devoid of clinically significant drug interactions. However, in vitro echinocandins have demonstrated varying inhibitory potential across a variety of drug transport proteins [18].

11.3 Amphotericin B

Drug interactions involving amphotericin B formulations are summarized in Table 11.1.

11.3.1 Amphotericin B Interactions Involving Synergistic/ Additive Nephrotoxicity

Amphotericin B is commonly used in patients who are severely immunocompromised and/or at high risk for renal failure and electrolyte disturbances. Amphotericin B causes nephrotoxicity via direct cytotoxicity to the renal tubules, which impairs proximal and distal reabsorption of electrolytes. Indirectly it reduces renal blood flow that causes ischemic damage and reduces glomerular filtration [1]. Significant amphotericin B-drug interactions involve concomitantly administered nephrotoxic and/or renally eliminated drugs with a narrow therapeutic index (i.e., aminoglycosides, cyclosporine, foscarnet, tacrolimus, etc.). These interactions produce additive or synergistic nephrotoxicity or result in the accumulation of renally eliminated drugs (i.e., 5-FC) to toxic concentrations that produce secondary extrarenal adverse effects (Table 11.1).

Table 11.1 Drug interactions caused by amphotericin B formulations

Interaction	Drugs	Comments
Additive/synergistic effects		
Direct or indirect nephrotoxicity[a]	Cyclosporine Tacrolimus Aminoglycosides	Monitor Scr, BUN, electrolytes, consider lipid amphotericin B formulations or other antifungal agents
Fluid and electrolyte disturbance (i.e., water retention, hypokalemia, hypomagnesemia)	Thiazide and loop diuretics aminoglycosides, corticosteroids	Monitor Scr, BUN, electrolytes. Supplement electrolytes as needed
Secondary non-renal toxicity		
Myelosuppression	5-Flucytosine (5-FC)	Effect due to diminished renal clearance of 5-FC secondary to amphotericin B-associated nephrotoxicity
Miscellaneous electrolyte disturbances		
Increase cardiac automaticity and inhibition of Na^+-K^+ ATPase pump	Digoxin	Effects secondary to amphotericin B-induced hypokalemia

[a]Any nephrotoxic drug that affects afferent and/or efferent arterioles can potentiate amphotericin B nephrotoxicity

11.3.1.1 Clinical Importance of Amphotericin B Interactions Involving Concomitantly Administered Nephrotoxic Drugs

In addition to increasing the risk of nephrotoxicity or extrarenal toxicities, amphotericin B interactions with concomitantly administered nephrotoxic drugs further complicate the use of additional renally eliminated medications. This often compels clinicians to switch to alternative drug therapies and/or empirically reduce medication doses.

11.3.1.2 Management of Amphotericin B Interactions with Concomitantly Administered Nephrotoxic Drugs

Amphotericin B-drug interactions are somewhat unavoidable and should be managed by limiting the risk or severity of these interactions. For example, although the lipid amphotericin B formulations may also cause nephrotoxicity with concomitant nephrotoxic drugs, they should be used in patients with or at high risk for nephrotoxicity [76, 77]. Depending upon the case, other intravenous, non-nephrotoxic antifungal agents (i.e., caspofungin, fluconazole, isavuconazole) should also be considered. Both voriconazole and posaconazole are available intravenously, but each contain SBECD, which can accumulate in patients with diminished renal function [34]. Although data suggest that accumulation of SBECD does not increase the risk of acute kidney injury at human doses, animal studies have demonstrated renal vacuolization, which has led labeling limitations in patients with a creatinine clearance less 50 mL/min [11, 12, 34]. However, SBECD is readily removed by hemodialysis and CRRT, but accumulation may still occur with repeated dosing, and enteral formulations are still preferred in these patients. In a typical 70 kg patient, maintenance doses of intravenous posaconazole contain about one third the amount of SBECD compared to voriconazole and may be more advantageous in patients with a creatinine clearance of 30 mL/min or more [34, 35, 78]. Intravenous isavuconazole is devoid of any solubilizing agent, and its pharmacokinetics were not influenced by renal dysfunction which may make it the most ideal intravenous anti-mold agent in patients with renal dysfunction [13, 14, 79].

11.3.2 Amphotericin B Interactions Involving Renally Eliminated, Narrow Therapeutic Index Drugs with Extrarenal Toxicity Including 5-FC

Amphotericin B reduces the CL_R of renally eliminated drugs with a narrow therapeutic index causing their accumulation and increasing the risk of extrarenal toxicities. For example, amphotericin B-associated nephrotoxicity can cause accumulation of 5-FC, leading to myelosuppression, hepatic necrosis, and diarrhea associated with its elevated plasma concentrations that occur with reduced renal function [5]. In addition, amphotericin B and 5-FC therapy can augment the myelosuppressive or cytotoxic effects of other drugs (i.e., zidovudine, ganciclovir) patients may be receiving. Often the clinical importance of the amphotericin B and 5-FC interaction is often outweighed by the efficacy of this combination in the treatment of cryptococcal meningitis [80].

11.3.2.1 Management of Amphotericin B Interactions with Renally Eliminated, Narrow Therapeutic Index Drugs with Extrarenal Toxicity

Often, in the treatment of cryptococcal meningitis, concomitant 5-FC therapy is unavoidable, so renal function and 5-FC blood concentrations should be monitored. Consensus guidelines for the treatment of cryptococcosis support the use of 5-FC in combination with amphotericin B with close monitoring for myelosuppression [80]. Ideally, 5-FC blood concentrations should be maintained between 25 and 100 µg/mL [81]. Several 5-FC dosing nomograms for patients with renal dysfunction exist, but they should not be used unless the renal dysfunction is chronic in nature and only used cautiously in elderly patients [82].

11.4 Systemic Azoles

11.4.1 Interaction Mechanisms

Drug interactions involving the systemic azoles primarily affect the pharmacokinetic processes of the concomitantly administered drug(s). However, in select circumstances the systemic azole or both drugs can be affected. As discussed above, all systemic azoles undergo oxidative CYP-mediated metabolism in the liver, and several also undergo significant CYP metabolism in the intestine. Itraconazole, voriconazole, posaconazole, and the active isavuconazole being more lipophilic are more extensively metabolized to polar metabolites than fluconazole. Thus, they possess a greater potential for CYP-mediated interactions [8].

11.4.1.1 Interactions Affecting Solubility and Absorption (pH Interactions)

Drug dissolution rate determines the intestinal lumen drug concentration available for absorption [83]. Therefore, intraluminal pH indirectly affects absorption. Weakly basic drugs such as itraconazole and posaconazole suspension dissolve more slowly at higher pH, whereas weakly acidic drugs dissolve faster at higher pH. The delayed release formulation of posaconazole is formulated to dissolve in the higher pH environments of the intestines [37]. Itraconazole and posaconazole are highly lipophilic weak bases with high pKa values, and their dissolution and subsequent absorption are optimal at pH 1–4 [84, 85]. Fluconazole and voriconazole are also weak bases, with lower pKa values, and thus their dissolution is unaffected by increases in gastric pH [8, 86, 87]. Likewise, isavuconazonium sulfate is not affected by alterations in gastric pH. Formulated as a triazolium salt linked to an aminocarboxyl moiety, it is a highly water-soluble prodrug that is stable at low pH (pH 1–4) which allows it to be solubilized in the gastrointestinal environment [13, 88].

11.4.1.2 Interactions Affecting CYP-Mediated Biotransformation

As described above, to varying degrees all azoles are CYP substrates. Moreover, the excretion of itraconazole, voriconazole, posaconazole, and isavuconazole from the body in the feces or bile requires their extensive conversion to hydrophilic metabolites. In contrast, fluconazole, being more hydrophilic, requires less biotransformation and is primarily eliminated unchanged in urine [8]. All the azoles are CYP inhibitors, but their affinities for specific isoforms differ. As CYP inhibitors, the systemic azoles generally exhibit rapidly reversible binding [89]. As reviewed in Chap. 2, this type of binding to CYP by an inhibitor or its metabolite results in either competitive or noncompetitive inhibition [89, 90]. The azoles, particularly itraconazole, primarily exert competitive inhibition, but fluconazole, voriconazole, and posaconazole also demonstrate noncompetitive or mixed-type inhibition of CYP [89–92]. Whether isavuconazole also exhibits noncompetitive or mixed-type inhibition has not been described.

As discussed in Chap. 2, isoforms of the CYP3A subfamily are the most abundant CYPs in the human liver and intestine and catalyze the metabolism of numerous xenobiotics. However, CYP3A5, which is expressed polymorphically in the intestine and liver, may contribute substantially to overall CYP3A activity [93]. CYP3A5 is present in more than 33% of the population [94]. All the systemic azoles inhibit CYP3A4, which is extensively expressed in the liver and intestine [8]. In general, all triazoles are weak inhibitors of CYP3A5 [8, 63, 94]. CYP3A5 catalyzes the biotransformation of isavuconazole but not other azoles [8, 46]. Therefore, the polymorphic expression of CYP3A5 may contribute to the observed interindividual variability in inhibition of CYP3A-mediated metabolism of other drugs [93–96].

The itraconazole stereoisomers inhibit CYP3A4 and weakly inhibit CYP3A5 [49, 97]. The three itraconazole metabolites circulate at concentrations sufficient to inhibit CYP3A4 and contribute to drug interactions involving itraconazole [49, 97]. Following multiple itraconazole dosing, 40–50% of overall CYP3A inhibition is attributed to hydroxyitraconazole (major) and N-desalkyl-itraconazole (minor) [49]. Fluconazole undergoes minimal CYP-mediated metabolism; it inhibits CYP3A4, albeit much more weakly than other systemic azoles [98]. Fluconazole inhibition of CYP3A5 is also much weaker than its effects on CYP3A4 [93]. However, fluconazole is a comparatively stronger inhibitor of several other isoforms (i.e., CYP2C9 and CYP2C19) [98]. Fluconazole binds noncompetitively to CYP, and in vivo it circulates largely as free drug. Given this, greater CYP inhibition may occur with higher fluconazole doses. Even though only a small percentage of fluconazole undergoes CYP-mediated metabolism, this percentage may greatly increase in the presence of a potent CYP inducer. Fluconazole also interacts with conjugative enzymes involved in glucuronidation and is a moderate inhibitor of UGT1A1 and UGT2B7 [24, 99]. Voriconazole is a potent competitive inhibitor of CYP2B6, CYP2C9, and CYP2C19 [92, 100]. In addition voriconazole is a potent competitive and noncompetitive CYP3A4 inhibitor [92, 100]. Therefore, it has the potential to interact with many medicines. Voriconazole inhibits CYP3A5 much more weakly than it does CYP3A4 [93]. Although very little posaconazole is metabolized by CYP, it is a moderate inhibitor of CYP3A4. However, compared to other systemic azoles, posaconazole inhibits CYP3A4 less significantly. Additionally it has no effect on the activity of CYP2C8/9, CYP1A2, CYP2D6, or CYP2E1 [101]. In vitro, preclinical studies using pooled human liver microsomes and phase I healthy volunteer data suggest that isavuconazole is a moderate inhibitor of CYP3A4 [46, 64]. Additional phase I studies in healthy volunteers did not demonstrate substantial inhibitory or induction effects on CYP1A2, CYP2C8, CYP2D6, or CYP2C19 [64]. The preclinical in vitro data demonstrate that isavuconazole may be an inducer of CYP3A4/5, CYP2B6, CYP2C8, and CYP2C9, but not CYP2C19; however, in healthy volunteers isavuconazole only demonstrated clinically significant induction on CYP2B6. Furthermore, isavuconazole only minimally inhibited warfarin and influenced R-warfarin more so than S-warfarin indicating minimal inhibitory effects on CYP2C9 [102]. Preclinical in vitro studies in human liver microsomes also demonstrate that isavuconazole is a weak inhibitor of UGT1A1, UGT1A9, and UGT2B7 as well [18]. The clinical impact of these interactions needs further study.

11.4.1.3   Interactions Affecting P-glycoprotein (P-gp)-Mediated Efflux and Other Transporters

As discussed in Chap. 3, transport proteins are important determinants of drug disposition and effects. In particular, transporters of the ABC transporter protein family are involved in unidirectional efflux of drugs across cellular membranes in vital organs or at blood-tissue barriers in the body [103]. For example, efflux transporters, including P-gp, BCRP, and MRP2, are localized to the apical membrane of enterocytes and mediate the bioavailability of orally administered substrates [103]. Similarly when drugs reach the basolateral membrane of hepatocytes, uptake transporters including OATP1B1 and organic cation transporters (OCT) like OCT1 mediate uptake into hepatocytes [103]. Efflux transporters including P-gp, MRP2, and BSEP are located in the canalicular membrane of hepatocytes and mediate transport of drugs and/or their metabolites into bile [103]. Uptake and transport proteins are expressed in the kidney and mediate renal drug clearance [103]. Finally, at blood-tissue barriers of the brain and placenta, uptake and transport including

OATP1A2, OATP2B1, P-gp, BCRP and several isoforms of MRP mediate distribution to sensitive tissues and organs [103].

The systemic azoles vary in how they interact with various transport proteins, but such interactions can be somewhat predicted based upon chemical structure and physicochemical properties. Itraconazole, posaconazole, and isavuconazole have comparable chemical structures and physicochemical properties, and all possess similar potentials to inhibit ABC transporters [18]. In contrast, the structural analogues fluconazole and voriconazole do not inhibit the ABC transporters to a significant extent [18].

Itraconazole is a substrate and a strong inhibitor of P-gp [15, 17, 18, 104, 105]. In addition it is a potent and strong inhibitor of BCRP and BSEP [16, 18]. In vitro, posaconazole is a substrate and strong inhibitor of P-gp and a strong inhibitor of BCRP at high concentrations [18, 60]. Isavuconazole is not a substrate of P-gp or other major transport proteins, but to varying degrees, it can act as an inhibitor of certain transporters [106]. In vitro data indicate that isavuconazole strongly inhibits P-gp and BCRP [18]. However, data using probe substrates in healthy volunteers indicate isavuconazole weakly inhibits P-gp, OCT1, OCT2, and multidrug and toxin extrusion protein-1 (MATE1) [106]. In vivo confirmation of in vitro inhibition of BCRP has not been adequately demonstrated due to the lack of a sensitive probe substrate for BCRP [106]. In vivo studies also indicate that isavuconazole does not inhibit OATP1B1 [106]. The clinical impact of interactions between isavuconazole and transport proteins needs further study. Fluconazole may be a P-gp substrate, but it is not a P-gp inhibitor, and voriconazole is neither a substrate nor an inhibitor of P-gp [16–18, 104, 105, 107]. In addition, it neither inhibits other ABC transporters involved in drug disposition (e.g., MRP1–5, BCRP, and BSEP) [18].

11.5 Drug Interactions Involving Itraconazole

Drug interactions involving itraconazole are summarized in Tables 11.2, 11.3, and 11.4.

11.5.1 Itraconazole Interactions Involving Gastric pH

Based on the physicochemical properties of itraconazole, the dissolution and subsequent absorption of its capsule form depend on gastric pH, retention time, and the fat content of a meal and are optimal in acidic gastric conditions [10, 84, 149]. Increased gastric pH does not affect the absorption of the oral solution [23]. H_2-receptor antagonists, proton pump inhibitors, and antacids reduce the exposure, C_{max}, or oral availability of the itraconazole capsule up to 67% [23, 85, 138–140, 150].

Table 11.2 Itraconazole interactions affecting CYP-mediated biotransformation of other drugs

Drug	Effect on drug (% change)	Inhibition site	Reference(s)
"Statins"			
Simvastatin	↑C_{max} (175%); ↑$AUC_{(0-\infty)}$ (417%); ↑$t_{1/2}$ (25%)	Hepatic CYP3A; perhaps intestinal CYP3A or P-gp	[108]
Atorvastatin	↑C_{max} (20–38%); ↑$AUC_{(0-\infty)}$ (150–231%); ↑$t_{1/2}$ (29–190%)	Hepatic CYP3A; perhaps intestinal CYP3A or P-gp	[109, 110]
Fluvastatin	None		[149]
Pravastatin			[108, 109]
Rosuvastatin			[111]
Pitavastatin			[112]
Benzodiazepines			
Midazolam (oral)		Hepatic CYP3A; perhaps intestinal CYP3A	[113]
+ ITZ day 1	↑C_{max} (75%); ↑$AUC_{(0-\infty)}$ (242%); ↑$t_{1/2}$ (104%)		
+ ITZ day 6	↑C_{max} (151%); ↑$AUC_{(0-\infty)}$ (564%); ↑$t_{1/2}$ (259%)		
Midazolam (i.v.)		Hepatic CYP3A	[113]
+ ITZ day 4	↓CL (69%); ↑$t_{1/2}$ (141%)		
Triazolam	↑C_{max} (41–76%); ↑T_{max} (11–94%);	Hepatic CYP3A; perhaps intestinal CYP3A	[114]
	↑$AUC_{(0-\infty)}$ (210–348%); ↑ $t_{1/2}$ (155–210%)		
Diazepam	↑$AUC_{(0-\infty)}$ (31.8%); ↑$t_{1/2}$ (34%)	Hepatic CYP3A	[115]
Estazolam			
Bromazepam			
Temazepam	None		[116–118]
Oxazepam			
Other anxiolytics, sedatives, and hypnotics			
Buspirone	↑ C_{max} (1240%); ↑$AUC_{(0-\infty)}$ (1815%),	Hepatic CYP3A; perhaps intestinal CYP3A	[119]
Zolpidem	None		[120, 121]
Antipsychotic agents			
Haloperidol	↑C_{max} (14%); ↑$AUC_{(0-\infty)}$ (82%); ↑$t_{1/2}$ (115%); ↓CL/F (33%)	Hepatic CYP3A	[122, 123]
Clozapine	None		[124]
Calcineurin inhibitors			
Cyclosporine	↑C_{ss} (80%) (range 24–149%)	Hepatic CYP3A; perhaps intestinal CYP3A/P-gp	[125]
Tacrolimus	↑C_{ss} (83%) (range 49–117%)	Hepatic CYP3A; perhaps intestinal CYP3A/P-gp	[125]
Corticosteroids			
Methylprednisolone			
Oral	↑C_{max} (57–87%); ↑$AUC_{(0-\infty)}$ (279%); ↑$t_{1/2}$ (71–132%);	Hepatic and intestinal CYP3A4/P-gp	[126, 127]

(continued)

Table 11.2   (continued)

Drug	Effect on drug (% change)	Inhibition site	Reference(s)
i.v.	$\uparrow AUC_{(0-\infty)}$ (143%); $\uparrow t_{1/2}$ (129%): $\downarrow CL$ (62%); $\downarrow Vd$ (15%)	Hepatic CYP; perhaps biliary P-gp	[128]
Dexamethasone		Primarily hepatic CYP3A4	[129]
Oral	$\uparrow AUC_{(0-\infty)}$ (269%); $\uparrow C_{max}$ (58%); $\uparrow t_{1/2}$ (172%); F (14.7%)		
i.v.	$\uparrow AUC_{(0-\infty)}$ (223%); $\uparrow t_{1/2}$ (197%); $\downarrow CL$ (69%)		
Prednisolone	$\uparrow C_{max}$ (2–14%); $\uparrow t_{1/2}$ (14–29%); $\uparrow AUC_{(0-\infty)}$ (24%)	Hepatic and intestinal CYP3A4	[127, 130]
Budesonide (inh)	$\uparrow C_{max}$ (64%); $\uparrow AUC_{(0-\infty)}$ (321%); $\uparrow t_{1/2}$ (287%); $\uparrow T_{max}$ (150%)	Hepatic and intestinal CYP3A4	[131]
Fluticasone (inh)	Increased plasma fluticasone levels by 2.5-fold	Hepatic CYP3A4; perhaps intestinal CYP3A4	[132]
Calcium channel blockers			
Felodipine	$\uparrow C_{max}$ (675%); $\uparrow AUC_{(0-\infty)}$ (534%); $\uparrow t_{1/2}$ (71%)	Hepatic and intestinal CYP3A4	[133]
Miscellaneous			
Oxybutynin	$\uparrow C_{max}$ (89%); $\uparrow AUC_{(0-t)}$ (85%)	Hepatic and intestinal CYP3A4	[134]
Busulfan	$\uparrow C_{ss}$ (25%); $\uparrow CL/F$ (20%)	Hepatic CYP3A; perhaps intestinal CYP3A4	[135]
Meloxicam	$\downarrow AUC_{(0-72)}$ (37%); $\downarrow C_{max}$ (63%); $\uparrow t_{1/2}$ (55%); $\uparrow T_{max}$ (500%)	Perhaps intestinal CYP3A4 and intestinal transport	[136]
Oxycodone		Hepatic and intestinal CYP3A4	[137]
Oral	$\uparrow C_{max}$ (43%); $\uparrow AUC_{(0-\infty)}$ (125%); $\uparrow t_{1/2}$ (48%); $\downarrow CL/F$ (58%); $\uparrow F$ (49%)		
i.v.	$\downarrow CL$ (33%); $\uparrow AUC^{(0-\infty)}$ (51%); $\uparrow t^{1/2}$ (44%)		

11.5.1.1   Clinical Importance of Itraconazole Interactions Involving Gastric pH Interactions

Reduced itraconazole absorption can lead to therapeutic failure. However, elevated gastric pH is unavoidable in patients who require high-dose corticosteroid therapy (i.e., transplant recipients) [139]. In these instances the oral solution may be preferred.

Table 11.3 Interactions that induce itraconazole biotransformation or inhibit its absorption

Drug	Effect on itraconazole (% change)	Mechanism	Reference(s)
Gastric pH			
Modifiers			
Famotidine	$\downarrow C_{max}$ (30–52%); $\downarrow C_{min}$ (35%); $AUC_{(0-48)}$(51%)	↑ gastric pH and ↓ absorption	[138, 139]
Omeprazole			
Itraconazole caps	$\downarrow C_{max}$ (67%); $\downarrow AUC_{(0-24)}$ (65%); $\uparrow T_{max}$ (27%)	↑ gastric pH and ↓ absorption	[140]
Itraconazole soln	None		[23]
Didanosine			
Enteric-coated formulation	None		[87]
Inducing agents			
Phenobarbital	Subtherapeutic serum concentrations	CYP3A induction	[141]
Carbamazepine	Undetectable serum concentrations	CYP3A induction	[141]
Phenytoin	$\downarrow C_{max}$ (83%); $\downarrow AUC_{(0-\infty)}$ (93%); $\downarrow T_{1/2}$ (82%); $\uparrow CL/F$ (1384%)	Hepatic and intestinal CYP3A induction	[142]
Rifampin	$\downarrow C_{max}$ (67%); $\downarrow AUC_{(0-24)}$ (67%); $\downarrow T_{max}$ (35%)	CYP3A induction	[143]
Nevirapine	$\downarrow C^{max}$ (38%); $\downarrow AUC^{(0-96)}$ (61%); $\downarrow AUC^{(0-\infty)}$ (62%); $\downarrow t^{1/2}$ (31%)	Hepatic CYP3A4 induction;P-gp induction	[144]

Table 11.4 Itraconazole interactions affecting P-gp-mediated transport of other drugs

Drug	Effect on drug (% Change)	Inhibition Site	Reference(s)
Cardiac			
Glycoside			
Digoxin	$\uparrow C_{max}$ (31%); $\uparrow AUC_{(0-\infty)}$ (68%); $\uparrow t_{1/2}$ (38%); $\downarrow CL_R$ (20%)	Renal P-gp; possibly hepatic/biliary P-gp	[145]
Alkaloids			
Quinidine	$\uparrow C_{max}$ (32–59%); $\uparrow t_{1/2}$ (35–67%); $\uparrow AUC_{(0-\infty)}$ (142%); $\uparrow T_{max}$ (150%);$\downarrow CL_R$ (49–60%)		
Quinidine metabolites		Inhibition of hepatic CYP3A; renal P-gp	[146, 147]
3-hydroxyquinidine	$\downarrow AUC_{(0-24)}$ (78%); $\downarrow CL_{(partial)}$ (84%)		
N-oxide	$\downarrow CL_{(partial)}$ (73%)		
Miscellaneous drugs			
Morphine	$\uparrow C^{max}$ (25%); $\uparrow AUC^{(0-9)}$ (27%);$\uparrow AUC^{(0-48)}$ (19%)	Inhibition of intestinal P-gp	[148]

11.5.1.2 Management of Itraconazole Interactions Involving Gastric pH Interactions

In patients requiring acid suppression therapy and short courses of itraconazole, the solution should be employed. The solution is somewhat dilute (20 mL per dose) and not very palatable which may be impractical for protracted courses of therapy. In such cases, alternative antifungal agents should be considered. If no suitable alternative agent exists, the itraconazole capsule can be used with routine therapeutic drug monitoring to document adequate oral availability [10].

11.5.2 Itraconazole Interactions Affecting CYP-Mediated Biotransformation of Other Drugs

11.5.2.1 The 3-Hydroxy-3 Methylglutaryl (HMG) Coenzyme: A Reductase Inhibitor (The "Statins")

Following oral administration the systemic availability of the statins is determined by a series of interactions with CYP (phase I) and conjugative (phase II) enzymes as well as ABC transport and export proteins [151]. Simvastatin and atorvastatin are metabolized by CYP3A4; fluvastatin is primarily metabolized by CYP2C9, with CYP3A4 and CYP2C8 marginally contributing [149, 152]. Although pitavastatin is a substrate of CYP2C9 and CYP2C8, it is minimally metabolized by these isoforms and instead undergoes lactonization via UGTA1/1A3 [152]. Likewise, pravastatin and rosuvastatin are negligibly metabolized by CYP and excreted primarily in the urine as unchanged drug [108, 111]. All statins are substrates of OATP1B1, which mediates hepatocyte uptake from the portal circulation [151, 152]. In addition, most statins are P-gp substrates, which modulate absorption from the intestine into the portal circulation [151]. Lastly, all stains have affinity for MRP2, BSEP, and BCRP [151]. Itraconazole inhibits P-gp, BSEP, BCRP, and CYP3A4; thus, attributing its interactions with certain statins solely to CYP3A4 inhibition is difficult [15–18, 104, 107].

Itraconazole co-administration with certain statins can elevate their systemic concentrations, which may result in rare, but severe, life-threatening toxicities [109, 153]. Itraconazole significantly increases the exposure and C_{max} of simvastatin, but it affects atorvastatin pharmacokinetics considerably less [108–110]. In contrast, itraconazole has no significant effect on fluvastatin or pravastatin [108, 109, 149]. Although the effects were not significant, itraconazole produced modest increases in rosuvastatin plasma concentrations and slightly reduced pitavastatin C_{max} and AUC [111, 112]. Nearly all studies identifying an interaction between itraconazole and a statin were performed before the role of transport proteins in statin disposition was fully realized. Nonetheless, itraconazole is not an OATP1B1 inhibitor; thus, its interaction with simvastatin results from inhibition of intestinal and hepatic CYP3A4, whereas its interaction with atorvastatin also may involve BCRP [108,

154, 155]. Neither pravastatin nor pitavastatin is a CYP3A4 substrate, but both are P-gp substrates; thus, the lack of a significant interaction with these statins also suggests itraconazole's effects on simvastatin, and atorvastatin is primarily a result of CYP3A4 inhibition [108, 112, 151, 152]. Rosuvastatin is neither a substrate of P-gp nor CYP3A4; thus, its modest interaction with itraconazole may reflect inhibition of intestinal BCRP [18, 151, 152, 155].

11.5.2.2   Benzodiazepines

Itraconazole co-administration with triazolam, midazolam, or diazepam produces significant pharmacokinetic interactions that enhance their pharmacologic effects [113, 114, 116–118]. The most notable alterations are observed with triazolam and midazolam, which are metabolized only by CYP3A4 [113, 114]. Following oral co-administration with itraconazole, the interaction increases both the triazolam and midazolam systemic availability and decreases their clearance (CL), leading to significant changes in exposure, C_{max}, T_{max}, and $t_{1/2}$ [113, 114]. Itraconazole does not affect the steady-state volume of distribution (V_{ss}) of i.v. midazolam, but it substantially reduces its plasma CL as reflected by a prolongation in $t_{1/2}$ [113].The effect of itraconazole on the CL of triazolam cannot be determined due to the lack of an i.v. formulation for this benzodiazepine [114]. The itraconazole-triazolam interaction occurs even if triazolam is administered up to 24 h after itraconazole and can persist for several days after discontinuing the azole [114]. This persistence is due to the itraconazole metabolites [47, 49, 156]. The N-desalkyl-itraconazole metabolite has a much longer half-life than the other metabolites or itraconazole, and along with the hydroxyl metabolite, it contributes substantially to CYP3A4 inhibition [49, 156].

The interaction with either benzodiazepines can occur with a single or multiple itraconazole doses and produces long-lasting pharmacological effects, including prolonged amnesia, significantly reduced psychomotor performance, and severe sedation [113, 114]. Given the nonlinear stereoselective sequential itraconazole metabolism and the prolonged elimination of it and its metabolites, the interaction will likely be greater and more prolonged with repeated or increased itraconazole doses [113, 114, 156]. Additionally, the variability in benzodiazepine interactions may be explained in part by the CYP3A5 genotype [94, 95]. As discussed previously, unlike many CYP3A4 substrates, of the azoles only isavuconazole is a CYP3A5 substrate, and the other azoles only weakly inhibit this isoform [93–96]. Also, in some individuals, CYP3A5 may represent the majority of hepatic CYP3A activity [94, 95]. Thus, in homozygous expressers of the wild-type genotype (CYP3A5*1/*1), the magnitude of itraconazole-midazolam interaction is less because the CYP3A5 pathway can compensate for the lack of midazolam metabolism caused by the itraconazole-mediated inhibition of the CYP3A4 pathway [94]. However, among homozygous (CYP3A5*3/*3) or heterozygous (CYP3A5*1/*3) expressers of the variant-type genotype, CYP3A5 activity is lacking or diminished, respectively, and they have relatively more CYP3A4 activity than homozygous expressers of the wild-type genotype [93–96]. Thus, in those expressing the variant-type genotype, the magnitude of interaction will be greater because there is less CYP3A5 to compensate for the lack of midazolam metabolism caused by the azole-mediated inhibition of the CYP3A4 pathway [94].

Diazepam undergoes minimal first-pass metabolism and is primarily metabolized by CYP2C19 and CYP3A4 [115, 157]. Concomitant itraconazole produces a small yet statistically significant increase in diazepam exposure and slightly prolongs its $t_{1/2}$, but it does not enhance the pharmacological effects of this benzodiazepine [115]. Estazolam is a short-acting triazolobenzodiazepine derivative that is extensively metabolized by CYP3A4 [117]. Itraconazole inhibits estazolam metabolism in vitro, but clinically its co-administration did not alter the pharmacokinetics or enhance the effects of estazolam [117, 158]. The lack of interaction in vivo may have resulted from using a low dose (100 mg/day) of itraconazole in the interaction study [117, 158]. Itraconazole does not affect the pharmacokinetics or enhance the effects of benzodiazepines that are not appreciably metabolized by CYP3A4 (i.e., bromazepam, temazepam, oxazepam) [116, 118].

11.5.2.3   Other Anxiolytics, Sedatives, and Hypnotics

Buspirone undergoes extensive first-pass metabolism via CYP3A4 [119]. Itraconazole co-administration significantly increases buspirone exposure, C_{max}, which moderately enhances its pharmacological effects [119]. However, the interaction does not alter the buspirone $t_{1/2}$, which suggests the interaction involves intestinal CYP3A4 inhibition [119]. The imidazopyridine hypnotic agent zolpidem is a substrate of CYP3A4 and to a lesser extent CYP1A2 [120]. It undergoes minimal first-pass metabolism and possesses good oral availability. While co-administered, itraconazole slightly reduces zolpidem clearance, but it does not appreciably affect zolpidem pharmacodynamics [120, 121].

11.5.2.4   Antipsychotic Agents

Haloperidol undergoes first-pass metabolism but has good oral availability. The drug is hepatically metabolized by CYP2D6 and CYP3A4 [122]. Several CYP2D6 alleles (CYP2D6*3, *4, and *5) significantly influence haloperidol disposition, while others (i.e., CYP2D6*10) only moderately influence haloperidol disposition [122]. Itraconazole significantly increases plasma concentrations of haloperidol and its metabolite (reduced haloperidol) and augments its neurological side effects [122, 123]. The CYP3A4 inhibition by itraconazole enhances the contribution of minor CYP2D6 alleles to haloperidol metabolism [122]. CYP3A4 has only a minor influence on the disposition of the atypical antipsychotic agent clozapine; thus its pharmacokinetics or activity is unaffected by itraconazole co-administration [124].

11.5.2.5 Calcineurin Inhibitors and mTOR Inhibitors

Pharmacokinetic interactions between tacrolimus or cyclosporine and the azoles are well known. Regardless of route of administration, itraconazole increases cyclosporine concentrations 40–200% and "trough" (C_{min}) tacrolimus concentrations up to sevenfold [97, 125, 159–162]. As described above with midazolam, the variability in the itraconazole-tacrolimus interaction may also be explained in part by the CYP3A5 genotype. A small study in Japanese allogeneic hematopoietic stem cell transplant recipients demonstrated that orally administered tacrolimus concentrations were largely unchanged in a homozygous expresser of the wild-type genotype (CYP3A5*1/*1) during the co-administration of itraconazole. In contrast, among patients with the variant CYP3A5*3/*3 or CYP3A5*1/*3 alleles, tacrolimus concentrations increased significantly after the start of itraconazole co-administration [97]. This study also comparatively demonstrated that interaction between oral cyclosporine and itraconazole is not influenced by CYP3A5 polymorphisms [97]. The interaction between itraconazole and the calcineurin inhibitors persists due to the itraconazole metabolites [49, 156]. Itraconazole has been reported to interact with sirolimus in several cases [163, 164]. An anecdotal observation from a large population pharmacokinetic analysis of everolimus demonstrated that itraconazole co-administration in a single patient reduced everolimus clearance 74% [165].

11.5.2.6 Corticosteroids

Itraconazole inhibits the metabolism of oral or i.v. methylprednisolone (i.e., two- to threefold increases in exposure, C_{max}, and $t_{1/2}$) [126–128]. The interaction can reduce morning plasma cortisol concentration 80%–90% [126–128]. The metabolism of methylprednisolone is complex, and these data suggest CYP3A4 is primarily involved [126–128].

Dexamethasone is also a CYP3A4 substrate, and itraconazole increases its systemic exposure after i.v. or oral administration, approximately three- and fourfold, respectively [129]. The interaction can also significantly reduce morning plasma cortisol concentrations. In contrast, itraconazole co-administration increases prednisolone exposure and $t_{1/2}$ 13–30% but produces only minimal changes in prednisolone C_{max} or morning plasma cortisol concentrations [127, 130].

Itraconazole also interacts with inhaled corticosteroids, which depending on the inhalation device and patient technique are absorbed into the systemic circulation and undergo hepatic metabolism or can be inadvertently swallowed and undergo intestinal and/or hepatic metabolism [131, 166]. Oral itraconazole significantly inhibits the metabolism of inhaled budesonide and leads to 1.5–4-fold increases in exposure, C_{max}, and $t_{1/2}$ [131]. This interaction enhances the adrenal suppressive effects of budesonide and should be considered when co-administering other similar corticosteroids with itraconazole [131, 132, 167–169].

11.5.2.7   Calcium Channel Blockers

Felodipine is a CYP3A4 substrate that undergoes extensive first-pass metabolism [133]. Itraconazole co-administration increases felodipine exposure approximately sixfold, C_{max} eightfold, and $t_{1/2}$ approximately twofold [133]. These pharmacokinetic changes significantly reduce systolic and diastolic blood pressure and increase heart rate [133].

11.5.2.8   Miscellaneous Drugs

Itraconazole interacts with several other medicines including oxybutynin (increases exposure and C_{max}) [134] and busulfan (increases steady-state busulfan concentrations and lowers apparent oral clearance (CL/F)) [135]. An initial case report has noted that concomitant itraconazole therapy also substantially enhances warfarin's effect; however, two subsequent reports have demonstrated minimal effects of itraconazole on warfarin [170–172]. The pharmacologically active S-warfarin enantiomer is a CYP2C9 substrate, while the R-warfarin is primarily a CYP3A4 substrate. Itraconazole does not inhibit CYP2C9 activity, except at supratherapeutic concentrations, but it is an inhibitor CYP3A4 at standard concentrations. CYP3A4 inhibition should only influence R-warfarin, and this may not result in a clinically significant anticoagulant effect [172]. Given this, itraconazole may only affect warfarin in patients who are PM of CYP2C9 or in patients who have concomitant CYP2C9 inhibitors which would need to be determined on a case-by-case basis.

   In healthy volunteers oral itraconazole co-administration significantly reduced the exposure and C_{max} and delayed the absorption of the selective COX-2 inhibitor meloxicam [136]. The mechanism behind this interaction is unknown but may involve inhibition of meloxicam absorption [136]. Oxycodone undergoes extensive hepatic metabolism via CYP3A4 to noroxycodone (inactive) and via CYP2D6 to oxymorphone (active), which are further metabolized to inactive metabolites via CYP3A4 [173]. Itraconazole co-administration reduces i.v. and oral oxycodone CL and prolongs its $t_{1/2}$, resulting in increased pharmacological effects [137, 174]. When i.v. and oral oxycodone are given with paroxetine (CYP2D6 inhibitor), there is a reduction in the formation of oxymorphone, an increase in the formation of noroxycodone, and negligible effects on parent oxycodone pharmacokinetics [174, 175]. The increase in noroxycodone following CYP2D6 inhibition suggests that oxycodone metabolism gets compensated for by CYP3A4 [174, 175]. However, when the combination of paroxetine (CYP2D6 inhibitor) and itraconazole (CYP3A4 inhibitor) is co-administered with i.v. and oral oxycodone, there is an increase in parent oxycodone levels, with a more pronounced increases with oral oxycodone [174, 175]. The larger increases in parent oxycodone following oral administration are likely a result of itraconazole inhibition of both hepatic and intestinal CYP3A4 [174, 175].

11.5.2.9 Clinical Importance of Itraconazole Interactions Affecting CYP Biotransformation of Other Drugs

Many interactions involving itraconazole are clinically important. Myopathy (skeletal muscle toxicity) is a potentially severe side effect of elevated statin concentrations that can progress to rhabdomyolysis [151]. The incidence of rhabdomyolysis associated with the CYP3A4-metabolized statins is substantially greater than that of those not metabolized by CYP3A4 [176]. When statins are administered with potent CYP3A4 inhibitors like itraconazole, the risk of rhabdomyolysis associated with their use increases significantly [108–110, 151, 177]. Concomitant itraconazole use may also increase the risk of dose-dependent adverse effects (i.e., hepatotoxicity) associated with the CYP3A4-metabolized statins [178].

Co-administration of itraconazole with midazolam, triazolam, or buspirone severely impairs intellectual capacity and psychomotor skills even when low doses of these benzodiazepines (particularly midazolam and triazolam) are used for prolonged periods. The interaction between the azoles like itraconazole and the calcineurin/mTOR inhibitors is largely unavoidable and if not properly managed can lead to calcineurin−/mTOR-associated nephrotoxicity. The interaction between itraconazole and corticosteroids administered orally, i.v., or via inhalation can produce significant suppression of endogenous cortisol production. Multiple cases of Cushing's syndrome or adrenal insufficiency reported in the literature are attributed to itraconazole co-administration with either fluticasone or budesonide [169].

Itraconazole co-administration with felodipine produces clinically significant cardiovascular effects. In addition, the pharmacological effects of oxycodone may be increased, particularly the oral dosage form. The interaction between itraconazole and warfarin is likely not clinically significant in most cases, but under the right conditions (CYP2C9 inhibition or PM of CYP2C9), it could lead to a significant bleeding event [170].

11.5.2.10 Management of Itraconazole Interactions Affecting CYP-Mediated Biotransformation of Other Drugs

Patients receiving simvastatin or atorvastatin with itraconazole should be closely monitored for clinical and laboratory signs of skeletal muscle toxicity (myalgia, arthralgia, CK elevations) and hepatotoxicity (transaminase elevations). Depending on the degree of lipid-lowering effects needed, non-CYP3A4-metabolized statins (pravastatin and rosuvastatin) are alternatives for patients receiving concurrent itraconazole therapy.

Itraconazole and its metabolites are slowly eliminated; therefore, the interaction between triazolam and midazolam cannot be avoided by adjusting itraconazole dosing [17, 49]. The benzodiazepines that are not appreciably metabolized by CYP3A4 are alternatives to triazolam and midazolam for patients receiving concurrent itraconazole therapy. Other alternatives include diazepam, estazolam, and zolpidem.

Management of the itraconazole-calcineurin inhibitor interaction necessitates monitoring calcineurin inhibitor blood concentrations, adjusting calcineurin inhibitor dosages, or switching antifungal therapy. Calcineurin inhibitor doses should be empirically reduced at the onset of itraconazole co-administration, and blood concentrations should be obtained before, during, and after azole use and dosages adjusted accordingly. The use of itraconazole is not recommended with the mTOR inhibitors, and other azole antifungals should be considered.

In patients requiring concomitant itraconazole and oral or i.v. corticosteroid therapy, prednisolone should be considered for immunosuppressive or anti-inflammatory agent. If patients are receiving dexamethasone or methylprednisolone (dosed chronically or as pulse therapy), corticosteroid dose reductions may be needed during concomitant itraconazole therapy.

Co-administration of itraconazole with felodipine or other chemically related calcium channel blockers should be avoided given the considerable clinical significance of the interaction. If these combinations cannot be avoided, then the dose of the calcium channel blocker should be reduced, and the patient's heart rate and blood pressure should be closely monitored until stable. The interaction between itraconazole and meloxicam results in limited meloxicam activity due to decreased absorption, at least during the first 48–72 h [136]. The combination of itraconazole and warfarin should necessitate close clinical monitoring of warfarin. Contemporary data indicate that this interaction is not significant, but some case reports have suggested otherwise [170–172]. If antifungal therapy is needed, an amphotericin B formulation or an echinocandin should be used.

11.5.3 Interactions That Induce Itraconazole Biotransformation

Phenytoin, phenobarbital, carbamazepine, rifampin, efavirenz, and nevirapine are CYP3A4 inducers. Itraconazole co-administration with these agents results in a pharmacokinetic interaction that markedly reduces its serum concentrations [141–144, 179]. The onset of induction varies with each drug and may not be detectable for up to 2 weeks [141–144, 179]. After discontinuation of these agents, induction may persist for up to 2 weeks [141–144, 179].

11.5.3.1 Clinical Importance of Interactions That Induce Itraconazole Biotransformation

Interactions between CYP3A inducers and itraconazole lead to undetectable or subtherapeutic serum itraconazole concentrations, which can result in therapeutic failure.

11.5.3.2 Management of Interactions That Induce Itraconazole Biotransformation

These interactions likely cannot be circumvented by increasing the itraconazole dose. If possible, these combinations should be avoided. However, this is often not possible, especially in HIV patients receiving rifampin or rifabutin. In these cases, if alternative antifungal therapy cannot be used, then itraconazole serum concentrations and the patient's clinical condition should be closely monitored for therapeutic failure. If alternative antifungal agents cannot be used, then antimycobacterial regimens without rifampin or rifabutin should be considered. Similarly, gabapentin, levetiracetam, lamotrigine, or other nonenzyme-inducing antiepileptic drugs may represent alternatives.

11.5.4 Itraconazole Interactions Affecting P-glycoprotein-Mediated Efflux of Other Drugs

11.5.4.1 Digoxin

Digoxin is not a CYP substrate, undergoes little hepatic metabolism, and is renally eliminated primarily as unchanged drug, through P-gp-mediated renal tubular secretion [145, 180–183]. The reduced P-gp-mediated efflux causes decreases in CL_R and increases in serum digoxin concentrations [145, 180–183]. Therefore, the interaction results from inhibition of intestinal P-gp and/or inhibition of P-gp-mediated digoxin renal secretion by itraconazole [18, 181, 182].

11.5.4.2 Quinidine

Quinidine is primarily metabolized by CYP3A4 to form 3-hydroxyquinidine and CYP2C9 and perhaps CYP3A4 to form quinidine N-oxide [146, 147]. Quinidine is also actively secreted by the renal tubules, which most likely involves P-gp. Itraconazole co-administration significantly increases quinidine exposure 2.5-fold and C_{max} nearly twofold, prolongs elimination $t_{1/2}$, and significantly reduces its CL_R [146, 147]. Itraconazole co-administration also significantly reduces the partial CL of both metabolites [147]. This interaction likely results from inhibition of intestinal and hepatic CYP3A4 metabolism and P-gp-mediated tubular secretion of quinidine by itraconazole [146, 147, 184].

11.5.4.3   Vinca Alkaloids and Opiates

Itraconazole reduces CYP3A4 metabolism and P-gp efflux of vincristine. The subsequent accumulation and distribution of vincristine produce neurological toxicities (seizures, paresthesia, sensory deficits, muscle weakness, neuropathy), gastrointestinal disturbances (abdominal pain/distention, constipation, ileus) hyponatremia, and SIADH [185]. Itraconazole also interacts to a similar degree with vinblastine [186]. Itraconazole produces subtle increases in oral morphine plasma concentrations but does not alter its pharmacological effects [148]. The interaction probably involves inhibition of intestinal P-gp [148].

11.5.4.4   Clinical Importance of Itraconazole Interactions Involving P-glycoprotein-Mediated Efflux of Other Drugs

Case reports document that interactions between itraconazole and digoxin or the vinca alkaloids are clinically significant [145, 181–183, 185–187]. Quinidine has a relatively narrow therapeutic index, and elevated concentrations can produce life-threatening toxicity. Therefore, the interaction is considered clinically significant [146].

11.5.4.5   Management of Itraconazole Interactions Involving P-glycoprotein-Mediated Efflux of Other Drugs

Patients receiving itraconazole and digoxin should be questioned about symptoms of digoxin toxicity and have their serum digoxin concentrations closely monitored [183]. Similarly, plasma quinidine concentrations should be closely monitored in patients receiving quinidine and itraconazole [146]. Due to the severity of the interaction, itraconazole or any azole should not be co-administered with vincristine- or vinblastine-containing regimens. If a vinca alkaloid is started in a patient receiving an azole, the azole should be discontinued [185–187].

11.6   Interactions Involving Fluconazole

Drug interactions involving fluconazole are summarized in Table 11.5.

Table 11.5 Fluconazole interactions affecting CYP-mediated biotransformation of other drugs

Drug	Effect on drug (% change)	Inhibition site	Reference(s)
Statins			
Fluvastatin	↑C_{max} (44%); ↑$AUC_{(0-\infty)}$ (84%); $t_{1/2}$ (80%)	Hepatic CYP2C9	[188]
Pravastatin	None		[188]
Rosuvastatin	None		[189]
Benzodiazepines			
Midazolam (oral)			
+ FCZ po day 1	↑C_{max} (130–150%); ↑$AUC_{(0-\infty)}$ (251–273%); ↑$t_{1/2}$ (71–123%)	Hepatic and intestinal CYP3A4	[113, 190]
+ FCZ po day 6	↑C_{max} (74%); ↑$AUC_{(0-\infty)}$ (259%); ↑$t_{1/2}$ (71%)	Hepatic and intestinal CYP3A4	[113]
+ FCZ iv day 1	↑C_{max} (79%); ↑T_{max} (100%); ↑$AUC_{(0-\infty)}$ (244%); ↑$t_{1/2}$ (123%)	Hepatic CYP3A4	[190]
Midazolam (i.v.)		Hepatic CYP3A4	[155]
+ FCZ po day 4	↓CL (51%); ↑$t_{1/2}$ (52%)		
α-OH-midazolam		Hepatic CYP3A4	[190]
+ FCZ po day 1	↓C_{max} (19%); ↑$AUC_{(0-17)}$ (50%); ↑$t_{1/2}$ (142%); ↓ratio (54%)		
+ FCZ iv day 1	↓C_{max} (10%); ↑$AUC_{(0-17)}$ (56%); ↑$t_{1/2}$ (157%); ↓ratio (56%)		
Triazolam		Hepatic and possibly intestinal CYP3A4	
+ FCZ po 50 mg	↓C_{max} (47%); ↑$AUC_{(0-\infty)}$ (63%); ↑$t_{1/2}$ (29%); ↑T_{max} (15%)		[191]
+ FCZ po 100 mg	↓C_{max} (25–40%); ↑$AUC_{(0-\infty)}$ (105–145%); ↑$t_{1/2}$ (77–83); ↑T_{max} (11–92%)		[191, 192]
+ FCZ po 200 mg	↓C_{max} (133%); ↑$AUC_{(0-\infty)}$ (342%); ↑$t_{1/2}$ (126%); ↑T_{max} (54%)		[191]
Diazepam	↑$AUC_{(0-48)}$ (51%); ↑$AUC_{(0-\infty)}$ (174%); ↑$t_{1/2}$ (135%); ↓CL/F (59%)	Hepatic CYP2C19 and CYP3A4	[193]
N-Desmethyldiazepam	↓C_{48hr} (60%); ↓$AUC_{(0-48)}$ (70%); ↓AUC_{ratio} (71%)	Hepatic CYP2C19	[193]
Calcineurin inhibitors			
Cyclosporine			
Day 4	↑C_{max} (39%); ↑C_{min} (38%); ↑AUC (87%); ↓CL (18%)	Hepatic intestinal CYP3A and P-gp	[194]

(continued)

Table 11.5 (continued)

Drug	Effect on drug (% change)	Inhibition site	Reference(s)
Tacrolimus	C^{min}, $AUC^{(0-12)}$ similar pre- and post-fluconazole with 40% ↓dose		[195]
Anticonvulsants			
Phenytoin	↑C_{min} (≈25%); ↑$AUC_{(0-24)}$ (75%)	Hepatic CYP3A4	[196]
Anticoagulants			
Warfarin	Inhibits *S-warfarin* metabolic pathway ≈ 70%	Hepatic CYP2C9	[98, 177]
Miscellaneous drugs			
Fentanyl	↓CL (17%); ↓norfentanyl $AUC_{(0-\infty)}$ (56%); ↓Ratio (67%)	Hepatic CYP3A4	[197]
Alfentanil		Hepatic CYP3A4	[198]
FCZ po	↑$AUC_{(0-10)}$ (96%); ↑$t_{1/2}$ (67%); ↓CL (54%); ↓V_{ss} (19%)		
FCZ i.v.	↑$AUC_{(0-10)}$ (107%); ↑$t_{1/2}$ (80%); ↓CL (58%); ↓V_{ss} (19%)		
Methadone	↑$AUC_{(0-24)}$ (35%); ↑C_{max} (27%); ↑C_{min} (48%);↓CL/F (24%)	Hepatic CYP	[199]
Cyclophosphamide	↑$AUC_{(0-24)}$ (79%); ↑C_{max} (33–36%)	Hepatic CYP	[200, 201]
4-OH-cyclophosphamide	↓C_{max} (33–36%)		
Nevirapine	↑$AUC_{(0-8)}$ (29%); ↑C_{max} (28%); ↓CL/F (22%)	Hepatic CYP3A4	[202]
Ibuprofen (*S-enantiomer*)	↑$AUC^{(0-24)}$ (83%); ↑C^{max} (16%); ↑$t^{1/2}$ (34%)	Hepatic CYP2C9	[203]

11.6.1   Fluconazole Interactions Affecting CYP-Mediated Biotransformation of Other Drugs

11.6.1.1   The 3-Hydroxy-3 Methylglutaryl (HMG) Coenzyme: A Reductase Inhibitor (The "Statins")

Fluconazole significantly increases fluvastatin exposure, C_{max}, and $t_{1/2}$ [188]. Fluconazole does not inhibit ABC transporters involved in drug disposition; thus, this interaction results from hepatic CYP2C9 inhibition [18, 155]. Pravastatin, pitavastatin, and rosuvastatin are not appreciably metabolized by CYP2C9 or CYP2C19; thus, fluconazole does not significantly affect their pharmacokinetics [153, 188, 189]. Case reports suggest even a moderate CYP3A4 inhibitor like fluconazole can inhibit the metabolism of CYP3A-metabolized statins (simvastatin, atorvastatin) [204–207].

11.6.1.2   Benzodiazepines

Fluconazole co-administration with triazolam or midazolam significantly alters the pharmacokinetics and enhances the pharmacological effect of both [113, 190–192]. Fluconazole increases the oral availability and decreases the CL (i.e., increases exposure, C_{max}, and $t_{1/2}$,) of both benzodiazepines [113, 190–192]. Fluconazole has no effect on V_{ss} of i.v. midazolam but does substantially reduce its plasma CL thereby prolonging its $t_{1/2}$ [113]. The interaction significantly enhances and prolongs the pharmacological effects of these benzodiazepines [113, 190–192]. The inhibition of CYP-mediated midazolam metabolism is greater with orally rather than i.v. administered fluconazole [190]. Like itraconazole, CYP3A5 genotype also influences the extent and interindividual variability of the midazolam-fluconazole interaction [47, 95]. The effects of fluconazole on midazolam did not increase with repeated dosing [113, 190], but with increasing doses, the extent of the interaction with triazolam increased accordingly [192]. Fluconazole significantly increases diazepam exposure, most likely by inhibiting CYP2C19-catalyzed formation of its primary metabolite N-desmethyldiazepam, but the interaction minimally changes its pharmacological effects [193].

11.6.1.3   Calcineurin Inhibitors and mTOR Inhibitors

Fluconazole interacts with calcineurin inhibitors in a dose-related manner, with interactions occurring at higher fluconazole ($\geq$200 mg) doses [91, 194, 195, 208–212]. The maximum effect occurs approximately 4 days after starting fluconazole [208, 213]. The magnitude of the interaction is influenced by the route of fluconazole administration and is less with i.v. dosing [195, 210]. Such differences may also be related to CYP3A5 genotype. Like other azoles, fluconazole interacts with tacrolimus, and the interaction is influenced by the CYP3A5 genotype. A small study examining the influence of the variant CYP3A5 alleles on tacrolimus exposure and dose before, during, and after fluconazole administration observed that the magnitude of the interaction is less in heterozygous expressers (CYP3A5*3/*1) compared to homozygous expressers (CYP3A5*3/*3) [214]. In addition, the study illustrated that CYP3A5 non-expressers are more frequently exposed to supratherapeutic trough tacrolimus concentrations during treatment with fluconazole [214]. Data also show that hepatic and intestinal CYP3A5 expression and activity may be higher among heterozygous expressers [214]. A case report describes a significant interaction between fluconazole and sirolimus [215]. The interaction manifests rapidly and results in toxic sirolimus concentrations [215]. Similarly, a case report describing the management of a pharmacokinetic drug interaction between everolimus and fluconazole suggests that the dose of everolimus should be reduced to avoid overexposure and that reduction is comparatively less with fluconazole than other triazoles (e.g., voriconazole) [216].

11.6.1.4 Phenytoin

In two healthy volunteer studies, fluconazole significantly increased phenytoin exposure and C_{min} [196, 217]. In the multidose study, the phenytoin dose and its duration were limited which demonstrated no appreciable effect on fluconazole pharmacokinetics, but in practice it likely will [196].

11.6.1.5 Warfarin

Therapeutic plasma fluconazole concentrations exceed its in vitro inhibitory constant for CYP2C9-mediated warfarin metabolism [98]. Therefore, fluconazole interacts with warfarin in a predictable manner. Fluconazole inhibits S-warfarin metabolism approximately 70% and R-warfarin metabolism by 45%. The inhibition of S-warfarin results in a 38% increase in the INR in previously stabilized patients [98, 177]. A single 150 mg dose of fluconazole can increase INR levels in excess of 8 days [218].

11.6.1.6 Miscellaneous Drugs

Oral fluconazole (400 mg) significantly decreased fentanyl plasma CL and the exposure of its primary active metabolite, norfentanyl [197]. The interaction did not affect fentanyl V_{ss} or $t_{1/2}$, which suggests it was due to inhibition of CYP3A-mediated norfentanyl formation [197]. Oral or i.v. fluconazole significantly reduces alfentanil CL and nearly doubles its $t_{1/2}$ [198]. The increased alfentanil concentrations were associated with enhanced pharmacological effects [198]. Fluconazole (200 mg daily) reduced methadone CL/F and increased its exposure in patients receiving a mean daily methadone dose of 55 mg, without clinically enhancing its pharmacological effect [199]. Contrary to this, a single case report demonstrated a significant increase in methadone's pharmacological effect with concomitant i.v. fluconazole [219].

Cyclophosphamide undergoes extensive metabolism including one pathway involving activation by several CYPs including CYP2C9 and CYP3A4, which produces the cytotoxic alkylating agent 4-hydroxycyclophosphamide. Fluconazole reduces cyclophosphamide CL and increases its $t_{1/2}$ in children [200]. Data also indicate that fluconazole increases cyclophosphamide exposure and C_{max} and reduces 4-hydroxycyclophosphamide C_{max} [201]. In HIV patients, the co-administration of fluconazole and rifabutin increased the rifabutin C_{max} 91% and the AUC 76% [220]. This interaction also extended to the primary rifabutin metabolite, 25-O-desacetylrifabutin, in which the C_{max} and C_{min} were increased by 3.6-fold and 2.3-fold, respectively [220]. Drug interactions involving antiretroviral agents are discussed in Chaps. 13 and 14. However, fluconazole (200 mg three times per week) co-administration significantly increases nevirapine exposure [202].

Ibuprofen is a chiral compound, and the pharmacologically active S-enantiomer, which produces most of its analgesic effect, is metabolized primarily by CYP2C9. Fluconazole significantly increases the S-enantiomer C_{max}, exposure, and its $t_{1/2}$ [203].

11.6.1.7 Clinical Importance of Fluconazole Interactions Affecting CYP-Mediated Biotransformation of Other Drugs

While fluconazole is generally regarded as a safe medication, interactions involving this azole and benzodiazepines, calcineurin inhibitors, mTOR inhibitors, nevirapine, phenytoin, and warfarin are clinically significant. Most of these interactions can lead to prolonged changes in the pharmacological effects and toxicity of the victim drugs. Some interactions with fluconazole if not recognized can produce fatalities [221]. Other interactions, like the fluconazole-cyclophosphamide interaction, may limit toxicities associated with cyclophosphamide regimens [201]. In certain patients, the potential for fluconazole to be co-administered with rifabutin is high; thus, clinicians must be aware of rifabutin toxicities (uveitis, flu-like symptoms, and liver enzymes) [220]. Ibuprofen is a component of many over-the-counter products and is largely overlooked as having CYP450 drug interactions. Patients should know that fluconazole co-administration may increase the risk of concentration-dependent ibuprofen toxicity (i.e., renal, cardiovascular, or gastrointestinal adverse effects) [203].

11.6.1.8 Management of Fluconazole Interactions Affecting CYP-Mediated Biotransformation of Other Drugs

Like other azoles, the interaction between fluconazole and triazolam, or midazolam, cannot be circumvented even with low doses of fluconazole, and these combinations should be avoided [113, 190–192]. In patients receiving fluconazole, temazepam, oxazepam, or lorazepam may be alternatives to triazolam and midazolam as these agents are not appreciably metabolized by CYP3A4. Even though fluconazole co-administration significantly reduces the clearance of diazepam, the interaction does not enhance its pharmacological effects [193]. Similarly, zolpidem clearance is decreased by 20% in the presence of fluconazole which does not influence the sedative effects of zolpidem [121]. Despite these interactions, diazepam and zolpidem could also represent alternatives to midazolam and triazolam.

When used with the calcineurin inhibitors, low doses of fluconazole (<200 mg) may produce less significant interactions; however, using these doses may not be possible [194, 208]. No matter what dose of fluconazole is being used, therapeutic drug monitoring of the calcineurin inhibitors is essential to maintain therapeutic concentrations and reduce potential toxicities. Phenytoin serum concentrations should be monitored with the addition of fluconazole therapy [196]. If the two are used together for prolonged times, clinicians should monitor for breakthrough fungal infections due to the known CYP450 induction effects of phenytoin. The interaction between fluconazole and warfarin cannot be avoided. The interaction occurs at any dose, with single doses, and the interaction may persist for >7 days [48, 98, 218, 222, 223]. Termination of this interaction requires fluconazole discontinuation and perhaps infusion of fresh-frozen plasma, vitamin K, or other therapeutic modalities to reverse excessive anticoagulation. In cases where antifungal therapy is needed, isavuconazole, amphotericin B, echinocandins, and perhaps itraconazole may be preferred [102, 172, 177].

When using fluconazole and fentanyl concomitantly, respiratory depression may occur if the fentanyl dose is not reduced and the patients are not monitored closely [197, 221]. The use of rifabutin therapy in *Mycobacterium avium* infections is often unavoidable; therefore, careful monitoring will be necessary [220]. Similar to fentanyl, ibuprofen exposure is significantly increased by fluconazole; however, the toxicities of ibuprofen may not be immediately apparent or clinically evident [203]. Ideally, a lower ibuprofen dose should be used in patients receiving fluconazole, particularly with long-term co-administration. Fluconazole reduces phenytoin clearance; therefore, therapeutic drug monitoring of phenytoin and monitoring for phenytoin toxicities are needed [196, 224]. If the fluconazole-phenytoin interaction persists, there is a potential for fluconazole induction by phenytoin [225, 226]. If the induction effects cannot be overcome by increasing the fluconazole dose or the patient is not responding to fluconazole therapy, then echinocandins or amphotericin B formulations should be considered.

11.6.2 Interactions That Induce Fluconazole Biotransformation

Although fluconazole undergoes minimal metabolism, co-administration with CYP3A4 inducers markedly reduces its exposure [227].

11.6.2.1 Clinical Importance of Interactions That Induce Fluconazole Biotransformation

Rifampin co-administration with fluconazole produces a clinically significant interaction [227]. Without adjusting the fluconazole dose, the resulting induction leads to undetectable or subtherapeutic serum fluconazole concentrations that could lead to therapeutic failure [228, 229].

11.6.2.2 Management of Interactions That Induce Fluconazole Biotransformation

Often the induction of fluconazole CYP-mediated metabolism cannot be overcome by increasing its dose. However, in patients receiving rifampin, the dose of fluconazole should be doubled [227, 229].

11.6.3 Fluconazole Interactions Affecting Conjugative Biotransformation of Other Drugs

In humans UGT2B7 catalyzes zidovudine metabolism to its major metabolite, zidovudine glucuronide [230]. Fluconazole, a moderate UGT2B7 inhibitor, co-administered at 400 mg daily dose significantly decreased zidovudine CL/F and formation of zidovudine glucuronide, which increased zidovudine exposure, C_{max}, and $t_{1/2}$ [52, 230].

11.6.3.1 Clinical Importance and Management of Fluconazole Interactions Affecting Conjugative Biotransformation of Other Drugs

The clinical significance of the fluconazole-zidovudine interaction is undetermined. Patients receiving this combination should be monitored for zidovudine toxicity [231].

11.7 Interactions Involving Voriconazole

Drug interactions involving voriconazole are summarized in Tables 11.6 and 11.7.

11.7.1 Voriconazole Interactions Involving Gastric pH and Motility

Voriconazole co-administration with high-fat meals reduces the absolute bioavailability by 22% and C_{max} by 34% [11].

Table 11.6 Voriconazole interactions affecting CYP-mediated biotransformation of other drugs

Drug	Effect on drug (% Change)	Inhibition site	Reference(s)
Benzodiazepines			
Midazolam (oral)		Hepatic and intestinal CYP3A4	[232]
+ VCZ po	$\uparrow C_{max}$ (259%); $\uparrow AUC_{(0-\infty)}$ (840%) $\uparrow t_{1/2}$ (252%); $\downarrow CL/F$ (91%)		
α-OH-midazolam			
+ VCZ po	$\downarrow C_{max}$ (6%); $\uparrow AUC_{(0-\infty)}$ (149%); $\downarrow$Ratio(77%)		
Midazolam (i.v.)		Hepatic CYP3A4	[232]
+ VCZ po	$\uparrow AUC_{(0-\infty)}$ (253%); $\downarrow CL$ (72%); $\uparrow t_{1/2}$ (196%)		
α-OH-midazolam			
+ VCZ po	$\downarrow C_{max}$ (18%); $\uparrow AUC_{(0-\infty)}$ (68%); $\uparrow T_{max}$ (168%); $\downarrow$Ratio (54%)		
Diazepam	$\uparrow AUC_{(0-48)}$ (39%); $\uparrow AUC_{(0-\infty)}$ (123%); $\uparrow t_{1/2}$ (97%); $\downarrow CL/F$ (47%)	Hepatic CYP2C19 and CYP3A4	[193]
N-desmethyldiazepam	$\downarrow C_{48hr}$ (48%); $\downarrow AUC_{(0-48)}$ (64%); $\downarrow AUC_{ratio}$ (71%)	Hepatic CYP2C19	[193]
Calcineurin inhibitors			
Cyclosporine	$\uparrow C_{min}$ (248%); $\uparrow AUC_{(0-12)}$ (70%)	Hepatic and intestinal CYP3A and P-gp	[233–235]
Tacrolimus	$\uparrow C_{min}$ great than predicted	Hepatic intestinal CYP3A	[234, 236–238]
mTOR inhibitors			
Sirolimus	No formal PK studies; required ~90% sirolimus dose reduction		[239–241]
Everolimus	No formal PK studies; required 65%–80% everolimus dose reduction		[216, 242, 243]
Anagesics and anti-inflamatory agents			
Ibuprofen (*S-enantiomer*)	$\uparrow AUC^{(0-24)}$ (103%); $\uparrow C^{max}$ (19%); $\uparrow t^{1/2}$ (33%)	Hepatic CYP2C9	[203]
Alfentanil	$\uparrow AUC_{(0-10)}$ (264%); $\uparrow AUC_{(0-\infty)}$ (444%) $\uparrow t_{1/2}$ (340%); $\downarrow CL$ (85%); $\downarrow V_{ss}$ (28%)	Hepatic CYP3A4	[244]
Fentanyl	$\downarrow CL$ (24%); $\uparrow AUC_{(0-\infty)}$ (39%);	Hepatic CYP3A4	[197]
Norfentanyl	$\downarrow AUC_{(0-\infty)}$ (56%); $\downarrow$Ratio (67%)		
Methadone	$\uparrow AUC_{(0-24)}$ (44%); $\uparrow C_{max}$ (30%)	Hepatic CYP2B6, 3A, 2C9, and 2C19	[245]
Oxycodone	$\uparrow C_{max}$ (69%); $\uparrow AUC_{(0-\infty)}$ (257%); $\downarrow CL/F$ (71%); $\uparrow t_{1/2}$ (102%)	Hepatic and intestinal CYP3A4 Hepatic CYP2D6	[246–248]

(continued)

Table 11.6 (continued)

Drug	Effect on drug (% Change)	Inhibition site	Reference(s)
Noroxycodone	$\downarrow C_{max}$ (87%); $\downarrow AUC_{(0-\infty)}$ (67%); $\downarrow$Ratio(92%); $\uparrow t_{1/2}$ (106%)		
Oxymorphone	$\uparrow C_{max}$ (104%); $\uparrow AUC_{(0-\infty)}$ (597%); $\uparrow$Ratio(100%); $\uparrow t_{1/2}$ (541%)		
Noroxymorphone	$\downarrow C^{max}$ (88%); $\downarrow AUC^{(0-\infty)}$ (49%); $\downarrow$Ratio(87%); $\uparrow t^{1/2}$ (218%)		
Buprenorphine	$\uparrow C_{max}$ (37%); $\uparrow AUC_{(0-\infty)}$ (80%);$\downarrow$CL/F (42%); $\uparrow t_{1/2}$ (39%)	Hepatic CYP3A4	[249]
Meloxicam	$\uparrow AUC^{(0-72)}$ (46%); $\uparrow t^{1/2}$ (50%)	Hepatic CYP2C9 and CYP3A4	[136]
Diclofenac	$\uparrow AUC^{(0-\infty)}$ (77%); $\uparrow C^{max}$ (114%)	Hepatic CYP2C9/19 and CYP3A4	[250]
Etoricoxib	$\uparrow AUC^{(0-\infty)}$ (49%)	Hepatic CYP3A	[251]
Miscellaneous drugs			
Warfarin	Inhibits *S-warfarin* metabolic pathway $\approx 41\%$	Hepatic CYP2C9	[252]
Phenytoin	$\uparrow AUC^{(0-24)}$ (80%); $\uparrow C_{max}$ (70%)	Hepatic CYP2C9 and CYP3A4	[253]
Efavirenz (400 mg/day)		Hepatic CYP2B6	[254]
+ VCZ 200 mg BID	$\uparrow AUC_{(0-24)}$ (44%); $\uparrow C_{max}$ (37%)		
Efavirenz (300 mg/day)[a]		Less inhibition with	[255]
+ VCZ 300 mg Q12 hr	$\downarrow AUC_{(0-24)}$ (8%); $\downarrow C_{max}$ (18%)	Lower efavirenz dose and	
+ VCZ 400 mg Q12 hr	$\uparrow AUC^{(0-24)}$ (6%); $\downarrow C^{max}$ (10%)	Higher voriconazole dose	
Ritonavir			
400 mg BID	None		
100mg B.I.D	$\downarrow AUC_{(0-12)}$ (18%); $\downarrow C_{max}$ (30%); $\downarrow C_{min}$ (22%)	Unknown	[256]

[a]Values compared to efavirenz 600 mg/day

11.7.1.1 Clinical Importance of Voriconazole Interactions Involving Gastric pH or Motility Interactions

When considered alone, the clinical significance of the impact of food on voriconazole disposition is minimal. However, given the large variation in voriconazole pharmacokinetics and its narrow therapeutic index for efficacy and toxicity, reduction in absolute bioavailability may be significant in select patients.

Table 11.7 Interactions that induce voriconazole biotransformation or inhibit its absorption

Drug	Effect on voriconazole (% change)	Comments	Reference(s)
Food			
Day 7	↓AUC$^{(0-12)}$ (28%); ↓C$_{max}$ (34%); ↑T$_{max}$ (73%); F (22%)	Delayed absorption; Decreased bioavailability	[11]
Phenytoin		CYP 3A4, 2C9/19 induction	[253]
+ VCZ 200 mg BID	↓AUC$^{(0-12)}$ (64%); ↓C$_{max}$ (39%)		
+ VCZ 400 mg BID[a]	↑AUC$^{(0-12)}$ (39%); ↑C$_{max}$ (34%)	Doubling dose compensated induction	
Ritonavir (chronic dose study)		CYP2C19 and CYP2C9 induction	[256]
400 mg BID	↓AUC$^{(0-12)}$ (84%); ↓C$_{max}$ (66%)		
100 mg BID	↓AUC$^{(0-12)}$ (27%); ↓C$_{max}$ (16%)		
Ritonavir (acute dose study)	↑AUC$^{(0-\infty)}$ (354%); ↑C$_{max}$ (17%); ↓CL/F (43%)	CYP 3A4 inhibition	[257]
Efavirenz (400 mg/ day)		CYP2C19/9 and CYP3A4 induction;	[254]
+ VCZ 200 mg BID	↓AUC$_{(0-24)}$ (78%); ↓C$_{max}$ (62%)	Greater effect with standard 600 mg dose?	
Efavirenz (300 mg/ day)[b]		Less induction with	[255]
+ VCZ 300 mg Q12 hr	↓AUC$_{(0-12)}$ (48%); ↓C$_{max}$ (27%)	Lower efavirenz dose and	
(VCZ-N-oxide)	None	Higher voriconazole dose	
+ VCZ 400 mg Q12 hr	↑AUC$^{(0-24)}$ (4.5%); ↑C^{max} (28%)		
(VCZ-N-oxide)	↑AUC$^{(0-24)}$ (41%); ↑C^{max} (45%)		
Flucloxacillin	↓Voriconazole concentrations	CYP3A4 and possible CYP2C8/9 induction	[258]

[a]Values compared to voriconazole 200 mg
[b]Values compared to voriconazole 400 mg/day

11.7.1.2 Management of Voriconazole Interactions Involving Gastric pH Interactions and Motility

The voriconazole-food interaction can be managed by separating the doses from meals or by therapeutic monitoring. Separating the voriconazole dose by more than 1 h pre or post a meal should maintain its high oral bioavailability [11]. Additionally, routine voriconazole therapeutic drug monitoring and close clinical monitoring could be employed in these instances.

11.7.2 Voriconazole Interactions Affecting CYP-Mediated Biotransformation of Other Drugs

11.7.2.1 Benzodiazepines

Voriconazole co-administration also increases oral midazolam exposure ninefold, C_{max} 3.5-fold, and bioavailability 2.7-fold, respectively [232]. Voriconazole profoundly enhanced the pharmacological effects of oral midazolam, more so than intravenous midazolam. As discussed previously, the variability in benzodiazepine interactions may be explained in part by the CYP3A5 genotype [94, 95]. Voriconazole co-administration significantly increases diazepam exposure but did not enhance the pharmacological effects of this benzodiazepine [193].

11.7.2.2 Calcineurin Inhibitors and mTOR Inhibitors

Voriconazole co-administration increases cyclosporine exposure 1.7-fold and C_{min} 1.7–2.5-fold [259]. Clinically, the mean cyclosporine concentration to dose ratio increased 1.8-fold with concomitant voriconazole [233]. Similar to cyclosporine, the median increase in tacrolimus concentrations (concentration/dose ratios) in the presence of voriconazole is 1.3-fold to 1.5-fold [159, 236, 259]. However, this interaction demonstrates considerable variability, which has now been at least in part attributed to genetic polymorphisms in CYP2C19 and CYP3A5 and the presence of concomitant CYP inhibitors [93, 233, 237, 260–265]. Whether administered orally or via i.v., voriconazole interacts with the calcineurin inhibitors with considerable interpatient variability, but oral voriconazole leads to more pronounced interactions [234, 236–238, 260]. Additionally, voriconazole serum concentrations do not seem to influence the extent of the tacrolimus interaction [236]. In several case series, concomitant sirolimus and voriconazole required sirolimus dose reductions of ~90% to maintain adequate sirolimus concentrations [239–241]. In three separate case reports, the addition of voriconazole to everolimus therapy resulted in significant increases in everolimus concentrations [216, 242, 243]. The culmination of these reports indicates that everolimus dose reductions of 65–80% are necessary to maintain therapeutic everolimus concentrations in the presence of voriconazole [216, 242, 243]. Additionally, two of these reports were able to compare the influence of fluconazole and posaconazole compared to voriconazole [216, 242]. Voriconazole increased everolimus concentrations threefold more than that of fluconazole and twofold that of posaconazole [216, 242].

11.7.2.3 Analgesics and Anti-inflammatory Agents

Voriconazole co-administration significantly decreases the mean alfentanil plasma CL, increases exposure (sixfold), and prolongs its $t_{1/2}$ [244]. Similar to fluconazole, oral voriconazole significantly increases fentanyl exposure and decreases its CL and norfentanyl exposure [197]. Voriconazole significantly increases the pharmacologically active R-methadone exposure by 47.2% and C_{max} by 30.7% [245]. Voriconazole co-administration with oxycodone can significantly decrease first-pass metabolism of oxycodone by inhibiting its CYP3A4-mediated metabolism [246]. This can produce compensatory changes in oxymorphone pharmacokinetics mediated by CYP2D6, which voriconazole does not inhibit [246]. The compensatory activity of CYP2D6 only modestly enhances the pharmacological effects of oxycodone; however, CYP2D6 exhibits genetic polymorphisms; and poor metabolizer phenotypes would be at an increased risk for oxycodone toxicities [246–248]. Voriconazole also demonstrated an inhibitory effect on buprenorphine metabolism. Concomitant administration of buprenorphine and voriconazole resulted in an 87% increase in buprenorphine exposure and a 37% increase in its C_{max} [249]. The resultant increase in buprenorphine exposure and C_{max} is most likely due to inhibition of CYP3A4 by voriconazole [249]. Co-administration of voriconazole and ibuprofen 400 mg resulted in a twofold increase in S-(+)-ibuprofen exposure and prolonged its $t_{1/2}$ by 43% [203]. Voriconazole exhibited little influence on R-(−)-ibuprofen [203]. Additionally, the magnitude of the S-(+)-ibuprofen changes was directly correlated with voriconazole concentrations [203]. Ibuprofen is a racemic mixture containing the physiologically active S-(+)-ibuprofen and physiologically inactive R-(−)-ibuprofen. S-(+)-ibuprofen is metabolized via CYP2C9 and R-(−)-ibuprofen is metabolized by CYP2C8 [266]. Given this, the effects of voriconazole on S-(+)-ibuprofen are a direct result of voriconazole's CYP2C9 inhibition, while its negligible effects on R-(−)-ibuprofen are due the lack of CYP2C8 inhibition of voriconazole [266]. Similar to ibuprofen, meloxicam and diclofenac are extensively metabolized by hepatic CYP2C9 and to a lesser extent by CYP3A4 and CYP2C19 (diclofenac) [136, 250]. Voriconazole co-administration markedly increases both meloxicam and diclofenac exposure. However, the interaction with voriconazole prolongs meloxicam's $t_{1/2}$, without affecting its C_{max}, while its effects on diclofenac increased T_{max} and did not change its $t_{1/2}$ [136, 250]. The differing pharmacokinetics of meloxicam and diclofenac suggest that the meloxicam interaction primarily occurs in the liver, while the diclofenac interaction occurs in the intestine [136, 250]. Voriconazole oral gel co-administration can moderately increase single-dose etoricoxib exposure, C_{max}, and $t_{1/2}$ [251].

11.7.2.4 Miscellaneous Drugs

Voriconazole interacts with several other medicines including warfarin, phenytoin, and efavirenz.

Voriconazole co-administration significantly enhances the pharmacological effects of warfarin [252]. The interaction can increase prothrombin time by 100% from baseline and can persist for up to 6 days [252]. The effects of voriconazole on the anticoagulant effect of warfarin are most likely due to the more pronounced stereoselective inhibition of the more pharmacologically active warfarin enantiomer, S-warfarin [98]. Steady-state plasma phenytoin C_{max} and exposure increase by 67% and 81%, respectively, following repeated administration of oral voriconazole (400 mg twice daily for 10 days) [253]. The phenytoin (a substrate of CYP2C9 and CYP2C19) pharmacokinetic changes are most likely a result of voriconazole CYP2C19 and CYP2C9 inhibition [92, 100]. However, as discussed below, this interaction is bi-directional, and phenytoin induces voriconazole metabolism [225, 226, 253].

Voriconazole demonstrates variable drug interactions in patients receiving antiretroviral agents. The resultant interactions and their magnitude are complex and dependent on genetic polymorphism phenotypes, voriconazole dose (i.e., concentration), and the co-administered antiretroviral drug and their concentrations. The variable nature of these interactions makes them difficult to predict, so close clinical monitoring is warranted in all patients [267]. Thorough reviews of antiretroviral drug interactions can be found in Chaps. 13 and 14. In healthy volunteers, co-administration of voriconazole (200 mg twice daily) with efavirenz (400 mg daily) moderately increased efavirenz exposure (43%) and C_{max} (37%) [254]. The efavirenz-voriconazole interaction is likely due to voriconazole inhibition of CYP2B6, which is subject to genetic polymorphisms [92, 268]. However, as discussed below, this interaction is bi-directional, and the effect of efavirenz induction on voriconazole metabolism is more pronounced [254, 255]. Voriconazole had no apparent effect on steady-state high-dose (400 mg twice daily) ritonavir exposure but did slightly reduce C_{max} of low dose (100 mg twice daily). The mechanism of this effect is not clear [256].

Interactions between voriconazole and statins have only been reported in case reports [205]. However, based upon its CYP metabolism and lack of interactions with ABC transporters, voriconazole can potentially inhibit CYP3A4-mediated first-pass metabolism of simvastatin and atorvastatin and the CYP2C9-mediated hepatic metabolism of fluvastatin.

11.7.2.5 Clinical Importance of Voriconazole Interactions Affecting CYP-Mediated Biotransformation of Other Drugs

Voriconazole interactions with benzodiazepines, the calcineurin inhibitors, mTOR inhibitors, opioids, warfarin, and phenytoin are clinically significant. Voriconazole increases and prolongs the effects of commonly used hypnotic doses of oral midazolam to the extent that its pharmacological effects are no longer considered "short acting" [232]. The use of voriconazole with oral midazolam should be avoided. The interaction between i.v. midazolam and oral voriconazole is also significant [232]. If high doses or continuous infusions of i.v. midazolam are co-administered with voriconazole, the doses should be adjusted and the patients should be monitored closely [232].

The impact of voriconazole on cyclosporine and tacrolimus pharmacokinetics are qualitatively similar and are likely to become clinically significant if appropriate empiric dose modifications are not made or if enhanced therapeutic drug monitoring is not employed [233–236, 238]. Additionally, voriconazole co-administration with everolimus and sirolimus requires 65%–80% and 90% reductions in everolimus and sirolimus doses, respectively [216, 240–243]. With all these interactions, oral voriconazole may have a more pronounced effect than the i.v. version [238].

Voriconazole co-administration with ibuprofen, meloxicam, and diclofenac may increase the risk of concentration-dependent toxicities, including renal, cardiovascular, or gastrointestinal adverse effects [203]. The interaction between voriconazole and alfentanil is probably only significant when larger alfentanil doses are given either by intermittent bolus or continuous infusion. In these cases, extubation procedures may be delayed, more nausea and vomiting may be observed, and respiratory depression can occur. In regard to other opiates, voriconazole increases the exposure of fentanyl, methadone, oxycodone and buprenorphine. The voriconazole-fentanyl interaction is similar to that of fluconazole and fentanyl [197]. In regard to methadone, voriconazole increases the R-methadone (pharmacologically active) exposure to a lesser extent than S-methadone, and therefore methadone dose reductions may or may not be required [245]. The effect of voriconazole on oral oxycodone is most likely due to intestinal inhibition of CYP3A4 and less oxycodone first-pass metabolism [246]. While this type of interaction can be significant, oxycodone also undergoes metabolism via CYP2D6 and can at least partially compensate for the reduction in CYP3A4 metabolism which mitigates this interaction [174, 175]. Voriconazole can increase the exposure of buprenorphine via CYP3A4 inhibition [249]. The resultant effect may lead to more pronounced buprenorphine analgesic effects and toxicities. Given the danger of prolonged and excessive anticoagulation, the voriconazole-warfarin interaction is clinically significant [252]. Likewise, the increased phenytoin concentrations in the presence of voriconazole may result in phenytoin toxicities, particularly with long-term exposure [253].

11.7.2.6 Management of Voriconazole Interactions Affecting CYP-Mediated Biotransformation of Other Drugs

The use of oral midazolam with voriconazole should be avoided, or substantially lower doses of midazolam should be used. To manage the voriconazole-cyclosporine and tacrolimus interactions, 50% dose reductions have been recommended [269]. However, given the substantial interpatient variability associated with this interaction, dosage adjustments should be individualized [159, 270]. The concomitant use of voriconazole and sirolimus is contraindicated. However, the use of this combination may be unavoidable; therefore, the current literature suggests a sirolimus dose reduction of 90% may be needed to maintain therapeutic concentrations [240, 241, 270]. Similarly, the co-administration of voriconazole and everolimus would require a 65–80% everolimus dose reduction. No matter which immunosuppressant agent is being used, there is a clinically significant interaction with voriconazole co-administration. In studies where empiric dose reductions (as stated above) were employed at the outset of voriconazole administration, patients were more adequately maintained with the respective therapeutic range for the agent being used [216, 234, 236, 240–243, 270].

Caution should be exercised when using alfentanil with voriconazole. Alfentanil dosage adjustments are not needed if only small bolus alfentanil doses are administered during voriconazole treatment. However, patients receiving larger alfentanil doses as repetitive bolus or continuous infusion may require 70–90% reductions in alfentanil dosage for the maintenance of analgesia [244]. The effect of voriconazole on fentanyl results in a 1.4-fold increase in fentanyl exposure [197]. This exposure may lead to increased incidence of sedation and even respiratory depression. A similar increase was seen in R-methadone exposure. Methadone dose reductions and close clinical monitoring may be necessary [245]. In patients receiving voriconazole and oxycodone, there is a modest increase in oxycodone exposure and a potential for increased toxicities [246]. The modest increase is oxycodone results because of compensatory CYP2D6 oxycodone metabolism. However, CYP2D6 is subject to genetic polymorphisms, and more severe oxycodone-induced adverse effects may been seen in poor metabolizer phenotypes or with concomitant CYP2D6 inhibitors [246–248]. In patients receiving warfarin who require voriconazole therapy, the empiric warfarin dose reductions should be considered, and individual dose reductions should be according to INR values [252].

11.7.3   Interactions That Induce Voriconazole Biotransformation

Co-administration of voriconazole with CYP inducers (i.e., phenytoin, ritonavir, efavirenz) can significantly reduce its serum concentrations, which could lead to therapeutic failure [25, 42, 225, 226, 253–256, 271, 272].

11.7.3.1   Phenytoin

The interaction between voriconazole and phenytoin is bi-directional. Initially, repeated administration of oral voriconazole increases steady-state phenytoin concentrations and exposure [250]. However, phenytoin (300 mg/day) co-administration for 2 weeks significantly reduces steady-state voriconazole C_{max}, and exposure, by 50% and 70%, respectively [253]. This interaction persists for up to 12 h post-dose [253].

11.7.3.2 Antiretroviral Agents

Voriconazole plasma concentrations increase with acute co-administration of rito-navir, particularly with higher doses (400 mg) of ritonavir and the effect is most pronounced in CYP2C19 PM phenotype patients [257]. This increase in voricon-azole results from CYP3A4 inhibition by ritonavir [257]. However, with chronic co-administration, ritonavir significantly reduces voriconazole exposure in a dose-dependent fashion [256, 272–275]. This interaction likely results from ritonavir induction of CYP2C19/2C9 [267, 275].

Efavirenz (400 mg daily) co-administration with voriconazole (200 mg twice daily) decreases voriconazole exposure and C_{max} [254]. The interaction is caused by efavirenz induction of CYP3A4, and possibly CYP2C19 or CYP2C9 [254, 255]. See Chaps. 13 and 14 for a comprehensive review of antiretroviral drug interactions.

11.7.3.3 Miscellaneous Drugs

Voriconazole may also be induced by co-administration with phenobarbital, carba-mazepine, rifampin, or other CYP inducers, but data from well-controlled studies describing these interactions are lacking. Recently, flucloxacillin co-administration with voriconazole resulted in subtherapeutic concentrations of voriconazole [258]. This interaction occurred in a single case report and needs more complete investiga-tion, but flucloxacillin is not readily recognized as a major source of drug interactions.

11.7.3.4 Clinical Importance of Interactions That Induce Voriconazole Biotransformation

Interactions that reduce voriconazole serum concentrations are clinically significant because they can precipitate therapeutic failure of voriconazole.

11.7.3.5 Management of Interactions That Induce Voriconazole Biotransformation

In many cases, given the magnitude of the interaction, induction of voriconazole cannot be completely overcome by increasing the voriconazole dose or reducing the dose of the CYP inducer. Therefore, the concomitant use of certain drugs (rifabutin, rifampin, phenobarbital, and carbamazepine) is contraindicated. While doubling the voriconazole dose may compensate for the effect of phenytoin on plasma voricon-azole levels in healthy volunteers [253], this may not work in clinical practice [226]. A similar doubling of the voriconazole dose balanced the efavirenz induction effects while also requiring a 25% decrease in the efavirenz dose [255]. The clinical effectiveness of these dose adjustments is unknown. Co-administration of flucloxacillin and voriconazole may warrant therapeutic drug monitoring of voriconazole [258].

11.8 Interactions Involving Posaconazole

Drug interactions involving posaconazole are summarized in Tables 11.8 and 11.9.

11.8.1 Posaconazole Interactions Involving Gastric pH and Motility

Posaconazole suspension co-administered with esomeprazole significantly reduced posaconazole C_{max} and exposure, clearly indicating this formulation is impacted by changes in pH [28]. Moreover, regardless of fat content, co-administration of solid or liquid food significantly increases posaconazole suspension systemic availability [30, 32, 84]. Increases in gastric emptying caused by metoclopramide may result in clinically insignificant reductions in C_{max} and exposure [28]. The posaconazole delayed release tablet formulation was designed to overcome the effects changes in gastric pH have on the absorption of the oral suspension. Consequently, delayed release tablets are unaffected by food, pH or GI motility agents [37–39, 287]. Compared with oral suspension, under optimal conditions, solid state dosage formulations of posaconazole demonstrated a 35–43% and 38%-43 increase in C_{max} and AUC, respectively. Additionally, the total CL of these solid dosage formulations was decreased by 35–40% [39]. Single daily doses of delayed release posaconazole tablets demonstrate linear pharmacokinetics from 200 to 400 mg, with an approximate three-fold accumulation following multiple doses. Twice daily dosing of posaconazole delayed release tablets (200 mg 2×/day) produced similar total posaconazole exposures compared to 400 mg daily, but the accumulation ratio was five-fold higher after multiple doses [287]. Prophylactic posaconazole delayed release tablets at 300 mg 2×/day on day 1 followed by 300 mg daily thereafter resulted in average steady-state concentrations of 1460 ng/mL in high-risk patients. Ninety-seven percent of these patients achieved concentrations >500 and <2500 ng/mL [288]. Compared to the oral suspension, the delayed release tablets consistently produce higher and more reliable steady-state concentrations in patients. Additionally, the mean steady-state concentrations of delayed release tablets are similar to the intravenous formulation using the same dosing scheme [35, 288]. These concentrations exceed the currently recommend threshold for both treatment and prophylaxis of invasive fungal infections [289–291].

Table 11.8 Posaconazole interactions affecting CYP-mediated biotransformation of other drugs

Drug	Effect on drug (% change)	Inhibition site	Reference(s)
Benzodiazepines			
Midazolam (oral)		Hepatic CYP3A; perhaps intestinal CYP3A	[276, 277]
+ PCZ 200 mg BID	$\uparrow C^{max}$ (120%); $\uparrow AUC_{(0-\infty)}$ (398%); $\uparrow t_{1/2}$ (112%); $\downarrow CL/F$ (81%)		
+ PCZ 400 mg BID	$\uparrow C_{max}$ (133%); $\uparrow AUC_{(0-\infty)}$ (426%); $\uparrow t_{1/2}$ (162%); $\downarrow CL/F$ (82%)		
Midazolam (i.v.)		Hepatic CYP3A	[276]
+ PCZ 200 mg BID	$\uparrow C_{max}$ (30%); $\uparrow AUC_{(0-\infty)}$ (342%); $\uparrow t_{1/2}$ (130%); $\downarrow CL/F$ (76%)		
+ PCZ 400 mg BID	$\uparrow C_{max}$ (68%); $\uparrow AUC_{(0-\infty)}$ (523%); $\uparrow t_{1/2}$ (130%); $\downarrow CL/F$ (83%)		
Calcineurin inhibitors			
Cyclosporine (CSA)*	$\uparrow Cmax$ (5%); $\uparrow AUC_{(tau)}$(33%); $\uparrow t_{1/2}$ (21%); $\downarrow CL/F$ (25%)	Hepatic CYP3A4	[278]
Tacrolimus	$\uparrow C^{max}$ (114%); $\uparrow AUC_{(0-\infty)}$ (323%); $\uparrow t_{1/2}$ (24%); $\downarrow CL/F$ (80%)	Hepatic CYP3A; perhaps intestinal CYP3A/P-gp	[279]
mTOR inhibitors			
Sirolimus	$\uparrow C_{max}$ (537%); $\uparrow AUC_{(0-\infty)}$ (690%); $\uparrow t_{1/2}$ (52%); $\downarrow CL/F$ (89%); $\downarrow Vd/F$ (80%)	Hepatic CYP3A4	[280]
Everolimus	3.8-fold increase trough concentrations	Hepatic CYP3A4	[242]
Antiretrovirals			
Atazanavir		Hepatic CYP3A4	[281]
Atazanavir + PCZ	$\uparrow C_{max}$ (115%); $\uparrow AUC_{(0-24)}$ (67%); $\uparrow t_{1/2}$ (88%); $\downarrow CL/F$ (81%); $\downarrow Vd/F$ (56%)		
Atazanavir/RTV			
Atazanavir + PCZ + RTV	$\uparrow C_{max}$ (47%); T_{max} (200%) $\uparrow AUC_{(0-24)}$ (140%); $\downarrow CL/F$(60%);		
RTV + atazanavir		Hepatic CYP3A4	[281]
RTV + atazanavir + PCZ	$\uparrow C_{max}$ (27%); $\uparrow AUC_{(0-24)}$ (63%); $\uparrow t^{1/2}$ (30%); $\downarrow CL/F$ (51%) $\downarrow Vd/F$ (31%)		
Fosamprenavir + PCZ			
Fosamprenavir + RTV	$\downarrow C_{max}$ (36%); $\downarrow C_{min}$(86%); $\downarrow Vd/F$ (39%); $\downarrow AUC_{(0-12)}$ (65%); CL/(183%)	Hepatic CYP3A4	[282]

(continued)

Table 11.8 (continued)

Drug	Effect on drug (% change)	Inhibition site	Reference(s)
Miscellaneous			
Phenytoin	↑C_{max} (24%); ↑$AUC_{(0-24)}$ (25%);↑bioavailability 15.5%	Considered clinically insignificant; Mechanism unknown	[283]
Simvastatin (40 mg)			
+ PCZ (50 mg daily)	↑C_{max} (643%); ↑$AUC_{(0-\infty)}$ (436%)	Intestinal and hepatic CYP3A4	[277]
+PCZ (100 mg daily)	↑C_{max} (837%); ↑$AUC_{(0-\infty)}$ (862%)		
+PCZ (200 mg daily)	↑C_{max} (1042%); ↑$AUC_{(0-\infty)}$ (907%)		
Simvastatin acid			
+ PCZ (50 mg daily)	↑C_{max} (452%); ↑$AUC_{(0-\infty)}$ (399%)	Intestinal and hepatic CYP3A4	[277]
+ PCZ (100 mg daily)	↑C_{max} (817%); ↑$AUC_{(0-\infty)}$ (584%)		
+ PCZ (200 mg daily)	↑C_{max} (851%); ↑$AUC_{(0-\infty)}$ (645%)		
Topiramate			
+ PCZ (200 mg Q.I.D.)	No formal PK studies ↑Topiramate serum concentrations ~fivefold	Likely intestinal and/or hepatic CYP3A4	[284]
Inhaled fluticasone			
+PCZ (200 mg T.I.D.)	No PCZ concentrations Drug-induced adrenal insufficiency after 12 months	Hepatic CYP3A4	[285]

Table 11.9 Interactions that induce posaconazole biotransformation or inhibit its absorption

Drug	Effect on posaconazole (% change)	Mechanism	Reference(s)
Gastric pH			
Modifiers			
Esomeprazole	↓C_{max} (43%); AUC (37%)	↑ gastric pH and ↓ absorption	[28]
Inducing agents			
Efavirenz	↓C_{max} (40%); ↓$AUC_{(0-24)}$ (46%);	Induction of UGT-mediated	[281]
	↑CL/F (99%)	Glucuronidation	
Phenytoin	↓C_{max} (44%); ↓$AUC_{(0-24)}$ (52%);	Mechanism unknown	[283]
	↑CL/F(90%)	Possibly induction of CYP3A4 or UGT1A4	
Rifampin	↓58%–80% in PCZ serum concentrations	CYP3A4; UGT1A4; P-gp	[286]

[a]Values are normalized to dose

Furthermore, the delayed release tablets produce significantly higher posaconazole concentrations and less pharmacokinetic variability compared to oral suspension even in its most optimal conditions [37–39, 287].

11.8.1.1 Clinical Importance of Posaconazole Interactions Involving Gastric pH Interactions

Reduced posaconazole absorption may lead to therapeutic failure. However, elevated gastric pH interactions with posaconazole are unavoidable in certain patients. The delayed release tablet formulation of posaconazole alleviates the food, pH and co-medications restrictions that limited the oral suspension, however in patients that cannot swallow (i.e., severe mucositis) the oral suspension may still be necessary. The availability of an i.v. formulation of posaconazole also provides an additional option for patients who cannot tolerate either oral dosage form.

11.8.1.2 Management of Posaconazole Interactions Involving Gastric pH Interactions

Posaconazole oral suspension interactions involving alterations in gastric pH may be managed by administering it in divided doses; with or after a meal, or with a nutritional supplement, or an acidic beverage [27, 28, 33]. Delayed release posaconazole tablets are not affected by food, pH of GI motility agents and this would be the preferred dosage formulation in most patients [38, 39, 287]. Switching patients from oral suspension to the delayed release tablet formulation has resulted in improved posaconazole concentrations that meet or exceed the current threshold for treatment and prophylaxis of invasive fungal infections [290–292].

11.8.2 Posaconazole Interactions Affecting CYP-Mediated Biotransformation of Other Drugs

11.8.2.1 Benzodiazepines

Posaconazole significantly inhibits CYP3A metabolism of intravenous or oral midazolam [276, 277]. Posaconazole oral suspension (200 or 400 mg BID) co-administration significantly increases oral midazolam (2 mg) exposure with dose-dependent increases in posaconazole exposure with numerically similar decrease in CL [276, 277]. Likewise, oral midazolam C_{max}, $t_{1/2}$, and T_{max} are all increased in a dose-dependent manner [276, 277]. Similar changes were seen when posaconazole oral suspension (200 or 400 mg BID) was co-administered with i.v. midazolam [276]. Posaconazole, regardless of dosage form likely interacts with other benzodiazepines that are CYP3A4 substrates (triazolam, alprazolam, etc.), but data regarding such interactions are lacking.

11.8.2.2 Calcineurin Inhibitors and mTOR Inhibitors

Posaconazole significantly interacts with the calcineurin inhibitors. The magnitude of the interaction between cyclosporine and systemic azoles is similar (~50% increase in cyclosporine concentrations), but its onset varies between azoles [233, 269, 278, 279]. In patients receiving cyclosporine and posaconazole oral suspension prophylaxis (200 mg 3×/day), the cyclosporine concentrations significantly increased over the 30 day period [278]. The mean cyclosporine concentration to dose ratio also significantly increased over this same period. The results of this interaction demonstrated a 50% increase in cyclosporine concentrations, however these increases did not require a cyclosporine dose adjustment until day 14 and with increased doses being needed through day 30 [278]. Given this, a 50% dose reduction in cyclosporine is likely needed, but due to the delay in this interaction, this dose adjustment should be made based upon individual cyclosporine concentrations and gradually changed over the period of the interaction [269, 278, 279, 293]. A significant interaction (increased exposure, C_{max}, $t_{1/2}$, and reduced CL/F) between posaconazole suspension and single-dose tacrolimus has also been reported [279]. The results of this study indicate that the tacrolimus dose should be decreased by a factor of 3, which is 25% more than the reduction needed for sirolimus [269, 294].

Posaconazole suspension (400 mg twice daily) significantly increased the single dose C_{max}, exposure and $t_{1/2}$ of sirolimus [280]. In addition, the interaction reduced sirolimus apparent volume of distribution (Vd/F), and CL/F, 80% and 88%, respectively [280]. Clinically, posaconazole oral suspension (200 mg 3×/day) increased the sirolimus concentration/dose ratio by 2.7-fold. An empiric 50% reduction in the sirolimus dose is recommended when posaconazole is co-administered [295, 296]. Lastly, a drug interaction modeling study suggests that the drug interaction potential of posaconazole may increase due to the increased systemic concentrations, improved bioavailability and pharmacokinetic profile associated with the delayed release tablets [297]. The investigation predicted a greater and more rapid interaction between the two drugs. The 50% inhibitory concentration for posaconazole was 0.68 mcg/mL, which is readily achievable with the new tablet formulation and suggests that an 80% reduction in the sirolimus dose may be needed when using delayed release posaconazole tablets [297]. The interaction is likely due to posaconazole inhibition of CYP3A-mediated sirolimus metabolism [280, 295–297]. Whether the delayed release tablets increase the drug interaction potential of posaconazole clinically remains to be seen. Furthermore, the contribution of P-gp inhibition by posaconazole to this interaction is unknown. In a single case report involving everolimus and sequential use of voriconazole and then posaconazole, both agents increased everolimus concentrations [242]. Posaconazole oral suspension (400 mg 2×/day) resulted in a 3.8-fold increase in everolimus trough concentrations, which was 50% less than the increase seen with voriconazole and everolimus co-administration [242].

11.8.2.3 Phenytoin

One parallel-designed interaction study demonstrated a bi-directional interaction between posaconazole suspension and phenytoin. Posaconazole suspension co-administration produced modest, but not statistically significant, increase in steady-state phenytoin C_{max} (24%), exposure (25%), and relative bioavailability (15.5%) which are not considered clinically significantly [283]. However, this study used a small number of healthy volunteers who did not serve as their own controls and received substandard doses of unmarketed posaconazole solid dosage forms (200 mg per day) and phenytoin (200 mg per day) [283]. Whether these limitations impacted the magnitude of the observed interaction or whether the interaction occurs with marketed dosage forms of posaconazole is unclear.

11.8.2.4 Miscellaneous Drugs

Posaconazole suspension interacts with several other medicines including atazanavir and ritonavir. Drug interactions involving antiretroviral agents are discussed in detail in Chaps. 13 and 14. However, healthy volunteers in part 1 of a two-part crossover study received the protease inhibitor atazanavir alone and then co-administered with either ritonavir or posaconazole. In addition, subjects received all three concomitantly [281]. Atazanavir and ritonavir are CYP3A4 substrates and inhibitors. Posaconazole suspension co-administration (400 mg twice daily for 7 days) also increased atazanavir exposure, C_{max}, and $t_{1/2}$. In addition, the interaction reduced atazanavir Vd/F and CL/F via CYP3A4 inhibition [281]. However, because both ritonavir and posaconazole inhibit CYP3A4, when all three were administered together, no additional increases in the concentrations and exposure of atazanavir were observed compared with ritonavir and atazanavir administration together [281]. In this study, posaconazole suspension co-administration modestly increased ritonavir exposure and C_{max} compared with ritonavir and atazanavir administration alone [281]. Similar to the above results, posaconazole suspension co-administration with fosamprenavir modestly increased amprenavir concentrations [282]. When fosamprenavir plus posaconazole was compared to fosamprenavir plus ritonavir, the fosamprenavir plus posaconazole exposure was decreased by 2.8-fold as a result of an increase in fosamprenavir CL [282]. The result of this interaction demonstrates that posaconazole when given as a suspension is a less potent inhibitor of CYP3A4 compared to ritonavir and that posaconazole cannot eliminate or replace the need for ritonavir boosting of fosamprenavir [282].

Steady-state plasma rifabutin exposure and C_{max} increase 72% and 31%, respectively, following repeated administration of an unmarketed posaconazole tablet (200 mg once daily for 10 days) in healthy volunteers [298].

Posaconazole suspension significantly increased the steady-state concentrations of simvastatin and simvastatin acid (primary metabolite) [277]. When posaconazole suspension (50 mg, 100 mg, and 200 mg daily) was administered with simvastatin 40 mg, the simvastatin C_{max} increased 6.4-fold to 10.4-fold, and its exposure was also significantly increased 5.4-fold to 9-fold [277]. Simvastatin clearance was also significantly reduced by a similar magnitude. The degree of simvastatin acid pharmacokinetic changes were parallel to those of simvastatin [277]. In a single case report, posaconazole suspension significantly increased the concentration of topiramate, but not concurrent valproic acid [284]. The case report lacked baseline topiramate concentrations, but compared to topiramate pharmacokinetic data, the patient's topiramate level was increased by over fivefold [284]. Removal of posaconazole resulted in a decrease in topiramate levels. Concurrent valproic acid concentrations remained stable, and posaconazole concentrations were not reported [284]. The primary mechanism of this interaction is believed to be from CYP3A4 inhibition, but inhibition of P-gp may also be involved [284]. A case report described drug-induced Cushing's syndrome associated with posaconazole and inhaled fluticasone [285]. Concomitant administration of inhaled fluticasone and posaconazole suspension resulted in adrenal insufficiency after 12 months; however, prior to posaconazole therapy, this patient received prophylactic itraconazole (with inhaled fluticasone) for 7 years without incident [285]. While no azole levels were reported with either drug, the lack of an interaction with itraconazole therapy may have been due to reduced absorption and therefore insufficient itraconazole concentrations to inhibit hepatic CYP3A4 [166].

11.8.2.5 Clinical Importance of Posaconazole Interactions Affecting CYP-Mediated Biotransformation of Other Drugs

Posaconazole interactions with midazolam, the calcineurin and mTOR inhibitors, and simvastatin are clinically significant. Similar to the other azoles, the posaconazole-calcineurin and mTOR inhibitor interactions may cause adverse events or toxicities when clinicians fail to properly monitor blood concentrations and make dosage adjustments accordingly. In the case of midazolam and simvastatin, the use of alternative agents that are not metabolized by CYP3A4 (i.e., temazepam and pravastatin, respectively) may be more prudent during posaconazole co-administration, particularly when the posaconazole treatment duration is prolonged. These interactions illustrate that even drugs like posaconazole that are minimally metabolized by CYP3A4 can potently inhibit this important metabolic enzyme. Clinicians may miss or confuse this point and mistakenly believe that because posaconazole is a poor CYP3A4 substrate, it will be relatively devoid of drug interactions.

11.8.2.6 Management of Posaconazole Interactions Affecting CYP-Mediated Biotransformation of Other Drugs

Data regarding the management of patients receiving benzodiazepines, other than midazolam, and concomitant posaconazole therapy are lacking. Clinicians should consider empirical dose adjustments and monitoring of benzodiazepine adverse events in patients receiving midazolam or other benzodiazepines that are metabolized by CYP3A4 (e.g., triazolam, alprazolam). Alternatively, non-CYP 3A4 metabolized benzodiazepines (e.g., temazepam, lorazepam) could be used when clinically appropriate.

The interaction between the azoles and calcineurin or mTOR inhibitors is well known and should be avoidable. Management of these interactions necessitates therapeutic drug monitoring, adjusting, or substituting calcineurin or mTOR inhibitor therapy. Empirically derived dose adjustments are a good starting point to manage these interactions. A small retrospective study in lung transplant recipients suggests the posaconazole interaction may be safely managed by empirically reducing the tacrolimus dose by a factor of 3, with subsequent tapering to a mean of 2 mg daily [294]. In patients receiving sirolimus and posaconazole oral suspension, a 50% dose reduction is recommended, but more recently modeling data suggest an 80% dosage reduction may be required with the delayed release tablet formulation [295–297]. Empiric cyclosporine dosage reductions of 50% or more maintained therapeutic drug concentrations more effectively than dosage reductions of less than 50% [278]. Such dosage reductions may need to be gradually implemented as the full extent of this interaction may not manifest until day 14 or after [278]. The combination of posaconazole and everolimus demonstrated a significant effect on everolimus concentrations; however, this was a single case report; no definitive dosing recommendations can be established [242]. Despite these empiric dose reduction recommendations, therapeutic drug monitoring of the calcineurin and mTOR inhibitors should be performed before, during, and after all azole use. Any dose adjustment should be based upon the objective results of these blood concentration data.

11.8.3   Interactions That Induce Posaconazole Biotransformation

Posaconazole co-administration with CYP inducers (i.e., phenytoin, ritonavir, efavirenz) can significantly reduce its serum concentrations and potentially lead to therapeutic failure. As discussed above, one study demonstrated that posaconazole interacts with phenytoin. Despite the previously addressed limitations of that study, phenytoin co-administration significantly reduced steady-state posaconazole exposure and C_{max}. There was also a 57% reduction in posaconazole $t_{1/2}$ and a 90% increase in its steady-state CL [283]. Rifampin reduced posaconazole serum concentration by 58–80% in a single case report [286]. Interactions between marketed posaconazole formulations and rifabutin have not been reported, but an interaction similar to that observed with rifampin is likely to occur. Efavirenz co-administration decreased posaconazole exposure and C_{max} [281]. Posaconazole undergoes glucuronidation via UGT1A4, and phenytoin, rifabutin, rifampin, and efavirenz all induce UGT activity. Therefore, all these interactions are believed to result from or be due in part to induction of UGT-mediated posaconazole glucuronidation [281,

283, 286, 298]. Co-administration with fosamprenavir resulted in a 29% decrease in posaconazole AUC corresponding to a 29% increase in posaconazole clearance [282]. The magnitude of this interaction is less than that seen with other UGT inducers and may reflect differences in induction of UGT1A4 [281–283, 298]. Secondly, fosamprenavir may induce P-gp, but the cause of this interaction is unknown [282].

11.8.3.1 Clinical Importance of Interactions That Induce Posaconazole Biotransformation

Interactions that induce posaconazole biotransformation may precipitate therapeutic failure and are therefore potentially clinically significant. In addition, these interactions are often bi-directional and may increase the risk of toxicity associated with the inducer.

11.8.3.2 Management of Interactions That Induce Posaconazole Biotransformation

Because these interactions are bi-directional, increased plasma concentrations of phenytoin, rifampin, efavirenz, and fosamprenavir should be expected when they are co-administered with posaconazole. Although frequent monitoring for adverse events and toxicity is recommended, if possible avoid these combinations due to the decreased posaconazole exposure and subsequent risk for therapeutic failure.

11.9   Interactions Involving Isavuconazole

Drug interactions involving isavuconazole are summarized in Tables 11.10 and 11.11.

11.9.1   Isavuconazole Interactions Involving Gastric pH and Motility

Isavuconazole is the active moiety that is formed after gut and plasma esterase cleavage of isavuconazonium. In healthy volunteers, isavuconazole was unaffected by the presence of food [301]. Additionally, isavuconazole co-administration with esomeprazole 40 mg resulted in negligible effects on isavuconazole C_{max} or AUC [301]. The effects of prokinetic agents on isavuconazole absorption are unknown.

Table 11.10 Isavuconazole interactions affecting CYP-mediated biotransformation of other drugs

Drug	Effect on drug (% change)	Inhibition site	Reference(s)
Benzodiazepines			
Midazolam	$\uparrow C_{max}$ (69%); $\uparrow T_{max}$ (63%); $\uparrow AUC_{(0-\infty)}$ (103%); $\downarrow Cl/F$ (50%)	Intestinal and hepatic CYP3A4	[46]
Calcineurin			
Inhibitors			
Cyclosporine	$\uparrow C_{max}$ (7%); $\uparrow AUC_{(0-\infty)}$ (29%); $\downarrow Cl/F$ (21%)	Hepatic CYP3A4	[299]
Tacrolimus	$\uparrow C_{max}$ (39%); $\uparrow T_{max}$ (100%); $\uparrow AUC_{(0-\infty)}$ (122%); $\downarrow Cl/F$ (57%)	Intestinal and hepatic CYP3A4	[299]
mTOR inhibitors			
Sirolimus	$\uparrow C_{max}$ (67%); $\downarrow T_{max}$ (20%); $\uparrow AUC_{(0-\infty)}$ (77%); $\downarrow Cl/F$ (46%)	Intestinal and hepatic CYP3A4	[299]
Anticoagulants			
Warfarin		Intestinal and hepatic CYP3A4	[102]
S-warfarin	$\downarrow C_{max}$ (12%); $\uparrow T_{max}$ (150%); $\uparrow AUC_{(0-\infty)}$ (11%)		
R-warfarin	$\downarrow C_{max}$ (8%); $\uparrow T_{max}$ (131%); $\uparrow AUC_{(0-\infty)}$ (20%)		
Miscellaneous			
Mycophenolate acid	$\downarrow C_{max}$ (10%); $\uparrow AUC_{(0-\infty)}$ (39%); $\downarrow Cl/F$ (26%)	UGT	[299]
Prednisolone	$\downarrow C_{max}$ (4%); $\uparrow T_{max}$ (33%); $\uparrow AUC_{(0-\infty)}$ (8%)	Intestinal and hepatic CYP3A4	[299]
Dextromethorphan	$\uparrow C_{max}$ (18%); $\uparrow AUC_{(0-\infty)}$ (19%); $\downarrow Cl/F$ (47%)		[64]
Buproprion	$\downarrow C_{max}$ (42%); $\downarrow AUC_{(0-\infty)}$ (40%); $\uparrow Cl/F$ (78%)	Induction of CYP2B6	[64]
Methadone		Induction of CYP2B6	[64]
S-Methadone	$\uparrow C_{max}$ (<1%); $\downarrow AUC_{(0-\infty)}$ (32%); $\uparrow Cl/F$ (59%)		
R-Methadone	$\uparrow C_{max}$ (4%); $\downarrow AUC_{(0-\infty)}$ (10%); $\uparrow Cl/F$ (11%)		
Lopinavir/ritonavir		Possibly induction of CYP3A4	[300]
Lopinavir	$\downarrow C_{max}$ (19%); $\downarrow AUC_{(tau)}$ (19%);		
Ritonavir	$\downarrow C_{max}$ (33%); $\downarrow AUC_{(tau)}$ (27%);		

11.9.1.1 Clinical Importance of Isavuconazole Interactions Involving Gastric pH and Motility Interactions

Co-administration with food or esomeprazole did not influence the absorption of isavuconazole. It is unknown if GI motility enhancers will effect isavuconazole absorption.

Table 11.11 Interactions that affect isavuconazole biotransformation or absorption

Drug	Effect on isavuconazole (% change)	Mechanism	Reference(s)
Miscellaneous			
Lopinavir/ritonavir		Inhibition of intestinal and hepatic CYP3A4	[300]
Isavuconazole			
(100 mg TID, 100 mg daily)	$\uparrow C_{max}$ (94%);$\uparrow AUC_{(tau)}$ (112%);		
Isavuconazole			
(200 mg TID, 200 mg daily)	$\uparrow C_{max}$ (74%);$\uparrow AUC_{(tau)}$ (96%);		
Ketoconazole	$\uparrow C_{max}$ (10%); $\uparrow T_{max}$ (50%); $\uparrow AUC_{(0-\infty)}$ (450%); $\downarrow Cl/F$ (80%)	Inhibition of hepatic CYP3A4	[46]
Rifampin	$\downarrow C_{max}$ (75%);$\downarrow AUC_{(0-\infty)}$ (97%);$\uparrow Cl/F$ (864%)	Induction of intestinal and hepatic CYP3A4	[46]

11.9.1.2 Management of Isavuconazole Interactions Involving Gastric pH and Motility Interactions

The evidence to date indicate isavuconazole can be administered without regard to food or acid-suppressing agents [301]. Its co-administration with prokinetic agents should be avoided or maximally separated until formal pharmacokinetic studies are performed.

11.9.2 Isavuconazole Interactions Affecting CYP-Mediated Biotransformation of Other Drugs

11.9.2.1 Benzodiazepines

Isavuconazole co-administered in therapeutic doses to healthy volunteers delayed single-dose midazolam (3 mg) T_{max} by 0.6 h; increased the C_{max} and AUC 69% and 103%, respectively; and reduced CL by approximately 50% [46].

11.9.2.2 Calcineurin Inhibitors and mTOR Inhibitors

Consistent with other azoles, isavuconazole increases the exposure of cyclosporine, sirolimus, and tacrolimus [299]. The magnitude of these interactions varies with cyclosporine being the least and tacrolimus being the most affected. In healthy volunteers, cyclosporine C_{max} and AUC are increased by 7% and 29%, respectively, while the clearance is decreased by 21% [299]. Sirolimus pharmacokinetics were studied in healthy volunteers with and without isavuconazole [299]. Sirolimus C_{max} and AUC increased 67% and 77%, respectively, while its clearance decreased by 46% with co-administered isavuconazole [299]. In healthy volunteers co-administration of isavuconazole increased tacrolimus C_{max} and exposure by 39% and 122%, respectively, while decreasing its clearance by 56% [299]. A single case report of co-administered isavuconazole and tacrolimus found a similar increase in tacrolimus exposure that continued to increase over time [302].

11.9.2.3 Warfarin

Warfarin 20 mg as a single dose was given to 20 healthy volunteers with and without isavuconazole 200 mg 3x/day for 2 days followed by 200 mg daily. The pharmacokinetics of both warfarin enantiomers, S-warfarin, and R-warfarin were measured. The mean S-warfarin and R-warfarin C_{max} were 12% and 8% lower, respectively, and AUCs were increased 11% and 20%, respectively [102]. Isavuconazole co-administration with warfarin results in clinically insignificant changes in warfarin pharmacokinetics and anticoagulant effects [102]. As expected, the R-warfarin enantiomer is more influenced by isavuconazole, most likely as a result of CYP3A4 inhibition.

11.9.2.4 Miscellaneous Drugs

The pharmacokinetics of mycophenolate acid were evaluated in 21 healthy volunteers with and without the presence of isavuconazole [299]. Overall exposure of mycophenolate acid was increased by 39% with a 26% decrease in clearance. However, the mycophenolate acid C_{max} was decreased by 10%. Mycophenolate acid phenyl glucuronide, the primary metabolite of mycophenolate acid, exposure, and C_{max} were decreased in the presence of isavuconazole [299]. The changes in the primary metabolite pharmacokinetics demonstrate mild inhibition of UGT by isavuconazole, which is consistent with previous reports [62, 63, 299].

The effect of isavuconazole on the disposition of prednisolone, ethinyl estradiol, and norethindrone has been explored in two healthy volunteer pharmacokinetic studies [46, 299]. Isavuconazole co-administration produced clinically insignificant changes in the pharmacokinetics of ethinyl estradiol, norethindrone, and prednisolone [46, 299].

In the presence of isavuconazole, bupropion C_{max} and AUC were decreased by 31% and 40%, respectively, while bupropion clearance increased by 78% [64]. These changes in bupropion concentrations and exposure are consistent with isavuconazole being a weak inducer of CYP2B6. Although CYP2B6 displays a wide variety of genetic polymorphisms, which may influence the magnitude of this interaction, genotypic testing for CYP2B6 was not performed in this study [303]. Methadone is formulated as a racemic mixture of S-methadone and R-methadone, with R-methadone being the pharmacologically active enantiomer [303].

Isavuconazole co-administration with a single dose of methadone decreased S-methadone exposure more than R-methadone (32% versus 10%, respectively) [64]. These results are consistent with the stereoselective metabolism that methadone undergoes. Metabolism of the pharmacologically active enantiomer R-methadone is catalyzed by CYP2C19, whereas S-methadone metabolism is preferentially catalyzed by CYP2B6. CYP3A4 plays only a minor role in the metabolism of methadone [64, 303, 304]. Thus, the difference in enantiomer exposures reflects weak induction of CYP2B6 by isavuconazole and its negligible effects on CYP2C19 metabolism [303, 304].

Multiple dose isavuconazole increased the Cmax and AUC of dextromethorphan by 18% and 19%, respectively. Dextromethorphan is a CYP2D6 and CYP3A4 substrate, but these effects most likely result from isavuconazole inhibition of CYP3A4 [64].

11.9.2.5 Clinical Importance of Isavuconazole Interactions Affecting CYP-Mediated Biotransformation of Other Drugs

While the increase in midazolam exposure produced by isavuconazole co-administration is clinically significant, the magnitude of this interaction is less than that seen with itraconazole, voriconazole, and posaconazole [46, 113, 115, 232, 276, 277] and similar to changes observed with fluconazole co-administration [113, 190]. Regardless, the interaction is sufficient enough to enhance the pharmacodynamic effects of midazolam, and individual patient response should be monitored. The inhibitory effects of isavuconazole on the metabolism of cyclosporine, sirolimus, and tacrolimus are consistently less than that of other azoles [299]. Nonetheless, like the interaction between isavuconazole and midazolam, the interaction with each agent is clinically significant and requires management to avoid adverse outcomes. In addition to the moderate CYP3A4 inhibition demonstrated above, isavuconazole is also a weak inducer of CYP2B6 and may decrease concentrations of drugs metabolized by this enzyme [62, 64]. The induction of CYP2B6 metabolism of bupropion by isavuconazole may produce subtherapeutic concentrations and clinical failures of bupropion [64]. Although isavuconazole decreased the pharmacologically inactive S-methadone exposure by CYP2B6 induction, this isoform may contribute to the variability in methadone dose-response relationships [64]. The clinical significance of this interaction is unknown.

11.9.2.6 Management of Isavuconazole Interactions Affecting CYP-Mediated Biotransformation of Other Drugs

The isavuconazole interaction with midazolam produces effects similar in magnitude to those observed when fluconazole and midazolam are co-administered and can be managed by reducing midazolam based on clinical response [46, 113, 190]. The moderate increases in cyclosporine exposure caused by isavuconazole co-administration do not require the dose of the calcineurin inhibitor be reduced. However, using these two drugs together requires cyclosporine concentrations be closely monitored and the patient assessed for signs of toxicity. In contrast, given the magnitude of the interaction caused by isavuconazole co-administration, empiric reductions in the sirolimus dose of at least 30% may be warranted. However, formal recommendations do not exist, and until they are established, early and more frequent sirolimus therapeutic drug monitoring should be performed. Similarly, when co-administered with isavuconazole, empiric reductions in the dose of tacrolimus of at least 50% may be warranted, and its concentrations should be closely monitored [299]. Additionally, further tacrolimus dose reductions may be needed with continued co-administration [302]. Isavuconazole minimally affected the pharmacokinetics of a single dose of warfarin [102]. Isavuconazole inhibited warfarin metabolism in a stereoselective manner, primarily via inhibition of CYP3A4 but not CYP2C9 [102]. Because CYP2C9 catalyzes metabolism of the pharmacologically active S-enantiomer, the effects of warfarin anticoagulation were considered clinically insignificant [102]. However, because the effects of isavuconazole on multiple doses of warfarin are unknown, INR should be closely monitored [102]. Isavuconazole produced increased mycophenolate acid exposure most likely by reducing the formation of the primary phenyl glucuronide metabolite via UGT inhibition [299]. This increase in the mycophenolate acid exposure should not require empiric dose reductions; however, in the absence of routine therapeutic drug monitoring, close clinical monitoring for mycophenolate acid toxicities (i.e., complete blood cell count with differential) is recommended [299]. Patients receiving concomitant bupropion and isavuconazole may require higher doses of bupropion due to isavuconazole's induction of CYP2B6. However, bupropion CYP2B6 metabolism is highly variable due to significant genetic polymorphisms; thus empiric increases in its dose are discouraged; instead dosing should be guided by patient response [64]. Even though the effects of isavuconazole co-administration are primarily limited to the pharmacologically inactive S-methadone, as an weak inducer of CYP2B6, it may cause additional variation in methadone dose-response relationships [303].

11.9.3   Interactions Affecting CYP-Mediated Biotransformation of Isavuconazole

The pharmacokinetics of isavuconazole were assessed with co-administered fixed-dose lopinavir-ritonavir (400 mg/100 mg) twice daily [300]. At 50% of the standard isavuconazole dose (100 mg three times a day for 2 days, then 100 mg daily), the mean C_{max} and mean AUC increased by 94% and 112%, respectively [300]. When isavuconazole was administered at the standard therapeutic dose with lopinavir-ritonavir, the mean C_{max} increased by 74% and the mean AUC increased by 96%. The isavuconazole T_{max} was unchanged in either analysis. This study also demonstrated a bi-directional induction effect of isavuconazole on the pharmacokinetics of lopinavir and ritonavir.

Cyclosporine, sirolimus, and tacrolimus all influenced the Cmax of isavuconazole but have little impact on isavuconazole's exposure ($< 12\%$); therefore, the clinical significance of these interactions is negligible [299].

11.9.3.1 Clinical Importance of Interactions Affecting CYP-Mediated Biotransformation of Isavuconazole

Co-administration of isavuconazole with ritonavir significantly increased its exposure by a factor of 2 [46, 300]. Ritonavir may also increase CYP protein transcription, which could have lessened the magnitude of this interaction [46, 300]. When these drugs are co-administered, close clinical monitoring for isavuconazole toxicities is necessary, and isavuconazole therapeutic drug monitoring may also be warranted as clinically available.

11.9.4 CYP-Mediated Interactions That Induce Isavuconazole Biotransformation

The pharmacokinetics of isavuconazole (400 mg on day 1, then 100 mg daily) alone and with concomitant rifampin (600 mg daily) were compared in healthy volunteers [46]. Rifampin co-administration increased isavuconazole clearance nearly tenfold, leading to decreases in its C_{max} and exposure by 75% and 97%, respectively [46].

11.9.4.1 Clinical Importance of CYP-Mediated Interactions That Induce Isavuconazole Biotransformation

Rifampin significantly decreased isavuconazole exposure and C_{max} [46]. These effects will result in isavuconazole treatment failures if no other measures are taken. Similarly, strong inducers of CYP3A4 and/or CYP3A5 are also likely to decrease isavuconazole exposure and compromise its efficacy; however, agents other than rifampin have not been studied to date.

The changes in isavuconazole disposition caused by rifampin co-administration are clinically significant as they would result in antifungal treatment failures if no other measures are taken [46]. However, interaction studies with CYP inducers other than rifampin have not been yet been published.

11.9.4.2 Management of CYP-Mediated Interactions That Induce Isavuconazole Biotransformation

Therefore, concomitant administration of isavuconazole and rifampin should be avoided. Similarly, co- administration of other inducers of CYP3A4 and perhaps CYP3A5 should also be avoided until these agents are formally studied.

11.9.5 Isavuconazole Interactions Affecting Non-CYP-Mediated Biotransformation of Other Drugs

The effects of isavuconazole on the pharmacokinetics of atorvastatin (OATP1B1 and OATP1B2, BCRP, P-gp, and CYP3A4 substrate), digoxin (P-gp substrate), metformin (organic cation transporter 1 (OCT1), OCT2 and multidrug and toxin extrusion protein (MATE)-1 substrate), and methotrexate (organic anion transporter 1 and 3 (OAT1 and OAT3) substrate) were studied in healthy volunteers [106, 151].

Isavuconazole increased atorvastatin exposure by 37% as a result of a 35% reduction in its clearance. The C_{max} and T_{max} of atorvastatin were largely unchanged by isavuconazole [106]. Atorvastatin is a substrate of OATP1B1/2, P-gp, BRCP, and CYP3A4 [106, 151]. Isavuconazole did not influence repaglinide, a substrate of OATP1B1, which suggests the underlying mechanism for this interaction does not involve OATP1B1 [64]. Although, as discussed below, isavuconazole is a P-gp inhibitor in vivo, the C_{max} of atorvastatin was only modestly changed (3%), suggesting that P-gp plays a minor role in atorvastatin disposition and is not the underlying mechanism of this interaction. Isavuconazole inhibits BCRP in vitro; however, on the interaction study with atorvastatin and, as described below with methotrexate, a BCRP substrate, these data suggest that isavuconazole is not an inhibitor of this ABC transporter [18, 106]. Thus, the isavuconazole-atorvastatin interaction is best explained by CYP3A4 inhibition [106]. The C_{max}, T_{max}, and exposure of digoxin were increased by 25%, 50%, and 33%, respectively, while total digoxin CL was reduced by 26% when co-administered with isavuconazole [106]. Isavuconazole inhibits P-gp at clinically achievable concentrations with 50% inhibitory concentrations (IC_{50}) similar to itraconazole. However, the magnitude of the isavuconazole-digoxin interaction is less than that observed with itraconazole [18, 145]. Isavuconazole increased total metformin exposure by 50% as well as C_{max} by 23% [106]. However, the drug transporter responsible for this change was not addressed. By comparison, when metformin was co-administered with cimetidine and pyrimethamine (known OCT1, OCT2, and MATE1 inhibitors), metformin concentrations were increased but varied considerably between agents [103]. Thus, it is likely that one or more of these drug transporters are responsible for this interaction, but the extent which each transporter contributes to the interaction needs further investigation. Methotrexate pharmacokinetics were largely unaffected by isavuconazole, but the C_{max} and exposure of its inactive and toxic metabolite (7-hydroxymethotrexate) were increased by 15% and 29%, respectively [106]. Methotrexate disposition is complex and in vitro and animal data suggest it may be governed by a variety of membrane transporters including, but not limited to, OATP1A2, OATP1B1, OATP1B3, OAT1, OAT3, MRP2, MRP3, and BCRP [106, 305–307]. Studies to assess whether isavuconazole inhibits OATP1A2 using a specific substrate in vivo are lacking. Durmus et al. [308] demonstrated in transgenic mice that OATP1A2 may have more of a role in methotrexate disposition via the renal tubule system; however, the lack of a significant effect of isavuconazole on methotrexate exposure suggests that it is not an inhibitor of this transport protein [106]. Similarly, as discussed previously, isavuconazole did not influence the pharmacokinetics of repaglinide, which is an OATP1B1 substrate. Therefore, isavuconazole seemingly does not influence OATP1B1. Additionally, the lack of a significant effect on methotrexate, an OATP1B3, OAT1, and OAT3 substrate, also indicates that isavuconazole does not inhibit these drug transport proteins [106, 308]. Isavuconazole also did not alter the pharmacokinetics of methotrexate or atorvastatin, both of which are BCRP substrates thus signifying that isavuconazole does not inhibit BCRP in vivo [106]. It should be noted that neither methotrexate nor atorvastatin is a specific substrate for BCRP; therefore, further data is needed to completely rule out the possibility of this interaction. Lastly, MRP2 and MRP3 have demonstrated a role in controlling methotrexate disposition in murine studies [308]. However, an in vitro study demonstrated that isavuconazole did not influence these transport proteins, but substrate-specific data is lacking [18]. Given the absence of an effect of isavuconazole on methotrexate pharmacokinetics, the increased exposure and C_{max} of the 7-hydroxymethotrexate, a toxic metabolite, with concomitant isavuconazole are puzzling and unknown [106, 305–308]. Given the role this metabolite has in causing methotrexate toxicity, further study is needed to understand and characterize the underlying mechanism of this interaction.

11.9.5.1 Clinical Importance of Isavuconazole Interactions Affecting Non-CYP-Mediated Biotransformation of Other Drugs

The increase in atorvastatin exposure caused by isavuconazole CYP3A inhibition could increase the risk of statin-associated skeletal muscle toxicity [106]. Similarly, given the narrow therapeutic index of digoxin, co-administration with isavuconazole results in clinically significant toxicity [106]. The overall increase in metformin exposure during isavuconazole co-administration is numerically larger; however, metformin dosing varies widely and its therapeutic index is large; thus, the interaction is likely not clinically significant and may result in enhanced blood glucose control [106]. Isavuconazole does not affect methotrexate pharmacokinetics, but it does affect the exposure of its primary circulating metabolite, which may result in increased methotrexate toxicity [106, 307].

11.9.5.2 Management of Isavuconazole Interactions Affecting Non-CYP-Mediated Biotransformation of Other Drugs

The increase in atorvastatin exposure will likely further enhance its lipid-lowering effects without clinically significant adverse effects. However, creatinine phospho-kinase levels should be monitored, and during concomitant isavuconazole therapy, patient counseling should address early warning signs of myopathy. Similarly, in most circumstances, isavuconazole co-administration may enhance metformin's blood glucose-lowering potential. However, in certain patients, like those with reduced renal clearance, metformin exposure can increase and further increases the risk lactic acidosis and rhabdomyolysis. Thus, in patients with reduced creatinine clearance, close monitoring of serum creatinine and assessment for early signs and symptoms of lactic acidosis and/or rhabdomyolysis may be warranted. If prolonged co-administration is anticipated, dose reductions of metformin or substituting vori-conazole or posaconazole should be considered. Due to digoxin's narrow therapeu-tic index and ability to routinely monitor drug concentrations, close clinical monitoring of digoxin levels is recommended when this agent is co-administered with isavuconazole. However, voriconazole does not inhibit P-gp and may be a therapeutic alternative to isavuconazole [105]. Isavuconazole increased 7-hydroxymethotrexate exposure; the clinical significance of this interaction is unknown, but this may lead to liver and renal damage associated with methotrexate therapy. Thus, when isavuconazole and methotrexate are used together, close clini-cal monitoring for methotrexate toxicity will be needed [307]. Alternatively, in vitro voriconazole did not demonstrate any inhibitory potential for OATP or BCRP and may be an alternative to isavuconazole with methotrexate therapy [18].

11.10 Echinocandins

11.10.1 Interaction Mechanisms

The echinocandins (caspofungin, micafungin, anidulafungin) have very few reported drug-drug interactions associated with their use. The echinocandins are not significant CYP inhibitors; however, they do inhibit a variety of ABC transporters including P-gp, BCRP, MRPs, and BSEP to varying extents [18]. Though the mech-anisms behind the few reported interactions have not been fully elucidated, the data suggest inhibition of ABC and/or OATP transporters may be an underlying cause.

11.10.1.1 Calcineurin Inhibitors

Early studies involving very few patients raised concerns about the potential for the co-administration of caspofungin and cyclosporine to produce additive or synergistic hepatotoxicity. The cause of this interaction is unclear. Using pharmacokinetic data from published trials, investigators found no evidence to support inhibition of OATP1B1 as a plausible mechanism underlying this interaction [309]. Experience with this combination has demonstrated that the combination is well-tolerated, and these concerns have dissipated. There are few published data describing drug interactions with micafungin. Micafungin does interact with cyclosporine, but the interaction varies in magnitude from producing nonsignificant inhibition of cyclosporine CYP metabolism to, in rare cases, a clinically significant increase in cyclosporine concentrations [310, 311]. Clinical data suggest that micafungin does not significantly interact with tacrolimus [312, 313]. Anidulafungin exposure is increased by 22% in the presence of cyclosporine, which is not considered clinically relevant, and cyclosporine pharmacokinetics are unchanged [314]. Anidulafungin does not interact with tacrolimus [315].

11.10.1.2 Rifampin

Co-administration of rifampin produces inhibitory and induction effects on caspofungin disposition, with an overall effect being slight induction at steady-state [316]. In the initial days of concomitant therapy, rifampin produces a transient increase in caspofungin plasma concentration [316]. This part of the interaction occurs during caspofungin's β distribution phase, which is the driving process behind the decline in its serum concentrations within 24 h of administration [316]. Caspofungin is an OATP1B1 substrate, and rifampin is a substrate, inhibitor, and inducer of this transport protein [67]. Therefore, the transient increases in caspofungin serum concentrations, observed in the initial days with concomitant rifampin, are most likely a result of rifampin's inhibition of caspofungin's OATP1B1-mediated uptake into hepatocytes and other tissues [66, 67, 316]. In vitro data using primary human hepatocytes demonstrate that rifampin can cause significant upregulation of OATP1B1 mRNA at clinically achievable concentrations (10 µg) [317, 318]. Thus, the decline in caspofungin C_{min} observed after 2 weeks of co-administered rifampin likely results from induction of OATP1B1 [316].

11.10.1.3 Clinical Importance of Interactions Involving Echinocandins

The interaction between cyclosporine and micafungin is significant only in those individuals with a very high cyclosporine CL/F. These individuals cannot be identified by obtaining a single cyclosporine blood concentration.

It is unlikely that the initial transient increase in serum caspofungin concentrations produced by rifampin co-administration is clinically important. However, the ongoing decline in caspofungin concentrations as the therapy continued could precipitate therapeutic failure [316].

11.10.1.4 Management of Interactions Involving Echinocandins

When micafungin is co-administered with cyclosporine, it is difficult to identify patients who will have a clinically significant interaction. Therefore, in these patients, careful monitoring of cyclosporine blood concentrations and dosage adjustment as needed are recommended upon initiating or discontinuing micafungin therapy.

A reduction in caspofungin dose is not necessary for the transient elevation in caspofungin plasma concentrations when rifampin and caspofungin are initiated on the same study day. However, when rifampin is added to caspofungin therapy, an increase in the daily caspofungin maintenance dose from 50 to 70 mg should be considered [316].

11.11 Summary

The myriad of potential drugs that antifungal agents can interact with is daunting and can be confusing. Antifungal agents differ markedly in their pharmacokinetic properties and in how they interact with other medicines. The clinical relevance of antifungal-drug interactions varies substantially. While certain interactions with antifungal agents are benign and result in little or no untoward clinical outcomes, others can produce significant toxicity or compromise efficacy if not properly managed. However, certain antifungal-drug interactions produce significant toxicity or compromise efficacy to such an extent that they cannot be managed, and the particular combination of antifungal and interacting medicine should be avoided. The amphotericin B formulations interact with other medicines by reducing their renal elimination or producing additive toxicities. Among the several classes of antifungal agents, the triazole class (fluconazole, itraconazole, voriconazole, posaconazole, and isavuconazole) produce the most drug interactions, primarily because of their ability to inhibit CYP. As a class these agents inhibit several CYP isoforms including CYP2C9, CYP2C19, and CYP3A4. Certain triazoles also interact with transport proteins, and depending on the dosage form, the absorption of some can be altered by changes in gastric pH. Therefore, collectively, triazoles interact with a vast array of medicines, and the degree of interaction is often agent specific. While their potential to interact with other drugs is vast, the most clinically significant interactions involving the triazoles involve benzodiazepines and anxiolytics, immunosuppressants (i.e., calcineurin inhibitors, mTOR inhibitors, and corticosteroids), the "statins," certain types of calcium channel blockers, phenytoin, and warfarin. The echinocandins have the lowest propensity to interact with other medicines.

References

1. Gallis HA, Drew RH, Pickard WW (1990) Amphotericin B: 30 years of clinical experience. Rev Infect Dis 12:308–329
2. Bekersky I, Fielding RM, Dressler DE et al (2002) Plasma protein binding of amphotericin B and pharmacokinetics of bound versus unbound amphotericin B after administration of intravenous liposomal amphotericin B (AmBisome) and amphotericin B deoxycholate. Antimicrob Agents Chemother 46:834–840
3. Bekersky I, Fielding RM, Dressler DE et al (2002) Pharmacokinetics, excretion, and mass balance of liposomal amphotericin B (AmBisome) and amphotericin B deoxycholate in humans. Antimicrob Agents Chemother 46:828–833
4. Wong-Beringer A, Jacobs RA, Guglielmo BJ (1998) Lipid formulations of amphotericin B: clinical efficacy and toxicities. Clin Infect Dis 27:603–618
5. Daneshmend TK, Warnock DW (1983) Clinical pharmacokinetics of systemic antifungal drugs. Clin Pharmacokinet 8:17–42
6. Stamm AM, Diasio RB, Dismukes WE et al (1987) Toxicity of amphotericin B plus flucytosine in 194 patients with cryptococcal meningitis. Am J Med 83:236–242
7. Bennett JE, Dismukes WE, Duma RJ et al (1979) A comparison of amphotericin B alone and combined with flucytosine in the treatment of cryptoccal meningitis. N Engl J Med 301:126–131
8. Gubbins PO (2011) Triazole antifungal agents drug-drug interactions involving hepatic cytochrome P450. Expert Opin Drug Metab Toxicol 7:1411–1429. https://doi.org/10.1517/17425 255.2011.627854
9. Barone JA, Moskovitz BL, Guarnieri J et al (1998) Enhanced bioavailability of itraconazole in hydroxypropyl-ß-cyclodextrin solution versus capsules in healthy volunteers. Antimicrob Agents Chemother 42:1862–1865
10. Poirier JM, Cheymol G (1998) Optimisation of itraconazole therapy using target drug concentrations. Clin Pharmacokinet 35:461–473. https://doi.org/10.2165/00003088-199835060-00004
11. Theuretzbacher U, Ihle F, Derendorf H (2006) Pharmacokinetic/pharmacodynamic profile of voriconazole. Clin Pharmacokinet 45:649–663. https://doi.org/10.2165/00003088-200645070-00002
12. Guarascio AJ, Slain D (2015) Review of the new delayed-release oral tablet and intravenous dosage forms of posaconazole. Pharmacotherapy 35:208–219. https://doi.org/10.1002/phar.1533
13. Rybak JM, Marx KR, Nishimoto AT et al (2015) Isavuconazole: pharmacology, pharmacodynamics, and current clinical experience with a new triazole antifungal agent. Pharmacotherapy 35:1037–1051. https://doi.org/10.1002/phar.1652
14. Slain D, Cleary JD (2015) Isavuconazonium sulfate: a novel antifungal agent. Curr Fungal Infect Rep 9:302–313
15. Wang EJ, Lew K, Casciano CN et al (2002) Interaction of common azole antifungals with P glycoprotein. Antimicrob Agents Chemother 46:160–165
16. Gupta A, Unadkat JDD, Mao QC (2007) Interactions of azole antifungal agents with the human breast cancer resistance protein (BCRP). J Pharm Sci 96:3226–3235. https://doi.org/10.1002/jps.20963
17. Wacher VJ, Wu CY, Benet LZ (1995) Overlapping substrate specificities and tissue distribution of cytochrome P450 3A and P-glycoprotein: implications for drug delivery and activity in cancer chemotherapy. Mol Carcinog 13:129–134
18. Lempers VJC, Van Den Heuvel JJMW, Russel FGM et al (2016) Inhibitory potential of antifungal drugs on atp-binding cassette transporters p-glycoprotein, MRP1 to MRP5, BCRP, and BSEP. Antimicrob Agents Chemother 60:3372–3379
19. Wacher VJ, Silverman JA, Zhang Y, Benet LZ (1998) Role of P-glycoprotein and cytochrome P450 3A in limiting oral absorption of peptides and peptidomimetics. J Pharm Sci 87:1322–1330. https://doi.org/10.1021/js980082d

20. Heykants J, Van Peer A, Van de Velde V et al (1989) The clinical pharmacokinetics of itraconazole: an overview. Mycoses 32:67–87. https://doi.org/10.1111/j.1439-0507.1989.tb02296.x
21. Van De Velde VJS, Van Peer AP, Heykants JJP et al (1996) Effect of food on the pharmacokinetics of a new hydroxypropyl-β-cyclodextrin formulation of itraconazole. Pharmacotherapy 16:424–428
22. Barone JA, Moskovitz BL, Guarnieri J et al (1998) Food interaction and steady-state pharmacokinetics of itraconazole oral solution in healthy volunteers. Pharmacotherapy 18:295–301
23. Johnson MD, Hamilton CD, Drew RH et al (2003) A randomized comparative study to determine the effect of omeprazole on the peak serum concentration of itraconazole oral solution. J Antimicrob Chemother 51:453–457
24. Debruyne D, Ryckelynck JP (1993) Clinical pharmacokinetics of fluconazole. Clin Pharmacokinet 24:10–27. https://doi.org/10.2165/00003088-199324010-00002
25. Johnson LB, Kauffman CA (2003) Voriconazole: a new triazole antifungal agent. Clin Infect Dis 36:630–637. https://doi.org/10.1086/367933
26. Courtney R, Pai S, Laughlin M et al (2003) Pharmacokinetics, safety, and tolerability of oral posaconazole administered in single and multiple doses in healthy adults. Antimicrob Agents Chemother 47:2788–2795
27. Ezzet F, Wexler D, Courtney R et al (2005) Oral bioavailability of posaconazole in fasted healthy subjects: comparison between three regimens and basis for clinical dosage recommendations. Clin Pharmacokinet 44:211–220. https://doi.org/10.2165/00003088-200544020-00006
28. Krishna G, Moton A, Ma L et al (2009) Pharmacokinetics and absorption of posaconazole oral suspension under various gastric conditions in healthy volunteers. Antimicrob Agents Chemother 53:958–966. https://doi.org/10.1128/AAC.01034-08
29. Krishna G, Ma L, Vickery D et al (2009) Effect of varying amounts of a liquid nutritional supplement on the pharmacokinetics of posaconazole in healthy volunteers. Antimicrob Agents Chemother 53:4749–4752
30. Sansone-Parsons A, Krishna G, Calzetta A et al (2006) Effect of a nutritional supplement on posaconazole pharmacokinetics following oral administration to healthy volunteers. Antimicrob Agents Chemother 50:1881–1883
31. Dolton MJ, Brüggemann RJM, Burger DM, McLachlan AJ (2014) Understanding variability in posaconazole exposure using an integrated population pharmacokinetic analysis. Antimicrob Agents Chemother 58:6879–6885. https://doi.org/10.1128/AAC.03777-14
32. Courtney R, Radwanski E, Lim J, Laughlin M (2004) Pharmacokinetics of posaconazole coadministered with antacid in fasting or nonfasting healthy men. Antimicrob Agents Chemother 48:804–808
33. Walravens J, Brouwers J, Spriet I et al (2011) Effect of pH and comedication on gastrointestinal absorption of posaconazole: monitoring of intraluminal and plasma drug concentrations. Clin Pharmacokinet 50:725–734
34. Luke DR, Tomaszewski K, Damle B, Schlamm HT (2010) Review of the basic and clinical pharmacology of sulfobutylether-beta-cyclodextrin (SBECD). J Pharm Sci 99:3291–3301. https://doi.org/10.1002/jps.22109
35. Maertens J, Cornely OA, Ullmann AJ et al (2014) Phase 1B study of the pharmacokinetics and safety of posaconazole intravenous solution in patients at risk for invasive fungal disease. Antimicrob Agents Chemother 58:3610–3617. https://doi.org/10.1128/AAC.02686-13
36. Wiederhold NP (2015) Pharmacokinetics and safety of posaconazole delayed-release tablets for invasive fungal infections. Clin Pharmacol Adv Appl 8:1–8. https://doi.org/10.2147/CPAA.S60933
37. Hens B, Corsetti M, Brouwers J, Augustijns P (2016) Gastrointestinal and systemic monitoring of posaconazole in humans after fasted and fed state administration of a solid dispersion. J Pharm Sci 105:2904–2912. https://doi.org/10.1016/j.xphs.2016.03.027
38. Kraft WK, Chang PS, Van Iersel MLPS et al (2014) Posaconazole tablet pharmacokinetics: lack of effect of concomitant medications altering gastric pH and gastric motility in healthy subjects. Antimicrob Agents Chemother 58:4020–4025

39. Krishna G, Ma L, Martinho M, O'Mara E (2012) Single-dose phase I study to evaluate the pharmacokinetics of posaconazole in new tablet and capsule formulations relative to oral suspension. Antimicrob Agents Chemother 56:4196–4201. https://doi.org/10.1128/AAC.00222-12

40. Schmitt-Hoffmann A, Roos B, Spickermann J et al (2009) Effect of mild and moderate liver disease on the pharmacokinetics of isavuconazole after intravenous and oral administration of a single dose of the prodrug BAL8557. Antimicrob Agents Chemother 53:4885–4890

41. Nagappan V, Deresinski S (2007) Posaconazole: a broad-spectrum triazole antifungal agent. Clin Infect Dis 45:1610–1617

42. Schwartz S, Milatovic D, Thiel E (1997) Successful treatment of cerebral aspergillosis with a novel triazole (voriconazole) in a patient with acute leukaemia. Br J Haematol 97:663–665

43. Wiederhold NP, Pennick GJ, Dorsey SA et al (2014) A reference laboratory experience of clinically achievable voriconazole, posaconazole, and itraconazole concentrations within the bloodstream and cerebral spinal fluid. Antimicrob Agents Chemother 58:424–431. https://doi.org/10.1128/AAC.01558-13

44. Wiederhold NP, Kovanda L, Najvar LK et al (2016) Isavuconazole is effective for the treatment of experimental cryptococcal meningitis. Antimicrob Agents Chemother 60:5600–5603. https://doi.org/10.1128/AAC.00229-16

45. Peixoto D, Gagne LS, Hammond SP et al (2014) Isavuconazole treatment of a patient with disseminated mucormycosis. J Clin Microbiol 52:1016–1019. https://doi.org/10.1128/JCM.03176-13

46. Townsend R, Dietz A, Hale C et al (2017) Pharmacokinetic evaluation of CYP3A4-mediated drug-drug interactions of isavuconazole with rifampin, ketoconazole, midazolam, and ethinyl estradiol/norethindrone in healthy adults. Clin Pharmacol Drug Dev 6:44–53. https://doi.org/10.1002/cpdd.285

47. Isoherranen N, Kunze KL, Allen KE et al (2004) Role of itraconazole metabolites in CYP3A4 inhibition. Drug Metab Dispos 32:1121–1131

48. Kunze KL, Trager WF (1996) Warfarin-fluconazole. III. A rational approach to management of a metabolically based drug interaction. Drug Metab Dispos 24:429–435

49. Templeton IE, Thummel KE, Kharasch ED et al (2008) Contribution of itraconazole metabolites to inhibition of CYP3A4 in vivo. Clin Pharmacol Ther 83:77–85

50. Brammer KW, Coakley AJ, Jezequel SG, Tarbit MH (1991) The disposition and metabolism of [14C]fluconazole in humans. Drug Metab Dispos 19:764–767

51. Locuson CW, Hutzler JM, Tracy TS (2007) Visible spectra of type II cytochrome P450-drug complexes: evidence that "incomplete" heme coordination is common. Drug Metab Dispos 35:614–622. https://doi.org/10.1124/dmd.106.012609

52. Bourcier K, Hyland R, Kempshall S et al (2010) Investigation into UDP-glucuronosyltransferase (UGT) enzyme kinetics of imidazole- and triazole-containing antifungal drugs in human liver microsomes and recombinant UGT enzymes. Drug Metab Dispos 38:923–929. https://doi.org/10.1124/dmd.109.030676

53. Hyland R, Jones BC, Smith DA (2003) Identification of the cytochrome P450 enzymes involved in the N-oxidation of voriconazole. Drug Metab Dispos 31:540–547

54. Sabo JA, Abdel-Rahman SM (2000) Voriconazole: a new triazole antifungal. Ann Pharmacother 34:1032–1043. https://doi.org/10.1345/aph.19237

55. Goldstein JA (2001) Clinical relevance of genetic polymorphisms in the human CYP2C subfamily. Br J Clin Pharmacol 52:349–355

56. Niinuma Y, Saito T, Takahashi M et al (2014) Functional characterization of 32 CYP2C9 allelic variants. Pharmacogenomics J 14:107–114. https://doi.org/10.1038/tpj.2013.22

57. Lee CR, Goldstein JA, Pieper JA (2002) Cytochrome P450 2C9 polymorphisms: a comprehensive review of the in-vitro and human data. Pharmacogenetics 12:251–263

58. Mikus G, Scholz IM, Weiss J (2011) Pharmacogenomics of the triazole antifungal agent voriconazole. Pharmacogenomics 12:861–872. https://doi.org/10.2217/pgs.11.18

59. Yanni SB, Annaert PP, Augustijns P et al (2008) Role of flavin-containing monooxygenase in oxidative metabolism of voriconazole by human liver microsomes. Drug Metab Dispos 36:1119–1125

60. Krieter P, Flannery B, Musick T et al (2004) Disposition of posaconazole following single-dose oral administration in healthy subjects. Antimicrob Agents Chemother 48:3543–3551. https://doi.org/10.1128/AAC.48.9.3543-3551.2004

61. Ghosal A, Hapangama N, Yuan Y et al (2004) Identification of human UDP-glucuronosyltransferase enzyme(s) responsible for the glucuronidation of posaconazole (Noxafil). Drug Metab Dispos 32:267–271. https://doi.org/10.1124/dmd.32.2.267

62. Murrell D, Bossaer JB, Carico R et al (2017) Isavuconazonium sulfate: a triazole prodrug for invasive fungal infections. Int J Pharm Pract 25:18–30. https://doi.org/10.1111/ijpp.12302

63. Yu J, Zhou Z, Owens KH et al (2017) What can be learned from recent new drug applications? A systematic review of drug interaction data for drugs approved by the U.S. FDA in 2015. Drug Metab Dispos 45:86–108. https://doi.org/10.1124/dmd.116.073411

64. Yamazaki T, Desai A, Goldwater R et al (2017) Pharmacokinetic effects of isavuconazole coadministration with the cytochrome P450 enzyme substrates bupropion, repaglinide, caffeine, dextromethorphan, and methadone in healthy subjects. Clin Pharmacol Drug Dev 6:54–65. https://doi.org/10.1002/cpdd.281

65. Chen SC-A, Slavin MA, Sorrell TC (2011) Echinocandin antifungal drugs in fungal infections: a comparison. Drugs 71:11–41. https://doi.org/10.2165/11585270-000000000-00000

66. Stone JA, Xu X, Winchell GA et al (2004) Disposition of caspofungin: role of distribution in determining pharmacokinetics in plasma. Antimicrob Agents Chemother 48:815–823

67. Sandhu P, Lee W, Xu X et al (2005) Hepatic uptake of the novel antifungal agent caspofungin. Drug Metab Dispos 33:676–682. https://doi.org/10.1124/dmd.104.003244

68. Yanni SB, Augustijns PF, Benjamin DK et al (2010) In vitro investigation of the hepatobiliary disposition mechanisms of the antifungal agent micafungin in humans and rats. Drug Metab Dispos 38:1848–1856. https://doi.org/10.1124/dmd.110.033811

69. Cappelletty D, Eiselstein-McKitrick K (2007) The echinocandins. Pharmacotherapy 27:369–388. https://doi.org/10.1592/phco.27.3.369

70. Stone JA, Holland SD, Wickersham PJ et al (2002) Single- and multiple-dose pharmacokinetics of caspofungin in healthy men. Antimicrob Agents Chemother 46:739–745

71. Balani SK, Xu X, Arison BH et al (2000) Metabolites of caspofungin acetate, a potent antifungal agent, in human plasma and urine. Drug Metab Dispos 28:1274–1278

72. Hebert MF, Smith HE, Marbury TC et al (2005) Pharmacokinetics of micafungin in healthy volunteers, volunteers with moderate liver disease, and volunteers with renal dysfunction. J Clin Pharmacol 45:1145–1152. 45/10/1145 [pii]\r. https://doi.org/10.1177/0091270005279580

73. Joseph JM, Jain R, Danziger LH (2007) Micafungin: a new echinocandin antifungal. Pharmacotherapy 27:53–67. https://doi.org/10.1592/phco.27.1.53

74. Damle BD, Dowell JA, Walsky RL et al (2009) In vitro and in vivo studies to characterize the clearance mechanism and potential cytochrome P450 interactions of anidulafungin. Antimicrob Agents Chemother 53:1149–1156

75. Estes KE, Penzak SR, Calis KA, Walsh TJ (2009) Pharmacology and antifungal properties of anidulafungin, a new echinocandin. Pharmacotherapy 29:17–30

76. White MH, Bowden RA, Sandler ES et al (1998) Randomized, double-blind clinical trial of amphotericin B colloidal dispersion vs. amphotericin B in the empirical treatment of fever and neutropenia. Clin Infect Dis 27:296–302

77. Ringden O, Andstrom E, Remberger M et al (1994) Safety of liposomal amphotericin B (AmBisome) in 187 transplant recipients treated with cyclosporin. Bone Marrow Transplant 14(Suppl 5):S10–S14

78. Morris AA, Mueller SW, Rower JE et al (2015) Evaluation of sulfobutylether-β-cyclodextrin exposure in a critically ill patient receiving intravenous posaconazole while undergoing continuous venovenous hemofiltration. Antimicrob Agents Chemother 59:6653–6656. https://doi.org/10.1128/AAC.01493-15

79. Maertens JA, Raad II, Marr KA et al (2016) Isavuconazole versus voriconazole for primary treatment of invasive mould disease caused by Aspergillus and other filamentous fungi (SECURE): a phase 3, randomised-controlled, non-inferiority trial. Lancet 387:760–769. https://doi.org/10.1016/S0140-6736(15)01159-9

80. Perfect JR, Dismukes WE, Dromer F et al (2010) Clinical practice guidelines for the management of cryptococcal disease: 2010 update by the Infectious Diseases Society of America. Clin Infect Dis 50:291–322. https://doi.org/10.1086/649858
81. Viviani MA (1995) Flucytosine—what is its future? J Antimicrob Chemother 35:241–244. https://doi.org/10.1093/jac/35.2.241
82. Chung D-K, Koenig MG (1971) Reversible cardiac enlargement during treatment with amphotericin B and hydrocortisone: report of three cases. Am Rev Respir Dis 103:831–841. https://doi.org/10.1164/arrd.1971.103.6.831
83. Fleisher D, Li C, Zhou Y et al (1999) Drug, meal and formulation interactions influencing drug absorption after oral administration. Clinical implications. Clin Pharmacokinet 36:233–254
84. Courtney R, Wexler D, Radwanski E et al (2004) Effect of food on the relative bioavailability of two oral formulations of posaconazole in healthy adults. Br J Clin Pharmacol 57:218–222
85. Lange D, Pavao JH, Wu J, Klausner M (1997) Effect of a cola beverage on the bioavailability of itraconazole in the presence of H2 blockers. J Clin Pharmacol 37:535–540
86. Wood N, Tan K, Purkins L et al (2003) Effect of omeprazole on the steady-state pharmacokinetics of voriconazole. Br J Clin Pharmacol 56(Suppl 1):56–61
87. Damle B, Hess H, Kaul S, Knupp C (2002) Absence of clinically relevant drug interactions following simultaneous administration of didanosine-encapsulated, enteric-coated bead formulation with either iltraconazole or fluconazole. Biopharm Drug Dispos 23:59–66
88. Ohwada J, Tsukazaki M, Hayase T et al (2003) Design, synthesis and antifungal activity of a novel water soluble prodrug of antifungal triazole. Bioorg Med Chem Lett 13:191–196. https://doi.org/10.1016/s0960-894x(02)00892-2
89. Thummel KE, Wilkinson GR (1998) In vitro and in vivo drug interactions involving human CYP3A. Annu Rev Pharmacol Toxicol 38:389–430. https://doi.org/10.1146/annurev.pharmtox.38.1.389
90. Foti RS, Dalvie DK (2016) Cytochrome P450 and non–cytochrome P450 oxidative metabolism: contributions to the pharmacokinetics, safety, and efficacy of xenobiotics. Drug Metab Dispos 44:1229–1245. https://doi.org/10.1124/dmd.116.071753
91. Omar G, Whiting PH, Hawksworth GM et al (1997) Ketoconazole and fluconazole inhibition of the metabolism of cyclosporin A by human liver in vitro. Ther Drug Monit 19:436–445
92. Jeong S, Nguyen PD, Desta Z (2009) Comprehensive in vitro analysis of voriconazole inhibition of eight cytochrome P450 (CYP) enzymes: major effect on CYPs 2B6, 2C9, 2C19, and 3A. Antimicrob Agents Chemother 53:541–551
93. Yamazaki H, Nakamoto M, Shimizu M et al (2010) Potential impact of cytochrome P450 3A5 in human liver on drug interactions with triazoles. Br J Clin Pharmacol 69:593–597. https://doi.org/10.1111/j.1365-2125.2010.03656.x
94. Yu K-S, Cho JJ-Y, Jang II-J et al (2004) Effect of the CYP3A5 genotype on the pharmacokinetics of intravenous midazolam during inhibited and induced metabolic states. Clin Pharmacol Ther 76:104–112. https://doi.org/10.1016/j.clpt.2004.03.009
95. Shirasaka Y, Chang S-Y, Grubb MF et al (2013) Effect of CYP3A5 expression on the inhibition of CYP3A-catalyzed drug metabolism: impact on modeling CYP3A-mediated drug-drug interactions. Drug Metab Dispos 41:1566–1574. https://doi.org/10.1124/dmd.112.049940
96. Lee S, Kim B-H, Nam W-S et al (2012) Effect of CYP2C19 polymorphism on the pharmacokinetics of voriconazole after single and multiple doses in healthy volunteers. J Clin Pharmacol 52:195–203
97. Nara M, Takahashi N, Miura M et al (2013) Effect of itraconazole on the concentrations of tacrolimus and cyclosporine in the blood of patients receiving allogeneic hematopoietic stem cell transplants. Eur J Clin Pharmacol 69:1321–1329. https://doi.org/10.1007/s00228-013-1471-2
98. Black DJ, Kunze KL, Wienkers LC et al (1996) Warfarin-fluconazole II – a metabolically based drug interaction: in vivo studies. Drug Metab Dispos 24:422–428
99. Seo K-A, Kim H-J, Jeong ES et al (2014) In vitro assay of six UGT isoforms in human liver microsomes, using cocktails of probe substrates and LC-MS/MS. Drug Metab Dispos 42:1803–1810. https://doi.org/10.1124/dmd.114.058818

100. Niwa T, Shiraga T, Takagi A (2005) Effect of antifungal drugs on cytochrome P450 (CYP) 2C9, CYP2C19, and CYP3A4 activities in human liver microsomes. Biol Pharm Bull 28:1805–1808

101. Wexler D, Courtney R, Richards W et al (2004) Effect of posaconazole on cytochrome P450 enzymes: a randomized, open-label, two-way crossover study. Eur J Pharm Sci 21:645–653

102. Desai A, Yamazaki T, Dietz AJ et al (2016) Pharmacokinetic and pharmacodynamic evaluation of the drug-drug interaction between isavuconazole and warfarin in healthy subjects. Clin Pharmacol Drug Dev 6(1):86–92. https://doi.org/10.1002/cpdd.283

103. König J, Müller F, Fromm MF (2013) Transporters and drug-drug interactions: important determinants of drug disposition and effects. Pharmacol Rev 65:944–966. https://doi.org/10.1124/pr.113.007518

104. Ferté J (2000) Analysis of the tangled relationships between P-glycoprotein-mediated multidrug resistance and the lipid phase of the cell membrane. Eur J Biochem 267:277–294

105. Purkins L, Wood N, Kleinermans D, Nichols D (2003) Voriconazole does not affect the steady-state pharmacokinetics of digoxin. Br J Clin Pharmacol 56(Suppl 1):45–50

106. Yamazaki T, Desai A, Goldwater R et al (2017) Pharmacokinetic interactions between isavuconazole and the drug transporter substrates atorvastatin, digoxin, metformin, and methotrexate in healthy subjects. Clin Pharmacol Drug Dev 6:76–85

107. Eytan GD, Regev R, Oren G, Assaraf YG (1996) The role of passive transbilayer drug movement in multidrug resistance and its modulation. J Biol Chem 271:12897–12902. https://doi.org/10.1074/jbc.271.22.12897

108. Neuvonen PJ, Kantola T, Kivistö KT (1998) Simvastatin but not pravastatin is very susceptible to interaction with the CYP3A4 inhibitor itraconazole. Clin Pharmacol Ther 63:332–341

109. Mazzu AL, Lasseter KC, Shamblen EC et al (2000) Itraconazole alters the pharmacokinetics of atorvastatin to a greater extent than either cerivastatin or pravastatin. Clin Pharmacol Ther 68:391–400

110. Kantola T, Kivistö KT, Neuvonen PJ (1998) Effect of itraconazole on the pharmacokinetics of atorvastatin. Clin Pharmacol Ther 64:58–65. https://doi.org/10.1016/S0009-9236(98)90023-6

111. Cooper K (2003) Effect of itraconazole on the pharmacokinetics of rosuvastatin. Clin Pharmacol Ther 73:322–329. https://doi.org/10.1016/S0009-9236(02)17633-8

112. Nakagawa S, Gosho M, Inazu Y, Hounslow N (2013) Pitavastatin concentrations are not increased by CYP3A4 inhibitor itraconazole in healthy subjects. Clin Pharmacol Drug Dev 2:195–200. https://doi.org/10.1002/cpdd.19

113. Olkkola KT, Ahonen J, Neuvonen PJ (1996) The effects of the systemic antimycotics, itraconazole and fluconazole, on the pharmacokinetics and pharmacodynamics of intravenous and oral midazolam. Anesth Analg 82:511–516

114. Neuvonen PJ, Varhe A, Olkkola KT (1996) The effect of ingestion time interval on the interaction between itraconazole and triazolam. Clin Pharmacol Ther 60:326–331

115. Ahonen J, Olkkola KT, Neuvonen PJ (1996) The effect of the antimycotic itraconazole on the pharmacokinetics and pharmacodynamics of diazepam. Fundam Clin Pharmacol 10:314–318

116. Ahonen J, Olkkola KT, Neuvonen PJ (1996) Lack of effect of antimycotic itraconazole on the pharmacokinetics or pharmacodynamics of temazepam. Ther Drug Monit 18:124–127

117. Otsuji Y, Okuyama N, Aoshima T et al (2002) No effect of itraconazole on the single oral dose pharmacokinetics and pharmacodynamics of estazolam. Ther Drug Monit 24:375–378

118. Oda M, Kotegawa T, Tsutsumi K et al (2003) The effect of itraconazole on the pharmacokinetics and pharmacodynamics of bromazepam in healthy volunteers. Eur J Clin Pharmacol 59:615–619

119. Kivistö KT, Lamberg TS, Kantola T, Neuvonen PJ (1997) Plasma buspirone concentrations are greatly increased by erythromycin and itraconazole. Clin Pharmacol Ther 62:348–354

120. Luurila H, Kivistö KT, Neuvonen PJ (1998) Effect of itraconazole on the pharmacokinetics and pharmacodynamics of zolpidem. Eur J Clin Pharmacol 54:163–166. https://doi.org/10.1007/s002280050439

121. Greenblatt DJ, von Moltke LL, Harmatz JS et al (1998) Kinetic and dynamic interaction study of zolpidem with ketoconazole, itraconazole, and fluconazole. Clin Pharmacol Ther 64:661–671. https://doi.org/10.1016/S0009-9236(98)90057-1

122. Park J-YY, Shon J-HH, Kim K-AA et al (2006) Combined effects of itraconazole and CYP2D6*10 genetic polymorphism on the pharmacokinetics and pharmacodynamics of haloperidol in healthy subjects. J Clin Psychopharmacol 26:135–142

123. Yasui N, Kondo T, Otani K et al (1999) Effects of itraconazole on the steady-state plasma concentrations of haloperidol and its reduced metabolite in schizophrenic patients: in vivo evidence of the involvement of CYP3A4 for haloperidol metabolism. J Clin Psychopharmacol 19:149–154

124. Raaska K, Neuvonen PJ (1998) Serum concentrations of clozapine and N-desmethylclozapine are unaffected by the potent CYP3A4 inhibitor itraconazole. Eur J Clin Pharmacol 54:167–170

125. Leather H, Boyette RM, Tian L, Wingard JR (2006) Pharmacokinetic evaluation of the drug interaction between intravenous itraconazole and intravenous tacrolimus or intravenous cyclosporin a in allogeneic hematopoietic stem cell transplant recipients. Biol Blood Marrow Transplant 12:325–334

126. Varis T, Kaukonen KM, Kivistö KT, Neuvonen PJ (1998) Plasma concentrations and effects of oral methylprednisolone are considerably increased by itraconazole. Clin Pharmacol Ther 64:363–368

127. Lebrun-Vignes B, Archer VC, Diquet B et al (2001) Effect of itraconazole on the pharmacokinetics of prednisolone and methylprednisolone and cortisol secretion in healthy subjects. Br J Clin Pharmacol 51:443–450

128. Varis T, Kivisto KT, Backman JT, Neuvonen PJ (1999) Itraconazole decreases the clearance and enhances the effects of intravenously administered methylprednisolone in healthy volunteers. Pharmacol Toxicol 85:29–32

129. Varis T, Kivistö KT, Backman JT, Neuvonen PJ (2000) The cytochrome P450 3A4 inhibitor itraconazole markedly increases the plasma concentrations of dexamethasone and enhances its adrenal-suppressant effect. Clin Pharmacol Ther 68:487–494

130. Varis T, Kivistö KT, Neuvonen PJ (2000) The effect of itraconazole on the pharmacokinetics and pharmacodynamics of oral prednisolone. Eur J Clin Pharmacol 56:57–60. https://doi.org/10.1007/s002280050720

131. Raaska K, Niemi M, Neuvonen M et al (2002) Plasma concentrations of inhaled budesonide and its effects on plasma cortisol are increased by the cytochrome P4503A4 inhibitor itraconazole. Clin Pharmacol Ther 72:362–369

132. Naef R, Schmid C, Hofer M et al (2007) Itraconazole comedication increases systemic levels of inhaled fluticasone in lung transplant recipients. Respiration 74:418–422

133. Jalava KM, Olkkola KT, Neuvonen PJ (1997) Itraconazole greatly increases plasma concentrations and effects of felodipine. Clin Pharmacol Ther 61:410–415

134. Lukkari E, Juhakoski A, Aranko K, Neuvonen PJ (1997) Itraconazole moderately increases serum concentrations of oxybutynin but does not affect those of the active metabolite. Eur J Clin Pharmacol 52:403–406

135. Buggia I, Zecca M, Alessandrino EP et al (1996) Itraconazole can increase systemic exposure to busulfan in patients given bone marrow transplantation. GITMO (Gruppo Italiano Trapianto di Midollo Osseo). Anticancer Res 16:2083–2088

136. Hynninen VV, Olkkola KT, Bertilsson L et al (2009) Voriconazole increases while itraconazole decreases plasma meloxicam concentrations. Antimicrob Agents Chemother 53:587–592

137. Saari TI, Grönlund J, Hagelberg NM et al (2010) Effects of itraconazole on the pharmacokinetics and pharmacodynamics of intravenously and orally administered oxycodone. Eur J Clin Pharmacol 66:387–397. https://doi.org/10.1007/s00228-009-0775-8

138. Lim SG, Sawyerr AM, Hudson M et al (1993) Short report: the absorption of fluconazole and itraconazole under conditions of low intragastric acidity. Aliment Pharmacol Ther 7:317–321

139. Kanda Y, Kami M, Matsuyama T et al (1998) Plasma concentration of itraconazole in patients receiving chemotherapy for hematological malignancies: the effect of famotidine on the absorption of itraconazole. Hematol Oncol 16:33–37

140. Jaruratanasirikul S, Sriwiriyajan S (1998) Effect of omeprazole on the pharmacokinetics of itraconazole. Eur J Clin Pharmacol 54:159–161

141. Bonay M, Jonville-Bera AP, Diot P et al (1993) Possible interaction between phenobarbital, carbamazepine and itraconazole. Drug Saf 9:309–311

142. Ducharme MP, Slaughter RL, Warbasse LH et al (1995) Itraconazole and hydroxyitraconazole serum concentrations are reduced more than tenfold by phenytoin*. Clin Pharmacol Ther 58:617–624. https://doi.org/10.1016/0009-9236(95)90017-9

143. Baciewicz AM, Chrisman CR, Finch CK, Self TH (2013) Update on rifampin, rifabutin, and rifapentine drug interactions. Curr Med Res Opin 29:1–12

144. Jaruratanasirikul S, Sriwiriyajan S (2007) Pharmacokinetic study of the interaction between itraconazole and nevirapine. Eur J Clin Pharmacol 63:451–456

145. Jalava KM, Partanen J, Neuvonen PJ (1997) Itraconazole decreases renal clearance of digoxin. Ther Drug Monit 19:609–613

146. Kaukonen KM, Olkkola KT, Neuvonen PJ (1997) Itraconazole increases plasma concentrations of quinidine. Clin Pharmacol Ther 62:510–517

147. Damkier P, Hansen LL, Brøsen K (1999) Effect of diclofenac, disulfiram, itraconazole, grapefruit juice and erythromycin on the pharmacokinetics of quinidine. Br J Clin Pharmacol 48:829–838

148. Heiskanen T, Backman JT, Neuvonen M et al (2008) Itraconazole, a potent inhibitor of P-glycoprotein, moderately increases plasma concentrations of oral morphine. Acta Anaesthesiol Scand 52:1319–1326. https://doi.org/10.1111/j.1399-6576.2008.01739.x

149. Kivistö KT, Kantola T, Neuvonen PJ (1998) Different effects of itraconazole on the pharmacokinetics of fluvastatin and lovastatin. Br J Clin Pharmacol 46:49–53. https://doi.org/10.1046/j.1365-2125.1998.00034.x

150. Lohitnavy M, Lohitnavy O, Thangkeattiyanon O, Srichai W (2005) Reduced oral itraconazole bioavailability by antacid suspension. J Clin Pharm Ther 30:201–206

151. Kellick KA, Bottorff M, Toth PP (2014) A clinician's guide to statin drug-drug interactions. J Clin Lipidol 8:S30–S46. https://doi.org/10.1016/j.jacl.2014.02.010

152. Bellosta S, Corsini A (2012) Statin drug interactions and related adverse reactions. Expert Opin Drug Saf 11:933–946. https://doi.org/10.1517/14740338.2012.712959

153. Neuvonen PJ, Niemi M, Backman JT (2006) Drug interactions with lipid-lowering drugs: mechanisms and clinical relevance. Clin Pharmacol Ther 80:565–581

154. Ishigam M, Uchiyama M, Kondo T et al (2001) Inhibition of in vitro metabolism of simvastatin by itraconazole in humans and prediction of in vivo drug-drug interactions. Pharm Res 18:622–631

155. Elsby R, Hilgendorf C, Fenner K (2012) Understanding the critical disposition pathways of statins to assess drug-drug interaction risk during drug development: it's not just about OATP1B1. Clin Pharmacol Ther 92:584–598. https://doi.org/10.1038/clpt.2012.163;10.1038/clpt.2012.163

156. Templeton I, Peng C-C, Thummel KE et al (2010) Accurate prediction of dose-dependent CYP3A4 inhibition by itraconazole and its metabolites from in vitro inhibition data. Clin Pharmacol Ther 88:499–505. https://doi.org/10.1038/clpt.2010.119

157. Ono S, Hatanaka T, Miyazawa S et al (1996) Human liver microsomal diazepam metabolism using cDNA-expressed cytochrome P450s: role of CYP2B6, 2C19 and the 3A subfamily. Xenobiotica 26:1155–1166. https://doi.org/10.3109/00498259609050260

158. Miura M, Otani K, Ohkubo T (2005) Identification of human cytochrome P450 enzymes involved in the formation of 4-hydroxyestazolam from estazolam. Xenobiotica 35:455–465

159. Mori T, Aisa Y, Kato J et al (2009) Drug interaction between oral solution itraconazole and calcineurin inhibitors in allogeneic hematopoietic stem cell transplantation recipients: an association with bioavailability of oral solution itraconazole. Int J Hematol 90:103–107

160. Florea NR, Capitano B, Nightingale CH et al (2003) Beneficial pharmacokinetic interaction between cyclosporine and itraconazole in renal transplant recipients. Transplant Proc 35:2873–2877

161. Kramer MR, Amital A, Fuks L, Shitrit D (2011) Voriconazole and itraconazole in lung transplant recipients receiving tacrolimus (FK 506): efficacy and drug interaction. Clin Transpl 25:E163–E167. https://doi.org/10.1111/j.1399-0012.2010.01373.x

162. Billaud EM, Guillemain R, Tacco F, Chevalier P (1998) Evidence for a pharmacokinetic interaction between itraconazole and tacrolimus in organ transplant patients. Br J Clin Pharmacol 46:271–272. https://doi.org/10.1046/j.1365-2125.1998.00784.x
163. Said A, Garnick JJ, Dieterle N et al (2006) Sirolimus-itraconazole interaction in a hematopoietic stem cell transplant recipient. Pharmacotherapy 26:289–295
164. Kuypers DR, Claes K, Evenepoel P et al (2005) Drug interaction between itraconazole and sirolimus in a primary renal allograft recipient. Transplantation 79:737
165. Kovarik JM, Hsu CH, McMahon L et al (2001) Population pharmacokinetics of everolimus in de novo renal transplant patients: impact of ethnicity and comedications. Clin Pharmacol Ther 70:247–254
166. Edsbäcker S, Johansson C-J (2006) Airway selectivity: an update of pharmacokinetic factors affecting local and systemic disposition of inhaled steroids. Basic Clin Pharmacol Toxicol 98:523–536. https://doi.org/10.1111/j.1742-7843.2006.pto_355.x
167. Bolland MJ, Bagg W, Thomas MG et al (2004) Cushing's syndrome due to interaction between inhaled corticosteroids and itraconazole. Ann Pharmacother 38:46–49
168. Gilchrist FJ, Cox KJ, Rowe R et al (2013) Itraconazole and inhaled fluticasone causing hypothalamic-pituitary-adrenal axis suppression in adults with cystic fibrosis. J Cyst Fibros 12:399–402
169. Daveluy A, Raignoux C, Miremont-Salamé G et al (2009) Drug interactions between inhaled corticosteroids and enzymatic inhibitors. Eur J Clin Pharmacol 65:743–745. https://doi.org/10.1007/s00228-009-0653-4
170. Yeh J, Soo SC, Summerton C, Richardson C (1990) Potentiation of action of warfarin by itraconazole. BMJ 301:669
171. Miura M, Takahashi N, Kanno S et al (2011) Drug interaction of (S)-warfarin, and not (R)-warfarin, with itraconazole in a hematopoietic stem cell transplant recipient. Clin Chim Acta 412(21-22):2002–2006. https://doi.org/10.1016/j.cca.2011.06.035
172. Yamamoto H, Habu Y, Yano I et al (2014) Comparison of the effects of azole antifungal agents on the anticoagulant activity of warfarin. Biol Pharm Bull 37:1990–1993
173. Lalovic B, Phillips B, Risler LL et al (2004) Quantitatlve contribution of CYP2D6 and CYP3A to oxycodone metabolism in human liver and intestinal microsomes. Drug Metab Dispos 32:447–454
174. Grönlund J, Saari TI, Hagelberg NM et al (2010) Exposure to oral oxycodone is increased by concomitant inhibition of CYP2D6 and 3A4 pathways, but not by inhibition of CYP2D6 alone. Br J Clin Pharmacol 70:78–87. https://doi.org/10.1111/j.1365-2125.2010.03653.x
175. Grönlund J, Saari TI, Hagelberg NM et al (2011) Effect of inhibition of cytochrome P450 enzymes 2D6 and 3A4 on the pharmacokinetics of intravenous oxycodone. Clin Drug Investig 31:143–153. https://doi.org/10.2165/11539950-000000000-00000
176. Law M, Rudnicka AR (2006) Statin safety: a systematic review. Am J Cardiol 97:52C–60C. https://doi.org/10.1016/j.amjcard.2005.12.010
177. Lomaestro BM, Piatek MA (1998) Update on drug interactions with azole antifungal agents. Ann Pharmacother 32:915–928
178. Jacobsen W, Kirchner G, Hallensleben K et al (1999) Comparison of cytochrome P-450-dependent metabolism and drug interactions of the 3-hydroxy-3-methylglutaryl-CoA reductase inhibitors lovastatin and pravastatin in the liver. Drug Metab Dispos 27:173–179
179. Jaruratanasirikul S, Sriwiriyajan S (1998) Effect of rifampicin on the pharmacokinetics of itraconazole in normal volunteers and AIDS patients. Eur J Clin Pharmacol 54:155–158
180. de Lannoy IAM, Silverman M (1992) The MDR1 gene product, P-glycoprotein, mediates the transport of the cardiac glycoside, digoxin. Biochem Biophys Res Commun 189:551–557
181. Alderman CP, Allcroft PD (1997) Digoxin-itraconazole interaction: possible mechanisms. Ann Pharmacother 31:438–440
182. Partanen J, Jalava KM, Neuvonen PJ (1996) Itraconazole increases serum digoxin concentration. Pharmacol Toxicol 79:274–276
183. Sachs MK, Blanchard LM, Green PJ (1993) Interaction of itraconazole and digoxin. Clin Infect Dis 16:400–403

184. De Lannoy IA, Koren G, Klein J et al (1992) Cyclosporin and quinidine inhibition of renal digoxin excretion: evidence for luminal secretion of digoxin. Am J Phys 263:F613–F622
185. Bermúdez M, Fuster JL, Llinares E et al (2005) Itraconazole-related increased vincristine neurotoxicity: case report and review of literature. J Pediatr Hematol Oncol 27:389–392
186. Bashir H, Motl S, Metzger ML et al (2006) Itraconazole-enhanced chemotherapy toxicity in a patient with Hodgkin lymphoma. J Pediatr Hematol Oncol 28:33–35
187. Moriyama B, Henning SA, Leung J et al (2012) Adverse interactions between antifungal azoles and vincristine: review and analysis of cases. Mycoses 55:290–297
188. Kantola T, Backman JT, Niemi M et al (2000) Effect of fluconazole on plasma fluvastatin and pravastatin concentrations. Eur J Clin Pharmacol 56:225–229
189. Cooper KJ, Martin PD, Dane AL et al (2002) The effect of fluconazole on the pharmacokinetics of rosuvastatin. Eur J Clin Pharmacol 58:527–531
190. Ahonen J, Olkkola KT, Neuvonen PJ (1997) Effect of route of administration of fluconazole on the interaction between fluconazole and midazolam. Eur J Clin Pharmacol 51:415–419. https://doi.org/10.1007/s002280050223
191. Varhe A, Olkkola KT, Neuvonen PJ (1996) Effect of fluconazole dose on the extent of fluconazole-triazolam interaction. Br J Clin Pharmacol 42:465–470
192. Varhe A, Olkkola KT, Neuvonen PJ (2003) Fluconazole, but not terbinafine, enhances the effects of triazolam by inhibiting its metabolism. Br J Clin Pharmacol 41:319–323. https://doi.org/10.1046/j.1365-2125.1996.03189.x
193. Saari TI, Laine K, Bertilsson L et al (2007) Voriconazole and fluconazole increase the exposure to oral diazepam. Eur J Clin Pharmacol 63:941–949
194. Sud K, Singh B, Krishna VS et al (1999) Unpredictable cyclosporin-fluconazole interaction in renal transplant recipients. Nephrol Dial Transplant 14:1698–1703
195. Mihara A, Mori T, Aisa Y et al (2008) Greater impact of oral fluconazole on drug interaction with intravenous calcineurin inhibitors as compared with intravenous fluconazole. Eur J Clin Pharmacol 64:89–91
196. Blum RA, Wilton JH, Hilligoss DM et al (1991) Effect of fluconazole on the disposition of phenytoin. Clin Pharmacol Ther 49:420–425
197. Saari TI, Laine K, Neuvonen M et al (2008) Effect of voriconazole and fluconazole on the pharmacokinetics of intravenous fentanyl. Eur J Clin Pharmacol 64:25–30
198. Palkama VJ, Isohanni MH, Neuvonen PJ, Olkkola KT (1998) The effect of intravenous and oral fluconazole on the pharmacokinetics and pharmacodynamics of intravenous alfentanil. Anesth Analg 87:190–194
199. Cobb MN, Desai J, Brown LS et al (1998) The effect of fluconazole on the clinical pharmacokinetics of methadone. Clin Pharmacol Ther 63:655–662
200. Yule SM, Walker D, Cole M et al (1999) The effect of fluconazole on cyclophosphamide metabolism in children. Drug Metab Dispos 27:417–421
201. Upton A, McCune JS, Kirby KA et al (2007) Fluconazole coadministration concurrent with cyclophosphamide conditioning may reduce regimen-related toxicity postmyeloablative hematopoietic cell transplantation. Biol Blood Marrow Transplant 13:760–764
202. Wakeham K, Parkes-Ratanshi R, Watson V et al (2009) Co-administration of fluconazole increases nevirapine concentrations in HIV-infected Ugandans. J Antimicrob Chemother 65:316–319
203. Hynninen VV, Olkkola KT, Leino K et al (2006) Effects of the antifungals voriconazole and fluconazole on the pharmacokinetics of S-(+)- and R-(-)-ibuprofen. Antimicrob Agents Chemother 50:1967–1972
204. Shaukat A, Benekli M, Vladutiu GD et al (2003) Simvastatin-fluconazole causing rhabdomyolysis. Ann Pharmacother 37:1032–1035
205. Cool RM, Gulbis AM (2013) Rhabdomyolysis after concomitant use of simvastatin and voriconazole in an allogeneic stem cell transplant patient. J Pharm Technol 29:135–138
206. Hazin R, Abuzetun JY, Suker M, Porter J (2008) Rhabdomyolysis induced by simvastatin-fluconazole combination. J Natl Med Assoc 100:444–446

207. Kahri J, Valkonen M, Bäcklund T et al (2005) Rhabdomyolysis in a patient receiving atorvastatin and fluconazole. Eur J Clin Pharmacol 60:905–907. https://doi.org/10.1007/s00228-004-0858-5
208. López-Gil JA (1993) Fluconazole-cyclosporine interaction: a dose-dependent effect? Ann Pharmacother 27:427–430
209. Canafax DM, Graves NM, Hilligoss DM et al (1991) Interaction between cyclosporine and fluconazole in renal allograft recipients. Transplantation 51:1014–1018
210. Osowski CL, Dix SP, Lin LS et al (1996) Evaluation of the drug interaction between intravenous high-dose fluconazole and cyclosporine or tacrolimus in bone marrow transplant patients. Transplantation 61:1268–1272
211. Salouage I, Trabelsi S, Charfi R et al (2012) Voriconazole and fluconazole in kidney transplant recipient receiving tacrolimus: a case of drug interaction. Fundam Clin Pharmacol 26:87–88
212. Kawazoe H, Takiguchi Y, Tanaka H et al (2006) Change of the blood concentration of tacrolimus after the switch from fluconazole to voriconazole in patients receiving allogeneic hematopoietic stem cell transplantation. Biol Pharm Bull 29:2528–2531
213. Toda F, Tanabe K, Ito S et al (2002) Tacrolimus trough level adjustment after administration of fluconazole to kidney recipients. Transplant Proc 1735:1733–1735
214. Kuypers DR, de Jonge H, Naesens M, Vanrenterghem Y (2008) Effects of CYP3A5 and MDR1 single nucleotide polymorphisms on drug interactions between tacrolimus and fluconazole in renal allograft recipients. Pharmacogenet Genomics 18:861–868
215. Cervelli MJ (2002) Fluconazole-sirolimus drug interaction. Transplantation 74:1477–1478
216. Pea F, Baccarani U, Tavio M et al (2008) Pharmacokinetic interaction between everolimus and antifungal triazoles in a liver transplant patient. Ann Pharmacother 42:1711–1716
217. Touchette M, Chandrasekar P, Milad M, Edwards D (1992) Contrasting effects of fluconazole and ketoconazole on phenytoin and testosterone disposition in man. Br J Clin Pharmacol 34:75–78
218. Turrentine MA (2006) Single-dose fluconazole for vulvovaginal candidiasis: impact on prothrombin time in women taking warfarin. Obstet Gynecol 107:310–313. https://doi.org/10.1097/01.AOG.0000196722.13403.33
219. Tarumi Y, Pereira J, Watanabe S (2002) Methadone and fluconazole: respiratory depression by drug interaction. J Pain Symptom Manag 23:148–153
220. Jordan MK, Polis MA, Kelly G et al (2000) Effects of fluconazole and clarithromycin on rifabutin and 25-O- desacetylrifabutin pharmacokinetics. Antimicrob Agents Chemother 44:2170–2172
221. Hallberg P, Martén L, Wadelius M (2006) Possible fluconazole – fentanyl interaction – a case report. Eur J Clin Pharmacol 62:491–492. https://doi.org/10.1007/s00228-006-0120-4
222. Gericke KR (2003) Possible interaction between warfarin and fluconazole. Pharmacotherapy 13:508–509
223. Seaton TL, Celum CL, Black DJ (1990) Possible potentiation of warfarin by fluconazole. DICP 24:1177–1178
224. Cadle RM, Zenon GJ, Rodriguez-Barradas MC, Hamill RJ (1994) Fluconazole-induced symptomatic phenytoin toxicity. Ann Pharmacother 28:191–195
225. Spriet I, Meersseman P, Meersseman W et al (2010) Increasing the dose of voriconazole compensates for enzyme induction by phenytoin. Br J Clin Pharmacol 69:701–702. https://doi.org/10.1111/j.1365-2125.2010.03650.x
226. Alffenaar J-WC, van der Elst KCM, Uges DRA et al (2009) Phenytoin-induced reduction of voriconazole serum concentration is not compensated by doubling the dosage. Br J Clin Pharmacol 68(3):462. https://doi.org/10.1111/j.1365-2125.2009.03464.x
227. Apseloff G, Hilligoss DM, Gardner MJ et al (1991) Induction of fluconazole metabolism by rifampin: in vivo study in humans. J Clin Pharmacol 31:358–361
228. Nicolau DP, Crowe HM, Nightingale CH, Quintiliani R (1995) Rifampin-fluconazole interaction in critically ill patients. Ann Pharmacother 29:994–996

229. Ayudhya DPN, Thanompuangseree N, Tansuphaswadikul S (2004) Effect of rifampicin on the pharmacokinetics of fluconazole in patients with AIDS. Clin Pharmacokinet 43:725–732. https://doi.org/10.2165/00003088-200443110-00003

230. Barbier O, Turgeon D, Girard C et al (2000) 3'-azido-3'-deoxythimidine (AZT) is glucuronidated by human UDP-glucuronosyltransferase 2B7 (UGT2B7). Drug Metab Dispos 28:497–502

231. Sahai J, Gallicano K, Pakuts A, Cameron DW (1994) Effect of fluconazole on zidovudine pharmacokinetics in patients infected with human immunodeficiency virus. J Infect Dis 169:1103–1107

232. Saari TI, Laine K, Leino K et al (2006) Effect of voriconazole on the pharmacokinetics and pharmacodynamics of intravenous and oral midazolam. Clin Pharmacol Ther 79:362–370

233. Kikuchi T, Mori T, Yamane A et al (2012) Variable magnitude of drug interaction between oral voriconazole and cyclosporine A in recipients of allogeneic hematopoietic stem cell transplantation. Clin Transpl 26:E544–E548. https://doi.org/10.1111/ctr.12016

234. Mori T, Aisa Y, Kato J et al (2009) Drug interaction between voriconazole and calcineurin inhibitors in allogeneic hematopoietic stem cell transplant recipients. Bone Marrow Transplant 44:371–374. https://doi.org/10.1038/bmt.2009.38

235. Park SJ, Song I-S, Kang SW et al (2012) Pharmacokinetic effect of voriconazole on cyclosporine in the treatment of aspergillosis after renal transplantation. Clin Nephrol 78:412–418

236. Mori T, Kato J, Yamane A et al (2012) Drug interaction between voriconazole and tacrolimus and its association with the bioavailability of oral voriconazole in recipients of allogeneic hematopoietic stem cell transplantation. Int J Hematol 95:564–569. https://doi.org/10.1007/s12185-012-1057-2

237. Tintillier M, Kirch L, Goffin E et al (2005) Interaction between voriconazole and tacrolimus in a kidney-transplanted patient. Nephrol Dial Transplant 20:664–665. https://doi.org/10.1093/ndt/gfh593

238. Spriet I, Grootaert V, Meyfroidt G et al (2013) Switching from intravenous to oral tacrolimus and voriconazole leads to a more pronounced drug-drug interaction. Eur J Clin Pharmacol 69:737–738. https://doi.org/10.1007/s00228-012-1365-8

239. Saad AH, DePestel DD, Carver PL (2006) Factors influencing the magnitude and clinical significance of drug interactions between azole antifungals and select immunosuppressants. Pharmacotherapy 26:1730–1744

240. Ceberio I, Dai K, Devlin SM et al (2015) Safety of voriconazole and sirolimus coadministration after allogeneic hematopoietic SCT. Bone Marrow Transplant 50:438–443. https://doi.org/10.1038/bmt.2014.286

241. Marty FM, Lowry CM, Cutler CS et al (2006) Voriconazole and sirolimus coadministration after allogeneic hematopoietic stem cell transplantation. Biol Blood Marrow Transplant 12:552–559. https://doi.org/10.1016/j.bbmt.2005.12.032

242. Billaud EM, Antoine C, Berge M et al (2009) Management of metabolic cytochrome P450 3A4 drug-drug interaction between everolimus and azole antifungals in a renal transplant patient. Clin Drug Investig 29:481–486

243. Lecefel C, Eloy P, Chauvin B et al (2015) Worsening pneumonitis due to a pharmacokinetic drug-drug interaction between everolimus and voriconazole in a renal transplant patient. J Clin Pharm Ther 40:119–120. https://doi.org/10.1111/jcpt.12234

244. Saari TI, Laine K, Leino K et al (2006) Voriconazole, but not terbinafine, markedly reduces alfentanil clearance and prolongs its half-life. Clin Pharmacol Ther 80:502–508. https://doi.org/10.1016/j.clpt.2006.07.008

245. Liu P, Foster G, LaBadie R et al (2007) Pharmacokinetic interaction between voriconazole and methadone at steady state in patients on methadone therapy. Antimicrob Agents Chemother 51:110–118

246. Hagelberg NM, Nieminen TH, Saari TI et al (2009) Voriconazole drastically increases exposure to oral oxycodone. Eur J Clin Pharmacol 65:263–271

247. Hagelberg NM, Nieminen TH, Saari TI et al (2011) Interaction of oxycodone and voriconazole – a case series of patients with cancer pain supports the findings of randomised

498 J. R. Amsden and P. O. Gubbins

controlled studies with healthy subjects. Eur J Clin Pharmacol 67:863–864. https://doi.org/10.1007/s00228-010-0969-0

248. Watanabe M, Homma M, Momo K et al (2011) Effects of voriconazole co-administration on oxycodone-induced adverse events: a case in the retrospective survey. Eur J Clin Pharmacol 67:859–861. https://doi.org/10.1007/s00228-010-0968-1

249. Fihlman M, Hemmila T, Hagelberg NM et al (2016) Voriconazole more likely than posaconazole increases plasma exposure to sublingual buprenorphine causing a risk of a clinically important interaction. Eur J Clin Pharmacol 72:1363–1371. https://doi.org/10.1007/s00228-016-2109-y

250. Hynninen VV, Olkkola KT, Leino K et al (2007) Effect of voriconazole on the pharmacokinetics of diclofenac. Fundam Clin Pharmacol 21:651–656

251. Hynninen VV, Olkkola KT, Neuvonen PJ, Laine K (2009) Oral voriconazole and miconazole oral gel produce comparable effects on the pharmacokinetics and pharmacodynamics of etoricoxib. Eur J Clin Pharmacol 65:89–95

252. Purkins L, Wood N, Kleinermans D, Nichols D (2003) Voriconazole potentiates warfarin-induced prothrombin time prolongation. Br J Clin Pharmacol 56:24–29

253. Purkins L, Wood N, Ghahramani P et al (2003) Coadministration of voriconazole and phenytoin: pharmacokinetic interaction, safety, and toleration. Br J Clin Pharmacol 56:37–44. https://doi.org/10.1046/j.1365-2125.2003.01997.x

254. Liu P, Foster G, LaBadie RR et al (2008) Pharmacokinetic interaction between voriconazole and efavirenz at steady state in healthy male subjects. J Clin Pharmacol 48:73–84

255. Damle B, LaBadie R, Crownover P, Glue P (2008) Pharmacokinetic interactions of efavirenz and voriconazole in healthy volunteers. Br J Clin Pharmacol 65:523–530

256. Liu P, Foster G, Gandelman K et al (2007) Steady-state pharmacokinetic and safety profiles of voriconazole and ritonavir in healthy male subjects. Antimicrob Agents Chemother 51:3617–3626

257. Mikus G, Schöwel V, Drzewinska M et al (2006) Potent cytochrome P450 2C19 genotype-related interaction between voriconazole and the cytochrome P450 3A4 inhibitor ritonavir. Clin Pharmacol Ther 80:126–135

258. Kennedy B, Larcombe R, Chaptini C, Gordon DL (2015) Interaction between voriconazole and flucloxacillin during treatment of disseminated Scedosporium apiospermum infection. J Antimicrob Chemother 70:2171–2173. https://doi.org/10.1093/jac/dkv069

259. Romero AJ, Le Pogamp P, Nilsson LG, Wood N (2002) Effect of voriconazole on the pharmacokinetics of cyclosporine in renal transplant patients. Clin Pharmacol Ther 71:226–234

260. Venkataramanan R, Zang S, Gayowski T, Singh N (2002) Voriconazole inhibition of the metabolism of tacrolimus in a liver transplant recipient and in human liver microsomes. Antimicrob Agents Chemother 46:3091–3093

261. Niwa T, Imagawa Y, Yamazaki H (2014) Drug interactions between nine antifungal agents and drugs metabolized by human cytochromes P450. Curr Drug Metab 15:651–679

262. Iwamoto T, Monma F, Fujieda A et al (2011) Hepatic drug interaction between tacrolimus and lansoprazole in a bone marrow transplant patient receiving voriconazole and harboring CYP2C19 and CYP3A5 heterozygous mutations. Clin Ther 33:1077–1080. https://doi.org/10.1016/j.clinthera.2011.07.006

263. Shao B, Ma Y, Li Q et al (2016) Effects of cytochrome P450 3A4 and non-genetic factors on initial voriconazole serum trough concentrations in hematological patients with different cytochrome P450 2C19 genotypes. Xenobiotica:1–28. https://doi.org/10.1080/00498254.2016.1271960

264. Hynninen VV, Olkkola KT, Bertilsson L et al (2008) Effect of terbinafine and voriconazole on the pharmacokinetics of the antidepressant venlafaxine. Clin Pharmacol Ther 83:342–348

265. Imamura CK, Furihata K, Okamoto S, Tanigawara Y (2016) Impact of cytochrome P450 2C19 polymorphisms on the pharmacokinetics of tacrolimus when coadministered with voriconazole. J Clin Pharmacol 56:408–413. https://doi.org/10.1002/jcph.605

266. Mazaleuskaya LL, Theken KN, Gong L et al (2015) PharmGKB summary: ibuprofen pathways. Pharmacogenet Genomics 25:96–106. https://doi.org/10.1097/FPC.0000000000000113

267. Vadlapatla RK, Patel M, Paturi DK et al (2014) Clinically relevant drug-drug interactions between antiretrovirals and antifungals. Expert Opin Drug Metab Toxicol 10:561–580. https://doi.org/10.1517/17425255.2014.883379

268. Ogburn ET, Jones DR, Masters AR et al (2010) Efavirenz primary and secondary metabolism in vitro and in vivo: identification of novel metabolic pathways and cytochrome P450 2A6 as the principal catalyst of efavirenz 7-hydroxylation. Drug Metab Dispos 38:1218–1229. https://doi.org/10.1124/dmd.109.031393

269. Dodds-Ashley E (2010) Management of drug and food interactions with azole antifungal agents in transplant recipients. Pharmacotherapy 30:842–854

270. Peksa GD, Schultz K, Fung HC (2015) Dosing algorithm for concomitant administration of sirolimus, tacrolimus, and an azole after allogeneic hematopoietic stem cell transplantation. J Oncol Pharm Pract 21:409–415. https://doi.org/10.1177/1078155214539825

271. Muijsers RB, Goa KL, Scott LJ (2002) Voriconazole: in the treatment of invasive aspergillosis. Drugs 62:2655–2656. doi: 621810 [pii]

272. Becker A, Sifaoui F, Gagneux M et al (2015) Drug interactions between voriconazole, darunavir/ritonavir and tenofovir/emtricitabine in an HIV-infected patient treated for Aspergillus candidus lung abscess. Int J STD AIDS 26:672–675. https://doi.org/10.1177/0956462414549035

273. Aouri M, Decosterd LA, Buclin T et al (2012) Drug interactions between voriconazole, darunavir/ritonavir and etravirine in an HIV-infected patient with Aspergillus pneumonia. AIDS 26:776–778

274. Toy J, Giguère P, Kravcik S, la Porte CJL (2011) Drug interactions between voriconazole, darunavir/ritonavir and etravirine in an HIV-infected patient with Aspergillus pneumonia. AIDS 25:541–542

275. Zhu L, Brüggemann RJ, Uy J et al (2017) CYP2C19 genotype-dependent pharmacokinetic drug interaction between voriconazole and ritonavir-boosted atazanavir in healthy subjects. J Clin Pharmacol 57:235–246. https://doi.org/10.1002/jcph.798

276. Krishna G, Moton A, Ma L et al (2009) Effects of oral posaconazole on the pharmacokinetic properties of oral and intravenous midazolam: a phase I, randomized, open-label, crossover study in healthy volunteers. Clin Ther 31:286–298

277. Krishna G, Ma L, Prasad P et al (2012) Effect of posaconazole on the pharmacokinetics of simvastatin and midazolam in healthy volunteers. Expert Opin Drug Metab Toxicol 8:1–10

278. Sánchez-Ortega I, Vázquez L, Montes C et al (2012) Effect of posaconazole on cyclosporine blood levels and dose adjustment in allogeneic blood and marrow transplant recipients. Antimicrob Agents Chemother 56:6422–6424. https://doi.org/10.1128/AAC.01489-12

279. Sansone-Parsons A, Krishna G, Martinho M et al (2007) Effect of oral posaconazole on the pharmacokinetics of cyclosporine and tacrolimus. Pharmacotherapy 27:825–834. https://doi.org/10.1592/phco.27.6.825

280. Moton A, Ma L, Krishna G et al (2009) Effects of oral posaconazole on the pharmacokinetics of sirolimus. Curr Med Res Opin 25:701–707

281. Krishna G, Moton A, Ma L et al (2009) Effects of oral posaconazole on the pharmacokinetics of atazanavir alone and with ritonavir or with efavirenz in healthy adult volunteers. J Acquir Immune Defic Syndr 51:437–444

282. Brüggemann RJM, van Luin M, Colbers EPH et al (2010) Effect of posaconazole on the pharmacokinetics of fosamprenavir and vice versa in healthy volunteers. J Antimicrob Chemother 65:2188–2194. https://doi.org/10.1093/jac/dkq280

283. Krishna G, Sansone-Parsons A, Kantesaria B (2007) Drug interaction assessment following concomitant administration of posaconazole and phenytoin in healthy men. Curr Med Res Opin 23:1415–1422

284. Marriott D, Levy R, Doyle T, Ray J (2009) Posaconazole-induced topiramate toxicity. Ann Intern Med 151:143

285. Pilmis B, Coignard-Biehler H, Jullien V et al (2013) Iatrogenic cushing's syndrome induced by posaconazole. Antimicrob Agents Chemother 57:5727–5728. https://doi.org/10.1128/AAC.00416-13

286. Hohmann C, Kang EM, Jancel T (2010) Rifampin and posaconazole coadministration leads to decreased serum posaconazole concentrations. Clin Infect Dis 50:939–940. https://doi.org/10.1086/650740
287. Krishna G, Ma L, Martinho M et al (2012) A new solid oral tablet formulation of posaconazole: a randomized clinical trial to investigate rising single- and multiple-dose pharmacokinetics and safety in healthy volunteers. J Antimicrob Chemother 67:2725–2730
288. Duarte RF, Lopez-Jimenez J, Cornely OA et al (2014) Phase 1b study of new posaconazole tablet for prevention of invasive fungal infections in high-risk patients with neutropenia. Antimicrob Agents Chemother 58:5758–5765. https://doi.org/10.1128/AAC.03050-14
289. Pham AN, Bubalo JS, Lewis JS (2016) Comparison of posaconazole serum concentrations from haematological cancer patients on posaconazole tablet and oral suspension for treatment and prevention of invasive fungal infections. Mycoses 59:226–233. https://doi.org/10.1111/myc.12452
290. Durani U, Tosh PK, Barreto JN et al (2015) Retrospective comparison of posaconazole levels in patients taking the delayed-release tablet versus the oral suspension. Antimicrob Agents Chemother 59:4914–4918. https://doi.org/10.1128/AAC.00496-15
291. Cumpston A, Caddell R, Shillingburg A et al (2015) Superior serum concentrations with posaconazole delayed-release tablets compared to suspension formulation in hematological malignancies. Antimicrob Agents Chemother 59:4424–4428. https://doi.org/10.1128/AAC.00581-15
292. Jung DS, Tverdek FP, Kontoyiannis DP (2014) Switching from posaconazole suspension to tablets increases serum drug levels in leukemia patients without clinically relevant hepatotoxicity. Antimicrob Agents Chemother 58:6993–6995
293. Lempers VJC, Martial LC, Schreuder MF et al (2015) Drug-interactions of azole antifungals with selected immunosuppressants in transplant patients: strategies for optimal management in clinical practice. Curr Opin Pharmacol 24:38–44. https://doi.org/10.1016/j.coph.2015.07.002
294. Berge M, Chevalier P, Benammar M et al (2009) Safe management of tacrolimus together with posaconazole in lung transplant patients with cystic fibrosis. Ther Drug Monit 31:396–399
295. Cho E, Chan H, Nguyen HM et al (2015) Management of drug interaction between posaconazole and sirolimus in patients who undergo hematopoietic stem cell transplant. Pharmacotherapy 35:578–585. https://doi.org/10.1002/phar.1600
296. Kubiak DW, Koo S, Hammond SP et al (2012) Safety of posaconazole and sirolimus coadministration in allogeneic hematopoietic stem cell transplants. Biol Blood Marrow Transplant 18:1462–1465. https://doi.org/10.1016/j.bbmt.2012.04.015
297. Petitcollin A, Crochette R, Tron C et al (2016) Increased inhibition of cytochrome P450 3A4 with the tablet formulation of posaconazole. Drug Metab Pharmacokinet 31:389–393. https://doi.org/10.1016/j.dmpk.2016.05.001
298. Krishna G, Parsons A, Kantesaria B, Mant T (2007) Evaluation of the pharmacokinetics of posaconazole and rifabutin following co-administration to healthy men. Curr Med Res Opin 23:545–552
299. Groll AH, Desai A, Han D et al (2017) Pharmacokinetic assessment of drug-drug interactions of isavuconazole with the immunosuppressants cyclosporine, mycophenolic acid, prednisolone, sirolimus, and tacrolimus in healthy adults. Clin Pharmacol Drug Dev 6:76–85. https://doi.org/10.1002/cpdd.284
300. Yamazaki T, Desai A, Han D et al (2017) Pharmacokinetic interaction between isavuconazole and a fixed-dose combination of lopinavir 400 mg/ritonavir 100 mg in healthy subjects. Clin Pharmacol Drug Dev 6:93–101. https://doi.org/10.1002/cpdd.282
301. Schmitt-Hoffmann A, Desai A, Kowalski D et al (2016) Isavuconazole absorption following oral administration in healthy subjects is comparable to intravenous dosing, and is not affected by food, or drugs that alter stomach pH. Int J Clin Pharmacol Ther 54:572–580
302. Kim T, Jancel T, Kumar P, Freeman AF (2015) Drug-drug interaction between isavuconazole and tacrolimus: a case report indicating the need for tacrolimus drug-level monitoring. J Clin Pharm Ther 40:609–611

303. Meyer MR, Maurer HH (2011) Absorption, distribution, metabolism and excretion pharmacogenomics of drugs of abuse. Pharmacogenomics 12:215–233. https://doi.org/10.2217/pgs.10.171

304. Totah RA, Sheffels P, Roberts T et al (2008) Role of CYP2B6 in stereoselective human methadone metabolism. Anesthesiology 108:363–374. https://doi.org/10.1097/ALN.0b013e3181642938

305. Badagnani I, Castro RA, Taylor TR et al (2006) Interaction of methotrexate with organic-anion transporting polypeptide 1A2 and its genetic variants. J Pharmacol Exp Ther 318:521–529. https://doi.org/10.1124/jpet.106.104364

306. van de Steeg E, van der Kruijssen CMM, Wagenaar E et al (2009) Methotrexate pharmacokinetics in transgenic mice with liver-specific expression of human organic anion-transporting polypeptide 1B1 (SLCO1B1). Drug Metab Dispos 37:277–281. https://doi.org/10.1124/dmd.108.024315

307. Vlaming MLH, van Esch A, van de Steeg E et al (2011) Impact of abcc2 [multidrug resistance-associated protein (MRP) 2], abcc3 (MRP3), and abcg2 (breast cancer resistance protein) on the oral pharmacokinetics of methotrexate and its main metabolite 7-hydroxymethotrexate. Drug Metab Dispos 39:1338–1344. https://doi.org/10.1124/dmd.111.038794

308. Durmus S, van Hoppe SS, Schinkel AH (2016) The impact of organic anion-transporting polypeptides (OATPs) on disposition and toxicity of antitumor drugs: insights from knockout and humanized mice. Drug Resist Updat 27:72–88. https://doi.org/10.1016/j.drup.2016.06.005

309. Stader F, Wuerthwein G, Groll AH et al (2015) Physiology-based pharmacokinetics of caspofungin for adults and paediatrics. Pharm Res 32:2029–2037. https://doi.org/10.1007/s11095-014-1595-9

310. Inoue Y, Saito T, Ogawa K et al (2012) Drug interactions between micafungin at high doses and cyclosporine A in febrile neutropenia patients after allogeneic hematopoietic stem cell transplantation. Int J Clin Pharmacol Ther 50:831–837

311. Hebert MF, Townsend RW, Austin S et al (2005) Concomitant cyclosporine and micafungin pharmacokinetics in healthy volunteers. J Clin Pharmacol 45:954–960

312. Hebert MF, Blough DK, Townsend RW et al (2005) Concomitant tacrolimus and micafungin pharmacokinetics in healthy volunteers. J Clin Pharmacol 45:1018–1024

313. Fukuoka N, Imataki O, Ohnishi H et al (2010) Micafungin does not influence the concentration of tacrolimus in patients after allogeneic hematopoietic stem cell transplantation. Transplant Proc 42:2725–2730

314. Dowell JA, Stogniew M, Krause D et al (2005) Assessment of the safety and pharmacokinetics of anidulafungin when administered with cyclosporine. J Clin Pharmacol 45:227–233. https://doi.org/10.1177/0091270004270146

315. Dowell JA, Stogniew M, Krause D et al (2007) Lack of pharmacokinetic interaction between anidulafungin and tacrolimus. J Clin Pharmacol 47:305–314

316. Stone JA, Migoya EM, Hickey L et al (2004) Potential for interactions between caspofungin and nelfinavir or rifampin. Antimicrob Agents Chemother 48:4306–4314

317. Williamson B, Dooley KE, Zhang Y et al (2013) Induction of influx and efflux transporters and cytochrome P450 3A4 in primary human hepatocytes by rifampin, rifabutin, and rifapentine. Antimicrob Agents Chemother 57:6366–6369. https://doi.org/10.1128/AAC.01124-13

318. Burhan E, Ruesen C, Ruslami R et al (2013) Isoniazid, rifampin, and pyrazinamide plasma concentrations in relation to treatment response in indonesian pulmonary tuberculosis patients. Antimicrob Agents Chemother 57:3614–3619. https://doi.org/10.1128/AAC.02468-12

Chapter 12
Drug-Drug Interactions of Antimalarial Drugs

Waheed A. Adedeji, Tunde Balogun, Fatai A. Fehintola, and Gene D. Morse

12.1 Introduction

Malaria is a cause of substantial morbidity and mortality. Malaria-associated deaths remain very high at about 450,000 annually [1]. Human populations within the tropical and subtropical countries across Africa, the Americas, and Asia are at risk of the disease, though with varying extent of endemicity even within the same country. An estimated 3.2 billion people across 95 countries and territories are at risk of malaria with estimated 214 million malaria cases in 2015 [1]. Malaria was known to account for considerably higher morbidity and mortality rates prior to the institution of malaria control measures, notably the use of long-lasting insecticidal nets (LLINs), presumptive treatment of malaria in "at-risk" populations, and the prompt treatment of acute uncomplicated malaria using artemisinin-based combination

W. A. Adedeji
Department of Clinical Pharmacology, University College Hospital,
Ibadan, Oyo State, Nigeria

T. Balogun
Department of Clinical Pharmacology and Therapeutics, College of Medical Sciences,
University of Maiduguri, Maiduguri, Borno State, Nigeria

F. A. Fehintola (✉)
Department of Clinical Pharmacology, University College Hospital,
Ibadan, Oyo State, Nigeria

Department of Pharmacology and Therapeutics, University of Ibadan,
Ibadan, Oyo State, Nigeria
e-mail: fehintolaf@com.ui.edu.ng; fentolamine@yahoo.com; fentolamine@gmail.com

G. D. Morse
School of Pharmacy and Pharmaceutical Sciences, Center for Integrated Global Biomedical
Sciences, University at Buffalo, Buffalo, NY, USA

© Springer International Publishing AG, part of Springer Nature 2018   503
M. P. Pai et al. (eds.), *Drug Interactions in Infectious Diseases: Antimicrobial
Drug Interactions*, Infectious Disease,
https://doi.org/10.1007/978-3-319-72416-4_12

therapy (ACT). Malaria-associated morbidity and mortality figures recorded in 2015 were improvements, respectively, 37% and 65% from the year 2000 rates [1].

Malaria is caused by a protozoan of genus *Plasmodium*, and five species, namely, *falciparum, malariae, ovale, vivax*, and *knowlesi* are known to naturally infect humans. Of these five species, *P. falciparum* is the most prevalent species in sub-Saharan Africa, accounting for virtually all malaria-related morbidity and mortality in the tropical Africa [2]. Malaria transmission is driven by infected species of *Anopheles* mosquito where the sexual reproductive stage as well as sporogony takes place. Human infections begin with the inoculation of sporozoite-laden salivary secretion by the infected mosquito during a blood meal. The initial development of the sporozoites occurs in the liver and is usually completed in about 7–10 days for the species of *P. falciparum, P. ovale*, and P. *vivax*, but *P. malariae* requires about 15 days. Some of the sporozoites of *P. ovale* and *P. vivax* can remain dormant as *hypnozoites* and when "awakened" resume hepatic schizogony and subsequent invasion of erythrocytes causing the malaria *relapse*. Merozoites that result from hepatic schizogony invade the erythrocytes to begin the erythrocytic schizogony [3, 4]. Parasite development within the erythrocytes subsequently produces schizonts and releases merozoites and some cytokines resulting in fever and other malaria-associated symptoms. Anemia is, at least, partly a result of red-cell lysis and dys-erythropoiesis [5].

Prevention of malaria may be achieved through chemoprophylaxis and vector control and, perhaps, by immunization. Various classes of drugs for the treatment of malaria exist (Tables 12.1 and 12.2), and, due to drug resistance in malaria, drug combinations from at least two different classes are currently the standard of care [6].

Table 12.1 Chemical classes[a] of antimalarial drugs with examples

Classes	Examples
4-aminoquinolines	Amodiaquine, chloroquine
8-aminoquinolines	Pamaquine, primaquine, tafenoquine
Antibacterials	Clindamycin, doxycycline, tetracycline, co-trimoxazole[b]
Artemisinin derivatives	Artemether, artemotil, artesunate, Arteether, dihydroartemisinin
Biguanides	Chlorproguanil, proguanil
Diaminopyrimidines	Pyrimethamine, trimethoprim
Phenanthrene methanol	Halofantrine, lumefantrine
Quinoline-methanol	Mefloquine, quinidine, quinine
Quinone	Atovaquone
Sulfonamides/sulfones	Dapsone, sulfadoxine

[a]Arranged in alphabetical order
[b]Co-trimoxazole is a co-formulation of sulfamethoxazole and trimethoprim

Table 12.2 Classification of antimalarial drugs based on the parasite life cycle

Classes	Examples
Blood schizonticides	Amodiaquine, artemisinins, chloroquine, mefloquine, pyronaridine, piperaquine, quinine
Tissue schizonticides	Primaquine, tafenoquine
Gametocides	Artemisinins, primaquine
Sporontocide	Pyrimethamine

12.2 Current Status of Malaria Chemotherapy

Until early this millennium, chloroquine and other schizonticides were used as monotherapy in the treatment of acute uncomplicated malaria [7, 8]. However, more than half a century ago, reports of chloroquine-resistant malaria from Southeast Asia and South America emerged, and it has subsequently assumed a global phenomenon; involving to a varying extent, all other drugs hitherto are used as monotherapy resulting in increased morbidity and mortality [9–13].

The initial response to stemming malaria-associated morbidity and mortality due to drug resistance was the development and introduction of such drugs as halofantrine and mefloquine [14–19]. Although both mefloquine and quinine remain substantially efficacious and safe even till date, in some countries on the continent of Africa, neither is currently used as monotherapy [20].

The artemisinins are derived from *Artemisia annua* L., a Chinese herbal treatment for fevers that has been used for many centuries [21]. Artemether, artesunate, and other drugs in this group contain the sesquiterpene lactone ring and are rapidly schizonticidal. Artesunate given intravenously has become the standard of care for severe or complicated malaria [20].

Treatment of acute uncomplicated malaria requires combination therapy administered orally. Ideally the combined drugs should have similar pharmacokinetics and pharmacodynamics and should display no enhanced adverse effects [22]. Depending on the resistance pattern in a given geographical region, a variety of combinations may be efficacious, but artemisinin-based combination therapy (ACT) is the mostly preferred [20].

In most countries of sub-Saharan Africa, the commonly used ACTs include artemether-lumefantrine, artesunate-amodiaquine, artesunate-mefloquine, and dihydroartemisinin-piperaquine.

12.2.1 Artemisinin Derivatives

Artemisinin is a sesquiterpene lactone endoperoxide derived from the weed sweet wormwood (*Artemisia annua*). Although the medicinal value of this plant has been known for over 2000 years, its active ingredient, *qinghaosu* (i.e., artemisinin), was first isolated in 1972 [23, 24]. Semisynthetic derivatives such as artesunate, artemether, dihydroartemisinin, and arteether, with improved potency and bioavailability, are also available for treatment of malaria. Artemisinins possess activity against all species of malaria parasites that infect humans [20, 25].

12.2.1.1   Artemether

Artemether, the methyl ether of dihydroartemisinin, is more lipid soluble than artemisinin or artesunate. Artemether is usually administered in combination with lumefantrine at a ratio of 1:6. Artemether is rapidly metabolized to dihydroartemisinin by various cytochrome P450 (CYPs) enzymes including CYP3A4 and CYP2B6. Both artemether and dihydroartemisinin are rapidly eliminated [20].

12.2.1.2   Artesunate

Artesunate is the hemisuccinate ester of artemisinin. Its antimalarial activity is due largely to its active metabolite, dihydroartemisinin (DHA).

12.2.1.3   Dihydroartemisinin

Dihydroartemisinin is the main active metabolite of artemisinin derivatives and is also available as a drug on its own.

12.2.1.4   Arteether

Arteether is an ethyl ether semisynthetic derivative of artemisinin; it is also known as artemotil. It exists in two isomers (α- and β-isomers) with both isomers having antimalarial activities. It has a relatively long elimination half-life when compared to other artemisinins such as artesunate or dihydroartemisinin (> 20 h vs < 1 h or ~ 2 h) [26].

12.2.2   Aminoquinolines: 4-Aminoquinolines

12.2.2.1   Chloroquine

Chloroquine (CQ) is a 4-aminoquinoline that has been used extensively for the treatment and prevention of malaria since the end of World War II. It is a blood schizonticide, a weak base that has strong affinity for various tissues and organs.

12.2.2.2   Amodiaquine

Amodiaquine hydrochloride is readily absorbed from the gastrointestinal tract. It is rapidly metabolized in the liver to the marginally less active metabolite, desethyl-amodiaquine, which accounts for nearly all of the antimalarial activity as it has substantially long half-life, respectively, 5 h and 9–18 days [20].

12.2.3   Aminoquinolines: 8-Aminoquinolines

12.2.3.1   Primaquine

Primaquine (PQ) is an 8-aminoquinoline developed from a large series of quinoline derivatives during World War II while searching for potent and less toxic 8-aminoquinoline antimalarial drugs [27].

12.3   Antimalarial Drug Interactions

The introduction of combination therapy in the treatment of malaria has imposed the need for careful considerations of both intragroup and intergroup drug(−drug) interactions. Drug interactions involving antimicrobials become especially important whenever there is substantial overlap in the epidemiology of respective conditions. For example, malaria and tuberculosis are rampant in the tropical and subtropical countries, and both are responsible for considerable burden [28]. Another important overlap of epidemiology exists between the malaria and HIV/AIDS demanding caution in order to avoid adverse drug interactions [29]. Probability of positive pharmacodynamic interaction between co-trimoxazole, an antibacterial, and chloroquine, the first-line antimalarial drug of the 1960s through the 1980s in most sub-Saharan African countries, is of note [30].

Both the anti-Tb and combination antiretroviral therapy (cART) are normally given for a prolonged period whereas the ACT requires a short course, usually not more than 3 days. Predictably, the pharmacokinetic interactions between either anti-Tb or cART with antimalarial drugs, particularly when it involves enzyme induction, would mostly situate the latter as the object drug, given the needed time for the evolution of induction of enzyme activity. Alteration of pharmacokinetic parameters of any drug by the other(s) may result in inadequate concentrations and ineffective treatment and/or toxic concentrations and drug-induced disorders. Expectedly, genetically imposed pharmacokinetic differences, for example, CYP polymorphisms, will further modulate the drug-environment interactions. Thus, these scenarios may be sources of increased morbidity and/or mortality, from the original diseases and/or adverse drug interactions.

The following paragraphs attempt to delve into the antimalarial, particularly, ACT-anti-Tb drug interactions, as ACT-cART interactions have been addressed elsewhere in the text.

12.3.1   Antimalarial-Antituberculosis Drug Interactions

The use of combination of drugs has remained the standard of care for the treatment of tuberculosis for decades, having been seen as a vital approach to improving treatment outcome of the disease. Rifampicin, a rifamycin, being a potent enzyme inducer, is well known for its pharmacokinetic drug-drug interactions [28]. Isoniazid, pyrazinamide, and ethambutol, together with streptomycin, are some of the other components of the 6- or more-month regimens used in the treatment of tuberculosis [31]. In desperate situations as multidrug-resistant (MDR) or extensively drug-resistant tuberculosis (XDR Tb), other drugs such as amikacin, second- and newer-generations of quinolones (e.g., ofloxacin or moxifloxacin), capreomycin, or ethionamide are used as salvage drugs [32]. Increasing number and/or complexities of antituberculosis regimens expose patients to varying degrees of drug-drug interactions.

ACT-anti-Tb drug interactions may be viewed as adverse, if concomitant use results in treatment failure of either (or both) condition(s) or there is enhancement of known adverse effects of either group of drugs. Pharmacokinetic interactions may involve absorption, distribution, and elimination of drugs; notably, malaria can alter all of these processes in an infected individual [28]. Theoretically, reduced exposure to ACT due to enhanced metabolism of its components by the enzyme-inducing rifamycins may predispose to poor treatment outcome of acute uncomplicated malaria in a tuberculosis patient. Clinically significant pharmacokinetic interactions between ACTs and anti-Tb drugs may also involve the drug transporters, particularly, the ubiquitous P-glycoprotein (P-gp) or multidrug resistance-associated protein (MRP) transporters [33]. In the following discussion for convenience, the commonly used ACTs (and few other antimalarial drugs) are considered, in the main, as the object drugs.

12.3.1.1   ACT-Anti-Tb Drug Interactions

Artemether-Lumefantrine and Anti-Tb Drugs

Rifamycins including rifampicin, rifabutin, and rifapentine are well known for their capacity to induce various isozymes of cytochrome P450 superfamily (CYPs); consequently, they reduce respective concentrations of artemether and lumefantrine. It is to be noted that the enzyme-inducing capacity of rifabutin is considerably less compared to that of rifampicin and may be preferred whenever necessary in, for example, HIV-infected individuals on protease inhibitors [28, 31]. Notable CYP enzyme activity induced by rifampicin includes 3A4/5, 1A2, 2C8, 2C9, and 2C18/19. Rifampicin also induces phase 2 enzymes, uridine-diphosphate glucuro-nosyltransferase (UGT), and drug transporters including P-glycoprotein (P-gp) and multidrug resistance-associated protein (MRP) [28, 34]. In vitro models have revealed that artemether is metabolized by CYPs, 1A2, 2B6, 2C9, and 3A4, but the biotransformation of artemether is mediated by the cytochrome P450 isozymes CYP1A2, CYP2B6, and CYP3A4 in humans [35]. The antimalarial activity of arte-mether depends largely on the formation of its active metabolite, dihydroartemis-inin (DHA); the phase 2 biotransformation is catalyzed by UGTs resulting in the final inactive glucuronide conjugates. Ultimately, there is substantial lowering of concentrations of artemether, its active metabolite, DHA, and lumefantrine when administered to individuals on chronic medication with rifampicin-containing anti-Tb drugs [36].

In view of the fact that both rifampicin and isoniazid are used throughout the 6-month or more duration of anti-Tb treatment, and that the latter agent inhibits CYPs, potential for counterbalancing the enzyme-inducing activity of rifampicin is theoretically possible. However, the few available studies suggest that the resultant effect is the reduced exposure produced by rifampicin-mediated enzyme induction [36]. In a study by Lamorde et al., rifampicin-containing anti-Tb therapy resulted in lower concentrations of artemether, DHA, and lumefantrine, prompting the conclu-sion that artemether-lumefantrine should not be coadministered with rifampicin [36].

Artesunate-Amodiaquine and Anti-Tb Drugs

Artesunate-amodiaquine is another commonly prescribed ACT for the treatment of acute uncomplicated malaria. As described previously, artesunate is a prodrug and its antimalarial activity is largely due to its primarily esterase- and CYP2A6-mediated conversion to its active metabolite, DHA. The phase 2 reaction and final inactivation of artesunate is catalyzed by uridine-diphosphate glucuronosyltransfer-ases (UGTs – 1A9 & 2B7). Rifampicin and, probably to lower extent, other rifamy-cins induce UGTs and P-glycoprotein (efflux) transport protein and will consequently result in reduced exposure to the artemisinins [28, 36, 37]. Amodiaquine, on the other hand, is a CYP2C8 substrate yielding desethylamodiaquine as the major metabolite with significant antimalarial activity and considerably longer half-life [38]. Concomitant medication of artesunate-amodiaquine with rifampicin-containing antituberculosis drugs would appear to present complicated pharmacoki-netic interactions that are yet to be fully elucidated.

The antiplasmodial activity of rifampicin had been documented, and, when con-sidered with its hepatotoxic effect shared with amodiaquine, the pharmacodynamic interactions would seem to have further confound the requisite treatment of these diseases when they co-exist [39, 40]. Amodiaquine-induced hepatotoxicity has been linked to the formation of a reactive species quinone-imine which subsequently covalently binds to cell structures to provoke immunological response and cell lysis

[38]. The mechanism(s) of rifampicin–/rifamycin-induced hepatotoxicity remain(s) to be fully elucidated though it may be idiosyncratic and unpredictable resulting in diffuse hepatocyte necrosis [41]. In view of the enhanced adverse hepatotoxic effect that may be exhibited by both arms of antimalarial (amodiaquine-induced) and anti-Tb drugs (Rifampicin and INH), there is an urgent need for relevant assessment of their concomitant use to bridge the information gap.

Artesunate-Mefloquine and Anti-Tb Drugs

Mefloquine in combination with artesunate is an important ACT employed in the treatment of acute uncomplicated malaria and is widely used in Southeast Asia. Mefloquine is also used as monotherapy in the chemoprevention of malaria [42, 43]. Rifampicin induces mefloquine metabolism resulting in reduced area under the concentration-time curve (AUC) and increased clearance [44] as CYP3A isozyme-mediated reaction results in the formation of two inactive metabolites of mefloquine. The coadministration of the two drugs should better be avoided [28].

12.3.1.2   Other Antimalarial Drugs and Anti-Tb Drugs

Atovaquone-Proguanil-Rifamycin Interaction

Atovaquone-proguanil combination is mostly used for treatment of *P. vivax* malaria and as prophylactic option for malaria [45]. While CYP3A4 metabolizes atovaquone, proguanil is metabolized by CYP2C19 to active metabolite, cycloguanil. Coadministration of atovaquone with rifampicin and rifabutin resulted in significant reduction in the AUC of atovaquone [28]. Similarly, since rifampicin is an inducer of CYP2C19, it is expected that proguanil concentrations will be reduced. However, relevant clinical data should be obtained to provide firm basis for appropriate clinical consultation and recommendations.

Quinine-Rifamycin Interaction

Quinine metabolism is catalyzed mainly by CYP3A4 and CYP2C19, and it possesses inhibitory activity against CYP2D6 and enterocyte P-glycoprotein. Rifampicin being a potent inducer of many CYPs and transporter proteins has been found to reduce exposure to quinine though the former enhances antimalarial activity of the latter when used concomitantly [46]. The enhanced antimalarial activity notwithstanding more assessment is required given the theoretical increased exposure to rifampicin and possible untoward effects. Caution has been advised when quinine has to be used with the rifamycins [28, 46].

Chloroquine-Rifamycin Interaction

Chloroquine still has relevance especially in the treatment of *P. vivax* malaria. Chloroquine metabolism is catalyzed by CYPs 3A4/5 and 2C8 to desethylchloroquine and bisdesethylchloroquine [47, 48]. In infected mice treated with rifampicin-chloroquine, delayed parasite clearance and an increased recrudescence were observed [49]. However, this is yet to be supported by human data.

Antifolate Antimalarial Drugs-Anti-Tb Drug Interactions

Isoniazid and sulfonamide components of antimalarial drugs, for example, sulfadoxine, undergo acetylation catalyzed by NAT-2 enzyme. Theoretically, coadministration of the two drugs may result in competitive inhibition of each other's metabolism. However, definitive recommendations can only await relevant data from pharmacokinetic and pharmacodynamic studies of interactions between sulfadoxine (a partner drug in the sulfadoxine-pyrimethamine) and isoniazid.

Co-trimoxazole (sulfamethoxazole-trimethoprim), an antifolate antibacterial, is known to possess substantial antimalarial activity, and its sulfonamide component undergoes acetylation like isoniazid [30]. It is commonly used in HIV/AIDS patients especially for prophylaxis and treatment of *Pneumocystis jirovecii* or PJP (formerly PCP) infection; thus there may be need for its concomitant use with cART and anti-Tb drugs, including isoniazid. Sulfadoxine-pyrimethamine and co-trimoxazole cannot be used together as both have similar antimalarial activity, and there may be enhanced adverse hematological effect, folate deficiency anemia [50, 51]. Genton et al. evaluated co-trimoxazole in co-formulation with rifampicin and isoniazid in the treatment of acute uncomplicated malaria and concluded that the combination has similar efficacy and tolerability profile with mefloquine or quinine in combination with sulfadoxine-pyrimethamine [52]. The relative antiplasmodial activity of rifampicin would be expected to have played a role, but its N-acetyltransferase-inhibitory activity may also have contributed to the observed positive pharmacodynamic interaction of antimalarial-anti-Tb drugs [28, 39].

References

1. WHO (2016) Malaria. www.who.int/mediacentre/factsheets/fs094/en/. Accessed Nov 2016
2. WHO (2014) World malaria report 2014. World Health Organization, Geneva
3. Harrison TR, Wilson JD (1991) Harrison's principles of internal medicine, vol 2, 12th edn. McGraw-Hill, New York, pp 782–788
4. Krotoski WA (1985) Discovery of the hypnozoite and a new theory of malarial relapse. Trans R Soc Trop Med Hyg 79(1):1–11
5. Kai OK, Roberts DJ (2008) The pathophysiology of malarial anaemia: where have all the red cells gone? BMC Med 6:24. https://doi.org/10.1186/1741-7015-6-24

6. Bosman A, Mendis KN (2007) A major transition in malaria treatment: the adoption and deployment of artemisinin-based combination therapies. Am J Trop Med Hyg 77(6 Suppl): 193–197
7. Mwai L, Ochong E, Abdirahman A, Kiara SM, Ward S, Kokwaro G, Sasi P, Marsh K, Borrmann S, Mackinnon M, Nzila A (2009) Chloroquine resistance before and after its withdrawal in Kenya. Malar J 8:106. https://doi.org/10.1186/1475-2875-8-106
8. WHO (2001) Antimalarial drug combination therapy, Report of a WHO technical consultation. World Health Organization, Geneva
9. Fogh S, Jepsen S, Effersoe P (1979) Chloroquine-resistant Plasmodium falciparum malaria in Kenya. Trans R Soc Trop Med Hyg 73(2):228–229
10. Weniger BG, Blumberg RS, Campbell CC, Jones TC, Mount DL, Friedman SM (1982) High-level chloroquine resistance of Plasmodium falciparum malaria acquired in Kenya. N Engl J Med 307(25):1560–1562. https://doi.org/10.1056/nejm198212163072506
11. Ekanem O, Weisfeld J, Salako LA, Nahlen BL, Ezedinachi E, Walker O, Breman JG, Laoye O, Hedberg K (1990) Sensitivity of Plasmodium falciparum to chloroquine and sulfadoxine/pyrimethamine in Nigerian children. Bull World Health Organ 68(1):45
12. Trape JF, Pison G, Preziosi MP, Enel C, Desgrees du Lou A, Delaunay V, Samb B, Lagarde E, Molez JF, Simondon F (1998) Impact of chloroquine resistance on malaria mortality. C R Acad Sci III 321(8):689–697
13. Howard DH, Scott RD, Packard R, Jones D (2003) The global impact of drug resistance. Clin Infect Dis 36(Supplement_1):S4–S10
14. Martin SK, Oduola AM, Milhous WK (1987) Reversal of chloroquine resistance in Plasmodium falciparum by verapamil. Science (New York, NY) 235(4791):899–901
15. Salako LA, Sowunmi A, Walker O (1990) Evaluation of the clinical efficacy and safety of halofantrine in falciparum malaria in Ibadan, Nigeria. Trans R Soc Trop Med Hyg 84(5):644–647
16. Sowunmi A, Oduola AM (1995) Open comparison of mefloquine, mefloquine/sulfadoxine/pyrimethamine and chloroquine in acute uncomplicated falciparum malaria in children. Trans R Soc Trop Med Hyg 89(3):303–305
17. Oduola AM, Omitowoju GO, Gerena L, Kyle DE, Milhous WK, Sowunmi A, Salako LA (1993) Reversal of mefloquine resistance with penfluridol in isolates of Plasmodium falciparum from south-west Nigeria. Trans R Soc Trop Med Hyg 87(1):81–83
18. Bitonti AJ, Sjoerdsma A, McCann PP, Kyle DE, Oduola AM, Rossan RN, Milhous WK, Davidson DE Jr (1988) Reversal of chloroquine resistance in malaria parasite Plasmodium falciparum by desipramine. Science (New York, NY) 242(4883):1301–1303
19. Sowunmi A, Oduola AM, Ogundahunsi OA, Falade CO, Gbotosho GO, Salako LA (1997) Enhanced efficacy of chloroquine-chlorpheniramine combination in acute uncomplicated falciparum malaria in children. Trans R Soc Trop Med Hyg 91(1):63–67
20. WHO (2015) Guidelines for the treatment of malaria. World Health Organization, Geneva
21. White NJ (1994) Artemisinin: current status. Trans R Soc Trop Med Hyg 88(Suppl 1):S3–S4
22. White N (1999) Antimalarial drug resistance and combination chemotherapy. Philos Trans R Soc Lond Ser B Biol Sci 354(1384):739–749. https://doi.org/10.1098/rstb.1999.0426
23. Klayman DL (1985) Qinghaosu (artemisinin): an antimalarial drug from China. Science (New York, NY) 228(4703):1049–1055
24. Hien TT, White NJ (1993) Qinghaosu. Lancet (London, England) 341(8845):603–608
25. White NJ (1997) Assessment of the pharmacodynamic properties of antimalarial drugs in vivo. Antimicrob Agents Chemother 41(7):1413–1422
26. de Vries PJ, Dien TK (1996) Clinical pharmacology and therapeutic potential of artemisinin and its derivatives in the treatment of malaria. Drugs 52(6):818–836
27. Nodiff EA, Chatterjee S, Musallam HA (1991) Antimalarial activity of the 8-aminoquinolines. Prog Med Chem 28:1–40
28. Sousa M, Pozniak A, Boffito M (2008) Pharmacokinetics and pharmacodynamics of drug interactions involving rifampicin, rifabutin and antimalarial drugs. J Antimicrob Chemother 62(5):872–878. https://doi.org/10.1093/jac/dkn330

29. German P, Greenhouse B, Coates C, Dorsey G, Rosenthal PJ, Charlebois E, Lindegardh N, Havlir D, Aweeka FT (2007) Hepatotoxicity due to a drug interaction between amodiaquine plus artesunate and efavirenz. Clin Infect Dis Off Publ Infect Dis Soc Am 44(6):889–891. https://doi.org/10.1086/511882

30. Fehintola FA, Adedeji AA, Sowunmi A (2002) Comparative efficacy of chloroquine and cotrimoxazole in the treatment of acute uncomplicated falciparum malaria in Nigerian children. Cent Afr J Med 48(9–10):101–105

31. Brunton L, Lazo J, Parker K (2005) Goodman & Gilman's the pharmacological basis of therapeutics, 11th edn. McGraw-Hill Education, New York

32. WHO (2016) WHO treatment guidelines for drug-resistant tuberculosis 2016 update. World Health Organization, Geneva

33. Elsherbiny D (2008) Pharmacokinetic drug-drug interactions in the management of malaria, HIV and tuberculosis. Acta Universitatis Upsaliensis, Uppsala

34. Chen J, Raymond K (2006) Roles of rifampicin in drug-drug interactions: underlying molecular mechanisms involving the nuclear pregnane X receptor. Ann Clin Microbiol Antimicrob 5(1):3

35. Ali S, Najmi MH, Tarning J, Lindegardh N (2010) Pharmacokinetics of artemether and dihydroartemisinin in healthy Pakistani male volunteers treated with artemether-lumefantrine. Malar J 9:275. https://doi.org/10.1186/1475-2875-9-275

36. Lamorde M, Byakika-Kibwika P, Mayito J, Nabukeera L, Ryan M, Hanpithakpong W, Lefevre G, Back DJ, Khoo SH, Merry C (2013) Lower artemether, dihydroartemisinin and lumefantrine concentrations during rifampicin-based tuberculosis treatment. AIDS (London, England) 27(6):961–965. https://doi.org/10.1097/QAD.0b013e32835cae3b

37. Baciewicz AM, Chrisman CR, Finch CK, Self TH (2013) Update on rifampin, rifabutin, and rifapentine drug interactions. Curr Med Res Opin 29(1):1–12. https://doi.org/10.1185/030079 95.2012.747952

38. Gil JP (2012) The pharmacogenetics of the antimalarial amodiaquine. In: Sanoudou D (ed) Clinical applications of pharmacogenetics, 1st edn. InTech, Rijeca

39. Pukrittayakamee S, Viravan C, Charoenlarp P, Yeamput C, Wilson RJ, White NJ (1994) Antimalarial effects of rifampin in Plasmodium vivax malaria. Antimicrob Agents Chemother 38(3):511–514

40. Zhang Y, Vermeulen NP, Commandeur JN (2017) Characterization of human cytochrome P450 mediated bioactivation of amodiaquine and its major metabolite N-desethylamodiaquine. Br J Clin Pharmacol 83(3):572–583. https://doi.org/10.1111/bcp.13148

41. Saukkonen JJ, Cohn DL, Jasmer RM, Schenker S, Jereb JA, Nolan CM, Peloquin CA, Gordin FM, Nunes D, Strader DB (2006) An official ATS statement: hepatotoxicity of antituberculosis therapy. Am J Respir Crit Care Med 174(8):935–952

42. Schlagenhauf P, Adamcova M, Regep L, Schaerer MT, Rhein HG (2010) The position of mefloquine as a 21st century malaria chemoprophylaxis. Malar J 9:357. https://doi.org/10.1186/1475-2875-9-357

43. Lalloo DG, Hill DR (2008) Preventing malaria in travellers. BMJ: Br Med J 336(7657):1362

44. Ridtitid W, Wongnawa M, Mahatthanatrakul W, Chaipol P, Sunbhanich M (2000) Effect of rifampin on plasma concentrations of mefloquine in healthy volunteers. J Pharm Pharmacol 52(10):1265–1269

45. Bloechliger M, Schlagenhauf P, Toovey S, Schnetzler G, Tatt I, Tomianovic D, Jick SS, Meier CR (2014) Malaria chemoprophylaxis regimens: a descriptive drug utilization study. Travel Med Infect Dis 12(6 Pt B):718–725. https://doi.org/10.1016/j.tmaid.2014.05.006

46. Pukrittayakamee S, Prakongpan S, Wanwimolruk S, Clemens R, Looareesuwan S, White NJ (2003) Adverse effect of rifampin on quinine efficacy in uncomplicated falciparum malaria. Antimicrob Agents Chemother 47(5):1509–1513

47. Kim KA, Park JY, Lee JS, Lim S (2003) Cytochrome P450 2C8 and CYP3A4/5 are involved in chloroquine metabolism in human liver microsomes. Arch Pharm Res 26(8):631–637

48. Projean D, Baune B, Farinotti R, Flinois JP, Beaune P, Taburet AM, Ducharme J (2003) In vitro metabolism of chloroquine: identification of CYP2C8, CYP3A4, and CYP2D6 as the main isoforms catalyzing N-desethylchloroquine formation. Drug Metab Dispos 31(6):748–754

49. Hou LJ, Raju SS, Abdulah MS, Nor NM, Ravichandran M (2004) Rifampicin antagonizes the effect of choloroquine on chloroquine-resistant Plasmodium berghei in mice. Jpn J Infect Dis 57(5):198–202

50. Peters PJ, Thigpen MC, Parise ME, Newman RD (2007) Safety and toxicity of sulfadoxine/pyrimethamine: implications for malaria prevention in pregnancy using intermittent preventive treatment. Drug Saf 30(6):481–501

51. Fehintola FA, Adedeji AA, Tambo E, Fateye BB, Happi TC, Sowunmi A (2004) Cotrimoxazole in the treatment of acute uncomplicated falciparum malaria in nigerian children : a controlled clinical trial. Clin Drug Investig 24(3):149–155

52. Genton B, Mueller I, Betuela I, Casey G, Ginny M, Alpers MP, Reeder JC (2006) Rifampicin/cotromixazole/isoniazid versus Mefloquine or quinine + sulfadoxine-pyrimethamine in malaria – a randomized trial. PLoS Clin Trials 1(8):e38. https://doi.org/10.1371/journal.pctr.0010038

Chapter 13
Antiprotozoal and Anthelmintic Agents

Tony K. L. Kiang, Kyle John Wilby, and Mary H. H. Ensom

13.1 Introduction

As access to medications for both infectious and noninfectious diseases improves worldwide, the potential for clinically significant drug interactions in endemic regions of parasitic disease increases [1]. Combination chemotherapy is mainstay practice in the management of certain parasitic diseases. In malaria, such a strategy is dictated by a requirement to combine the aim of effective chemotherapy with the wish to minimize the emergence of drug resistance [2–4]. The control of lymphatic filariasis and onchocerciasis involves combinations geared at reducing transmission as a prelude to elimination of diseases posing huge socioeconomic problems [5, 6]. Consequently, combination therapy is associated with many pharmacokinetic and pharmacodynamic implications.

The current treatment regimens recommended for most types of malaria in most populations consist of an artemisinin agent in combination with a longer half-life partner agent. This strategy is known as artemisinin-based combination therapy (ACT) [3]. Several arguments favor use of these combinations in the treatment of malaria. Synergy among drugs or the potentiation of their individual effects is the

T. K. L. Kiang
Faculty of Pharmacy and Pharmaceutical Sciences, University of Alberta,
Edmonton, AB, Canada
e-mail: tkiang@ualberta.ca

K. J. Wilby
College of Pharmacy, Qatar University, Doha, Qatar
e-mail: kjw@qu.edu.qa

M. H. H. Ensom (✉)
Faculty of Pharmaceutical Sciences, University of British Columbia, Vancouver, BC, Canada

Children's and Women's Health Centre of British Columbia, Vancouver, BC, Canada
e-mail: ensom@mail.ubc.ca

© Springer International Publishing AG, part of Springer Nature 2018   515
M. P. Pai et al. (eds.), *Drug Interactions in Infectious Diseases: Antimicrobial Drug Interactions*, Infectious Disease,
https://doi.org/10.1007/978-3-319-72416-4_13

reason for use in the treatment of an individual patient. Resistance of parasites is a main reason why drugs are being combined and advocated for on a population level. Development of resistance as a result of drug pressure depends upon numerous factors [3]. Among them, genetic determinants include mutation frequency and the number of mutations required for expression of resistance. Single point mutations may confer resistance to inhibitors of dihydrofolate reductase in *Plasmodium falciparum*. Combination with other drugs is advantageous when the number of genes required to express resistance is increased, for example, with combinations such as sulfadoxine with pyrimethamine. The survival and selection of resistant parasites may additionally depend on the pharmacodynamics of the component drugs. Parasitemia should ideally be reduced rapidly in order to decrease the opportunity for the development of mutations and the likelihood that parasites will survive under drug pressure. Rapid reduction of the parasite burden in patients with malaria and the relatively short terminal elimination half-life of the artemisinin drugs lead to little or no selective pressure, yet parasites may not fully be eliminated and recrudescence may take place. The benefits of the artemisinin drugs are better realized when combined with other, longer half-life drugs. However, these longer half-life agents lead to residual levels after elimination of the artemisinin component, therefore exposing vulnerability to selective development of resistance [7].

Advantages of combination therapy need to be viewed alongside the increased probability of drug-drug interactions. Throughout the last few decades, the clinical pharmacokinetics and metabolism of many antiparasitic agents have been elucidated, particularly the role of drug-metabolizing enzymes, notably cytochromes P450, and drug transporter proteins [8–13]. Moreover, the pharmacodynamic properties of antiparasitic agents are becoming a research focus and enable predictions of clinically significant drug interactions [14, 15]. This is true for drug interactions between antiparasitic combinations, as well as between antiparasitic agents and other drug classes. Specifically, the use of antiparasitic agents in combination with agents used to treat other endemic diseases (i.e., human immunodeficiency virus, tuberculosis, cardiovascular diseases) may pose risk to achieving optimal patient outcomes in terms of both efficacy and safety [16].

Equally important to drug-drug interactions is the consequence of dietary change on pharmacokinetics. Patients normally take drugs with meals unless advised to the contrary. Diets may differ substantially between developed and underdeveloped countries where diseases susceptible to antiparasitic agents are most prevalent. Failure to understand the nature of any food effects may lead to a poor clinical outcome and/or unacceptable adverse effects [17, 18].

Finally, many drug-drug interactions can be postulated on the basis of common pathways of metabolism among combinations of therapeutic agents. While it is outside the scope of this chapter to deal with all possible effects, they will be highlighted with regard to available evidence pertaining to the most commonly co-administered agents.

13.2 Interactions with Food

Bioavailability of orally ingested antiprotozoal and anthelmintic drugs can be affected by food. The nature of such interactions is complex and may be influenced by quantity of food ingested as well as its composition. Moreover, both pharmacokinetics and pharmacodynamics can be affected. Co-ingested food may change bioavailability of orally administered drugs and alter their dose-response relationships. Consequently, food can unintentionally reduce or increase pharmacodynamic effects of antiprotozoal or anthelmintic drugs, potentially resulting in therapeutic failure or increased toxicity. Influence of drug formulation on interactions with food is predictable from knowledge of gastric function, with solutions and suspensions less susceptible to food interactions than solid formulations and enteric-coated drugs more susceptible, as retention of capsule in the stomach delays drug release [17].

13.2.1 Food Interactions with Antiparasitic Drugs

13.2.1.1 Antimalarial Agents

Halofantrine

Halofantrine is one of the three classes of arylaminoalcohols identified in the United States as potential antimalarial agents by the World War II Chemotherapy Programme. It is a blood schizonticide with selective activity against intraerythrocytic asexual stages of plasmodia. Bioavailability of halofantrine is low with wide intra- and inter-subject variability [19, 20]. Absorption of halofantrine may increase dramatically when taken with food. Both maximal concentration (Cmax) and area under the curve (AUC) of halofantrine and desbutyl-halofantrine are increased by an order of magnitude after administration of a 250 mg dose of halofantrine hydrochloride with a fatty meal [20]. Studies in dogs and rats have shown that clearance of halofantrine is influenced by composition of plasma lipoproteins and may help to explain dramatic changes in circulating plasma concentrations in the postprandial state [21, 22]. The most serious effects of halofantrine relate to QTc prolongation, torsades de pointes, or sudden cardiac death [23–28]. These events led to the curtailment of halofantrine as a frontline antimalarial agent. Because halofantrine is primarily metabolized by cytochrome P450 (CYP) 3A4 [29], it can be subjected to food interactions, notably grapefruit juice, that can affect intrinsic clearance of this enzyme. Effects of grapefruit juice on pharmacokinetics and pharmacodynamics of halofantrine have been documented in healthy subjects, where co-administration (250 mL juice) has been shown to increase exposure of halofantrine by 2.8-fold with a concomitant elevation in QTc interval, compared to controls [30].

Table 13.1 The five artemisinin combination therapies (ACTs) listed below are those currently recommended by WHO

ACT	Proprietary brand	Manufacturer	Date of introduction
Artemether-lumefantrine (AL)	Coartem® Riamet®	Novartis	2006
Artesunate-mefloquine (AS + MQ)	N/A	N/A	N/A
Artesunate-sulfadoxine/ pyrimethamine (AS + SP)	N/A	N/A	N/A
Artesunate-amodiaquine (AS + AQ)	Coarsucam®	Sanofi-Aventis	2008
Dihydroartemisinin-piperaquine (DHA + PPQ)	Artekin®	Chong Qing Holley, Sigma-Tau	2008

Details of their manufacturer and brand name are given where appropriate. Those marked N/A are not yet available as co-formulations

Artemisinin, Its Derivatives, and Partner Drugs

Artemisinin (qinghaosu) was introduced into clinical practice in the 1980s. Subsequently, semisynthetic derivatives were developed, and these have been used in some tropical countries since the early 1990s. Artemisinin derivatives available on the market today (artesunate, artemether, and arteether) are often used in combination with other antimalarials with different mechanisms of action. The World Health Organization (WHO) recommends combining (short-acting) artemisinins with longer-acting antimalarials such as lumefantrine, amodiaquine, mefloquine, piperaquine, and sulfadoxine-pyrimethamine [31–33].

Currently available ACTs are included in Table 13.1 and described in detail below. For such ACT combinations to be effective, the parasite biomass must be reduced sufficiently by the shorter-acting artemisinin derivative, so that chances of mutation to the other, more slowly eliminated drug are greatly reduced. Artemisinin derivatives are the most active of available antimalarial compounds and produce fractional reduction in parasite biomass of approximately 10^4 per asexual cycle. As a result, 3 days of treatment, which involves two cycles, usually produce a 10^8-fold reduction in biomass, leaving a maximum of 10^5 parasites for other slower-acting antimalarial drugs (e.g., lumefantrine, amodiaquine, mefloquine, piperaquine, or sulfadoxine-pyrimethamine) to clear. This reduces considerably exposure of the parasite population to ACT regimens, thereby reducing chance of an escape-resistant mutant arising from the infection [15].

Artemether and Lumefantrine

Artemether-lumefantrine (Coartem™) was one of the first ACTs marketed to treat acute uncomplicated *Plasmodium falciparum* malaria. Food, especially dietary fat, may enhance oral availability of artemether and lumefantrine [34, 35]. Administration of artemether and lumefantrine to healthy volunteers at the same time as a high-fat meal increases bioavailability of both drugs by 2- and 16-fold, respectively, when compared with the fasted state. Healthy adult subjects administered artemether-lumefantrine (80 mg/480 mg, single oral dose) had higher exposures of lumefantrine when co-ingested with milk or maize porridge with oil compared to fasted or oil-free subjects [36]. This may be particularly important given reduced food intake of many patients in the acute phase of malaria. A double-blind trial of patients with uncomplicated malaria in Thailand demonstrated that extent and variability of lumefantrine absorption improved alongside clinical recovery as normal food intake was resumed [37]. Pediatric patients in Mali and Niger with severe malnutrition exhibited lower concentrations of lumefantrine compared to those with normal nutrition [38], although no difference in therapeutic response as measured by clinical and parasitic load was observed. Moreover, data were also available on relative lumefantrine exposures in African children with uncomplicated *Plasmodium falciparum* malaria receiving full treatment doses of artemether-lumefantrine within a randomized, investigator-blinded phase III trial based on their consumption of foods [39]. Specifically, lumefantrine plasma concentration increased by 55–100% (depending on tablet formulation) when given with a meal, and its bioavailability increased by 27–65% when co-administered with milk. However, despite increased lumefantrine exposure, presence (or absence) of food with artemether-lumefantrine treatment did not affect ultimate clinical outcome, with 99% overall cure observed in these children.

These observations prompt the question as to how much dietary fat is necessary to achieve plasma concentrations of lumefantrine that would affect total parasite clearance when it is used in combination with artemether. A population model developed from lumefantrine concentration measurements in a crossover study in healthy volunteers receiving different volumes of soya milk or with no milk demonstrated that 36 mL of soya milk (containing 1.2 g of fat) was associated with 90% of lumefantrine exposure obtained with 500 mL of regular milk (16 g fat) [40]. Fat intake in sub-Saharan countries is approximately 15–30 g/day during breastfeeding, greater than 10 g/day in the postweaning phase, and upward of 30–60 g/day in a normal diet, supporting the view that typical fat intake should support optimal absorption, hence therapeutic concentration, of lumefantrine. This is corroborated by a trial of 957 patients in Uganda receiving artemether-lumefantrine in a hospital under supervision with a meal containing 23 g fat or unsupervised at home after the first dose with advice to take drug with a meal or breast milk, when no difference in cure rates was observed [41, 42]. Consistent with other studies discussed above, lumefantrine plasma concentrations were higher in the supervised group, but adequate pharmacological concentrations of lumefantrine were obtained from home food consumption to achieve good clinical response [42, 43]. Overall, it may be concluded that a small amount of dietary fat may be necessary to ensure adequate absorption of lumefantrine and that standard African diets or breast milk are sufficient to fulfill this need. However, it is important that patients maintain normal food or milk intake during drug administration and resume intake quickly once able to do so to minimize variability with drug absorption. Van Agtmael and colleagues [44, 45] have demonstrated that grapefruit juice (350 mL) increases exposure of artemether (single oral dose of 100 mg) and its metabolite (dihydroartemisinin) by ~twofold in healthy subjects in an interaction that likely involves inhibition by grapefruit juice of CYP3A4, the predominant enzyme responsible for deactivation of artemether [46]. Theoretically, grapefruit juice could conceivably reduce recrudescence with artemether monotherapy by enhancing effective plasma concentrations. This has been demonstrated in preclinical studies of experimental infection with *Schistosoma mansoni* in mice, where co-administration of grapefruit juice with a lower dose (150 mg/kg) of artemether achieved similar protection of host animal compared to a higher dose of artemether (300 mg/kg) from damage induced by schistosomal infection, indicating significant effects of grapefruit juice on pharmacodynamics of artemether [47]. These observations, however, remain to be shown in humans.

Artesunate-Amodiaquine

Artesunate-amodiaquine is one of the ACTs currently recommended by the WHO (Table 13.1) and adopted as first-line treatment in many African countries [48, 49]. Relative to the fasting state, administration of the fixed-dose combination after a high-fat breakfast resulted in a statistically significant increase in circulating concentrations of amodiaquine and desethyl-amodiaquine and a decrease in blood concentrations of artesunate and dihydroartemisinin [50]. These observations were also evident in healthy male subjects administered a single oral dose of artesunate-amodiaquine with a high-fat breakfast compared to the fasting state [49]. Subjects taking artesunate-amodiaquine with a high-fat breakfast exhibited higher amodiaquine and desethyl-amodiaquine exposures but lower artesunate and dihydroartemisinin maximum blood concentrations compared to fasted individuals. One might hypothesize that altered pharmacokinetics of amodiaquine and artesunate may enhance toxicity or reduce efficacy, but alterations in pharmacological effects from this food-drug interaction remain to be verified.

Piperaquine

Piperaquine (PQ) is a *bis*-quinoline antimalarial drug that was first synthesized in the 1950s. It was seen as less toxic than chloroquine, and its efficacy against chloroquine-resistant strains of *Plasmodium falciparum* led to widespread distribution in China and Indochina in the 1970s. With emergence of piperaquine-resistant parasites, its use declined, but continuing search for suitable partner drugs prompted a renewed interest in piperaquine [51]. Sim and colleagues [52] investigated oral bioavailability of piperaquine with food relative to the fasting state and discovered a 1.2-fold increase after a high-fat meal. Side effects (i.e., changes in postural blood pressure, QTc interval, serum glucose, and other biochemical and hematological indices) were similar in fasting and fed states. Hai and colleagues [53] also found no significant difference in drug exposure between fed and fasting subjects after administration of piperaquine with dihydroartemisinin with or without a standard Vietnamese meal [53]. Similar observations were reported by Annerberg and colleagues [54] in adult subjects diagnosed with uncomplicated falciparum malaria, where no significant differences in exposure of piperaquine were observed in fasted subjects compared to patients provided with 200 mL of milk (~6.4 g of fat). Likewise, population pharmacokinetic analysis in Thai patients with uncomplicated *Plasmodium falciparum* malaria also indicated no effects of food intake on piperaquine oral bioavailability [55]. In contrast, in healthy subjects, a high-fat meal appears to increase exposure of piperaquine (~threefold) and dihydroartemisinin (~1.4-fold) after a single-dose combination regimen [56]. The apparent discrepancy in the latter study raises the question of whether degree of drug-food interaction might be dependent on amount of fat content (which has yet to be studied systematically) and cautions against co-administration of piperaquine-dihydroartemisinin with high-fat meals to minimize potential adverse effects.

Mefloquine

Mefloquine is a chiral quinoline-methanol active against asexual forms of the species of *Plasmodium* that infect humans. Mefloquine is poorly water soluble, and extent of its absorption in healthy volunteers is increased modestly when taken with food [57]. In Vietnamese patients with *Plasmodium falciparum* malaria, mefloquine exposures were similar between those co-administered a low-fat versus high-fat meal, although the sample population ($N = 6$) was relatively small. Because its oral bioavailability is relatively high, drug-food interaction is likely clinically insignificant.

Atovaquone

Atovaquone is a hydroxyl-naphthoquinone with broad-spectrum antiprotozoal activity initially selected for development as an antimalarial agent on the basis of potent activity against drug-resistant strains of *Plasmodium falciparum* in vitro. Atovaquone was subsequently found to be active against a number of other microorganisms including *Pneumocystis carinii* and *Toxoplasma gondii*. Studies on in vitro potentiation of atovaquone by other antimalarial drugs revealed evidence of marked synergistic activity with proguanil stimulated subsequent clinical evaluation of these two drugs, culminating in development of a fixed-dose combination for treatment and prevention of malaria. Food increases bioavailability of atovaquone in healthy male subjects administered the tablet formulation by threefold (toast with 28 g butter) or 3.9-fold (toast with 56 g butter) [58]. In patients with HIV, target concentrations for treatment of *Pneumocystis carinii* pneumonia are more consistently reached when atovaquone is administered with food or a nutrition supplement with a moderate fat content [59, 60]. Based on these data, atovaquone is recommended to be administered with food. Findings of some of the major investigations into effect of food on pharmacokinetics of antimalarial agents are summarized in Table 13.2.

Table 13.2 Key food-drug interactions with antimalarial agents

Antimalarial agent(s)	Interaction with food	Effect on drug	[a]Reference(s)
Halofantrine (HF)	Cmax of HF ↑ AUC of HF ↑ *Grapefruit juice*: ↑ AUC	Possible ↑ in QT prolongation with food or grapefruit juice	[19–21, 29, 30]
Artemether (ARM) / lumefantrine (LUM)	*High fat*: Bioavailability of ARM and LUM ↑ 200–1600%	Effects of pharmacokinetic alterations on pharmacodynamic effects not known	[34–36]
	Milk: Bioavailability of LUM ↑ 157% *Pancakes*: Bioavailability of LUM ↑ 274% Grapefruit juice: F of LUM ↑ 200%		[40] [35]
Artesunate (ARTS) /amodiaquine (AQ)	*High fat*: AUC of AQ/desethyl-AQ ↑ after high-fat breakfast	Effects of pharmacokinetic alterations on pharmacodynamic effects [not known]	[44, 45]
	AUC of ARTS and dihydroartemisinin ↓		[49, 50]
Piperaquine (PIP)	Inconsistent findings between studies on the effects of food on exposure of PIP. High-fat meals may ↑ AUC	Effects of pharmacokinetic alterations on pharmacodynamic effects not known	[52]
Mefloquine (MQ)	*Food*: AUC of MQ↑	No clinical relevance as MQ already has high F	[57]
Atovaquone (ATQ)	*Fatty meal*: Bioavailaiblity of ATQ ↑ 200–300%	More consistent target concentrations of ATQ achieved in *Pneumocystis carinii*	[58]

Standard abbreviations for *AUC* area under the curve, *Cmax* maximal concentration are used. If no additional information is available concerning the effect or its clinical importance, this is indicated by N/A.
[a]Note that the references given are the key source in each case. The reader is referred to the text of the chapter for more detailed information.

13.2.1.2 Anthelmintics

Benzimidazoles

Albendazole and mebendazole are benzimidazole carbamates with a broad spectrum of anthelmintic activity. While poor absorption may be advantageous for therapy of helminth infections located in the gut lumen, successful treatment of tissue helminth infections, such as hydatid disease or neurocysticercosis, requires a sufficient concentration of drug to reach site of infection. Despite low and variable bioavailability of benzimidazoles, albendazole bioavailability can be increased significantly when taken with fatty meals or grapefruit juice [61–63]. Based on these observations, it is recommended that albendazole be co-administered with a meal to increase absorption. However, little data are available on food interactions with mebendazole.

Ivermectin

Ivermectin is a potent antiparasitic drug from the macrocyclic lactone family, the most powerful agents against a broad spectrum of ecto- and endoparasites. It was used exclusively in veterinary medicine due to its high efficacy and wide margin of safety until 1987 when it was introduced into human use for the treatment of onchocerciasis [64, 65]. Since then, it has been used in combination with albendazole (ABZ) and diethylcarbamazine (DEC) for treatment of onchocerciasis and lymphatic filariasis [6]. The effects of food on pharmacokinetics of ivermectin have not been well characterized; co-ingestion of alcoholic drinks, however, is not recommended. Although the mechanism of interaction remains to be established, in healthy volunteers given ivermectin orally (150 µg/kg), plasma concentrations were significantly higher when ivermectin was co-administered with beer (750 mL) than with an equivalent volume of water (66). It has been hypothesized that increased ivermectin concentration in the presence of alcohol may result in manifestation of adverse effects, and the unpleasant pharmacological interaction may serve as the basis for an alcohol abstinence regimen [66].

Praziquantel

Praziquantel is a pyrazino-isoquinoline whose potent anthelmintic activity against all *Schistosoma* species and the majority of other trematodes and cestodes was seen as a major advance in medical parasitology. Food increases bioavailability of praziquantel, as demonstrated in fed healthy volunteers where both Cmax and AUC were two to three times higher compared to fasted individuals [67]. Another study showed that meals high in fat and carbohydrate increased AUC by 180% and 271%, respectively, after a single oral dose (1800 mg) [68]. These data were consistent with a further study involving healthy Sudanese volunteers where meals with high

Table 13.3 Key food-drug interactions with anthelmintics*

Albendazole (ALB)	*Fatty meal and grapefruit juice*: AUC of ALB sulfoxide↑	Potentially ↑ chemosterilant properties vs. systemic parasites	[61–63]
Ivermectin (IVM)	*Beer vs. water*: AUC of IVM ↑	n/a	[71]
Praziquantel (PZQ)	*Fat*: AUC of PZQ ↑ 180% *Carbohydrate*: AUC PZQ ↑ 271% *High-oil meal*: 134% *Low-oil meal*: 174% *Grapefruit juice*: 190%	n/a	[67–70]

Standard abbreviations for *AUC* area under the curve, *Cmax* maximal concentration are used. Values in square parentheses refer to circulating concentrations of a particular drug. If no additional information is available concerning the effect or its clinical importance, this is indicated by N/A.
*Note that the references given are the key source in each case. The reader is referred to the text of the chapter for more detailed information.

and low oil contents were associated with mean AUC praziquantel values that were 134% and 174%, respectively, of those during fasting [69]. Grapefruit juice (250 mL) increased AUC and Cmax of praziquantel after a single oral dose in healthy male volunteers ($N = 18$) [70]. Findings of some major investigations into effect of food, on the pharmacokinetics of anthelmintics are summarized in Table 13.3.

13.3   Antimalarial Pharmacokinetic and Pharmacodynamic Drug-Drug Interactions

13.3.1   4-Aminoquinolines

13.3.1.1   Amodiaquine

Amodiaquine has been used in treatment of malaria for over 40 years having once been considered as a successor to chloroquine in East Africa. The use of amodiaquine in prophylaxis was ended due to unacceptable incidences of agranulocytosis and hepatotoxicity [72–74]. CYP2C8 is primarily responsible for the metabolism of amodiaquine and exclusively catalyzes formation of desethyl-amodiaquine, a well-established marker reaction for the enzyme [75] indicating a potential interaction with co-substrates, although CYP2C8, CYP1A1, CYP1B1, CYP2D6, and CYP3A4 may also play a minor role [76–78]. Ketoconazole, an inhibitor of CYP3A4, was associated with decreased formation of desethyl-amodiaquine in human liver microsomes [79]. The prominent role of CYP2C8 in amodiaquine metabolism has been further demonstrated in vitro where expressed CYP2C8*2 Supersomes™ exhibited decreased intrinsic clearance compared to wild-type control [80]. More recently, amodiaquine received a new lease on life as a partner drug with artesunate, where a pharmacokinetic interaction has been observed such that total AUC for dihydroarte-misinin and desethyl-amodiaquine was significantly reduced when compared with equivalent parameters from the individual drugs [81]. Artesunate is rapidly con-verted to dihydroartemisinin, suggesting its principal role is as a prodrug for the former. Dihydroartemisinin is largely glucuronidated, principally by 5'-diphospho-(UDP)-glucuronosyltransferase (UGT) 1A9 and UGT2B7, suggesting that common CYP isoenzymes are not involved in metabolism of either drug, thereby pointing to some other, as yet poorly understood, mechanism for the interaction [82, 83]. In human studies, steady-state nevirapine (in combination therapy containing zidovu-dine and lamivudine) significantly decreased AUC of amodiaquine (29%) and desethyl-amodiaquine (33%), when HIV-infected, but malaria-free, patients were given artesunate/amodiaquine (200 mg/600 mg daily for 3 days) [84]. The mecha-nisms of interaction, however, remain to be determined because none of these anti-retrovirals are known to inhibit CYP2C8 (Table 13.4).

From a pharmacodynamic perspective, amodiaquine shows synergy with arte-misinins [104], quinine, retinol, atovaquone, and atorvastatin [105, 106]. No addi-tive drug interactions were documented, yet antagonism was demonstrated with chloroquine [107] and methylene blue [108]. Clinically, amodiaquine-artesunate combination resulted in increased transaminases after 5 days combination therapy with efavirenz, which may limit use of co-administration of these agents [109].

13.3.1.2 Chloroquine

Chloroquine's antimalarial effects are mediated through the interference of nucleic acid synthesis. Although not recommended for the treatment or prophylaxis of *P. falciparum*, chloroquine may be considered an option for other forms of uncompli-cated malaria [33]. Magnesium trisilicate and kaolin caused a modest reduction in bioavailability of chloroquine. To avoid drug loss, it is suggested that chloroquine should not be administered with gastrointestinal medications of this type or that they should be separated by at least 4 h to reduce risk of adsorption to antacids or adsorbents [85, 86]. In vitro reaction phenotyping studies indicated that chloroquine is predominately metabolized by CYP2D6, CYP2C8, and CYP3A4 to form desethyl-chloroquine [76]. In vitro and in vivo, chloroquine is also a weak inhibitor of CYP2D6, and the clinical relevance of this metabolic characteristic to drug-drug interactions is yet to be established fully [110–112]. A small reduction was observed in Cmax for the fluoroquinolone ciprofloxacin when it was administered with chlo-roquine, but clinical significance of this observation is unknown [113]. However, chloroquine had little effects on pharmacokinetics of various other co-administered drugs (e.g., debrisoquine, chlorzoxazone, S-mephenytoin, ampicillin, chlorproma-zine, imipramine, azithromycin, antipyrine) in healthy volunteers [112, 114–119]. Likewise, little effects of co-administered drugs (e.g., cimetidine, ranitidine, imip-ramine, aspirin, azithromycin, promethazine, and chlorpheniramine) on exposure of chloroquine have been documented in healthy human subjects [118, 120–124].

Table 13.4 Key drug-drug interactions with antimalarial agents

Antimalarial agent(s)	Interaction	Effect on drug/importance	[a]Key reference(s)
4-aminoquinolines-Amodiaquine (AQ)	*In vitro*: AQ metabolism ↓ by ketoconazole or CYP2C8*2 Supersomes	Likely not clinically significant	[79, 84]
	In vivo: Nevirapine (+zidovudine/lamivudine) ↑AUC of AQ Bioavailability of CQ ↓ by antacids	N/A	[85, 86]
8-aminoquinolines-Primaquine (PQ)	Methemoglobin Little pharmacokinetic interactions observed in humans	Likely not clinically significant -avoid PQ/DAP combination	
Antifolates Biguanides Proguanil (PG) Chlorproguanil (CG) Sulfa drugs	*PG and cycloguanil*: AUC and Cmax of ↑ by cimetidine and omeprazole. AUC ↓ by efavirenz, lopinavir/ritonavir, atazanavir/ritonavir	N/A	[87, 88]
Sulfadoxine-pyrimethamine	Little pharmacokinetic interactions documented in the literature		
Atovaquone	Atovaquone: AUC ↓ by efavirenz, lopinavir/ritonavir, atazanavir/ritonavir		[87]
Artemisinin and derivatives Artemether/lumefantrine (ARM/LUM)	*ARM/ART*: Time dependent ↓ in [ARM] and ↑ in [DHA] repeat dosage of ARM	N/A	[89]
Artemisinin (ART)	Time dependent ↓ in [ART] on repeat dosage	Not clinically relevant	[90–92]
Artesunate (ARTS)	*Pyrimethamine*: Cmax ↑ and Vd ↓ with ARM	Not clinically relevant	[93–95]
Dihydroartemisinin (DHA)	*ART*: Cl$_{oral}$ of DHA ↓75% by ART AUC of ARM ↑ (140%) and AUC of LUM ↑ (70%) by ketoconazole and ↓ by rifampin	Increased activity of ARM and LUM	

Antimalarial agent(s)	Interaction	Effect on drug/importance	[a]Key reference(s)
Mefloquine (MQ)	PQ/sulfadoxine/ MQ: $t_{1/2}$ of MQ ↓	No clinical significance	[96]
	ARM/MQ: $t_{1/2}$ of ART unchanged; Vd of ART ↓ AUC of ART ↑	N/A	[97]
	ARTS/MQ: AUC and F of MQ ↑ from day 0 to day 2 when ARTS given for 3 days	Recovery from malaria leads to ↑ AUC	[98]
	ARM/MQ: AUC of MQ ↓ slightly after ARM	N/A	[99]
	Co-ARM (6 doses)/MQ: AUC of LUM ↓ 30–40%	N/A	[100–103]
	Cimetidine, ketoconazole: ↑ mefloquine AUC Rifampin: ↓ mefloquine AUC	N/A	

Standard abbreviations for *AUC* area under the curve, CL_{oral} apparent oral clearance, *Cmax* maximal concentration, *CYP* cytochrome, *DAP* dapsone, $t_{1/2}$ half-life, *Vd* volume of distribution are used. Values in square parentheses refer to circulating concentrations of a particular drug. If no additional information is available concerning the effect or its clinical importance, this is indicated by N/A.
[a]Note that the references given are the key source in each case. The reader is referred to the text of the chapter for more detailed information.

These data suggest that clinically significant pharmacokinetic interactions are likely rarely observed with chloroquine in humans.

Pharmacodynamic interactions are less established with chloroquine. There is evidence of antagonism between choloroquine and amodiaquine [107], artemisinins [107], atorvastatin [107], mefloquine [107], methylene blue [125], omeprazole [126], quinine [107, 126], and sulfadoxine-pyrimethamine [107]. Additive interactions were found with artemisinins [127, 128], azithromycin [129, 130], cepharanthine [131], and methylene blue [108]. Synergy was found between chloroquine and azithromycin [130], cepharanthine [131], and retinol [106].

13.3.2   8-Aminoquinolines

13.3.2.1   Primaquine

Hepatic biotransformation of primaquine and metabolites is partly mediated by cytochromes P450 [132]. The principal plasma metabolite is carboxyprimaquine [133, 134]. Clinically, the most significant interactions would be those that facilitate the formation or accumulation of toxic metabolites [79]. In vitro reaction phenotyping studies have provided support that CYP2D6 and CYP3A4 act as primary enzymes in the conversion to carboxyprimaquine [135]. Specifically, in human liver microsomes, carboxyprimaquine formation is inhibited by ketoconazole, a potent CYP3A4 inhibitor, but little effects of quinine, artemether, artesunate, halofantrine, or chloroquine were observed [136]. Despite being a substrate for CYP2D6 and CYP3A4, little has been documented in the literature to support metabolism-mediated drug interactions affecting pharmacokinetics of primaquine (or effects of primaquine on pharmacokinetics of co-administered drugs). In humans, a small decrease in Cmax and AUC of carboxyprimaquine was observed after co-administration with quinine [137], but this is probably of little clinical relevance. Co-administration with mefloquine had little effect on elimination of primaquine or its main metabolite carboxyprimaquine in healthy Thai male adults [137]. Likewise, in healthy subjects, a single oral dose of primaquine did not alter the pharmacokinetics of a single dose of antipyrine [119]. These studies support overall lack of significant pharmacokinetic interactions observed with primaquine in humans.

Little data exist regarding pharmacodynamic interactions with primaquine. One study found additive properties when combined with methylene blue [108], while another study found an additive to synergistic interaction when co-administered with azithromycin [130]. Adverse effects of primaquine in combination with artemether were insignificant [138].

13.3.3 Antifolates

Combination of inhibitors of folate synthesis takes advantage of their synergism and observation that different genes contribute to the resistance phenotype, thus reducing likelihood that resistant strains will be selected. Unfortunately, resistance to these drugs has developed in most endemic areas of the world, but degree of resistance varies [139].

13.3.3.1 Biguanides

Proguanil and chlorproguanil are biguanides that inhibit dihydrofolate reductase (DHFR) which is an enzyme involved in the folate-thymidylate pathway. Proguanil is a prodrug, as it is rapidly transformed in the liver to the DHFR inhibitor cycloguanil. Metabolism of proguanil is possibly mediated by CYP1A2, CYP3A4, and/or CYP2C19 as demonstrated in in vitro reaction phenotyping studies [140–144]. Although role of CYP2C19 has been clearly established by various investigators [140, 143, 144], inconsistent findings have been reported for CYP1A2 and CYP3A4. A genetic polymorphism in CYP2C19 enzyme, with up to 20% poor metabolizers in Asian and African populations, has been demonstrated [145–147]. Poor metabolizers have reduced plasma concentrations of cycloguanil during prophylaxis, and this could conceivably contribute to prophylactic failure in this group, but large inter-subject variability and role of CYP3A4 or CYP1A2 mean there is no clear association [141, 142, 146]. Chlorproguanil is a chloro derivative of proguanil and intrinsically more active. A similar difference in activity exists with regard to its active metabolite, chlorcycloguanil, when compared with cycloguanil [148]. Despite known metabolic properties that can potentially mediate drug-drug interactions, available human literature has indicated little effects of proguanil on co-administered drugs. In contrast, concurrent drugs have been shown to affect pharmacokinetics of proguanil. Steady-state efavirenz, lopinavir/ritonavir, or atazanavir/ritonavir has been demonstrated to decrease exposure of proguanil in HIV-infected individuals, compared to pharmacokinetic parameters obtained in healthy subjects [87]; these effects may have been mediated by inductive properties of these drugs toward CYP enzymes known to metabolize proguanil. The H_2-receptor antagonist cimetidine and proton pump inhibitor omeprazole significantly increased Cmax and AUC of proguanil (with corresponding decreases in Cmax and AUC of cycloguanil in the case of cimetidine), which also suggests potential inhibitory effects of omeprazole and cimetidine toward CYP isoenzymes [88, 149].

Pharmacodynamic interactions with proguanil have been assessed through numerous studies. Synergy with atovaquone was first demonstrated in a study co-administering the two agents with artemisinin [150]. However, synergy was later discovered to the atovaquone-proguanil pair alone [151]. One other study demonstrated a synergistic pharmacodynamic interaction between proguanil and monodebutyl-benflumetol [152]. Proguanil is active against *P. falciparum*, and future studies should further assess its role as a recommended alternative agent.

13.3.3.2 Sulfonamides

Sulfonamides are inhibitors of DHPS (dihydropteroate synthase) and historically have been used extensively in combination with inhibitors of DHFR, notably sulfadoxine, in prevention and treatment of malaria [139]. Chiefly, sulfadoxine is usually used in combination with pyrimethamine (Fansidar™). Few pharmacokinetic interactions of clinical importance involving sulfadoxine and pyrimethamine have been documented in humans, probably because sulfadoxine undergoes minimal liver metabolism. Even though pyrimethamine undergoes extensive hepatic biotransformation, relatively little is known of the exact metabolic enzymes involved. From a pharmacodynamic perspective, the sulfadoxine-pyrimethamine combination showed antagonism with chloroquine and is not recommended for use [107].

13.3.4 Atovaquone

The combination of proguanil and atovaquone was originally developed to combat multidrug-resistant falciparum malaria [127]. Atovaquone undergoes minimal oxidation but extensive conjugation in humans [153]. Overall, there is no effect of co-administration of proguanil on pharmacokinetics of atovaquone [154, 155]. In healthy Caucasians, pharmacokinetics of proguanil, cycloguanil, and atovaquone are unaffected by the combination. In patients with *P. falciparum* malaria, pharmacokinetics of proguanil with atovaquone was comparable with healthy volunteers treated with proguanil alone. In HIV-infected subjects, trimethoprim-sulfamethoxazole had little effects on pharmacokinetics atovaquone [60]. However, steady-state efavirenz, lopinavir/ritonavir, and atazanavir/ritonavir decreased exposure of atovaquone from a single oral dose of atovaquone/proguanil (250 mg/100 mg) by 75%, 74%, and 44%, respectively [87]. Because atovaquone undergoes minimal oxidative metabolism, it might be postulated that inductive effects of these antiretrovirals toward phase II conjugative enzymes could be the mechanism behind this pharmacokinetic interaction. Furthermore, atovaquone decreases the oral clearance of zidovudine, leading to a $35 \pm 23\%$ increase in its plasma AUC. Clinical significance of this is not known, and presently no dose modification is recommended [156]. Atovaquone and indinavir might exhibit mutual pharmacokinetic interactions, where indinavir AUC is decreased and atovaquone AUC is increased [157]. These changes, however, are relatively small and unlikely clinically significant. Moreover, clinical studies have shown higher plasma indinavir in Thai patients with much lower body weight, and given toxicity of indinavir at higher doses, dosage adjustments are not indicated for ritonavir-boosted indinavir when given with atovaquone or the atovaquone-proguanil combination (Malarone™).

Pharmacodynamic interactions with atovaquone are documented. Antagonism between combinations of atovaquone and methylene blue was previously reported [108]. However, enhanced efficacy was shown when atovaquone was combined with cepharanthine [131]. From a safety perspective, a case report showed a temporal relationship between atovaquone and increased international normalized ratio for a patient taking warfarin [158]. Therefore, co-administration of these agents should be cautioned.

13.3.5 Artemisinin and Derivatives

Artemisinin drugs rapidly reduce parasite burden. Because of short half-lives of most artemisinin derivatives, recrudescence occurs after monotherapy. In combination with other drugs, rapidly acting artemisinin may help to minimize selection pressure [3, 4, 46]. Pharmacokinetics of the combination of artemether and lumefantrine (Coartem™) are comparable to pharmacokinetics of the individual agents. A time-dependent decline in artemether concentration and corresponding increase in concentration of dihydroartemisinin were observed, possibly due to autoinduction [89]. When combined with artemether, C_{max} of pyrimethamine was increased significantly, and volume of distribution was reduced slightly, although these changes are likely not clinically significant [93]. As demonstrated in in vitro reaction phenotyping studies, artemisinin is primarily metabolized by CYP2B6 [159] and is capable of inhibiting CYP1A2 [160]. Artesunate undergoes bioactivation in formation of dihydroartemisinin in a reaction likely predominately mediated by CYP2A6 [76]. Artemether also undergoes activation to form dihydroartemisinin but in a reaction mediated primarily by CYP3A4 [161]. In contrast, the reactive metabolite dihydroartemisinin is not further oxidized but rather undergoes subsequent conjugation by phase II uridine UGT1A9 and UGT2B7 [83]. However, additional metabolic pathways might be possible and remain to be investigated. After repeated doses, plasma concentrations of artemisinin decline steadily, with a six- to sevenfold reduction of AUC after 6 days of daily administration [90, 91]. It is likely that artemisinin induces its own metabolism but the exact mechanism has not yet been elucidated. This time-dependent decline of artemisinin also occurs after rectal administration, which suggests that site of induction is hepatic [92]. This effect has also been observed with artemether and dihydroartemisinin after oral administration of artesunate [162] and probably contributes to recrudescence. Time-dependent pharmacokinetics of artemisinin suggests that artemisinin is a selective inducer of drug metabolism, in a reaction most likely mediated by CYP2B6 [159, 163]. In addition to autoinduction, a few pharmacokinetic drug-drug interactions mediated by or affecting artemisinin derivatives have been reported in the literature. Steady-state rifampin is shown to decrease exposure of artemether, dihydroartemisinin, and lumefantrine in HIV-infected subjects without malaria [94], whereas the administration of artemether-lumefantrine with ketoconazole, a potent inhibitor of CYP3A4, increased the AUC for both artemether (2.4-fold) and lumefantrine (1.7-fold) [95] in

reactions likely mediated by induction and inhibition of CYP3A4, respectively. In contrast, artemisinin had little effects on the pharmacokinetics of caffeine [160] and coumarin [164] but decreased exposure of nicotine (46%) in healthy volunteers. Although studies reporting clinical pharmacokinetic interactions involving artemisinin derivatives are relatively scarce (antiretroviral-artemisinin derivative drug interactions to be discussed below), metabolic properties of these compounds (discussed above) indicate that metabolism-associated drug-drug interactions, mediated by CYP or UGT enzymes, can be manifested in clinical situations but may be reasonably predicted.

Pharmacodynamic interactions between artemisinin agents and partner drugs are well established. A Cochrane review identified 50 studies assessing efficacy of ACTs against *P. falciparum*. Findings showed notable efficacy for each ACT regimen assessed, with failure rates of <10% [165]. Numerous other interactions have also been reported. Antagonism was demonstrated with artemisinin agents and cepharanthine [131], chloroquine [107], and ketoconazole [166]. Additive interactions were reported with amphotericin B [167], azithromycin [130, 168], chloroquine [128], clindamycin [169], clotrimazole [167], methylene blue [108], and omeprazole [126]. Finally, synergy was demonstrated with amodiaquine [104], atovaquone [105, 170], chalcones [167], clindamycin [169], doxycycline [171], mefloquine [127, 172, 173], methylene blue [125], pyronaridine [174], quinine [127], retinol plus mefloquine [173], and triclosan [166]. Many of these additive and synergistic interactions should be further explored as potential combination or adjunctive regimens for malaria treatment.

13.3.6   Artemisinin Derivatives and Antiviral Interactions

There is extensive overlap in patient populations likely infected with both malaria and HIV, which may lead to increased probability of polypharmacy and drug interactions [175]. In general, protease inhibitors have potential to increase exposure of artemisinin derivatives, whereas non-nucleoside reverse transcriptase inhibitors tend to have opposite effects. In contrast, artemisinin derivatives have fewer effects on pharmacokinetics of antiretroviral agents, suggesting more significant impact on malarial treatment when these drugs are combined. In healthy volunteers or HIV-infected subjects, steady-state lopinavir/ritonavir (400 mg/100 mg) increased exposure of lumefantrine but had little effects on artemether [176, 177]. To the contrary, steady-state darunavir/ritonavir (80 mg/480 mg) increased exposure of lumefantrine but decreased that of artemether [178]. These findings are consistent with inhibitory effects of protease inhibitors toward CYP3A4, which is responsible for metabolism of lumefantrine. However, it is unclear why concentrations of artemether, which is also metabolized primarily by CYP3A4, remain unchanged (or decreased) in presence of protease inhibitors. Contribution of alternative metabolic pathways and mixed inhibitory/inductive properties of protease inhibitors might be responsible for these observations.

With respect to non-nucleoside reverse transcriptase inhibitors, steady-state etravirine decreased both single-dose and steady-state artemether concentrations (with corresponding decreases in dihydroartemisinin and lumefantrine concentrations) in healthy subjects [178]. Efavirenz, when used in combination with zidovudine/lamivudine, also had similar effects (i.e., decreased artemether/lumefantrine/dihydroartemisinin exposures) in HIV-infected patients, but when used alone, it did not change pharmacokinetics of artemether/lumefantrine in healthy volunteers [179, 180]. Available data for nevirapine are inconsistent as it has been shown to decrease artemether/dihydroartemisinin exposure in one study [180] but increase lumefantrine exposure in another [181]. Most of these effects can be explained by inductive effects of etravirine, efavirenz, and nevirapine toward CYP3A4, the primary enzyme responsible for metabolism of artemether and lumefantrine. Discrepancy in data obtained with efavirenz and nevirapine might be attributed to design differences between studies and possibility of alternative, not yet identified, metabolic pathways involved in the drug interaction. Please see Kiang et al. [175] for a detailed, systematic discussion of antimalarial-antiviral drug-drug interactions.

13.3.7 Cinchona Alkaloids

Quinine is transformed into 3-hydroxyquinine principally by CYP3A4 [182] and to a minor, but significant, extent by CYP2C19 [76]. Co-administration with rifampicin (CYP3A4 inducer) and cigarette smoking (CYP1A inducer) each increases metabolic clearance of quinine [183, 184], confirming the role of CYP3A4 and suggesting some contribution of CYP1A in the elimination of quinine [183]. Fortunately, among other antimalarials, there are few potent inducers or inhibitors of CYP3A4, and clinically relevant interactions revolving around enzyme induction and inhibition are unlikely. The diastereoisomer of quinine, quinidine, used as an antimalarial in North America, is the more potent inhibitor of CYP2D6 in vivo [185]. Theoretically, drug-drug interactions would be expected to be more problematic; but, in context of antimalarial therapy, relatively few have been observed. In preclinical studies, a number of drugs, such as phenobarbital, that induce cytochromes P450 increase quinine clearance [186]. Clearance of quinine is inhibited by cimetidine and ciprofloxacin, inhibitors of cytochrome P450 isoenzymes [187, 188]. In healthy subjects, steady-state nevirapine and lopinavir/ritonavir significantly reduced exposure of quinine (likely mediated by CYP3A4 induction) but had opposite effects on exposure of 3-hydroxyquinine [189, 190]. In contrast, ritonavir itself increased exposure of quinine (340%) and reduced formation of 3-OH quinine (90%), in a reaction likely mediated by ritonavir's potent inhibitory effects toward CYP3A4. Clinical significance of these pharmacokinetic interactions, however, remains to be determined in the patient population. Quinine is known to inhibit metabolism of phenobarbital and carbamazepine but not phenytoin [191] and to reduce clearance of flecainide [192] although not excessively [193]. Despite significant effects on pharmacokinetics of quinine from co-administration of some antiretroviral agents (see above), quinine had little effects on ritonavir or lopinavir exposure in healthy subjects [189, 190]. In contrast, quinine and quinidine may increase plasma concentrations of digoxin although the magnitude of effect is less for quinine than for quinidine [194, 195]. Clearance of quinidine is unaltered in smokers [196]. Elimination of quinidine is markedly increased by phenobarbital, phenytoin, and rifampin but reduced by inhibitors of cytochrome P450 isoenzymes, such as cimetidine, amiodarone, verapamil, and erythromycin [197, 198]. Pharmacokinetic interactions between phenobarbital and quinine have not been noted in patients with cerebral malaria where both drugs are commonly co-administered without evidence of toxicity [198, 199].

13.3.8 Mefloquine

Mefloquine has a very long terminal elimination half-life [8]. This may increase selection pressure and is probably the principal reason why resistance developed soon after its introduction in Thailand. A combination of mefloquine and artemisinin derivatives is effective against multidrug-resistant parasites. Moreover, it is argued that use of the combination will also delay development of mefloquine resistance where monotherapy has not been used. Mefloquine is metabolized by CYP3A4 to carboxy-mefloquine [136]. While co-administration of mefloquine has no effect on pharmacokinetics of primaquine or carboxyprimaquine in healthy Thai males [137], primaquine, but not sulfadoxine-pyrimethamine, may reduce half-life of mefloquine [96]. Mefloquine co-administration to patients with uncomplicated falciparum malaria increased AUC of artemisinin and reduced apparent volume of distribution and clearance without affecting half-life. Because mefloquine and artemisinin exhibit distinct metabolism characteristics, mechanisms of interaction for this observation remain to be investigated [97]. Mefloquine did not affect pharmacokinetics of dihydroartemisinin in healthy Thai males or patients with malaria (and vice versa) [200, 201]. To the contrary, interaction between other artemisinin derivatives and mefloquine may be more conflicting: In Thai children with falciparum malaria, AUC of mefloquine on day 0 was lower than AUC on day 2 when artesunate was given for 3 days. Rather than a pharmacokinetic interaction, the authors suggested that recovery from malaria was the main cause of the increased bioavailability [98]. However, in a study comparing adult Thai patients and healthy volunteers, malaria was shown to slow meloflloquine's rate of absorption without any effects on AUC [202]. A study in patients with uncomplicated malaria revealed AUC of mefloquine to be slightly reduced when given 24 h after artemether [99]. In a further study of healthy volunteers, no interaction between artemether and quinine, mefloquine, or primaquine was observed [138]. Co-artemether (40 mg artemether +480 mg lumefantrine) given in six doses over 60 h following a 1000 mg dose of mefloquine elicited a significant decrease (30–40%) in plasma lumefantrine concentrations compared with lumefantrine alone. However, the authors considered that clinical effects were unlikely to be influenced by the interaction [100].

Pharmacokinetics of mefloquine can be altered by co-administered drugs. In healthy volunteers, cimetidine and ketoconazole have been shown to increase exposure of mefloquine, whereas rifampin had opposite effects [101–103]. In contrast, co-administration of metoclopramide, ampicillin, tetracycline, and ritonavir had little influence on mefloquine exposure [203–205]. In the same study by Khaliq et al., mefloquine also did not affect pharmacokinetics of ritonavir. These data suggest some potential of mefloquine to interact with co-administered drugs in humans, but clinical significance of these pharmacokinetic alterations remains to be studied.

From a pharmacodynamic perspective, mefloquine showed additive and synergistic behavior with methylene blue [108, 125], while antagonism was demonstrated with cepharanthine [131].

The findings of some of the major investigations into drug-drug interactions among antimalarials and co-administered drugs are summarized in Table 13.4.

13.4 Anthelmintic Drug-Drug Interactions

13.4.1 Albendazole

Albendazole is converted in vivo into albendazole sulfoxide, the systemically active form of the drug, and albendazole sulfone, which is inactive, in sequential sulfoxidation reactions. CYP3A4 and flavin monooxygenases have been implicated as primary catalysts in these reactions [206, 207]. Plasma concentrations of albendazole are increased in presence of grapefruit juice, in a reaction likely mediated by CYP3A4 inhibition [63]. In the same study, cimetidine with grapefruit juice decreased exposure of albendazole, compared to grapefruit juice alone, potentially indicating a pH-dependent absorption of albendazole [63]. Other drugs have been shown to affect pharmacokinetics of albendazole: dexamethasone and praziquantel (increased exposure); phenytoin, phenobarbital, carbamazepine, and ritonavir (decreased exposure); ivermectin and azithromycin (no change in exposure), as summarized in [208]. However, not all of these pharmacokinetic interactions can be explained by albendazole's known metabolic characteristics (i.e., CYP3A4 being the primary metabolic enzyme). Involvement of alternative metabolic pathways and clinical significance of these interactions should be further explored.

Albendazole has a chiral center, and formation of albendazole (−) sulfoxide appears to be dependent upon on P450 isoenzymes, whereas formation of albendazole (+) sulfoxide is dependent upon flavin monooxygenases. Subsequent oxidation to albendazole sulfone is wholly dependent on P450 enzymes [207]. Albendazole, although a substrate of CYP3A4, is neither a substrate nor an inhibitor of P-glycoprotein (P-gp) or breast cancer-resistant protein, BCRP/ABCG2. Accordingly, interactions between albendazole and P-gp substrates or inhibitors are unlikely to be clinically important [209, 210].

13.4.2 Ivermectin

Ivermectin and other macrocyclic lactones are highly lipophilic molecules and therefore widely distributed in the body [211, 212]. The antiparasitic activities of ivermectin and other macrocyclic lactones are related to the presence of effective concentrations for a suitable length of time in the systemic circulation and in target tissues [213, 214]. As demonstrated in human liver microsomes, ivermectin is primarily metabolized by CYP3A4 converting the drug to at least ten metabolites, most of them hydroxylated and demethylated derivatives [215]. These data correspond to information obtained in healthy volunteers where a number of radioactive metabolites were reported after oral administration of [14]C- ivermectin [216]. Ivermectin is both a substrate and inhibitor of P-gp [215] and has been demonstrated to inhibit P-gp, ABCC1, ABCC2, and ABCC3 activities [217, 218]. Preclinical studies have indicated that it is a potential inducer of several cytochrome P450 subfamilies including CYP1A, CYP2B, and CYP3A, but inductive effects of ivermectin in humans remain to be determined [219]. Most literature on ivermectin pharmacokinetic interactions have been reported in animal models. In healthy volunteers, levamisole increased bioavailability of ivermectin, but the study used only historical controls [220]. Likewise, no significant changes in pharmacokinetics of ivermectin were observed when healthy Thai subjects were administered the combination of albendazole, praziquantel, and ivermectin, but this study also lacked proper controls [221]. In contrast, concurrent administration of azithromycin and ivermectin/albendazole appeared to increase exposure of ivermectin (31%), the mechanism of which remains to be determined as azithromycin is not known to cause significant CYP3A4 inhibition [222]. Clinical relevance of this positive pharmacokinetic interaction remains to be studied.

13.4.3 Praziquantel

The commercial preparation of praziquantel is a racemate composed of R (−) and S (+) isomers of which only (−) enantiomer has antischistosomal activity [223]. The isomers do, however, have similar toxicity. Orally administered praziquantel is rapidly absorbed, measurable amounts appearing in blood as early as 15 min after dosing with peak levels occurring after 1–2 h. Maximum plasma concentrations after standard dose of 40 mg/kg show wide interindividual variations in the range of 200–2000 ng/mL. Praziquantel undergoes pronounced liver first-pass metabolism with rapid disappearance from the circulation, plasma half-life ranging between 1 and 3 h. Elimination occurs through urine and feces and is more than 80% complete after 24 h [224]. The principal enzyme responsible for oxidation of praziquantel in humans appears to be CYP3A4 [225] although systematic reaction phenotyping studies are still required to determine relative contribution of other CYP enzymes. Bioavailability of praziquantel is increased by simultaneous administration of

Table 13.5 Key drug-drug interactions with anthelmintics

Anthelmintic agent	Interaction	Effect on drug/importance	[a]Key reference(s)
Albendazole (ABZ)	[ABZ] ↑ by grapefruit juice, dexamethasone, praziquantel. ABZ exposure ↓ by phenytoin, phenobarbital, carbamazepine, and ritonavir.	Clinical significance not known	[63]
Ivermectin (IVM)	IVM is a substrate for/inhibitor of P-glycoprotein and substrate for CYP3A4 Azithromycin: IVM AUC ↑ 31%	Clinical significance not known	[222, 234]
Praziquantel (PZQ)	*Cimetidine*: PZQ AUC ↑ 100%. *Ketoconazole*: [PZQ] ↑ 100%. *Rifampicin*: [PZQ] ↓ Albendazole: [PZQ] ↑	Enhanced effectiveness in cysticercosis Some dose adjustment may be needed	[226, 227] [229–231]

Standard abbreviations for area under the curve (AUC) and cytochrome (CYP) are used. Values in square parentheses refer to circulating concentrations of a particular drug. If no additional information is available concerning the effect or its clinical importance, this is indicated by N/A.
[a]Note that the references given are the key source in each case. The reader is referred to the text of the chapter for more detailed information.

substances that inhibit cytochrome P450 activities, e.g., cimetidine leads to a 100% increase in humans [226, 227]. For this reason, cimetidine has been used in combination with praziquantel, especially in treatment of neurocysticercosis, where high concentrations are required. Chloroquine similarly decreases praziquantel's bioavailability to a significant extent [228]. Ketoconazole, a CYP3A inhibitor, has been observed to double plasma concentration of praziquantel in humans, while rifampin, an inducer of drug metabolism, has been reported dramatically to reduce its concentration, and dose adjustment upon co-administration has been recommended [229, 230]. In healthy volunteers, co-administration of albendazole was shown to increase exposure of praziquantel, in a reaction that may be mediated by CYP3A4 [231]. In contrast, praziquantel has not conclusively been characterized in relation to its effects on drug transporters. Available in vitro studies indicate that praziquantel may act as an inhibitor, but not substrate, of P-gp enzymes [232, 233]; thus, it may mediate pharmacokinetic interactions via transporter inhibition, although clinical significance of this theoretical effect remains to be determined. Findings of some major investigations into drug-drug interactions among anthelmintics and co-administered drugs are summarized in Table 13.5.

13.5 Conclusions

Combination chemotherapy is commonly indicated for treatment of malaria and certain anthelmintic infections. Limited clinical drug-drug interaction data are available in the literature, and there are few documented interactions that can be considered clinically significant. This chapter has summarized relevant clinical pharmacokinetic and pharmacodynamic drug interactions associated with commonly used agents today. Where possible, hepatic drug metabolism characteristics are provided to explain mechanisms of observed interactions. Future studies should also focus on establishing relationships between significant pharmacokinetic interactions with pharmacodynamic effects. Likewise, further emphasis on common comorbid conditions such as HIV infection and tuberculosis, in which polypharmacy and likelihood of drug-drug interactions are significant, would be more impactful.

Acknowledgment The important work of Geoffrey Edwards B.Sc. Ph.D. to a previous version of this chapter is acknowledged.

References

1. Mabey D, Peeling RW, Ustianowski A, Perkins MD (2004) Diagnostics for the developing world. Nat Rev Microbiol 2:231–240
2. White NJ (1997) Assessment of the pharmacodynamic properties of antimalarial drugs in vivo. Antimicrob Agents Chemother 41:1413–1422
3. White N (1999) Antimalarial drug resistance and combination chemotherapy. Philos Trans R Soc Lond Ser B Biol Sci 354:739–749
4. Eastman RT, Fidock DA (2009) Artemisinin-based combination therapies: a vital tool in efforts to eliminate malaria. Nat Rev Microbiol 7:864–874
5. Ottesen EA, Ismail MM, Horton J (1999) The role of albendazole in programmes to eliminate lymphatic filariasis. Parasitol Today 15:382–386
6. Molyneux DH, Bradley M, Hoerauf A, Kyelem D, Taylor MJ (2003) Mass drug treatment for lymphatic filariasis and onchocerciasis. Trends Parasitol 19:516–522
7. Lin JT, Juliano JJ, Wongsrichanalai C (2010) Drug-resistant malaria: the era of ACT. Curr Infect Dis Rep 12:165–173
8. White NJ (1985) Clinical pharmacokinetics of antimalarial drugs. Clin Pharmacokinet 10:187–215
9. Edwards G, Winstanley PA, Ward SA (1994) Clinical pharmacokinetics in the treatment of tropical diseases. Some applications and limitations. Clin Pharmacokinet 27:150–165
10. Edwards G, Breckenridge AM (1988) Clinical pharmacokinetics of anthelmintic drugs. Clin Pharmacokinet 15:67–93
11. Krishna S, White NJ (1996) Pharmacokinetics of quinine, chloroquine and amodiaquine. Clinical implications. Clin Pharmacokinet 30:263–299
12. Winstanley P (2003) The contribution of clinical pharmacology to antimalarial drug discovery and development. Br J Clin Pharmacol 55:464–468
13. Kerb R, Fux R, Morike K et al (2009) Pharmacogenetics of antimalarial drugs: effect on metabolism and transport. Lancet Infect Dis 9:760–774
14. White NJ (2013) Pharmacokinetic and pharmacodynamic considerations in antimalarial dose optimization. Antimicrob Agents Chemother 57:5792–5807
15. Edwards G, Krishna S (2004) Pharmacokinetic and pharmacodynamic issues in the treatment of parasitic infections. Eur J Clin Microbiol Infect Dis 23:233–242
16. Kiang TKL, Wilby KJ, Ensom MHH (2015) Clinical pharmacokinetic and pharmacodynamic drug interactions associated with antimalarials. Springer International Publishing AG,Cham. ISBN 978-3-319-10527-7 (eBook). https://doi.org/10.1007/978-3-319-10527-7

17. Winstanley PA, Orme ML (1989) The effects of food on drug bioavailability. Br J Clin Pharmacol 28:621–628
18. Schmidt LE, Dalhoff K (2002) Food-drug interactions. Drugs 62:1481–1502
19. Broom C (1989) Human pharmacokinetics of halofantrine hydrochloride. Parasitol Today (Suppl):15–19
20. Milton KA, Edwards G, Ward SA, Orme ML, Breckenridge AM (1989) Pharmacokinetics of halofantrine in man: effects of food and dose size. Br J Clin Pharmacol 28:71–77
21. Brocks DR, Ramaswamy M, MacInnes AI, Wasan KM (2000) The stereoselective distribution of halofantrine enantiomers within human, dog, and rat plasma lipoproteins. Pharm Res 17:427–431
22. Brocks DR, Wasan KM (2002) The influence of lipids on stereoselective pharmacokinetics of halofantrine: important implications in food-effect studies involving drugs that bind to lipoproteins. J Pharm Sci 91:1817–1826
23. Castot A, Rapoport P, Le Coz P, Prolonged QT (1993) Interval with halofantrine. Lancet 341:1541
24. Karbwang J, Na Bangchang K, Bunnag D, Harinasuta T, Laothavorn P (1993) Cardiac effect of halofantrine. Lancet 342:501
25. Monlun E, Le Metayer P, Szwandt S et al (1995) Cardiac complications of halofantrine: a prospective study of 20 patients. Trans R Soc Trop Med Hyg 89:430–433
26. Monlun E, Pillet O, Cochard JF, Favarel Garrigues JC, le Bras M, Prolonged QT (1993) Interval with halofantrine. Lancet 341:1541–1542
27. Nosten F, ter Kuile FO, Luxemburger C et al (1993) Cardiac effects of antimalarial treatment with halofantrine. Lancet 341:1054–1056
28. Toivonen L, Viitasalo M, Siikamaki H, Raatikka M, Pohjola-Sintonen S (1994) Provocation of ventricular tachycardia by antimalarial drug halofantrine in congenital long QT syndrome. Clin Cardiol 17:403–404
29. Baune B, Flinois JP, Furlan V et al (1999) Halofantrine metabolism in microsomes in man: major role of CYP 3A4 and CYP 3A5. J Pharm Pharmacol 51:419–426
30. Charbit B, Becquemont L, Lepere B, Peytavin G, Funck-Brentano C (2002) Pharmacokinetic and pharmacodynamic interaction between grapefruit juice and halofantrine. Clin Pharmacol Ther 72:514–523
31. Ilett KF, Batty KT (2006) Artemisinin and its derivatives. In: Yu VL, Edwards G, PS MK, Peloquin C, Morse GD (eds) Antimicrobial therapy and vaccines. ESun Technologies, LLC, Pittsburg, pp 981–1002
32. Djimde A, Lefevre G (2009) Understanding the pharmacokinetics of Coartem. Malar J 8(Suppl 1):S4. https://doi.org/10.1186/1475-2875-8-S1-S4
33. WHO (2015) Guidelines for the treatment of malaria, 3rd edn. World Health Organization, Geneva, Switzerland, pp 1–316
34. White NJ, van Vugt M, Ezzet F (1999) Clinical pharmacokinetics and pharmacodynamics and pharmacodynamics of artemether-lumefantrine. Clin Pharmacokinet 37:105–125
35. Ezzet F, van Vugt M, Nosten F, Looareesuwan S, White NJ (2000) Pharmacokinetics and pharmacodynamics of lumefantrine (benflumetol) in acute falciparum malaria. Antimicrob Agents Chemother 44:697–704
36. Mwebaza N, Jerling M, Gustafsson LL et al (2013) Comparable lumefantrine oral bioavailability when co-administered with oil-fortified maize porridge or milk in healthy volunteers. Basic Clin Pharmacol Toxicol 113:66–72
37. Ezzet F, Mull R, Karbwang J (1998) Population pharmacokinetics and therapeutic response of CGP 56697 (artemether + benflumetol) in malaria patients. Br J Clin Pharmacol 46:553–561
38. Denoeud-Ndam L, Dicko A, Baudin E et al (2016) Efficacy of artemether-lumefantrine in relation to drug exposure in children with and without severe acute malnutrition: an open comparative intervention study in Mali and Niger. BMC Med 14:167
39. Borrmann S, Sallas WM, Machevo S et al (2010) The effect of food consumption on lumefantrine bioavailability in African children receiving artemether-lumefantrine crushed or dis-

persible tablets (Coartem) for acute uncomplicated plasmodium falciparum malaria. Tropical Med Int Health 15:434–441

40. Ashley EA, Stepniewska K, Lindegardh N et al (2007) How much fat is necessary to optimize lumefantrine oral bioavailability? Tropical Med Int Health 12:195–200

41. Premji ZG, Abdulla S, Ogutu B et al (2008) The content of African diets is adequate to achieve optimal efficacy with fixed-dose artemether-lumefantrine: a review of the evidence. Malar J 7:244. https://doi.org/10.1186/1475-2875-7-244

42. Piola P, Fogg C, Bajunirwe F et al (2005) Supervised versus unsupervised intake of six-dose artemether-lumefantrine for treatment of acute, uncomplicated plasmodium falciparum malaria in Mbarara, Uganda: a randomised trial. Lancet 365:1467–1473

43. Checchi F, Piola P, Fogg C et al (2006) Supervised versus unsupervised antimalarial treatment with six-dose artemether-lumefantrine: pharmacokinetic and dosage-related findings from a clinical trial in Uganda. Malar J 5:59

44. van Agtmael MA, Gupta V, van der Graaf CA, van Boxtel CJ (1999) The effect of grapefruit juice on the time-dependent decline of artemether plasma levels in healthy subjects. Clin Pharmacol Ther 66:408–414

45. van Agtmael MA, Gupta V, van der Wosten TH, Rutten JP, van Boxtel CJ (1999) Grapefruit juice increases the bioavailability of artemether. Eur J Clin Pharmacol 55:405–410

46. German PI, Aweeka FT (2008) Clinical pharmacology of artemisinin-based combination therapies. Clin Pharmacokinet 47:91–102

47. El-Lakkany NM (2004) Seif el-din SH, Badawy AA, Ebeid FA. Effect of artemether alone and in combination with grapefruit juice on hepatic drug-metabolising enzymes and biochemical aspects in experimental Schistosoma Mansoni. Int J Parasitol 34:1405–1412

48. Zwang J, Olliaro P, Barennes H et al (2009) Efficacy of artesunate-amodiaquine for treating uncomplicated falciparum malaria in sub-Saharan Africa: a multi-centre analysis. Malar J 8:203. https://doi.org/10.1186/1475-2875-8-203

49. Fitoussi S, Thang C, Lesauvage E et al (2009) Bioavailability of a co-formulated combination of amodiaquine and artesunate under fed and fasted conditions. A randomised, open-label crossover study. Arzneimittelforschung 59:370–376

50. Stepniewska K, Taylor W, Sirima SB et al (2009) Population pharmacokinetics of artesunate and amodiaquine in African children. Malar J 8:200. https://doi.org/10.1186/1475-2875-8-200

51. Davis TM, Hung TY, Sim IK, Karunajeewa HA, Ilett KF (2005) Piperaquine: a resurgent antimalarial drug. Drugs 65:75–87

52. Sim IK, Davis TM, Ilett KF (2005) Effects of a high-fat meal on the relative oral bioavailability of piperaquine. Antimicrob Agents Chemother 49:2407–2411

53. Hai TN, Hietala SF, Van Huong N, Ashton M (2008) The influence of food on the pharmacokinetics of piperaquine in healthy Vietnamese volunteers. Acta Trop 107:145–149

54. Annerberg A, Lwin KM, Lindegardh N et al (2011) A small amount of fat does not affect piperaquine exposure in patients with malaria. Antimicrob Agents Chemother 55:3971–3976

55. Tarning J, Lindegardh N, Lwin KM et al (2014) Population pharmacokinetic assessment of the effect of food on piperaquine bioavailability in patients with uncomplicated malaria. Antimicrob Agents Chemother 58:2052–2058

56. Reuter SE, Evans AM, Shakib S et al (2015) Effect of food on the pharmacokinetics of piperaquine and dihydroartemisinin. Clin Drug Investig 35:559–567

57. Crevoisier C, Handschin J, Barre J, Roumenov D, Kleinbloesem C (1997) Food increases the bioavailability of mefloquine. Eur J Clin Pharmacol 53:135–139

58. Rolan PE, Mercer AJ, Weatherley BC et al (1994) Examination of some factors responsible for a food-induced increase in absorption of atovaquone. Br J Clin Pharmacol 37:13–20

59. Dixon R, Pozniak AL, Watt HM, Rolan P, Posner J (1996) Single-dose and steady-state pharmacokinetics of a novel microfluidized suspension of atovaquone in human immunodeficiency virus-seropositive patients. Antimicrob Agents Chemother 40:556–560

60. Falloon J, Sargent S, Piscitelli SC et al (1999) Atovaquone suspension in HIV-infected volunteers: pharmacokinetics, pharmacodynamics, and TMP-SMX interaction study. Pharmacotherapy 19:1050–1056

61. Awadzi K, Hero M, Opoku NO et al (1994) The chemotherapy of onchocerciasis XVII. A clinical evaluation of albendazole in patients with onchocerciasis; effects of food and pre-treatment with ivermectin on drug response and pharmacokinetics. Trop Med Parasitol 45:203–208

62. Lange H, Eggers R, Bircher J (1988) Increased systemic availability of albendazole when taken with a fatty meal. Eur J Clin Pharmacol 34:315–317

63. Nagy J, Schipper HG, Koopmans RP, Butter JJ, Van Boxtel CJ, Kager PA (2002) Effect of grapefruit juice or cimetidine coadministration on albendazole bioavailability. Am J Trop Med Hyg 66:260–263

64. Lindley D (1987) Merck's new drug free to WHO for river blindness programme. Nature 329:752

65. Steel JW (1993) Pharmacokinetics and metabolism of avermectins in livestock. Vet Parasitol 48:45–57

66. Roche DJ, Yardley MM, Lunny KF et al (2016) A pilot study of the safety and initial efficacy of ivermectin for the treatment of alcohol use disorder. Alcohol Clin Exp Res 40:1312–1320

67. Homeida M, Leahy W, Copeland S, Ali MM, Harron DW (1994) Pharmacokinetic interaction between praziquantel and albendazole in Sudanese men. Ann Trop Med Parasitol 88:551–559

68. Castro N, Medina R, Sotelo J, Jung H (2000) Bioavailability of praziquantel increases with concomitant administration of food. Antimicrob Agents Chemother 44:2903–2904

69. Mandour ME, el Turabi H, Homeida MM et al (1990) Pharmacokinetics of praziquantel in healthy volunteers and patients with schistosomiasis. Trans R Soc Trop Med Hyg 84:389–393

70. Castro N, Jung H, Medina R, Gonzalez-Esquivel D, Lopez M, Sotelo J (2002) Interaction between grapefruit juice and praziquantel in humans. Antimicrob Agents Chemother 46:1614–1616

71. Shu EN, Onwujekwe EO, Okonkwo PO (2000) Do alcoholic beverages enhance availability of ivermectin? Eur J Clin Pharmacol 56:437–438

72. Hatton CS, Peto TE, Bunch C et al (1986) Frequency of severe neutropenia associated with amodiaquine prophylaxis against malaria. Lancet 1:411–414

73. Larrey D, Castot A, Pessayre D et al (1986) Amodiaquine-induced hepatitis. A report of seven cases. Ann Intern Med 104:801–803

74. Neftel KA, Woodtly W, Schmid M, Frick PG, Fehr J (1986) Amodiaquine induced agranulo-cytosis and liver damage. Br Med J (Clin Res Ed) 292:721–723

75. Walsky RL, Obach RS, Gaman EA, Gleeson JP, Proctor WR (2005) Selective inhibition of human cytochrome P4502C8 by montelukast. Drug Metab Dispos 33:413–418

76. Li XQ, Bjorkman A, Andersson TB, Gustafsson LL, Masimirembwa CM (2003) Identification of human cytochrome P(450)s that metabolise anti-parasitic drugs and predictions of in vivo drug hepatic clearance from in vitro data. Eur J Clin Pharmacol 59:429–442

77. Li XQ, Bjorkman A, Andersson TB, Ridderstrom M, Masimirembwa CM (2002) Amodiaquine clearance and its metabolism to N-desethylamodiaquine is mediated by CYP2C8: a new high affinity and turnover enzyme-specific probe substrate. J Pharmacol Exp Ther 300:399–407

78. Gil JP (2008) Amodiaquine pharmacogenetics. Pharmacogenomics 9:1385–1390

79. Giao PT, de Vries PJ (2001) Pharmacokinetic interactions of antimalarial agents. Clin Pharmacokinet 40:343–373

80. Parikh S, Ouedraogo JB, Goldstein JA, Rosenthal PJ, Kroetz DL (2007) Amodiaquine metab-olism is impaired by common polymorphisms in CYP2C8: implications for malaria treat-ment in Africa. Clin Pharmacol Ther 82:197–203

81. Orrell C, Little F, Smith P et al (2008) Pharmacokinetics and tolerability of artesunate and amodiaquine alone and in combination in healthy volunteers. Eur J Clin Pharmacol 64:683–690

82. Maggs JL, Madden S, Bishop LP, O'Neill PM, Park BK (1997) The rat biliary metabolites of dihydroartemisinin, an antimalarial endoperoxide. Drug Metab Dispos 25:1200–1204

83. Ilett KF, Ethell BT, Maggs JL et al (2002) Glucuronidation of dihydroartemisinin in vivo and by human liver microsomes and expressed UDP-glucuronosyltransferases. Drug Metab Dispos 30:1005–1012

84. Scarsi KK, Fehintola FA, Ma Q et al (2014) Disposition of amodiaquine and desethylamodiaquine in HIV-infected Nigerian subjects on nevirapine-containing antiretroviral therapy. J Antimicrob Chemother 69:1370–1376

85. McElnay JC, Mukhtar HA, D'Arcy PF, Temple DJ, Collier PS (1982) The effect of magnesium trisilicate and kaolin on the in vivo absorption of chloroquine. J Trop Med Hyg 85:159–163

86. McElnay JC, Sidahmed AM, D'Arcy PF (1982) Examination of the chloroquine-kaolin drug absorption interaction using the buccal partitioning model. J Clin Hosp Pharm 7:269–273

87. van Luin M, Van der Ende ME, Richter C et al (2010) Lower atovaquone/proguanil concentrations in patients taking efavirenz, lopinavir/ritonavir or atazanavir/ritonavir. AIDS 24:1223–1226

88. Kolawole JA, Mustapha A, Abdul-Aguye I, Ochekpe N, Taylor RB (1999) Effects of cimetidine on the pharmacokinetics of proguanil in healthy subjects and in peptic ulcer patients. J Pharm Biomed Anal 20:737–743

89. van Agtmael MA, Cheng-Qi S, Qing JX, Mull R, van Boxtel CJ (1999) Multiple dose pharmacokinetics of artemether in Chinese patients with uncomplicated falciparum malaria. Int J Antimicrob Agents 12:151–158

90. Hassan Alin M, Ashton M, Kihamia CM, Mtey GJ, Bjorkman A (1996) Multiple dose pharmacokinetics of oral artemisinin and comparison of its efficacy with that of oral artesunate in falciparum malaria patients. Trans R Soc Trop Med Hyg 90:61–65

91. Ashton M, Hai TN, Sy ND et al (1998) Artemisinin pharmacokinetics is time-dependent during repeated oral administration in healthy male adults. Drug Metab Dispos 26:25–27

92. Ashton M, Nguyen DS, Nguyen VH et al (1998) Artemisinin kinetics and dynamics during oral and rectal treatment of uncomplicated malaria. Clin Pharmacol Ther 63:482–493

93. Tan-ariya P, Na-Bangchang K, Ubalee R, Thanavibul A, Thipawangkosol P, Karbwang J (1998) Pharmacokinetic interactions of artemether and pyrimethamine in healthy male Thais. Southeast Asian J Trop Med Public Health 29:18–23

94. Lamorde M, Byakika-Kibwika P, Mayito J et al (2013) Lower artemether, dihydroartemisinin and lumefantrine concentrations during rifampicin-based tuberculosis treatment. AIDS 27:961–965

95. Lefevre G, Carpenter P, Souppart C, Schmidli H, McClean M, Stypinski D (2002) Pharmacokinetics and electrocardiographic pharmacodynamics of artemether-lumefantrine (Riamet) with concomitant administration of ketoconazole in healthy subjects. Br J Clin Pharmacol 54:485–492

96. Karbwang J (1990) Back DJ, Bunnag D, Breckenridge AM. Pharmacokinetics of mefloquine in combination with sulfadoxine-pyrimethamine and primaquine in male Thai patients with falciparum malaria. Bull World Health Organ 68:633–638

97. Hassan A, Lin M, Ashton M, Kihamia CM (1996) Clinical efficacy and pharmacokinetics of artemisinin monotherapy and in combination with mefloquine in patients with falciparum malaria. Br J Clin Pharmacol 41:592

98. Price R, Simpson JA, Teja-Isavatharm P et al (1999) Pharmacokinetics of mefloquine combined with artesunate in children with acute falciparum malaria. Antimicrob Agents Chemother 43:341–346

99. Na-Bangchang K, Karbwang J, Molunto P, Banmairuroi V, Thanavibul A (1995) Pharmacokinetics of mefloquine, when given alone and in combination with artemether, in patients with uncomplicated falciparum malaria. Fundam Clin Pharmacol 9:576–582

100. Lefevre G, Bindschedler M, Ezzet F, Schaeffer N, Meyer I, Thomsen MS (2000) Pharmacokinetic interaction trial between co-artemether and mefloquine. Eur J Pharm Sci 10:141–151

101. Kolawole JA, Mustapha A, Abudu-Aguye I, Ochekpe N (2000) Mefloquine pharmacokinetics in healthy subjects and in peptic ulcer patients after cimetidine administration. Eur J Drug Metab Pharmacokinet 25:165–170

102. Ridtitid W, Wongnawa M, Mahatthanatrakul W, Raungsri N, Sunbhanich M (2005) Ketoconazole increases plasma concentrations of antimalarial mefloquine in healthy human volunteers. J Clin Pharm Ther 30:285–290

103. Ridtitid W, Wongnawa M, Mahatthanatrakul W, Chaipol P, Sunbhanich M (2000) Effect of rifampin on plasma concentrations of mefloquine in healthy volunteers. J Pharm Pharmacol 52:1265–1269

104. Gupta S, Thapar MM, Wernsdorfer WH, Bjorkman A (2002) Vitro interactions of artemisinin with atovaquone, quinine, and mefloquine against plasmodium falciparum. Antimicrob Agents Chemother 46:1510–1515

105. Mariga ST, Gil JP, Wernsdorfer WH, Bjorkman A (2005) Pharmacodynamic interactions of amodiaquine and its major metabolite desethylamodiaquine with artemisinin, quinine and atovaquone in plasmodium falciparum in vitro. Acta Trop 93:221–231

106. Ley B, Wernsdorfer G, Frank C, Sirichaisinthop J, Congpuong K, Wernsdorfer WH (2008) Pharmacodynamic interaction between 4-aminoquinolines and retinol in Plasmodium falciparum in vitro. Wien Klin Wochenschr 120(19–20 Suppl 4):74–79

107. Stahel E, Druilhe P, Gentilini M (1988) Antagonism of chloroquine with other antimalarials. Trans R Soc Trop Med Hyg 82:221

108. Garavito G, Bertani S, Rincon J et al (2007) Blood schizontocidal activity of methylene blue in combination with antimalarials against plasmodium falciparum. Parasite 14:135–140

109. German P, Greenhouse B, Coates C et al (2007) Hepatotoxicity due to a drug interaction between amodiaquine plus artesunate and efavirenz. Clin Infect Dis 44:889–891

110. Lancaster DL, Adio RA, Tai KK, Simooya OO, Broadhead GD, Tucker GT et al (1990) Inhibition of metoprolol metabolism by chloroquine and other antimalarial drugs. J Pharm Pharmacol 42:267–271

111. Masimirembwa CM, Hasler JA, Johansson I (1995) Inhibitory effects of antiparasitic drugs on cytochrome P450 2D6. Eur J Clin Pharmacol 48:35–38

112. Simooya OO, Sijumbil G, Lennard MS, Tucker GT (1998) Halofantrine and chloroquine inhibit CYP2D6 activity in healthy Zambians. Br J Clin Pharmacol 45:315–317

113. Ilo CE, Ilondu NA, Okwoli N et al (2006) Effect of chloroquine on the bioavailability of ciprofloxacin in humans. Am J Ther 13:432–435

114. Adedoyin A, Frye RF, Mauro K, Branch RA (1998) Chloroquine modulation of specific metabolizing enzymes activities: investigation with selective five drug cocktail. Br J Clin Pharmacol 46:215–219

115. Masimirembwa CM, Gustafsson LL, Dahl ML, Abdi YA, Hasler JA (1996) Lack of effect of chloroquine on the debrisoquine (CYP2D6) and S-mephenytoin (CYP2C19) hydroxylation phenotypes. Br J Clin Pharmacol 41:344–346

116. Ali HM (1985) Reduced ampicillin bioavailability following oral coadministration with chloroquine. J Antimicrob Chemother 15:781–784

117. Makanjuola RO, Dixon PA, Oforah E (1988) Effects of antimalarial agents on plasma levels of chlorpromazine and its metabolites in schizophrenic patients. Trop Geogr Med 40:31–33

118. Cook JA, Randinitis EJ, Bramson CR, Wesche DL (2006) Lack of a pharmacokinetic interaction between azithromycin and chloroquine. Am J Trop Med Hyg 74:407–412

119. Back DJ, Purba HS, Park BK, Ward SA, Orme ML (1983) Effect of chloroquine and primaquine on antipyrine metabolism. Br J Clin Pharmacol 16:497–502

120. Ette EI, Brown-Awala A, Essien EE (1987) Effect of ranitidine on chloroquine disposition. Drug Intell Clin Pharm 21:732–734

121. Ette EI, Brown-Awala EA, Essien EE (1987) Chloroquine elimination in humans: effect of low-dose cimetidine. J Clin Pharmacol 27:813–816
122. Onyeji CO, Toriola TA, Ogunbona FA (1993) Lack of pharmacokinetic interaction between chloroquine and imipramine. Ther Drug Monit 15:43–46
123. Raina RK, Bano G, Amla V, Kapoor V, Gupta KL (1993) The effect of aspirin, paracetamol and analgin on pharmacokinetics of chloroquine. Indian J Physiol Pharmacol 37:229–231
124. Gbotosho GO, Happi CT, Sijuade A, Ogundahunsi OA, Sowunmi A, Oduola AM (2008) Comparative study of interactions between chloroquine and chlorpheniramine or promethazine in healthy volunteers: a potential combination-therapy phenomenon for resuscitating chloroquine for malaria treatment in Africa. Ann Trop Med Parasitol 102:3–9
125. Dormoi J, Pascual A, Briolant S et al (2012) Proveblue (methylene blue) as an antimalarial agent: in vitro synergy with dihydroartemisinin and atorvastatin. Antimicrob Agents Chemother 56:3467–3469
126. Skinner-Adams T, Davis TM (1999) Synergistic in vitro antimalarial activity of omeprazole and quinine. Antimicrob Agents Chemother 43:1304–1306
127. Gupta S, Thapar MM, Mariga ST, Wernsdorfer WH, Bjorkman A (2002) Plasmodium falciparum: in vitro interactions of artemisinin with amodiaquine, pyronaridine, and chloroquine. Exp Parasitol 100:28–35
128. Kyavar L, Rojanawatsirivet C, Kollaritsch H, Wernsdorfer G, Sirichaisinthop J, Wernsdorfer WH (2006) In vitro interaction between artemisinin and chloroquine as well as desbutylbenflumetol in Plasmodium vivax. Wien Klin Wochenschr 118(19–20 Suppl 3):62–69
129. Pereira MR, Henrich PP, Sidhu AB et al (2011) Vivo and in vitro antimalarial properties of azithromycin-chloroquine combinations that include the resistance reversal agent amlodipine. Antimicrob Agents Chemother 55:3115–3124
130. Ohrt C, Willingmyre GD, Lee P, Knirsch C, Milhous W (2002) Assessment of azithromycin in combination with other antimalarial drugs against plasmodium falciparum in vitro. Antimicrob Agents Chemother 46:2518–2524
131. Desgrouas C, Dormoi J, Chapus C, Ollivier E, Parzy D, Taudon N (2014) In vitro and in vivo combination of cepharanthine with anti-malarial drugs. Malar J 13:90. https://doi.org/10.1186/1475-2875-13-90
132. Constantino L, Paixao P, Moreira R, Portela MJ, Do Rosario VE, Iley J (1999) Metabolism of primaquine by liver homogenate fractions. Evidence for monoamine oxidase and cytochrome P450 involvement in the oxidative deamination of primaquine to carboxyprimaquine. Exp Toxicol Pathol 51:299–303
133. Baker JK, McChesney JD, Hufford CD, Clark AM (1982) High-performance liquid chromatographic analysis of the metabolism of primaquine and the identification of a new mammalian metabolite. J Chromatogr 230:69–77
134. Mihaly GW, Ward SA, Edwards G, Orme ML, Breckenridge AM (1984) Pharmacokinetics of primaquine in man: identification of the carboxylic acid derivative as a major plasma metabolite. Br J Clin Pharmacol 17:441–446
135. Jin X, Pybus BS, Marcsisin R et al (2014) An LC-MS based study of the metabolic profile of primaquine, an 8-aminoquinoline antiparasitic drug, with an in vitro primary human hepatocyte culture model. Eur J Drug Metab Pharmacokinet 39:139–146
136. Bangchang KN, Karbwang J (1992) Back DJ. Primaquine metabolism by human liver microsomes: effect of other antimalarial drugs. Biochem Pharmacol 44:587–590
137. Edwards G, McGrath CS, Ward SA et al (1993) Interactions among primaquine, malaria infection and other antimalarials in Thai subjects. Br J Clin Pharmacol 35:193–198
138. Na-Bangchang K, Karbwang J, Ubalee R, Thanavibul A, Saenglertsilapachai S (2000) Absence of significant pharmacokinetic and pharmacodynamic interactions between artemether and quinoline antimalarials. Eur J Drug Metab Pharmacokinet 25:171–178
139. Nzila A (2006) The past, present and future of antifolates in the treatment of plasmodium falciparum infection. J Antimicrob Chemother 57:1043–1054

140. Birkett DJ, Rees D, Andersson T, Gonzalez FJ, Miners JO, Veronese ME (1994) Vitro proguanil activation to cycloguanil by human liver microsomes is mediated by CYP3A isoforms as well as by S-mephenytoin hydroxylase. Br J Clin Pharmacol 37:413–420

141. Skjelbo E, Mutabingwa TK, Bygbjerg I, Nielsen KK, Gram LF, Broosen K (1996) Chloroguanide metabolism in relation to the efficacy in malaria prophylaxis and the S-mephenytoin oxidation in Tanzanians. Clin Pharmacol Ther 59:304–311

142. Rasmussen BB, Nielsen TL, Brosen K (1998) Fluvoxamine inhibits the CYP2C19-catalysed metabolism of proguanil in vitro. Eur J Clin Pharmacol 54:735–740

143. AH L, Shu Y, Huang SL, Wang W, Ou-Yang DS, Zhou HH (2000) Vitro proguanil activation to cycloguanil is mediated by CYP2C19 and CYP3A4 in adult Chinese liver microsomes. Acta Pharmacol Sin 21:747–752

144. Coller JK, Somogyi AA, Bochner F (1999) Comparison of (S)-mephenytoin and proguanil oxidation in vitro: contribution of several CYP isoforms. Br J Clin Pharmacol 48:158–167

145. Helsby NA, Edwards G, Breckenridge AM, Ward SA (1993) The multiple dose pharmacokinetics of proguanil. Br J Clin Pharmacol 35:653–656

146. Kaneko A, Bergqvist Y, Taleo G, Kobayakawa T, Ishizaki T, Bjorkman A (1999) Proguanil disposition and toxicity in malaria patients from Vanuatu with high frequencies of CYP2C19 mutations. Pharmacogenetics 9:317–326

147. Watkins WM, Mberu EK, Nevill CG, Ward SA, Breckenridge AM, Koech DK (1990) Variability in the metabolism of proguanil to the active metabolite cycloguanil in healthy Kenyan adults. Trans R Soc Trop Med Hyg 84:492–495

148. Veenendaal JR, Edstein MD, Rieckmann KH (1988) Pharmacokinetics of chlorproguanil in man after a single oral dose of Lapudrine. Chemotherapy 34:275–283

149. Funck-Brentano C, Becquemont L, Lenevu A, Roux A, Jaillon P, Beaune P (1997) Inhibition by omeprazole of proguanil metabolism: mechanism of the interaction in vitro and prediction of in vivo results from the in vitro experiments. J Pharmacol Exp Ther 280:730–738

150. Canfield CJ, Pudney M, Gutteridge WE (1995) Interactions of atovaquone with other antimalarial drugs against plasmodium falciparum in vitro. Exp Parasitol 80:373–381

151. Thapar MM, Gupta S, Spindler C, Wernsdorfer WH, Bjorkman A (2003) Pharmacodynamic interactions among atovaquone, proguanil and cycloguanil against plasmodium falciparum in vitro. Trans R Soc Trop Med Hyg 97:331–337

152. Raffelsberger J, Wernsdorfer G, Sirichaisinthop J, Kollaritsch H, Congpuong K, Wernsdorfer WH (2008) Pharmacodynamic interaction between monodesbutyl-benflumetol and artemisinin as well as proguanil in Plasmodium falciparum in vitro. Wien Klin Wochenschr 120(19–20 Suppl 4):90–94

153. Rolan PE, Mercer AJ, Tate E, Benjamin I, Posner J (1997) Disposition of atovaquone in humans. Antimicrob Agents Chemother 41:1319–1321

154. Gillotin C, Mamet JP, Veronese L (1999) Lack of a pharmacokinetic interaction between atovaquone and proguanil. Eur J Clin Pharmacol 55:311–315

155. Hussein Z, Eaves CJ, Hutchinson DB, Canfield CJ (1996) Population pharmacokinetics of proguanil in patients with acute P. Falciparum malaria after combined therapy with atovaquone. Br J Clin Pharmacol 42:589–597

156. Lee BL, Tauber MG, Sadler B, Goldstein D, Chambers HF (1996) Atovaquone inhibits the glucuronidation and increases the plasma concentrations of zidovudine. Clin Pharmacol Ther 59:14–21

157. Khoo S (2005) Back D, Winstanley P. The potential for interactions between antimalarial and antiretroviral drugs. AIDS 19:995–1005

158. Hidalgo K, Lyles A, Dean SR (2011) A potential interaction between warfarin and atovaquone. Ann Pharmacother 45:e3

159. Svensson US, Ashton M (1999) Identification of the human cytochrome P450 enzymes involved in the in vitro metabolism of artemisinin. Br J Clin Pharmacol 48:528–535

160. Bapiro TE, Egnell AC, Hasler JA, Masimirembwa CM (2001) Application of higher through-put screening (HTS) inhibition assays to evaluate the interaction of antiparasitic drugs with cytochrome P450s. Drug Metab Dispos 29:30–35

161. Grace JM, Aguilar AJ, Trotman KM, Peggins JO, Brewer TG (1998) Metabolism of beta-arteether to dihydroqinghaosu by human liver microsomes and recombinant cytochrome P450. Drug Metab Dispos 26:313–317

162. Khanh NX, de Vries PJ, Ha LD, van Boxtel CJ, Koopmans R, Kager PA (1999) Declining concentrations of dihydroartemisinin in plasma during 5-day oral treatment with artesunate for falciparum malaria. Antimicrob Agents Chemother 43:690–692

163. Svensson US, Ashton M, Trinh NH et al (1998) Artemisinin induces omeprazole metabolism in human beings. Clin Pharmacol Ther 64:160–167

164. Asimus S, Elsherbiny D, Hai TN et al (2007) Artemisinin antimalarials moderately affect cytochrome P450 enzyme activity in healthy subjects. Fundam Clin Pharmacol 21:307–316

165. Sinclair D, Zani B, Donegan S, Olliaro P, Garner P (2009) Artemisinin-based combination therapy for treating uncomplicated malaria. Cochrane Database Syst Rev 3:CD007483

166. Mishra LC, Bhattacharya A, Bhasin VK (2007) Antiplasmodial interactions between artemis-inin and triclosan or ketoconazole combinations against blood stages of plasmodium falci-parum in vitro. Am J Trop Med Hyg 76:497–501

167. Bhattacharya A, Mishra LC, Bhasin VK (2008) Vitro activity of artemisinin in combination with clotrimazole or heat-treated amphotericin B against plasmodium falciparum. Am J Trop Med Hyg 78:721–728

168. Noedl H, Krudsood S, Leowattana W et al (2007) Vitro antimalarial activity of azithromycin, artesunate, and quinine in combination and correlation with clinical outcome. Antimicrob Agents Chemother 51:651–656

169. Ramharter M, Noedl H, Winkler H et al (2003) Vitro activity and interaction of clindamy-cin combined with dihydroartemisinin against plasmodium falciparum. Antimicrob Agents Chemother 47:3494–3499

170. Lutgendorf C, Rojanawatsirivet C, Wernsdorfer G, Sirichaisinthop J, Kollaritsch H, Wernsdorfer WH (2006) Pharmacodynamic interaction between atovaquone and other antimalarial compounds against Plasmodium falciparum in vitro. Wien Klin Wochenschr 118(19–20 Suppl 3):70–76

171. Sponer U, Prajakwong S, Wiedermann G, Kollaritsch H, Wernsdorfer G, Wernsdorfer WH (2002) Pharmacodynamic interaction of doxycycline and artemisinin in plasmodium falci-parum. Antimicrob Agents Chemother 46:262–264

172. Bwijo B, Alin MH, Abbas N, Wernsdorfer W, Bjorkman A (1997) Efficacy of artemisinin and mefloquine combinations against plasmodium falciparum. In vitro simulation of in vivo pharmacokinetics. Tropical Med Int Health 2:461–467

173. Kerschbaumer G, Wernsdorfer G, Wiedermann U, Congpuong K, Sirichaisinthop J, Wernsdorfer WH (2010) Synergism between mefloquine and artemisinin and its enhancement by retinol in plasmodium falciparum in vitro. Wien Klin Wochenschr 122(Suppl 3):57–60

174. Vivas L, Rattray L, Stewart L et al (2008) Anti-malarial efficacy of pyronaridine and artesu-nate in combination in vitro and in vivo. Acta Trop 105:222–228

175. Kiang TK, Wilby KJ, Ensom MH (2014) Clinical pharmacokinetic drug interactions associ-ated with artemisinin derivatives and HIV-antivirals. Clin Pharmacokinet 53:141–153

176. German P, Parikh S, Lawrence J et al (2009) Lopinavir/ritonavir affects pharmacokinetic exposure of artemether/lumefantrine in HIV-uninfected healthy volunteers. J Acquir Immune Defic Syndr 51:424–429

177. Byakika-Kibwika P, Lamorde M, Okaba-Kayom V et al (2012) Lopinavir/ritonavir sig-nificantly influences pharmacokinetic exposure of artemether/lumefantrine in HIV-infected Ugandan adults. J Antimicrob Chemother 67:1217–1223

178. Kakuda TN, DeMasi R, van Delft Y, Mohammed P (2013) Pharmacokinetic interaction between etravirine or darunavir/ritonavir and artemether/lumefantrine in healthy volunteers: a two-panel, two-way, two-period, randomized trial. HIV Med 14:421–429

179. Huang L, Parikh S, Rosenthal PJ et al (2012) Concomitant efavirenz reduces pharmaco-kinetic exposure to the antimalarial drug artemether-lumefantrine in healthy volunteers. J Acquir Immune Defic Syndr 61:310–316

180. Byakika-Kibwika P, Lamorde M, Mayito J et al (2012) Significant pharmacokinetic interactions between artemether/lumefantrine and efavirenz or nevirapine in HIV-infected Ugandan adults. J Antimicrob Chemother 67:2213–2221

181. Kredo T, Mauff K, Van der Walt JS et al (2011) Interaction between artemether-lumefantrine and nevirapine-based antiretroviral therapy in HIV-1-infected patients. Antimicrob Agents Chemother 55:5616–5623

182. Zhao XJ, Yokoyama H, Chiba K, Wanwimolruk S, Ishizaki T (1996) Identification of human cytochrome P450 isoforms involved in the 3-hydroxylation of quinine by human live microsomes and nine recombinant human cytochromes P450. J Pharmacol Exp Ther 279:1327–1334

183. Wanwimolruk S, Wong SM, Zhang H, Coville PF, Walker RJ (1995) Metabolism of quinine in man: identification of a major metabolite, and effects of smoking and rifampicin pretreatment. J Pharm Pharmacol 47:957–963

184. Wanwimolruk S, Wong SM, Coville PF, Viriyayudhakorn S, Thitiarchakul S (1993) Cigarette smoking enhances the elimination of quinine. Br J Clin Pharmacol 36:610–614

185. Muralidharan G, Hawes EM, McKay G, Midha KK (1991) Quinine is a more potent inhibitor than quinidine in rat of the oxidative metabolic routes of methoxyphenamine which involve debrisoquine 4-hydroxylase. Xenobiotica 21:1441–1450

186. Krishna S (2006) Quinine and quinidine. In: Yu VL, Edwards G, PS MK, Peloquin C, Morse GD (eds) Antimicrobial therapy and vaccines, Antimicrobial agents, vol 2. ESun Technologies, LLC, Pittsburg, pp 1167–1194

187. Wanwimolruk S, Sunbhanich M, Pongmarutai M, Patamasucon P (1986) Effects of cimetidine and ranitidine on the pharmacokinetics of quinine. Br J Clin Pharmacol 22:346–350

188. Adegbola AJ, Soyinka JO, Adeagbo BA, Iqbinoba SI, Nathaniel TI (2016) Alteration of the disposition of quinine in healthy volunteers after concurrent ciprofloxacin administration. Am J Ther 23:e398–e404

189. Soyinka JO, Onyeji CO (2010) Alteration of pharmacokinetics of proguanil in healthy volunteers following concurrent administration of efavirenz. Eur J Pharm Sci 39:213–218

190. Nyunt MM, Lu Y, El-Gasim M, Parsons TL, Petty BG, Hendrix CW (2012) Effects of ritonavir-boosted lopinavir on the pharmacokinetics of quinine. Clin Pharmacol Ther 91:889–895

191. Amabeoku GJ, Chikuni O, Akino C, Mutetwa S (1993) Pharmacokinetic interaction of single doses of quinine and carbamazepine, phenobarbitone and phenytoin in healthy volunteers. East Afr Med J 70:90–93

192. Suphakawanich W, Thithapandha A (1987) Inhibition of hepatic drug metabolism by quinine. Asia Pacific. J Pharmacol 2:241–247

193. Munafo A, Reymond-Michel G, Biollaz J (1990) Altered flecainide disposition in healthy volunteers taking quinine. Eur J Clin Pharmacol 38:269–273

194. Hager WD, Fenster P, Mayersohn M et al (1979) Digoxin-quinidine interaction pharmacokinetic evaluation. N Engl J Med 300:1238–1241

195. Wandell M, Powell JR, Hager WD et al (1980) Effect of quinine on digoxin kinetics. Clin Pharmacol Ther 28:425–430

196. Edwards DJ, Axelson JE, Visco JP (1987) Van every S, slaughter RL, Lalka D. Lack of effect of smoking on the metabolism and pharmacokinetics of quinidine in patients. Br J Clin Pharmacol 23:351–354

197. Spinler SA, Cheng JW, Kindwall KE, Charland SL (1995) Possible inhibition of hepatic metabolism of quinidine by erythromycin. Clin Pharmacol Ther 57:89–94

198. White NJ, Looareesuwan S, Phillips RE, Chanthavanich P, Warrell DA (1988) Single dose phenobarbitone prevents convulsions in cerebral malaria. Lancet 2:64–66

199. Winstanley PA, Newton CR, Pasvol G et al (1992) Prophylactic phenobarbitone in young children with severe falciparum malaria: pharmacokinetics and clinical effects. Br J Clin Pharmacol 33:149–154

200. Na-Bangchang K, Tippanangkosol P, Ubalee R, Chaovanakawee S, Saenglertsilapachai S, Karbwang J (1999) Comparative clinical trial of four regimens of dihydroartemisinin-mefloquine in multidrug-resistant falciparum malaria. Tropical Med Int Health 4:602–610

201. Na-Bangchang K, Tippawangkosol P, Thanavibul A, Ubalee R, Karbwang J (1999) Pharmacokinetic and pharmacodynamic interactions of mefloquine and dihydroartemisinin. Int J Clin Pharmacol Res 19:9–17

202. Boudreau EF, Fleckenstein L, Pang LW et al (1990) Mefloquine kinetics in cured and recrudescent patients with acute falciparum malaria and in healthy volunteers. Clin Pharmacol Ther 48:399–409

203. Na Bangchang K, Karbwang J, Bunnag D, Harinasuta T (1991) Back DJ. The effect of metoclopramide on mefloquine pharmacokinetics. Br J Clin Pharmacol 32:640–641

204. Karbwang J, Na Bangchang K (1991) Back DJ, Bunnag D. Effect of ampicillin on mefloquine pharmacokinetics in Thai males. Eur J Clin Pharmacol 40:631–633

205. Khaliq Y, Gallicano K, Tisdale C, Carignan G, Cooper C, McCarthy A (2001) Pharmacokinetic interaction between mefloquine and ritonavir in healthy volunteers. Br J Clin Pharmacol 51:591–600

206. Rawden HC, Kokwaro GO, Ward SA, Edwards G (2000) Relative contribution of cytochromes P-450 and flavin-containing monooxygenases to the metabolism of albendazole by human liver microsomes. Br J Clin Pharmacol 49:313–322

207. Souhaili-El Amri H, Mothe O, Totis M et al (1988) Albendazole sulfonation by rat liver cytochrome P-450c. J Pharmacol Exp Ther 246:758–764

208. Pawluk SA, Roels CA, Wilby KJ, Ensom MHA (2015) Review of pharmacokinetic drug-drug interactions with the anthelmintic medications albendazole and mebendazole. Clin Pharmacokinet 54:371–383

209. Merino G, Jonker JW, Wagenaar E et al (2005) Transport of anthelmintic benzimidazole drugs by breast cancer resistance protein (BCRP/ABCG2). Drug Metab Dispos 33:614–618

210. Merino G, Alvarez AI, Prieto JG, Kim RB (2002) The anthelminthic agent albendazole does not interact with p-glycoprotein. Drug Metab Dispos 30:365–369

211. Krishna DR, Klotz U (1993) Determination of ivermectin in human plasma by high-performance liquid chromatography. Arzneimittelforschung 43:609–611

212. Scott EW, McKellar QA (1992) The distribution and some pharmacokinetic parameters of ivermectin in pigs. Vet Res Commun 16:139–146

213. Lanusse C, Lifschitz A, Virkel G et al (1997) Comparative plasma disposition kinetics of ivermectin, moxidectin and doramectin in cattle. J Vet Pharmacol Ther 20:91–99

214. McKellar QA, Benchaoui HA (1996) Avermectins and milbemycins. J Vet Pharmacol Ther 19:331–351

215. Zeng Z, Andrew NW, Arison BH, Luffer-Atlas D, Wang RW (1998) Identification of cytochrome P4503A4 as the major enzyme responsible for the metabolism of ivermectin by human liver microsomes. Xenobiotica 28:313–321

216. Fink DW, Porras AG (1989) Pharmacokinetics of ivermectin in animals and humans. In: Campbell WC (ed) Ivermectin and abamectin. Springer, New York, pp 113–130

217. Lespine A, Dupuy J, Orlowski S et al (2006) Interaction of ivermectin with multidrug resistance proteins (MRP1, 2 and 3). Chem Biol Interact 159:169–179

218. Pouliot JF, L'Heureux F, Liu Z, Prichard RK, Georges E (1997) Reversal of P-glycoprotein-associated multidrug resistance by ivermectin. Biochem Pharmacol 53:17–25

219. Skalova L, Szotakova B, Machala M et al (2001) Effect of ivermectin on activities of cytochrome P450 isoenzymes in mouflon (Ovis Musimon) and fallow deer (Dama Dama). Chem Biol Interact 137:155–167

220. Awadzi K, Edwards G, Opoku NO et al (2004) The safety, tolerability and pharmacokinetics of levamisole alone, levamisole plus ivermectin, and levamisole plus albendazole, and their efficacy against Onchocerca volvulus. Ann Trop Med Parasitol 98:595–614

221. Na-Bangchang K, Kietinun S, Pawa KK, Hanpitakpong W, Na-Bangchang C, Lazdins J (2006) Assessments of pharmacokinetic drug interactions and tolerability of albendazole, praziquantel and ivermectin combinations. Trans R Soc Trop Med Hyg 100:335–345

222. Amsden GW, Gregory TB, Michalak CA, Glue P, Knirsch CA (2007) Pharmacokinetics of azithromycin and the combination of ivermectin and albendazole when administered alone and concurrently in healthy volunteers. Am J Trop Med Hyg 76:1153–1157

223. MH W, Wei CC, ZY X et al (1991) Comparison of the therapeutic efficacy and side effects of a single dose of levo-praziquantel with mixed isomer praziquantel in 278 cases of schistosomiasis japonica. Am J Trop Med Hyg 45:345–349

224. Giorgi M, Salvatori AP, Soldani G et al (2001) Pharmacokinetics and microsomal oxidation of praziquantel and its effects on the P450 system in three-month-old lambs infested by Fasciola Hepatica. J Vet Pharmacol Ther 24:251–259

225. Godawska-Matysik A, Kiec-Kononowicz K (2006) Biotransformation of praziquantel by human cytochrome p450 3A4 (CYP 3A4). Acta Pol Pharm 63:381–385

226. Jung H, Medina R, Castro N, Corona T, Sotelo J (1997) Pharmacokinetic study of praziquantel administered alone and in combination with cimetidine in a single-day therapeutic regimen. Antimicrob Agents Chemother 41:1256–1259

227. Metwally A, Bennett JL, Botros S, Ebeid F (1995) Effect of cimetidine, bicarbonate and glucose on the bioavailability of different formulations of praziquantel. Arzneimittelforschung 45:516–518

228. Masimirembwa CM, Naik YS, Hasler JA (1994) The effect of chloroquine on the pharmacokinetics and metabolism of praziquantel in rats and in humans. Biopharm Drug Dispos 15:33–43

229. Ridtitid W, Ratsamemonthon K, Mahatthanatrakul W, Wongnawa M (2007) Pharmacokinetic interaction between ketoconazole and praziquantel in healthy volunteers. J Clin Pharm Ther 32:585–593

230. Ridtitid W, Wongnawa M, Mahatthanatrakul W, Punyo J, Sunbhanich M (2002) Rifampin markedly decreases plasma concentrations of praziquantel in healthy volunteers. Clin Pharmacol Ther 72:505–513

231. Lima RM, Ferreira MA (2011) De Jesus Ponte Carvalho TM, et al. Albendazole-praziquantel interaction in healthy volunteers: kinetic disposition, metabolism and enantioselectivity. Br J Clin Pharmacol 71:528–535

232. Gonzalez-Esquivel D, Rivera J, Castro N, Yepez-Mulia L, Jung Cook H (2005) Vitro characterization of some biopharmaceutical properties of praziquantel. Int J Pharm 295:93–99

233. Hayeshi R, Masimirembwa C, Mukanganyama S, Ungell AL (2006) The potential inhibitory effect of antiparasitic drugs and natural products on P-glycoprotein mediated efflux. Eur J Pharm Sci 29:70–81

234. Edwards G (2003) Ivermectin: does P-glycoprotein play a role in neurotoxicity? Filaria J 2(Suppl 1):S8

Index

A

Abacavir (ABC), 257, 283
  hepatic alcohol dehydrogenase, 301
  modern ART regimens, 298
  NRTIs, 303–305
  pharmacokinetic parameters, 310
  and stavudine distribution, 300
  tipranavir/ritonavir, 308
  UGT1A1, 301
Acetaminophen, 179–180
Acid-neutralizing agents, 23
Acid-suppressive agents, 1, 2, 22–23
  antacids, 23
  ranitidine, famotidine and omeprazole,
    22, 23
Acinetobacter anitratum, 11
Acinetobacter baumannii, 33
Acquired rifamycin resistance (ARR),
    237–238, 243
ACT-anti-Tb drug interactions
  artemether-lumefantrine, 508, 509
  artesunate-amodiaquine, 509, 510
  artesunate-mefloquine, 510
Acute kidney injury (AKI), 20, 21
Acyclovir, 384, 395
Adamantane drugs, 402
Adedeji, W.A., 505
Adefovir dipivoxil, 359
Adenosine, 298, 307
Adjusted odds ratios (AORs), 99, 118
Adrenal insufficiency, 452
Albendazole (ABZ), 523, 535
Albert, K.S., 192
Allopurinol therapy, 2, 3
Alvimopan, 108

Amantadine, 402, 403
Aminoglycosides, 141, 151–153, 158, 186, 240
  amphotericin B, 201
  chemotherapeutic agents, 203
  cyclosporine, 203
  diuretics, 204, 205
  ethacrynic acid, 204
  furosemide, 204, 205
  indomethacin, 202
  in vitro inactivation, 5
  in vivo inactivation, 3–5
  neuromuscular blocking agents, 201, 202
  penicillins, 3
  in serum samples, 5
  synergy, 5, 6
  vancomycin, 205, 206
Aminoquinolines
  CQ, 506
  PQ, 507
Amiodarone, 367
Amodiaquine (AQ) hydrochloride, 507, 520,
    522, 524, 525, 528
Amoxicillin-clavulanate (A-C), 112, 113
Amphotericin B, 201, 438–440
  adverse effects, 432
  distribution, 432
  elimination, 432
  ergosterol, 431
  extrarenal toxicity, 5-FC
    cryptococcal meningitis, 440
    management, 440
    myelosuppression, hepatic necrosis and
      diarrhea, 439
  synergistic/additive nephrotoxicity
    electrolytes, 438

© Springer International Publishing AG, part of Springer Nature 2018   551
M. P. Pai et al. (eds.), *Drug Interactions in Infectious Diseases: Antimicrobial
Drug Interactions*, Infectious Disease,
https://doi.org/10.1007/978-3-319-72416-4

Printed in the United States
By Bookmasters